Advances in Geometry and Lie Algebras from Supergravity

Theoretical and Mathematical Physics

The series founded in 1975 and formerly (until 2005) entitled *Texts and Monographs in Physics* (TMP) publishes high-level monographs in theoretical and mathematical physics. The change of title to *Theoretical and Mathematical Physics* (TMP) signals that the series is a suitable publication platform for both the mathematical and the theoretical physicist. The wider scope of the series is reflected by the composition of the editorial board, comprising both physicists and mathematicians.
The books, written in a didactic style and containing a certain amount of elementary background material, bridge the gap between advanced textbooks and research monographs. They can thus serve as basis for advanced studies, not only for lectures and seminars at graduate level, but also for scientists entering a field of research.

Editorial Board

More information about this series at http://www.springer.com/series/720

Pietro Giuseppe Fré

Advances in Geometry and Lie Algebras from Supergravity

Springer

Pietro Giuseppe Fré
Pietra Marazzi, Alessandria
Italy

ISSN 1864-5879 ISSN 1864-5887 (electronic)
Theoretical and Mathematical Physics
ISBN 978-3-030-09005-0 ISBN 978-3-319-74491-9 (eBook)
https://doi.org/10.1007/978-3-319-74491-9

Softcover re-print of the Hardcover 1st edition 2018

Printed on acid-free paper

This Springer imprint is published by the registered company Springer International Publishing AG part of Springer Nature
The registered company address is: Gewerbestrasse 11, 6330 Cham, Switzerland

This book is dedicated to my beloved daughter Laura, to my darling wife Olga and to my young son Vladimir. My deep feelings of love are hereby transmitted to the three of them, with all the hopes and the worries of an old father who stubbornly believes that culture and science are just one thing and encode, for mankind, the unique escape route from global disasters.

Preface

This book forms a twin pair with another book [1] by the same author, which is of a different, historically oriented character, while the character of the present volume is thoroughly mathematical. These two pieces of work constitute a twin pair since, notwithstanding their different profile and contents, they arise from the same vision and pursue complementary goals.

The vision, extensively discussed in [1], consists of the following main conceptual assessments:

1. Our current understanding of the Fundamental Laws of Nature is based on a coherent, yet provisional, set of five meta-theoretical principles, listed by me as (A)–(E) and dubbed the current *episteme*. This episteme is of genuine geometrical nature and can be viewed as the current evolutionary state of Einstein's ideas concerning the geometrization of physics.
2. *Geometry* and *Symmetry* are inextricably entangled, and their current conception is the result of a long process of abstraction, traced back in [1], which was historically determined and makes sense only within the *Analytic System of Thought* of Western Civilization, started by the ancient Greeks.
3. The evolution of *Geometry and Symmetry Theory* in the last forty years has been deeply and very much constructively influenced by *Supersymmetry/Supergravity* and the allied constructions of *Strings and Branes*.
4. Further advances in Theoretical Physics cannot be based simply on the Galilean Method of interrogating first Nature and then formulating a testable theory that explains the observed phenomena. As stated in [1], one ought to interrogate also *Human Thought*, by this meaning frontier-line mathematics concerned with geometry and symmetry in order to find there the threads of so far unobserved correspondences, reinterpretations, and renewed conceptions.

The complementary pursued goals are:

(a) In the case of book [1]

- the historical and conceptual analysis of the process mentioned in point (2) of the above list which led to the current episteme.

- the philosophical argumentation, on historical basis, of the assessment made in point (4) of the above list.

(b) In the case of the present book, the mathematical full-fledged illustration of the main developments in geometry and symmetry theory that occurred under the fertilizing influence of *Supersymmetry/Supergravity* and that would be inconceivable without the latter.

In view of this, it is reasonable to quote from the ample discussion presented in [1] the summary of the current *episteme* as I understand it. There I say what follows.

The Episteme

As a theoretical physicist, I consider myself very fortunate to have witnessed, in my own lifetime, the following series of experimental discoveries:

1. The detection of the $W^{\pm}$ and Z particles, definitely confirming that fundamental non-gravitational interactions can be described by gauge theories.
2. The detection of the Brout Englert Higgs boson, definitely confirming that gauge theories can be spontaneously broken by scalar fields falling into non-symmetric extrema of some potential.
3. The direct detection of gravitational waves emitted in the coalescence of two compact stars (black holes or neutron stars) which not only confirms the general structure of General Relativity, but directly tests the dynamics encoded in Einstein Equations, namely in a set of purely geometrical differential equations.

Trying to summarize the implications for the *episteme* of the last thirty-three years of experimental physics, we can say the following.

Leaving apart the issue of *quantization* that we can generically identify with the *functional path integral over classical configurations*, we have, within our Western Analytic System of Thought, a rather simple and universal scheme of interpretation of the Fundamental Interactions and of the Fundamental Constituents of Matter based on the following few principles:

(A) The categorical reference frame is provided by Field Theory defined by some action $\mathscr{A} = \int_{\mathscr{M}} \mathscr{L}(\Phi, \partial\Phi)$ where $\mathscr{L}(\Phi, \partial\Phi)$ denotes some Lagrangian depending on a set of fields $\Phi(x)$.

(B) All fundamental interactions are described by *connections* $\mathbf{A}$ on *principal fiber bundles* $P(\mathrm{G}, \mathscr{M})$ where G is a Lie group and the base manifold $\mathscr{M}$ is some space-time in $d = 4$ or in higher dimensions.

(C) All the fields Φ describing fundamental constituents are *sections* of *vector bundles* $B(\mathrm{G}, V, \mathscr{M})$, associated with the principal one $P(\mathrm{G}, \mathscr{M})$ and determined by the choice of suitable *linear representations* $D(\mathrm{G}) : V \to V$ of the structural group G.

(D) The spin-zero particles described by scalar fields ϕ^I have the additional feature of admitting nonlinear interactions encoded in a scalar potential $\mathscr{V}(\phi)$ for whose choice general principles, supported by experimental confirmation, have not yet been determined.

(E) Gravitational interactions are special among the others and universal since they deal with the *tangent bundle* $T\mathcal{M} \rightarrow \mathcal{M}$ to space-time. The relevant connection is in this case the Levi-Civita connection (or some of its generalization with torsion) which is determined by a metric g on $\mathcal{M}$.

A quick look at the list of principles (A)–(E) immediately reveals that, notwithstanding their simplicity and unifying power, they can be only provisional. There are still too many ad hoc choices which strongly demand some deeper unifying principle able to predict them from above. Most prominent among these choices are those of the structural group G, of the representations $D(\mathrm{G})$ and of the potential $\mathscr{V}(\phi)$, the latter choice including also, in some extended sense, the determination of quark and lepton masses. What I have described in the above way is described in the physical literature of the last forty years as the problem of *grand unification* or of *super unification*.

Supersymmetry-Inspired Trends in Geometry and Group Theory

In the same forty years, an enormously extended set of developments have taken place in the quest for unification, starting from the new idea of *Supersymmetry* which, as the word reveals, is an extension of the notion of *Symmetry*, meaning by that Lie algebras. The reason why Supersymmetry, which leads to the fields of *Supergravity*, *Superstrings,* and *Brane Physics,* entrains so many structural and ramified implications is because it tackles with one of the most fundamental and, in my opinion, not yet fully penetrated, principles of physics, namely the distinction among fermion and bosons, intertwined, by means of the spin–statistics theorem with Lie algebra theory, the distinction between two groups of representations, the vector and the spinor ones, being a distinctive property of the $\mathfrak{so}(n)$ Lie algebras, unexisting for the others.

The largest part of the developments mentioned above, related with Supergravity/Superstrings, have a distinctive geometrical/algebraic basis. Entire chapters of algebraic geometry and of algebraic topology have been integrated by these developments into the fabrics of theoretical physics, while some new geometries have been introduced into the fabrics of mathematics. Furthermore, the very way to analyze and interpret mathematical structures is sometimes redirected by the influence of Supergravity/Superstrings. Two or three examples suffice to illustrate what I mean. Exceptional Lie algebras that, up to the mid-1960s were considered by the majority of physicists like mathematical curiosities, have been promoted to the role of primary actors on the stage of the *Superworld*. Special Kähler geometries, never defined by pure mathematicians have by now entered, with full rights, the mathematical club, revealing their relation with other geometries, already introduced by mathematicians, like HyperKähler geometry and quaternionic Kähler geometry. The notions of momentum map, Kähler, and HyperKähler quotients find a deep interpretation in the context of supersymmetric field theories and connect with some of the most brilliant mathematical achievements of the last few decades like the Kronheimer construction of ALE manifolds.

The Topics of the Present Volume and Its Mission

Relying on the above arguments and explanations, I can now more appropriately restate the topic of the present book, which is *the scope of Group Theory, of the Differential Geometry of Coset Manifolds and of various issues in Special Geometries* as they have been promoted and assessed under the influence of current research in Supergravity.

In line with above the statements, it goes without saying that the education of present time physicists, in particular theoretical, but not only, should include, from a very early stage of their student career, a ground course in the basic *Mathematics of Symmetry*, namely in group theory, discrete and Lie groups being equally essential, and in the fundaments of differential geometry. Such course should be mathematically precise, yet more focused on the fundamental mathematical ideas than on the task of mathematical rigorous proofs. Furthermore, it should provide explicit constructions and train the student in the art of explicit calculations, especially those implemented on computers. To such a task is devoted the textbook [2] which was recently published.

Repeating my words in a slightly different form, I think that what is currently practiced in the whole world as Fundamental Physics or Mathematics is based on the Greek view of the *episteme* and it is meaningful only inside the Analytic System of Thought founded by the ancient Greeks. To recuperate a full conscience of this fact is mandatory in order to continue on the difficult but exciting path we are confronted with.

The twin pair of which this book is a member, together with the more introductory textbook [2], is viewed by the author as his limited, humble contribution to the promotion of a new season of more scholarly teaching of *physical mathematics*.

Spes, ultima dea.

Turin, Italy
November 2017

Pietro Giuseppe Fré

References

1. P.G. Fré, *A Conceptual History of Symmetry from Plato to the Superworld* (Springer, Berlin, 2018)
2. P.G. Fré, A.M. FedotoBv, *Groups and Manifolds, Lectures for Physicists with examples in Mathematica* (De Gruyter, Berlin, 2018)

Acknowledgements

With great pleasure, I would like to thank my collaborators and dear friends Pietro Antonio Grassi, Ugo Bruzzo, Alexander Sorin, and Mario Trigiante for several discussions, continuous encouragement, and many precious consultations. It is also my pleasure to acknowledge important discussions with my great distinguished friend Sergio Ferrara, in relation with the several common papers that are reflected in the texture of the present book quite ubiquitously. Similarly, I express my gratitude to the Editors of Springer-Verlag, in particular to Maria Bellantone for their continuous assistance, constructive criticism, and suggestions.

My thoughts, while finishing the writing of this long essay that occurred in Moscow, were frequently directed to my late parents, whom I miss very much and I will never forget. To them, I also express my gratitude for all what they taught me in their life, in particular to my father who, with his own example, introduced me, since my childhood, to the great satisfaction and deep suffering of writing books.

Contents

Chapter 1
Finite Groups and Lie Algebras: The ADE Classification and Beyond

Tout ce qu'on invente est vrai, soi-en sure. La poésie est une chose aussi précise que la géométrie.
Gustave Flaubert, letter to Louise Colet

The geometrical structures, mostly motivated by supergravity, that are considered in this book are strongly related with the theory of symmetric spaces and of Lie Algebras, the exceptional ones being of utmost relevance in this context.

At various stages of the here considered constructions also the finite groups play an important role and, among them, those that are crystallographic in certain dimensions. This is not too much surprising since there exists a profound relation among the classification of simple, simply-laced, complex Lie Algebras and the classification of finite subgroups of the three-dimensional rotation group, the so named ADE classification.[1]

This ADE correspondence, known for a long time, finds a deeper and fertile interpretation in the McKay correspondence, to be explained in Chap. 8, that is crucial for the Kronheimer construction of ALE-manifolds as HyperKähler quotients. This construction is reviewed in Chap. 8. The McKay correspondence admits a generalization to finite subgroups $\Gamma \subset \mathrm{SU(n)}$, in particular for the case $n = 3$, which has a significant role to play in the context of the *gravity/gauge correspondence* and in

[1]The ADE classification, according to the name frequently utilized in the physical literature, is based on a diophantine inequality that we spell out in the sequel of the present chapter. It encompasses in just one scheme the classification of several different types of mathematical objects:

1. the finite rotation groups,
2. the simple simply-laced Lie complex Lie algebras,
3. the locally Euclidean gravitational instantons,
4. the singularities $\mathbb{C}^2/\Gamma$,
5. the modular invariant partition functions of $2D$-conformal field theories.

P. G. Frè, *Advances in Geometry and Lie Algebras from Supergravity*, Theoretical and Mathematical Physics,
https://doi.org/10.1007/978-3-319-74491-9_1

the theory of D-branes and M-branes. This we will illustrate in Chap. 8 relying in a decisive way on the constructions known as (Hyper)Kähler quotients.

We anticipate here the discussion of the joint ADE classification of binary extensions $\Gamma_b \subset \mathrm{SU}(2)$ of finite subgroups $\Gamma \subset \mathrm{SO}(3)$ and of simple, simply-laced, complex Lie algebras, since some of the involved notions happen to be needed in other chapters previous to the last one.

For an identical reason we present in this chapter also the full-fledged theory of the simple group $\mathrm{L}_{168} \equiv \mathrm{PSL}(2, \mathbb{Z}_7)$ which fits into the discussion of crystallographic groups and plays a relevant role not only in Chap. 8, but in other chapters.

In the third part of this chapter, relying on the Dynkin and on the root system language, we consider a particular splitting, named by us *golden*, of the Lie algebras that appear in later supergravity constructions and, focusing on the relevant instances of the $\mathfrak{g}_{(2,2)}$ and $\mathfrak{f}_{(4,4)}$ cases, we explicitly construct the fundamental representations of the two corresponding Lie groups.

Let me note that the explicit construction of the exceptional Lie algebras is addressed at a paedagogical level in [1], with general group theoretical aims. Furthermore in that book the algorithmic details of the construction are presented and a guide is provided to the use of the special MATHEMATICA codes that have been devoted to such a task. In the present book the emphasis is on the *special geometries* introduced by supergravity. Within such a context exceptional Lie algebras play a quite relevant role and several aspects of their structure seem just devised to satisfy the constraints imposed by supersymmetry at various levels. It is in the light of these considerations that the construction $\mathfrak{g}_{(2,2)}$ and $\mathfrak{f}_{(4,4)}$ is reviewed here. In particular the *golden splitting* turns out to be fundamental for the discussion of the c-map and of the $c^\star$-map, firstly addressed in Sect. 4.3 and then systematically reviewed in Chap. 5.

1.1 The ADE Classification of the Finite Subgroups of SU(2)

We start with the ADE classification of platonic groups. This classification is encoded in the possible solutions of a diophantine equation that we presently derive. To this effect we begin with some preliminaries.

Let us start by considering the homomorphism:

$$\omega : \quad \mathrm{SU}(2) \rightarrow \mathrm{SO}(3) \tag{1.1.1}$$

between the group SU(2) of unitary 2×2 matrices, each of which can be written as follows

$$\mathrm{SU}(2) \ni \mathscr{U} = \begin{pmatrix} \alpha & \mathrm{i}\beta \\ \mathrm{i}\bar{\beta} & \bar{\alpha} \end{pmatrix} \tag{1.1.2}$$

in terms of two complex numbers α, β satisfying the constraint:

$$|\alpha|^2 + |\beta|^2 = 1 \tag{1.1.3}$$

and the group SO(3) of 3×3 of orthogonal matrices with unit determinant:

$$\mathcal{O} \in \mathrm{SO}(3) \Leftrightarrow \mathcal{O}^T \mathcal{O} = \mathbf{1} \quad \text{and} \quad \det \mathcal{O} = 1 \tag{1.1.4}$$

The homomorphism ω can be explicitly constructed utilizing the so-named triplet σ^x of hermitian Pauli matrices:

$$\sigma^1 = \begin{pmatrix} 0 & 1 \\ 1 & 0 \end{pmatrix} \; ; \; \sigma^2 = \begin{pmatrix} 0 & -\mathrm{i} \\ \mathrm{i} & 0 \end{pmatrix} \; ; \; \sigma^3 = \begin{pmatrix} 1 & 0 \\ 0 & -1 \end{pmatrix} \tag{1.1.5}$$

Using the above we can define:

$$\mathcal{H} = \sum_{x=1}^{3} h_x \, \sigma^x \tag{1.1.6}$$

where h_x is a three-vector with real components. The matrix $\mathcal{H} = \mathcal{H}^\dagger$ is hermitian by construction and we have:

$$\mathrm{Tr}\left[\mathcal{H}^2\right] = \sum_{x=1}^{3} h_x^2 \tag{1.1.7}$$

Consider next the following matrix transformed by means of an SU(2) element:

$$\begin{aligned} \tilde{\mathcal{H}} &= \mathcal{U}^\dagger \, \mathcal{H} \, \mathcal{U} = \tilde{h}_x \, \sigma^x \\ \tilde{h}_x &= \mathcal{O}_x{}^y \, h_y \end{aligned} \tag{1.1.8}$$

The first line of equation (1.1.8) can be written since the Pauli matrices form a complete basis for the space of 2×2 hermitian traceless matrices. The second line can be written since the matrix $\tilde{\mathcal{H}}$ depends linearly on the matrix $\mathcal{H}$. Next we observe that because of its definition the matrix $\tilde{\mathcal{H}}$ has the following property:

$$\mathrm{Tr}\left[\tilde{\mathcal{H}}^2\right] = \sum_{x=1}^{3} \tilde{h}_x^2 = \sum_{x=1}^{3} h_x^2 \tag{1.1.9}$$

This implies that the matrix $\mathcal{O}_x{}^y$ is orthogonal and, by definition it is the image of $\mathcal{U}$ through the homomorphism ω. We can write an explicit formula for the matrix elements $\mathcal{O}_x{}^y$ in terms of $\mathcal{U}$:

$$\forall \, \mathcal{U} \in \mathrm{SU}(2) : \omega[\mathcal{U}] = \mathcal{O} \in \mathrm{SO}(3) \; / \; \mathcal{O}_x{}^y = \tfrac{1}{2} \mathrm{Tr}\left[\mathcal{U}^\dagger \, \sigma_x \, \mathcal{U} \, \sigma^y\right] \tag{1.1.10}$$

which follows from the trace-orthogonality of the Pauli matrices $\frac{1}{2}\,\mathrm{Tr}\left[\sigma^y \, \sigma_x\right] = \delta_x^y$.

We named the map defined above a homomorphism rather than an isomorphism since it has a non trivial kernel of order two. Indeed the following two SU(2) matrices constitute the kernel of ω since they are both mapped into the identity element of SO(3).

$$\begin{aligned} \ker\omega &= \left\{ \mathbf{e} = \begin{pmatrix} 1 & 0 \\ 0 & 1 \end{pmatrix}, \mathscr{Z} = \begin{pmatrix} -1 & 0 \\ 0 & -1 \end{pmatrix} \right\} \\ \mathbf{1} &= \omega\,[\mathbf{e}] = \omega\,[\mathscr{Z}] \end{aligned} \tag{1.1.11}$$

We will now obtain the classification of all finite subgroups of SU(2) that are binary extensions of SO(3) finite subgroups. We collectively name G^b_{2n} such subgroups denoting by $2n$ their necessarily even order. Through the isomorphism ω each of them maps into a finite subgroup $\mathrm{G_n} \subset \mathrm{SO(3)}$, whose order is just n because of the two-dimensional kernel mentioned above:

$$\omega\left[\mathrm{G}^b_{2n}\right] = \mathrm{G_n} \tag{1.1.12}$$

1.1.1 The Argument Leading to the Diophantine Equation

We begin by considering one parameter subgroups of SO(3). These are singled out by a rotation axis, namely by a point on the two–sphere S^2. Explicitly let us consider a solution (ℓ, m, n) to the sphere equation (Fig. 1.1):

$$\ell^2 + m^2 + n^2 = 1 \tag{1.1.13}$$

The triplet of real numbers (ℓ, m, n) parametrize the direction of a possible rotation angle. The generator of infinitesimal rotations around such an axis is given by the following matrix

$$A_{\ell,m,n} = \begin{pmatrix} 0 & -n & m \\ n & 0 & -\ell \\ -m & \ell & 0 \end{pmatrix} = -A^T_{\ell,m,n} \tag{1.1.14}$$

which being antisymmetric belongs to the SO(3) Lie algebra. The matrix A has the property that $A^3 = -A$ and explicitly we have:

$$A^2_{\ell,m,n} = \begin{pmatrix} -1+\ell^2 & \ell\,m & \ell\,n \\ \ell\,m & -1+m^2 & m\,n \\ \ell\,n & m\,n & -1+n^2 \end{pmatrix} \tag{1.1.15}$$

Hence a finite element of the group SO(3) corresponding to a rotation of an angle θ around this axis is given by:

$$\mathscr{O}_{(\ell,m,n)} = \exp[\theta\,A_{\ell,m,n}] = \mathbf{1} + \sin\theta\,A_{\ell,m,n} + (1-\cos\theta)\,A^2_{\ell,m,n} \tag{1.1.16}$$

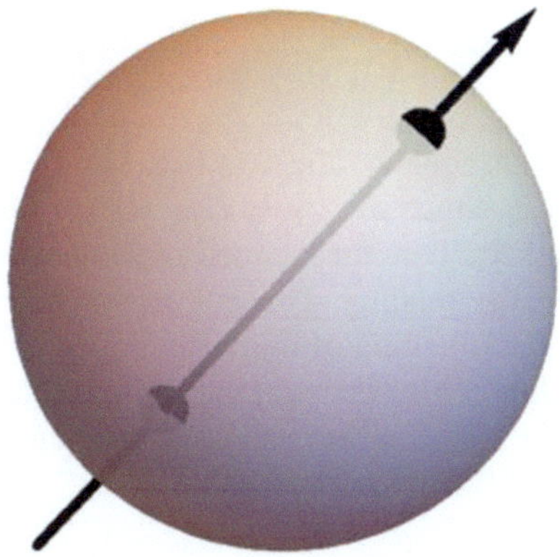

Fig. 1.1 Every element of the rotation group $\mathscr{O}_{(\ell,m,n)} \in \mathrm{SO}(3)$ corresponds to a rotation around some axis $\mathbf{a} = \{\ell, m, n\}$. On the surface of the two-sphere $\mathbb{S}^2$ this rotation has two fixed points, a North Pole and a South Pole that do not rotate to any other point. The rotation $\mathscr{O}_{(\ell,m,n)}$ is the image, under the homomorphism ω of either one of 2×2 - matrices $\mathscr{U}^{\pm}_{\ell,m,n}$ that, acting on the space $\mathbb{C}^2$, admit two eigenvectors $\mathbf{z}_1$ and $\mathbf{z}_2$. The one-dimensional complex spaces $p_{1,2} \equiv \lambda_{1,2}\mathbf{z}_{1,2}$ are named the two poles of the unitary rotation

Setting

$$\lambda = \ell \sin\frac{\theta}{2} \quad ; \quad \mu = m \sin\frac{\theta}{2} \quad ; \quad \nu = n \sin\frac{\theta}{2} \quad ; \quad \rho = \cos\frac{\theta}{2} \tag{1.1.17}$$

the corresponding SU(2) finite group elements, realizing the double covering are:

$$\mathscr{U}^{\pm}_{\ell,m,n} = \pm \begin{pmatrix} \rho + \mathrm{i}\nu & \mu - \mathrm{i}\lambda \\ -\mu - \mathrm{i}\lambda & \rho - \mathrm{i}\nu \end{pmatrix} \tag{1.1.18}$$

namely we have:

$$\omega\left[\mathscr{U}^{\pm}_{\ell,m,n}\right] = \mathscr{O}_{(\ell,m,n)} \tag{1.1.19}$$

We can now consider the argument that leads to the ADE classification of the finite subgroups of SU(2). Let us consider the action of the SU(2) matrices on $\mathbb{C}^2$. A generic $\mathscr{U} \in \mathrm{SU}(2)$ acts on a $\mathbb{C}^2$-vector $\mathbf{z} = \begin{pmatrix} z_1 \\ z_2 \end{pmatrix}$ by usual matrix multiplication $\mathscr{U}\,\mathbf{z}$. Each element $\mathscr{U} \in \mathrm{SU}(2)$ has two eigenvectors $\mathbf{z}_1$ and $\mathbf{z}_2$, such that

$$\begin{aligned} \mathscr{U}\,\mathbf{z}_1 &= \exp[\mathrm{i}\theta]\,\mathbf{z}_1 \\ \mathscr{U}\,\mathbf{z}_2 &= \exp[-\mathrm{i}\theta]\,\mathbf{z}_2 \end{aligned} \tag{1.1.20}$$

where θ is some (half)-rotation angle. Namely for each $\mathscr{U} \in \mathrm{SU}(2)$ we can find an orthogonal basis where $\mathscr{U}$ is diagonal and given by:

$$\mathscr{U} = \begin{pmatrix} \exp[\mathrm{i}\theta] & 0 \\ 0 & \exp[-\mathrm{i}\theta] \end{pmatrix} \tag{1.1.21}$$

for some angle θ. Then let us consider the rays $\{\lambda\, \mathbf{z}_1\}$ and $\{\mu\, \mathbf{z}_2\}$ where $\lambda, \mu \in \mathbb{C}$ are arbitrary complex numbers. Since $\mathbf{z}_1 \cdot \mathbf{z}_2 = \mathbf{z}_1^\dagger \mathbf{z}_2 = 0$ it follows that each element of SU(2) singles out two rays, hereafter named *poles* that are determined one from the other by the orthogonality relation. This concept of pole is the basic item in the argument leading to the classification of finite rotation groups.

Let $H \subset \mathrm{SO}(3)$ be a finite, discrete subgroup of the rotation group and let $\hat{H} \subset \mathrm{SU}(2)$ be its pre-image in SU(2) with respect to the homomorphism ω. Then the order of H is some positive integer number:

$$|H| = n \in \mathbb{N} \tag{1.1.22}$$

The total number of poles associated with H is:

$$\#\,\text{of poles} = 2n - 2 \tag{1.1.23}$$

since $n-1$ is the number of elements in H that are different from the identity. Let us then adopt the notation

$$p_i \equiv \{\lambda \mathbf{z}_i\} \tag{1.1.24}$$

for the pole or ray singled out by the eigenvector $\mathbf{z}_i$. We say that two poles are equivalent if there exists an element of the group H that maps one into the other:

$$p_i \sim p_j \quad \text{iff} \quad \exists \gamma \in H \,/\, \gamma p_i = p_j \tag{1.1.25}$$

Let us distribute the poles p_i into orbits under the action of the group H:

$$\mathscr{Q}_\alpha = \left\{p_1^\alpha, \ldots, p_{m_\alpha}^\alpha\right\} \quad ; \quad \alpha = 1, \ldots, r \tag{1.1.26}$$

and name m_α the cardinality of the orbit class $\mathscr{Q}_\alpha$, namely the number of poles it contains. Hence we have assumed that there are r orbits and that each orbit $\mathscr{Q}_\alpha$ contains m_α elements.

Each pole $p \in \mathscr{Q}_\alpha$ has a stability subgroup $K_p \subset H$:

$$\forall h \in K_p \quad : \quad h\, p = p \tag{1.1.27}$$

that is finite, abelian and cyclic of order k_α. Indeed it must be finite since it is a subgroup of a finite group, it must be abelian since in the basis $\mathbf{z}_1, \mathbf{z}_2$ the SU(2) matrices that preserve the poles $\lambda \mathbf{z}_1$ and $\mu \mathbf{z}_2$ are, of the form (1.1.21) and therefore it is cyclic of some order. The H group can be decomposed into cosets according to the subgroup K_p:

$$H = K_p + v_1\, K_p + \cdots + v_{m_\alpha}\, K_p \quad m_\alpha \in \mathbb{N} \tag{1.1.28}$$

Consider now an element $x_i \in v_i K_p$ belonging to one of the cosets and define the group conjugate to K_p through x_i:

$$K_{(xp)_i} = x_i \, K_p \, x_i^{-1} \tag{1.1.29}$$

Each element $h \in K_{(xp)_i}$ admits a pole p_x:

$$h \, p_x = p_x \tag{1.1.30}$$

that is given by:

$$p_x = x_i \, p \tag{1.1.31}$$

since

$$h \, p_x = x \, h_p x \, x^{-1} \, p = x \, h_p \, p = x \, p = p_x \tag{1.1.32}$$

Hence the set of poles $\left\{p, v_1 \, p, v_2 \, p, \ldots v_{m_\alpha} \, p\right\}$ are equivalent forming an orbit. Each of them has a stability group K_{p_i} conjugate to K_p which implies that all K_{p_i} are finite of the same order:

$$\forall v_i \, p \quad |K_{p_i}| = k_\alpha \tag{1.1.33}$$

By this token we have proven that in each orbit $\mathcal{Q}_\alpha$ the stability subgroups of each element are isomorphic, and cyclic of the same order k_α which is a property of the orbit. Hence we must have:

$$\forall \mathcal{Q}_\alpha \quad ; \quad k_\alpha \, m_\alpha = n \tag{1.1.34}$$

The total number of poles we have in the orbit $\mathcal{Q}_\alpha$ (counting coincidences) is:

$$\#\,\text{of poles in the orbit } \mathcal{Q}_\alpha = m_\alpha \, (k_\alpha - 1) \tag{1.1.35}$$

since the number of elements in K_p different from the identity is $k_\alpha - 1$. Hence we find

$$2n - 2 = \sum_{\alpha=1}^{r} m_\alpha \, (k_\alpha - 1) \tag{1.1.36}$$

Dividing by n we obtain:

$$2\left(1 - \frac{1}{n}\right) = \sum_{\alpha=1}^{r} \left(1 - \frac{1}{k_\alpha}\right) \tag{1.1.37}$$

We consider next the possible solutions to the diophantine equation (1.1.37) and to this effect we rewrite it as follows:

$$r + \frac{2}{n} - 2 = \sum_{\alpha=1}^{r} \frac{1}{k_\alpha} \tag{1.1.38}$$

We observe that $k_\alpha \geq 2$. Indeed each pole admits at least two group elements that keep it fixed, the identity and the non trivial group element that defines it by diagonalization. Hence we have the bound:

$$r + \frac{2}{n} - 2 \leq \frac{r}{2} \tag{1.1.39}$$

which implies:

$$r \leq 4 - \frac{4}{n} \quad \Rightarrow \quad r = 1, 2, 3 \tag{1.1.40}$$

On the other hand we also have $k_\alpha \leq n$ so that:

$$r + \frac{2}{n} - 2 \geq \frac{r}{n} \quad \Rightarrow \quad r\left(1 - \frac{1}{n}\right) \geq 2\left(1 - \frac{1}{n}\right) \quad \Rightarrow \quad r \geq 2 \tag{1.1.41}$$

Therefore there are only two possible cases:

$$r = 2 \quad \text{or} \quad r = 3 \tag{1.1.42}$$

Let us now consider the solutions of the diophantine equation (1.1.39) and identify the finite rotation groups and their binary extensions.

Taking into account the conclusion (1.1.42) we have two cases.

1.1.2 Case $r = 2$: The Infinite Series of Cyclic Groups $\mathfrak{a}_n$

Choosing $r = 2$, the diophantine equation (1.1.38) reduces to:

$$\frac{2}{n} = \frac{1}{k_1} + \frac{1}{k_2} \tag{1.1.43}$$

Since we have $k_{1,2} \leq n$, the only solution of (1.1.43) is $k_1 = k_2 = n$, with n arbitrary. Since the order of the cyclic stability subgroup of the two poles coincides with the order of the full group H it follows that H itself is a cyclic subgroup of SU(2) of order n. We name it $\Gamma_b[n, n, 1]$. The two orbits are given by the two eigenvectors of the unique cyclic group generator:

$$\mathscr{A} \in \mathrm{SU}(2) \quad : \quad \mathscr{Z} \equiv \mathscr{A}^n \tag{1.1.44}$$

The finite subgroup of SU(2), isomorphic to the abstract group $\mathbb{Z}_{2n}$ is composed by the following $2n$ elements:

$$\mathbb{Z}_{2n} \sim \Gamma_b[n, n, 1] = \left\{\mathbf{1}, \mathscr{A}, \mathscr{A}^2, \ldots, \mathscr{A}^{n-1}, \mathscr{Z}, \mathscr{Z}\mathscr{A}, \mathscr{A}^2, \ldots, \mathscr{Z}\mathscr{A}^{n-1}\right\} \tag{1.1.45}$$

Under the homomorphism ω, the SU(2)-element $\mathscr{Z}$ maps into the identity and both $\mathscr{A}$ and $\mathscr{Z}\,\mathscr{A}$ map into the same 3×3 orthogonal matrix A $\in$ SO(3) with the property $\mathrm{A}^n = \mathbf{1}$. Hence we have:

$$\omega\left[\Gamma_b[n,n,1]\right] = \Gamma[n,n,1] \sim \mathbb{Z}_n \tag{1.1.46}$$

In conclusion we can define the cyclic subgroups of SO(3) and their binary extensions in SU(2) by means of the following presentation in terms of generators and relations:

$$\mathfrak{a}_n \quad \Leftrightarrow \quad \begin{cases} \Gamma_b[n,n,1] = (\mathscr{A},\mathscr{Z} \mid \mathscr{A}^n = \mathscr{Z} \ ; \ \mathscr{Z}^2 = \mathbf{1}) \\ \Gamma[n,n,1] = \qquad (\mathrm{A} \mid \mathrm{A}^n = \mathbf{1}) \end{cases} \tag{1.1.47}$$

The nomenclature $\mathfrak{a}_n$ introduced in the above equation is just for future comparison. As we will see, in the ADE-classification of simply laced Lie algebras the case of cyclic groups corresponds to that of $\mathfrak{a}_n$ algebras.

1.1.3 Case $r = 3$ and its Solutions

In the $r = 3$ case the Diophantine equation becomes:

$$\frac{1}{k_1} + \frac{1}{k_2} + \frac{1}{k_3} = 1 + \frac{2}{n} \tag{1.1.48}$$

In order to analyze its solutions in a unified way and inspired by the above case it is convenient to introduce the following notations:

$$\mathscr{R} = 1 + \sum_{\alpha}^{r} k_\alpha \tag{1.1.49}$$

and consider the abstract groups, that turn out to be of finite order, associated with each triple of integers $\{k_1, k_2, k_3\}$ satisfying (1.1.48) and defined by the following presentation:

$$\begin{aligned} &\Gamma_b\,[k_1,k_2,k_3] = \left(\mathscr{A},\mathscr{B},\mathscr{Z} \mid (\mathscr{A}\mathscr{B})^{k_1} = \mathscr{A}^{k_2} = \mathscr{B}^{k_3} = \mathscr{Z}\,;\ \mathscr{Z}^2 = \mathbf{1}\right) \\ &\Gamma\,[k_1,k_2,k_3] = \left(\mathrm{A},\mathrm{B} \mid (\mathrm{AB})^{k_1} = \mathrm{A}^{k_2} = \mathrm{B}^{k_3} = \mathbf{1}\right) \end{aligned} \tag{1.1.50}$$

We will see that the finite subgroups of SU(2) are indeed isomorphic to the above defined abstract groups $\Gamma_b\,[k_1,k_2,k_3]$ and that their image under the homomorphism ω are isomorphic to $\Gamma\,[k_1,k_2,k_3]$.

1.1.3.1 The Solution $(k, 2, 2)$ and the Dihedral Groups Dih_k

One infinite class of solutions of the diophantine equation (1.1.48) is given by

$$\{k_1, k_2, k_3\} = \{k, 2, 2\} \quad ; \quad 2 < k \in \mathbb{Z} \tag{1.1.51}$$

The corresponding subgroups of SU(2) and SO(3) are:

$$\mathrm{Dih}_k \quad \Leftrightarrow \quad \begin{cases} \Gamma_b[k,2,2] = (\mathscr{A}, \mathscr{B}, \mathscr{Z} \mid (\mathscr{A}\mathscr{B})^k = \mathscr{A}^2 = \mathscr{B}^2 = \mathscr{Z}\,; \\ \qquad\qquad\qquad\qquad \mathscr{Z}^2 = \mathbf{1}) \\ \Gamma[k,2,2] = \qquad (\mathrm{A}, \mathrm{B} \mid (\mathrm{AB})^k = \mathrm{A}^2 = \mathrm{B}^2 = \mathbf{1}) \end{cases} \tag{1.1.52}$$

whose structure we illustrate next.

$\Gamma_b[k, 2, 2] \simeq \mathrm{Dih}_k^b$ is the binary dihedral subgroup. Its order is

$$|\,\mathrm{Dih}_k^b\,| = 4\,k \tag{1.1.53}$$

and it contains a cyclic subgroup of order k that we name K. Its index in Dih_k^b is two. The elements of Dih_k^b that are not in K are of period equal to two since $k_2 = k_3 = 2$. Altogether the elements of the dihedral group are the matrices given below:

$$F_l = \begin{pmatrix} e^{il\pi/k} & 0 \\ 0 & e^{-il\pi/k} \end{pmatrix} \quad ; \quad (l = 0, 1, 2, \ldots, 2k-1)$$

$$G_l = \begin{pmatrix} 0 & i\,e^{-il\pi/k} \\ i\,e^{il\pi/k} & 0 \end{pmatrix} \quad ; \quad (l = 0, 1, 2, \ldots, 2k-1)$$

In terms of them the generators are identified as follows:

$$F_0 = 1 \quad ; \quad F_1\,G_0 = \mathscr{A} \quad ; \quad F_k = \mathscr{Z} \quad ; \quad G_0 = \mathscr{B}\,. \tag{1.1.54}$$

There are exactly $\mathscr{R} = k + 3$ conjugacy classes

1. K_e contains only the identity F_0
2. K_Z contains the central extension $\mathscr{Z}$
3. $K_{G\,even}$ contains the elements $G_{2\nu}$ $(\nu = 1, \ldots, k-1)$
4. $K_{G\,odd}$ contains the elements $G_{2\nu+1}$ $(\nu = 1, \ldots, k-1)$
5. the $k-1$ classes K_{F_μ}: each of these classes contains the pair of elements F_μ and $F_{2k-\mu}$ for $(\mu = 1, \ldots, k-1)$.

Correspondingly the group Dih_k^b admits $k + 3$ irreducible representations, 4 of which are 1-dimensional while $k-1$ are 2-dimensional. We name them as follows:

Table 1.1 Character table of the group Dih_k^b

.	KE	KZ	KG_e	KG_o	KF_1	$\cdots$	KF_{k-1}
DE	1	1	1	1	1	$\cdots$	1
DZ	1	1	-1	-1	1	$\cdots$	1
DG_e	1	$(-1)^k$	i^k	$-i^k$	$(-1)^1$	$\cdots$	$(-1)^{k-1}$
DG_o	1	$(-1)^k$	$-i^k$	i^k	$(-1)^1$	$\cdots$	$(-1)^{k-1}$
DF_1	2	$(-2)1$	0	0	$2\,Cos\,\frac{\pi}{k}$	$\cdots$	$2\,Cos\,\frac{(k-1)\pi}{k}$
$\vdots$	$\vdots$	$\vdots$	$\vdots$	$\vdots$	$\vdots$	$\ddots$	$\vdots$
DF_{k-1}	2	$(-2)^{k-1}$	0	0	$2\,Cos\,\frac{(k-1)\pi}{k}$	$\cdots$	$2\,Cos\,\frac{(k-1)^2\pi}{k}$

$$\begin{cases} D_e\ ;\ D_Z\ ;\ D_{G\,even}\ ;\ D_{G\,odd}\ ; \text{1-dimensional} \\ D_{F_1}\ ; \ldots ;\ D_{F_{k-1}} \qquad\qquad ; \text{2-dimensional} \end{cases} \tag{1.1.55}$$

The combinations of the $\mathbf{C}^2$ vector components (z_1, z_2) that transform in the four 1-dimensional representations are easily listed:

$$\begin{aligned} D_e &\longrightarrow |z_1|^2 + |z_2|^2 \\ D_Z &\longrightarrow z_1\, z_2 \\ D_{G\,even} &\longrightarrow z_1^k + z_2^k \\ D_{G\,odd} &\longrightarrow z_1^k - z_2^k\,. \end{aligned} \tag{1.1.56}$$

The matrices of the $k-1$ two-dimensional representations are obtained in the following way. In the DF_s representation, $s = 1, \ldots k-1$, the generator $\mathscr{A}$, namely the group element F_1, is represented by the matrix F_s. The generator $\mathscr{B}$ is instead represented by $(i)^{s-1} G_0$ and the generator $\mathscr{Z}$ is given by F_{sk}, so that:

$$\begin{aligned} DF_s\,(F_j) &= F_{sj} \\ DF_s\,(G_j) &= (i)^{s-1} G_{sj}\,. \end{aligned} \tag{1.1.57}$$

The character table is immediately obtained and it is displayed in Table 1.1.[2] This concludes the discussion of the binary dihedral groups.

[2]In finite group-theory the square matrix $\chi_i^\mu \equiv \mathrm{Tr}\left(D_\mu\,(\gamma_i)\right)$ where $\mu = 1, 2, \ldots, r+1$ labels the irreducible representations of a group Γ and $i = 1, \ldots, r+1$ labels the conjugacy classes $\mathscr{C}^i$ of Γ-group elements, ($\gamma_i \in \mathscr{C}^i$ is any representative of the class) is named the character table and plays a fundamental, central role.

1.1.3.2 The Three Isolated Solutions Corresponding to the Tetrahedral, Octahedral and Icosahedral Groups

There remain three isolated solutions of the Diophantine equation (1.1.48), namely:

$$\{k_1, k_2, k_3\} = \{3, 3, 2\} \tag{1.1.58}$$
$$\{k_1, k_2, k_3\} = \{4, 3, 2\} \tag{1.1.59}$$
$$\{k_1, k_2, k_3\} = \{5, 3, 2\} \tag{1.1.60}$$

They respectively correspond to the tetrahedral T_{12}, octahedral O_{24} and icosahedral I_{60} groups and to their binary extensions, namely:

$$\Gamma[3, 3, 2] \simeq \mathrm{T}_{12} \tag{1.1.61}$$
$$\Gamma[4, 3, 2] \simeq \mathrm{O}_{24} \tag{1.1.62}$$
$$\Gamma[5, 3, 2] \simeq \mathrm{I}_{60} \tag{1.1.63}$$

As their name reveals these three groups have, 12,24 and 60 elements, respectively. The corresponding binary extensions have 24,48 and 120 elements respectively. With a procedure completely analogous to the one utilized in the case of the dihedral groups we might reconstruct all these elements and organize them into conjugacy classes. We do not do this explicitly; in the next section, while discussing crystallographic groups, we will rather study in full detail the example of the octahedral group O_{24} and we will do that starting from the three-dimensional realization in SO(3).

1.1.4 Summary of the ADE Classification of Finite Rotation Groups

Here we prepare the stage for the illustration of the deep and surprising relation, already anticipated, between the platonic classification of finite rotation groups and that of simple Lie algebras. To this effect let us consider Fig. 1.2 and diagrams of the sort there displayed. Such diagrams are named Dynkin diagrams and obtain a well-defined interpretation while studying root spaces and the classification of simple Lie Algebras. For the time being let us note that Dynkin diagrams such as that in Fig. 1.2 are characterized by three-integer numbers $\{k_1, k_2, k_3\}$, denoting the lengths of three chains of dots, linked one to the other and departing from a central node which belongs to each of the three chains. In the case one of the number k_α is equal to one (say k_3), the corresponding chain disappears and we are left with a simple chain of length $k_1 + k_2 - 1$. In Sect. 1.5 we will see that the admissible Dynkin diagrams

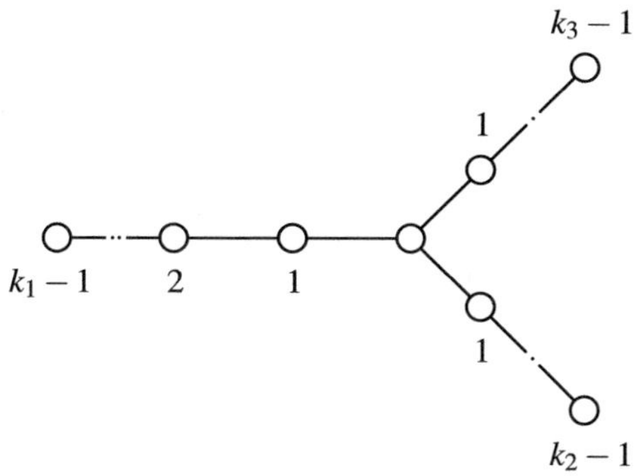

	Simple Lie Algebras	Finite subgroups of $\Gamma_b \subset \mathrm{SU}(2)$
r	number of simple chains in the Dynkin diagram	# of different types of group–element orders present in $\Gamma \equiv \omega[\Gamma_b]$
k_α	$k_\alpha - 1$ = lengths of the simple chains in the Dynkin diagram	group-element orders in $\Gamma \equiv$ $\left(\mathrm{A,B} \mid (\mathrm{AB})^{k_1} = \mathrm{A}^{k_2} = \mathrm{B}^{k_3} = \mathbf{1}\right)$
$\mathscr{R} - 1 \equiv \sum_{\alpha=1}^{r}(k_\alpha - 1)$	$\mathscr{R}$ = rank of the Lie algebra	$\mathscr{R} + 1$ = # of conjugacy classes in Γ_b

Fig. 1.2 Interpretation of the solutions of the same Diophantine equation in the case of finite subgroups of $\Gamma_b \subset \mathrm{SU}(2)$ and of simply laced Lie algebras

with one node are those and only those where the numbers $\{k_1, k_2, k_3\}$ satisfy the diophantine equation (1.1.48). Hence each solution of that equation has a double interpretation: it singles out a finite rotation group and labels a simple Lie algebra. The anticipated correspondence is the following one:

$$\Gamma[\ell, \ell, 1] \simeq \mathbb{Z}_\ell \Leftrightarrow \mathfrak{a}_\ell \tag{1.1.64}$$

$$\Gamma[\ell, 2, 2] \simeq \mathrm{Dih}_\ell \Leftrightarrow \mathfrak{d}_\ell \tag{1.1.65}$$

$$\Gamma[3, 3, 2] \simeq \mathrm{T}_{12} \Leftrightarrow \mathfrak{e}_6 \tag{1.1.66}$$

$$\Gamma[4, 3, 2] \simeq \mathrm{O}_{24} \Leftrightarrow \mathfrak{e}_7 \tag{1.1.67}$$

$$\Gamma[5, 3, 2] \simeq \mathrm{I}_{60} \Leftrightarrow \mathfrak{e}_8 \tag{1.1.68}$$

where $\mathfrak{a}_\ell$ is the Lie algebra associated with the Lie group $\mathrm{SL}(\ell+1,\mathbb{C})$, $\mathfrak{d}_\ell$ is the Lie algebra associated with the Lie group $\mathrm{SO}(2\ell,\mathbb{C})$, and $\mathfrak{e}_{6,7,8}$ are the Lie algebras of three exceptional Lie groups of dimensions 78, 133 and 248, respectively. A very important concept, in Lie Algebra theory is that of *rank* that is the maximal number of mutually commuting and diagonalizable elements of the algebra. As we see from Fig. 1.2, the rank has a counterpart in the binary extension of the corresponding finite rotation group: it is the number of non trivial conjugacy classes of the group, except the class of the identity element. The property of Lie algebras that in Dynkin diagrams there are no nodes with more than three converging lines corresponds on the finite rotation group side to the property that in such a group there are at most three different types of group-element orders.

A further challenging reinterpretation of the ADE-classification will be discussed later on and regards the construction of the so called ALE-manifolds, that are four-dimensional spaces with a self-dual curvature and asymptotic flatness. On their turn such manifolds are in relation with certain finite polynomial rings also classified by the same diophantine equation (see Chap. 8).

1.2 Lattices and Crystallographic Groups

In this section we consider the finite rotation groups from the point of view of crystallography, namely as groups of automorphisms of certain lattices. To this effect we need first to introduce the very notion of lattice and then introduce the notion of crystallographic group.

1.2.1 Lattices

We begin by fixing our notations for space and momentum lattices that define an n-torus T^n endowed with a flat metric structure, namely with a symmetric positive definite inner product.[3]

Let us consider the standard $\mathbb{R}^n$ manifold and introduce a basis of n linearly independent n-vectors that are not necessarily orthogonal to each other and of equal length:

$$\mathbf{w}_\mu \in \mathbb{R}^n \quad \mu = 1, \ldots n \tag{1.2.1}$$

Any vector in $\mathbb{R}$ can be decomposed along such a basis and we have:

$$\mathbf{r} = r^\mu \mathbf{w}_\mu \tag{1.2.2}$$

[3]*A clarification for mathematicians:* a metric on T^n is an inner product on the tangent spaces $T_p\,(\mathrm{T^n})$ for each $p \in \mathrm{T}^n$. In physical jargon we identify the inner product on $T_p\,(\mathrm{T^n})$ with the manifold metric since the metric coefficients $g_{\mu\nu}$ are the same for all $p \in \mathrm{T}^n$.

The flat (constant) metric on $\mathbb{R}^n$ is defined by:

$$g_{\mu\nu} = \langle \mathbf{w}_\mu \, , \, \mathbf{w}_\nu \rangle \tag{1.2.3}$$

where $\langle \, , \, \rangle$ denotes the standard Euclidean scalar product. The space lattice Λ consistent with the metric (1.2.3) is the free abelian group (with respect to sum) generated by the n basis vectors (1.2.1), namely:

$$\mathbb{R}^n \ni \mathbf{q} \in \Lambda \Leftrightarrow \mathbf{q} = q^\mu \, \mathbf{w}_\mu \quad \text{where} \quad q^\mu \in \mathbb{Z} \tag{1.2.4}$$

The dual lattice $\Lambda^\star$ is defined by the property:

$$\mathbf{p} \in \Lambda^\star \subset \mathbb{R}^n \Leftrightarrow \langle \mathbf{p} \, , \, \mathbf{q} \rangle \in \mathbb{Z} \quad \forall \mathbf{q} \in \Lambda \tag{1.2.5}$$

A basis for the dual lattice is provided by a set of n *dual vectors* $\mathbf{e}^\mu$ defined by the relations[4]:

$$\langle \mathbf{w}_\mu \, , \, \mathbf{e}^\nu \rangle = \delta^\nu_\mu \tag{1.2.6}$$

so that

$$\forall \mathbf{p} \in \Lambda^\star \quad \mathbf{p} = p_\mu \, \mathbf{e}^\mu \quad \text{where} \quad p_\mu \in \mathbb{Z} \tag{1.2.7}$$

1.2.2 Crystallographic Groups and the Bravais Lattices for $n = 3$

Every lattice Λ yields a metric g and every metric g singles out an isomorphic copy $\mathrm{SO_g(3)}$ of the continuous rotation group SO(n), which leaves it invariant:

$$M \in \mathrm{SO_g(n)} \quad \Leftrightarrow \quad M^T \, g \, M = g \tag{1.2.8}$$

By definition $\mathrm{SO_g(n)}$ is the conjugate of the standard SO(n) in $\mathrm{GL(n, \mathbb{R})}$:

$$\mathrm{SO_g(n)} = \mathscr{S} \, \mathrm{SO(n)} \, \mathscr{S}^{-1} \tag{1.2.9}$$

with respect to the matrix $\mathscr{S} \in \mathrm{GL(n, \mathbb{R})}$ which reduces the metric g to the Kronecker delta:

$$\mathscr{S}^T \, g \, \mathscr{S} = \mathbf{1} \tag{1.2.10}$$

Notwithstanding this a generic lattice Λ is not invariant with respect to any proper subgroup of the rotation group $\mathrm{G} \subset \mathrm{SO_g(n)} \equiv \mathrm{SO(n)}$. Indeed by invariance of the lattice one understands the following condition:

---

[4]In the sequel for the scalar product of two vectors we utilize also the equivalent shorter notation $\mathbf{a} \cdot \mathbf{b} = \langle \mathbf{a} \, , \mathbf{b} \rangle$.

$$\forall \gamma \in \mathrm{G} \;\; \text{and} \;\; \forall \mathbf{q} \in \Lambda \; : \qquad \gamma \cdot \mathbf{q} \in \Lambda \tag{1.2.11}$$

For $n = 3$ lattices that have a non trivial symmetry group $\mathrm{G} \subset \mathrm{SO}(3)$ are those relevant to Solid State Physics and Crystallography. There are 14 of them grouped in 7 classes that were already classified in the XIX century by Bravais. The symmetry group G of each of these Bravais lattices Λ_B is necessarily one of the well known finite subgroups of the three-dimensional rotation group O(3). In the language universally adopted by Chemistry and Crystallography for each Bravais lattice Λ_B the corresponding invariance group $\mathrm{G_B}$ is named the *Point Group*.

According to a standard nomenclature the 7 classes of Bravais lattices are respectively named *Triclinic, Monoclinic, Orthorhombic, Tetragonal, Rhombohedral, Hexagonal and Cubic*. Such classes are specified by giving the lengths of the basis vectors $\mathbf{w}_\mu$ and the three angles between them, in other words, by specifying the 6 components of the metric (1.2.3).

In general we have the following

Definition 1.2.1 An abstract group Γ is named crystallographic in n-dimensions if there exists an n-dimensional lattice Λ_n with basis vectors $\mathbf{w}_\mu$ such that:

1. there is a isomorphism:
$$\omega \, : \, \Gamma \, \rightarrow \, H \, \subset \mathrm{SO_g}(n) \tag{1.2.12}$$
where $\mathrm{SO_g}(n)$ is the conjugate of the n-dimensional group rotation group respecting a metric g (see Eq. (1.2.10)
2. the metric g is that defined by the basis vectors of the lattice Λ_n (see Eq. (1.2.3)
3. all elements of H are $n \times n$ matrices with integer valued entries.

This is equivalent to the statement that Γ has an orthogonal action in $\mathbb{R}^n$ and preserves the lattice Λ_n.

When a group Γ is crystallographic with respect to a given n-dimensional lattice Λ_n we say that is the *Point Group* of Λ_n.

1.2.3 The Proper Point Groups

Restricting one's attention to $n = 3$, it was shown in the classical crystallographic literature that the proper point groups that appear in the 7 lattice classes are either the cyclic groups $\mathbb{Z}_h$ with $h = 2, 3, 4$ or the dihedral groups Dih_k with $k = 3, 4, 6$ or the tetrahedral group T_{12} or the octahedral group O_{24}. Indeed the $n = 3$ crystallographic point groups are, by definition, finite subgroups of the rotation group, hence they must fall in the ADE-classification. Yet not every finite rotation group is crystallographic. For instance there is no lattice that is invariant under the icosahedral group and in general in a $n = 3$ point group there are no elements with orders different from $2, 3, 4, 6$.

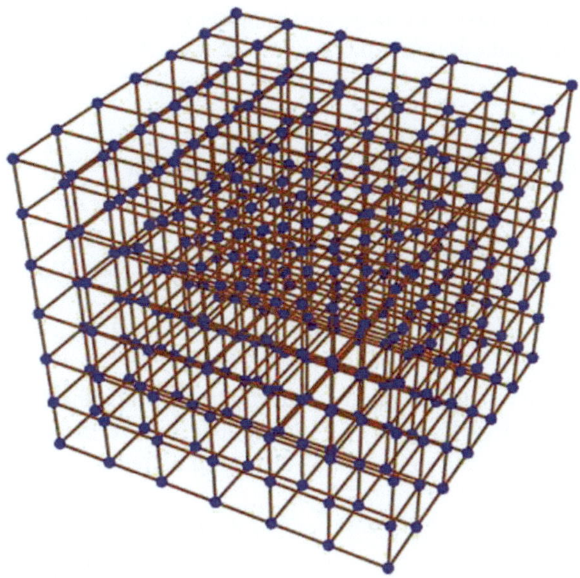

Fig. 1.3 A view of the self-dual cubic lattice

In this section, for the sake of illustration by means of a well structured example, we restrict our attention to the largest possible point group, namely that of the cubic lattice which has O_{24} symmetry.

1.2.4 *The Cubic Lattice and Its Point Group*

Let us now consider, within the general frame presented above the cubic lattice.

The cubic lattice is displayed in Fig. 1.3.

The basis vectors of the cubic lattice Λ_{cubic} are:

$$\mathbf{w}_1 = \{1, 0, 0\} \quad ; \quad \mathbf{w}_2 = \{0, 1, 0\} \quad ; \quad \mathbf{w}_3 = \{0, 0, 1\} \tag{1.2.13}$$

which implies that the metric is just the Kronecker delta:

$$g_{\mu\nu} = \delta_{\mu\nu} \tag{1.2.14}$$

and the basis vectors $\mathbf{e}^\mu$ of the dual lattice $\Lambda^\star_{cubic}$ coincide with those of the lattice Λ. Hence the cubic lattice is self-dual:

$$\mathbf{w}_\mu = \mathbf{e}^\mu \quad \Rightarrow \quad \Lambda_{cubic} = \Lambda^\star_{cubic} \tag{1.2.15}$$

The subgroup of the proper rotation group which maps the cubic lattice into itself is the octahedral group O whose order is 24.

1.2.5 *The Octahedral Group* $O_{24} \sim S_4$

Abstractly the octahedral Group $O_{24} \sim S_{24}$ is isomorphic to the symmetric group of permutations of 4 objects. It is defined by the following generators and relations:

$$\mathrm{A}, \mathrm{B} \quad : \quad \mathrm{A}^3 = \mathbf{e} \;\; ; \quad \mathrm{B}^2 = \mathbf{e} \;\; ; \quad (\mathrm{B}\,\mathrm{A})^4 = \mathbf{e} \tag{1.2.16}$$

Since O_{24} is a finite, discrete subgroup of the three-dimensional rotation group, any $\gamma \in O_{24} \subset SO(3)$ of its 24 elements can be uniquely identified by its action on the coordinates x, y, z, as it is displayed below:

e	$1_1 = \{x, y, z\}$		$4_1 = \{-x, -z, -y\}$	
	$2_1 = \{-y, -z, x\}$		$4_2 = \{-x, z, y\}$	
	$2_2 = \{-y, z, -x\}$	C_4	$4_3 = \{-y, -x, -z\}$	
	$2_3 = \{-z, -x, y\}$		$4_4 = \{-z, -y, -x\}$	
C_2	$2_4 = \{-z, x, -y\}$		$4_5 = \{z, -y, x\}$	
	$2_5 = \{z, -x, -y\}$		$4_6 = \{y, x, -z\}$	(1.2.17)
	$2_6 = \{z, x, y\}$		$5_1 = \{-y, x, z\}$	
	$2_7 = \{y, -z, -x\}$		$5_2 = \{-z, y, x\}$	
	$2_8 = \{y, z, x\}$	C_5	$5_3 = \{z, y, -x\}$	
	$3_1 = \{-x, -y, z\}$		$5_4 = \{y, -x, z\}$	
C_3	$3_2 = \{-x, y, -z\}$		$5_5 = \{x, -z, y\}$	
	$3_3 = \{x, -y, -z\}$		$5_6 = \{x, z, -y\}$	

As one sees from the above list the 24 elements are distributed into 5 conjugacy classes mentioned in the first column of the table. The relation between the abstract and concrete presentation of the octahedral group is obtained by identifying in the list (1.2.17) the generators A and B mentioned in Eq. (1.2.16). Explicitly we have:

$$\mathrm{A} = 2_8 = \begin{pmatrix} 0 & 1 & 0 \\ 0 & 0 & 1 \\ 1 & 0 & 0 \end{pmatrix} \;\; ; \quad \mathrm{B} = 4_6 = \begin{pmatrix} 0 & 1 & 0 \\ 1 & 0 & 0 \\ 0 & 0 & -1 \end{pmatrix} \tag{1.2.18}$$

All other elements are reconstructed from the above two using the multiplication table of the group which is displayed below:

	1_1	2_1	2_2	2_3	2_4	2_5	2_6	2_7	2_8	3_1	3_2	3_3	4_1	4_2	4_3	4_4	4_5	4_6	5_1	5_2	5_3	5_4	5_5	5_6
1_1	1_1	2_1	2_2	2_3	2_4	2_5	2_6	2_7	2_8	3_1	3_2	3_3	4_1	4_2	4_3	4_4	4_5	4_6	5_1	5_2	5_3	5_4	5_5	5_6
2_1	2_1	2_5	2_4	3_3	3_2	1_1	3_1	2_6	2_3	2_7	2_2	2_8	5_3	4_4	5_6	4_6	5_4	4_2	4_1	4_3	5_1	5_5	4_5	5_2
2_2	2_2	2_6	2_3	1_1	3_1	3_3	3_2	2_5	2_4	2_8	2_1	2_7	4_5	5_2	5_5	5_4	4_6	4_1	4_2	5_1	4_3	5_6	5_3	4_4
2_3	2_3	3_2	1_1	2_2	2_8	2_7	2_1	3_3	3_1	2_4	2_6	2_5	4_6	5_1	5_3	5_6	4_1	4_5	5_2	4_2	5_5	4_4	4_3	5_4
2_4	2_4	3_1	3_3	2_1	2_7	2_8	2_2	1_1	3_2	2_3	2_5	2_6	5_4	4_3	4_5	5_5	4_2	5_3	4_4	4_1	5_6	5_2	5_1	4_6
2_5	2_5	1_1	3_2	2_8	2_2	2_1	2_7	3_1	3_3	2_6	2_4	2_3	5_1	4_6	5_2	4_2	5_5	4_4	5_3	5_6	4_1	4_5	5_4	4_3
2_6	2_6	3_3	3_1	2_7	2_1	2_2	2_8	3_2	1_1	2_5	2_3	2_4	4_3	5_4	4_4	4_1	5_6	5_2	4_5	5_5	4_2	5_3	4_6	5_1
2_7	2_7	2_3	2_6	3_1	1_1	3_2	3_3	2_4	2_5	2_1	2_8	2_2	5_2	4_5	4_2	5_1	4_3	5_6	5_5	5_4	4_6	4_1	4_4	5_3
2_8	2_8	2_4	2_5	3_2	3_3	3_1	1_1	2_3	2_6	2_2	2_7	2_1	4_4	5_3	4_1	4_3	5_1	5_5	5_6	4_6	5_4	4_2	5_2	4_5
3_1	3_1	2_8	2_7	2_6	2_5	2_4	2_3	2_2	2_1	1_1	3_3	3_2	5_6	5_5	4_6	5_3	5_2	4_3	5_4	4_5	4_4	5_1	4_2	4_1
3_2	3_2	2_7	2_8	2_5	2_6	2_3	2_4	2_1	2_2	3_3	1_1	3_1	5_5	5_6	5_4	4_5	4_4	5_1	4_6	5_3	5_2	4_3	4_1	4_2
3_3	3_3	2_2	2_1	2_4	2_3	2_6	2_5	2_8	2_7	3_2	3_1	1_1	4_2	4_1	5_1	5_2	5_3	5_4	4_3	4_4	4_5	4_6	5_6	5_5
4_1	4_1	5_4	4_6	4_5	5_3	5_2	4_4	5_1	4_3	5_5	5_6	4_2	1_1	3_3	2_8	2_6	2_3	2_2	2_7	2_5	2_4	2_1	3_1	3_2
4_2	4_2	4_6	5_4	5_3	4_5	4_4	5_2	4_3	5_1	5_6	5_5	4_1	3_3	1_1	2_7	2_5	2_4	2_1	2_8	2_6	2_3	2_2	3_2	3_1
4_3	4_3	5_3	5_2	5_6	4_2	5_5	4_1	4_5	4_4	4_6	5_1	5_4	2_6	2_4	1_1	2_8	2_7	3_1	3_2	2_2	2_1	3_3	2_5	2_3
4_4	4_4	4_2	5_5	5_1	5_4	4_6	4_3	5_6	4_1	5_2	4_5	5_3	2_8	2_1	2_6	1_1	3_2	2_5	2_3	3_1	3_3	2_4	2_2	2_7
4_5	4_5	5_6	4_1	4_6	4_3	5_1	5_4	4_2	5_5	5_3	4_4	5_2	2_2	2_7	2_4	3_2	1_1	2_3	2_5	3_3	3_1	2_6	2_8	2_1
4_6	4_6	4_4	4_5	4_1	5_5	4_2	5_6	5_2	5_3	4_3	5_4	5_1	2_3	2_5	3_1	2_1	2_2	1_1	3_3	2_7	2_8	3_2	2_4	2_6
5_1	5_1	4_5	4_4	5_5	4_1	5_6	4_2	5_3	5_2	5_4	4_3	4_6	2_5	2_3	3_3	2_7	2_8	3_2	3_1	2_1	2_2	1_1	2_6	2_4
5_2	5_2	4_1	5_6	4_3	4_6	5_4	5_1	5_5	4_2	4_4	5_3	4_5	2_7	2_2	2_5	3_3	3_1	2_6	2_4	3_2	1_1	2_3	2_1	2_8
5_3	5_3	5_5	4_2	5_4	5_1	4_3	4_6	4_1	5_6	4_5	5_2	4_4	2_1	2_8	2_3	3_1	3_3	2_4	2_6	1_1	3_2	2_5	2_7	2_2
5_4	5_4	5_2	5_3	4_2	5_6	4_1	5_5	4_4	4_5	5_1	4_6	4_3	2_4	2_6	3_2	2_2	2_1	3_3	1_1	2_8	2_7	3_1	2_3	2_5
5_5	5_5	4_3	5_1	4_4	5_2	5_3	4_5	4_6	5_4	4_1	4_2	5_6	3_2	3_1	2_2	2_4	2_5	2_8	2_1	2_3	2_6	2_7	3_3	1_1
5_6	5_6	5_1	4_3	5_2	4_4	4_5	5_3	5_4	4_6	4_2	4_1	5_5	3_1	3_2	2_1	2_3	2_6	2_7	2_2	2_4	2_5	2_8	1_1	3_3

(1.2.19)

This observation is important in relation with representation theory. Any linear representation of the group is uniquely specified by giving the matrix representation of the two generators $\mathrm{A} = 2_8$ and $\mathrm{S} = 4_6$.

The Solvable Structure of O_{24}

The group O_{24} is solvable since there exists the following chain of normal subgroups:

$$\mathrm{O}_{24} \ \triangleright \ \mathrm{N}_{12} \ \triangleright \ \mathrm{N}_4 \tag{1.2.20}$$

where the mentioned subgroups are given by the following lists of elements:

$$\mathrm{N}_{12} \equiv \{1_1, 2_1, 2_2, \ldots, 2_8, 3_1, 3_2, 3_3\} \tag{1.2.21}$$

$$\mathrm{N}_4 \equiv \{1_1, 3_1, 3_2, 3_3\} \tag{1.2.22}$$

The group N_4 is abelian and we have:

$$\mathrm{N}_4 \ \sim \ \mathbb{Z}_2 \times \mathbb{Z}_2 \tag{1.2.23}$$

since all of its elements are of order two. This abstract structure allows for an a priori determination of all the irreducible representations, simply starting from the multiplication table. Yet because of the interpretation of O_{24} as made of proper rotations in three dimensions, its five irreps can also be constructed directly with some ingenuity. This is what we do in the next section.

1.2.6 *Irreducible Representations of the Octahedral Group*

There are five conjugacy classes in O_{24} and therefore according to theory there are five irreducible representations of the same group, that we name D_i, $i = 1, \dots, 5$. They have dimensions:

$$\dim D_1 = 1 \;;\; \dim D_2 = 1 \;;\; \dim D_3 = 2 \;;\; \dim D_4 = 3 \;;\; \dim D_5 = 4 \tag{1.2.24}$$

Let us briefly describe them.

1.2.6.1 D_1: The Identity Representation

The identity representation which exists for all groups is that one where to each element of O we associate the number 1

$$\forall \gamma \in O_{24} \;:\; D_1(\gamma) = 1 \tag{1.2.25}$$

Obviously the character of such a representation is[5]:

$$\chi_1 = \{1, 1, 1, 1, 1\} \tag{1.2.26}$$

1.2.6.2 D_2: The Quadratic Vandermonde Representation

The representation D_2 is also one-dimensional. It is constructed as follows. Consider the following polynomial of order six in the coordinates of a point in $\mathbb{R}^3$ or T^3:

$$\mathfrak{V}(x, y, z) = (x^2 - y^2)\,(x^2 - z^2)\,(y^2 - z^2) \tag{1.2.27}$$

As one can explicitly check under the transformations of the octahedral group listed in Eq. (1.2.17) the polynomial $\mathfrak{V}(x, y, z)$ is always mapped into itself modulo an overall sign. Keeping track of such a sign provides the form of the second one-dimensional representation whose character is explicitly calculated to be the following one:

$$\chi_1 = \{1, 1, 1, -1, -1\} \tag{1.2.28}$$

1.2.6.3 D_3: The Two-Dimensional Representation

The representation D_3 is two-dimensional and it corresponds to a homomorphism:

---

[5]Here as elsewhere we utilize the notion of group-characters for which we refer the reader to standard textbooks on finite group theory as [2].

$$D_3 \;:\quad \mathrm{O}_{24} \;\rightarrow\; \mathrm{SL}(2,\mathbb{Z}) \tag{1.2.29}$$

which associates to each element of the octahedral group a 2×2 integer valued matrix of determinant one. The homomorphism is completely specified by giving the two matrices representing the two generators:

$$D_3(\mathrm{A}) \;=\; \begin{pmatrix} 0 & 1 \\ -1 & -1 \end{pmatrix} \quad ; \quad D_3(\mathrm{B}) \;=\; \begin{pmatrix} 0 & 1 \\ 1 & 0 \end{pmatrix} \tag{1.2.30}$$

The character vector of D_2 is easily calculated from the above information and we have:

$$\chi_3 \;=\; \{2, -1, 2, 0, 0\} \tag{1.2.31}$$

1.2.6.4 D_4: The Three-Dimensional Defining Representation

The three dimensional representation D_4 is simply the defining representation, where the generators A and B are given by the matrices in Eq. (1.2.18).

$$D_4(\mathrm{A}) \;=\; \mathrm{A} \quad ; \quad D_4(\mathrm{B}) \;=\; \mathrm{B} \tag{1.2.32}$$

From this information the characters are immediately calculated and we get:

$$\chi_3 \;=\; \{3, 0, -1, -1, 1\} \tag{1.2.33}$$

1.2.6.5 D_5: The Three-Dimensional Unoriented Representation

The three dimensional representation D_5 is simply that where the generators A and B are given by the following matrices:

$$D_5(\mathrm{A}) \;=\; \begin{pmatrix} 0 & 1 & 0 \\ 0 & 0 & 1 \\ 1 & 0 & 0 \end{pmatrix} \quad ; \quad D_5(\mathrm{B}) \;=\; \begin{pmatrix} 0 & 1 & 0 \\ 1 & 0 & 0 \\ 0 & 0 & 1 \end{pmatrix} \tag{1.2.34}$$

From this information the characters are immediately calculated and we get:

$$\chi_5 \;=\; \{3, 0, -1, 1, -1\} \tag{1.2.35}$$

The table of characters is summarized in Table 1.2.

Table 1.2 Character table of the proper octahedral group

Irrep	Class	$\{\mathbf{e},1\}$	$\{C_2,8\}$	$\{C_3,3\}$	$\{C_4,6\}$	$\{C_5,6\}$
D_1,	$\chi_1 =$	1	1	1	1	1
D_2,	$\chi_2 =$	1	1	1	-1	-1
D_3,	$\chi_3 =$	2	-1	2	0	0
D_4,	$\chi_4 =$	3	0	-1	-1	1
D_5,	$\chi_5 =$	3	0	-1	1	-1

1.3 A Simple Crystallographic Point-Group in 7-Dimensions

In the previous section we analyzed the possible crystallographic point groups in our familiar three-dimensional Euclidean space.

Summarizing our discussion we point out some group–theoretical features that follow from the ADE classification, combined with the further compatibility constraints which emerge when you impose the crystallographic condition that a lattice should be left invariant by the action of the Point Group:

(a) The Point Group $\mathfrak{P}$ must be a finite rotation group in $d = 3$ hence it must belong to the list:

$$\mathfrak{P} \in \{\mathbb{Z}_k\,,\,\mathrm{Dih}_k\,,\,\mathrm{T}_{12}\,,\,\mathrm{O}_{24}\,,\,\mathrm{I}_{60}\} \tag{1.3.1}$$

(b) The order of any element $\gamma \in \mathfrak{P}$ belonging to the Point Group must be in the range 2, 3, 4, 6

The intersection of these two conditions leads to the conclusion that:

$$\mathfrak{P} \in \left\{\mathbb{Z}_{2,3,4,6}\,,\,\mathrm{Dih}_{3,4,6}\,,\,\mathrm{T}_{12}\,,\,\mathrm{O}_{24}\right\} \tag{1.3.2}$$

The classification of Bravais lattices, which is responsible for so many chemical-physical properties of matter, is essentially encoded in Eq. (1.3.2). In this list of candidate Point Groups there is no simple one which is non abelian. They are all either solvable or abelian and this implies that their irreducible representations can be constructed by means of an induction algorithm starting from the one-dimensional irreps of their largest normal abelian subgroup. A simple group which occurs in the ADE classification is the icosahedral group I_{60} which is isomorphic to the simple alternating group A_5 (the even permutations of 5 objects). It is barred out by the crystallographic condition because it contains elements of order 5.

Under many respects this is the analogue of what happens with algebraic equations. The algebraic equations of order 2, 3, 4 are always solvable by radicals since their Galois group is solvable. In degree $d \geq 5$ the generic equation is not solvable because the Galois group is generically not solvable.

A natural question arises at this point. Is the condition b) on the possible orders of the Point Group elements intrinsic to the crystallographic constraint in any dimension or it is a specific feature of $d = 3$?

The correct answer to the above question is the second option and in this section we show a counterexample of a crystallographic group in 7-dimensions that has group elements of order 7. Not only that. Ours is an example of a simple non abelian crystallographic Point Group!

It is quite remarkable that the analogue of the ADE classification of finite rotation groups in $d > 5$ is so far non existing up to the knowledge of this author. Even less is known about higher dimensional crystallographic groups.

It is philosophically quite challenging to imagine what Chemistry, Geology and even Molecular Biology and Genetics might be in a world where the Point Group is a simple non abelian group!

1.3.1 The Simple Group L_{168}

The finite group:

$$\mathrm{L}_{168} \equiv \mathrm{PSL}(2, \mathbb{Z}_7) \tag{1.3.3}$$

is the second smallest simple group after the alternating group A_5 which has 60 elements and coincides with the symmetry group of the regular icosahedron or dodecahedron. As anticipated by its given name, L_{168} has 168 elements: they can be identified with all the possible 2×2 matrices with determinant one whose entries belong to the finite field $\mathbb{Z}_7$, counting them up to an overall sign. In projective geometry, L_{168} is classified as a *Hurwitz group* since it is the automorphism group of a Hurwitz Riemann surface, namely a surface of genus g with the maximal number $84\,(g-1)$ of conformal automorphisms.[6] The Hurwitz surface pertaining to the Hurwitz group L_{168} is the Klein quartic [4], namely the locus $\mathscr{K}_4$ in $\mathbb{P}_2(\mathbb{C})$ cut out by the following quartic polynomial constraint on the homogeneous coordinates $\{x, y, z\}$:

$$x^3\, y + y^3\, z + z^3\, x = 0 \tag{1.3.4}$$

Indeed $\mathscr{K}_4$ is a genus $g = 3$ compact Riemann surface and it can be realized as the quotient of the hyperbolic Poincaré plane $\mathbb{H}_2$ by a certain group Γ that acts freely on $\mathbb{H}_2$ by isometries.

The L_{168} group, which is also isomorphic to $\mathrm{GL}(3, \mathbb{Z}_2)$, has received a lot of attention in Mathematics and it has important applications in algebra, geometry, and number theory: for instance, besides being associated with the Klein quartic, L_{168} is the automorphism group of the Fano plane [5].

[6] Hurwitz's automorphisms theorem proved in 1893 [3] states that the order $|\mathscr{G}|$ of the group $\mathscr{G}$ of orientation-preserving conformal automorphisms, of a compact Riemann surface of genus $g > 1$ admits the following upper bound $|\mathscr{G}| \leq 84(g-1)$.

The reason why we consider L_{168} in this section is associated with another property of this finite simple group which was proved fifteen years ago in [6], namely:

$$\mathrm{L}_{168} \subset \mathfrak{g}_{2(-14)} \tag{1.3.5}$$

This means that L_{168} is a finite subgroup of the compact form of the exceptional Lie group $\mathfrak{g}_2$ and the 7-dimensional fundamental representation of the latter is irreducible upon restriction to L_{168}.

The key reason to consider L_{168} in this section is that it happens to be crystallographic in $d = 7$, the preserved lattice being the root lattice of either the simple Lie algebra $\mathfrak{a}_7$ or, even more inspiringly, of the exceptional Lie algebra $\mathfrak{e}_7$. Hence L_{168} is a subgroup of the $\mathfrak{e}_7$ Weyl group. Because of the role of $\mathfrak{e}_7$ in supergravity related special geometries we will come back to it in the sequel. Here we are interested in its properties in order to illustrate the case of a **simple crystallographic non abelian group**.

1.3.2 *Structure of the Simple Group* $\mathbf{L}_{168} = \mathbf{PSL}(2, \mathbb{Z}_7)$

For the reasons outlined above we consider the simple group (1.3.3) and its crystallographic action in $d = 7$. The Hurwitz simple group L_{168} is abstractly presented as follows[7]:

$$\mathrm{L}_{168} = \left(R, S, T \;\|\; R^2 = S^3 = T^7 = RST = (TSR)^4 = \mathbf{e}\right) \tag{1.3.6}$$

and, as its name implicitly advocates, it has order 168:

$$|\,\mathrm{L}_{168}\,| = 168 \tag{1.3.7}$$

The elements of this simple group are organized in six conjugacy classes according to the scheme displayed below:

Conjugacy class	$\mathscr{C}_1$	$\mathscr{C}_2$	$\mathscr{C}_3$	$\mathscr{C}_4$	$\mathscr{C}_5$	$\mathscr{C}_6$
representative of the class	$\mathbf{e}$	R	S	TSR	T	SR
order of the elements in the class	1	2	3	4	7	7
number of elements in the class	1	21	56	42	24	24

(1.3.8)

As one sees from the above table (1.3.8) the group contains elements of order 2, 3, 4 and 7 and there are two inequivalent conjugacy classes of elements of the highest order. According to the general theory of finite groups, there are 6 different irreducible representations of dimensions 1, 6, 7, 8, 3, 3, respectively. The character table of the

[7]In the rest of this section we follow closely the results obtained by the present author in a recent paper [7].

group L_{168} can be found in the mathematical literature, for instance in the book [2]. It reads as follows:

Representation	$\mathscr{C}_1$	$\mathscr{C}_2$	$\mathscr{C}_3$	$\mathscr{C}_4$	$\mathscr{C}_5$	$\mathscr{C}_6$
$D_1\,[L_{168}]$	1	1	1	1	1	1
$D_6\,[L_{168}]$	6	2	0	0	-1	-1
$D_7\,[L_{168}]$	7	-1	1	-1	0	0
$D_8\,[L_{168}]$	8	0	-1	0	1	1
$DA_3\,[L_{168}]$	3	-1	0	1	$\frac{1}{2}\left(-1+i\sqrt{7}\right)$	$\frac{1}{2}\left(-1-i\sqrt{7}\right)$
$DB_3\,[L_{168}]$	3	-1	0	1	$\frac{1}{2}\left(-1-i\sqrt{7}\right)$	$\frac{1}{2}\left(-1+i\sqrt{7}\right)$

(1.3.9)

Soon we will retrieve it by constructing explicitly all the irreducible representations.

1.3.3 The 7-Dimensional Irreducible Representation

For our purposes the most interesting representations are the real 7 dimensional and the complex three dimensional ones. The properties of these irreps are the very reason to consider the group L_{168} in the present context.

As for the 7-dimensional irrep the following three statements are true:

1. The 7-dimensional irreducible representation is crystallographic since all elements $\gamma \in L_{168}$ are represented by integer valued matrices $D_7(\gamma)$ in a basis of vectors that span a lattice, namely the root lattice Λ_{root} of the $\mathfrak{a}_7$ simple Lie algebra.
2. The 7-dimensional irreducible representation provides an immersion $L_{168} \hookrightarrow SO(7)$ since its elements preserve the symmetric Cartan matrix of $\mathfrak{a}_7$:

$$\forall\gamma \in L_{168} \quad : \quad D_7^T(\gamma)\,\mathscr{C}\,D_7(\gamma) = \mathscr{C}$$
$$\mathscr{C}_{i,j} = \alpha_i \cdot \alpha_j \qquad (i, j = 1 \ldots, 7) \tag{1.3.10}$$

defined in terms of the simple roots α_i whose standard construction in terms of the unit vectors ε_i of $\mathbb{R}^8$ is recalled below[8]:

$$\begin{aligned} &\alpha_1 = \varepsilon_1 - \varepsilon_2 \ ; \ \alpha_2 = \varepsilon_2 - \varepsilon_3 = \ ; \ \alpha_3 = \varepsilon_3 - \varepsilon_4 \\ &\alpha_4 = \varepsilon_4 - \varepsilon_5 \ ; \ \alpha_5 = \varepsilon_5 - \varepsilon_6 = \ ; \ \alpha_6 = \varepsilon_6 - \varepsilon_7 \\ &\alpha_7 = \varepsilon_7 - \varepsilon_8 \end{aligned} \tag{1.3.11}$$

[8]We refer the reader to Sect. 1.5 for the explicit form of the Cartan matrices associated with $\mathfrak{a}_\ell$ algebras.

3. Actually the 7-dimensional representation defines an embedding $\mathrm{L}_{168} \hookrightarrow \mathfrak{g}_2 \subset \mathrm{SO}(7)$ since there exists a three-index antisymmetric tensor ϕ_{ijk} satisfying the relations of the octonionic structure constants[9] that is preserved by all the matrices $D_7(\gamma)$:

$$\forall \gamma \in \mathrm{L}_{168} \quad : \quad D_7(\gamma)_{ii'}\, D_7(\gamma)_{jj'}\; D_7(\gamma)_{kk'}\, \phi_{i'j'k'} \;=\; \phi_{ijk} \tag{1.3.12}$$

Let us prove the above statements. It suffices to write the explicit form of the generators R, S and T in the crystallographic basis of the considered root lattice:

$$\mathbf{v} \in \Lambda_{\text{root}} \quad \Leftrightarrow \quad \mathbf{v} = n_i\, \alpha_i \quad n_i \in \mathbb{Z} \tag{1.3.13}$$

Explicitly if we set:

$$\begin{gathered}
\mathscr{R} = \begin{pmatrix} 0 & 0 & 0 & 0 & 0 & 0 & -1 \\ 0 & 0 & 0 & 0 & 0 & -1 & 0 \\ 0 & 0 & -1 & 1 & 0 & -1 & 0 \\ 0 & -1 & 0 & 1 & 0 & -1 & 0 \\ 0 & -1 & 0 & 1 & -1 & 0 & 0 \\ 0 & -1 & 0 & 0 & 0 & 0 & 0 \\ -1 & 0 & 0 & 0 & 0 & 0 & 0 \end{pmatrix} \; ; \; \mathscr{S} = \begin{pmatrix} 0 & 0 & 0 & 0 & 0 & 0 & -1 \\ 1 & 0 & 0 & 0 & 0 & 0 & -1 \\ 1 & 0 & 0 & -1 & 1 & 0 & -1 \\ 1 & 0 & -1 & 0 & 1 & 0 & -1 \\ 1 & 0 & -1 & 0 & 1 & -1 & 0 \\ 1 & 0 & -1 & 0 & 0 & 0 & 0 \\ 1 & -1 & 0 & 0 & 0 & 0 & 0 \end{pmatrix} \\
\mathscr{T} = \begin{pmatrix} 0 & 0 & 0 & 0 & 0 & -1 & 1 \\ 1 & 0 & 0 & 0 & 0 & -1 & 1 \\ 0 & 1 & 0 & 0 & 0 & -1 & 1 \\ 0 & 0 & 1 & 0 & 0 & -1 & 1 \\ 0 & 0 & 0 & 1 & 0 & -1 & 1 \\ 0 & 0 & 0 & 0 & 1 & -1 & 1 \\ 0 & 0 & 0 & 0 & 0 & 0 & 1 \end{pmatrix}
\end{gathered} \tag{1.3.14}$$

we find that the defining relations of L_{168} are satisfied:

$$\mathscr{R}^2 \;=\; \mathscr{S}^3 \;=\; \mathscr{T}^7 \;=\; \mathscr{R}\mathscr{S}\mathscr{T} \;=\; (\mathscr{T}\mathscr{S}\mathscr{R})^4 \;=\; \mathbf{1}_{7\times 7} \tag{1.3.15}$$

and furthermore we have:

$$\mathscr{R}^{\mathrm{T}}\mathscr{C}\mathscr{R} \;=\; \mathscr{S}^{\mathrm{T}}\mathscr{C}\mathscr{S} \;=\; \mathscr{T}^{\mathrm{T}}\mathscr{C}\mathscr{T} \;=\mathscr{C} \tag{1.3.16}$$

where the explicit form of the $\mathfrak{a}_7$ Cartan matrix is recalled below:

[9] For the history of quaternions and octonions I refer the reader to my book [8].

$$\mathscr{C} = \begin{pmatrix} 2 & -1 & 0 & 0 & 0 & 0 & 0 \\ -1 & 2 & -1 & 0 & 0 & 0 & 0 \\ 0 & -1 & 2 & -1 & 0 & 0 & 0 \\ 0 & 0 & -1 & 2 & -1 & 0 & 0 \\ 0 & 0 & 0 & -1 & 2 & -1 & 0 \\ 0 & 0 & 0 & 0 & -1 & 2 & -1 \\ 0 & 0 & 0 & 0 & 0 & -1 & 2 \end{pmatrix} \tag{1.3.17}$$

This proves statements (1) and (2).

In order to prove statement (3) we proceed as follows. In $\mathbb{R}^7$ we consider the antisymmetric three-index tensor ϕ_{ABC} that, in the standard orthonormal basis, has the following components:

$$\begin{aligned} \phi_{1,2,6} &= \tfrac{1}{6} \\ \phi_{1,3,4} &= -\tfrac{1}{6} \\ \phi_{1,5,7} &= -\tfrac{1}{6} \\ \phi_{2,3,7} &= \tfrac{1}{6} \quad ; \text{ all other components vanish} \\ \phi_{2,4,5} &= \tfrac{1}{6} \\ \phi_{3,5,6} &= -\tfrac{1}{6} \\ \phi_{4,6,7} &= -\tfrac{1}{6} \end{aligned} \tag{1.3.18}$$

This tensor satisfies the algebraic relations of octonionic structure constants, namely [10]:

$$\phi_{ABM}\,\phi_{CDM} = \frac{1}{18}\delta^{AB}_{CD} + \frac{2}{3}\Phi_{ABCD} \tag{1.3.19}$$

$$\phi_{ABC} = -\frac{1}{6}\varepsilon_{ABCPQRS}\,\Phi_{ABCD} \tag{1.3.20}$$

and the subgroup of SO(7) which leaves ϕ_{ABC} invariant is, by definition, the compact section $\mathfrak{g}_{(2,-14)}$ of the complex $\mathfrak{g}_2$ Lie group (see for instance [9]). A particular matrix that transforms the standard orthonormal basis of $\mathbb{R}^7$ into the basis of simple roots α_i is the following one:

$$\mathfrak{M} = \begin{pmatrix} \sqrt{2} & -\frac{1}{\sqrt{2}} & 0 & 0 & 0 & 0 & 0 \\ 0 & -\frac{1}{\sqrt{2}} & \sqrt{2} & -\frac{1}{\sqrt{2}} & 0 & 0 & 0 \\ 0 & 0 & 0 & -\frac{1}{\sqrt{2}} & \sqrt{2} & -\frac{1}{\sqrt{2}} & 0 \\ 0 & 0 & 0 & 0 & 0 & -\frac{1}{\sqrt{2}} & \sqrt{2} \\ 0 & -\frac{1}{\sqrt{2}} & 0 & \frac{1}{\sqrt{2}} & 0 & -\frac{1}{\sqrt{2}} & 0 \\ 0 & 0 & 0 & -\frac{1}{\sqrt{2}} & 0 & 0 & 0 \\ 0 & \frac{1}{\sqrt{2}} & 0 & 0 & 0 & -\frac{1}{\sqrt{2}} & 0 \end{pmatrix} \tag{1.3.21}$$

[10] In this equation the indices of the $\mathfrak{g}_2$-invariant tensor are denoted with capital letter of the Latin alphabet, as it was the case in the quoted literature on weak $\mathfrak{g}_2$-structures. In the following we will use lower case latin letters, the upper Latin letters being reserved for $d = 8$.

since:

$$\mathfrak{M}^T \, \mathfrak{M} = \mathscr{C} \tag{1.3.22}$$

Defining the transformed tensor:

$$\varphi_{ijk} \equiv \left(\mathfrak{M}^{-1}\right)_i^{\ I} \left(\mathfrak{M}^{-1}\right)_j^{\ J} \left(\mathfrak{M}^{-1}\right)_k^{\ K} \phi_{IJK} \tag{1.3.23}$$

we can explicitly verify that:

$$\begin{aligned}
\varphi_{ijk} &= (\mathscr{R})_i^{\ p} \, (\mathscr{R})_j^{\ q} \, (\mathscr{R})_k^{\ r} \, \varphi_{pqr} \\
\varphi_{ijk} &= (\mathscr{S})_i^{\ p} \, (\mathscr{S})_j^{\ q} \, (\mathscr{S})_k^{\ r} \, \varphi_{pqr} \\
\varphi_{ijk} &= (\mathscr{T})_i^{\ p} \, (\mathscr{T})_j^{\ q} \, (\mathscr{T})_k^{\ r} \, \varphi_{pqr}
\end{aligned} \tag{1.3.24}$$

Hence, being preserved by the three-generators $\mathscr{R}$,$\mathscr{S}$ and $\mathscr{T}$, the antisymmetric tensor φ_{ijk} is preserved by the entire discrete group L_{168} which, henceforth, is a subgroup of $\mathfrak{g}_{(2,-14)} \subset \mathrm{SO}(7)$, as it was shown by intrinsic group theoretical arguments in [6]. The other representations of the group L_{168} were explicitly constructed about ten years ago by Pierre Ramond and his younger collaborators in [10]. They are completely specified by giving the matrix form of the three generators R, S, T satisfying the defining relations (1.3.6).

1.3.4 The 3-Dimensional Complex Representations

The two three dimensional irreducible representations are complex and they are conjugate to each other. It suffices to give the form of the generators for one of them. The generators of the conjugate representation are the complex conjugates of the same matrices.

Setting:

$$\rho \equiv e^{\frac{2i\pi}{7}} \tag{1.3.25}$$

we have the following form for the representation **3**:

$$D[R]_3 = \begin{pmatrix} \frac{i(\rho^2-\rho^5)}{\sqrt{7}} & \frac{i(\rho-\rho^6)}{\sqrt{7}} & \frac{i(\rho^4-\rho^3)}{\sqrt{7}} \\ \frac{i(\rho-\rho^6)}{\sqrt{7}} & \frac{i(\rho^4-\rho^3)}{\sqrt{7}} & \frac{i(\rho^2-\rho^5)}{\sqrt{7}} \\ \frac{i(\rho^4-\rho^3)}{\sqrt{7}} & \frac{i(\rho^2-\rho^5)}{\sqrt{7}} & \frac{i(\rho-\rho^6)}{\sqrt{7}} \end{pmatrix}$$

$$D[S]_3 = \begin{pmatrix} \frac{i(\rho^3-\rho^6)}{\sqrt{7}} & \frac{i(\rho^3-\rho)}{\sqrt{7}} & \frac{i(\rho-1)}{\sqrt{7}} \\ \frac{i(\rho^2-1)}{\sqrt{7}} & \frac{i(\rho^6-\rho^5)}{\sqrt{7}} & \frac{i(\rho^6-\rho^2)}{\sqrt{7}} \\ \frac{i(\rho^5-\rho^4)}{\sqrt{7}} & \frac{i(\rho^4-1)}{\sqrt{7}} & \frac{i(\rho^5-\rho^3)}{\sqrt{7}} \end{pmatrix}$$

$$D[T]_3 = \begin{pmatrix} -ie^{\frac{3i\pi}{14}} & 0 & 0 \\ 0 & -ie^{-\frac{i\pi}{14}} & 0 \\ 0 & 0 & -e^{-\frac{i\pi}{7}} \end{pmatrix} \tag{1.3.26}$$

1.3.5 The 6-Dimensional Representation

Introducing the following short-hand notation:

$$\begin{aligned} c_n &= \cos\left[\frac{2\pi}{7} n\right] \\ s_n &= \sin\left[\frac{2\pi}{7} n\right] \end{aligned} \tag{1.3.27}$$

The generators of the group L_{168} in the 6-dimensional irreducible representation can be explicitly written as it is displayed below:

$$D[R]_6 = \left(\begin{array}{c|c|c|c|c|c} \frac{c_3-1}{\sqrt{2}} & \frac{c_2-1}{\sqrt{2}} & \frac{c_1-1}{\sqrt{2}} & c_3-c_1 & c_1-c_2 & c_2-c_3 \\ \hline \frac{c_2-1}{\sqrt{2}} & \frac{c_1-1}{\sqrt{2}} & \frac{c_3-1}{\sqrt{2}} & c_2-c_3 & c_3-c_1 & c_1-c_2 \\ \hline \frac{c_1-1}{\sqrt{2}} & \frac{c_3-1}{\sqrt{2}} & \frac{c_2-1}{\sqrt{2}} & c_1-c_2 & c_2-c_3 & c_3-c_1 \\ \hline c_3-c_1 & c_2-c_3 & c_1-c_2 & \frac{c_1-1}{\sqrt{2}} & \frac{c_2-1}{\sqrt{2}} & \frac{c_3-1}{\sqrt{2}} \\ \hline c_1-c_2 & c_3-c_1 & c_2-c_3 & \frac{c_2-1}{\sqrt{2}} & \frac{c_3-1}{\sqrt{2}} & \frac{c_1-1}{\sqrt{2}} \\ \hline c_2-c_3 & c_1-c_2 & c_3-c_1 & \frac{c_3-1}{\sqrt{2}} & \frac{c_1-1}{\sqrt{2}} & \frac{c_2-1}{\sqrt{2}} \end{array}\right)$$

$$D[S]_6 = \left(\begin{array}{c|c|c|c|c|c} \frac{(c_3-1)\rho^2}{\sqrt{2}} & \frac{(c_2-1)\rho^4}{\sqrt{2}} & \frac{(c_1-1)\rho}{\sqrt{2}} & (c_3-c_1)\rho^3 & (c_1-c_2)\rho^5 & (c_2-c_3)\rho^6 \\ \hline \frac{(c_2-1)\rho^2}{\sqrt{2}} & \frac{(c_1-1)\rho^4}{\sqrt{2}} & \frac{(c_3-1)\rho}{\sqrt{2}} & (c_2-c_3)\rho^3 & (c_3-c_1)\rho^5 & (c_1-c_2)\rho^6 \\ \hline \frac{(c_1-1)\rho^2}{\sqrt{2}} & \frac{(c_3-1)\rho^4}{\sqrt{2}} & \frac{(c_2-1)\rho}{\sqrt{2}} & (c_1-c_2)\rho^3 & (c_2-c_3)\rho^5 & (c_3-c_1)\rho^6 \\ \hline (c_3-c_1)\rho^2 & (c_2-c_3)\rho^4 & (c_1-c_2)\rho & \frac{(c_1-1)\rho^3}{\sqrt{2}} & \frac{(c_2-1)\rho^5}{\sqrt{2}} & \frac{(c_3-1)\rho^6}{\sqrt{2}} \\ \hline (c_1-c_2)\rho^2 & (c_3-c_1)\rho^4 & (c_2-c_3)\rho & \frac{(c_2-1)\rho^3}{\sqrt{2}} & \frac{(c_3-1)\rho^5}{\sqrt{2}} & \frac{(c_1-1)\rho^6}{\sqrt{2}} \\ \hline (c_2-c_3)\rho^2 & (c_1-c_2)\rho^4 & (c_3-c_1)\rho & \frac{(c_3-1)\rho^3}{\sqrt{2}} & \frac{(c_1-1)\rho^5}{\sqrt{2}} & \frac{(c_2-1)\rho^6}{\sqrt{2}} \end{array}\right)$$

$$D[T]_6 = (D[R]_6 \cdot D[S]_6)^{-1} \tag{1.3.28}$$

1.3.6 The 8-Dimensional Representation

Utilizing the same notations as before we can write the matrix form of the generators also in the irreducible 8-dimensional representation.

$$
D[R]_8 = \begin{pmatrix}
2-2c_1 & 0 & 2c_1+2c_2-4c_3 & 2-2c_2 & 0 & 2-2c_3 & 0 & 2\sqrt{3}c_1-2\sqrt{3}c_2 \\
0 & -2c_1+4c_2-2 & 0 & 0 & 2c_2-4c_3+2 & 0 & 4c_1-2c_3-2 & 0 \\
2c_1+2c_2-4c_3 & 0 & -c_1+2c_2-c_3 & -4c_1+2c_2+2c_3 & 0 & 2c_1-4c_2+2c_3 & 0 & \sqrt{3}c_1-\sqrt{3}c_3 \\
2-2c_2 & 0 & -4c_1+2c_2+2c_3 & 2-2c_3 & 0 & 2-2c_1 & 0 & 2\sqrt{3}c_2-2\sqrt{3}c_3 \\
0 & 2c_2-4c_3+2 & 0 & 0 & 4c_1-2c_3-2 & 0 & 2c_1-4c_2+2 & 0 \\
2-2c_3 & 0 & 2c_1-4c_2+2c_3 & 2-2c_1 & 0 & 2-2c_2 & 0 & 2\sqrt{3}c_3-2\sqrt{3}c_1 \\
0 & 4c_1-2c_3-2 & 0 & 0 & 2c_1-4c_2+2 & 0 & -2c_2+4c_3-2 & 0 \\
2\sqrt{3}c_1-2\sqrt{3}c_2 & 0 & \sqrt{3}c_1-\sqrt{3}c_3 & 2\sqrt{3}c_2-2\sqrt{3}c_3 & 0 & 2\sqrt{3}c_3-2\sqrt{3}c_1 & 0 & c_1-2c_2+c_3
\end{pmatrix}
$$

$$
D[S]_8 = \begin{pmatrix}
c_1 & s_1 & 0 & 0 & 0 & 0 & 0 & 0 \\
-s_1 & c_1 & 0 & 0 & 0 & 0 & 0 & 0 \\
0 & 0 & 1 & 0 & 0 & 0 & 0 & 0 \\
0 & 0 & 0 & c_3 & s_3 & 0 & 0 & 0 \\
0 & 0 & 0 & -s_3 & c_3 & 0 & 0 & 0 \\
0 & 0 & 0 & 0 & 0 & c_2 & s_2 & 0 \\
0 & 0 & 0 & 0 & 0 & -s_2 & c_2 & 0 \\
0 & 0 & 0 & 0 & 0 & 0 & 0 & 1
\end{pmatrix}
$$

$$
D[T]_8 = (D[R]_8 \cdot D[S]_8)^{-1} \tag{1.3.29}
$$

1.3.7 The Proper Subgroups of L_{168}

From the complexity of the other irreps, in relation with the simplicity of the 7-dimensional one, it is already clear that this latter should be considered the natural defining representation. The crystallographic nature of the group in $d = 7$ has already been stressed and we will have more to say about it in Chap. 7. Next we introduce the

$\mathfrak{a}_7$ weight lattice which, by definition, is just the dual of the root lattice. Explicitly

$$\Lambda_w \ni \mathbf{w} = n_i \lambda^i \quad : \quad n^i \in \mathbb{Z} \tag{1.3.30}$$

is spanned by the simple weights that are implicitly defined by the relations:

$$\lambda^i \cdot \alpha_j = \delta^i_j \quad \Rightarrow \quad \lambda^i = \left(\mathscr{C}^{-1}\right)^{ij} \alpha_j \tag{1.3.31}$$

Since the group L_{168} is crystallographic on the root lattice, by necessity it is crystallographic also on the weight lattice. Given the generators of the group L_{168} in the basis of simple roots we obtain the same in the basis of simple weights through the following transformation:

$$\mathscr{R}_w = \mathscr{C}\,\mathscr{R}\,\mathscr{C}^{-1} \quad ; \quad \mathscr{S}_w = \mathscr{C}\,\mathscr{S}\,\mathscr{C}^{-1} \quad ; \quad \mathscr{T}_w = \mathscr{C}\,\mathscr{T}\,\mathscr{C}^{-1} \tag{1.3.32}$$

Explicitly we find:

$$\mathscr{R}_W = \begin{pmatrix} 0 & 0 & 0 & 0 & 0 & 0 & -1 \\ 0 & 0 & 0 & -1 & -1 & -1 & 0 \\ 0 & 0 & -1 & 0 & 0 & 0 & 0 \\ 0 & 0 & 1 & 1 & 1 & 0 & 0 \\ 0 & 0 & 0 & 0 & -1 & 0 & 0 \\ 0 & -1 & -1 & -1 & 0 & 0 & 0 \\ -1 & 0 & 0 & 0 & 0 & 0 & 0 \end{pmatrix} \quad ; \quad \mathscr{S}_W = \begin{pmatrix} -1 & -1 & -1 & -1 & -1 & -1 & -1 \\ 1 & 1 & 1 & 1 & 0 & 0 & 0 \\ 0 & 0 & 0 & -1 & 0 & 0 & 0 \\ 0 & 0 & 0 & 1 & 1 & 1 & 0 \\ 0 & 0 & 0 & 0 & 0 & -1 & 0 \\ 0 & 0 & -1 & -1 & -1 & 0 & 0 \\ 0 & -1 & 0 & 0 & 0 & 0 & 0 \end{pmatrix} \tag{1.3.33}$$

$$\mathscr{T}_w = \begin{pmatrix} -1 & -1 & -1 & -1 & -1 & -1 & 0 \\ 1 & 0 & 0 & 0 & 0 & 0 & 0 \\ 0 & 1 & 0 & 0 & 0 & 0 & 0 \\ 0 & 0 & 1 & 0 & 0 & 0 & 0 \\ 0 & 0 & 0 & 1 & 0 & 0 & 0 \\ 0 & 0 & 0 & 0 & 1 & 0 & 0 \\ 0 & 0 & 0 & 0 & 0 & 1 & 1 \end{pmatrix} \tag{1.3.34}$$

Given the weight basis, which is useful in several constructions, let us conclude our survey of the remarkable simple group L_{168} by a brief discussion of its subgroups, none of which, obviously, is normal.

L_{168} contains maximal subgroups only of index 8 and 7, namely of order 21 and 24. The order 21 subgroup G_{21} is the unique non-abelian group of that order and abstractly it has the structure of the semidirect product $\mathbb{Z}_3 \ltimes \mathbb{Z}_7$. Up to conjugation there is only one subgroup G_{21} as we have explicitly verified with the computer. On the other hand, up to conjugation, there are two different groups of order 24 that are both isomorphic to the octahedral group O_{24}.

1.3.7.1 The Maximal Subgroup G_{21}

The group G_{21} has two generators $\mathscr{X}$ and $\mathscr{Y}$ that satisfy the following relations:

$$\mathscr{X}^3 = \mathscr{Y}^7 = \mathbf{1} \quad ; \quad \mathscr{X}\mathscr{Y} = \mathscr{Y}^2\mathscr{X} \tag{1.3.35}$$

The organization of the 21 group elements into conjugacy classes is displayed below:

ConjugacyClass	C_1	C_2	C_3	C_4	C_5
representative of the class	e	$\mathscr{Y}$	$\mathscr{X}^2\mathscr{Y}\mathscr{X}\mathscr{Y}^2$	$\mathscr{Y}\mathscr{X}^2$	$\mathscr{X}$
order of the elements in the class	1	7	7	3	3
number of elements in the class	1	3	3	7	7

(1.3.36)

As we see there are five conjugacy classes which implies that there should be five irreducible representations the square of whose dimensions should sum up to the group order 21. The solution of this problem is:

$$21 = 1^2 + 1^2 + 1^2 + 3^2 + 3^2 \tag{1.3.37}$$

and the corresponding character table is mentioned below:

0	e	$\mathscr{Y}$	$\mathscr{X}^2\mathscr{Y}\mathscr{X}\mathscr{Y}^2$	$\mathscr{Y}\mathscr{X}^2$	$\mathscr{X}$
$D_1\,[G_{21}]$	1	1	1	1	1
$DX_1\,[G_{21}]$	1	1	1	$-(-1)^{1/3}$	$(-1)^{2/3}$
$DY_1\,[G_{21}]$	1	1	1	$(-1)^{2/3}$	$-(-1)^{1/3}$
$DA_3\,[G_{21}]$	3	$\frac{1}{2}i\left(i+\sqrt{7}\right)$	$-\frac{1}{2}i\left(-i+\sqrt{7}\right)$	0	0
$DB_3\,[G_{21}]$	3	$-\frac{1}{2}i\left(-i+\sqrt{7}\right)$	$\frac{1}{2}i\left(i+\sqrt{7}\right)$	0	0

(1.3.38)

In the weight-basis the two generators of the G_{21} subgroup of L_{168} can be chosen to be the following matrices and this fixes our representative of the unique conjugacy class:

$$\mathscr{X} = \begin{pmatrix} 1 & 1 & 1 & 1 & 1 & 1 & 1 \\ 0 & 0 & 0 & 0 & 0 & 0 & -1 \\ 0 & -1 & -1 & -1 & -1 & -1 & 0 \\ 0 & 1 & 1 & 1 & 0 & 0 & 0 \\ 0 & 0 & -1 & -1 & 0 & 0 & 0 \\ 0 & 0 & 1 & 1 & 1 & 0 & 0 \\ 0 & 0 & 0 & -1 & -1 & 0 & 0 \end{pmatrix} \quad \mathscr{Y} = \begin{pmatrix} 0 & 1 & 1 & 0 & 0 & 0 & 0 \\ 0 & 0 & 0 & 1 & 1 & 1 & 1 \\ 0 & 0 & -1 & -1 & -1 & -1 & -1 \\ 0 & 0 & 1 & 1 & 0 & 0 & 0 \\ -1 & -1 & -1 & -1 & 0 & 0 & 0 \\ 1 & 1 & 1 & 1 & 1 & 0 & 0 \\ 0 & 0 & 0 & 0 & 0 & 1 & 0 \end{pmatrix} \tag{1.3.39}$$

1.3.7.2 The Maximal Subgroups O_{24A} and O_{24B}

As we know from Sect. 1.2.5, the octahedral group O_{24} has two generators $S \equiv B$ and $T \equiv A$ that satisfy the following relations:

$$S^2 = T^3 = (ST)^4 = \mathbf{1} \tag{1.3.40}$$

The 24 elements are organized in five conjugacy classes according to the scheme displayed below:

Conjugacy Class	C_1	C_2	C_3	C_4	C_5
representative of the class	e	T	$STST$	S	ST
order of the elements in the class	1	3	2	2	4
number of elements in the class	1	8	3	6	6

(1.3.41)

The irreducible representations of O_{24} were explicitly constructed in Sect. 1.2.6. We repeat here the corresponding character table mentioning also a standard representative of each conjugacy class:

0	e	T	$STST$	S	ST
$D_1\,[O_{24}]$	1	1	1	1	1
$D_2\,[O_{24}]$	1	1	1	−1	−1
$D_3\,[O_{24}]$	2	−1	2	0	0
$D_4\,[O_{24}]$	3	0	−1	−1	1
$D_5\,[O_{24}]$	3	0	−1	1	−1

(1.3.42)

By computer calculations we have verified that there are just two disjoint conjugacy classes of O_{24} maximal subgroups in L_{168} that we have named A and B, respectively. We have chosen two standard representatives, one for each conjugacy class, that we have named O_{24A} and O_{24B} respectively. To fix these subgroups it suffices to mention the explicit form of the their generators in the weight basis.

For the group O_{24A}, we chose:

$$T_A = \begin{pmatrix} 1 & 1 & 1 & 1 & 1 & 1 & 1 \\ 0 & 0 & 0 & 0 & 0 & 0 & -1 \\ 0 & -1 & -1 & -1 & -1 & -1 & 0 \\ 0 & 1 & 1 & 1 & 0 & 0 & 0 \\ 0 & 0 & -1 & -1 & 0 & 0 & 0 \\ 0 & 0 & 1 & 1 & 1 & 0 & 0 \\ 0 & 0 & 0 & -1 & -1 & 0 & 0 \end{pmatrix} \quad S_A = \begin{pmatrix} 0 & 0 & 0 & 1 & 1 & 1 & 0 \\ 0 & 0 & 0 & 0 & -1 & -1 & 0 \\ -1 & -1 & -1 & -1 & 0 & 0 & 0 \\ 1 & 1 & 0 & 0 & 0 & 0 & 0 \\ 0 & 0 & 1 & 1 & 1 & 1 & 1 \\ 0 & -1 & -1 & -1 & -1 & -1 & -1 \\ 0 & 1 & 1 & 1 & 1 & 0 & 0 \end{pmatrix} \tag{1.3.43}$$

For the group O_{24B}, we chose:

$$T_B = \begin{pmatrix} 1 & 1 & 1 & 1 & 0 & 0 & 0 \\ 0 & -1 & -1 & -1 & 0 & 0 & 0 \\ 0 & 1 & 1 & 1 & 1 & 0 & 0 \\ 0 & 0 & -1 & -1 & -1 & 0 & 0 \\ 0 & 0 & 1 & 1 & 1 & 1 & 0 \\ 0 & 0 & 0 & -1 & -1 & -1 & 0 \\ 0 & 0 & 0 & 1 & 1 & 1 & 1 \end{pmatrix} \quad S_B = \begin{pmatrix} 0 & 0 & 1 & 1 & 1 & 0 & 0 \\ -1 & -1 & -1 & -1 & -1 & 0 & 0 \\ 1 & 1 & 1 & 1 & 1 & 1 & 1 \\ 0 & 0 & 0 & 0 & 0 & 0 & -1 \\ 0 & -1 & -1 & -1 & -1 & -1 & 0 \\ 0 & 1 & 1 & 1 & 0 & 0 & 0 \\ 0 & 0 & 0 & -1 & 0 & 0 & 0 \end{pmatrix} \tag{1.3.44}$$

1.3.7.3 The Tetrahedral Subgroup $T_{12} \subset O_{24}$

Every octahedral group O_{24} has, up to O_{24}-conjugation, a unique tetrahedral subgroup T_{12} whose order is 12. The abstract description of the tetrahedral group is provided by the following presentation in terms of two generators:

$$T_{12} = \left(s, t \,\middle|\, s^2 = t^3 = (st)^3 = 1\right) \tag{1.3.45}$$

The 12 elements are organized into four conjugacy classes as displayed below:

Classes	C_1	C_2	C_3	C_4
standard representative	1	s	t	t^2s
order of the elements in the conjugacy class	1	2	3	3
number of elements in the conjugacy class	1	3	4	4

(1.3.46)

We do not display the character table since we will not use it. The two tetrahedral subgroups $T_{12A} \subset O_{24A}$ and $T_{12B} \subset O_{24B}$ are not conjugate under the big group L_{168}. Hence we have two conjugacy classes of tetrahedral subgroups of L_{168}.

1.3.7.4 The Dihedral Subgroup $Dih_3 \subset O_{24}$

Every octahedral group O_{24} has a dihedral subgroup Dih_3 whose order is 6. The abstract description of the dihedral group Dih_3 is provided by the following presentation in terms of two generators:

$$Dih_3 = \left(A, B \,\middle|\, A^3 = B^2 = (BA)^2 = 1\right) \tag{1.3.47}$$

The 6 elements are organized into three conjugacy classes as displayed below:

ConjugacyClasses	C_1	C_2	C_3
standard representative of the class	1	A	B
order of the elements in the class	1	3	2
number of elements in the class	1	2	3

(1.3.48)

We do not display the character table since we will not use it. Differently from the case of the tetrahedral subgroups the two dihedral subgroups $\mathrm{Dih}_{3A} \subset \mathrm{O}_{24A}$ and $\mathrm{Dih}_{3B} \subset \mathrm{O}_{24B}$ turn out to be conjugate under the big group L_{168}. Actually there is just one L_{168}-conjugacy class of dihedral subgroups Dih_3.

1.3.7.5 Enumeration of the Possible Subgroups and Orbits

In $d = 3$ the orbits of the octahedral group acting on the cubic lattice are the vertices of regular geometrical figures. Since L_{168} has a crystallographic action on the mentioned 7-dimensional weight lattice, its orbits $\mathscr{O}$ in $\Lambda_{\mathbf{w}}$ correspond to the analogue regular geometrical figures in $d = 7$. Every orbit is in correspondence with a coset G/H where G is the big group and H one of its possible subgroups. Indeed H is the stability subgroup of an element of the orbit.

Since the maximal subgroups of L_{168} are of index 7 or 8 we can have subgroups $\mathrm{H} \subset \mathrm{L}_{168}$ that are either G_{21} or O_{24} or subgroups thereof. Furthermore, as we know, the order $|\mathrm{H}|$ of any subgroup $\mathrm{H} \subset \mathrm{G}$ must be a divisor of $|\mathrm{G}|$. Hence we conclude that

$$|\mathrm{H}| \in \{1, 2, 3, 4, 6, 7, 8, 12, 21, 24\} \tag{1.3.49}$$

Correspondingly we might have L_{168}-orbits $\mathscr{O}$ in the weight lattice Λ_w, whose length is one of the following nine numbers:

$$\ell_{\mathscr{O}} \in \{168, 84, 56, 42, 28, 24, 21, 14, 8, 7\} \tag{1.3.50}$$

Combining the information about the possible group orders (1.3.49) with the information that the maximal subgroups are of index 8 or 7, we arrive at the following list of possible subgroups H (up to conjugation) of the group L_{168}:

Order (24) Either $\mathrm{H} = \mathrm{O}_{24A}$ or $\mathrm{H} = \mathrm{O}_{24B}$.
Order (21) The only possibility is $\mathrm{H} = \mathrm{G}_{21}$.
Order (12) The only possibilities are $\mathrm{H} = \mathrm{T}_{12A}$ or $\mathrm{H} = \mathrm{T}_{12B}$ where T_{12} is the tetrahedral subgroup of the octahedral group O_{24}.
Order (8) Either $\mathrm{H} = \mathbb{Z}_2 \times \mathbb{Z}_2 \times \mathbb{Z}_2$ or $\mathrm{H} = \mathbb{Z}_2 \times \mathbb{Z}_4$.
Order (7) The only possibility is $\mathbb{Z}_7$.
Order (6) Either $\mathrm{H} = \mathbb{Z}_2 \times \mathbb{Z}_3$ or $\mathrm{H} = \mathrm{Dih}_3$, where Dih_3 denotes the dihedral subgroup of index 3 of the octahedral group O_{24}.
Order (4) Either $\mathrm{H} = \mathbb{Z}_2 \times \mathbb{Z}_2$ or $\mathrm{H} = \mathbb{Z}_4$.
Order (3) The only possibility is $\mathrm{H} = \mathbb{Z}_3$
Order (2) The only possibility is $\mathrm{H} = \mathbb{Z}_2$.

1.3.7.6 Synopsis of the L_{168} Orbits in the Weight Lattice Λ_w

In [7], the author of this book has presented his results, obtained by means of computer calculations, on the orbits of the considered simple group acting on the $\mathfrak{a}_7$ weight lattice. They are briefly summarized below

1. Orbits of length 8 (one parameter **n**; stability subgroup $H^s = G_{21}$)
2. Orbits of length 14 (two types A & B) (one parameter **n**; stability subgroup $H^s = T_{12A,B}$)
3. Orbits of length 28 (one parameter **n** ; stability subgroup $H^s = \mathrm{Dih}_3$)
4. Orbits of length 42 (one parameter **n**; stability subgroup $H^s = \mathbb{Z}_4$))
5. Orbits of length 56 (three parameters **n,m,p**; stability subgroup $H^s = \mathbb{Z}_3$)
6. Orbits of length 84 (three parameters **n,m,p**; stability subgroup $H^s = \mathbb{Z}_2$)
7. Generic orbits of length 168 (seven parameters; stability subgroup $H^s = \mathbf{1}$)

As we already said, the above list is in some sense the 7-dimensional analogue of Platonic solids. It is only in some sense, since it is a complete classification for the group L_{168} yet we are not aware of a classification of the other crystallographic subgroups of SO(7), if any.

Notwithstanding this ignorance, the piece of knowledge we have summarized above is already impressively complicated and demonstrates how even flat geometry becomes more sophisticated in higher dimensions.

The next natural question is why just $d = 7$ should attract our geometrical attention. There are several reasons for the number 7. They are probably all related to each other:

1. The possible division algebras are $\mathbb{R}$, $\mathbb{C}$, $\mathbb{H}$, $\mathbb{O}$, the real numbers, the complex numbers, the quaternions and the octonions. The corresponding number of imaginary units are 0, 1, 3, 7. The automorphisms groups of these division algebras are 1, U(1), SU(2), $\mathfrak{g}_{2(-14)}$.
2. The spheres that are globally parallelizable are $\mathbb{S}^1$,$\mathbb{S}^3$,$\mathbb{S}^7$.
3. The manifolds of restricted holonomy are the complex ones, the Kähler ones, the quaternionic ones, that exist in all dimensions $d = 2n$, respectively $d = 4n$, and then, just in $d = 7$, we have the $\mathfrak{g}_2$ manifolds and in $d = 8$ we have the Spin(7) manifolds.
4. Seven are the dimensions that one has to compactify in order to step down from the 11-dimensional M-theory to our $d = 4$ space-time and many solutions of the theory naturally perform the splitting $11 = 4 + 7$.

1.4 The General Form of a Simple Lie Algebra and the Root Systems

Every simple Lie algebra $\mathbb{G}$ of dimension $n = 2m + r$ can be described in a compact and quite inspiring way. There exists an abelian subalgebra (the Cartan subalgebra CSA) made of elements whose adjoint action is fully diagonalizable and whose dimension $\ell < n$ is named the *rank* of $\mathbb{G}$. A basis of generators spanning the CSA is usually denoted by H_i ($i =, \ldots, \ell$). The remaining $2m$ generators, denoted E^α are in one-to-one correspondence with a set Δ of vectors α living in an ℓ-dimensional Euclidean space that are named the *roots*. The set Δ is dubbed a *root space* and it is formally defined as follows:

Definition 1.4.1 A root space Δ of rank ℓ is a finite set of vectors $\{\alpha\}$, named the roots and defined in an Euclidean space of dimension ℓ, that satisfy the following properties. If $\alpha, \beta \in \Delta$ are two roots, then the following two statements are true:

1. $2\,\frac{(\alpha\,,\beta)}{(\alpha\,,\alpha)} \in \mathbb{Z}$
2. $\sigma_\alpha(\beta) \equiv \beta - 2\,\alpha\,\frac{(\alpha\,,\beta)}{(\alpha\,,\alpha)} \in \Delta$ is also a root.

The vector $\sigma_\alpha(\beta)$ defined above is named the reflection of β with respect to α and the second part of the definition can be reformulated by saying that any root system Δ is invariant under reflection with respect to any of its elements.

Utilizing these notations and the advocated notion of root system we have:

Theorem 1.4.1 *The commutation relations of a* complex simple Lie algebra *take necessarily the following general form:*

$$\begin{aligned} \left[H_i\,,\,H_j\right] &= 0 \\ \left[H_i\,,\,E^\alpha\right] &= \alpha_i\,E^\alpha \\ \left[E^\alpha\,,\,E^{-\alpha}\right] &= \alpha^i\,H_i \\ \left[E^\alpha\,,\,E^\beta\right] &= N(\alpha,\beta)\,E^{\alpha+\beta} \quad \textit{if} \quad \alpha+\beta \in \Delta \\ \left[E^\alpha\,,\,E^\beta\right] &= 0 \quad \textit{if} \quad \alpha+\beta \notin \Delta \end{aligned} \tag{1.4.1}$$

where $N(\alpha, \beta)$ is a coefficient that has to be determined using Jacobi identities.

From now on we can associate to every complex simple Lie algebra its root system Δ. Furthermore each root system singles out a well-defined finite group, named the *Weyl group* that is obtained combining together the reflections with respect to all the roots.

Definition 1.4.2 Let Δ be a root system in dimension ℓ. The Weyl group of Δ, denoted $\mathscr{W}(\Delta)$ is the finite group generated by the reflections σ_α, $\forall\alpha \in \Delta$.

Since for any two vectors $\mathbf{v}, \mathbf{w} \in \mathbb{E}$ we have:

$$(\sigma_\alpha(\mathbf{v})\,,\,\sigma_\alpha(\mathbf{w})) = (\mathbf{v}\,,\,\mathbf{w}) \tag{1.4.2}$$

it follows that the Weyl group, which is finite, is always a subgroup of the rotation group in ℓ dimensions:

$$\mathscr{W}(\Delta) \subset \mathrm{SO}(\ell) \tag{1.4.3}$$

1.4.1 The Cartan Matrices

The main token in the classification of root systems is provided by the Cartan matrices, which we presently define. We begin with the notion of simple roots.

Definition 1.4.3 Given a root system $\Delta \subset \mathbb{E}^\ell$ in an Euclidean space of dimension ℓ, a set Δ of exactly ℓ roots is named a simple root basis if:

1. Δ is a basis for the entire $\mathbb{E}^\ell$.
2. Every root $\alpha \in \Delta$ can be written as a linear combination of the elements α_i whose coefficients are either all positive or all negative integers

$$\alpha = \sum_{i=1}^{\ell} k^i \, \alpha_i \quad ; \quad k^i \in \begin{cases} \mathbb{Z}_+ \\ \text{or } \mathbb{Z}_- \end{cases} \tag{1.4.4}$$

The vectors α_i comprised in Δ are named the simple roots of Δ.

A rather simple, yet fundamental theorem establishes that every root system has a simple root basis $\alpha_1, \ldots, \alpha_\ell$. This being the case to every root system and hence to every complex Lie algebra we can associated the following $\ell \times \ell$ matrix:

$$C_{ij} =< \alpha_i \, , \, \alpha_j > \equiv 2 \, \frac{(\alpha_i \, , \, \alpha_j)}{(\alpha_j \, , \, \alpha_j)} \tag{1.4.5}$$

Another simple and constructive theorem shows that from the Cartan matrix one can retrieve the entire root system and hence the simple Lie algebra.

Having established that all possible irreducible root systems Δ are uniquely determined (up to isomorphisms) by the Cartan matrix, we can classify *all the complex simple Lie algebras* by classifying all possible Cartan matrices. This is the classification originally achieved by Killing and Cartan. Later on in the XXth century the theory of Cartan matrices of root systems and of the finite reflection groups associated with them was extensively developed by three mathematicians Hermann Weyl, Harold Coxeter and Evgenij Dynkin.

1.4.2 Dynkin Diagrams

Each Cartan matrix can be given a graphical representation in the following way. To each simple root α_i we associate a circle ◯ as in Fig. 1.4 and then we link the ith circle with the jth circle by means of a line which is *simple*, *double* or *triple* depending on whether

$$< \alpha_i \, , \, \alpha_j >< \alpha_j \, , \, \alpha_i >= 4 \cos^2 \theta_{ij} = \begin{cases} 1 \\ 2 \\ 3 \end{cases} \tag{1.4.6}$$

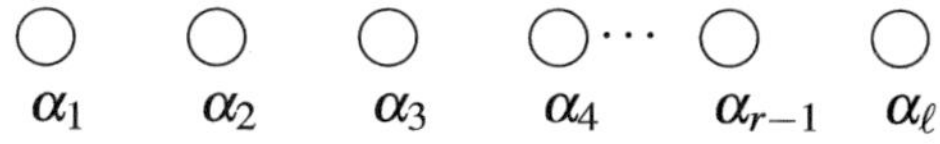

Fig. 1.4 The simple roots α_i are represented by circles

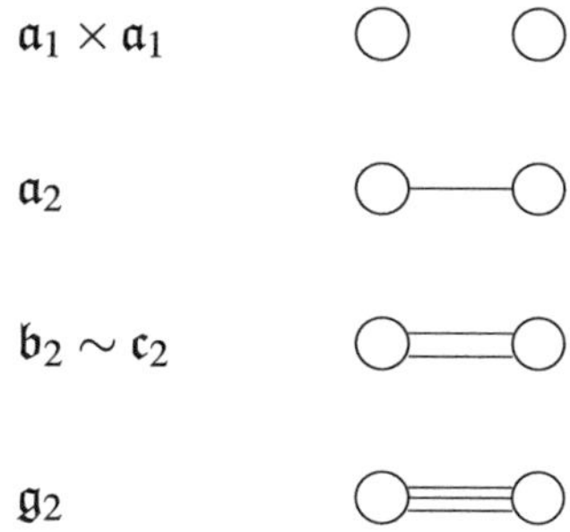

Fig. 1.5 The four possible Coxeter graphs with two vertices

having denoted θ_{ij} the angle between the two simple roots α_i and α_j. The corresponding graph is named a **Coxeter graph**.

If we consider the simplest case of two-dimensional Cartan matrices we have the four possible Coxeter graphs depicted in Fig. 1.5 Given a Coxeter graph if it is *simply laced*, namely if there are only simple lines, then all the simple roots appearing in such a graph have the same *length* and the corresponding Cartan matrix is completely identified. On the other hand if the Coxeter graph involves double or triple lines, then, in order to identify the corresponding Cartan matrix, we need to specify which of the two roots sitting at the end points of each multiple line is the *long* root and which is the *short* one. This can be done by associating an arrow to each multiple line. By convention we decide that this *arrow points* in the direction of the *short root*. A Coxeter graph equipped with the necessary arrows is named a **Dynkin diagram**. Applying this convention to the case of the Coxeter graphs of Fig. 1.5 we obtain the result displayed in Fig. 1.6. The one-to-one correspondence between the Dynkin diagram and the associated Cartan matrix is illustrated by considering in some detail the case B_2 of Fig. 1.6. By definition of the Cartan matrix we have:

$$\begin{aligned} 2\,\frac{(\alpha_1\,,\,\alpha_2)}{(\alpha_2\,,\,\alpha_2)} &= 2\,\frac{|\alpha_1|}{|\alpha_2|}\,\cos\theta = -2 \\ 2\,\frac{(\alpha_2\,,\,\alpha_1)}{(\alpha_1\,,\,\alpha_1)} &= 2\,\frac{|\alpha_2|}{|\alpha_1|}\,\cos\theta = -1 \end{aligned} \tag{1.4.7}$$

so that we conclude:

$$|\alpha_1|^2 = 2\,|\alpha_2|^2 \tag{1.4.8}$$

which shows that α_1 is a long root, while α_2 is a short one. Hence the arrow in the Dynkin diagram pointing towards the short root α_2 tells us that the matrix elements C_{12} is -2 while the matrix element C_{21} is -1. It happens the opposite in the example C_2.

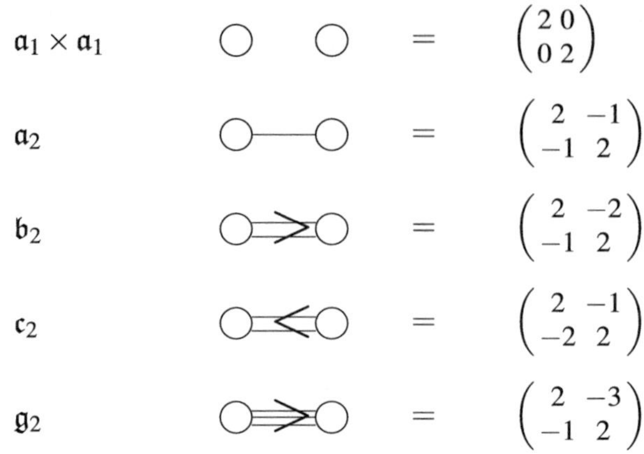

Fig. 1.6 The distinct Cartan matrices in two dimensions (and therefore the simple Algebras in rank two) correspond to the Dynkin diagrams displayed above. We have distinguished a $\mathfrak{b}_2$ and a $\mathfrak{c}_2$ matrix since they are the limiting case for $\ell = 2$ of two series of Cartan matrices the $\mathfrak{b}_\ell$ and the $\mathfrak{c}_\ell$ series that for $\ell > 2$ are truly different. However $\mathfrak{b}_2$ is the transposed of $\mathfrak{c}_2$ so that they correspond to isomorphic algebras obtained one from the other by renaming the two simple roots $\alpha_1 \leftrightarrow \alpha_2$

Fig. 1.7 The Dynkin diagrams of the four infinite families of classical simple algebras

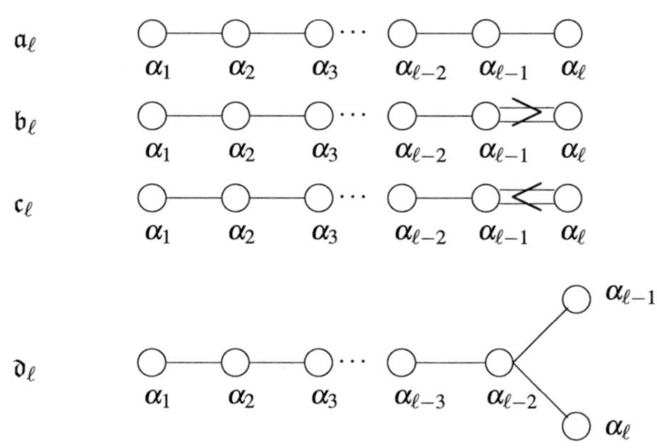

1.5 The Classification Theorem

Having clarified the notation of Dynkin diagrams the basic classification theorem of *complex simple Lie algebras* is the following:

Theorem 1.5.1 *If Δ is an irreducible system of roots of rank ℓ then its Dynkin diagram is either one of those shown in Fig. 1.7 or for special values of ℓ is one of those shown in Fig. 1.8. There are no other irreducible root systems besides these ones.*

This fundamental theorem encoding the classification of complex simple Lie algebras is proved in many textbooks and a proof, essentially based on that of [11], is provided in the same notations of the present book in Chap. 7 of [1]. Of that proof we report

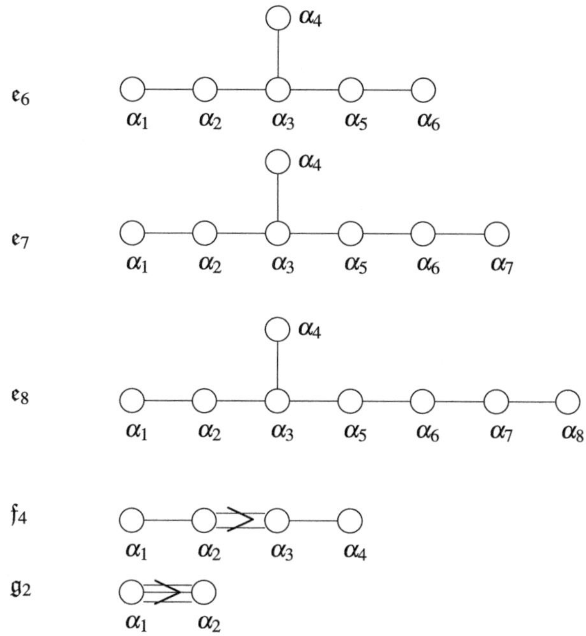

Fig. 1.8 The Dynkin diagrams of the five exceptional algebras

here only the crucial segment that leads to the diophantine Eq. (1.1.48) and shows the ADE isomorphism between the classification of simply laced Lie algebras and of finite Kleinian subgroups of SU(2).

The strategy of the proof, which is organized in ten steps is based on the introduction of a set of vectors:

$$\mathscr{U} = \{\varepsilon_1\,,\,\varepsilon_2\,,\,\ldots\,,\varepsilon_\ell\} \tag{1.5.1}$$

that satisfy the following three conditions:

$$\begin{aligned} (\varepsilon_i\,,\,\varepsilon_i) &= 1 \\ \left(\varepsilon_i\,,\,\varepsilon_j\right) &\leq 0 \quad i \neq j \\ 4\,\left(\varepsilon_i\,\varepsilon_j\right)^2 &= 0, 1, 2, 3 \quad i \neq j \end{aligned} \tag{1.5.2}$$

Such a system of vectors is named *admissible*. It is clear that each admissible system of vectors singles out a Coxeter graph Γ. Indeed the vectors ε_i correspond to the simple roots α_i divided by their norm:

$$\varepsilon_i = \frac{\alpha_i}{\sqrt{|\alpha_i|^2}} \tag{1.5.3}$$

The task is that of classifying all connected Coxeter graphs.

In the first eight steps of the proof one establishes that there is a set of prohibited Coxeter subgraphs that are those displayed in Fig. 1.9.

In this way, apart from the Coxeter graph of the $\mathfrak{g}_2$ Lie algebra (see Fig. 1.8), which is admissible, one is left with the candidate graphs displayed in Figs. 1.10 and 1.11.

In step 9 one considers the graphs of the type shown in Fig. 1.10 and utilizing the properties of Euclidean geometry one establishes that there are only two solutions namely:

$$\begin{array}{llll} p=2 & ; q=2 \Rightarrow & \mathfrak{f}_4 & \text{Dynkin diagram} \\ p=\ell\in\mathbb{N} & ; q=1 \Rightarrow & \mathfrak{b}_\ell \;\; or \;\; \mathfrak{c}_\ell & \text{Dynkin diagrams} \end{array} \qquad (1.5.4)$$

The first solution leads to the Dynkin diagram of the exceptional Lie algebra $\mathfrak{f}_4$, while the second solution leads to the two infinite series of classical Lie algebras $\mathfrak{b}_\ell$ and $\mathfrak{c}_\ell$.

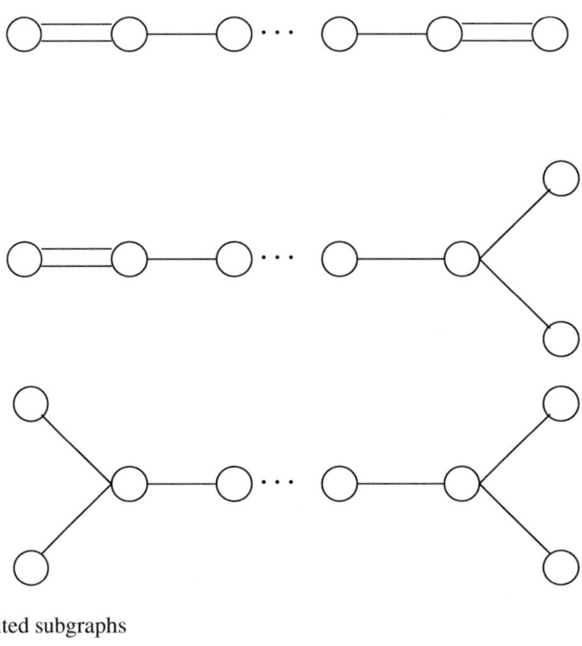

Fig. 1.9 Prohibited subgraphs

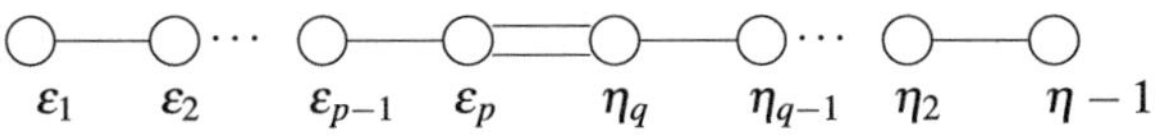

Fig. 1.10 Coxeter graph with a double link that is preceded by a simple chain of length p and followed by a simple chain of length q

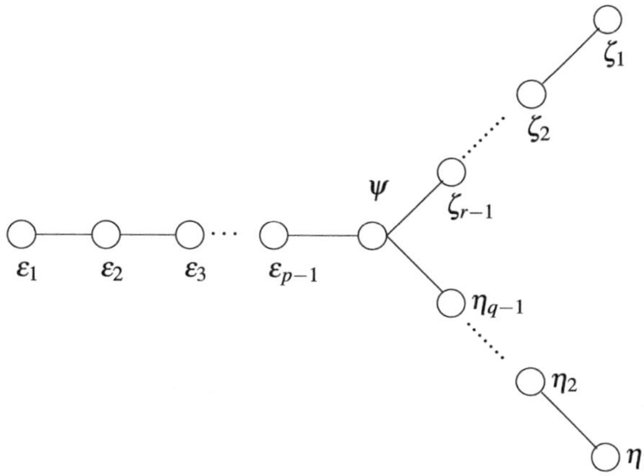

Fig. 1.11 Coxeter graph with a node. The unit vector in the node is named ψ while the unit vectors along the three simple lines departing from the node are respectively named $\varepsilon_1, \ldots, \varepsilon_{p-1}, \eta_1, \ldots, \eta_{q-1}, \zeta_1, \ldots, \zeta_{r-1}$. The graph is characterized by the three integer numbers p, q, r that denote the lengths of the three simple lines departing from the node

Finally in step 10 one considers the Coxeter graphs of the type shown in Fig. 1.11. The claim is that the only possible solutions are:

$$(p, q, r) = \begin{cases} (\ell, 1, 1) & \Rightarrow A_\ell \text{ Dynkin diagrams} \quad \ell \in \mathbb{N} \\ (\ell - 2, 2, 2) & \Rightarrow \mathfrak{d}_\ell \text{ Dynkin diagrams } 4 \leq \ell \in \mathbb{N} \\ (3, 3, 2) & \Rightarrow \mathfrak{e}_6 \text{ Dynkin diagram} \\ (4, 3, 2) & \Rightarrow \mathfrak{e}_7 \text{ Dynkin diagram} \\ (5, 3, 2) & \Rightarrow \mathfrak{e}_8 \text{ Dynkin diagram} \end{cases} \tag{1.5.5}$$

To prove this statement we follow a strategy similar to that used in the proof of Step 9, namely we define the following three vectors:

$$\varepsilon = \sum_{i=1}^{p-1} i\, \varepsilon_i \quad ; \quad \eta = \sum_{i=1}^{q-1} i\, \eta_i \quad ; \quad \sum_{i=1}^{r-1} i\, \zeta_i \tag{1.5.6}$$

Clearly ε, η, ζ are mutually orthogonal and ψ, the vector in the node is not in the subspace generated by ε, η, ζ. Hence if in the linear span of $\{\psi, \varepsilon, \eta, \zeta\}$ we construct a vector γ that is orthogonal to $\{\varepsilon, \eta, \zeta\}$ we obtain that $(\gamma\,, \psi) \neq 0$. Normalizing this vector to 1 we can write:

$$\psi = (\psi\,, \gamma)\,\gamma + \frac{(\psi\,, \varepsilon)}{\sqrt{(\varepsilon\,, \varepsilon)}}\,\varepsilon + \frac{(\psi\,, \eta)}{\sqrt{(\eta\,, \eta)}}\,\eta + \frac{(\psi\,, \zeta)}{\sqrt{(\zeta\,, \zeta)}}\,\zeta \tag{1.5.7}$$

and we obtain:

$$(\psi\,, \psi) = 1 = (\psi\,, \gamma)^2 + \frac{(\psi\,, \varepsilon)^2}{(\varepsilon\,, \varepsilon)} + \frac{(\psi\,, \eta)^2}{(\eta\,, \eta)} + \frac{(\psi\,, \zeta)^2}{(\zeta\,, \zeta)} \tag{1.5.8}$$

that implies the inequality:

$$1 > \frac{(\psi \,,\, \varepsilon)^2}{(\varepsilon \,,\, \varepsilon)} + \frac{(\psi \,,\, \eta)^2}{(\eta \,,\, \eta)} + \frac{(\psi \,,\, \zeta)^2}{(\zeta \,,\, \zeta)} \tag{1.5.9}$$

By definition of the Coxeter graph in Fig. 1.11 we have:

$$\begin{aligned} (\psi \,,\, \varepsilon) = (p-1)\left(\varepsilon_{p-1} \,,\, \psi\right) \quad \Rightarrow \quad (\psi \,,\, \varepsilon)^2 &= \frac{(p-1)^2}{4} \\ (\varepsilon \,,\, \varepsilon) &= \frac{p(p-1)}{2} \end{aligned} \tag{1.5.10}$$

and similarly for the scalar products associated with the other chains. Inserting these results into the inequality of Eq. (1.5.9) we obtain the Diophantine inequality:

$$\frac{1}{p} + \frac{1}{q} + \frac{1}{r} > 1 \tag{1.5.11}$$

whose independent solutions are those displayed in Eq. (1.5.5). To this effect it is sufficient to note that Eq. (1.5.11) has an obvious permutational symmetry in the three numbers p, q, r. To avoid double counting of solutions we break this symmetry by setting $p \geq q \geq r$ and then we see that the only possibilities are those listed in Eq. (1.5.5).

Having concluded the above proof we can look back and compare the just obtained results with those summarized in Sect. 1.1.4. The anticipated correspondence between finite rotation subgroups and simply laced Lie algebras should now be clear: the profound meaning of the correspondence was displayed in Fig. 1.2. The rank of the Lie algebra $\mathscr{A}$ corresponds to the number of non trivial conjugacy classes of the finite group Γ, while the lengths k_α of the simple chains in the Dynkin diagram correspond to the order of the group generators. More implications of the correspondence will be unveiled in Chap. 8.

1.6 The Exceptional Lie Algebra $\mathfrak{g}_2$

It was Killing who, through his own classification of the root systems, first discovered the possible existence of the exceptional Lie algebras: yet their concrete existence was proved only later by Cartan who was able to construct the fundamental representation of all of them. In this section we study the smallest of the five exceptional algebras, namely, $\mathfrak{g}_2$ and we explicitly exhibit its fundamental representation which is 7-dimensional.

Our presentation is aimed not only at showing that $\mathfrak{g}_2$ exists but it also enlightens some features of its structure that will turn out to be general within a certain algebraic scheme that encompasses an entire set of classical and exceptional Lie algebras relevant for the *special geometries* implied by supergravity and superstrings.

Before the advent of supergravity, exceptional Lie algebras were viewed by physicists as some mathematical extravagance good only for a Dickensian *Old Curiosity Shop*. Supergravity quite surprisingly shew that all exceptional Lie algebras have a distinct and essential role to play in the connected web of gravitational theories that one obtains through dimensional reduction and coupling of matter multiplets in diverse dimensions. Furthermore there is an inner algebraic structure of the exceptional algebras, shared with other classical algebras that appears to be specially prepared to fit the geometrical yields of supersymmetry. This provides a new structural viewpoint motivated by physics that, in Weyl's spirit, encodes a deep truth, at the same time physical and mathematical, the distinction being somewhat irrelevant. The full-fledged span of the considerations first brought to the stage in this section will be fully appreciated by the reader when he will address Chap. 4 on special geometries and Chap. 5 on the theory of the Tits Satake projection. Let us next turn to the specific topic of the present section.

The complex Lie algebra $\mathfrak{g}_2(\mathbb{C})$ has rank two and it is defined by the 2×2 Cartan matrix encoded in the following Dynkin diagram:

$\mathfrak{g}_2$  $= \begin{pmatrix} 2 & -1 \\ -3 & 2 \end{pmatrix}$

The $\mathfrak{g}_2$ root system Δ consists of the following six positive roots plus their negatives:

$$\begin{array}{llll} \alpha_1 = \alpha_1 = (1,0) & ; \alpha_2 = \alpha_2 = \frac{\sqrt{3}}{2}(-\sqrt{3},1) \\ \alpha_3 = \alpha_1 + \alpha_2 = \frac{1}{2}(-1,\sqrt{3}) & ; \alpha_4 = 2\,\alpha_1 + \alpha_2 = \frac{1}{2}(1,\sqrt{3}) \\ \alpha_5 = 3\,\alpha_1 + \alpha_2 = \frac{\sqrt{3}}{2}(\sqrt{3},1) & ; \alpha_6 = 3\,\alpha_1 + 2\,\alpha_2 = (0,\sqrt{3}) \end{array} \tag{1.6.1}$$

The two fundamental weights are easily derived and have the following form:

$$\begin{aligned} \lambda^1 &= \left\{1, \sqrt{3}\right\} \\ \lambda^2 &= \left\{0, \frac{2}{\sqrt{3}}\right\} \end{aligned} \tag{1.6.2}$$

Simple roots, fundamental weights and the Weyl chamber are displayed in Fig. 1.12. Figure 1.13 instead displays the entire root system. The fundamental representation of the Lie algebra is identified as the one which admits the fundamental weight λ^1 as highest weight. Using the Weyl group symmetry and the α through λ string technique one derives all the weights of the 7-dimensional fundamental representation that are the following ones:

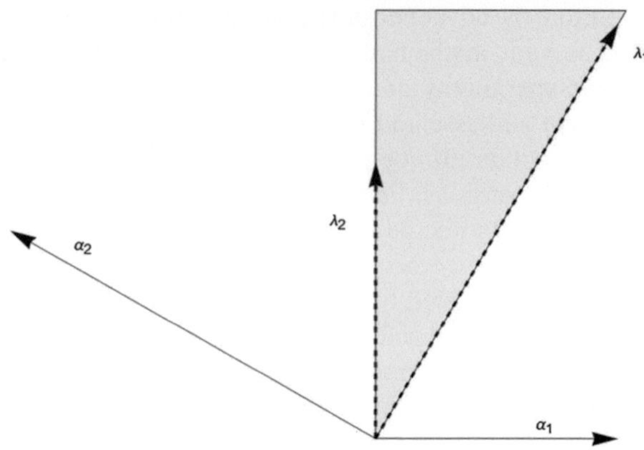

Fig. 1.12 The simple roots and the fundamental weights of the $\mathfrak{g}_2$ Lie algebra. The shaded region is the Weyl Chamber

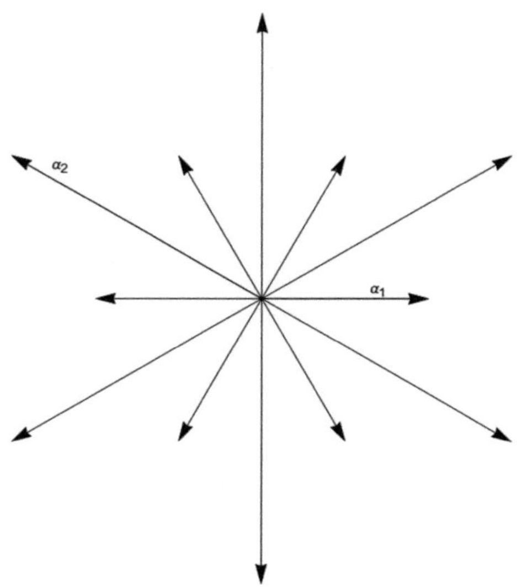

Fig. 1.13 The complete root system of the $\mathfrak{g}_2$ Lie algebra

Name		Dynk lab		Orth. comp.	mult.
Λ_1	$=$	$\{1, 0\}$	$\Rightarrow$	$\left\{1, \sqrt{3}\right\}$	1
Λ_2	$=$	$\{-1, 1\}$	$\Rightarrow$	$\left\{-1, -\frac{1}{\sqrt{3}}\right\}$	1
Λ_3	$=$	$\{2, -1\}$	$\Rightarrow$	$\left\{2, \frac{4}{\sqrt{3}}\right\}$	1
Λ_4	$=$	$\{0, 0\}$	$\Rightarrow$	$\{0, 0\}$	1
Λ_5	$=$	$\{-2, 1\}$	$\Rightarrow$	$\left\{-2, -\frac{4}{\sqrt{3}}\right\}$	1
Λ_6	$=$	$\{1, -1\}$	$\Rightarrow$	$\left\{1, \frac{1}{\sqrt{3}}\right\}$	1
Λ_7	$=$	$\{-1, 0\}$	$\Rightarrow$	$\left\{-1, -\sqrt{3}\right\}$	1

(1.6.3)

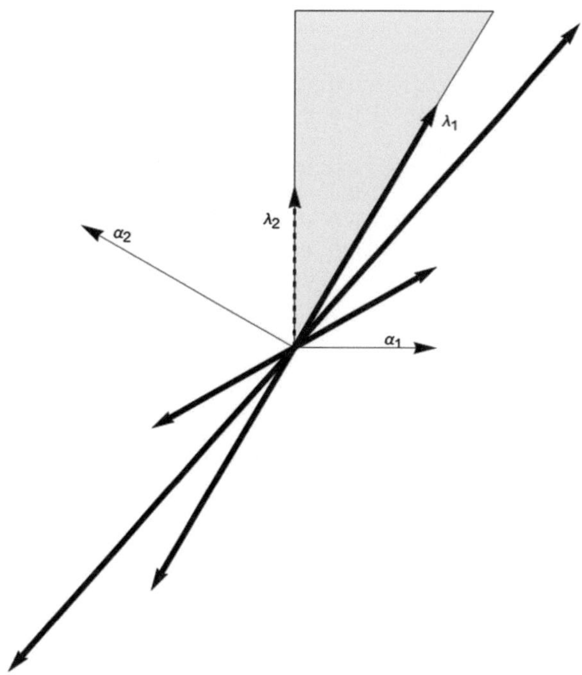

Fig. 1.14 The six non vanishing weights of the fundamental representation of the $\mathfrak{g}_2$ Lie algebra. The fundamental weight λ^1 is the highest weight of this representation

The six non-vanishing weights are displayed in Fig. 1.14

Given this information we are ready to derive the fundamental representation of the algebra. According to our general strategy we are supposed to construct 7×7 upper triangular matrices spanning the Borel subalgebra of the maximally split real section $\mathfrak{g}_{2(2)}$ of $\mathfrak{g}_2(\mathbb{C})$:

$$\mathrm{Bor}\,[\mathfrak{g}_2] = \mathrm{span}\,\{H_1, H_2, E^{\alpha_1}, E^{\alpha_2}, \ldots, E^{\alpha_6}\} \tag{1.6.4}$$

As for all maximally split algebras the Cartan generators H_i and the step operators E^α associated with each positive root α can be chosen completely real in all representations.

In the fundamental 7-dimensional representation the explicit form of the $\mathfrak{g}_{2(2)}$-generators with the above properties is presented hereby. Naming $\{H_1, H_2\}$ the Cartan generators along the two ortho-normal directions and adopting the standard Cartan–Weyl normalizations:

$$[E_\alpha, E_\alpha] = \alpha^i\, H_i \ , \ \ [H_i, E_\alpha] = \alpha^i\, E_\alpha \, . \tag{1.6.5}$$

we have:

$$H_1 = \begin{pmatrix} \frac{1}{2} & 0 & 0 & 0 & 0 & 0 & 0 \\ 0 & -\frac{1}{2} & 0 & 0 & 0 & 0 & 0 \\ 0 & 0 & 1 & 0 & 0 & 0 & 0 \\ 0 & 0 & 0 & 0 & 0 & 0 & 0 \\ 0 & 0 & 0 & 0 & -1 & 0 & 0 \\ 0 & 0 & 0 & 0 & 0 & \frac{1}{2} & 0 \\ 0 & 0 & 0 & 0 & 0 & 0 & -\frac{1}{2} \end{pmatrix} \; ; \; H_2 = \begin{pmatrix} \frac{\sqrt{3}}{2} & 0 & 0 & 0 & 0 & 0 & 0 \\ 0 & \frac{\sqrt{3}}{2} & 0 & 0 & 0 & 0 & 0 \\ 0 & 0 & 0 & 0 & 0 & 0 & 0 \\ 0 & 0 & 0 & 0 & 0 & 0 & 0 \\ 0 & 0 & 0 & 0 & 0 & 0 & 0 \\ 0 & 0 & 0 & 0 & 0 & -\frac{\sqrt{3}}{2} & 0 \\ 0 & 0 & 0 & 0 & 0 & 0 & -\frac{\sqrt{3}}{2} \end{pmatrix} \tag{1.6.6}$$

$$E_{\alpha_1} = \begin{pmatrix} 0 & \frac{1}{\sqrt{2}} & 0 & 0 & 0 & 0 & 0 \\ 0 & 0 & 0 & 0 & 0 & 0 & 0 \\ 0 & 0 & 0 & 1 & 0 & 0 & 0 \\ 0 & 0 & 0 & 0 & 1 & 0 & 0 \\ 0 & 0 & 0 & 0 & 0 & 0 & 0 \\ 0 & 0 & 0 & 0 & 0 & 0 & \frac{1}{\sqrt{2}} \\ 0 & 0 & 0 & 0 & 0 & 0 & 0 \end{pmatrix} \; ; \; E_{\alpha_2} = \begin{pmatrix} 0 & 0 & 0 & 0 & 0 & 0 & 0 \\ 0 & 0 & \sqrt{\frac{3}{2}} & 0 & 0 & 0 & 0 \\ 0 & 0 & 0 & 0 & 0 & 0 & 0 \\ 0 & 0 & 0 & 0 & 0 & 0 & 0 \\ 0 & 0 & 0 & 0 & 0 & -\sqrt{\frac{3}{2}} & 0 \\ 0 & 0 & 0 & 0 & 0 & 0 & 0 \\ 0 & 0 & 0 & 0 & 0 & 0 & 0 \end{pmatrix} \tag{1.6.7}$$

$$E_{\alpha_1+\alpha_2} = \begin{pmatrix} 0 & 0 & \frac{1}{\sqrt{2}} & 0 & 0 & 0 & 0 \\ 0 & 0 & 0 & -1 & 0 & 0 & 0 \\ 0 & 0 & 0 & 0 & 0 & 0 & 0 \\ 0 & 0 & 0 & 0 & 0 & -1 & 0 \\ 0 & 0 & 0 & 0 & 0 & 0 & \frac{1}{\sqrt{2}} \\ 0 & 0 & 0 & 0 & 0 & 0 & 0 \\ 0 & 0 & 0 & 0 & 0 & 0 & 0 \end{pmatrix} \; ; \; E_{2\alpha_1+\alpha_2} = \begin{pmatrix} 0 & 0 & 0 & -1 & 0 & 0 & 0 \\ 0 & 0 & 0 & 0 & \frac{1}{\sqrt{2}} & 0 & 0 \\ 0 & 0 & 0 & 0 & 0 & -\frac{1}{\sqrt{2}} & 0 \\ 0 & 0 & 0 & 0 & 0 & 0 & 1 \\ 0 & 0 & 0 & 0 & 0 & 0 & 0 \\ 0 & 0 & 0 & 0 & 0 & 0 & 0 \\ 0 & 0 & 0 & 0 & 0 & 0 & 0 \end{pmatrix} \tag{1.6.8}$$

$$E_{3\alpha_1+\alpha_2} = \begin{pmatrix} 0 & 0 & 0 & 0 & -\sqrt{\frac{3}{2}} & 0 & 0 \\ 0 & 0 & 0 & 0 & 0 & 0 & 0 \\ 0 & 0 & 0 & 0 & 0 & 0 & -\sqrt{\frac{3}{2}} \\ 0 & 0 & 0 & 0 & 0 & 0 & 0 \\ 0 & 0 & 0 & 0 & 0 & 0 & 0 \\ 0 & 0 & 0 & 0 & 0 & 0 & 0 \\ 0 & 0 & 0 & 0 & 0 & 0 & 0 \end{pmatrix} \; ; \; E_{3\alpha_1+2\alpha_2} = \begin{pmatrix} 0 & 0 & 0 & 0 & 0 & -\sqrt{\frac{3}{2}} & 0 \\ 0 & 0 & 0 & 0 & 0 & 0 & -\sqrt{\frac{3}{2}} \\ 0 & 0 & 0 & 0 & 0 & 0 & 0 \\ 0 & 0 & 0 & 0 & 0 & 0 & 0 \\ 0 & 0 & 0 & 0 & 0 & 0 & 0 \\ 0 & 0 & 0 & 0 & 0 & 0 & 0 \\ 0 & 0 & 0 & 0 & 0 & 0 & 0 \end{pmatrix} \tag{1.6.9}$$

1.7 A Golden Splitting for Quaternionic Algebras

In Chap. 4 we shall address the study of *special Kähler geometry* and of *quaternionic geometry* that are both implied by $\mathcal{N} = 2$ supersymmetry, the first applying to the scalars of vector multiplets, the second to the scalar of hypermultiplets. Furthermore we shall discuss a very interesting relation between such geometries that is named the *c*-map:

$$\mathfrak{c} - \text{map} \quad : \quad \mathcal{SK}_{2n} \rightarrow \mathcal{Q}_{4n+4} \tag{1.7.1}$$

where $\mathscr{S}\mathscr{K}_{2n}$ denotes a special Kähler manifold of $2n$-real dimension while $\mathscr{Q}_{4n+4}$ denotes a quaternionic Kähler manifold of $4n+4$ real dimension. For the definition and properties of such manifolds we refer the reader to later chapters. What is of interest to us here is that among the special Kähler and quaternionic manifolds there are also classes of homogeneous symmetric spaces G/H leading to a split of the Lie algebra:

$$\mathbb{G} = \mathbb{H} \oplus \mathbb{K} \tag{1.7.2}$$

into a subalgebra $\mathbb{H}$ and an orthogonal subspace $\mathbb{K}$. We refer the reader to Chap. 2 for the notion of coset manifolds and symmetric spaces: here we just focus on the fact that the existence of a c-map between two symmetric spaces implies the existence of a well-defined relation between two Lie algebras that we can respectively dub the *special Kählerian algebra* $\mathbb{U}_{\mathscr{S}\mathscr{K}}$ and the *quaternionic algebra* $\mathbb{U}_{\mathscr{Q}}$.[11] For reasons that the reader will fully appreciate in later chapters this relation is provided by the following decomposition of the adjoint representation of the quaternionic algebra $\mathbb{U}_{\mathscr{Q}}$ with respect to its special Kähler subalgebra:

$$\mathrm{adj}(\mathbb{U}_{\mathscr{Q}}) = \mathrm{adj}(\mathbb{U}_{\mathscr{S}\mathscr{K}}) \oplus \mathrm{adj}(\mathrm{SL}(2,\mathbb{R})_{\mathrm{E}}) \oplus \mathbf{W}_{(2,\mathbf{W})} \tag{1.7.3}$$

where $\mathbf{W}$ is a **symplectic** representation of $\mathbb{U}_{\mathscr{S}\mathscr{K}}$ in which the symplectic section of Special Geometry (to be defined in Chap. 4) transforms. Denoting the generators of $\mathbb{U}_{\mathscr{S}\mathscr{K}}$ by T^a, the generators of $\mathrm{SL}(2,\mathbb{R})_{\mathrm{E}}$, which is named the Ehlers subalgebra, by L^{x} and denoting by $\mathbf{W}^{i\alpha}$ the generators in $\mathbf{W}_{(2,\mathbf{W})}$, the commutation relations that correspond to the decomposition (1.7.3) have the following general form:

$$\begin{aligned}
&[T^a, T^b] = f^{ab}{}_c\, T^c \\
&[L^x_E, L^y_E] = f^{xy}{}_z\, L^z, \\
&[T^a, \mathbf{W}^{i\alpha}] = (\Lambda^a)^\alpha{}_\beta\, \mathbf{W}^{i\beta}, \\
&[L^x_E, \mathbf{W}^{i\alpha}] = (\lambda^x)^i{}_j\, \mathbf{W}^{j\alpha}, \\
&[\mathbf{W}^{i\alpha}, \mathbf{W}^{j\beta}] = \varepsilon^{ij}\, (K_a)^{\alpha\beta}\, T^a + \mathbf{C}^{\alpha\beta}\, k^{ij}_x\, L^x_E
\end{aligned} \tag{1.7.4}$$

where the 2×2 matrices $(\lambda^x)^i_j$, are the canonical generators of $\mathrm{SL}(2,\mathbb{R})$ in the fundamental, defining representation:

$$\lambda^3 = \begin{pmatrix} \frac{1}{2} & 0 \\ 0 & -\frac{1}{2} \end{pmatrix} \quad ; \quad \lambda^1 = \begin{pmatrix} 0 & \frac{1}{2} \\ \frac{1}{2} & 0 \end{pmatrix} \quad ; \quad \lambda^2 = \begin{pmatrix} 0 & \frac{1}{2} \\ -\frac{1}{2} & 0 \end{pmatrix} \tag{1.7.5}$$

while Λ^a are the generators of $\mathbb{U}_{\mathscr{S}\mathscr{K}}$ in the symplectic representation $\mathbf{W}$. By

[11] We name a Lie algebra $\mathbb{U}_{\mathscr{S}\mathscr{K}}$ special Kählerian if the corresponding Lie group $\mathrm{U}_{\mathscr{S}\mathscr{K}}$ modded by its maximal compact subgroup $\mathrm{H}_{\mathscr{S}\mathscr{K}}$ defines a symmetric coset space $\frac{\mathrm{U}_{\mathscr{S}\mathscr{K}}}{\mathrm{H}_{\mathscr{S}\mathscr{K}}}$ that is *special Kählerian*. Similarly we name a Lie algebra $\mathbb{U}_{\mathscr{Q}}$ quaternionic if the corresponding Lie group $\mathrm{U}_{\mathscr{Q}}$ modded by its maximal compact subgroup $\mathrm{H}_{\mathscr{Q}}$ defines a symmetric coset space $\frac{\mathrm{U}_{\mathscr{Q}}}{\mathrm{H}_{\mathscr{Q}}}$ that is *quaternionic*.

$$\mathbf{C}^{\alpha\beta} \equiv \left(\begin{array}{c|c} \mathbf{0}_{(n+1)\times(n+1)} & \mathbf{1}_{(n+1)\times(n+1)} \\ \hline -\mathbf{1}_{(n+1)\times(n+1)} & \mathbf{0}_{(n+1)\times(n+1)} \end{array}\right) \tag{1.7.6}$$

we denote the antisymmetric symplectic metric in $2n + 2$ dimensions, n being the complex dimension of the Special Kähler manifold $\frac{\mathbb{U}_{\mathscr{SK}}}{\mathbb{H}_{\mathscr{SK}}}$. The symplectic character of the representation $\mathbf{W}$ is asserted by the identity:

$$\Lambda^a \, \mathbf{C} + \mathbf{C} \, \left(\Lambda^a\right)^T = 0 \tag{1.7.7}$$

The fundamental doublet representation of $\mathrm{SL}(2, \mathbb{R})_{\mathrm{E}}$ is also symplectic and we have denoted by $\varepsilon^{ij} = \begin{pmatrix} 0 & 1 \\ -1 & 0 \end{pmatrix}$ the 2-dimensional symplectic metric, so that:

$$\lambda^x \, \varepsilon + \varepsilon \, \left(\lambda^x\right)^T = 0, \tag{1.7.8}$$

The matrices $(K_a)^{\alpha\beta} = (K_a)^{\beta\alpha}$ and $(k_x)^{ij} = \left(k_y\right)^{ji}$ are just symmetric matrices in one-to-one correspondence with the generators of $\mathbb{U}_{\mathscr{Q}}$ and $\mathrm{SL}(2, \mathbb{R})$, respectively. Implementing Jacobi identities we find the following relations:

$$K_a \Lambda^c + \Lambda^c K_a = f^{bc}{}_a K_b, \quad k_x \lambda^y + \lambda^y k_x = f^{yz}{}_x k_z,$$

which admit the unique solution:

$$K_a = c_1 \, \mathbf{g}_{ab} \, \Lambda^b \mathbf{C}, \quad ; \quad k_x = c_2 \, \mathbf{g}_{xy} \, \lambda^y \varepsilon \tag{1.7.9}$$

where $\mathbf{g}_{ab}$, $\mathbf{g}_{xy}$ are the Cartan-Killing metrics on the algebras $\mathbb{U}_{\mathscr{SK}}$ and $\mathrm{SL}(2, \mathbb{R})$, respectively and c_1 and c_2 are two arbitrary constants. These latter can always be reabsorbed into the normalization of the generators $\mathbf{W}^{i\alpha}$ and correspondingly set to one. Hence the algebra (1.7.4) can always be put into the following elegant form:

$$\begin{aligned}
&[T^a, T^b] = f^{ab}{}_c \, T^c \\
&[L^x, L^y] = f^{xy}{}_z \, L^z, \\
&[T^a, \mathbf{W}^{i\alpha}] = (\Lambda^a)^\alpha{}_\beta \, \mathbf{W}^{i\beta}, \\
&[L^x, \mathbf{W}^{i\alpha}] = (\lambda^x)^i{}_j \, \mathbf{W}^{j\alpha}, \\
&[\mathbf{W}^{i\alpha}, \mathbf{W}^{j\beta}] = \varepsilon^{ij} \, (\Lambda_a)^{\alpha\beta} \, T^a + \mathbf{C}^{\alpha\beta} \, \lambda_x^{ij} \, L^x
\end{aligned} \tag{1.7.10}$$

where we have used the convention that symplectic indices are raised and lowered with the symplectic metric, while adjoint representation indices are raised and lowered with the Cartan-Killing metric.

We name (1.7.10) the golden splitting of quaternionic Lie algebras and it is obviously an intrinsic property of certain Lie algebras that might have been discovered by Killing, Cartan or Weyl if they had searched for it, independently of any supersymmetry or dimensional reduction of supergravity theories. It is the algebraic basis of the c-map and it has far reaching geometrical consequences.

As we emphasized above, starting from Eq. (1.7.10) we can embark on the programme of classifying all pairs of Lie algebras $(\mathbb{U}_{\mathscr{Q}}, \mathbb{U}_{\mathscr{SK}})$ whose structure fits into such a presentation with the additional necessary constraint that the dimension $2n + 2$ of the representation $\mathbf{W}$ should be consistent with

$$2n = \dim [\mathbb{U}_{\mathscr{SK}}] - \dim [\mathbb{H}_{\mathscr{SK}}] \tag{1.7.11}$$

the subalgebra $\mathbb{H}_{\mathscr{SK}} \subset \mathbb{U}_{\mathscr{SK}}$ being compact.

The result of such a scanning leads to the classfication of all the homogeneous special Kähler manifolds and of their quaternionic images through the c-map, which will be presented in Chap. 5.

Here we illustrate the first example of the golden splitting with the case of the $\mathfrak{g}_2$ Lie algebra.

1.7.1 *The Golden Splitting of the Quaternionic Algebra* $\mathfrak{g}_2$

The Lie algebra $\mathfrak{g}_2$ is quaternionic since it contains two $\mathfrak{a}_1 \sim \mathfrak{sl}(2, \mathbb{C})$ subalgebras with respect to which the adjoint representation decomposes as follows:

$$\text{adj}\,[\mathfrak{g}] = (\text{adj}\,[\mathfrak{sl}(2,\mathbb{C})_E]\ ,\ \mathbf{1}) \oplus (\mathbf{1}\,,\ \text{adj}\,[\mathfrak{sl}(2,\mathbb{C})]) \oplus (\mathbf{2}\,,\ \mathbf{4}) \tag{1.7.12}$$

where $\mathbf{4}$, which is the present instance of the symplectic $\mathbf{W}$, denotes the $J = \frac{3}{2}$ irreducible representation of the Lie algebra $\mathfrak{so}(3, \mathbb{C}) \sim \mathfrak{sl}(2, \mathbb{C})$.

To show this we begin to analyse the $\mathbf{W}$-representation proving that it is symplectic. To this effect we find it convenient to restrict our attention to the maximally split real section of the algebra.

1.7.1.1 The $J = \frac{3}{2}$-Representation of SL(2, $\mathbb{R}$)

The group SL(2, $\mathbb{R}$) is also locally isomorphic to SO(1, 2) and the fundamental representation of the first corresponds to the spin $J = \frac{1}{2}$ of the latter. The spin $J = \frac{3}{2}$ representation is obviously four-dimensional and, in the SL(2, $\mathbb{R}$) language, it corresponds to a symmetric three-index tensor t_{abc}. Let us explicitly construct the 4×4 matrices of such a representation. This is easily done by choosing an order for the four independent components of the symmetric tensor t_{abc}. For instance we can identify the four axes of the representation with $t_{111}, t_{112}, t_{122}, t_{222}$. So doing, the image of the group element:

$$\mathfrak{A} = \begin{pmatrix} a & b \\ c & d \end{pmatrix} \quad ; \quad a\,d - b\,c = 1 \tag{1.7.13}$$

in the cubic symmetric tensor product representation is the following 4×4 matrix:

$$\mathscr{D}_3\,(\mathfrak{A}) = \begin{pmatrix} a^3 & 3a^2b & 3ab^2 & b^3 \\ a^2c & da^2+2bca & cb^2+2adb & b^2d \\ ac^2 & bc^2+2adc & ad^2+2bcd & bd^2 \\ c^3 & 3c^2d & 3cd^2 & d^3 \end{pmatrix} \tag{1.7.14}$$

By explicit evaluation we can easily check that:

$$\mathscr{D}_3^T\,(\mathfrak{A})\,\widehat{\mathbf{C}}_4\,\mathscr{D}_3\,(A) = \widehat{\mathbf{C}}_4 \quad \text{where} \quad \widehat{\mathbf{C}}_4 = \begin{pmatrix} 0 & 0 & 0 & 1 \\ 0 & 0 & -3 & 0 \\ 0 & 3 & 0 & 0 \\ -1 & 0 & 0 & 0 \end{pmatrix} \tag{1.7.15}$$

Since $\widehat{\mathbf{C}}_4$ is antisymmetric, Eq. (1.7.15) is already a clear indication that the triple symmetric representation defines a symplectic embedding. To make this manifest it suffices to change basis. Consider the matrix:

$$S = \begin{pmatrix} 0 & 1 & 0 & 0 \\ -\frac{1}{\sqrt{3}} & 0 & 0 & 0 \\ 0 & 0 & \frac{1}{\sqrt{3}} & 0 \\ 0 & 0 & 0 & 1 \end{pmatrix} \tag{1.7.16}$$

and define:

$$\Lambda\,(\mathfrak{A}) = S^{-1}\,D_3\,(\mathfrak{A})\,S \tag{1.7.17}$$

We can easily check that:

$$\Lambda^T\,(\mathfrak{A})\,\mathbf{C}_4\,\Lambda\,(\mathfrak{A}) = \mathbf{C}_4 \quad \text{where} \quad \mathbf{C}_4 = \begin{pmatrix} 0 & 0 & 1 & 0 \\ 0 & 0 & 0 & 1 \\ -1 & 0 & 0 & 0 \\ 0 & -1 & 0 & 0 \end{pmatrix} \tag{1.7.18}$$

So we have indeed constructed a standard symplectic embedding $\mathrm{SL}(2,\mathbb{R}) \mapsto \mathrm{Sp}(4,\mathbb{R})$ whose explicit form is the following:

$$\mathfrak{A} = \begin{pmatrix} a & b \\ c & d \end{pmatrix} \mapsto \left(\begin{array}{cc|cc} da^2+2bca & -\sqrt{3}a^2c & -cb^2-2adb & -\sqrt{3}b^2d \\ -\sqrt{3}a^2b & a^3 & \sqrt{3}ab^2 & b^3 \\ \hline -bc^2-2adc & \sqrt{3}ac^2 & ad^2+2bcd & \sqrt{3}bd^2 \\ -\sqrt{3}c^2d & c^3 & \sqrt{3}cd^2 & d^3 \end{array}\right) \equiv \Lambda\,(\mathfrak{A}) \tag{1.7.19}$$

The 2×2 blocks A, B, C, D of the 4×4 symplectic matrix Λ ($\mathfrak{A}$) are easily readable from Eq. (1.7.19).

1.7.1.2 Putting the $\mathfrak{g}_2$ Lie Algebra in the Quaternionic Form

Explicitly the $\mathfrak{g}_2$ Lie algebra can be cast into the form (1.7.4) in the following way.

First we single out the two relevant $\mathfrak{sl}(2, \mathbb{C})$ subalgebras. The Ehlers algebra is associated with the highest root and we have:

$$L_0^E = \frac{1}{\sqrt{3}} H_2 \quad ; \quad L_\pm^E = \sqrt{\frac{2}{3}} E^{\pm(3\alpha_1+2\alpha_2)} \tag{1.7.20}$$

while the special Kähler subalgebra $\mathbb{U}_{\mathscr{SK}} = \mathfrak{sl}(2, \mathbb{C})$ is associated with the first simple root orthogonal to the highest one and we have:

$$L_0 = H_1 \quad ; \quad L_\pm = \sqrt{2}\, E^{\pm\alpha_1} \tag{1.7.21}$$

Then we can arrange the remaining eight generators in the tensor $W^{i\beta}$ as follows:

$$\begin{aligned} W^{1M} &= \sqrt{\frac{2}{3}} \left(E^{\alpha_1+\alpha_2}, \, E^{\alpha_2}, \, E^{2\alpha_1+\alpha_2}, \, E^{3\alpha_1+\alpha_2}\right) \\ W^{2M} &= \sqrt{\frac{2}{3}} \left(-E^{-2\alpha_1-\alpha_2}, \, -E^{-3\alpha_1-\alpha_2}, \, E^{-\alpha_1-\alpha_2}, \, E^{-\alpha_2}\right) \end{aligned} \tag{1.7.22}$$

Calculating the commutators of W^{iM} with the generators of the two $\mathfrak{sl}(2)$ algebras we find:

$$\begin{aligned} \left[L_0^E, \begin{pmatrix} W^1 \\ W^2 \end{pmatrix}\right] &= \left(\begin{array}{c|c} \frac{1}{2}\mathbf{1} & 0 \\ \hline 0 & -\frac{1}{2}\mathbf{1} \end{array}\right) \begin{pmatrix} W^1 \\ W^2 \end{pmatrix} \\ \left[L_+^E, \begin{pmatrix} W^1 \\ W^2 \end{pmatrix}\right] &= \left(\begin{array}{c|c} 0 & 0 \\ \hline -\mathbf{1} & 0 \end{array}\right) \begin{pmatrix} W^1 \\ W^2 \end{pmatrix} \\ \left[L_-^E, \begin{pmatrix} W^1 \\ W^2 \end{pmatrix}\right] &= \left(\begin{array}{c|c} 0 & -\mathbf{1} \\ \hline 0 & 0 \end{array}\right) \begin{pmatrix} W^1 \\ W^2 \end{pmatrix} \end{aligned} \tag{1.7.23}$$

and:

$$\begin{aligned} \left[L_0, \begin{pmatrix} W^1 \\ W^2 \end{pmatrix}\right] &= -\left(\begin{array}{c|c} U_0 & 0 \\ \hline 0 & U_0 \end{array}\right) \begin{pmatrix} W^1 \\ W^2 \end{pmatrix} \\ \left[L_\pm, \begin{pmatrix} W^1 \\ W^2 \end{pmatrix}\right] &= -\left(\begin{array}{c|c} U_\pm & 0 \\ \hline 0 & U_\pm \end{array}\right) \begin{pmatrix} W^1 \\ W^2 \end{pmatrix} \end{aligned} \tag{1.7.24}$$

where:

$$U_0 = \begin{pmatrix} \frac{1}{2} & 0 & 0 & 0 \\ 0 & \frac{3}{2} & 0 & 0 \\ 0 & 0 & -\frac{1}{2} & 0 \\ 0 & 0 & 0 & -\frac{3}{2} \end{pmatrix}$$
$$U_+ = \begin{pmatrix} 0 & 0 & -2 & 0 \\ -\sqrt{3} & 0 & 0 & 0 \\ 0 & 0 & 0 & \sqrt{3} \\ 0 & 0 & 0 & 0 \end{pmatrix}$$
$$U_- = \begin{pmatrix} 0 & -\sqrt{3} & 0 & 0 \\ 0 & 0 & 0 & 0 \\ -2 & 0 & 0 & 0 \\ 0 & 0 & \sqrt{3} & 0 \end{pmatrix} \tag{1.7.25}$$

which are the generators of $\mathfrak{sl}(2,\mathbb{C})$ in the symplectic embedding (1.7.19) as it can be easily verified by considering the embedding of a group element infinitesimally closed to the identity:

$$\begin{pmatrix} a & b \\ c & d \end{pmatrix} = \begin{pmatrix} 1 + \frac{1}{2}\,\varepsilon_0 & \varepsilon_+ \\ \varepsilon_- & 1 - \frac{1}{2}\,\varepsilon_0 \end{pmatrix} \tag{1.7.26}$$

and collecting the matrix coefficients of the first order terms in ε_0 and $\varepsilon_\pm$.

1.7.2 Chevalley-Serre Basis

We utilize the case of the $\mathfrak{g}_2$ algebra to illustrate another canonical presentation of the Lie algebra commutation relations that is named the presentation in terms of Chevalley-Serre triples. It is the analogue for Lie algebras of the presentation of discrete groups through generators and relations and proves to be quite useful in several applications. Given a simple Lie algebra of rank r defined by its Cartan matrix C_{ij}, a Chevalley-Serre basis is given by r-triplets of generators:

$$(h_i\,,\,e_i\,,\,f_i) \quad ; \quad i = 1,\ldots r \tag{1.7.27}$$

such that the following commutation relations are satisfied:

$$\begin{aligned} \left[h_i\,,\,h_j\right] &= 0 \\ \left[h_i\,,\,e_j\right] &= C_{ij}\,e_j \\ \left[h_i\,,\,f_j\right] &= -C_{ij}\,f_j \\ \left[e_i\,,\,f_j\right] &= \delta_{ij}\,h_i \end{aligned}$$

$$\begin{aligned}\text{adj}\,[e_i]^{(C_{ji}+1)}\left(e_j\right) &= 0\\ \text{adj}\,[f_i]^{(C_{ji}+1)}\left(f_j\right) &= 0\end{aligned}\tag{1.7.28}$$

When such r-triplets are given the entire algebra is defined. Indeed all the other generators are constructed by commuting these ones modulo the relations (1.7.28). For simply-laced finite simple Lie algebras a Chevalley basis is easily constructed in terms of simple roots. Let α_i denote the simple roots, then it suffices to set:

$$(h_i\,,\,e_i\,,\,f_i) = \left(H_{\alpha_i}\,,\,E^{\alpha_i}\,,\,E^{-\alpha_i}\right)\tag{1.7.29}$$

where $H_{\alpha_i} \equiv \alpha_i \cdot H$ are the Cartan generator associated with the simple roots and $E^{\pm\alpha_i}$ are the step operators respectively associated with the simple roots and their negative.

1.7.2.1 The $\mathfrak{g}_2$ Lie Algebra in Terms of Chevalley Triples

Let us rewrite the commutation relations of the $\mathfrak{g}_{(2,2)}$ in terms of triples of Chevalley generators.

Since the algebra has rank two there are two fundamental triples of Chevalley generators:

$$(\mathscr{H}_1, e_1, f_1) \quad ; \quad (\mathscr{H}_2, e_2, f_2)\tag{1.7.30}$$

with the following commutation relations:

$$\begin{array}{llll}[\mathscr{H}_2, e_2] = 2e_2 & [\mathscr{H}_1, e_2] = -3e_2 & [\mathscr{H}_2, f_2] = -2f_2 & [\mathscr{H}_1, f_2] = 3f_2\\ [\mathscr{H}_2, e_1] = -e_1 & [\mathscr{H}_1, e_1] = 2e_1 & [\mathscr{H}_2, f_1] = f_1 & [\mathscr{H}_1, f_1] = -2f_1\\ [e_2, f_2] = \mathscr{H}_2 & [e_2, f_1] = 0 & [e_1, f_1] = \mathscr{H}_1 & [e_1, f_2] = 0\end{array}\tag{1.7.31}$$

The remaining basis elements are defined as follows:

$$\begin{array}{llll}e_3 = [e_1, e_2] & e_4 = \frac{1}{2}\,[e_1, e_3] & e_5 = \frac{1}{3}\,[e_4, e_1] & e_6 = [e_2, e_5]\\ f_3 = [f_2, f_1] & f_4 = \frac{1}{2}\,[f_3, f_1] & f_5 = \frac{1}{3}\,[f_1, f_4] & f_6 = [f_5, f_2]\end{array}\tag{1.7.32}$$

and satisfy the following Serre relations:

$$[e_2, e_3] = [e_5, e_1] = [f_2, f_3] = [f_5, f_1] = 0\tag{1.7.33}$$

The Chevalley form of the commutation relation is obtained from the standard Cartan Weyl basis introducing the following identifications:

$$
\begin{aligned}
e_1 &= \sqrt{2}E^{\alpha_1} \; ; \; e_2 = \sqrt{\tfrac{2}{3}}E^{\alpha_2} \\
e_3 &= \sqrt{2}E^{\alpha_3} \; ; \; e_4 = \sqrt{2}E^{\alpha_4} \\
e_5 &= \sqrt{\tfrac{2}{3}}E^{\alpha_5} \; ; \; e_6 = \sqrt{\tfrac{2}{3}}E^{\alpha_6} \\
f_1 &= \sqrt{2}E^{-\alpha_1} \; ; \; f_2 = \sqrt{\tfrac{2}{3}}E^{-\alpha_2} \\
f_3 &= \sqrt{2}E^{-\alpha_3} \; ; \; f_4 = \sqrt{2}E^{-\alpha_4} \\
f_5 &= \sqrt{\tfrac{2}{3}}E^{-\alpha_5} \; ; \; f_6 = \sqrt{\tfrac{2}{3}}E^{-\alpha_6}
\end{aligned}
\tag{1.7.34}
$$

and

$$
\mathscr{H}_1 = 2\alpha_1 \cdot H \quad ; \quad \mathscr{H}_2 = \tfrac{2}{3}\alpha_2 \cdot H \tag{1.7.35}
$$

1.8 The Lie Algebra $\mathfrak{f}_4$ and its Fundamental Representation

Another exceptional Lie algebra that is also quaternionic and will be of concern to us in the sequel is $\mathfrak{f}_4$. We consider it here and we construct its fundamental representation for later use. $\mathfrak{f}_4$ has rank $r = 4$ and we cannot visualize its root system as we easily did for the planar $\mathfrak{g}_2$ system. In any case the Lie algebra structure is codified by the Dynkin diagram presented in Fig. 1.15. We show how we can explicitly construct the fundamental and the adjoint representations of this exceptional, non simply laced Lie algebra.

Calling $y_{1,2,3,4}$ a basis of orthonormal vectors:

$$
y_i \cdot y_j = \delta_{ij} \tag{1.8.1}
$$

a possible choice of simple roots β_i which reproduces the Cartan matrix encoded in the Dynkin diagram (1.15) is the following:

$$
\begin{aligned}
\beta_1 &= -y_1 - y_2 - y_3 + y_4 \\
\beta_2 &= 2\,y_3 \\
\beta_3 &= y_2 - y_3 \\
\beta_4 &= y_1 - y_2
\end{aligned}
\tag{1.8.2}
$$

With this basis of simple roots the full root system composed of 48 vectors is given by:

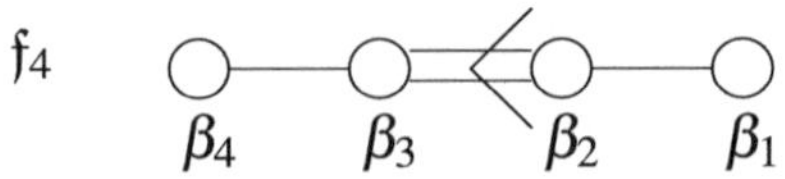

Fig. 1.15 The Dynkin diagram of F_4 and the labeling of simple roots

Table 1.3 List of the positive roots of the exceptional Lie algebra $\mathfrak{f}_4$. In this table the first column is the name of the root, the second column gives its decomposition in terms of simple roots, while the last column provides the component of the root vector in $\mathbb{R}^4$

$\beta_1 = \beta_1$	$= \{-1, -1, -1, 1\}$
$\beta_2 = \beta_2$	$= \{0, 0, 2, 0\}$
$\beta_3 = \beta_3$	$= \{0, 1, -1, 0\}$
$\beta_4 = \beta_4$	$= \{1, -1, 0, 0\}$
$\beta_5 = \beta_1 + \beta_2$	$= \{-1, -1, 1, 1\}$
$\beta_6 = \beta_2 + \beta_3$	$= \{0, 1, 1, 0\}$
$\beta_7 = \beta_3 + \beta_4$	$= \{1, 0, -1, 0\}$
$\beta_8 = \beta_1 + \beta_2 + \beta_3$	$= \{-1, 0, 0, 1\}$
$\beta_9 = \beta_2 + 2\beta_3$	$= \{0, 2, 0, 0\}$
$\beta_{10} = \beta_2 + \beta_3 + \beta_4$	$= \{1, 0, 1, 0\}$
$\beta_{11} = \beta_1 + \beta_2 + 2\beta_3$	$= \{-1, 1, -1, 1\}$
$\beta_{12} = \beta_1 + \beta_2 + \beta_3 + \beta_4$	$= \{0, -1, 0, 1\}$
$\beta_{13} = \beta_2 + 2\beta_3 + \beta_4$	$= \{1, 1, 0, 0\}$
$\beta_{14} = \beta_1 + 2\beta_2 + 2\beta_3$	$= \{-1, 1, 1, 1\}$
$\beta_{15} = \beta_1 + \beta_2 + 2\beta_3 + \beta_4$	$= \{0, 0, -1, 1\}$
$\beta_{16} = \beta_2 + 2\beta_3 + 2\beta_4$	$= \{2, 0, 0, 0\}$
$\beta_{17} = \beta_1 + 2\beta_2 + 2\beta_3 + \beta_4$	$= \{0, 0, 1, 1\}$
$\beta_{18} = \beta_1 + \beta_2 + 2\beta_3 + 2\beta_4$	$= \{1, -1, -1, 1\}$
$\beta_{19} = \beta_1 + 2\beta_2 + 3\beta_3 + \beta_4$	$= \{0, 1, 0, 1\}$
$\beta_{20} = \beta_1 + 2\beta_2 + 2\beta_3 + 2\beta_4$	$= \{1, -1, 1, 1\}$
$\beta_{21} = \beta_1 + 2\beta_2 + 3\beta_3 + 2\beta_4$	$= \{1, 0, 0, 1\}$
$\beta_{22} = \beta_1 + 2\beta_2 + 4\beta_3 + 2\beta_4$	$= \{1, 1, -1, 1\}$
$\beta_{23} = \beta_1 + 3\beta_2 + 4\beta_3 + 2\beta_4$	$= \{1, 1, 1, 1\}$
$\beta_{24} = 2\beta_1 + 3\beta_2 + 4\beta_3 + 2\beta_4$	$= \{0, 0, 0, 2\}$

$$\Delta_{\mathfrak{f}_4} \equiv \underbrace{\pm y_i \pm y_j}_{24 \text{ roots}} \; ; \; \underbrace{\pm y_i}_{8 \text{ roots}} \; ; \; \underbrace{\pm y_1 \pm y_2 \pm y_3 \pm y_4}_{16 \text{ roots}} \tag{1.8.3}$$

and one can list the positive roots by height as displayed in Table 1.3. Since the considered Lie algebra is not simply-laced the 24 positive roots split into two subsets of 12 long roots α^ℓ and 12 short roots α^s. They are displayed in Tables 1.4 and 1.5, respectively.

Calling Δ_ℓ and Δ_s the two subsets we have the following structure:

$$\forall \alpha^\ell, \beta^\ell \in \Delta_\ell : \alpha^\ell + \beta^\ell = \begin{Bmatrix} \text{not a root or} \\ \gamma^\ell \in \Delta_\ell \end{Bmatrix}$$

$$\forall \alpha^\ell \in \Delta_\ell \text{ and } \forall \beta^s \in \Delta_s : \alpha^\ell + \beta^s = \begin{Bmatrix} \text{not a root or} \\ \gamma^s \in \Delta_s \end{Bmatrix}$$

Table 1.4 The Δ_ℓ set of the 12 long positive roots in the $\mathfrak{f}_4$ root system

	$\mathfrak{f}_4$ root labels	$\mathfrak{f}_4$ root in eucl. basis	Root ordered by height
α_1^ℓ	$\{0, 1, 0, 0\}$	$2\,y_3$	β_2
α_2^ℓ	$\{1, 0, 0, 0\}$	$-y_1 - y_2 - y_3 + y_4$	β_1
α_3^ℓ	$\{1, 1, 0, 0\}$	$-y_1 - y_2 + y_3 + y_4$	β_3
α_4^ℓ	$\{0, 1, 2, 0\}$	$2\,y_2$	β_9
α_5^ℓ	$\{1, 1, 2, 0\}$	$-y_1 + y_2 - y_3 + y_4$	β_{11}
α_6^ℓ	$\{1, 2, 2, 0\}$	$-y_1 + y_2 + y_3 + y_4$	β_{14}
α_7^ℓ	$\{0, 1, 2, 2\}$	$2\,y_1$	β_{16}
α_8^ℓ	$\{1, 1, 2, 2\}$	$y_1 - y_2 - y_3 + y_4$	β_{18}
α_9^ℓ	$\{1, 2, 2, 2\}$	$y_1 - y_2 + y_3 + y_4$	β_{20}
α_{10}^ℓ	$\{1, 2, 4, 2\}$	$y_1 + y_2 - y_3 + y_4$	β_{22}
α_{11}^ℓ	$\{1, 3, 4, 2\}$	$y_1 + y_2 + y_3 + y_4$	β_{23}
α_{12}^ℓ	$\{2, 3, 4, 2\}$	$2\,y_4$	β_{24}

Table 1.5 The Δ_s set of 12 short positive roots in the $\mathfrak{f}_4$ root system

	$\mathfrak{f}_4$ root labels	$\mathfrak{f}_4$ root in eucl. basis	Root ordered by height
α_1^s	$\{0, 0, 0, 1\}$	$y_1 - y_2$	β_4
α_2^s	$\{0, 0, 1, 0\}$	$y_2 - y_3$	β_3
α_3^s	$\{0, 1, 1, 0\}$	$y_2 + y_3$	β_6
α_4^s	$\{0, 0, 1, 1\}$	$y_1 - y_3$	β_7
α_5^s	$\{1, 1, 1, 0\}$	$-y_1 + y_4$	β_8
α_6^s	$\{0, 1, 1, 1\}$	$y_1 + y_3$	β_{10}
α_7^s	$\{1, 1, 1, 1\}$	$-y_2 + y_4$	β_{12}
α_8^s	$\{0, 1, 2, 1\}$	$y_1 + y_2$	β_{13}
α_9^s	$\{1, 1, 2, 1\}$	$-y_3 + y_4$	β_{15}
α_{10}^s	$\{1, 2, 2, 1\}$	$y_3 + y_4$	β_{17}
α_{11}^s	$\{1, 2, 3, 1\}$	$y_2 + y_4$	β_{19}
α_{12}^s	$\{1, 2, 3, 2\}$	$y_1 + y_4$	β_{21}

$$\forall \alpha^s\,,\ \beta^s\ \in\ \Delta_s : \alpha^s + \beta^s\ =\ \left\{\begin{array}{l}\text{not a root or}\\ \gamma^s\ \in\ \Delta_s\ \text{ or}\\ \gamma^\ell\ \in\ \Delta_\ell\end{array}\right\} \tag{1.8.4}$$

The standard Cartan-Weyl form of the Lie algebra is as follows:

$$\left[\mathscr{H}_i\,,\ E^{\pm\beta}\right] = \pm\,\beta^i\ E^{\pm\beta_l} \tag{1.8.5}$$

$$\left[E^\beta\,,\ E^{-\beta}\right] =\ \beta\ \cdot\ \mathscr{H} \tag{1.8.6}$$

$$\left[E^{\beta}\,,\,E^{\gamma}\right]=\begin{cases}N_{\beta\gamma}\,E^{\beta+\gamma} & \text{if }\beta+\gamma\text{ is a root}\\ 0 & \text{if }\beta+\gamma\text{ is not a root}\end{cases} \tag{1.8.7}$$

where $N_{\beta\gamma}$ are numbers that can be worked constructing an explicit representation of the Lie algebra.

In the following three tables (1.8.8), (1.8.9), (1.8.10) we exhibit the values of $N_{\beta\gamma}$ for the $\mathfrak{f}_4$ Lie algebra.

α_1^ℓ	α_2^ℓ	α_3^ℓ	α_4^ℓ	α_5^ℓ	α_6^ℓ	α_7^ℓ	α_8^ℓ	α_9^ℓ	α_{10}^ℓ	α_{11}^ℓ	α_{12}^ℓ	
0	$-\sqrt{2}$	0	0	$-\sqrt{2}$	0	0	$-\sqrt{2}$	0	$\sqrt{2}$	0	0	α_1^ℓ
$\sqrt{2}$	0	0	$-\sqrt{2}$	0	0	$-\sqrt{2}$	0	0	0	$-\sqrt{2}$	0	α_2^ℓ
0	0	0	$-\sqrt{2}$	0	0	$-\sqrt{2}$	0	0	$-\sqrt{2}$	0	0	α_3^ℓ
0	$\sqrt{2}$	$\sqrt{2}$	0	0	0	0	$-\sqrt{2}$	$\sqrt{2}$	0	0	0	α_4^ℓ
$\sqrt{2}$	0	0	0	0	0	$\sqrt{2}$	0	$\sqrt{2}$	0	0	0	α_5^ℓ
0	0	0	0	0	0	$-\sqrt{2}$	$-\sqrt{2}$	0	0	0	0	α_6^ℓ
0	$\sqrt{2}$	$\sqrt{2}$	0	$-\sqrt{2}$	$\sqrt{2}$	0	0	0	0	0	0	α_7^ℓ
$\sqrt{2}$	0	0	$\sqrt{2}$	0	$\sqrt{2}$	0	0	0	0	0	0	α_8^ℓ
0	0	0	$-\sqrt{2}$	$-\sqrt{2}$	0	0	0	0	0	0	0	α_9^ℓ
$-\sqrt{2}$	0	$\sqrt{2}$	0	0	0	0	0	0	0	0	0	α_{10}^ℓ
0	$\sqrt{2}$	0	0	0	0	0	0	0	0	0	0	α_{11}^ℓ
0	0	0	0	0	0	0	0	0	0	0	0	α_{12}^ℓ

$N_{\alpha^\ell\beta^\ell}$ (1.8.8)

α_1^s	α_2^s	α_3^s	α_4^s	α_5^s	α_6^s	α_7^s	α_8^s	α_9^s	α_{10}^s	α_{11}^s	α_{12}^s	
0	$\sqrt{2}$	0	$-\sqrt{2}$	0	0	0	0	$\sqrt{2}$	0	0	0	α_1^ℓ
0	0	$-\sqrt{2}$	0	0	$-\sqrt{2}$	0	$\sqrt{2}$	0	0	0	0	α_2^ℓ
0	$-\sqrt{2}$	0	$\sqrt{2}$	0	0	0	$-\sqrt{2}$	0	0	0	0	α_3^ℓ
$\sqrt{2}$	0	0	0	0	0	$\sqrt{2}$	0	0	0	0	0	α_4^ℓ
$-\sqrt{2}$	0	0	0	0	$-\sqrt{2}$	0	0	0	0	0	0	α_5^ℓ
$\sqrt{2}$	0	0	$\sqrt{2}$	0	0	0	0	0	0	0	0	α_6^ℓ
0	0	0	0	$-\sqrt{2}$	0	0	0	0	0	0	0	α_7^ℓ
0	0	$\sqrt{2}$	0	0	0	0	0	0	0	0	0	α_8^ℓ
0	$\sqrt{2}$	0	0	0	0	0	0	0	0	0	0	α_9^ℓ
0	0	0	0	0	0	0	0	0	0	0	0	α_{10}^ℓ
0	0	0	0	0	0	0	0	0	0	0	0	α_{11}^ℓ
0	0	0	0	0	0	0	0	0	0	0	0	α_{12}^ℓ

$N_{\alpha^\ell\beta^s}$ (1.8.9)

α^s_1	α^s_2	α^s_3	α^s_4	α^s_5	α^s_6	α^s_7	α^s_8	α^s_9	α^s_{10}	α^s_{11}	α^s_{12}	
0	1	-1	0	-1	0	0	$\sqrt{2}$	$-\sqrt{2}$	$\sqrt{2}$	1	0	α^s_1
-1	0	$\sqrt{2}$	0	$\sqrt{2}$	1	-1	0	0	-1	0	$\sqrt{2}$	α^s_2
1	$-\sqrt{2}$	0	1	$-\sqrt{2}$	0	-1	0	1	0	0	$\sqrt{2}$	α^s_3
0	0	-1	0	1	$\sqrt{2}$	$\sqrt{2}$	0	0	1	$-\sqrt{2}$	0	α^s_4
1	$-\sqrt{2}$	$\sqrt{2}$	-1	0	1	0	1	0	0	0	$\sqrt{2}$	α^s_5
0	-1	0	$-\sqrt{2}$	-1	0	$\sqrt{2}$	0	1	0	$\sqrt{2}$	0	α^s_6
0	1	1	$-\sqrt{2}$	0	$-\sqrt{2}$	0	1	0	0	$\sqrt{2}$	0	α^s_7
$-\sqrt{2}$	0	0	0	-1	0	-1	0	$\sqrt{2}$	$\sqrt{2}$	0	0	α^s_8
$\sqrt{2}$	0	-1	0	0	-1	0	$-\sqrt{2}$	0	$-\sqrt{2}$	0	0	α^s_9
$-\sqrt{2}$	1	0	-1	0	0	0	$-\sqrt{2}$	$\sqrt{2}$	0	0	0	α^s_{10}
-1	0	0	$\sqrt{2}$	0	$-\sqrt{2}$	$-\sqrt{2}$	0	0	0	0	0	α^s_{11}
0	$-\sqrt{2}$	$-\sqrt{2}$	0	$-\sqrt{2}$	0	0	0	0	0	0	0	α^s_{12}

$\underbrace{\qquad\qquad}_{N_{\alpha^s\beta^s}}$

(1.8.10)

The ordering of long and short roots is that displayed in Tables: 1.4 and 1.5. The explicit determination of the tensor $N_{\alpha\beta}$ was performed via the explicit construction of the fundamental 26-dimensional representation of this Lie algebra which we describe in the next subsection.

1.8.1 Explicit Construction of the Fundamental and Adjoint Representation of $\mathfrak{f}_4$

The semisimple complex Lie algebra $\mathfrak{f}_4$ is defined by the Dynkin diagram in Fig. 1.15 and a set of simple roots corresponding to such diagram was provided in Eq. (1.8.2). A complete list of the 24 positive roots was given in Table 1.3. The roots were further subdivided into the set of 12 long roots and 12 short roots respectively listed in Tables 1.4 and 1.5. The adjoint representation of $\mathfrak{f}_4$ is 52-dimensional, while its fundamental representation is 26-dimensional. This dimensionality is true for all real sections of the Lie algebra but the explicit structure of the representation is quite different in each real section. Here we are interested in the maximally split real section $\mathfrak{f}_4$. For such a section we have a maximal, regularly embedded, subgroup $\mathfrak{so}(5,4) \subset \mathfrak{f}_{4(4)}$. The decomposition of the representations with respect to this particular subgroup is the essential instrument for their actual construction. For the adjoint representation we have the decomposition:

$$\underbrace{\mathbf{52}}_{\text{adj}\,\mathfrak{f}_{4(4)}} \overset{\mathfrak{so}(5,4)}{\Longrightarrow} \underbrace{\mathbf{36}}_{\text{adj}\,\mathfrak{so}(5,4)} \oplus \underbrace{\mathbf{16}}_{\text{spinor of }\mathfrak{so}(5,4)} \tag{1.8.11}$$

while for the fundamental one we have:

$$\underbrace{\mathbf{26}}_{\text{fundamental } \mathfrak{f}_{4(4)}} \stackrel{\mathfrak{so}(5,4)}{\Longrightarrow} \underbrace{\mathbf{9}}_{\text{vector of } \mathfrak{so}(5,4)} \oplus \underbrace{\mathbf{16}}_{\text{spinor of } \mathfrak{so}(5,4)} \oplus \underbrace{\mathbf{1}}_{\text{singlet of } \mathfrak{so}(5,4)} \tag{1.8.12}$$

In view of this, we fix our conventions for the $\mathfrak{so}(5,4)$ invariant metric as it follows

$$\eta_{AB} = \text{diag}\,\{+,+,+,+,+,-,-,-,-\} \tag{1.8.13}$$

and we perform an explicit construction of the 16×16 dimensional gamma matrices which satisfy the Clifford algebra

$$\{\Gamma_A\,,\,\Gamma_B\} = \eta_{AB}\,\mathbf{1} \tag{1.8.14}$$

and are *all completely real*. This construction is provided by the following tensor products:

$$\begin{aligned}
\Gamma_1 &= \sigma_1 \otimes \sigma_3 \otimes \mathbf{1} \otimes \mathbf{1} \\
\Gamma_2 &= \sigma_3 \otimes \sigma_3 \otimes \mathbf{1} \otimes \mathbf{1} \\
\Gamma_3 &= \mathbf{1} \otimes \sigma_1 \otimes \mathbf{1} \otimes \sigma_1 \\
\Gamma_4 &= \mathbf{1} \otimes \sigma_1 \otimes \sigma_1 \otimes \sigma_3 \\
\Gamma_5 &= \mathbf{1} \otimes \sigma_1 \otimes \sigma_3 \otimes \sigma_3 \\
\Gamma_6 &= \mathbf{1} \otimes \mathrm{i}\sigma_2 \otimes \mathbf{1} \otimes \mathbf{1} \\
\Gamma_7 &= \mathbf{1} \otimes \sigma_1 \otimes \mathrm{i}\,\sigma_2 \otimes \sigma_3 \\
\Gamma_8 &= \mathbf{1} \otimes \sigma_1 \otimes \mathbf{1} \otimes \mathrm{i}\sigma_2 \\
\Gamma_9 &= \mathrm{i}\,\sigma_2 \otimes \sigma_3 \otimes \mathbf{1} \otimes \mathbf{1}
\end{aligned} \tag{1.8.15}$$

where by σ_i we have denoted the standard Pauli matrices:

$$\sigma_1 = \begin{pmatrix} 0 & 1 \\ 1 & 0 \end{pmatrix} \quad ; \quad \sigma_2 = \begin{pmatrix} 0 & -\mathrm{i} \\ \mathrm{i} & 0 \end{pmatrix} \quad ; \quad \sigma_3 = \begin{pmatrix} 1 & 0 \\ 0 & -1 \end{pmatrix} \tag{1.8.16}$$

Moreover we introduce the C_+ charge conjugation matrix, such that:

$$\begin{aligned}
C_+ = (C_+)^T \quad &; \quad C_+^2 = \mathbf{1} \\
C_+\,\Gamma_A\,C_+ = (\Gamma_A)^T &
\end{aligned} \tag{1.8.17}$$

In the basis of Eq. (1.8.15) the explicit form of C_+ is given by:

$$C_+ = \mathrm{i}\,\sigma_2 \otimes \sigma_1 \otimes \mathrm{i}\,\sigma_2 \otimes \sigma_1 \tag{1.8.18}$$

Then we define the usual generators $J_{AB} = -J_{BA}$ of the pseudorthogonal algebra $\mathfrak{so}(5,4)$ satisfying the commutation relations:

$$[J_{AB}\,,\,J_{CD}] = \eta_{BC}\,J_{AD} - \eta_{AC}\,J_{BD} - \eta_{BD}J_{AC} + \eta_{AD}J_{BC} \tag{1.8.19}$$

and we construct the spinor and the vector representations by respectively setting:

$$J^{s}_{CD} = \frac{1}{4}\,[\Gamma_C\,,\,\Gamma_D] \quad ; \quad \left(J^{v}_{CD}\right)_A{}^{B} = \eta_{CA}\,\delta^B_D - \eta_{DA}\,\delta^B_C \tag{1.8.20}$$

In this way if v_A denote the components of a vector, ξ those of a real spinor and $\varepsilon^{AB} = -\varepsilon^{BA}$ are the parameters of an infinitesimal $\mathfrak{so}(5,4)$ rotation we can write the $\mathfrak{so}(5,4)$ transformation as follows:

$$\delta_{\mathfrak{so}(5,4)}\,v_A = 2\,\varepsilon_{AB}\,v^B \quad ; \quad \delta_{\mathfrak{so}(5,4)}\,\xi = \tfrac{1}{2}\,\varepsilon^{AB}\,\Gamma_{AB}\xi \tag{1.8.21}$$

where indices are raised and lowered with the metric (1.8.13). Furthermore we introduce the conjugate spinors via the position:

$$\overline{\xi} \equiv \xi^T\,C_+ \tag{1.8.22}$$

With these preliminaries, we are now in a position to write the explicit form of the 26-dimensional fundamental representation of $\mathfrak{f}_4$ and in this way to construct also its structure constants and hence its adjoint representation, which is our main goal.

According to Eq. (1.8.11) the parameters of an $\mathfrak{f}_4$ representation are given by an anti-symmetric tensor ε_{AB} and a spinor q. On the other hand a *vector* in the 26-dimensional representation is specified by a collection of three objects, namely a scalar ϕ, a vector v_A and a spinor ξ. The representation is constructed if we specify the $\mathfrak{f}_{4(4)}$ transformation of these objects. This is done by writing:

$$\delta_{F_{4(4)}}\begin{pmatrix}\phi\\ v_A\\ \xi\end{pmatrix} \equiv \left[\varepsilon^{AB}T_{AB} + \overline{q}\,Q\right]\begin{pmatrix}\phi\\ v_A\\ \xi\end{pmatrix} = \begin{pmatrix}\overline{q}\,\xi\\ 2\,\varepsilon_{AB}\,v^B + a\,\overline{q}\,\Gamma_A\,\xi\\ \frac{1}{2}\,\varepsilon^{AB}\,\Gamma_{AB}\,\xi - 3\,\phi\,q - \frac{1}{a}\,v^A\,\Gamma_A\,q\end{pmatrix} \tag{1.8.23}$$

where a is a numerical real arbitrary but non-null parameter. Equation (1.8.23) defines the generators T_{AB} and Q as 26×26 matrices and therefore completely specifies the fundamental representation of the Lie algebra $\mathfrak{f}_{4(4)}$. Explicitly we have:

$$T_{AB} = \left(\begin{array}{c|c|c} 0 & 0 & 0\\ \hline 0 & J^{v}_{AB} & 0\\ \hline 0 & 0 & J^{s}_{AB}\end{array}\right) \tag{1.8.24}$$

and

$$Q_\alpha = \left(\begin{array}{c|c|c} 0 & 0 & \delta^\beta_\alpha\\ \hline 0 & 0 & a\,(\Gamma_A)^\beta_\alpha\\ \hline -3\,\delta^\beta_\alpha & -\frac{1}{a}\,(\Gamma_B)^\beta_\alpha & 0\end{array}\right) \tag{1.8.25}$$

and the Lie algebra commutation relations are evaluated to be the following ones:

$$
\begin{aligned}
[T_{AB}\,,\, T_{CD}] &= \eta_{BC}\, T_{AD} - \eta_{AC}\, T_{BD} - \eta_{BD} T_{AC} + \eta_{AD} T_{BC} \\
[T_{AB}\,,\, Q] &= \tfrac{1}{2}\, \Gamma_{AB}\, Q \\
\left[Q_\alpha\,,\, Q_\beta\right] &= -\frac{1}{12}\, \left(C_+ \Gamma^{AB}\right)_{\alpha\beta}\, T_{AB}
\end{aligned}
\tag{1.8.26}
$$

Equation (1.8.26), together with Eqs. (1.8.15) and (1.8.17) provides an explicit numerical construction of the structure constants of the maximally split $\mathfrak{f}_{4(4)}$ Lie algebra. What we still have to do is to identify the relation between the tensorial basis of generators in Eq. (1.8.26) and the Cartan-Weyl basis in terms of Cartan generators and step operators. To this effect let us enumerate the 52 generators of $\mathfrak{f}_4$ in the tensorial representation according to the following table:

$\Omega_1 = T_{12}$	$\Omega_2 = T_{13}$	$\Omega_3 = T_{14}$	$\Omega_4 = T_{15}$
$\Omega_5 = T_{16}$	$\Omega_6 = T_{17}$	$\Omega_7 = T_{18}$	$\Omega_8 = T_{19}$
$\Omega_9 = T_{23}$	$\Omega_{10} = T_{24}$	$\Omega_{11} = T_{25}$	$\Omega_{12} = T_{26}$
$\Omega_{13} = T_{27}$	$\Omega_{14} = T_{28}$	$\Omega_{15} = T_{29}$	$\Omega_{16} = T_{34}$
$\Omega_{17} = T_{35}$	$\Omega_{18} = T_{36}$	$\Omega_{19} = T_{37}$	$\Omega_{20} = T_{38}$
$\Omega_{21} = T_{39}$	$\Omega_{22} = T_{45}$	$\Omega_{23} = T_{46}$	$\Omega_{24} = T_{47}$
$\Omega_{25} = T_{48}$	$\Omega_{26} = T_{49}$	$\Omega_{27} = T_{56}$	$\Omega_{28} = T_{57}$
$\Omega_{29} = T_{58}$	$\Omega_{30} = T_{59}$	$\Omega_{31} = T_{67}$	$\Omega_{32} = T_{68}$
$\Omega_{33} = T_{69}$	$\Omega_{34} = T_{78}$	$\Omega_{35} = T_{79}$	$\Omega_{36} = T_{89}$
$\Omega_{37} = Q_1$	$\Omega_{38} = Q_2$	$\Omega_{39} = Q_3$	$\Omega_{40} = Q_4$
$\Omega_{41} = Q_5$	$\Omega_{42} = Q_6$	$\Omega_{43} = Q_7$	$\Omega_{44} = Q_8$
$\Omega_{45} = Q_9$	$\Omega_{46} = Q_{10}$	$\Omega_{47} = Q_{11}$	$\Omega_{48} = Q_{12}$
$\Omega_{49} = Q_{13}$	$\Omega_{50} = Q_{14}$	$\Omega_{51} = Q_{15}$	$\Omega_{52} = Q_{16}$

(1.8.27)

Then, as Cartan subalgebra we take the linear span of the following generators:

$$
CSA \equiv \text{span}\,(\Omega_5\,,\, \Omega_{13}\,,\, \Omega_{20}\,,\, \Omega_{26}) \tag{1.8.28}
$$

and furthermore we specify the following basis:

$$
\begin{aligned}
\mathscr{H}_1 &= \Omega_5 + \Omega_{13} \;\; ; \; \mathscr{H}_2 = \Omega_5 - \Omega_{13} \\
\mathscr{H}_3 &= \Omega_{20} + \Omega_{26} \; ; \; \mathscr{H}_4 = \Omega_{20} - \Omega_{26}
\end{aligned}
\tag{1.8.29}
$$

With respect to this basis the step operators corresponding to the positive roots of $\mathfrak{f}_{4(4)}$ as ordered and displayed in Table 1.3 are those enumerated in Table 1.6. The steps operators corresponding to negative roots are obtained from those associate with positive ones via the following relation:

$$
E^{-\beta} = -\mathscr{C}\, E^{\beta}\, \mathscr{C} \tag{1.8.30}
$$

where the 26×26 symmetric matrix $\mathscr{C}$ is defined in the following way:

Table 1.6 Listing of the step operators corresponding to the positive roots of $\mathfrak{f}_4$

Name	Dynkin lab	Comp. root	Step operator E^{β} =
$\beta[1]$	$\{1,0,0,0\}$	$-y_1-y_2-y_3+y_4$	$(-\Omega_3-\Omega_8+\Omega_{23}-\Omega_{33})$
$\beta[2]$	$\{0,1,0,0\}$	$2\,y_3$	$(\Omega_{16}-\Omega_{21}+\Omega_{25}+\Omega_{36})$
$\beta[3]$	$\{0,0,1,0\}$	y_2-y_3	$(\Omega_{37}+\Omega_{39}+\Omega_{41}-\Omega_{43}+\Omega_{45}-\Omega_{47}+\Omega_{49}+\Omega_{51})$
$\beta[4]$	$\{0,0,0,1\}$	y_1-y_2	$(\Omega_{11}+\Omega_{28})$
$\beta[5]$	$\{1,1,0,0\}$	$-y_1-y_2+y_3+y_4$	$-\frac{1}{\sqrt{2}}(-\Omega_2+\Omega_7+\Omega_{18}+\Omega_{32})$
$\beta[6]$	$\{0,1,1,0\}$	y_2+y_3	$-\frac{1}{\sqrt{2}}(-\Omega_{38}-\Omega_{40}+\Omega_{42}-\Omega_{44}+\Omega_{46}-\Omega_{48}-\Omega_{50}-\Omega_{52})$
$\beta[7]$	$\{0,0,1,1\}$	y_1-y_3	$-(-\Omega_{37}-\Omega_{39}+\Omega_{41}-\Omega_{43}+\Omega_{45}-\Omega_{47}-\Omega_{49}-\Omega_{51})$
$\beta[8]$	$\{1,1,1,0\}$	$-y_1+y_4$	$-\frac{1}{2}(\Omega_{38}+\Omega_{40}-\Omega_{42}+\Omega_{44}+\Omega_{46}-\Omega_{48}-\Omega_{50}-\Omega_{52})$
$\beta[9]$	$\{0,1,2,0\}$	$2\,y_2$	$-\frac{1}{2}(\Omega_1+\Omega_6+\Omega_{12}-\Omega_{31})$
$\beta[10]$	$\{0,1,1,1\}$	y_1+y_3	$-\frac{1}{\sqrt{2}}(\Omega_{38}+\Omega_{40}+\Omega_{42}-\Omega_{44}+\Omega_{46}-\Omega_{48}+\Omega_{50}+\Omega_{52})$
$\beta[11]$	$\{1,1,2,0\}$	$-y_1+y_2-y_3+y_4$	$-\frac{1}{2\sqrt{2}}(\Omega_{10}+\Omega_{15}-\Omega_{24}+\Omega_{35})$
$\beta[12]$	$\{1,1,1,1\}$	$-y_2+y_4$	$-\frac{1}{2}(-\Omega_{38}-\Omega_{40}-\Omega_{42}+\Omega_{44}+\Omega_{46}-\Omega_{48}+\Omega_{50}+\Omega_{52})$
$\beta[13]$	$\{0,1,2,1\}$	y_1+y_2	$-\frac{1}{\sqrt{2}}(\Omega_4+\Omega_{27})$
$\beta[14]$	$\{1,2,2,0\}$	$-y_1+y_2+y_3+y_4$	$-\frac{1}{4}(-\Omega_9+\Omega_{14}+\Omega_{19}+\Omega_{34})$
$\beta[15]$	$\{1,1,2,1\}$	$-y_3+y_4$	$-\frac{1}{2}(\Omega_{22}-\Omega_{30})$
$\beta[16]$	$\{0,1,2,2\}$	$2\,y_1$	$-\frac{1}{2}(\Omega_1-\Omega_6+\Omega_{12}+\Omega_{31})$
$\beta[17]$	$\{1,2,2,1\}$	y_3+y_4	$-\frac{1}{2\sqrt{2}}(\Omega_{17}+\Omega_{29})$
$\beta[18]$	$\{1,1,2,2\}$	$y_1-y_2-y_3+y_4$	$-\frac{1}{2\sqrt{2}}(\Omega_{10}+\Omega_{15}+\Omega_{24}-\Omega_{35})$
$\beta[19]$	$\{1,2,3,1\}$	y_2+y_4	$-\frac{1}{2\sqrt{2}}(\Omega_{38}-\Omega_{40}+\Omega_{42}+\Omega_{44}+\Omega_{46}+\Omega_{48}+\Omega_{50}-\Omega_{52})$
$\beta[20]$	$\{1,2,2,2\}$	$y_1-y_2+y_3+y_4$	$-\frac{1}{4}(-\Omega_9+\Omega_{14}-\Omega_{19}-\Omega_{34})$
$\beta[21]$	$\{1,2,3,2\}$	y_1+y_4	$-\frac{1}{2\sqrt{2}}(-\Omega_{38}+\Omega_{40}+\Omega_{42}+\Omega_{44}+\Omega_{46}+\Omega_{48}-\Omega_{50}+\Omega_{52})$
$\beta[22]$	$\{1,2,4,2\}$	$y_1+y_2-y_3+y_4$	$-\frac{1}{4}(\Omega_3+\Omega_8+\Omega_{23}-\Omega_{33})$
$\beta[23]$	$\{1,3,4,2\}$	$y_1+y_2+y_3+y_4$	$-\frac{1}{4\sqrt{2}}(\Omega_2-\Omega_7+\Omega_{18}+\Omega_{32})$
$\beta[24]$	$\{2,3,4,2\}$	$2\,y_4$	$-\frac{1}{8}(\Omega_{16}+\Omega_{21}+\Omega_{25}-\Omega_{36})$

$$\mathscr{C} = \left(\begin{array}{c|c|c} \mathbf{1} & 0 & 0 \\ \hline 0 & \eta & 0 \\ \hline 0 & 0 & C_+ \end{array}\right) \tag{1.8.31}$$

A further comment is necessary about the normalizations of the step operators E^β which are displayed in Table 1.6. They have been fixed with the following criterion. Once we have constructed the algebra, via the generators (1.8.24),(1.8.25), we have the Lie structure constants encoded in Eq. (1.8.26) and hence we can diagonalize the adjoint action of the Cartan generators (1.8.29) finding which linear combinations of the remaining generators correspond to which root. Each root space is one-dimensional and therefore we are left with the task of choosing an absolute normalization for what we want to call the step operators:

$$E^\beta = \lambda_\beta \text{ (linear combination of } \Omega\text{.s)} \tag{1.8.32}$$

The values of λ_β are now determined by the following non trivial conditions:

1. The differences $\mathbb{H}^i = \left(E^{\beta_i} - E^{-\beta_i}\right)$ should close a subalgebra $\mathbb{H} \subset F_{4(4)}$, the maximal compact subalgebra $\mathfrak{su}(2)_R \oplus \mathfrak{usp}(6)$
2. The sums $\mathbb{K}^i = \frac{1}{\sqrt{2}}\left(E^{\beta_i} + E^{-\beta_i}\right)$ should span a 28-dimensional representation of $\mathbb{H}$, namely the aforementioned $\mathfrak{su}(2)_R \oplus \mathfrak{usp}(6)$

We arbitrarily choose the first four λ_β associated with simple roots and then all the others are determined. The result is that displayed in Table 1.6. Using the Cartan generators defined by Eqs. (1.8.29) and the step operators enumerated in Table 1.6 one can calculate the structure constants of $\mathfrak{f}_4$ in the Cartan-Weyl basis, namely:

$$\begin{aligned} \left[\mathscr{H}_i\,,\,\mathscr{H}_j\right] &= 0 \\ \left[\mathscr{H}_i\,,\,E^\beta\right] &= \beta^i\,E^\beta \\ \left[E^\beta\,,\,E^{-\beta}\right] &= \beta\cdot\mathscr{H} \\ \left[E^{\beta_i}\,,\,E^{\beta_j}\right] &= \mathscr{N}_{\beta_i,\beta_j}\,E^{\beta_i+\beta_j} \end{aligned} \tag{1.8.33}$$

in particular one obtains the explicit numerical value of the coefficients $\mathscr{N}_{\beta_i,\beta_j}$, which, as it is well known, are the only ones not completely specified by the components of the root vectors in the root system. The result of this computation, following from Eq. (1.8.26) is that encoded in Eqs. (1.8.8)–(1.8.10).

As a last point we can investigate the properties of the maximal compact subalgebra $\mathfrak{su}(2) \oplus \mathfrak{usp}(6) \subset \mathfrak{f}_{4(4)}$. As we know a basis of generators for this subalgebras is provided by:

$$H_i = \left(E^{\beta_i} - E^{-\beta_i}\right) \quad ; \quad (i = 1, \ldots, 24) \tag{1.8.34}$$

but it is not a priori clear which are the generators of $SU(2)_R$ and which of Usp(6). By choosing a basis of Cartan generators of the compact algebra and diagonalizing their adjoint action this distinction can be established. The generators of $SU(2)_R$ are the following linear combinations:

$$
\begin{aligned}
J_X &= \frac{1}{4\sqrt{2}}\left(H_1 - H_{14} + H_{20} - H_{22}\right) \\
J_Y &= \frac{1}{4\sqrt{2}}\left(H_5 + H_{11} - H_{18} + H_{23}\right) \\
J_Z &= \frac{1}{4\sqrt{2}}\left(-H_2 + H_9 - H_{16} - H_{24}\right)
\end{aligned} \tag{1.8.35}
$$

close the standard commutation relations:

$$
\left[J_i\,,\,J_j\right] = \varepsilon_{ijk}\,J_k \tag{1.8.36}
$$

and commute with all the generators of Usp(6). These latter are displayed as follows.

$$
\begin{aligned}
\mathscr{H}_1^{(Usp6)} &= -\tfrac{H_2}{2} - \tfrac{H_9}{2} + \tfrac{H_{16}}{2} - \tfrac{H_{24}}{2} \\
\mathscr{H}_2^{(Usp6)} &= -\tfrac{H_2}{2} + \tfrac{H_9}{2} + \tfrac{H_{16}}{2} + \tfrac{H_{24}}{2} \\
\mathscr{H}_3^{(Usp6)} &= \tfrac{H_2}{2} + \tfrac{H_9}{2} + \tfrac{H_{16}}{2} - \tfrac{H_{24}}{2}
\end{aligned} \tag{1.8.37}
$$

are the Cartan generators. On the other hand the nine pairs of generators which are rotated one into the other by the Cartans with eigenvalues equal to the roots of the compact algebra are the following ones

$W_1 = H_{10}$	$Z_1 = H_7$
$W_2 = H_4$	$Z_2 = -H_{13}$
$W_3 = H_6$	$Z_3 = -H_3$
$W_4 = -H_1 + H_{14} + H_{20} - H_{22}$	$Z_4 = -H_5 - H_{11} - H_{18} + H_{23}$
$W_5 = H_{21}$	$Z_5 = -H_8$
$W_6 = H_1 + H_{14} + H_{20} + H_{22}$	$Z_6 = H_5 - H_{11} - H_{18} - H_{23}$
$W_7 = -H_1 - H_{14} + H_{20} + H_{22}$	$Z_7 = H_5 - H_{11} + H_{18} + H_{23}$
$W_8 = H_{17}$	$Z_8 = H_{15}$
$W_9 = H_{12}$	$Z_9 = H_{19}$

(1.8.38)

The construction of the $\mathfrak{f}_4$ Lie algebra presented in this section was published in [12].

1.9 Conclusions for This Chapter

As the last example of the $\mathfrak{f}_4$ Lie algebra should have clearly illustrated, although deterministic and implicitly defined by the Dynkin diagram, the actual construction of exceptional Lie algebras is far from being a trivial matter and involves a series of strategies and long calculations that are best done by means of computer codes. One deals with large matrices that is difficult to display on paper and the best approach is to save the constructions in electronic libraries that can be utilized in subsequent calculations. It is not surprising that it took such a giant of mathematics as Elie

Cartan to explicitly construct the fundamental representations of the exceptional Lie algebras, especially at a time when computers were not available.

From another point of view, the existing mathematical literature usually presents the construction of Lie algebra representations in a very compact format that is not of too friendly use to physicists concerned with their application to the problems and the conceptions discussed in this book. As we stressed it is not only a question of convenience but also a conceptual one. There are in the architecture of Lie algebras and of their representations deep and significant aspects that are easily lost if you are not looking at them in the proper way, motivated by those questions that are posed by the various special geometries implied by supersymmetry.

The explicit construction of the exceptional and non exceptional Lie algebras in the light of supergravity is one of the motivations to write the present book. Our constructions are at many stages different from the conventional approaches of most text books [11, 13–15].

Similarly one can say about the issues in finite group theory that were reviewed in the first part of the present chapter. Although pertaining to classical topics in mathematics and retrievable with some considerable effort from various standard textbooks, the constructions we presented here are, in their form and in their spirit, original. The adopted viewpoint is motivated by the role of the considered mathematical structures in *supergravity inspired geometries* that, according to the ideas expressed in my other book [8], I deem not just one among their many possible applications, rather the manifestation of their deepest intrinsic meaning.

References

1. P.G. Fré, A. Fedotov, *Groups and Manifolds: Lectures for Physicists with Examples in Mathematica* (De Gruyter, Berlin, 2018)
2. G. James, M.W. Liebeck, *Representations and Characters of Groups* (Cambridge University Press, Cambridge, 2001)
3. A. Hurwitz, Uber algebraische gebilde mit eindeutigen transformationen in sich. Mathematische Annalen **41**, 3 (1893)
4. F. Klein, Ueber die transformation siebenter ordnung der elliptischen functionen. Mathematische Annalen **14**, 3 (1878)
5. G. Fano, *Sui postulati fondamentali della geometria proiettiva*, Giornale di Matematiche, 30 (1892)
6. R.C. King, F. Toumazet, B.G. Wybourne, A finite subgroup of the exceptional Lie group G2. J. Phys. A. Math. Gen. **32**, 8527–8537 (1999)
7. P. Fré, Supersymmetric M2-branes with Englert fluxes, and the simple group PSL(2, 7). Fortsch. Phys. **64**, 425–462 (2016)
8. P. G. Fré, *A Conceptual History of Symmetry from Plato to the Superworld* (Springer, Berlin, 2018)
9. P. Fre', M. Trigiante, Twisted tori and fluxes: a no go theorem for Lie groups of weak G(2) holonomy. Nucl. Phys. B **751**, 343–375 (2006)
10. C. Luhn, S. Nasri, P. Ramond, Simple finite non-Abelian flavor groups. J. Math. Phys. **48**, 123519 (2007). arXiv:0709.1447 [hep-th]
11. J.E. Humphreys, *Introduction to Lie Algebras and Representation Theory* (Springer, Berlin, 1972)

12. P. Fré, F. Gargiulo, K. Rulik, Cosmic billiards with painted walls in non-maximal supergravities: a worked out example. Nucl. Phys. B **737**, 1–48 (2006), https://doi.org/10.1016/j.nuclphysb.2005.10.023 [hep-th/0507256]
13. S. Helgason, *Differential Geometry and Symmetric Spaces* (Academic Press, Cambridge, 1962)
14. N. Jacobson, *Lie Algebras* (Courier Corporation, 1979)
15. V.S. Varadarajan, *Lie Groups, Lie Algebras, and their Representations* (Springer Science & Business Media, Berlin, 2013)

Chapter 2
Isometries and the Geometry of Coset Manifolds

The art of doing mathematics consists in finding that special case which contains all the germs of generality.

David Hilbert

2.1 Conceptual and Historical Introduction

The word isometry comes from the Greek word ή ισομετρία which means the equality of measures.

The origin of the modern concept of isometry is rooted in that of *congruence* of geometrical figures that Euclid never introduced explicitly, yet implicitly assumed when he proceeded to identify those triangles that can be superimposed one onto the other.

As I explained in my other book [1], it was indeed the question about what are the transformations that define such congruences what led Felix Klein to the Erlangen Programme. Klein understood that Euclidean congruences are based on the transformations of the Euclidean Group and he came to the idea that other geometries are based on different groups of transformations with respect to which we consider congruences.

Such a concept, however, would have been essentially empty without an additional element: the *metric*. The area and the volume of geometrical figures, the length of sides and the relative angles have to be measured in order to compare them. These measurements can be performed if and only if we have a metric g, in other words if the *substratum* of the considered geometry is a *Riemannian* or a *pseudo Riemannian* manifold $(\mathcal{M}, g)$.

Therefore the group of transformations which, according to the vision of the Erlangen Programme, defines a geometry, is the *group of isometries* $\mathrm{G}_{\mathrm{iso}}$ of a given Riemannian space $(\mathcal{M}, g)$, the elements of this group being diffeomorphisms:

P. G. Frè, *Advances in Geometry and Lie Algebras from Supergravity*, Theoretical and Mathematical Physics,
https://doi.org/10.1007/978-3-319-74491-9_2

$$\phi \quad : \quad \mathscr{M} \to \mathscr{M} \tag{2.1.1}$$

such that their pull-back on the metric form leaves it invariant:

$$\forall \phi \in \mathrm{G_{iso}} \quad : \quad \phi^\star \left[g_{\mu\nu}(x)\, dx^\mu\, dx^\nu \right] = g_{\mu\nu}(x)\, dx^\mu\, dx^\nu \tag{2.1.2}$$

Quite intuitively it becomes clear that the structure of $\mathrm{G_{iso}}$ is determined by the manifold $\mathscr{M}$ and by its metric g, so that the *Kleinian concept* of geometries is to be identified with that of Riemannian spaces $(\mathscr{M}, g)$.

A generic metric g has no isometries and hence there are no congruences to study. (Pseudo)-Riemannian manifolds with no isometry, or with few isometries, are relevant to several different problems pertaining to physics and also to other sciences, yet they are not in the vein of the Erlangen Programme, aiming at the classification of geometries in terms of groups. Hence we can legitimately ask ourselves the question whether such a programme can be ultimately saved, notwithstanding our discovery that a geometry, according to Klein's conception, is necessarily based on a (pseudo)-Riemannian manifold $(\mathscr{M}, g)$. The answer is obviously yes if we can invert the relation between the metric g and its isometry group $\mathrm{G_{iso}}$. Given a Lie group G can we construct the Riemannian manifold $(\mathscr{M}, g)$ which admits G as its own isometry group $\mathrm{G_{iso}}$? Indeed we can; the answers are also exhaustive if we add an additional request, that of transitivity.

Definition 2.1.1 A group G acting on a manifold $\mathscr{M}$ by means of diffeomorphisms:

$$\forall \gamma \in \mathrm{G} \quad \gamma : \mathscr{M} \to \mathscr{M} \tag{2.1.3}$$

has a transitive action if and only if

$$\forall p, q \in \mathscr{M} \quad , \quad \exists \gamma \in \mathrm{G} \,/\, \gamma(q) = p \tag{2.1.4}$$

If the Riemannian manifold $(\mathscr{M}, g)$ admits a transitive group of isometries it means that any point of $\mathscr{M}$ can be mapped into any other by means of a transformation that is an isometry. In this case the very manifold $\mathscr{M}$ and its metric g are completely determined by group theory: $\mathscr{M}$ is necessarily a *coset manifold* $\mathrm{G/H}$, namely the space of equivalence classes of elements of G with respect to multiplication (either on the right or on the left) by elements of a subgroup $\mathrm{H} \subset \mathrm{G}$. The metric g is induced on the equivalence classes by the Killing metric of the Lie algebra, defined on $\mathbb{G}$.

The present chapter, after a study of Killing vector fields, namely of the infinitesimal generators that realize the Lie algebra $\mathbb{G}$ of the isometry group, will be devoted to the geometry of coset manifolds. Among them particular attention will be given to the so named *symmetric spaces* characterized by an additional reflection symmetry whose nature will become clear to the reader in the following sections.

2.1.1 Symmetric Spaces and Elie Cartan

The full-fledged classification of all symmetric spaces was the gigantic achievement of Élie Cartan. As the reader will appreciate in the sequel, the classification of symmetric spaces is at the same time a classification of the real forms of the complex Lie algebras and it is the conclusive step in the path initiated by Killing in his papers of 1888, 1889. At the same time the geometries of non-compact symmetric spaces can be formulated in terms of other quite interesting algebraic structures, the *normed solvable Lie algebras*. The class of these latter is wider than that of symmetric spaces and this provides a generalization path leading to a wider class of geometries, all of them under firm algebraic control. This will be the topic of the last two sections of the present chapter which is propaedeutical to the developments of the subsequent chapters.

2.1.2 Where and How Do Coset Manifolds Come into Play?

By now it should be clear to the reader that, just as we have the whole spectrum of linear representations of a Lie algebra $\mathbb{G}$ and of its corresponding Lie group G, in the same way we have the set of *non-linear representations*[1] of the same Lie algebra $\mathbb{G}$ and of the same Lie group G. These are encoded in all possible coset manifolds G/H with their associated G–invariant metrics.

Where and how do these geometries pop up?

The answer is that they appear at several different levels of analysis and in connection with different aspects of physical theories. Let us enumerate them and discover a conceptual hierarchy.

(A) A first context of utilization of coset manifolds G/H is in the quest for solutions of Einstein Equations in $d = 4$ or in higher dimensions. One is typically interested in space-times with a prescribed *isometry* and one tries to fit into the equations G/H metrics whose parameters depend on some residual coordinate like the time t in cosmology or the radius r in black-hole physics. The field equations of the theory reduce to few parameter differential equations in the residual space.

[1] *Clarification for mathematicians:* in the physical literature *linear representation* of a symmetry corresponds to the case where the fundamental fields spanning the theory transform in a linear representation of the considered Lie group G. The Lagrangian defining the considered theory is supposed to be invariant with respect to such transformations. On the other hand the wording *non-linear representation* is universally used when the fundamental fields of the theory are identified with the coordinates of a Riemannian manifold $\mathcal{M}$ on which the Lie group G acts as a group of isometries. Indeed in order to be a symmetry of the theory, the action of the group G must leave the lagrangian invariant and this implies the existence of an invariant metric g on $\mathcal{M}$. The metric g appears in the kinetic term of the fields.

(B) Another instance of utilization of coset manifolds is in the context of σ-models. In physical theories that include scalar fields $\phi^I(x)$ the kinetic term is necessarily of the following form:

$$\mathscr{L}_{kin} = \frac{1}{2}\gamma_{IJ}(\phi)\,\partial_\mu\phi^I(x)\,\partial_\nu\phi^J(x)\,g^{\mu\nu}(x) \tag{2.1.5}$$

where $g^{\mu\nu}(x)$ is the metric of space-time, while $\gamma_{IJ}(\phi)$ can be interpreted as the metric of some manifold $\mathscr{M}_{\text{target}}$ of which the fields ϕ^I are the coordinates and whose dimension is just equal to the number of scalar fields present in the theory. If we require the field theory to have some Lie Group symmetry G, either we have linear representations or non linear ones. In the first case the metric γ_{IJ} is constant and invariant under the linear transformations of G acting on the $\phi^I(x)$. In the second case the manifold $\mathscr{M}_{\text{target}} = \mathrm{G/H}$ is some coset of the considered group and $\gamma_{IJ}(\phi)$ is the corresponding G-invariant metric.

(C) In mathematics and sometimes in physics you can consider structures that depend on a continuous set of parameters, for instance the solutions of certain differential equations, like the self-duality constraint for gauge-field strengths or the Ricci-flat metrics on certain manifolds, or the algebraic surfaces of a certain degree in some projective spaces. The parameters corresponding to all the possible deformations of the considered structure constitute themselves a manifold $\mathscr{M}$ which typically has some symmetries and in many cases is actually a coset manifold. A typical example is provided by the so named Kummer surface K3 whose Ricci flat metric no one has so far constructed, yet we know a priori that it depends on 3×19 parameters that span the homogeneous space $\frac{\mathrm{SO}(3,19)}{\mathrm{SO}(3)\times\mathrm{SO}(19)}$.

(D) In many instances of field theories that include scalar fields there is a scalar potential term $\mathrm{V}(\phi)$ which has a certain group of symmetries G. The vacua of the theory, namely the set of extrema of the potential usually fill up a coset manifold G/H where $\mathrm{H} \subset \mathrm{G}$ is the residual symmetry of the vacuum configuration $\phi = \phi_0$.

2.1.3 The Deep Insight of Supersymmetry

In supersymmetric field theories, in particular in supergravities that are supersymmetric extensions of Einstein Gravity coupled to matter multiplets, all the uses listed above of coset manifolds do occur, but there is an additional ingredient whose consequences are very deep and far reaching for geometry: supersymmetry itself. Consistency with supersymmetry introduces further restrictions on the geometry of target manifolds $\mathscr{M}_{\text{target}}$ that are required to fall in specialized categories like *Kähler manifolds*, *special Kähler manifolds*, *quaternionic Kähler manifolds* and so on. These geometries, that we collectively dub *Special Geometries*, require the existence of complex structures and encompass both manifolds that do not have transitive groups of isometries and homogeneous manifolds G/H. In the second case, which is one of the main focuses of interest for the present essay, the combination of the special structures with the theory of Lie algebras produces new insights in homogenous geometries that would have been inconceivable outside the framework of supergravity. This is what we call the deep geometrical insight of supersymmetry. In this book we neither discuss the construction of supergravity theories, nor we derive the constraints imposed by supersymmetry on geometry. Our commitment is simply to present the vast wealth of geometrical lore that supergravity Occam's razor has introduced, or systematically reorganized, in the field of mathematics.

2.2 Isometries and Killing Vector Fields

The existence of continuous isometries is related with the existence of Killing vector fields. Here we explain the underlying mathematical theory which leads to the study of coset manifolds and symmetric spaces.

Suppose that the diffeomorphism considered in Eq. (2.1.1) is infinitesimally close to the identity[2]

$$x^{\mu} \rightarrow \phi^{\mu}(x) \simeq x^{\mu} + k^{\mu}(x) \tag{2.2.1}$$

The condition for this diffeomorphism to be an isometry, is a differential equation for the components of the vector field $\mathbf{k} = k^{\mu} \partial_{\mu}$ which immediately follows from (2.1.2):

$$\nabla_{\mu} k_{\nu} + \nabla_{\nu} k_{\mu} = 0 \tag{2.2.2}$$

Hence given a metric one can investigate the nature of its isometries by trying to solve the linear homogeneous Eq. (2.2.2) determining its general integral. The important point is that, if we have two Killing vectors $\mathbf{k}$ and $\mathbf{w}$ also their commutator $[\mathbf{k}, \mathbf{w}]$ is a Killing vector. This follows from the fact that the product of two finite isometries is also an isometry. Hence Killing vector fields form a finite dimensional Lie algebra $\mathbb{G}_{iso}$ and one can turn the question around. Rather then calculating the isometries of a given metric one can address the problem of constructing (pseudo)-Riemannian manifolds that have a prescribed isometry algebra. Due to the well established classification of semi-simple Lie algebras this becomes a very fruitful point of view.

---

[2] *Clarification for mathematicians*: in the physical literature it is universally utilized the following jargon which turns out to be very clear to readers with an education as physicists. A Lie group element $g \in \mathbb{G}$ is named *infinitesimally close to the identity* when its Taylor series expansion in terms of a parameter ε that parameterizes a one-dimensional subgroup $\mathcal{G} \subset G$ to which g belongs is truncated to the first order term: $g = e + \varepsilon\, \mathbf{g} + \mathcal{O}(\varepsilon^2)$. Clearly the coefficient $\mathbf{g}$ of the first order term is an element of the Lie algebra $\mathbb{G}$ of G. Applying this jargon to the case of the group of diffeomorphisms, by means of a *diffeomorphism infinitesimally close to the identity* we define a *vector field*, the Lie algebra of the diffeomorphism group being the Lie algebra of vector fields. In the case the considered infinitesimally close to identity diffeomorphism is an isometry, the corresponding vector field is named a *Killing vector field*.

In particular, also in view of the Cosmological Principle, one is interested in homogeneous spaces, namely in (pseudo)-Riemannian manifolds where each point of the manifold can be reached from a reference one by the action of an isometry.

Homogeneous spaces are identified with coset manifolds, whose differential geometry can be thoroughly described and calculated in pure Lie algebra terms.

2.3 Coset Manifolds

Coset manifolds are a natural generalization of group manifolds and play a very important, ubiquitous, role both in Mathematics and in Physics.

In group-theory (irrespectively whether the group G is finite or infinite, continuous or discrete) we have the concept of *coset space* G/H which is just the set of equivalence classes of elements $g \in \mathrm{G}$, where the equivalence is defined by right multiplication with elements $h \in \mathrm{H} \subset \mathrm{G}$ of a subgroup:

$$\forall g, g' \in \mathrm{G} \quad : \quad g \sim g' \ \ \text{iff} \ \ \exists h \in \mathrm{H} \ \setminus \ g\,h = g' \tag{2.3.1}$$

Namely two group elements are equivalent if and only if they can be mapped into each other by means of some element of the subgroup. The equivalence classes, which constitute the elements of G/H are usually denoted g H, where g is any representative of the class, namely any one of the equivalent G-group elements the class is composed of. The definition we have just provided by means of right multiplication can be obviously replaced by an analogous one based on left-multiplication. In this case we construct the coset H\G composed of *right lateral classes* H g while g H are named the *left lateral classes*. For non abelian groups G and generic subgroups H the left G/H and right H\G coset spaces have different not coinciding elements. Working with one or with the other definition is just a matter of conventions. We choose to work with *left classes*.

Coset manifolds arise in the context of Lie group theory when G is a Lie group and H is a Lie subgroup thereof. In that case the set of lateral classes g H can be endowed with a manifold structure inherited from the manifold structure of the parent group G. Furthermore on G/H we can construct *invariant metrics* such that all elements of the original group G are isometries of the constructed metric. As we show below, the curvature tensor of invariant metrics on coset manifolds can be constructed in purely algebraic terms starting from the structure constants of the G Lie algebra, by-passing all analytic differential calculations.

Coset manifolds are easily identified with *homogeneous spaces* which we presently define.

Definition 2.3.1 A Riemannian or pseudo-Riemannian manifold $\mathcal{M}_g$ is said to be homogeneous if it admits as an isometry the transitive action of a group G. A group acts transitively if any point of the manifold can be reached from any other by means of the group action.

A notable and very common example of such homogeneous manifolds is provided by the spheres $\mathbb{S}^n$ and by their non-compact generalizations, the pseudo-spheres $\mathbb{H}_{\pm}^{(n+1-m,m)}$. Let x^I denote the cartesian coordinates in $\mathbb{R}^{n+1}$ and let:

$$\eta_{IJ} = \operatorname{diag}\left(\underbrace{+,+\cdots,+}_{n+1-m},\underbrace{-,-,\cdots,-}_{m}\right) \tag{2.3.2}$$

be the coefficient of a non degenerate quadratic form with signature $(n+1-m, m)$:

$$\langle \mathbf{x}\,,\,\mathbf{x}\rangle_\eta \equiv x^I\,x^J\,\eta_{IJ} \tag{2.3.3}$$

We obtain a pseudo-sphere $\mathbb{H}_{\pm}^{(n+1-m,m)}$ by defining the algebraic locus:

$$\mathbf{x} \in \mathbb{H}_{\pm}^{(n+1-m,m)} \quad \Leftrightarrow \quad \langle \mathbf{x}\,,\,\mathbf{x}\rangle_\eta \equiv \pm 1 \tag{2.3.4}$$

which is a manifold of dimension n. The spheres $\mathbb{S}^n$ correspond to the particular case $\mathbb{H}_{+}^{n+1,0}$ where the quadratic form is positive definite and the sign in the right hand side of Eq. (2.3.4) is positive. Obviously with a positive definite quadratic form this is the only possibility.

All these algebraic loci are invariant under the transitive action of the group $\mathrm{SO(n+1, n+1-m)}$ realized by matrix multiplication on the vector $\mathbf{x}$ since:

$$\forall g \in \mathrm{G} \;:\; \langle \mathbf{x}\,,\,\mathbf{x}\rangle_\eta = \pm 1 \quad \Leftrightarrow \quad \langle g\,\mathbf{x}\,,\,g\,\mathbf{x}\rangle_\eta = \pm 1 \tag{2.3.5}$$

namely the group maps solutions of the constraint (2.3.4) into solutions of the same and, furthermore, all solutions can be generated starting from a standard reference vector:

$$\langle \mathbf{x}\,,\,\mathbf{x}\rangle_\eta = 0 \;\Rightarrow\; \exists g \in \mathrm{G} \setminus \mathbf{x} = g\,\mathbf{x}_0^{\pm} \tag{2.3.6}$$

where:

$$x_0^+ = \begin{pmatrix} 1\\ 0\\ \vdots\\ 0\\ \hline 0\\ 0\\ \vdots\\ 0 \end{pmatrix} \;;\; x_0^- = \begin{pmatrix} 0\\ 0\\ \vdots\\ 0\\ \hline 1\\ 0\\ \vdots\\ 0 \end{pmatrix} \tag{2.3.7}$$

the line separating the first $n+1-m$ entries from the last m. Equation (2.3.6) guarantees that the locus is invariant under the action of G, while Eq. (2.3.7) states that G is transitive.

Definition 2.3.2 In a homogeneous space $\mathcal{M}_g$, the subgroup $\mathrm{H}_p \subset \mathrm{G}$ which leaves a point $p \in \mathcal{M}_g$ fixed ($\forall h \in \mathrm{H}_p, h\,p = p$) is named the **isotropy subgroup** of the point. Because of the transitive action of G, any other point $p' = g\,p$ has an isotropy subgroup $\mathrm{H}_{p'} = g\,\mathrm{H}_p\,g^{-1}$ which is conjugate to H_p and therefore isomorphic to it.

It follows that, up to conjugation, the isotropy group of a homogeneous manifold $\mathcal{M}_g$ is unique and corresponds to an intrinsic property of such a space. It suffices to calculate the isotropy group H_0 of a conventional properly chosen reference point p_0: all other isotropy groups will immediately follow. For brevity H_0 will be just renamed H.

In our example of the spaces $\mathbb{H}_{\pm}^{(n+1-m,m)}$ the isotropy group is immediately derived by looking at the form of the vectors $\mathbf{x}_0^{\pm}$: all elements of G which rotate the vanishing entries of these vectors among themselves are clearly elements of the isotropy group. Hence we find:

$$\begin{aligned} \mathrm{H} &= \mathrm{SO(n, m)} && \text{for} \quad \mathbb{H}_{+}^{(n+1-m,m)} \\ \mathrm{H} &= \mathrm{SO(n+1, m-1)} && \text{for} \quad \mathbb{H}_{-}^{(n+1-m,m)} \end{aligned} \tag{2.3.8}$$

It is natural to label any point p of a homogeneous space by the parameters describing the G-group element which carries a conventional point p_0 into p. These parameters, however, are redundant: because of the H-isotropy there are infinitely many ways to reach p from p_0. Indeed, if g does that job, any other element of the lateral class $g\,\mathrm{H}$ does the same. It follows by this simple discussion that the homogeneous manifold $\mathcal{M}_g$ can be identified with the coset manifold G/H defined by the transitive group G divided by the isotropy group H.

Focusing once again on our example we find:

$$\mathbb{H}_{+}^{(n+1-m,m)} = \frac{\mathrm{SO(n+1-m, m)}}{\mathrm{SO(n-m, m)}} \quad ; \quad \mathbb{H}_{-}^{(n+1-m,m)} = \frac{\mathrm{SO(n+1-m, m)}}{\mathrm{SO(n+1-m, m-1)}} \tag{2.3.9}$$

In particular the spheres correspond to:

$$\mathbb{S}^n = \mathbb{H}_{+}^{(n+1,0)} = \frac{\mathrm{SO(n+1)}}{\mathrm{SO(n)}} \tag{2.3.10}$$

Other important examples, relevant for cosmology are:

$$\mathbb{H}_{+}^{(n+1,1)} = \frac{\mathrm{SO(n+1, 1)}}{\mathrm{SO(n, 1)}} \quad ; \quad \mathbb{H}_{-}^{(n+1,1)} = \frac{\mathrm{SO(n+1, 1)}}{\mathrm{SO(n+1)}} \tag{2.3.11}$$

The general classification of homogeneous (pseudo)-Riemannian spaces corresponds therefore to the classification of the coset manifolds G/H for all Lie groups G and for their closed Lie subgroups $\mathrm{H} \subset \mathrm{G}$.

The equivalence classes constituting the points of the coset manifold can be labeled by a set of d coordinates $y \equiv \{y^1, \ldots, y^d\}$ where:

$$d = \dim \frac{\mathrm{G}}{\mathrm{H}} \equiv \dim \mathrm{G} - \dim \mathrm{H} \tag{2.3.12}$$

There are of course many different ways of choosing the y-parameters since, just as in any other manifold, there are many possible coordinate systems. What is specific of coset manifolds is that, given any coordinate system y by means of which we label the equivalence classes, within each equivalence class we can choose a representative group element $\mathbb{L}(y) \in \mathrm{G}$. The choice must be done in such a way that $\mathbb{L}(y)$ should be a smooth function of the parameters y. Furthermore for different values y and y', the group elements $\mathbb{L}(y)$ and $\mathbb{L}(y')$ should never be equivalent, in other words no $h \in \mathrm{H}$ should exist such that $\mathbb{L}(y) = \mathbb{L}(y')\, h$. Under left multiplication by $g \in \mathrm{G}$, $\mathbb{L}(y)$ is in general carried into another equivalence class with coset representative $\mathbb{L}(y')$. Yet the g image of $\mathbb{L}(y)$ is not necessarily $\mathbb{L}(y')$: it is typically some other element of the same class, so that we can write:

$$\forall g \in G \quad : \quad g\,\mathbb{L}(y) = \mathbb{L}(y')\, h(g, y) \quad ; \quad h(g, y) \in \mathrm{H} \tag{2.3.13}$$

where we emphasized that the H-element necessary to map $\mathbb{L}(y')$ into the g-image of $\mathbb{L}(y)$, depends, in general both from the point y and from the chosen transformation g. Equation (2.3.13) is pictorially described in Fig. 2.1. For the spheres a possible set of coordinates y can be obtained by means of the stereographic projection whose conception is recalled in Fig. 2.2

As an other explicit example, which will be useful in the sequel, we consider the case of the Euclidean hyperbolic spaces $\mathbb{H}^{(n,1)}_{-}$ identified as coset manifolds in Eq. (2.3.11). In this case, to introduce a coset parametrization means to write a family of SO(n, 1) matrices $\mathbb{L}(\mathbf{y})$ depending smoothly on an n-component vector $\mathbf{y}$ in such a way that for different values of $\mathbf{y}$ such matrices cannot be mapped one in the other by means of right multiplication with any element h of the subgroup SO(n) $\subset$ SO(n, 1):

$$\mathrm{SO(n,1)} \supset \mathrm{SO(n)} \ni h = \left(\begin{array}{c|c} \mathcal{O} & 0 \\ \hline 0 & 1 \end{array}\right) \quad ; \quad \mathcal{O}^T \mathcal{O} = \mathbf{1}_{n\times n} \tag{2.3.14}$$

An explicit parametrization of this type can be written as follows:

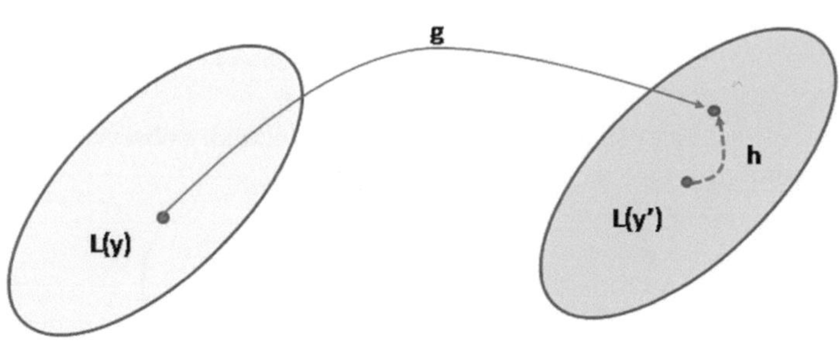

Fig. 2.1 Pictorial description of the action of the group G on the coset representatives

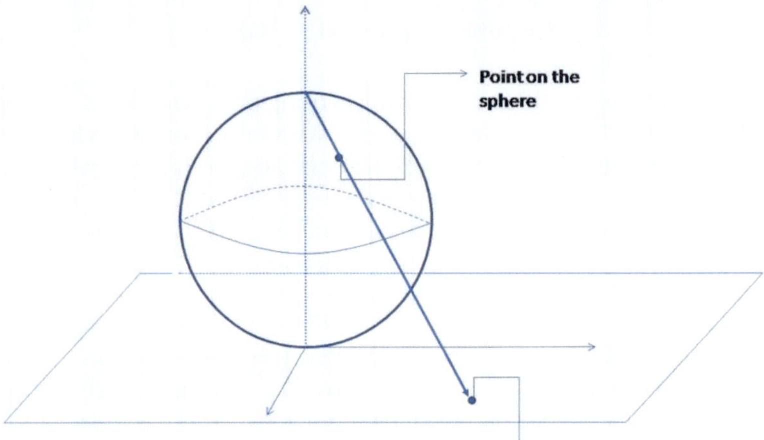

Fig. 2.2 The idea of the stereographic projection. Considering the $\mathbb{S}^n$ sphere immersed in $\mathbb{R}^{n+1}$, from the North-Pole $\{1, 0, 0, \ldots, 0\}$ one draws the line that goes through the point $p \in \mathbb{S}^n$ and considers the point $\pi(p) \in \mathbb{R}^n$ where such a line intersects the $\mathbb{R}^n$ plane tangent to sphere in the South Pole and orthogonal to the line that joins the North and the South Pole. The n-coordinates $\{y^1, \ldots, y^n\}$ of $\pi(p)$ can be taken as labels of an open chart in $\mathbb{S}^n$

$$\mathbb{L}(\mathbf{y}) = \left(\begin{array}{c|c} \mathbf{1}_{n\times n} + 2\,\frac{\mathbf{y}\,\mathbf{y}^T}{1-\mathbf{y}^2} & -2\,\frac{\mathbf{y}}{1-\mathbf{y}^2} \\ \hline -\,2\,\frac{\mathbf{y}^T}{1-\mathbf{y}^2} & \frac{1+\mathbf{y}^2}{1-\mathbf{y}^2} \end{array}\right) \tag{2.3.15}$$

where $\mathbf{y}^2 \equiv \mathbf{y}\cdot\mathbf{y}$ denotes the standard SO(n) invariant scalar product in $\mathbb{R}^n$. Why the matrices $\mathbb{L}(\mathbf{y})$ form a good parametrization of the coset? The reason is simple, first of all observe that:

$$\mathbb{L}(\mathbf{y})^T\,\eta\,\mathbb{L}(\mathbf{y}) = \eta \tag{2.3.16}$$

where

$$\eta = \text{diag}\,(+,\,+,\,\cdots,\,+,\,-) \tag{2.3.17}$$

This guarantees that $\mathbb{L}(\mathbf{y})$ are elements of SO(n, 1),secondly observe that the image $\mathbf{x}(\mathbf{y})$ of the standard vector $\mathbf{x}_0$ through $\mathbb{L}(\mathbf{y})$,

$$\mathbf{x}(\mathbf{y}) \equiv \mathbb{L}(\mathbf{y})\,\mathbf{x}_0 = \mathbb{L}(\mathbf{y})\begin{pmatrix} 0 \\ \vdots \\ 0 \\ \hline 1 \end{pmatrix} = \frac{1}{1-\mathbf{y}^2}\begin{pmatrix} 2\,y^1 \\ \vdots \\ 2y^n \\ \hline \frac{1+\mathbf{y}^2}{1-\mathbf{y}^2} \end{pmatrix} \tag{2.3.18}$$

lies, as it should, in the algebraic locus $\mathbb{H}_{-}^{(n,1)}$,

$$\mathbf{x}(\mathbf{y})^T \, \eta \, \mathbf{x}(\mathbf{y}) \; = \; -1 \tag{2.3.19}$$

and has n linearly independent entries (the first n) parameterized by $\mathbf{y}$. Hence the lateral classes can be labeled by y and this concludes our argument to show that (2.3.15) is a good coset parametrization. $\mathbb{L}(0) = \mathbf{1}_{(n+1)\times(n+1)}$ corresponds to the identity class which is usually named the *origin* of the coset.

2.3.1 The Geometry of Coset Manifolds

In order to study the geometry of a coset manifold G/H, the first important step is provided by the orthogonal decomposition of the corresponding Lie algebra, namely by

$$\mathbb{G} = \mathbb{H} \oplus \mathbb{K} \tag{2.3.20}$$

where $\mathbb{G}$ is the Lie algebra of G and the subalgebra $\mathbb{H} \subset \mathbb{G}$ is the Lie algebra of the subgroup H and where $\mathbb{K}$ denotes a vector space orthogonal to $\mathbb{H}$ with respect to the Cartan Killing metric[3] of $\mathbb{G}$. By definition of subalgebra we always have:

$$[\mathbb{H}\,,\,\mathbb{H}] \subset \mathbb{H} \tag{2.3.21}$$

while in general one has:

$$[\mathbb{H}\,,\,\mathbb{K}] \subset \mathbb{H} \oplus \mathbb{K} \tag{2.3.22}$$

Definition 2.3.3 Let G/H be a Lie coset manifold and let the orthogonal decomposition of the corresponding Lie algebra be as in Eq. (2.3.20). If the condition:

$$[\mathbb{H}\,,\,\mathbb{K}] \subset \mathbb{K} \tag{2.3.23}$$

applies, the coset G/H is named **reductive**.

Equation (2.3.23) has an obvious and immediate interpretation. The complementary space $\mathbb{K}$ forms a linear representation of the subalgebra $\mathbb{H}$ under its adjoint action within the ambient algebra $\mathbb{G}$.

Almost all of the "reasonable" coset manifolds which occur in various provinces of Mathematical Physics are reductive. Violation of *reductivity* is a sort of pathology whose study we can disregard in the scope of this book. We will consider only reductive coset manifolds.

---

[3] We assume that G is semi-simple so that the Cartan-Killing metric is non degenerate.

Definition 2.3.4 Let G/H be a reductive coset manifold. If in addition to (2.3.23) also the following condition:

$$[\mathbb{K}\,,\,\mathbb{K}] \subset \mathbb{H} \tag{2.3.24}$$

applies, then the coset manifold G/H is named a **symmetric space**.

Let T_A $(A = 1, \ldots, n)$ denote a complete basis of generators for the Lie algebra $\mathbb{G}$:

$$[T_A\,,\,T_B] \;=\; C^{\;C}_{AB}\,T_C \tag{2.3.25}$$

and T_i $(i = 1, \ldots, m)$ denote a complete basis for the subalgebra $\mathbb{H} \subset \mathbb{G}$. We also introduce the notation T_a $(a = 1, \ldots, n - m)$ for a set of generators that provide a basis of the complementary subspace $\mathbb{K}$ in the orthogonal decomposition (2.3.20). We nickname T_a the *coset generators*. Using such notations, Eq. (2.3.25) splits into the following three ones:

$$\left[T_j\,,\,T_k\right] = C^{i}_{\;jk}\,T_i \tag{2.3.26}$$
$$[T_i\,,\,T_b] = C^{a}_{\;ib}\,T_a \tag{2.3.27}$$
$$[T_b\,,\,T_c] = C^{i}_{\;bc}\,T_i \;+\; C^{a}_{\;bc}\,T_a \tag{2.3.28}$$

Equation (2.3.26) encodes the property of $\mathbb{H}$ of being a subalgebra. Equation (2.3.27) encodes the property of the considered coset of being reductive. Finally if in Eq. (2.3.28) we have $C^{a}_{\;bc} = 0$, the coset is not only reductive but also symmetric.

We will be able to provide explicit formulae for the Riemann tensor of reductive coset manifolds equipped with G-invariant metrics in terms of such structure constants. Prior to that we consider the infinitesimal transformation and the very definition of the Killing vectors with respect to which the metric has to be invariant.

2.3.1.1 Infinitesimal Transformations and Killing Vectors

Let us consider the transformation law (2.3.13) of the coset representative. For a group element g infinitesimally close to the identity, we have:

$$g \simeq 1 \,+\, \varepsilon^A\, T_A \tag{2.3.29}$$
$$h(y, g) \simeq 1 \,-\, \varepsilon_A\, W^i_A(y)\, T_i \tag{2.3.30}$$
$$y'^{\alpha} \simeq y^{\alpha} \,+\, \varepsilon^A\, k^{\alpha}_A \tag{2.3.31}$$

The induced h transformation in Eq. (2.3.13) depends in general on the infinitesimal G-parameters ε^A and on the point in the coset manifold y, as shown in Eq. (2.3.30). The y-dependent rectangular matrix $W^i_A(y)$ is usually named the $\mathbb{H}$-compensator. The shift in the coordinates y^α is also proportional to ε^A and the vector fields:

$$\mathbf{k}_A \;=\; k^{\alpha}_A(y)\,\frac{\partial}{\partial y^{\alpha}} \tag{2.3.32}$$

are named the *Killing vectors of the coset*. The reason for such a name will be justified when we will show that on G/H we can construct a (pseudo)-Riemannian metric which admits the vector fields (2.3.32) as generators of infinitesimal isometries. For the time being those in (2.3.32) are just a set of vector fields that, as we prove few lines below, close the Lie algebra of the group G.

Inserting Eqs. (2.3.29)–(2.3.31) into the transformation law (2.3.13) we obtain:

$$T_A\,\mathbb{L}(y) \;=\; \mathbf{k}_A\,\mathbb{L}(y) \;-\; W^i_A(y)\,\mathbb{L}(y)\,T_i \tag{2.3.33}$$

Consider now the commutator $g_2^{-1}\,g_1^{-1}g_2\,g_1$ acting on $\mathbb{L}(y)$. If both group elements $g_{1,2}$ are infinitesimally close to the identity in the sense of Eq. (2.3.29), then we obtain:

$$g_2^{-1}\,g_1^{-1}g_2\,g_1\,\mathbb{L}(y) \;\simeq\; \left(1 \;-\; \varepsilon_1^A\,\varepsilon_2^B\,[T_A\,,\,T_B]\right)\,\mathbb{L}(y) \tag{2.3.34}$$

By explicit calculation we find:

$$\begin{aligned}[][T_A\,,\,T_B]\,\mathbb{L}(y) &= T_A\,T_B\,\mathbb{L}(y) \;-\; T_B\,T_A\,\mathbb{L}(y)\\ &= [\mathbf{k}_A\,,\,\mathbf{k}_B]\,\mathbb{L}(y) - \left(\mathbf{k}_A\,W^i_B - \mathbf{k}_B\,W^i_A \;+\; 2\,C^i{}_{jk}\,W^j_A\,W^k_B\right)\,\mathbb{L}(y)\,T_i\end{aligned} \tag{2.3.35}$$

On the other hand, using the Lie algebra commutation relations we obtain:

$$[T_A\,,\,T_B]\,\mathbb{L}(y) \;=\; C^{\;C}_{AB}\,T_C\,\mathbb{L}(y) \;=\; C^{\;C}_{AB}\,\left(\mathbf{k}_C\,\mathbb{L}(y) \;-\; W^i_C\,\mathbb{L}(y)\,T_i\right) \tag{2.3.36}$$

By equating the right hand sides of Eqs. (2.3.35) and (2.3.36) we conclude that:

$$[\mathbf{k}_A\,,\,\mathbf{k}_B] = C^{\;C}_{AB}\,\mathbf{k}_C \tag{2.3.37}$$

$$\mathbf{k}_A\,W^i_B \;-\; \mathbf{k}_B\,W^i_A + 2\,C^i{}_{jk}\,W^j_A\,W^k_B = C^{\;C}_{AB}\,W^i_C \tag{2.3.38}$$

where we separately compared the terms with and without W's, since the decomposition of a group element into $\mathbb{L}(y)\,h$ is unique.

Equation (2.3.37) shows that the Killing vector fields defined above close the commutation relations of the $\mathbb{G}$-algebra.

Instead, Eq. (2.3.38) will be used to construct a consistent $\mathbb{H}$-covariant Lie derivative.

In the case of the spaces $\mathbb{H}^{(n,1)}_{-}$, which we choose as illustrative example, the Killing vectors can be easily calculated by following the above described procedure step by step. For later purposes we find it convenient to present such a calculation in a slightly more general set up by introducing the following coset representative that depends on a discrete parameter $\kappa = \pm 1$:

$$\mathbb{L}_\kappa(\mathbf{y}) = \left(\begin{array}{c|c} \mathbf{1}_{n\times n} + 2\,\mathbf{y}\,\mathbf{y}^T\,\frac{\kappa}{1+\kappa\mathbf{y}^2} & -2\,\frac{\mathbf{y}}{1+\kappa\mathbf{y}^2} \\ \hline 2\,\kappa\,\frac{\mathbf{y}^T}{1+\kappa\,\mathbf{y}^2} & \frac{1-\kappa\,\mathbf{y}^2}{1+\kappa\mathbf{y}^2} \end{array}\right) \tag{2.3.39}$$

An explicit calculation shows that:

$$\mathbb{L}_\kappa(\mathbf{y})^T \underbrace{\left(\begin{array}{cccc|c} 1 & 0 & \dots & 0 & 0 \\ 0 & 1 & \dots & 0 & 0 \\ \vdots & \vdots & \vdots & \vdots & \vdots \\ 0 & \dots & 0 & 1 & 0 \\ \hline 0 & \dots & 0 & 0 & \kappa \end{array}\right)}_{\eta_\kappa} \mathbb{L}_\kappa(\mathbf{y}) = \underbrace{\left(\begin{array}{cccc|c} 1 & 0 & \dots & 0 & 0 \\ 0 & 1 & \dots & 0 & 0 \\ \vdots & \vdots & \vdots & \vdots & \vdots \\ 0 & \dots & 0 & 1 & 0 \\ \hline 0 & \dots & 0 & 0 & \kappa \end{array}\right)}_{\eta_\kappa} \tag{2.3.40}$$

Namely $\mathbb{L}_{-1}(\mathbf{y})$ is an SO(n, 1) matrix, while $\mathbb{L}_1(\mathbf{y})$ is an SO(n + 1) group element. Furthermore defining, as in Eq. (2.3.18):

$$\mathbf{x}_\kappa(\mathbf{y}) \equiv \mathbb{L}_\kappa(\mathbf{y}) \left(\begin{array}{c} 0 \\ \vdots \\ 0 \\ \hline 1 \end{array}\right) \tag{2.3.41}$$

we find that:

$$\mathbf{x}_\kappa(\mathbf{y})^T\,\eta_\kappa\,\mathbf{x}_\kappa(\mathbf{y}) = \kappa \tag{2.3.42}$$

Hence by means of $\mathbb{L}_1(\mathbf{y})$ we parameterize the points of the n-sphere $\mathbb{S}^n$, while by means of $\mathbb{L}_{-1}(\mathbf{y})$ we parameterize the points of $\mathbb{H}_-^{(n,1)}$ named also the n-pseudo-sphere or the n-hyperboloid. In both cases the stability subalgebra is $\mathfrak{so}(n)$ for which a basis of generators is provided by the following matrices:

$$J_{ij} = \mathscr{I}_{ij} - \mathscr{I}_{ji} \quad ; \quad i, j = 1, \dots, n \tag{2.3.43}$$

having named:

$$\mathscr{I}_{ij} = \left(\begin{array}{cccc|c} 0 & \dots & \dots & 0 & 0 \\ 0 & \dots & 1 & 0 & 0 \;\}\,\text{i-th row} \\ 0 & \dots & \dots & 0 & 0 \\ \hline 0 & \dots & \underbrace{\dots}_{\text{j-th column}} & 0 & 0 \end{array}\right). \tag{2.3.44}$$

the $(n+1)\times(n+1)$ matrices whose only non vanishing entry is the ij-th one, equal to 1.

The commutation relations of the $\mathfrak{so}(n)$ generators are very simple. We have:

$$\left[J_{ij}\,,\,J_{k\ell}\right] = -\,\delta_{ik}\,J_{j\ell} + \delta_{jk}\,J_{i\ell} - \delta_{j\ell}\,J_{ik} + \delta_{i\ell}\,J_{jk} \qquad (2.3.45)$$

The coset generators can instead be chosen as the following matrices:

$$P_i = \left(\begin{array}{cccc|c} 0 & \dots & \dots & 0 & 0 \\ 0 & \dots & 0 & 0 & 1 \; \}\,\text{i-th row} \\ 0 & \dots & \dots & 0 & 0 \\ \hline 0 & \dots & \underbrace{-\kappa}_{\text{i-th column}} & 0 & 0 \end{array}\right) \qquad (2.3.46)$$

and satisfy the following commutation relations:

$$\left[J_{ij}\,,\,P_k\right] = -\,\delta_{ik}\,P_j + \delta_{jk}\,P_i \qquad (2.3.47)$$

$$\left[P_i\,,\,P_j\right] = -\kappa\,J_{ij} \qquad (2.3.48)$$

Equation (2.3.47) states that the generators P_i transform as an n-vector under $\mathfrak{so}(n)$ rotations (reductivity) while Eq. (2.3.48) shows that for both signs $\kappa = \pm 1$ the considered coset manifold is a symmetric space. Correspondingly we name $\mathbf{k}_{ij} = k^{\ell}_{ij}(y)\,\frac{\partial}{\partial y^{\ell}}$ the Killing vector fields associated with the action of the generators J_{ij}:

$$J_{ij}\,\mathbb{L}_{\kappa}(\mathbf{y}) = \mathbf{k}_{ij}\,\mathbb{L}_{\kappa}(\mathbf{y}) + \mathbb{L}_{\kappa}(\mathbf{y})\,J_{pq}\,W^{pq}_{ij}(y) \qquad (2.3.49)$$

while we name $\mathbf{k}_i = k^{\ell}_i(y)\,\frac{\partial}{\partial y^{\ell}}$ the Killing vector fields associated with the action of the generators P_i:

$$P_i\,\mathbb{L}_{\kappa}(\mathbf{y}) = \mathbf{k}_i\,\mathbb{L}_{\kappa}(\mathbf{y}) + \mathbb{L}_{\kappa}(\mathbf{y})\,J_{pq}\,W^{pq}_{i}(y) \qquad (2.3.50)$$

Resolving conditions (2.3.49) and (2.3.50) we obtain:

$$\mathbf{k}_{ij} = y_i\,\partial_j - y_j\,\partial_i \qquad (2.3.51)$$

$$\mathbf{k}_i = \tfrac{1}{2}\left(1 - \kappa\,\mathbf{y}^2\right)\partial_i + \kappa\,y_i\,\mathbf{y}\cdot\boldsymbol{\partial} \qquad (2.3.52)$$

The $\mathbb{H}$-compensators W^{pq}_i and W^{pq}_{ij} can also be extracted from the same calculation but since their explicit form is not essential for our future discussion we skip them.

2.3.1.2 Vielbeins, Connections and Metrics on G/H

Consider next the following 1-form defined over the reductive coset manifold G/H:

$$\Sigma(y) = \mathbb{L}^{-1}(y)\, \mathrm{d}\, \mathbb{L}(y) \tag{2.3.53}$$

which generalizes the Maurer Cartan form defined over the group manifold G. As a consequence of its own definition the 1-form Σ satisfies the equation:

$$0 = \mathrm{d}\Sigma + \Sigma \wedge \Sigma \tag{2.3.54}$$

which provides the clue to the entire (pseudo)-Riemannian geometry of the coset manifold. To work out this latter we start by decomposing Σ along a complete set of generators of the Lie algebra $\mathbb{G}$. According with the notations introduced in the previous subsection we put:

$$\Sigma = V^a\, T_a + \omega^i\, T_i \tag{2.3.55}$$

The set of $(n-m)$ 1-forms $V^a = V^a_\alpha(y)\, dy^\alpha$ provides a covariant frame for the cotangent bundle CT(G/H), namely a complete basis of sections of this vector bundle that transform in a proper way under the action of the group G. On the other hand $\omega = \omega^i\, T_i = \omega^i_\alpha(y)\, dy^\alpha\, T_i$ is called the $\mathbb{H}$-connection. Indeed ω turns out to be the 1-form of a bona-fide principal connection on the principal fiber bundle:

$$\mathscr{P}\left(\frac{\mathrm{G}}{\mathrm{H}}, \mathrm{H}\right) \quad : \quad \mathrm{G} \stackrel{\pi}{\rightarrow} \frac{\mathrm{G}}{\mathrm{H}} \tag{2.3.56}$$

which has the Lie group G as total space, the coset manifold G/H as base space and the closed Lie subgroup H ⊂ G as structural group. The bundle $\mathscr{P}\left(\frac{\mathrm{G}}{\mathrm{H}}, \mathrm{H}\right)$ is uniquely defined by the projection that associates to each group element $g \in \mathrm{G}$ the equivalence class $g\,\mathrm{H}$ it belongs to.

Introducing the decomposition (2.3.55) into the Maurer Cartan equation (2.3.54), this latter can be rewritten as the following pair of equations:

$$\mathrm{d}V^a + C^a_{\ ib}\, \omega^i \wedge V^b = -\tfrac{1}{2}\, C^a_{\ bc}\, V^b \wedge V^c \tag{2.3.57}$$

$$\mathrm{d}\omega^i + \tfrac{1}{2}\, C^i_{\ jk} \omega^j \wedge \omega^k = -\tfrac{1}{2}\, C^i_{\ bc}\, V^b \wedge V^c \tag{2.3.58}$$

where we have used the Lie algebra structure constants organized as in Eqs. (2.3.26)–(2.3.28).

Let us now consider the transformations of the 1-forms we have introduced.

Under left multiplication by a constant group element $g \in \mathrm{G}$ the 1-form $\Sigma(y)$ transforms as follows:

$$\begin{aligned}\Sigma(y') &= h(y,g)\,\mathbb{L}^{-1}(y)g^{-1}\,\mathrm{d}\left(gL(y)h^{-1}\right)\\ &= h(y,g)^{-1}\,\Sigma(y)\,h(y,g) + h(y,g)^{-1}\,\mathrm{d}h(y,g)\end{aligned} \tag{2.3.59}$$

where $y' = g.y$ is the new point in the manifold G/H whereto y is moved by the action of g. Projecting the above equation on the coset generators T_a we obtain:

$$V^a(y') = V^b(y)\,\mathscr{D}_b^{\ a}\,(h(y,g)) \tag{2.3.60}$$

where $\mathscr{D} = \exp[\mathfrak{D}_{\mathbb{H}}]$, having denoted by $\mathfrak{D}_{\mathbb{H}}$ the $(n-m)$ dimensional representation of the subalgebra $\mathbb{H}$ which occurs in the decomposition of the adjoint representation of $\mathbb{G}$:

$$\mathrm{adj}\,\mathbb{G} = \underbrace{\mathrm{adj}\mathbb{H}}_{=\mathfrak{A}_{\mathbb{H}}} \oplus \mathfrak{D}_{\mathbb{H}} \tag{2.3.61}$$

Projecting on the other hand on the $\mathbb{H}$-subalgebra generators T_i we get:

$$\omega(y') = \mathscr{A}\,[h(y,g)]\,\omega(y)\,\mathscr{A}^{-1}\,[h(y,g)] + \mathscr{A}\,[h(y,g)]\,\mathrm{d}\mathscr{A}^{-1}\,[h(y,g)] \tag{2.3.62}$$

where we have set:

$$\mathscr{A} = \exp[\mathfrak{A}_{\mathbb{H}}] \tag{2.3.63}$$

Considering a complete basis T_A of generators for the full Lie algebra $\mathbb{G}$, the adjoint representation is defined as follows:

$$\forall g \in \mathrm{G} \quad : \quad g^{-1}\,T_A\,g \equiv \mathrm{adj}(g)_A^{\ B}\,T_B \tag{2.3.64}$$

In the explicit basis of T_A generators the decomposition (2.3.61) means that, once restricted to the elements of the subgroup $\mathrm{H} \subset \mathrm{G}$, the adjoint representation becomes block-diagonal:

$$\forall h \in \mathrm{H} \quad : \quad \mathrm{adj}(h) = \left(\begin{array}{c|c} \mathscr{D}(h) & 0 \\ \hline 0 & \mathscr{A}(h) \end{array}\right) \tag{2.3.65}$$

Note that for such decomposition to hold true the coset manifold has to be reductive according to definition (2.3.23).

The infinitesimal form of Eq. (2.3.60) is the following one:

$$V^a(y+\delta y) - V^a(y) = -\,\varepsilon^A\,W_A^i(y)\,C^a_{\ ib}\,V^b(y) \tag{2.3.66}$$

$$\delta y^\alpha = \varepsilon^A\,k_A^\alpha(y) \tag{2.3.67}$$

for a group element $g \in \mathrm{G}$ very close to the identity as in Eq. (2.3.29).

Similarly the infinitesimal form of Eq. (2.3.62) is:

$$\omega^i(y+\delta y) - \omega^i(y) = -\,\varepsilon^A\left(C^i_{\ kj}\,W_A^k\,\omega^j + \mathrm{d}W_A^i\right) \tag{2.3.68}$$

2.3.1.3 Lie Derivatives

The Lie derivative of a tensor $T_{\alpha_1\ldots\alpha_p}$ along a vector field v^μ provides the change in shape of that tensor under an infinitesimal diffeomorphism:

$$y^\mu \mapsto y^\mu + v^\mu(y) \tag{2.3.69}$$

Explicitly one sets:

$$\begin{aligned}\ell_{\mathbf{v}}\, T_{\alpha_1\ldots\alpha_p}(y) = v^\mu\,\partial_\mu\, T_{\alpha_1\ldots\alpha_p} + (\partial_{\alpha_1} v^\gamma)\, T_{\gamma\alpha_2\ldots\alpha_p} + \cdots \\ + (\partial_{\alpha_p} v^\gamma)\, T_{\alpha_1\alpha_2\ldots\gamma}\end{aligned} \tag{2.3.70}$$

In the case of p-forms, namely of antisymmetric tensors the definition (2.3.70) of Lie derivative can be recast into a more intrinsic form using both the exterior differential d and the *contraction operator*.

Definition 2.3.5 Let $\mathscr{M}$ be a differentiable manifold and let $\Lambda_k\,(\mathscr{M})$ be the vector bundles of differential k-forms on $\mathscr{M}$, let $\mathbf{v} \in \Gamma(T\mathscr{M}, \mathscr{M})$ be a vector field. The contraction $\mathrm{i}_{\mathbf{k}}$ is a linear map:

$$\mathrm{i}_{\mathbf{k}} \quad : \quad \Lambda_k\,(\mathscr{M}) \to \Lambda_{k-1}\,(\mathscr{M}) \tag{2.3.71}$$

such that for any $\omega^{(k)} \in \Lambda_k\,(\mathscr{M})$ and for any set of $k-1$ vector fields $\mathbf{w}_1, \ldots, \mathbf{w}_{k-1}$, we have:

$$\mathrm{i}_{\mathbf{k}}\,\omega^{(k)}\,(\mathbf{w}_1, \ldots, \mathbf{w}_{k-1}) \equiv k\,\omega^{(k)}\,(\mathbf{v}, \mathbf{w}_1, \ldots, \mathbf{w}_{k-1}) \tag{2.3.72}$$

Then by going to components we can verify that the tensor definition (2.3.70) is equivalent to the following one:

Definition 2.3.6 Let $\mathscr{M}$ be a differentiable manifold and let $\Lambda_k\,(\mathscr{M})$ be the vector bundles of differential k-forms on $\mathscr{M}$, let $\mathbf{v} \in \Gamma(T\mathscr{M}, \mathscr{M})$ be a vector field. The Lie derivative $\ell_{\mathbf{v}}$ is a linear map:

$$\ell_{\mathbf{v}} \quad : \quad \Lambda_k\,(\mathscr{M}) \to \Lambda_k\,(\mathscr{M}) \tag{2.3.73}$$

such that for any $\omega^{(k)} \in \Lambda_k\,(\mathscr{M})$ we have:

$$\ell_{\mathbf{v}}\,\omega^{(k)} \equiv \mathrm{i}_{\mathbf{v}}\,\mathrm{d}\omega^{(k)} + d\,\mathrm{i}_{\mathbf{v}}\,\omega^{(k)} \tag{2.3.74}$$

On the other hand for vector fields the tensor definition (2.3.70) is equivalent to the following one.

Definition 2.3.7 Let $\mathscr{M}$ be a differentiable manifold and let $T\mathscr{M} \to \mathscr{M}$ be the tangent bundle, whose sections are the vector fields. Let $\mathbf{v} \in \Gamma(T\mathscr{M}, \mathscr{M})$ be a vector field. The Lie derivative $\ell_{\mathbf{v}}$ is a linear map:

$$\ell_{\mathbf{v}} \quad : \quad \Gamma(T\mathcal{M}, \mathcal{M}) \to \Gamma(T\mathcal{M}, \mathcal{M}) \tag{2.3.75}$$

such that for any $\mathbf{w} \in \Gamma(T\mathcal{M}, \mathcal{M})$ we have:

$$\ell_{\mathbf{v}}\,\mathbf{w} \equiv [\mathbf{v}\,,\,\mathbf{w}] \tag{2.3.76}$$

The most important properties of the Lie derivative, which immediately follow from its definition are the following ones:

$$\begin{aligned} [\ell_{\mathbf{v}}\,,\,\mathrm{d}] &= 0 \\ [\ell_{\mathbf{v}}\,,\,\ell_{\mathbf{w}}] &= \ell_{[\mathbf{v}\,,\,\mathbf{w}]} \end{aligned} \tag{2.3.77}$$

The first of the above equations states that the Lie derivative commutes with exterior derivative. This is just a consequence of the invariance of the exterior algebra of k-forms with respect to diffeomorphisms. On the other hand the second equation states that the Lie derivative provides an explicit representation of the Lie algebra of vector fields on tensors.

The Lie derivatives along the Killing vectors of the frames V^a and of the $\mathbb{H}$-connection ω^i introduced in the previous subsection are:

$$\ell_{\mathbf{v}_A}\,V^a = W^i_A\,C^a_{\ ib}\,V^b \tag{2.3.78}$$

$$\ell_{\mathbf{v}_A}\,\omega^i = -\left(\mathrm{d}W^i_A + C^i_{\ kj}\,W^k_A\,\omega^j\right) \tag{2.3.79}$$

This result can be interpreted by saying that, associated with every Killing vector $\mathbf{k}_A$ there is a an infinitesimal $\mathbb{H}$-gauge transformation:

$$\mathbf{W}_A = W^i_A(y)\,T_i \tag{2.3.80}$$

and that the Lie derivative of both V^a and ω^i along the Killing vectors is just such local gauge transformation pertaining to their respective geometrical type. The frame V^a is a section of an H-vector bundle and transforms as such, while ω^i is a connection and it transforms as a connection should do.

2.3.1.4 Invariant Metrics on Coset Manifolds

The result (2.3.78), (2.3.79) has a very important consequence which constitutes the fundamental motivation to consider coset manifolds. Indeed this result instructs us to construct G-invariant metrics on G/H, namely metrics that admit all the above discussed Killing vectors as generators of true isometries.

The argument is quite simple. We saw that the one-forms V^a transform in a linear representation $\mathfrak{D}_{\mathbb{H}}$ of the isotropy subalgebra $\mathbb{H}$ (and group H). Hence if τ_{ab} is a symmetric H-invariant constant two-tensor, by setting:

$$ds^2 = \tau_{ab} V^a \otimes V^b = \underbrace{\tau_{ab} V^a_\alpha(y) V^b_\beta(y)}_{g_{\alpha\beta}(y)} dy^\alpha \otimes dy^\beta \tag{2.3.81}$$

we obtain a metric for which all the above constructed Killing vectors are indeed Killing vectors, namely:

$$\ell_{\mathbf{k}_A} ds^2 = \tau_{ab} \left(\ell_{\mathbf{k}_A} V^a \otimes V^b + V^a \otimes \ell_{\mathbf{k}_A} V^b\right) \tag{2.3.82}$$

$$= \underbrace{\tau_{ab} \left([\mathfrak{D}_{\mathbb{H}}(W_A)]^a_{\ c}\, \delta^b_d + [\mathfrak{D}_{\mathbb{H}}(W_A)]^b_{\ c}\, \delta^a_d\right)}_{=0 \text{ by invariance}} V^c \otimes V^d$$

$$= 0 \tag{2.3.83}$$

The key point, in order to utilize the above construction, is the decomposition of the representation $\mathfrak{D}_{\mathbb{H}}$ into irreducible representations. Typically, for most common cosets, $\mathfrak{D}_{\mathbb{H}}$ is already irreducible. In this case there is just one invariant H-tensor τ and the only free parameter in the definition of the metric (2.3.81) is an overall scale constant. Indeed if τ_{ab} is an invariant tensor, any multiple thereof $\tau'_{ab} = \lambda\, \tau_{ab}$ is also invariant. In the case $\mathfrak{D}_{\mathbb{H}}$ splits into $\mathfrak{r}$ irreducible representations:

$$\mathfrak{D}_{\mathbb{H}} = \left(\begin{array}{c|c|c|c|c} \mathfrak{D}_1 & 0 & \dots & 0 & 0 \\ \hline 0 & \mathfrak{D}_2 & 0 & \dots & 0 \\ \hline \vdots & \vdots & \vdots & \vdots & \vdots \\ \hline 0 & \dots & 0 & \mathfrak{D}_{\mathfrak{r}-1} & 0 \\ \hline 0 & 0 & \dots & 0 & \mathfrak{D}_{\mathfrak{r}} \end{array}\right) \tag{2.3.84}$$

we have $\mathfrak{r}$ irreducible invariant tensors $\tau^{(i)}_{a_i b_i}$ in correspondence of such irreducible blocks and we can introduce $\mathfrak{r}$ independent scale factors:

$$\tau = \left(\begin{array}{c|c|c|c|c} \lambda_1\, \tau^{(1)} & 0 & \dots & 0 & 0 \\ \hline 0 & \lambda_2 \tau^{(2)} & 0 & \dots & 0 \\ \hline \vdots & \vdots & \vdots & \vdots & \vdots \\ \hline 0 & \dots & 0 & \lambda_{p-1} \tau^{(p-1)} & 0 \\ \hline 0 & 0 & \dots & 0 & \lambda_p\, \tau^{(p)} \end{array}\right) \tag{2.3.85}$$

Correspondingly we arrive at a continuous family of G-invariant metrics on G/H depending on $\mathfrak{r}$-parameters or, as it is customary to say in this context, of $\mathfrak{r}$ *moduli*. The number r defined by Eq. (2.3.84) is named the *rank of the coset manifold* G/H.

In this section we confine ourself to the most common case of rank one cosets ($\mathfrak{r} = 1$), assuming, furthermore, that the algebras $\mathbb{G}$ and $\mathbb{H}$ are both semi-simple. By an appropriate choice of basis for the coset generators T^a, the invariant tensor τ_{ab} can always be reduced to the form:

$$\tau_{ab} = \eta_{ab} = \text{diag}\left(\underbrace{+,+,\cdots,+}_{n_+},\underbrace{-,-,\cdots,-}_{n_-}\right) \tag{2.3.86}$$

where the two numbers n_+ and n_- sum up to the dimension of the coset:

$$n_+ + n_- = \dim\frac{\mathrm{G}}{\mathrm{H}} = \dim\mathbb{K} \tag{2.3.87}$$

and provide the dimensions of the two eigenspaces, $\mathbb{K}_\pm \subset \mathbb{K}$, respectively corresponding to real and pure imaginary eigenvalues of the matrix $\mathfrak{D}_{\mathbb{H}}(W)$ which represents a generic element W of the isotropy subalgebra $\mathbb{H}$.

Focusing on our example (2.3.39), that encompasses both the spheres and the pseudo-spheres, depending on the sign of κ, we find that:

$$n_+ = 0 \quad ; \quad n_- = n \tag{2.3.88}$$

so that in both cases ($\kappa = \pm 1$) the invariant tensor is proportional to a Kronecker delta:

$$\eta_{ab} = \delta_{ab} \tag{2.3.89}$$

The reason is that the subalgebra $\mathbb{H}$ is the compact $\mathfrak{so}(n)$, hence the matrix $\mathfrak{D}_{\mathfrak{H}}(W)$ is antisymmetric and all of its eigenvalues are purely imaginary.

If we consider cosets with non-compact isotropy groups, then the invariant tensor τ_{ab} develops a non trivial Lorentzian signature η_{ab}. In any case, if we restrict ourselves to rank one cosets, the general form of the metric is:

$$ds^2 = \lambda^2\,\eta_{ab}\,V^a \otimes V^b \tag{2.3.90}$$

where λ is a scale factor. This allows us to introduce the *vielbein*

$$E^a = \lambda\,V^a \tag{2.3.91}$$

and calculate the *spin connection* from the vanishing torsion equation:

$$0 = \mathrm{d}E^a - \omega^{ab} \wedge E^c\,\eta_{bc} \tag{2.3.92}$$

Using the Maurer Cartan equations (2.3.57)–(2.3.58), Eq. (2.3.92) can be immediately solved by:

$$\omega^{ab}\,\eta_{bc} \equiv \omega^a_{\ c} = \frac{1}{2\lambda}\,C^a_{\ cd}\,E^d + C^a_{\ ci}\,\omega^i \tag{2.3.93}$$

Inserting this in the definition of the curvature two-form

$$\mathfrak{R}^a_{\ b} = \mathrm{d}\omega^a_{\ b} - \omega^a_{\ c} \wedge \omega^c_{\ b} \tag{2.3.94}$$

allows to calculate the Riemann tensor defined by:

$$\mathfrak{R}^{a}{}_{b} = R^{a}{}_{bcd}\, E^{c} \wedge E^{d} \tag{2.3.95}$$

Using once again the Maurer Cartan equations (2.3.57)–(2.3.58), we obtain:

$$\begin{aligned} R^{a}{}_{bcd} = \frac{1}{\lambda^2}\Big(& -\tfrac{1}{4}\,\frac{1}{2\lambda}\, C^{a}{}_{be}\, C^{e}{}_{cd} - \tfrac{1}{8} C^{a}{}_{ec}\, C^{e}{}_{bd} + \tfrac{1}{8} C^{a}{}_{ed}\, C^{e}{}_{bc} \\ & - \tfrac{1}{2}\, C^{a}{}_{bi}\, C^{i}{}_{cd}\Big) \end{aligned} \tag{2.3.96}$$

which, as previously announced provides the expression of the Riemann tensor in terms of structure constants.

In the case of symmetric spaces $C^{a}{}_{be} = 0$ formula (2.3.96) simplifies to:

$$R^{a}{}_{bcd} = -\frac{1}{2\,\lambda^2}\, C^{a}{}_{bi}\, C^{i}{}_{cd} \tag{2.3.97}$$

2.3.1.5 For Spheres and Pseudo-spheres

In order to illustrate the structures presented in the previous section we consider the explicit example of the spheres and pseudo-spheres. Applying the outlined procedure to this case we immediately get:

$$\begin{aligned} E^{a} &= -\frac{2}{\lambda}\, \frac{\mathrm{d}y^{a}}{1+\kappa \mathbf{y}^2} \\ \omega^{ab} &= 2\,\frac{\kappa}{\lambda^2}\, E^{a} \wedge E^{b} \end{aligned} \tag{2.3.98}$$

This means that for spheres and pseudo-spheres the Riemann tensor is proportional to an antisymmetrized Kronecker delta:

$$R^{ab}{}_{cd} = \frac{\kappa}{\lambda^2}\, \delta^{[a}_{[c}\, \delta^{b]}_{d]} \tag{2.3.99}$$

2.4 The Real Sections of a Complex Lie Algebra and Symmetric Spaces

In the context of coset manifolds a very interesting class that finds important applications in supergravity and superstring theories is the following one:

$$\mathscr{M}_{\mathbb{G}_R} = \frac{\mathrm{G}_R}{\mathrm{H}_c} \tag{2.4.1}$$

where G_R is some semi-simple Lie group and $H_c \subset G_R$ is its maximal compact subgroup.The Lie algebra $\mathbb{H}_c$ of the denominator H_c is the maximal compact subalgebra $\mathbb{H} \subset \mathbb{G}_R$ which has typically rank $r_{compact} > r$. Denoting, as usual, by $\mathbb{K}$ the orthogonal complement of $\mathbb{H}_c$ in $\mathbb{G}_R$:

$$\mathbb{G}_R = \mathbb{H}_c \oplus \mathbb{K} \tag{2.4.2}$$

and defining as non compact rank or rank of the coset G_R/H the dimension of the non compact Cartan subalgebra:

$$r_{nc} = \text{rank}\,(G_R/H) \equiv \dim \mathscr{H}^{n.c.} \quad ; \quad \mathscr{H}^{n.c.} \equiv \text{CSA}_{\mathbb{G}(\mathbb{C})} \bigcap \mathbb{K} \tag{2.4.3}$$

we obtain that $r_{nc} < r$.

By definition the Lie algebra $\mathbb{G}_R$ is a real section of a complex semi-simple Lie algebra. Two universal instances of real sections of a simple Lie algebra $\mathbb{G}(\mathbb{C})$, are the *maximally split* and the *maximally compact real sections.*

The Maximally Split and the Maximal Compact Real Sections of a Simple Lie Algebra $\mathbb{G}(\mathbb{C})$.

Given the simple Lie algebra generators in the canonical Cartan-Weyl form: $T_A = \{H_i, E^\alpha, E^{-\alpha}\}$ the question is which restrictions on the imaginary and the real parts of the coefficients c^A of Lie algebra elements $c^A\, T_A$ can be introduced that are consistent with the Lie bracket and produce a real Lie algebra $\mathbb{G}_r$. Furthermore one would like to know how many such real sections do exist up to isomorphism.

Here we just introduce two real sections that are simply and universally defined for all simple Lie algebras:

(a) **The maximally split real section** $\mathbb{G}_{\mathbf{max}}$. This is defined by assuming that the allowed coefficients c^A are all real. In any linear representation of $\mathbb{G}_{\mathbf{max}}$ the matrices representing
$$T_A \equiv \{H_i\,,\, E^\alpha\,,\, E^{-\alpha}\} \tag{2.4.4}$$
are all *real*. From the representations of $\mathbb{G}_{\mathbf{max}}$, by taking linear combinations of the generators with complex coefficients one obtains all the linear representations of the complex Lie algebra $\mathbb{G}(\mathbb{C})$.

(b) **The maximally compact real section** $\mathbb{G}_c$. This real section, whose exponentiation produces a compact Lie group, is obtained by allowing linear combinations with real coefficients of the set of generators:
$$\mathfrak{T}_A \equiv \{\mathrm{i}\,H_i\,,\, \mathrm{i}\left(E^\alpha + E^{-\alpha}\right)\,,\, \left(E^\alpha - E^{-\alpha}\right)\} \tag{2.4.5}$$
In all linear representations of $\mathbb{G}_{\mathbf{c}}$ the matrices representing the generators $\mathfrak{T}_A$ are *anti-hermitian*.

One easily obtains the hermitian matrix representation of the generators $\mathfrak{T}_A$ from the real representation of the generators T_A and viceversa. It also follows that the matrices representing $E^{-\alpha}$ are the transposed of those representing E^α.

A very useful instrument in the explicit construction of matrix representations that has also important consequences for later developments is provided by the notion of Borel subalgebra. Starting from the Cartan-Weyl basis, if one consideres the subset of generators:

$$\mathrm{Bor}\,[\mathbb{G}] \;=\; \mathrm{span}\,\{H_i\,,\,E^\alpha\} \quad ; \quad \alpha > 0 \tag{2.4.6}$$

we see that it corresponds to a *solvable subalgebra of* $\mathbb{G}$. Hence every representation of Bor $[\mathbb{G}]$ can be put into an upper triangular form. This gives a powerful construction criterion for the fundamental representation. We just construct an upper triangular representation of the Bor $[\mathbb{G}]$ subalgebra and then we promote it to a representation of the full Lie algebra $\mathbb{G}$, by setting:

$$E^{-\alpha} \;=\; (E^\alpha)^T \tag{2.4.7}$$

Furthermore in view of the above discussion, the representations of the real sections $\mathbb{G}_{\mathbf{max}}$ and $\mathbb{G}_{\mathbf{c}}$ can be considered together and on that we rely in the following.

Classification of all the Real Sections

All other possible real sections are obtained by studying the available Cartan involutions of the complex Lie algebra. So consider:

Definition 2.4.1 Let:

$$\theta \;:\; \mathfrak{g} \to \mathfrak{g} \tag{2.4.8}$$

be a linear automorphism of the compact Lie algebra $\mathfrak{g} \;=\; \mathbb{G}_c$, where $\mathbb{G}_c$ is the maximal compact real section of a complex semi-simple Lie algebra $\mathbb{G}(\mathbb{C})$. By definition we have:

$$\forall \alpha, \beta \in \mathbb{R} \quad , \quad \forall X, Y \in \mathfrak{g} \quad : \quad \begin{cases} \theta\,(\alpha\, X + \beta\, Y) = \alpha\,\theta(X) \,+\, \beta\,\theta(Y) \\ \theta\,([X\,,\,Y]) = [\theta(X)\,,\,\theta(Y)] \end{cases} \tag{2.4.9}$$

If $\theta^2 \;=\; \mathbf{Id}$ then θ is named a Cartan involution of the Lie algebra $\mathfrak{g}$.

For any Cartan involution θ the possible eigenvalues are ± 1. This allows us to split the entire Lie algebra $\mathfrak{g}$ in two subspaces corresponding to the eigenvalues 1 and -1 respectively:

$$\mathfrak{g} \;=\; \mathfrak{H}_\theta \oplus \mathfrak{p}_\theta \tag{2.4.10}$$

One immediately realizes that:

$$\begin{aligned} [\mathfrak{H}_\theta\,,\,\mathfrak{H}_\theta] &\subset \mathfrak{H}_\theta \\ [\mathfrak{H}_\theta\,,\,\mathfrak{p}_\theta] &\subset \mathfrak{p}_\theta \\ [\mathfrak{p}_\theta\,,\,\mathfrak{p}_\theta] &\subset \mathfrak{H}_\theta \end{aligned} \tag{2.4.11}$$

Hence for any Cartan involution $\mathfrak{H}_\theta$ is a subalgebra and θ singles out a symmetric homogeneous compact coset manifold:

$$\mathscr{M}_\theta = \frac{\mathrm{G_c}}{\mathrm{H}_\theta} \quad \text{where} \quad \mathrm{H}_\theta \equiv \exp\left[\mathfrak{H}_\theta\right] \quad ; \qquad \mathrm{G_c} \equiv \exp\left[\mathfrak{g}\right] \tag{2.4.12}$$

The structure (2.4.11) has also another important consequence. If we define the vector space:

$$\mathfrak{g}^\star_\theta = \mathfrak{H}_\theta \oplus \mathfrak{p}^\star_\theta \quad ; \quad \mathfrak{p}^\star_\theta \equiv \mathrm{i}\mathfrak{p}_\theta \tag{2.4.13}$$

we see that $\mathfrak{g}^\star_\theta$ is closed under the Lie bracket and hence it is a Lie algebra. It is some real section of the complex Lie algebra $\mathbb{G}(\mathbb{C})$ and we can consider a new, generally non compact coset manifold:

$$\mathscr{M}^\star_\theta = \frac{\mathrm{G}^\star_\theta}{\mathrm{H}_\theta} \quad ; \quad \mathrm{H}_\theta \equiv \exp\left[\mathfrak{H}_\theta\right] \quad ; \qquad \mathrm{G}^\star_\theta \equiv \exp\left[\mathfrak{g}^\star_\theta\right] \tag{2.4.14}$$

An important theorem for which we refer the reader to classical textbooks [2–4][4] states that all real forms of a Lie algebra, up to isomorphism, are obtained in this way. Furthermore as part of the same theorem one has that θ can always be chosen in such a way that it maps the compact Cartan subalgebra into itself:

$$\theta \; : \; \mathscr{H}_c \; \rightarrow \; \mathscr{H}_c \tag{2.4.15}$$

This short discussion reveals that the classification of real forms of a complex Lie Algebra $\mathbb{G}(\mathbb{C})$ is in one-to-one correspondence with the classification of symmetric spaces, the complexification of whose Lie algebra of isometries is $\mathbb{G}(\mathbb{C})$. For this reason we discuss the real forms in the present chapter devoted to homogeneous coset manifolds.

Let us now consider the action of the Cartan involution on the Cartan subalgebra: $\mathscr{H}_c = \mathrm{span}\{\mathrm{i}H_i\}$ of the maximal compact section $\mathbb{G}_c$. Choosing a basis of $\mathscr{H}_c$ aligned with the simple roots:

$$\mathscr{H}_c = \mathrm{span}\left\{\mathrm{i}\, H_{\alpha_i}\right\} \tag{2.4.16}$$

we see that the action of the Cartan involution θ is by duality transferred to the simple roots α_i and hence to the entire root lattice. As a consequence we can introduce the notion of real and imaginary roots. One argues as follows.

We split the Cartan subalgebra into its compact and non compact subalgebras:

$$\begin{array}{rcl} \mathrm{CSA}_{\mathbb{G}_R} & = & \mathrm{i}\mathscr{H}^{comp} \oplus \mathscr{H}^{n.c.} \\ & & \quad\Updownarrow \qquad\quad \Updownarrow \\ \mathrm{CSA}_{\mathbb{G}_{\mathbf{max}}} & = & \mathscr{H}^{comp} \oplus \mathscr{H}^{n.c.} \end{array} \tag{2.4.17}$$

[4]The proof is also summarized in Appendix B of [5].

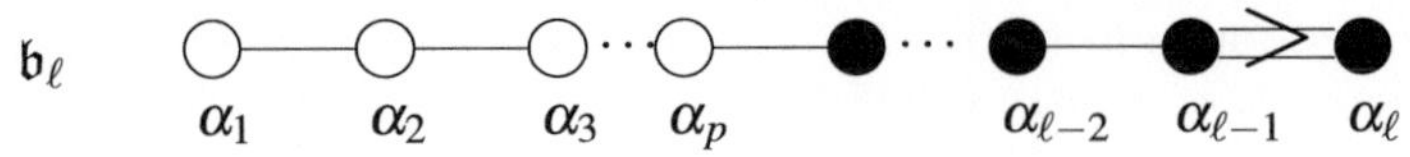

Fig. 2.3 The Tits–Satake diagram representing the real form $\mathfrak{so}(p, 2\ell - p + 1)$ of the complex $\mathfrak{so}(2\ell + 1)$ Lie algebra

defining:

$$\begin{aligned} h \in \mathscr{H}^{comp} &\Leftrightarrow \theta(h) = h \\ h \in \mathscr{H}^{n.c} &\Leftrightarrow \theta(h) \neq h \end{aligned} \tag{2.4.18}$$

Then every vector in the dual of the full Cartan subalgebra, in particular every root α can be decomposed into its parallel and its transverse part to $\mathscr{H}^{n.c.}$:

$$\alpha = \alpha_{||} \oplus \alpha_{\perp} \tag{2.4.19}$$

A root α is named *imaginary* if $\alpha_{||} = 0$. On the contrary a root α is called real if $\alpha_{\perp} = 0$. Generically a root is complex.

Given the original Dynkin diagram of a complex Lie algebra we can characterize a real section by mentioning which of the simple roots are imaginary. We do this by painting black the imaginary roots. The result is a Tits–Satake diagram like that in Fig. 2.3 which corresponds to the real Lie Algebra $\mathfrak{so}(p, 2\ell - p + 1)$ for $p > 2, \ell > 2$.

2.5 The Solvable Group Representation of Non-compact Coset Manifolds

Definition 2.5.1 A Riemannian space $(\mathscr{M}, g)$ is named **normal** if it admits a completely solvable[5] Lie group $\exp[\mathrm{Solv}(\mathscr{M})]$ of isometries that acts on the manifold in a simply transitive manner (i.e. for every 2 points in the manifold there is one and only one group element connecting them). The group $\exp[\mathrm{Solv}(\mathscr{M})]$ is generated by a so-called **normal metric Lie algebra**, $\mathrm{Solv}(\mathscr{M})$ that is a completely solvable Lie algebra endowed with a suitable, invariant Euclidean metric.

The main tool to classify and study homogeneous spaces of the type (2.4.14) is provided by a theorem [3] that states that if a Riemannian manifold $(\mathscr{M}, g)$ is normal, according to Definition 2.5.1, then it is metrically equivalent to the solvable group manifold

[5] A solvable Lie algebra s is completely solvable if the adjoint operation ad_X for all generators $X \in s$ has only real eigenvalues. The nomenclature of the Lie algebra is carried over to the corresponding Lie group in general in this chapter.

$$\begin{aligned} \mathscr{M} &\simeq \exp\left[\mathrm{Solv}(\mathscr{M})\right] \\ g\mid_{e\in\mathscr{M}} &= <,> \end{aligned} \tag{2.5.1}$$

where $<,>$ is a Euclidean metric defined on the normal solvable Lie algebra $\mathrm{Solv}(\mathscr{M})$. The key point is that non-compact coset manifolds of the form (2.4.14) are all normal. This is so because there is always, *for all real forms except the maximally compact one* a **solvable subalgebra** with the following features:

$$\begin{aligned} \mathrm{Solv}\left(\frac{\mathrm{G}_R}{\mathrm{H}_c}\right) &\subset \mathrm{G}_R \\ \dim\left[\mathrm{Solv}\left(\frac{\mathrm{G}_R}{\mathrm{H}_c}\right)\right] &= \dim\left(\frac{\mathrm{G}_R}{\mathrm{H}_c}\right) \\ \exp\left[\mathrm{Solv}\left(\frac{\mathrm{G}_R}{\mathrm{H}_c}\right)\right] &= \text{transitive on } \frac{\mathrm{G}_R}{\mathrm{H}_c} \end{aligned} \tag{2.5.2}$$

It is very easy to single out the appropriate solvable algebra in the case of the maximally split real form $\mathbb{G}_{\mathbf{max}}$. In that case, as we know, the maximal compact subalgebra has the following form:

$$\mathbb{H}_c = \mathrm{span}\left\{\left(E^\alpha - E^{-\alpha}\right)\right\} \quad ; \quad \forall\alpha\in\Delta_+ \tag{2.5.3}$$

The solvable algebra that does the required job is the Borellian subalgebra:

$$\mathrm{Bor}\left(\mathbb{G}_{\mathbf{max}}\right) \equiv \mathscr{H}\oplus\mathrm{span}\left(E^\alpha\right) \quad ; \quad \forall\alpha\in\Delta_+ \tag{2.5.4}$$

where $\mathscr{H}$ is the complete Cartan subalgebra and E^α are the step operators associated with the positive roots. That $\mathrm{Bor}\left(\mathbb{G}_{\mathbf{max}}\right)$ is a solvable Lie algebra follows from the canonical structure of Lie algebras displayed in Eq. (1.4.74). If you exclude the negative roots, you immediately see that the Cartan generators are not in the first derivative of the algebra. The second derivative excludes all the simple roots: the third derivative excludes the roots of height 2 and so on until you end up in a derivative that makes zero. Hence the Lie algebra is solvable. Furthermore it is obvious that any equivalence class of $\frac{\mathrm{G}_R}{\mathrm{H}_c}$ has a representative that is an element of the solvable Lie group $\exp\left[\mathrm{Bor}\left(\mathbb{G}_{\mathbf{max}}\right)\right]$. This is intuitive at the infinitesimal level from the fact that each element of the complementary space:

$$\mathbb{K} = \mathscr{H}\oplus\mathrm{span}\left[\left(E^\alpha + E^{-\alpha}\right)\right] \tag{2.5.5}$$

which generates the coset, can be uniquely rewritten as an element of $\mathrm{Bor}\left(\mathbb{G}_{\mathbf{max}}\right)$ plus an element of the subalgebra $\mathbb{H}_c$. At the finite level we will show later an exact formula which connects the solvable representative $\exp[\mathbf{s}]$ (with $\mathbf{s}\in\mathrm{Bor}$) to the orthogonal representative $\exp[\mathbf{k}]$ (with $\mathbf{k}\in\mathbb{K}$) of the same equivalence class. For the moment it suffices to understand that the action of the Borel group is transitive on the coset manifold, so that the coset manifold G_R/H_c is indeed normal and its metric can be obtained from the non degenerate Euclidean metric $<,>$ defined over $\mathrm{Bor}(G_{\mathbf{max}}) = \mathrm{Solv}\left(\frac{G_{\mathbf{max}}}{H_c}\right)$.

The example of the maximally split case clearly suggests what is the required solvable algebra for other normal forms. We have:

$$\mathrm{Solv}\left(\frac{G_R}{H_c}\right) = \mathscr{H}^{n.c.} \oplus \mathrm{span}\,(\mathscr{E}^{\alpha}) \quad ; \quad \forall \alpha \in \Delta_+ / \alpha_{\parallel} \neq 0 \tag{2.5.6}$$

where $\mathscr{H}^{n.c.}$ is the non-compact part of the Cartan subalgebra and $\mathscr{E}^{\alpha}$ denotes the combination of step operators pertaining to the positive roots α that appear in the real form $\mathbb{G}_R$ and the sum is extended only to those roots that are not purely imaginary. Indeed the step operators pertaining to imaginary roots are included into the maximal compact subalgebra that now is larger than the number of positive roots.

For any solvable group manifold with a non degenerate invariant metric[6] the differential geometry of the manifold is completely rephrased in algebraic language through the relation of the Levi-Civita connection and the *Nomizu operator* acting on the solvable Lie algebra. The latter is defined as

$$\mathbb{L} : \mathrm{Solv}\,(\mathscr{M}) \otimes \mathrm{Solv}\,(\mathscr{M}) \rightarrow \mathrm{Solv}\,(\mathscr{M}), \tag{2.5.7}$$
$$\forall X, Y, Z \in \mathrm{Solv}\,(\mathscr{M}) : 2 < \mathbb{L}_X Y, Z >=< [X, Y], Z > - < X, [Y, Z] > - < Y, [X, Z] > .$$

The *Riemann curvature operator* on this group manifold can be expressed as

$$\mathrm{Riem}(X, Y) = [\mathbb{L}_X, \mathbb{L}_Y] - \mathbb{L}_{[X,Y]}. \tag{2.5.8}$$

This implies that the covariant derivative explicitly reads:

$$\mathbb{L}_X Y = \Gamma^Z_{XY}\, Z \tag{2.5.9}$$

where

$$\Gamma^Z_{XY} = \frac{1}{2}\left(\langle Z, [X, Y]\rangle - \langle X, [Y, Z]\rangle - \langle Y, [X, Z]\rangle\right) \frac{1}{< Z, Z >} \qquad \forall X, Y, Z \in Solv \tag{2.5.10}$$

Equation (2.5.10) is true for any solvable Lie algebra, but in the case of **maximally non-compact, split algebras** we can write a general form for Γ^Z_{XY}, namely:

[6]See [6–12] for reviews on the solvable Lie algebra approach to supergravity scalar manifolds and the use of the Nomizu operator.

$$\begin{aligned}
\Gamma^i_{jk} &= 0 \\
\Gamma^i_{\alpha\beta} &= \tfrac{1}{2}\left(-\langle E_\alpha, \left[E_\beta, H^i\right]\rangle - \langle E_\beta, \left[E_\alpha, H^i\right]\rangle\right) = \tfrac{1}{2}\alpha^i \delta_{\alpha\beta} \\
\Gamma^\alpha_{ij} &= \Gamma^\alpha_{i\beta} = \Gamma^i_{j\alpha} = 0 \\
\Gamma^\alpha_{\beta i} &= \frac{1}{2}\left(\langle E^\alpha, \left[E_\beta, H_i\right]\rangle - \langle E_\beta, \left[H_i, E^\alpha\right]\rangle\right) = -\alpha_i\, \delta^\alpha_\beta \\
\Gamma^{\alpha+\beta}_{\alpha\beta} &= -\Gamma^{\alpha+\beta}_{\beta\alpha} = \frac{1}{2} N_{\alpha\beta} \\
\Gamma^\alpha_{\alpha+\beta\,\beta} &= \Gamma^\alpha_{\beta\,\alpha+\beta} = \frac{1}{2} N_{\alpha\beta}
\end{aligned} \tag{2.5.11}$$

where $N^{\alpha\beta}$ is defined by the commutator

$$\left[E_\alpha\,,\, E_\beta\right] = N_{\alpha\beta}\, E_{\alpha+\beta} \tag{2.5.12}$$

The explicit form (2.5.11) follows from the following choice of the non degenerate metric:

$$\begin{aligned}
\langle \mathscr{H}_i\,,\, \mathscr{H}_j\rangle &= 2\,\delta_{ij} \\
\langle \mathscr{H}_i\,,\, E_\alpha\rangle &= 0 \\
\langle E_\alpha\,,\, E_\beta\rangle &= \delta_{\alpha,\beta}
\end{aligned} \tag{2.5.13}$$

$\mathscr{H}_i \in$ CSA and E_α are the step operators associated with positive roots $\alpha \in \Delta_+$. For any other **non split case**, the Nomizu connection exists nonetheless although it does not take the form (2.5.11). It follows from Eq. (2.5.10) upon the choice of an invariant positive metric on *Solv* and the use of the structure constants of *Solv*.

2.5.1 The Tits Satake Projection: An Anticipation

Let us now come back to Eq. (2.4.19). Setting all $\alpha_\perp = 0$ corresponds to a projection:

$$\Pi_{TS} \quad : \quad \Delta_{\mathbb{G}} \mapsto \overline{\Delta} \tag{2.5.14}$$

of the original root system $\Delta_{\mathbb{G}}$ onto a new system of vectors living in an Euclidean space of dimension equal to the non compact rank r_{nc}. A priori this is not obvious, but it is nonetheless true that $\overline{\Delta}$, with only one exception, is by itself the root system of a simple Lie algebra $\mathbb{G}_{TS}$, the Tits–Satake subalgebra of $\mathbb{G}_R$:

$$\overline{\Delta} = \text{root system of } \mathbb{G}_{TS} \subset \mathbb{G}_R \tag{2.5.15}$$

The Tits–Satake subalgebra $\mathbb{G}_{TS} \subset \mathbb{G}_R$ is always the maximally non compact real section of its own complexification. For this reason, considering its maximal compact subalgebra $\mathbb{H}_{TS} \subset \mathbb{G}_{TS}$, we have a new smaller coset $\frac{G_{TS}}{H_{TS}}$, which is maximally split. What is the relation between the two solvable Lie algebras $\mathrm{Solv}\left(\frac{G}{H}\right)$ and $\mathrm{Solv}\left(\frac{G_{TS}}{H_{TS}}\right)$ is the natural question which arises. The explicit answer to this question and the systematic illustration of the geometrical relevance of the Tits Satake projection is the subject of an entire later chapter, namely Chap. 5. To appreciate the role of this projection we still have to introduce Kähler and Quaternionic geometry, special geometries and the c-map, all items that are the conspicuous contribution of Supergravity to Modern Geometry.

References

1. P.G. Fré, *A Conceptual History of Symmetry from Plato to the Superworld* (Springer, Berlin, 2018)
2. R. Gilmore, *Lie Groups, Lie Algebras, and Some of their Applications* (Courier Corporation, 2012)
3. S. Helgason, *Differential Geometry and Symmetric Spaces* (Academic Press, London, 1962)
4. A.W. Knapp, *Lie Groups Beyond an Introduction*, Progress in Mathematics (Springer Science & Business Media, Berlin, 2013)
5. M. Henneaux, D. Persson, P. Spindel, Spacelike singularities and hidden symmetries of gravity. Living Rev. Relativ. **11** (2008). https://doi.org/10.12942/lrr-2008-1 (arXiv:0710.1818 [hep-th])
6. P. G. Fré, *Gravity, A Geometrical Course*, volumes 1,2 (Springer Science & Business Media, Berlin, 2012)
7. P. Fré, F. Gargiulo, K. Rulik, Cosmic billiards with painted walls in non-maximal supergravities: a worked out example. Nucl. Phys. B **737** (2006), pp. 1–48. https://doi.org/10.1016/j.nuclphysb.2005.10.023 [hep-th/0507256]
8. P. Fré, F. Gargiulo, J. Rosseel, K. Rulik, M. Trigiante, A. Van Proeyen, Tits–Satake projections of homogeneous special geometries. Class. Quantum Gravity, **24** (2006), pp. 27–78. https://doi.org/10.1088/0264-9381/24/1/003 [hep-th/0606173]
9. L. Andrianopoli, R. D'Auria, S. Ferrara, U-duality and central charges in various dimensions revisited. Int. J. Mod. Phys. A **13**, 431–492 (1998)
10. L. Andrianopoli, R. D'Auria, S. Ferrara, P. Fré, M. Trigiante, R-R scalars, U-duality and solvable Lie algebras. Nucl. Phys. B **496**, 617–629 (1997). arXiv:hep-th/9611014
11. L. Andrianopoli, R. D'Auria, S. Ferrara, P. Fré, R. Minasian, M. Trigiante, Solvable Lie algebras in type IIA, type IIB and M-theories. Nucl. Phys. B **493**, 249–277 (1997). arXiv:hep-th/9612202
12. P. Fré, Gaugings and other supergravity tools of p-brane physics, *Proceedings of the Workshop on Latest Development in M-Theory, Paris, France, 1-9 Feb 2001* (2001). arXiv:hep-th/0102114

Chapter 3
Complex and Quaternionic Geometry

Mathematics, however, is, as it were, its own explanation; this, although it may seem hard to accept, is nevertheless true, for the recognition that a fact is so is the cause upon which we base the proof.

Girolamo Cardano

3.1 Imaginary Units and Geometry

Considering the possible types of numbers we have $\mathbb{R}$, $\mathbb{C}$, $\mathbb{H}$, or $\mathbb{O}$. This is a message for geometry. Keeping the fundamental idea that a geometrical space should be viewed as a manifold, constructed by means of an atlas of open charts, the local coordinates could be chosen not only as real numbers but also as complex, quaternionic or even octonionic numbers. Yet an important lesson is immediately learnt from the story told in my other book [1], twin of the present one: the possible numbers are, anyhow, division algebras over the reals, whose classification is due to Frobenius, so that the real structure remains the basis for everything.

This must be the same also in geometry. Manifolds of complex, quaternionic or octonionic type, if they exist, are, first of all, real manifolds. Their characterization as complex, quaternionic or octonionic must reside in some additional richer structure they are able to support. It is evident that this additional structure are the imaginary units, the same that provide the extensions of the field $\mathbb{R}$ to $\mathbb{C}$, $\mathbb{H}$ or $\mathbb{O}$.

Hence the conceptual path we have to follow starts revealing itself. We have to imagine what the imaginary units might be in the context of differential geometry. The catch is the relation $\mathbf{J}^2 = -\mathbf{1}$. How to reinterpret such a relation? It is rather natural to consider $\mathbf{J}$ as a map, in particular a linear map, and $\mathbf{1}$ as the identity map which always exists. We are almost there, the remaining question is *on which space does* $\mathbf{J}$ *act*? The answer is obvious since for linear maps we need vector spaces and if

P. G. Frè, *Advances in Geometry and Lie Algebras from Supergravity*, Theoretical and Mathematical Physics,
https://doi.org/10.1007/978-3-319-74491-9_3

we want to do things locally, point by point on the manifold, we need *vector bundles*. The universal vector-bundle that it is intrinsically associated with any manifold $\mathscr{M}$ is the tangent bundle $T\mathscr{M} \rightarrow \mathscr{M}$. Hence the imaginary units, that from now on we will name *complex structures*, are linear maps operating on sections of the tangent bundle that square to minus one.

Complex and quaternionic or hyper-complex geometries arise when a manifold admits one or more complex structures satisfying appropriate algebraic relations. This mixture of algebra and geometry leads to new classes of very interesting spaces:

(a) Complex Manifolds
(b) Complex Kähler Manifolds
(c) HyperKähler Manifolds
(d) Quaternionic Kähler Manifolds

that is the mission of the present chapter to define and illustrate.

Furthermore when we come to discuss the symmetries of such manifolds, namely their isometries, which is the main interest of this book, we discover that the presence of the complex-structures entrains a new very much challenging viewpoint on continuous symmetries. To the Killing vectors, thanks to the symplectic structures implied by the complex-structures we are able to associate *hamiltonian functions*, named *moment maps*. These moment maps open a vast playing ground for new constructions of high relevance both in Physics and Mathematics.

3.1.1 The Precognitions of Supersymmetry

Supersymmetric field-theories and in particular Supergravity have the remarkable property of an intrinsic precognition of geometric and algebraic structures. All classes of existing geometries found, in due time, their proper role within the frame of supersymmetric field theories. For instance Kähler Manifolds describe the most general coupling of scalar multiplets in $\mathscr{N} = 1$ rigid supersymetry, while HyperKähler Manifolds do the same for the rigid $\mathscr{N} = 2$ case (see [2] which will be extensively discussed in Chap. 8). Quaternionic Kähler Manifolds are the obligatory structure for the coupling of hypermultiplets to $\mathscr{N} = 2$ supergravity [3–5]. In these cases the precognition resides in algebraic relations that come from supersymmetry and, once duely interpreted, were shown to imply the mentioned geometry. In other, even more spectacular cases, the geometric structures required by supersymmetry were not yet available in the mathematical supermarkets when the corresponding supermultiplets were studied. They were just discovered by the physicists working in supergravity and now constitute new chapters of mathematics. These are the *Special Geometries* to which Chap. 4 is devoted.

Let us now turn to complex structures and their heritage.

3.2 Complex Structures on 2n-Dimensional Manifolds

Let $\mathscr{M}$ be a 2n-dimensional manifold, $T\mathscr{M}$ its tangent space and $T^*\mathscr{M}$ its cotangent space. Denoting by $\{\phi^\alpha\}$ $(\alpha = 1, \ldots, 2n)$ the $2n$ coordinates in a patch, a section $\mathbf{t} \in \Gamma(T\mathscr{M}, \mathscr{M})$ is represented by a linear differential operator:

$$\mathbf{t} = t^\alpha \boldsymbol{\partial}_\alpha \tag{3.2.1}$$

while a section in $T^*\mathscr{M}$ is a differential 1-form

$$\omega = d\phi^\alpha \omega_\alpha(\phi) \tag{3.2.2}$$

The contraction is an operation that to each vector field $\mathbf{t} \in \Gamma(T\mathscr{M}, \mathscr{M})$ associates a map

$$i_\mathbf{t} \; : \; T^*\mathscr{M} \longrightarrow \mathbb{C}^\infty\left(\mathscr{M}\right) \tag{3.2.3}$$

of 1-forms into 0-forms locally given by the following expression:

$$i_\mathbf{t}\, \omega = t^\alpha(\phi)\omega_\alpha(\phi) \tag{3.2.4}$$

In particular, if $\omega = df$ we have

$$i_\mathbf{t}\, df = t^\alpha \partial_\alpha f = \mathbf{t} f \tag{3.2.5}$$

The contraction is also canonically extended to higher forms:

$$\forall \mathbf{t} \in \Gamma(T\mathscr{M}, \mathscr{M}) \quad : \quad \begin{cases} i_\mathbf{t} \; : \; \Omega^p\left(\mathscr{M}\right) \longrightarrow \Omega^{p-1}\left(\mathscr{M}\right) \\ i_\mathbf{t}\, \omega = t^\alpha(\phi)\omega_{\alpha\beta_1\ldots\beta_{p-1}}(\phi)\, d\phi^{\beta_1} \wedge \cdots \wedge d\phi^{\beta_{p-1}} \end{cases} \tag{3.2.6}$$

Now we can consider a linear operator L acting on the tangent bundle $T\mathscr{M}$, or more precisely acting on $\Gamma(T\mathscr{M}, M)$:

$$\begin{aligned} & L : \Gamma(T\mathscr{M}, \mathscr{M}) \rightarrow \Gamma(T\mathscr{M}, \mathscr{M}) \\ & \forall \mathbf{t} \in \Gamma(T\mathscr{M}, \mathscr{M}) : \quad L\mathbf{t} \in \Gamma(T\mathscr{M}, \mathscr{M}) \\ & \forall \alpha, \beta \in \mathbb{C}, \;\; \forall \mathbf{t}_1, \mathbf{t}_2 \in \Gamma(T\mathscr{M}, \mathscr{M}) : \quad L(\alpha \mathbf{t}_1 + \beta \mathbf{t}_2) = \alpha L\mathbf{t}_1 + \beta L\mathbf{t}_2 \end{aligned} \tag{3.2.7}$$

In every local chart L is represented by a mixed tensor $L^\beta_\alpha(\phi)$ with one covariant index and one controvariant index such that

$$L\mathbf{t} = t^\alpha(\phi) L^\beta_\alpha(\phi) \boldsymbol{\partial}_\beta \tag{3.2.8}$$

Moreover the action of L is naturally pulled back on the cotangent space:

$$L : \Gamma(T\mathscr{M}^*, \mathscr{M}) \to \Gamma(T\mathscr{M}^*, \mathscr{M}) \tag{3.2.9}$$

by defining

$$i_{\mathbf{t}} L\omega = i_{L\mathbf{t}}\omega \tag{3.2.10}$$

which in a local chart yields

$$L\omega = d\phi^\alpha L_\alpha^\beta(\phi)\omega_\beta \tag{3.2.11}$$

Definition 3.2.1 A 2n-dimensional manifold $\mathscr{M}$ is called almost complex if it has an almost complex structure. An almost complex structure is a linear operator $J : \Gamma(T\mathscr{M}, \mathscr{M}) \to \Gamma(T\mathscr{M}, \mathscr{M})$ which satisfies the following property:

$$J^2 = -\,\mathbb{1} \tag{3.2.12}$$

In every local chart the operator J is represented by a tensor $J_\beta^\alpha(\phi)$ such that

$$J_\alpha^\beta(\phi) J_\beta^\gamma(\phi) = -\delta_\alpha^\gamma \tag{3.2.13}$$

and by a suitable change of basis at every point $p \in \mathscr{M}$ we can reduce J_α^β to the form

$$\begin{pmatrix} 0 & \mathbb{1} \\ -\mathbb{1} & 0 \end{pmatrix}$$

where $\mathbb{1}$ is the $n \times n$ unity matrix. A local frame where J takes the form (3.2.14) is called a "well-adapted" frame to the almost complex structure. Naming

$$\mathbf{e}_\alpha = \boldsymbol{\partial}_\alpha = \frac{\partial}{\partial \phi_\alpha} \tag{3.2.14}$$

the basis of the well-adapted frame we have

$$\begin{aligned} J\mathbf{e}_\alpha &= -\mathbf{e}_{\alpha+n} \quad if \quad \alpha \le n \\ J\mathbf{e}_\alpha &= \mathbf{e}_{\alpha-n} \quad if \quad \alpha > n \end{aligned} \tag{3.2.15}$$

At this point, introducing the index i with range $i = 1, \ldots, n$ we can define the complex vectors:

$$\begin{aligned} \mathbf{E}_i &= \mathbf{e}_i - i\mathbf{e}_{i+n} \\ \mathbf{E}_{i^*} &= \mathbf{e}_i + i\mathbf{e}_{i+n} \end{aligned} \tag{3.2.16}$$

and we obtain the following result:

$$\begin{aligned} J\mathbf{E}_i &= i\mathbf{E}_i \\ J\mathbf{E}_{i^*} &= -i\mathbf{E}_{i^*} \end{aligned} \tag{3.2.17}$$

The tangent vectors $\mathbf{E}_i$ are the partial derivatives along the complex coordinates:

$$z^i = \phi^i + i\phi^{i+n} \tag{3.2.18}$$

while $\mathbf{E}_{i^*}$ are the partial derivatives along the complex conjugate coordinates $\overline{z}^{i^*} = \phi^i - i\phi^{i+n}$:

$$\mathbf{E}_i = \boldsymbol{\partial}_i = \frac{\partial}{\partial z^i} \quad \mathbf{E}_{i^*} = \boldsymbol{\partial}_{i^*} = \frac{\partial}{\partial z^{i^*}} \tag{3.2.19}$$

This construction is the reason why J is called an almost complex structure: the existence of this latter guarantees that at every point $p \in \mathscr{M}$ we can replace the $2n$ real coordinates by n complex coordinates, corresponding to a well-adapted frame. Moreover every two well-adapted frames are related to each other by a coordinate transformation which is a holomorphic function of the corresponding complex coordinates. Indeed let

$$\phi^\alpha \to \phi^\alpha + \zeta^\alpha(\phi) \tag{3.2.20}$$

be an infinitesimal coordinate transformation connecting two well adapted frames. By definition this means

$$\partial_\alpha \zeta^\beta J^\gamma_\beta = J^\beta_\alpha \partial_\beta \zeta^\gamma \tag{3.2.21}$$

which is nothing but the Cauchy–Riemann equation for the real and imaginary parts of a holomorphic function. Hence Eq. (3.2.20) can be replaced by

$$z^i \to z^i + \zeta^i(z) \tag{3.2.22}$$

where $\zeta^i(z)$ is a holomorphic function of z^j. Conversely if $\mathscr{M}$ is a complex analytic manifold,[1] in every local chart $\{z^i\}$ we can set

$$\phi^\alpha = \mathrm{Re} z^i \ (\alpha \le n) \quad \phi^\alpha = \mathrm{Im} z^i \ (\alpha > n) \tag{3.2.23}$$

and we can define an almost complex structure J. Now let J act on $T^*(\mathscr{M})$. In a well-adapted frame we have

$$\begin{aligned} J dz^i &= i dz^i \\ J dz^{i^*} &= -i dz^{i^*} \end{aligned} \tag{3.2.24}$$

[1]Complex analytic manifold means a manifold whose transition functions in the intersection of two charts are holomorphic functions of the local coordinates.

Equation (3.2.24) characterize the holomorphic coordinates. More generally let $\{x^\alpha\}$ be a generic coordinate system (not necessarily well-adapted) and let $w(x)$ be a complex-valued function on the manifold $\mathscr{M}$: we say that w is holomorphic if it satisfies the equation:

$$Jdw = idw \tag{3.2.25}$$

which in the generic coordinate system $\{x^\alpha\}$ reads as follows:

$$J_\alpha^\beta \partial_\beta w(x) = i\partial_\alpha w(x) \tag{3.2.26}$$

As we have seen, at every point $p \in \mathscr{M}$, J can be reduced to the canonical form (3.2.14) by a suitable coordinate transformation: what is not guaranteed is whether J can be reduced to this canonical form in a whole open neighbourhood $\mathscr{U}_p$. This amounts to asking the question whether Eq. (3.2.26) admits n $\mathbb{C}$-linearly independent solutions in some open subset $\mathscr{U} \in \mathscr{U}_X$, where $\mathscr{U}_X$ is the domain of the considered local chart $\{x^\alpha\}$. If these solutions $w^i(x)$ exist we can consider them as the holomorphic coordinates in the neighbourhood $\mathscr{U}$, that is we can set

$$z^i = w^i(z) \tag{3.2.27}$$

In view of what we discussed before, the transition function between any two such coordinate systems is holomorphic. Hence if Eq. (3.2.25) is integrable, then a holomorphic coordinate system exists and any function ϕ on the manifold can be viewed as a function of z^i and z^{i^*}: $\phi = \phi(z, z^{i^*})$. In this case we have

$$\begin{aligned} d\phi &= \partial_i \phi dz^i + \partial_{i^*}\phi d\overline{z}^{i^*} \\ Jd\phi &= i(\partial_i\phi dz^i - \partial_{i^*}\phi dz^{i^*}) \end{aligned} \tag{3.2.28}$$

By taking the exterior derivative of Eq. (3.2.28) we obtain

$$dJ \wedge d\phi = -2i\partial_i\partial_{i^*}\phi dz^i \wedge dz^{i^*} \tag{3.2.29}$$

and we can verify the equation

$$(1 - J)dJ \wedge d\phi = 0 \tag{3.2.30}$$

which follows from

$$JdJ \wedge d\phi = -2i\partial_i\partial_{j^*}\phi Jdz^i \wedge Jdz^{j^*} = -2i\partial_i\partial_{j^*}\phi dz^i \wedge dz^{j^*} = dJ \wedge d\phi \tag{3.2.31}$$

Equation (3.2.30) is true in a holomorphic coordinate system and, being an exterior algebra statement, must be true in every coordinate system. In the real coordinate system Eq. (3.2.30) reads

$$T^\alpha_{\beta\gamma}\partial_\alpha\phi dx^\beta \wedge dx^\gamma = 0 \tag{3.2.32}$$

where the tensor

$$T^{\alpha}_{\beta\gamma} = \partial_{[\beta} J^{\alpha}_{\gamma]} - J^{\mu}_{\beta} J^{\nu}_{\gamma} \partial_{[\mu} J^{\alpha}_{\nu]} \tag{3.2.33}$$

is called the "torsion", or the Nienhuis tensor of the almost complex structure J^{α}_{β}. The vanishing of $T^{\alpha}_{\beta\gamma}$ is a necessary condition for the integrability of Eq. (3.2.26) and hence for the existence of a complex structure. It can be shown that it is also sufficient provided $T^{\alpha}_{\beta\gamma}$ is real analytic with respect to some real coordinate system.

3.3 Metric and Connections on Holomorphic Vector Bundles

In the previous section we considered the structure of complex manifolds. When both the base space and the standard fibre are complex manifolds we can refine the notion of fibre bundle by requiring that the transition function be locally holomorphic functions. In particular a very relevant concept, which plays a major role in our subsequent developments, is that of holomorphic vector bundle. For convenience we recall the complete definition that follows from the general definition of fibre-bundle.

Definition 3.3.1 Let $\mathcal{M}$ be a complex manifold and E be another complex manifold. A holomorphic vector bundle with total space E and base manifold $\mathcal{M}$ is given by a projection map:

$$\pi \; : \; E \longrightarrow \mathcal{M} \tag{3.3.1}$$

such that

(a) π is a holomorphic map of E onto $\mathcal{M}$

(b) Let $p \in \mathcal{M}$, then the fibre over p

$$E_p \; = \; \pi^{-1}(p) \tag{3.3.2}$$

is a complex vector space of dimension r. (The number r is called the rank of the vector bundle.)

(c) For each $p \in \mathcal{M}$ there is a neighbourhood U of p and a holomorphic homeomorphism

$$h \; : \; \pi^{-1}(U) \longrightarrow U \times \mathbb{C}^r \tag{3.3.3}$$

such that

$$h\left(\pi^{-1}(p)\right) \; = \; \{p\} \times \mathbb{C}^r \tag{3.3.4}$$

(The pair (U, h) is called a local trivialization.)

(d) The transition functions between two local trivializations (U_α, h_α) and (U_β, h_β):

$$h_\alpha \circ h_\beta^{-1} \; : \; (U_\alpha \cap U_\beta) \otimes \mathbb{C}^r \longrightarrow (U_\alpha \cap U_\beta) \otimes \mathbb{C}^r \tag{3.3.5}$$

induce holomorphic maps

$$g_{\alpha\beta} : (U_\alpha \cap U_\beta) \longrightarrow \mathrm{GL}(r, \mathbb{C}) \tag{3.3.6}$$

Let $E \longrightarrow \mathcal{M}$ be a holomorphic vector bundle of rank r and $U \subset \mathcal{M}$ an open subset of the base manifold. A frame over U is a set of r holomorphic sections $\{s_1, \ldots, s_r\}$ such that $\{s_1(z), \ldots, s_r(z)\}$ is a basis for $\pi^{-1}(z)$ for any $z \in U$. Let $f \equiv \{e_I(z)\}$ be a frame of holomorphic sections. Any other holomorphic section ξ is described by

$$\xi = \xi^I(z)\, e_I \tag{3.3.7}$$

where

$$\overline{\partial}\, \xi^I = d\overline{z}^{j^\star}\, \overline{\partial}_{j^\star}\, \xi^I = 0 \tag{3.3.8}$$

Given a holomorphic bundle with a frame of sections we can discuss metrics connections and curvatures, as we already did for the general case of bundles.

In general a connection θ is defined by introducing the covariant derivative of any section ξ

$$D\xi = d\xi + \theta\, \xi \tag{3.3.9}$$

where $\theta = \theta^I{}_J$, the connection coefficient, is an $r \times r$ matrix-valued 1-form. On a complex manifold this 1-form can be decomposed into its parts of holomorphic type (1, 0) and (0, 1), respectively:

$$\begin{aligned} \theta &= \theta^{(1,0)} + \theta^{(0,1)} \\ \theta^{(1,0)} &= dz^i\, \theta_i \\ \theta^{(0,1)} &= d\overline{z}^{i^\star}\, \theta_{i^\star} \end{aligned} \tag{3.3.10}$$

Let now a *fiber hermitian metric* h be defined on the holomorphic vector bundle. This is a sesquilinear form that yields the scalar product of any two holomorphic sections ξ and η at each point of the base manifold:

$$\langle\, \xi\, ,\, \eta\, \rangle_h \equiv \overline{\xi}^{I^\star}(\overline{z})\, \eta^J(z)\, h_{I^\star J}(z, \overline{z}) = \xi^\dagger\, h\, \eta \tag{3.3.11}$$

As it is evident from the above formula, the metric h is defined by means of the point-dependent hermitian matrix $h_{I^\star J}(z, \overline{z})$, which is requested to transform, from one local trivialization to another, with the inverses of the transition functions $g_{\alpha\beta}$ defined in Eq. (3.3.6). This is so because the scalar product $\langle\, \xi\, ,\, \eta\, \rangle_h$ is by definition an invariant (namely a scalar function globally defined on the manifold).

Definition 3.3.2 A hermitian metric for a complex manifold $\mathcal{M}$ is a hermitian fibre metric on the canonical tangent bundle $T\mathcal{M}$. In this case the transition functions $g_{\alpha\beta}$ are given by the jacobians of the coordinate transformations.

In general h is just a metric on the fibres and the transition functions are different objects from the Jacobian of the coordinate transformations. In any case, given a fibre metric on a holomorphic vector bundle we can introduce a canonical connection θ associated with it. It is defined by requiring that

$$\begin{array}{lll} (A) & d\,\langle \xi\,,\,\eta\,\rangle_h & = \langle\, D\xi\,,\,\eta\,\rangle_h + \langle \xi\,,\,D\eta\,\rangle_h \\ (B) & D^{(0,1)}\xi \equiv \left[\bar{\partial} + \theta^{(0,1)}\right]\xi = 0 & \end{array} \tag{3.3.12}$$

namely by demanding that the scalar product be invariant with respect to the parallel transport defined by θ and by requiring that the holomorphic sections be transported into holomorphic sections. Let f be a holomorphic frame. In this frame the canonical connection is given by

$$\theta(f) \;=\; h(f)^{-1}\,\partial\, h(f) \tag{3.3.13}$$

or, in other words, by

$$\theta^I{}_J \;=\; dz^i\, h^{IJ^\star}\,\partial_i\, h_{K^\star J} \tag{3.3.14}$$

In the particular case of a manifold metric (see Definition 3.3.2), where h is a fibre metric on the tangent bundle $T\mathscr{M}$, the general formula (3.3.14) provides the definition of the Levi-Civita connection:

$$dz^k\, \Gamma^i_{kj} \;=\; -\, g^{il^\star}\,\partial\, g_{l^\star j} \tag{3.3.15}$$

Given a connection we can compute its curvature by means of the standard formula $\Theta \;=\; d\theta + \theta \wedge \theta$. In the case of the above-defined canonical connection we obtain

$$\Theta\;(f) \;=\; \partial\,\theta + \bar{\partial}\,\theta + \theta \wedge \theta \;=\; \bar{\partial}\,\theta \tag{3.3.16}$$

This identity follows from $\partial\,\theta \;+\; \theta \;\wedge\; \theta \;=\; 0$, which is identically true for the canonical connection (3.3.13). Component-wise the curvature 2-form is given by

$$\Theta^I{}_J \;=\; \bar{\partial}_i\,\left(h^{IK^\star}\,\partial_j\, h_{K^\star J}\right)\, d\bar{z}^i \;\wedge\; dz^j \tag{3.3.17}$$

For the case of the Levi-Civita connection defined in Eq. (3.3.15) we find

$$\begin{aligned} \Gamma^i_j &= \Gamma^i_{kj} dz^k \\ \Gamma^i_{kj} &= -g^{i\ell^*}(\partial_j g_{k\ell^*}) \\ \Gamma^{i^*}_{j^*} &= \Gamma^{i^*}_{k^*\,j^*} d\bar{z}^{k^*} \\ \Gamma^{i^*}_{k^*\,j^*} &= -g^{i^*\ell}(\partial_{j^*} g_{k^*\ell}) \end{aligned} \tag{3.3.18}$$

for the connection coefficients and

$$
\begin{aligned}
\mathscr{R}^i_j &= \mathscr{R}^i_{jk^*\ell} d\bar{z}^{k^*} \wedge dz^\ell \\
\mathscr{R}^i_{jk^*\ell} &= \partial_{k^*} \Gamma^i_{j\ell} \\
\mathscr{R}^{i^*}_{j^*} &= \mathscr{R}^{i^*}_{j^*k\ell^*} dz^k \wedge d\bar{z}^{\ell^*} \\
\mathscr{R}^{i^*}_{j^*k\ell^*} &= \partial_k \Gamma^{i^*}_{j^*\ell^*}
\end{aligned}
\tag{3.3.19}
$$

for the curvature 2-form. The Ricci tensor has a remarkable simple expression:

$$
\mathscr{R}^n_{m^*} = \mathscr{R}^i_{m^*n\,i} = \partial_{m^*} \Gamma^i_{ni} = \partial_{m^*} \partial_n \, ln\,(\sqrt{g}) \tag{3.3.20}
$$

where $g = det|g_{\alpha\beta}| = (det|g_{ij^*}|)^2$.

3.4 Characteristic Classes and Elliptic Complexes

The cohomology[2] of differential forms on differentiable manifolds is named *de Rham cohomology*.[3] There are more general constructions of the same type. They are named *elliptic complexes*.

Elliptic complexes are associated with fibre-bundles and their general definition is provided below. To each elliptic complex we can associate a topological number that is named its *index*. On its turn the index of a complex can be calculated as the integral of certain polynomials in the curvature 2-forms of the connection that can be introduced on the corresponding principle bundle. These polynomials are named characteristic classes.

More precisely characteristic classes are maps from the ring $I^\star(\mathbb{G})$ of invariant polynomials on the Lie algebra $\mathbb{G}$ of the structural group of the bundle to the de Rham cohomology ring $H^\star(\mathscr{M})$ of its base manifold. They provide an intrinsic way of measuring the twisting, or deviation from triviality, of a fibre bundle. They are also an essential ingredient of the index theorems that express the difference of zero modes of an elliptic operator minus its adjoint precisely in terms of integrals of characteristic classes. Index theorems play a fundamental role in many physical problems. Characteristic classes are also needed in the definition of special geometries that we later consider. For this reason we devote the present section to their general discussion.

We begin by recalling the notion of de Rham cohomology groups. The differential forms of degree r on a k-dimensional manifold $\mathscr{M}$ are sections of a vector bundle, namely of the completely antisymmetrized tensor product $\Lambda^r\,(T^\star\mathscr{M})$ of the cotangent bundle $T^\star\mathscr{M}$, r times with itself. We name $\Omega^r = \Gamma\,(\mathscr{M}, \Lambda^r\,(T^\star\mathscr{M}))$ the

[2]For a pedagogical short introduction to cohomology theory I refer the reader to my book [6], Vol 1, Chap. 2.

[3]The development of de Rham cohomology and of characteristic classes is historically reviewed in the twin book to this one [1], within the general frame of the evolution of geometry in the XXth century.

space of sections of this bundle, namely the space of r-forms. The exterior derivative d provides a sequence of maps d_i:

$$\Omega^0(\mathscr{M}) \xrightarrow{d_0} \Omega^1(\mathscr{M}) \xrightarrow{d_1} \cdots \xrightarrow{d_{k-2}} \Omega^{k-1}(\mathscr{M}) \xrightarrow{d_{k-1}} \Omega^k(\mathscr{M}) \xrightarrow{d_k} 0 \quad (3.4.1)$$

where d_r is the exterior derivative acting on r-forms and producing $r+1$-forms as a result. The property of the exterior derivative $d^2 = 0$ implies that

$$d_i\, d_{i+1} = 0 \qquad \forall i = 0, \ldots, k \quad (3.4.2)$$

What we have just described is named the **de Rham complex** and provides the first and most prominent example of an elliptic complex. More generally we have

Definition 3.4.1 An elliptic complex $(E^\star, D)$ is a sequence of vector bundles $E_i \xrightarrow{\pi_i} \mathscr{M}$ constructed over the same base manifold and a sequence of Fredholm operators D_i mapping the sections of the ith bundle into those of the (i+1)th bundle:

$$\Gamma(\mathscr{M}, E_0) \xrightarrow{D_0} \Gamma(\mathscr{M}, E_1) \xrightarrow{D_1} \cdots \xrightarrow{D_{k-2}} \Gamma(\mathscr{M}, E_{k-1}) \xrightarrow{D_{k-1}} \Gamma(\mathscr{M}, E_k) \xrightarrow{D_k} 0 \quad (3.4.3)$$

such that

$$D_i\, D_{i+1} = 0 \qquad \forall i = 0, \ldots, k \quad (3.4.4)$$

A Fredholm operator is a differential operator of elliptic type with finite kernel and cokernel, as we discuss below. To each elliptic complex and to the de Rham complex in particular we can attach the notion of cohomology groups. The ith cohomology group is defined as follows:

$$H^i\left(E^\star, \mathscr{M}\right) = \frac{\ker D_i}{\operatorname{Im} D_{i-1}} \quad (3.4.5)$$

It is the space of sections of the ith bundle E_i satisfying $D_i\, s = 0$, modulo those of the form $s = D_{i-1}\, s'$. In the de Rham complex $H^r\,(\Omega^\star(\mathscr{M}))$ is the space of closed r-forms modulo exact forms. For any Fredholm operator D_i appearing in the elliptic complex (3.4.3) we denote $D_i^\dagger$ its adjoint, which is defined by

$$\begin{aligned} D_i^\dagger\ &: \Gamma(\mathscr{M}, E_{i+1}) \to \Gamma(\mathscr{M}, E_i) \\ (s', D_i s)_{E_{i+1}} &= (D_i^\dagger s', s)_{E_i} \end{aligned} \quad (3.4.6)$$

where $s \in \Gamma(\mathscr{M}, E_i)$, $s' \in \Gamma(\mathscr{M}, E_{i+1})$ and $(\ ,\ )_E$ denotes the fibre metric in the specified fibre. The laplacian operator is defined by

$$\begin{aligned} \Delta_i\ &: \Gamma(\mathscr{M}, E_i) \to \Gamma(\mathscr{M}, E_i) \\ \Delta_i &\equiv D_{i-1} D_{i-1}^\dagger + D_i^\dagger D_i \end{aligned} \quad (3.4.7)$$

The cohomology group $H^i(E^*, \mathcal{M})$ is isomorphic to the kernel of the operator Δ_i, so that we have

$$\dim H^i(E^*, D) = \dim \text{Harm}^i(E^*, D) \tag{3.4.8}$$

where by $\text{Harm}^i(E^*, D)$ we denote the vector space spanned by sections $h_i \in \Gamma(\mathcal{M}, E_i)$ which satisfy

$$\Delta_i h_i = 0\,. \tag{3.4.9}$$

Given a section $s_i \in \Gamma(\mathcal{M}, E_i)$ we can write the Hodge decomposition:

$$s_i = D_i s_{i-1} + D_i^\dagger s_{i+1} + h_i \tag{3.4.10}$$

where $s_{i\pm 1} \in \Gamma(\mathcal{M}, E_i)$.

Definition 3.4.2 Given an elliptic complex (E^*, D) we define the index of this complex by

$$\text{ind}\,(E^*, D) = \sum(-)^i \dim H^i(E^*, D) = \sum(-)^i \dim \ker \Delta_i \tag{3.4.11}$$

Equation (3.4.11), when specialized to the de Rham complex, gives the Euler characteristic of the base manifold:

$$\text{ind}\, d = \sum(-)^i \dim H^i(E^*, d) \equiv \chi(\mathcal{M}) = \sum(-)^i b^i \tag{3.4.12}$$

where b^i is the ith Betti number, equal, by definition, to the number of linearly independent harmonic i-forms. For a generic Fredholm operator $D : \Gamma(\mathcal{M}, E) \to \Gamma(\mathcal{M}, F)$ we can define the *analytical index* of D as

$$\text{ind} D = \dim \ker D - \dim \text{coker} D \tag{3.4.13}$$

To show the relation between Eqs. (3.4.11) and (3.4.13), we have to resume our discussion on Fredholm operators. Let $D : \Gamma(\mathcal{M}, E) \to \Gamma(\mathcal{M}, F)$ be an elliptic operator. The kernel of D is the following set of sections:

$$\ker D = \{s \in \Gamma(\mathcal{M}, E) | Ds = 0\}\,. \tag{3.4.14}$$

We define the cokernel of D by

$$\text{coker} D = \frac{\Gamma(\mathcal{M}, F)}{\text{Im} D} \tag{3.4.15}$$

We now state without proof the following theorem:

Theorem 3.4.1 *Let $D : \Gamma(\mathcal{M}, E) \to \Gamma(\mathcal{M}, F)$ be a Fredholm operator. Then*

$$\text{coker} D \sim \ker D^\dagger \tag{3.4.16}$$

Using Theorem 3.4.1 we immediately rewrite Eq. (3.4.11) as

$$\mathrm{ind} D = \dim \ker D - \dim \ker D^\dagger \tag{3.4.17}$$

Consider now the one-operator complex $\Gamma(\mathscr{M}, E) \xrightarrow{D} \Gamma(\mathscr{M}, F)$, which can also be written as

$$0 \xrightarrow{i} \Gamma(\mathscr{M}, E) \xrightarrow{D} \Gamma(\mathscr{M}, F) \xrightarrow{\phi} 0 \tag{3.4.18}$$

where i is the inclusion map (defined by $i(0) = 0$), and ϕ is a map from a generic section in $\Gamma(\mathscr{M}, F)$ into 0. Using Eq. (3.4.11) for the complex (3.4.18) we find

$$\dim \ker D - [\dim \Gamma(\mathscr{M}, F) - \dim \mathrm{Im} D] = \dim \ker D \; - \; \dim \mathrm{coker}\, D \tag{3.4.19}$$

The above equation shows the simple relation between the analytical index (3.4.13) and the index of the elliptic complex (3.4.11). Equation (3.4.13) provides an easy formula that is always recalled in physical literature. Moreover, given an elliptic complex, it is always possible to construct a Fredholm operator whose analytical index coincides with the index of the complex (E^*, D). Indeed if we define

$$E_+ = \oplus_i E_{2i}, \qquad E_- = \oplus_i E_{2i+1} \tag{3.4.20}$$

which are respectively called the even and the odd bundles and we consider the operators

$$D \equiv \oplus_i (D_{2i} + D^\dagger_{2i-1}) \quad D^\dagger \equiv \oplus_i (D_{2i+1} + D^\dagger_{2i}) \tag{3.4.21}$$

we easily verify that

$$\begin{aligned} D &: \Gamma(\mathscr{M}, E_+) \to \Gamma(\mathscr{M}, E_-) \\ D &: \Gamma(\mathscr{M}, E_-) \to \Gamma(\mathscr{M}, E_+) \end{aligned} \tag{3.4.22}$$

Next, if we define

$$\Delta_+ \equiv D^\dagger D = \oplus_i \Delta_{2i} \quad \Delta_- \equiv D D^\dagger = \oplus_i \Delta_{2i+1} \tag{3.4.23}$$

then we have

$$\mathrm{ind}(E_\pm, D) = \dim \ker \Delta_+ - \dim \ker \Delta_- = \sum (-)^i \dim \ker \Delta_i = \mathrm{ind}(E^*, D) \tag{3.4.24}$$

In general the index of an elliptic complex can be expressed by an integral over $\mathscr{M}$ of suitable characteristic classes. At the beginning of the present section we have defined characteristic classes as maps from the ring of invariant polynomials on the Lie algebra of the structural group to the de Rham cohomology group ring of the base manifold. Let us now go a little deeper on the meaning of this definition. Let

$\mathscr{M}(k, \mathbb{C})$ be the set of complex $k \times k$ matrices. We denote by $S^r(\mathscr{M}(k, \mathbb{C}))$ the vector space of symmetric r-linear $\mathbb{C}$-valued functions on $\mathscr{M}(k, \mathbb{C})$. A map

$$\hat{P} : \otimes_r \mathscr{M}(k, \mathbb{C}) \to \mathbb{C} \tag{3.4.25}$$

belongs to $S^r(\mathscr{M}(k, \mathbb{C}))$ if it satisfies, in addition to linearity in each entry, the symmetry

$$\hat{P}(a_1, \dots, a_i, \dots, a_j, \dots .a_r) = \hat{P}(a_1, \dots, a_j, \dots, a_i, \dots, a_r) \qquad \forall i, j \leq r \tag{3.4.26}$$

Consider now the formal sum

$$S^*(\mathscr{M}(k, \mathbb{C})) = \oplus_0^\infty S^r(\mathscr{M}(k, \mathbb{C})) \tag{3.4.27}$$

and define a product of $\hat{P} \in S^p(\mathscr{M}(k, \mathbb{C}))$ and $\hat{Q} \in S^q(\mathscr{M}(k, \mathbb{C}))$ by

$$\hat{P} \cdot \hat{Q}(a_1, \dots, a_{p+q}) = \frac{1}{(p+q)!} \sum_P \hat{P}(a_{P(1)}, \dots, a_{P(p)}) \hat{Q}(a_{P(p+1)}, \dots, a_{P(p+q)}) \tag{3.4.28}$$

where P denotes the permutation of the set $(1, \dots, p+q)$. $S^*(\mathscr{M}(k, \mathbb{C}))$ equipped with the product (3.4.28) is an algebra. If we now consider a Lie algebra $\mathbb{G} \in \mathscr{M}(k, \mathbb{C})$, and the corresponding simply connected Lie group $\mathscr{G} = \exp[\mathbb{G}]$, in full analogy with Eqs. (3.4.27) and (3.4.26), we can define the sum $S^*(\mathbb{G}) = \oplus_{r \geq 0} S^r(\mathbb{G})$. An element $\hat{P}(h_1, \dots, h_r) \in S^r(\mathbb{G})$ $(h_i \in \mathbb{G})$ is said to be invariant if, for any $g \in G$, it satisfies

$$\hat{P}(g^{-1} h_1 g, \dots, g^{-1} h_r g) = \hat{P}(h_1, \dots, h_r) \tag{3.4.29}$$

The set of invariant elements of $S^r(\mathbb{G})$ is denoted by $I^r(\mathbb{G})$. The product defined in (3.4.28) induces a natural multiplication

$$\cdot : I^p(\mathbb{G}) \otimes I^q(\mathbb{G}) \to I^{p+q}(\mathbb{G}) \tag{3.4.30}$$

The sum $I^* = \oplus_{r \geq 0} I^r(\mathbb{G})$ equipped with the product (3.4.30) is an algebra. The diagonal combination $P(h) = P(h, \dots, h)$ containing r-times the element $h \in \mathbb{G}$ is a polynomial of degree r, which is said to be an *invariant polynomial*. Let now $P(\mathscr{M}, G)$ be a principal bundle that has as structural group a Lie group $\mathscr{G}$ with Lie algebra $\mathbb{G}$. We extend the domain of invariant polynomials from $\mathbb{G}$ to $\mathbb{G}$-valued p-forms on $\mathscr{M}$. We define

$$\hat{P}(h_1 \omega_1, \dots, h_r \omega_r) \equiv \omega_1 \wedge \dots \wedge \omega_r \, \hat{P}(h_1, \dots, h_r) \tag{3.4.31}$$

where $h_i \in \mathbb{G}$, $\omega_i \in \Omega^{p_i}(\mathscr{M})$ $(i = 1 \dots r)$. The diagonal combination is now given by

$$P(h\omega) = \omega \wedge \dots \wedge \omega \, P(h) \tag{3.4.32}$$

where the wedge product of $\omega \in \Omega^p(\mathcal{M})$ is repeated r-times in (3.4.32). Consider now the curvature 2-form Θ associated with a connection in a complex fibre bundle. In the following we are particularly interested in invariant polynomials of the form $P(\Theta)$. We can state the following theorem (*Chern–Weil theorem*).

Theorem 3.4.2 *Let $P(\Theta)$ be an invariant polynomial in the curvature 2-form; then*
(i) $dP(\Theta) = 0$
(ii) Let Θ, Θ' be curvature 2-forms corresponding to different connections θ, θ' on the fibre bundle. Then the difference $P(\Theta) - P(\Theta')$ is exact.

This theorem proves that an invariant polynomial $P(\Theta)$ is closed and in general non-trivial. We can then associate to $P(\Theta)$ a cohomology class of $\mathcal{M}$. Moreover Theorem 3.4.2 ensures that this cohomology class is independent of the chosen connection. The cohomology class defined by $P(\Theta)$ is called a *characteristic class*. The characteristic class defined by an invariant polynomial P is denoted by $\chi_E(P)$, where E is the fibre bundle on which curvatures and connections are defined.

Theorem 3.4.3 *Let P be an invariant polynomial in $I^*(\mathbb{G})$ and E be a fibre bundle over $\mathcal{M}$, whose structural group $\mathcal{G}$ has $\mathbb{G}$ as Lie algebra. The map*

$$\chi_E : I^*(\mathbb{G}) \to H^*(\mathcal{M}) \tag{3.4.33}$$

defined by $P \to \chi_E(P)$ is a homomorphism.

Theorem 3.4.3 establishes a homomorphism, called the Chern–Weil homomorphism,[4] between the ring $I^*(\mathbb{G})$ and the de Rham cohomology ring $H^*(\mathcal{M})$, defined by

$$H^*(\mathcal{M}) = \oplus_r H^r(\mathcal{M}) \tag{3.4.34}$$

where H^r is the rth cohomolgy group. The Chern–Weil homomorphism is the fundamental instrument that allows one to relate the index of an elliptic complex with the integral of particular characteristic classes, through the so called *index theorem* (stated below in Eq. (3.4.56)). Before giving the statement of this theorem, due to Atiyah and Singer, we list some specific examples of characteristic classes, which will be useful in the following.

Definition 3.4.3 Given a complex vector bundle E equipped with a connection θ, whose fibre is $\mathbb{C}^r$, we can define its total Chern class $c(E, \Theta)$ as the following formal determinant:

$$c(E, \Theta) = det\left(\mathbf{1} + \frac{i}{2\pi}\Theta\right) \tag{3.4.35}$$

where Θ is the matrix-valued curvature 2-form.

The determinant is calculated with respect to the matrix indices. As it is well known, the determinant $det(\mathbf{1} + A)$ is a polynomial in the matrix elements of A and can

[4]The interesting history of the Chern–Weil homomorphism, independently discovered by the two great mathematicians in the years of World War II, is reported in the twin book [1].

be expanded in powers of A. Such an expansion of the total Chern class yields the definition of the individual Chern classes $c_k(E, \Theta)$. In particular, if we call $x_1, \ldots x_r$ the (formal) eigenvaules[5] 2-forms of the matrix $\frac{i}{2\pi}\Theta$ we easily find

$$\det\left(1 + \frac{i}{2\pi}\Theta\right) = \prod_{1}^{r}(1 + x_j) = 1 + (x_1 + \cdots + x_r) + (x_1 x_2 + \cdots + x_{r-1}x_r) + \cdots + (x_1 x_2 \cdots x_r) \quad (3.4.36)$$

so that, by writing

$$c(E, \Theta) = \sum_{k=0}^{r} c_k(E, \Theta) \quad (3.4.37)$$

we get

$$\begin{aligned} c_0 &= 1, \\ c_1 &= \frac{i}{2\pi}\operatorname{tr}(\Theta), \\ c_2 &= \frac{1}{8\pi^2}\left[\operatorname{tr}\left(\Theta^2\right) - (\operatorname{tr}\Theta)^2\right] \\ &\vdots \\ c_r &= \det\frac{i\Theta}{2\pi} \end{aligned} \quad (3.4.38)$$

where, for a generic form Ω, by Ω^n we mean the nth wedge product $\wedge^n\Omega$. A remarkable property of the Chern class is the following: given two complex vector bundles $E \xrightarrow{\pi} \mathcal{M}$, $F \xrightarrow{\pi'} \mathcal{M}$ we have

$$c(E \oplus F) = c(E) \wedge c(F) \quad (3.4.39)$$

Definition 3.4.4 Given a rank r vector bundle $E \xrightarrow{\pi} \mathcal{M}$ we define the total Chern character by

$$\operatorname{ch}(E, \Theta) = \operatorname{tr} exp\left(\frac{i\Theta}{2\pi}\right) = \sum_{l=1} \frac{1}{l!}\operatorname{tr}\left(\frac{i\Theta}{2\pi}\right)^j \quad (3.4.40)$$

and the jth Chern character by

[5] We stress the word "formal eigenvalues" because the correct framework to understand these eigenvalues is the "splitting principle", which, for convenience, is mentioned after the Eq. (2.7.59).

$$\mathrm{ch}_j(E, \Theta) = \frac{1}{j!}\mathrm{tr}\left(\frac{i\Theta}{2\pi}\right)^j \tag{3.4.41}$$

From now on, for notational convenience we refer to $\mathrm{ch}(E, \Theta)$ as $\mathrm{ch}\, E$ or $\mathrm{ch}\, \Theta$ indifferently (and similarly for the Chern class $c(E, \Theta)$). In terms of the eigenvectors x_j we get

$$\mathrm{ch}(\Theta) = \sum_{j=1}^{r}\left(1 + x_j + \frac{1}{2}x_j^2 + \cdots\right) \tag{3.4.42}$$

so that we can write

$$\begin{aligned}
\mathrm{ch}_0(\Theta) &= r \\
\mathrm{ch}_1(\Theta) &= c_1(\Theta) \\
\mathrm{ch}_2(\Theta) &= \frac{1}{2}[c_1^2(\Theta) - 2c_2(\Theta)]
\end{aligned} \tag{3.4.43}$$

Theorem 3.4.4 *Let E and F be two vector bundles over a manifold $\mathcal{M}$. The Chern character of $E \otimes F$ and $E \oplus F$ are given by*

$$\begin{aligned}
ch(E \otimes F) &= ch(E) \wedge ch(F) \\
ch(E \oplus F) &= ch(E) + ch(F)
\end{aligned} \tag{3.4.44}$$

Another useful characteristic class associated with a complex vector bundle is the **Todd class** defined by

$$\mathrm{Td}(\Theta) = \prod_{j=1}^{r}\frac{x_j}{1 - e^{-x_j}} \tag{3.4.45}$$

where x_j are the eigenvalues of the curvature 2-form $\frac{i}{2\pi}\Theta$. We obtain

$$\begin{aligned}
\mathrm{Td}(\Theta) &= 1 + \frac{1}{2}\sum_j x_j + \frac{1}{12}x_j^2 + \cdots \\
&= \prod_j\left(1 + \tfrac{1}{2}x_j + \sum_{k\geq 1}(-)^{k-1}\frac{B_k}{2k!}x_j^{2k}\right) \\
&= 1 + \frac{1}{2}c_1(\Theta) + \frac{1}{12}[c_1^2(\Theta) + c_2(\Theta)] + \cdots
\end{aligned} \tag{3.4.46}$$

where the numbers B_k appearing in Eq. (3.4.46) are the Bernoulli numbers.

Finally we define the **Euler class**. The characteristic classes previously introduced are naturally defined for complex vector bundles. On the other hand the Euler class can be defined for real vector bundles over an orientable Riemann manifold $\mathcal{M}$. In particular it is consistently defined for even rank real bundles, while it is zero for odd rank bundles. Given a rank k real bundle E it is useful to construct a complex vector bundle from E by a *complexification* procedure. The complexification of E is the bundle over $\mathcal{M}$ obtained by replacing the fibres $\mathbb{R}^k$ by $\mathbb{C}^k = (\mathbb{R} \oplus i\mathbb{R})^k$. We denote the complexification of E by $E^{\mathbb{C}}$. We can think of $E^{\mathbb{C}}$ as the following product

$$E^{\mathbb{C}} = E \otimes (\mathbb{R} \oplus i\mathbb{R}) \tag{3.4.47}$$

Complex vector bundles can also be complexified by converting them into real vector bundles and then complexifying the result. If the starting complex bundle has rank r, its complexification has rank $2r$. Notice that, given a complex vector bundle E, and denoting by $E_{\mathbb{R}}$ the underlying real bundle, we have

$$E_{\mathbb{R}}^{\mathbb{C}} = E_{\mathbb{R}} \otimes (\mathbb{R} + i\mathbb{R}) \sim E \oplus \overline{E} \tag{3.4.48}$$

where $\overline{E}$ denotes the conjugate complex bundle, defined by applying complex conjugation to the coordinates of the fibres $\mathbb{C}^r$ of E. Having outlined the complexification procedure for a real vector bundle, we define the Euler class through another typical characteristic class defined in real bundles: the Pontrjagin class. Let E be a real vector bundle of rank r over $\mathcal{M}$, the ith Pontrjagin class is defined as

$$p_i(E) = (-)^i c_{2i}(E^{\mathbb{C}}) \tag{3.4.49}$$

where $c_{2i}(E^{\mathbb{C}})$ is the $2i$th Chern class of the complexified bundle. The total Pontrjagin class is defined as

$$P(E) = 1 + p_1(E) + \cdots + p_{[r/2]} \tag{3.4.50}$$

where $[r/2]$ is the largest integer not greater than r. Consider now real vector bundles E of *even rank* over an orientable manifold $\mathcal{M}$. The Euler class is defined by

$$e^2(V) = p_{[r/2]} \tag{3.4.51}$$

The Euler class of a Whitney sum $E \oplus V$ is

$$e(E \oplus V) = e(E)e(V) \tag{3.4.52}$$

where we denote $c(E)c(V) = c(E) \wedge c(V)$. For a complex vector bundle the Pontrjagin and the Euler class are the Pontrjagin and the Euler class of the underlying real bundle. Since the eigenvalues of the curvature 2-form in the conjugate bundle are given by $-x_i$, we have

$$c(E^{\mathbb{C}}) = c(E \oplus \overline{E}) = c(E)c(\overline{E}) = \prod_{i=1}^{r}(1 + x_i)(1 - x_i) = \prod_{i=1}^{r}(1 - x_i^2) \tag{3.4.53}$$

so that

$$c_r(E^{\mathbb{C}}) = (-)^r x_1^2 \cdots x_r^2 \tag{3.4.54}$$

and (recalling that $E^{\mathbb{C}}$ has rank $2r$)

$$\begin{aligned} p_r(E) &= x_1^2 \cdots x_r^2 \\ e(E) &= x_1 x_2 \cdots x_r = c_r(E) \end{aligned} \tag{3.4.55}$$

We are now able to state the *Atiyah–Singer index theorem* in its full generality:

Theorem 3.4.5 *Given an elliptic complex (E^*, D) over an m-dimensional ($\dim_{\mathbb{R}}\mathcal{M} = m$) compact manifold $\mathcal{M}$ without a boundary, then*

$$ind(E^*, D) = (-)^{\frac{m(m+1)}{2}} \int_{\mathcal{M}} ch\left(\oplus_j (-)^j E_j\right) \frac{Td(T\mathcal{M}^{\mathbb{C}})}{e(T\mathcal{M})} \tag{3.4.56}$$

where $T\mathcal{M}$ is the tangent bundle over $\mathcal{M}$.

Let us now consider the application of the index theorem to some particular elliptic complexes. Consider an m-dimensional compact orientable manifold without boundaries and the elliptic de Rham complex:

$$\cdots \xrightarrow{d} \Omega^{r-1}(\mathcal{M})^{\mathbb{C}} \xrightarrow{d} \Omega^{r}(\mathcal{M})^{\mathbb{C}} \xrightarrow{d} \Omega^{r+1}(\mathcal{M})^{\mathbb{C}} \xrightarrow{d} \cdots \tag{3.4.57}$$

with $\Omega^r(\mathcal{M})^{\mathbb{C}} = \Gamma(\mathcal{M}, \wedge^r T^*\mathcal{M}^{\mathbb{C}})$, where we have complexified the forms to apply the Atiyah–Singer theorem. The analytical index is given by

$$\operatorname{ind} d = \sum_{r=0}^{m} (-)^r \dim_{\mathbb{C}} H^r(\mathcal{M}, \mathbb{C}) = \sum_{r=0}^{m} (-)^r \dim{}_{\mathbb{R}} H^r(\mathcal{M}, \mathbb{R}) = \chi(\mathcal{M}) \tag{3.4.58}$$

where $\chi(\mathcal{M})$ is the Euler characteristic of $\mathcal{M}$. Suppose $\mathcal{M}$ is even dimensional $m = 2l$. Equation 3.4.56 gives the following result for the de Rham index:

$$\operatorname{ind} d = (-)^{l(2l+1)} \int_{\mathcal{M}} \operatorname{ch}\left(\oplus_r^{2l} (-)^r \wedge^r T^*\mathcal{M}^{\mathbb{C}}\right) \frac{\operatorname{Td} T\mathcal{M}^{\mathbb{C}}}{e(T\mathcal{M})} \tag{3.4.59}$$

To compute $\operatorname{ch}\left(\oplus_r^m (-)^r \wedge^r T^*\mathcal{M}^{\mathbb{C}}\right)$ we employ the splitting principle. The splitting principle uses the fact that in order to prove an identity for characteristic classes, it is sufficient to prove it only for bundles which decompose into a sum of line bundles. Suppose that a fibre bundle F is a Whitney sum of n line bundles L_i; then

$$\wedge^p F = \oplus_{1 \le i_1 \cdots i_p \le n} \left(L_{i_1} \otimes \cdots \otimes L_{i_p}\right) \tag{3.4.60}$$

This means that

$$\mathrm{ch}(\wedge^p F) = \sum_{1\le i_1\cdots i_p\le n} \mathrm{ch}(L_{i_1})\mathrm{ch}(L_{i_2})\cdots\mathrm{ch}(L_{i_p}) \tag{3.4.61}$$

Since for any line bundle appearing in the Whitney sum $\mathrm{ch}(L_i) = e^{x_i}$, we finally get

$$\mathrm{ch}(\wedge^p F) = \sum_{1\le i_1\cdots i_p\le n} e^{x_{i_1}+\cdots+x_{i_p}} \tag{3.4.62}$$

Applying this result to $\oplus_r^m (-)^r \wedge^r T^*\mathcal{M}^{\mathbb{C}}$, and using the fact that taking the dual bundle merely changes the sign of x_i we get

$$\mathrm{ch}\oplus_r^m (-)^r \wedge^r T^*\mathcal{M}^{\mathbb{C}} = \prod_{i=1}^m (1-e^{-x_i})(T\mathcal{M}^{\mathbb{C}}) \tag{3.4.63}$$

Moreover we can write

$$\mathrm{Td}(T\mathcal{M}^{\mathbb{C}}) = \prod_{i=1}^m \frac{x_i}{1-e^{-x_i}}(T\mathcal{M}^{\mathbb{C}}) \tag{3.4.64}$$

Then the index of the de Rham complex is given by

$$\mathrm{ind}\, d = (-)^l \int_{\mathcal{M}} \frac{\prod_{i=1}^m x_i(T\mathcal{M}^{\mathbb{C}})}{e(T\mathcal{M})} = (-)^l \int_{\mathcal{M}} \frac{c_m(T\mathcal{M}^{\mathbb{C}})}{e(T\mathcal{M})} = \int_{\mathcal{M}} e(T\mathcal{M}) \tag{3.4.65}$$

where we have used

$$c_m(T\mathcal{M}^{\mathbb{C}}) = (-)^{m/2} e(T\mathcal{M}\oplus T\mathcal{M}) = (-)^l x_1^2\cdots x_m^2 = (-)^l e^2(T\mathcal{M})$$

By combining the results for the analytical index and for the Atiyah–Singer index (often referred to as the topological index), we get the Gauss–Bonnet theorem

$$\int_{\mathcal{M}} e(T\mathcal{M}) = \chi(\mathcal{M}) \tag{3.4.66}$$

For m odd, the de Rham index is zero. Let us consider now the application of the index theorem to the Dolbeault complex, which we are going to define below. Consider a complex manifold $\mathcal{M}$ with $dim_{\mathbb{C}}\mathcal{M} = m$. We denote by $T^{(1,0)}\mathcal{M}$ the tangent bundle spanned by the vectors $\{\partial/\partial z^\mu\}$ and by $T^{(0,1)}\mathcal{M}$ its complex conjugate. The space dual to $T^{(1,0)}\mathcal{M}$ is spanned by the 1-forms $\{dz^\mu\}$. We denote it by $T^{*(1,0)}\mathcal{M}$. The space $\Omega^r(\mathcal{M})^{\mathbb{C}}$ of complexified r-forms is decomposed as

$$\Omega^r(\mathcal{M})^{\mathbb{C}} = \oplus_{p+q=r}\Omega^{p,q}(\mathcal{M}) \tag{3.4.67}$$

where by $\Omega^{p,q}(\mathcal{M})$ we denote the space of (p, q) forms. The exterior derivative can be written as

$$d = dz^\mu \wedge \frac{\partial}{\partial z^\mu} + d\overline{z}^\mu \wedge \frac{\partial}{\partial \overline{z}^\mu} \tag{3.4.68}$$

It is immediate to verify that $\partial, \overline{\partial}$ satisfy the following relations:

$$\partial\overline{\partial} - \overline{\partial}\partial = \partial^2 = \overline{\partial}^2 = 0 \tag{3.4.69}$$

Moreover ∂ maps (p, q)-forms into $(p + 1, q)$-forms and $\overline{\partial}$ maps (p, q) forms into $(p, q + 1)$ forms. Let us consider the sequence

$$\cdots \overset{\overline{\partial}}{\rightarrow} \Omega^{(0,q)}(\mathcal{M}) \overset{\overline{\partial}}{\rightarrow} \Omega^{(0,q+1)}(\mathcal{M}) \overset{\overline{\partial}}{\rightarrow} \cdots \tag{3.4.70}$$

This sequence is called the **Dolbeault complex**. It can be shown that (3.4.70) defines an elliptic complex. The index theorem in this case gives

$$\text{ind}\,\overline{\partial} = \int_{\mathcal{M}} \text{ch}\left(\oplus_r (-)^r \wedge^r T^{*\,(0,1)}\mathcal{M}\right) \frac{\text{Td}T\mathcal{M}^{\mathbb{C}}}{e(T\mathcal{M})} \tag{3.4.71}$$

The left hand side of the above equation can be computed using the Eq. (3.4.13), so that

$$\text{ind}\,\overline{\partial} = \sum_{r=0}^{n} (-)^r h^{(0,r)} \tag{3.4.72}$$

where

$$h^{(0,r)} = \dim_{\mathbb{C}} H^{(0,r)}(\mathcal{M}) = \dim_{\mathbb{C}} \frac{\ker\overline{\partial}_r}{\text{im}\overline{\partial}_{r-1}} \tag{3.4.73}$$

is the complex dimension of the cohomology group $H^{(0,r)}$. The application of theorem (3.4.56) to this case is analogous to the one presented for the de Rham complex and gives

$$\sum_{r=0}^{n} (-)^r b^{(0,r)} = \int_{\mathcal{M}} \text{Td}(T^{(1,0)}\mathcal{M}) \tag{3.4.74}$$

In the Dolbeault complex the space $\Omega^{(0,r)}$ can be replaced by a tensor product bundle $\Omega^{(0,r)} \otimes V$, where V is a holomorphic vector bundle. In this case we define the following elliptic complex, named the **twisted Dolbeault complex**:

$$\cdots \overset{\overline{\partial}_V}{\rightarrow} \Omega^{(0,q)}(\mathcal{M}) \otimes V \overset{\overline{\partial}_V}{\rightarrow} \Omega^{(0,q+1)}(\mathcal{M}) \otimes V \overset{\overline{\partial}_V}{\rightarrow} \cdots \tag{3.4.75}$$

The Atiyah–Singer theorem for this particular complex reduces to the Hirzebruch–Riemann–Roch theorem:

$$\text{ind}\,\overline{\partial}_V = \int_{\mathcal{M}} \text{Td}\,(T^{(1,0)}\mathcal{M})\text{ch}(V) \tag{3.4.76}$$

In the case of complex dimension one, namely $dim_{\mathbb{C}}\mathcal{M} = 1$, we get

$$\text{ind}\,\overline{\partial}_V = \frac{1}{2}\text{dim}V \int_{\mathcal{M}} c_1(T^{(1,0)}\mathcal{M}) + \int_{\mathcal{M}} c_1(\mathcal{M}) \tag{3.4.77}$$

Since it can be shown that

$$\int_{\mathcal{M}} c_1(T^{(1,0)}\mathcal{M}) = \int_{\mathcal{M}} e(T\mathcal{M}) = 2(1-g) \tag{3.4.78}$$

where g is the genus of the base manifold, which in complex dimension one is nothing but a Riemann surface Σ_g, in this case we get

$$\text{ind}\,\overline{\partial}_V = \text{dim}V(1-g) + \int_{\Sigma_g} \frac{i\Theta}{2\pi} \tag{3.4.79}$$

In the general case of a complex manifold $\mathcal{M}$ of complex dimension n, the dimensions

$$h^{(p,q)} \stackrel{\text{def}}{=} \text{dim}_{\mathbb{C}}\, H^{(p,q)}\,(\mathcal{M}) \tag{3.4.80}$$

of the Dolbeault cohomology groups are named *Hodge numbers*.

3.5 Kähler Metrics

In the previous sections we have discussed the general notion of hermitian fibre metrics on holomorphic vector bundles and in particular of hermitian manifold metrics defined on the tangent bundle. In this section we introduce the more restricted concept of Kählerian metrics that plays a fundamental role in many applications.[6] The definition of the previous section Definition 3.3.2 can also be restated in the following way: a manifold metric g is a symmetric bilinear scalar valued functional on $\Gamma(T\mathcal{M},\mathcal{M}) \otimes \Gamma(T\mathcal{M},\mathcal{M})$

$$g\,:\,\Gamma(T\mathcal{M},\mathcal{M}) \otimes \Gamma(T\mathcal{M},\mathcal{M}) \;\rightarrow\; \mathbf{C}^{\infty}(\mathcal{M}) \tag{3.5.1}$$

In every coordinate system it is represented by the familiar symmetric tensor $g_{\alpha\beta}(x)$. Indeed we have

$$g(\mathbf{u},\mathbf{w}) = g_{\alpha\beta}u^{\alpha}w^{\beta} \tag{3.5.2}$$

[6] For Kähler's life, his relations with Chern and other outstanding mathematicians and for the conceptual development of Kähler metrics we refer the reader to the twin book [1].

where u^α, w^β are the components of the vector fields $\mathbf{u}$ and $\mathbf{w}$, respectively. In this language the hermiticity of the manifold metric g can be rephrased in the following way:

Definition 3.5.1 Let $\mathscr{M}$ be a 2n-dimensional manifold with an almost complex structure J. A metric g on $\mathscr{M}$ is called hermitian with respect to J if

$$g(J\mathbf{u}, J\mathbf{w}) = g(\mathbf{u}, \mathbf{w}) \tag{3.5.3}$$

Given a metric g and an almost complex structure J let us introduce the following differential 2-form K:

$$K(\mathbf{u}, \mathbf{w}) = \frac{1}{2\pi} g(J\mathbf{u}, \mathbf{w}) \tag{3.5.4}$$

The components $K_{\alpha\beta}$ of K are given by

$$K_{\alpha\beta} = g_{\gamma\beta} J^\gamma_\alpha \tag{3.5.5}$$

and by direct computation we can easily verify that:

Theorem 3.5.1 *g is hermitian if and only if K is anti-symmetric.*

Definition 3.5.2 A hermitian almost complex manifold is an almost complex manifold endowed with a hermitian metric g.

In a well-adapted basis we can write

$$g(u, w) = g_{ij} u^i w^j + g_{i^* j^*} u^{i^*} w^{j^*} + g_{\imath j^*} u^i w^{j^*} + g_{i^* j} u^i w^{j^*} \tag{3.5.6}$$

Reality of $g(u, w)$ implies

$$\begin{aligned} g_{ij} &= (g_{i^* j^*})^* \\ g_{i^* j} &= \left(g_{i j^*}\right)^\star \end{aligned} \tag{3.5.7}$$

symmetry ($g(u, w) = g(w, u)$) yields

$$\begin{aligned} g_{ij} &= g_{ji} \\ g_{j^* i} &= g_{i j^*} \end{aligned} \tag{3.5.8}$$

while the hermiticity condition gives

$$g_{ij} = g_{i^* j^*} = 0 \tag{3.5.9}$$

Finally in the well-adapted basis the 2-form K associated to the hermitian metric g can be written as

$$K = \frac{i}{2\pi} g_{ij^\star} dz^i \wedge d\overline{z}^{j^*} \tag{3.5.10}$$

Definition 3.5.3 A hermitian metric on a complex manifold $\mathscr{M}$ is called a Kähler metric if the associated 2-form K is closed:

$$dK = 0 \tag{3.5.11}$$

A hermitian complex manifold endowed with a Kähler metric is called a Kähler manifold.

Equation (3.5.11) is a differential equation for g_{ij^*} whose general solution in any local chart is given by the following expression:

$$g_{ij^*} = \partial_i \partial_{j^*} \mathscr{K} \tag{3.5.12}$$

where $\mathscr{K} = \mathscr{K}^* = \mathscr{K}(z, z^*)$ is a real function of z^i, z^{i^*}. The function $\mathscr{K}$ is called the Kähler potential and it is defined only up to the real part of a holomorphic function $f(z)$. Indeed one sees that

$$\mathscr{K}'(z, z^{i^*}) = \mathscr{K}(z, z^{i^*}) + f(z) + f^*(z^*) \tag{3.5.13}$$

give rise to the same metric g_{ij^*} as $\mathscr{K}$. The transformation (3.5.13) is called a Kähler transformation. The differential geometry of a Kähler manifold is described by Eqs. (3.3.18) and (3.3.19) with g_{ij^*} given by (3.5.12). Kähler geometry is that implied by $\mathscr{N} = 1$ supersymmetry for the scalar multiplets [7].

3.6 Hypergeometry

Next we turn our attention to the geometry that emerges when the manifold admits three complex structures satisfying the quaternionic algebra first discovered by Hamilton. To this effect the prerequisite is that the dimension of the manifold should be a multiple of 4. This is precisely what happens in supersymmetry when we consider the so called $\mathscr{N} = 2$ hypermultiplets. Each of them contains 4 real scalar fields and, at least locally, they can be regarded as the four components of a quaternion. The locality caveat is, in this case, very substantial because global quaternionic coordinates can be constructed only occasionally even on those manifolds that are denominated quaternionic in the mathematical literature [2, 3]. Anyhow, what is important is that, in the hypermultiplet sector, the scalar manifold $\mathscr{QM}$ has dimension multiple of four:

$$\dim_{\mathbf{R}} \mathscr{QM} = 4m \equiv 4\,\#\,\text{of hypermultiplets} \tag{3.6.1}$$

and, in some appropriate sense, it has a quaternionic structure.

We name *Hypergeometry* that pertaining to the hypermultiplet sector, irrespectively whether we deal with global or local $\mathcal{N} = 2$ theories. Yet there are two kinds of hypergeometries. Supersymmetry requires the existence of a principal SU(2)-bundle

$$\mathcal{SU} \longrightarrow \mathcal{QM} \tag{3.6.2}$$

The bundle $\mathcal{SU}$ is **flat** in the *rigid supersymmetry case* while its curvature is proportional to the Kähler forms in the *local case*.

These two versions of hypergeometry were already known in mathematics prior to their use [2–5, 8–10] in the context of $\mathcal{N} = 2$ supersymmetry and are identified as:

$$\begin{aligned} \text{rigid hypergeometry} &\equiv \text{HyperKähler geometry.} \\ \text{local hypergeometry} &\equiv \text{Quaternionic Kähler geometry} \end{aligned} \tag{3.6.3}$$

3.6.1 *Quaternionic Kähler, Versus HyperKähler Manifolds*

Both a Quaternionic Kähler or a HyperKähler manifold $\mathcal{QM}$ is a $4m$-dimensional real manifold endowed with a metric h:

$$ds^2 = h_{uv}(q)dq^u \otimes dq^v \quad ; \quad u, v = 1, \ldots, 4m \tag{3.6.4}$$

and three complex structures

$$(J^x) \; : \; T(\mathcal{QM}) \longrightarrow T(\mathcal{QM}) \qquad (x = 1, 2, 3) \tag{3.6.5}$$

that satisfy the quaternionic algebra

$$J^x J^y = -\delta^{xy}\, \mathbb{1} + \varepsilon^{xyz} J^z \tag{3.6.6}$$

and respect to which the metric is hermitian:

$$\forall \mathbf{X}, \mathbf{Y} \in T\mathcal{QM} \; : \quad h\left(J^x\mathbf{X}, J^x\mathbf{Y}\right) = h\left(\mathbf{X}, \mathbf{Y}\right) \qquad (x = 1, 2, 3) \tag{3.6.7}$$

From Eq. (3.6.7) it follows that one can introduce a triplet of 2-forms

$$K^x = K^x_{uv} dq^u \wedge dq^v \; ; \; K^x_{uv} = h_{uw}(J^x)^w_v \tag{3.6.8}$$

that provides the generalization of the concept of Kähler form occurring in the complex case. The triplet K^x is named the *HyperKähler* form. It is an SU(2) Lie-algebra valued 2-form in the same way as the Kähler form is a U(1) Lie-algebra valued 2-form. In the complex case the definition of Kähler manifold involves the statement that the Kähler 2-form is closed. At the same time in Hodge–Kähler manifolds the Kähler 2-form can be identified with the curvature of a line-bundle which in the case of rigid supersymmetry is flat. Similar steps can be taken also here and lead to two possibilities: either HyperKähler or Quaternionic Kähler manifolds.

Let us introduce a principal SU(2)-bundle $\mathcal{SU}$ as defined in Eq. (3.6.2). Let ω^x denote a connection on such a bundle. To obtain either a HyperKähler or a Quaternionic Kähler manifold we must impose the condition that the HyperKähler 2-form is covariantly closed with respect to the connection ω^x:

$$\nabla K^x \equiv dK^x + \varepsilon^{xyz}\omega^y \wedge K^z = 0 \tag{3.6.9}$$

The only difference between the two kinds of geometries resides in the structure of the $\mathcal{SU}$-bundle.

Definition 3.6.1 A HyperKähler manifold is a $4m$-dimensional manifold with the structure described above and such that the $\mathcal{SU}$-bundle is **flat**

Defining the $\mathcal{SU}$-curvature by:

$$\Omega^x \equiv d\omega^x + \frac{1}{2}\varepsilon^{xyz}\omega^y \wedge \omega^z \tag{3.6.10}$$

in the HyperKähler case we have:

$$\Omega^x = 0 \tag{3.6.11}$$

Viceversa

Definition 3.6.2 A Quaternionic Kähler manifold is a $4m$-dimensional manifold with the structure described above and such that the curvature of the $\mathcal{SU}$-bundle is proportional to the HyperKähler 2-form

Hence, in the quaternionic case we can write:

$$\Omega^x = \lambda\, K^x \tag{3.6.12}$$

where λ is a non vanishing real number.

As a consequence of the above structure the manifold $\mathcal{QM}$ has a holonomy group of the following type:

$$\begin{aligned}\mathrm{Hol}(\mathcal{QM}) &= \mathrm{SU}(2) \otimes \mathrm{H} \quad \text{(Quaternionic Kähler)}\\ \mathrm{Hol}(\mathcal{QM}) &= \mathbb{1} \otimes \mathrm{H} \quad \text{(HyperKähler)}\\ \mathrm{H} &\subset \mathrm{Sp}(2\mathrm{m}, \mathbb{R})\end{aligned} \tag{3.6.13}$$

In both cases, introducing flat indices $\{A, B, C = 1, 2\}\{\alpha, \beta, \gamma = 1, \ldots, 2m\}$ that run, respectively, in the fundamental representation of SU(2) and of $\mathrm{Sp}(2m, \mathbb{R})$, we can find a vielbein 1-form

$$\mathcal{U}^{A\alpha} = \mathcal{U}_u^{A\alpha}(q)dq^u \tag{3.6.14}$$

such that

$$h_{uv} = \mathcal{U}_u^{A\alpha}\mathcal{U}_v^{B\beta}\mathbb{C}_{\alpha\beta}\varepsilon_{AB} \tag{3.6.15}$$

where $\mathbb{C}_{\alpha\beta} = -\mathbb{C}_{\beta\alpha}$ and $\varepsilon_{AB} = -\varepsilon_{BA}$ are, respectively, the flat Sp(2m) and Sp(2) $\sim$ SU(2) invariant metrics. The vielbein $\mathcal{U}^{A\alpha}$ is covariantly closed with respect to the SU(2)-connection ω^z and to some Sp($2m, \mathbb{R}$)-Lie Algebra valued connection $\Delta^{\alpha\beta} = \Delta^{\beta\alpha}$:

$$\begin{aligned}\nabla\mathcal{U}^{A\alpha} &\equiv d\mathcal{U}^{A\alpha} + \frac{i}{2}\omega^x(\varepsilon\sigma_x\varepsilon^{-1})^A{}_B \wedge \mathcal{U}^{B\alpha}\\ &+ \Delta^{\alpha\beta} \wedge \mathcal{U}^{A\gamma}\mathbb{C}_{\beta\gamma} = 0\end{aligned} \tag{3.6.16}$$

where $(\sigma^x)_A{}^B$ are the standard Pauli matrices. Furthermore $\mathcal{U}^{A\alpha}$ satisfies the reality condition:

$$\mathcal{U}_{A\alpha} \equiv (\mathcal{U}^{A\alpha})^* = \varepsilon_{AB}\mathbb{C}_{\alpha\beta}\mathcal{U}^{B\beta} \tag{3.6.17}$$

Equation (3.6.17) defines the rule to lower the symplectic indices by means of the flat symplectic metrics ε_{AB} and $\mathbb{C}_{\alpha\beta}$. More specifically we can write a stronger version of Eq. (3.6.15) [7]:

$$(\mathcal{U}_u^{A\alpha}\mathcal{U}_v^{B\beta} + \mathcal{U}_v^{A\alpha}\mathcal{U}_u^{B\beta})\mathbb{C}_{\alpha\beta} = h_{uv}\varepsilon^{AB} \tag{3.6.18}$$

We have also the inverse vielbein $\mathcal{U}^u_{A\alpha}$ defined by the equation

$$\mathcal{U}^u_{A\alpha}\mathcal{U}_v^{A\alpha} = \delta^u_v \tag{3.6.19}$$

Flattening a pair of indices of the Riemann tensor $\mathcal{R}^{uv}{}_{ts}$ we obtain

$$\mathcal{R}^{uv}{}_{ts}\mathcal{U}_u^{\alpha A}\mathcal{U}_v^{\beta B} = -\frac{\mathrm{i}}{2}\Omega^x_{ts}\varepsilon^{AC}(\sigma_x)_C{}^B\mathbb{C}^{\alpha\beta} + \mathbb{R}^{\alpha\beta}_{ts}\varepsilon^{AB} \tag{3.6.20}$$

where $\mathbb{R}^{\alpha\beta}_{ts}$ is the field strength of the Sp(2m) connection:

$$d\Delta^{\alpha\beta} + \Delta^{\alpha\gamma} \wedge \Delta^{\delta\beta}\mathbb{C}_{\gamma\delta} \equiv \mathbb{R}^{\alpha\beta} = \mathbb{R}^{\alpha\beta}_{ts}dq^t \wedge dq^s \tag{3.6.21}$$

Equation (3.6.20) is the explicit statement that the Levi Civita connection associated with the metric h has a holonomy group contained in SU(2) $\otimes$ Sp(2m). Consider now Eqs. (3.6.6), (3.6.8) and (3.6.12). We easily deduce the following relation:

$$h^{st}K^x_{us}K^y_{tw} = -\delta^{xy}h_{uw} + \varepsilon^{xyz}K^z_{uw} \tag{3.6.22}$$

that holds true both in the HyperKähler and in the quaternionic case. In the latter case, using Eqs. (3.6.12), (3.6.22) can be rewritten as follows:

$$h^{st}\,\Omega^x_{us}\,\Omega^y_{tw} = -\lambda^2\delta^{xy}h_{uw} + \lambda\varepsilon^{xyz}\,\Omega^z_{uw} \tag{3.6.23}$$

Equation (3.6.23) implies that the intrinsic components of the curvature 2-form Ω^x yield a representation of the quaternion algebra. In the HyperKähler case such a representation is provided only by the HyperKähler form. In the quaternionic case we can write:

$$\Omega^x_{A\alpha,B\beta} \equiv \Omega^x_{uv}\,\mathcal{U}^u_{A\alpha}\,\mathcal{U}^v_{B\beta} = -i\lambda\mathbb{C}_{\alpha\beta}(\sigma_x)_A{}^C\varepsilon_{CB} \tag{3.6.24}$$

Alternatively Eq. (3.6.24) can be rewritten in an intrinsic form as

$$\Omega^x = -\mathrm{i}\,\lambda\mathbb{C}_{\alpha\beta}(\sigma_x)_A{}^C\varepsilon_{CB}\,\mathcal{U}^{\alpha A}\wedge\mathcal{U}^{\beta B} \tag{3.6.25}$$

whence we also get:

$$\frac{i}{2}\Omega^x(\sigma_x)_A{}^B = \lambda\,\mathcal{U}_{A\alpha}\wedge\mathcal{U}^{B\alpha} \tag{3.6.26}$$

3.7 Moment Maps

The conception of moment maps has its root in Hamiltonian mechanics where the time-derivative of any dynamical variable can be represented by the Poisson bracket of that variable with the hamiltonian. More generally the action of any vector field $\mathbf{t}$ on functions defined over the phase-space $\mathcal{M}$ can be represented as the Poisson bracket of that function with a generalized hamiltonian $\mathcal{H}_{\mathfrak{t}}$ which is associated with the vector field:

$$\begin{aligned}\mathbf{t} &\equiv t^i(p,q)\frac{\partial}{\partial q^i} + t_i(p,q)\frac{\partial}{\partial p_i}\\ \mathbf{t}\,\mathfrak{f}(p,q) &= \{\mathfrak{f}\,,\,\mathcal{H}_{\mathfrak{t}}\}\end{aligned} \tag{3.7.1}$$

The moment map is the map:

$$\begin{aligned}\mu\,:\,\Gamma[T\mathcal{M},\mathcal{M}] &\to \mathbb{C}[\mathcal{M}]\\ \mu[\mathbf{t}] &= \mathcal{H}_{\mathfrak{t}}\end{aligned} \tag{3.7.2}$$

which to every vector field associates its proper hamiltonian.

In the present geometrical context, conceptually very much different from that of dynamical systems which are of no concern to us in this book, the focus is on the moment-maps of Killing vectors, associated with isometries of the manifold $\mathcal{M}$. The symplectic structure which allows for the definition of Poisson-like brackets is provided by the presence of the complex-structure leading to closed or covariantly closed 2-forms, the Kähler or the HyperKähler ones. Our generalized hamiltonians or simply *moment-maps* have another important role to play. On one hand they appear as constructive items in supergravity lagrangians with gauge-symmetries, on the other, purely mathematical side, they are the building blocks in a general procedure, the *Kähler or HyperKähler quotient* which allows to construct non trivial Kähler or HyperKähler manifolds starting from simple trivial ones.

In Chap. 8 we plan to exemplify such constructions with the derivation of ALE-manifolds by means of HyperKḧaler quotients. Here we just begin with the general definitions of holomorphic and tri-holomorphic moment maps.

3.7.1 The Holomorphic Moment Map on Kähler Manifolds

The concept of holomorphic moment map applies to all Kähler manifolds, not necessarily special. Indeed it can be constructed just in terms of the Kähler potential without advocating any further structure. In this subsection we review its properties and definition, as usual in order to fix conventions, normalizations and notations.

Let $g_{ij^\star}$ be the Kähler metric of a Kähler manifold $\mathscr{M}$ and let us assume that $g_{ij^\star}$ admits a non trivial group of continuous isometries $\mathscr{G}$ generated by Killing vectors $k^i_{\mathbf{I}}$ ($\mathbf{I} = 1, \ldots, \dim \mathscr{G}$) that define the infinitesimal variation of the complex coordinates z^i under the group action:

$$z^i \to z^i + \varepsilon^{\mathbf{I}} k^i_{\mathbf{I}}(z) \tag{3.7.3}$$

Let $k^i_{\mathbf{I}}(z)$ be a basis of holomorphic Killing vectors for the metric $g_{ij^\star}$. Holomorphicity means the following differential constraint:

$$\partial_{j^*} k^i_{\mathbf{I}}(z) = 0 \leftrightarrow \partial_j k^{i^*}_{\mathbf{I}}(\overline{z}) = 0 \tag{3.7.4}$$

while the generic Killing equation (suppressing the gauge index $\mathbf{I}$):

$$\nabla_\mu k_\nu + \nabla_\mu k_\nu = 0 \tag{3.7.5}$$

in holomorphic indices reads as follows:

$$\nabla_i k_j + \nabla_j k_i = 0 \; ; \; \nabla_{i^*} k_j + \nabla_j k_{i^*} = 0 \tag{3.7.6}$$

where the covariant components are defined as $k_j = g_{ji^*} k^{i^*}$ (and similarly for k_{i^*}).

The vectors $k^i_{\mathbf{I}}$ are generators of infinitesimal holomorphic coordinate transformations $\delta z^i = \varepsilon^{\mathbf{I}} k^i_{\mathbf{I}}(z)$ which leave the metric invariant. In the same way as the metric is the derivative of a more fundamental object, the Killing vectors in a Kähler manifold are the derivatives of suitable prepotentials. Indeed the first of Eq. (3.7.6) is automatically satisfied by holomorphic vectors and the second equation reduces to the following one:

$$k^i_{\mathbf{I}} = i g^{ij^*} \partial_{j^*} \mathscr{P}_{\mathbf{I}}, \quad \mathscr{P}^*_{\mathbf{I}} = \mathscr{P}_{\mathbf{I}} \tag{3.7.7}$$

In other words if we can find a real function $\mathscr{P}^{\mathbf{I}}$ such that the expression $i g^{ij^*} \partial_{j^*} \mathscr{P}_{(\mathbf{I})}$ is holomorphic, then Eq. (3.7.7) defines a Killing vector.

The construction of the Killing prepotential can be stated in a more precise geometrical fashion through the notion of *moment map*. Let us review this construction.

Consider a Kählerian manifold $\mathscr{M}$ of real dimension $2n$. Consider an isometry group $\mathscr{G}$ acting on $\mathscr{M}$ by means of Killing vector fields $\overrightarrow{X}$ which are holomorphic with respect to the complex structure J of $\mathscr{M}$; then these vector fields preserve also the Kähler 2-form

$$\left.\begin{array}{l} \mathscr{L}_{\overrightarrow{X}} g = 0 \leftrightarrow \nabla_{(\mu} X_{\nu)} = 0 \\ \mathscr{L}_{\overrightarrow{X}} J = 0 \end{array}\right\} \Rightarrow 0 = \mathscr{L}_{\overrightarrow{X}} K = i_{\overrightarrow{X}} dK + d(i_{\overrightarrow{X}} K) = d(i_{\overrightarrow{X}} K) \tag{3.7.8}$$

Here $\mathscr{L}_{\overrightarrow{X}}$ and $i_{\overrightarrow{X}}$ denote respectively the Lie derivative along the vector field $\overrightarrow{X}$ and the contraction (of forms) with it.

If $\mathscr{M}$ is simply connected, $d(i_{\overrightarrow{X}} K) = 0$ implies the existence of a function $\mathscr{P}_{\overrightarrow{X}}$ such that

$$-\frac{1}{2} d\mathscr{P}_{\overrightarrow{X}} = i_{\overrightarrow{X}} K \tag{3.7.9}$$

The function $\mathscr{P}_{\overrightarrow{X}}$ is defined up to a constant, which can be arranged so as to make it equivariant:

$$\overrightarrow{X} \mathscr{P}_{\overrightarrow{Y}} = \mathscr{P}_{[\overrightarrow{X}, \overrightarrow{Y}]} \tag{3.7.10}$$

$\mathscr{P}_{\overrightarrow{X}}$ constitutes then a *moment map*. This can be regarded as a map

$$\mathscr{P} : \mathscr{M} \longrightarrow \mathbb{R} \otimes \mathbb{G}^* \tag{3.7.11}$$

where $\mathbb{G}^*$ denotes the dual of the Lie algebra $\mathbb{G}$ of the group $\mathscr{G}$. Indeed let $x \in \mathbb{G}$ be the Lie algebra element corresponding to the Killing vector $\overrightarrow{X}$; then, for a given $m \in \mathscr{M}$

$$\mu(m) : x \longrightarrow \mathscr{P}_{\overrightarrow{X}}(m) \in \mathbb{R} \tag{3.7.12}$$

is a linear functional on $\mathbb{G}$. If we expand $\overrightarrow{X} = a^{\mathbf{I}} k_{\mathbf{I}}$ in a basis of Killing vectors $k_{\mathbf{I}}$ such that

$$[k_{\mathbf{I}}, k_{\mathbf{L}}] = f_{\mathbf{IL}}{}^{\mathbf{K}} k_{\mathbf{K}} \tag{3.7.13}$$

we have also

$$\mathscr{P}_{\overrightarrow{X}} = a^{\mathbf{I}} \mathscr{P}_{\mathbf{I}} \tag{3.7.14}$$

In the following we use the shorthand notation $\mathscr{L}_{\mathbf{I}}$, $i_{\mathbf{I}}$ for the Lie derivative and the contraction along the chosen basis of Killing vectors $k_{\mathbf{I}}$.

From a geometrical point of view the prepotential, or moment map, $\mathscr{P}_{\mathbf{I}}$ is the Hamiltonian function providing the Poissonian realization of the Lie algebra on the Kähler manifold. This is just another way of stating the already mentioned *equivariance*. Indeed the very existence of the closed 2-form K guarantees that every Kähler space is a symplectic manifold and that we can define a Poisson bracket.

Consider Eq. (3.7.7). To every generator of the abstract Lie algebra $\mathbb{G}$ we have associated a function $\mathscr{P}_{\mathbf{I}}$ on $\mathscr{M}$; the Poisson bracket of $\mathscr{P}_{\mathbf{I}}$ with $\mathscr{P}_{\mathbf{J}}$ is defined as follows:

$$\{\mathscr{P}_{\mathbf{I}}, \mathscr{P}_{\mathbf{J}}\} \equiv 4\pi K(\mathbf{I}, \mathbf{J}) \tag{3.7.15}$$

where $K(\mathbf{I}, \mathbf{J}) \equiv K(\mathbf{k}_{\mathbf{I}}, \mathbf{k}_{\mathbf{J}})$ is the value of K along the pair of Killing vectors.

In Ref. [4] the following lemma was proved:

Lemma 3.1 *The following identity is true:*

$$\{\mathscr{P}_{\mathbf{I}}, \mathscr{P}_{\mathbf{J}}\} = f_{\mathbf{IJ}}{}^{\mathbf{L}}\mathscr{P}_{\mathbf{L}} + C_{\mathbf{IJ}} \tag{3.7.16}$$

where $C_{\mathbf{IJ}}$ is a constant fulfilling the cocycle condition

$$f_{\mathbf{IM}}{}^{\mathbf{L}}C_{\mathbf{LJ}} + f_{\mathbf{MJ}}{}^{\mathbf{L}}C_{\mathbf{LI}} + f_{\mathbf{JI}}{}^{\mathbf{L}}C_{\mathbf{LM}} = 0 \tag{3.7.17}$$

If the Lie algebra $\mathbb{G}$ has a trivial second cohomology group $H^2(\mathbb{G}) = 0$, then the cocycle $C_{\mathbf{IJ}}$ is a coboundary; namely we have

$$C_{\mathbf{IJ}} = f_{\mathbf{IJ}}{}^{\mathbf{L}}C_{\mathbf{L}} \tag{3.7.18}$$

where $C_{\mathbf{L}}$ are suitable constants. Hence, assuming $H^2(\mathbb{G}) = 0$ we can reabsorb $C_{\mathbf{L}}$ in the definition of $\mathscr{P}_{\mathbf{I}}$:

$$\mathscr{P}_{\mathbf{I}} \to \mathscr{P}_{\mathbf{I}} + C_{\mathbf{I}} \tag{3.7.19}$$

and we obtain the stronger equation

$$\{\mathscr{P}_{\mathbf{I}}, \mathscr{P}_{\mathbf{J}}\} = f_{\mathbf{IJ}}{}^{\mathbf{L}}\mathscr{P}_{\mathbf{L}} \tag{3.7.20}$$

Note that $H^2(\mathbb{G}) = 0$ is true for all semi-simple Lie algebras. Using Eqs. (3.7.16), (3.7.20) can be rewritten in components as follows:

$$\frac{i}{2}g_{ij^*}(k_{\mathbf{I}}^i k_{\mathbf{J}}^{j^*} - k_{\mathbf{J}}^i k_{\mathbf{I}}^{j^*}) = \frac{1}{2}f_{\mathbf{IJ}}{}^{\mathbf{L}}\mathscr{P}_{\mathbf{L}} \tag{3.7.21}$$

Equation (3.7.21) is identical with the equivariance condition in Eq. (3.7.10).

Finally let us recall the explicit general way of solving Eq. (3.7.9) obtaining the real valued function $\mathscr{P}_{\mathbf{I}}$ which satisfies Eq. (3.7.7). In terms of the Kähler potential $\mathscr{K}$ we have:

$$\mathscr{P}_{\mathbf{I}}{}^x = -\frac{i}{2}\left(k_{\mathbf{I}}^i\partial_i\mathscr{K} - k_{\mathbf{I}}^{\bar{\imath}}\partial_{\bar{\imath}}\mathscr{K}\right) + \mathrm{Im}(f_{\mathbf{I}})\,, \tag{3.7.22}$$

where $f_{\mathbf{I}} = f_{\mathbf{I}}(z)$ is a holomorphic transformation on the line-bundle, defining a compensating Kähler transformation:

$$k^i_{\mathbf{I}} \partial_i \mathcal{K} + k^{\bar{\imath}}_{\mathbf{I}} \partial_{\bar{\imath}} \mathcal{K} = -f_{\mathbf{I}}(z) - \overline{f}_{\mathbf{I}}(\bar{z}) \,. \tag{3.7.23}$$

3.7.2 The Triholomorphic Moment Map on Quaternionic Manifolds

Next, following closely the original derivation of [4, 11] let us turn to a discussion of the triholomorphic isometries of the manifold $\mathcal{QM}$ associated with hypermultiplets. In $D = 4$ supergravity the manifold of hypermultiplet scalars $\mathcal{QM}$ is a Quaternionic Kähler manifold and we can gauge only those of its isometries that are triholomorphic and that either generate an abelian group $\mathcal{G}$ or are *suitably realized* as isometries also on the special manifold $\widehat{\mathcal{SK}}_n$. This means that on $\mathcal{QM}$ we have Killing vectors:

$$\mathbf{k}_{\mathbf{I}} = k^u_{\mathbf{I}} \frac{\boldsymbol{\partial}}{\partial q^u} \tag{3.7.24}$$

satisfying the same Lie algebra as the corresponding Killing vectors on $\widehat{\mathcal{SK}}_n$. In other words

$$\mathfrak{K}_{\mathbf{I}} = \hat{k}^i_{\mathbf{I}} \boldsymbol{\partial}_i + \hat{k}^{i^*}_{\mathbf{I}} \boldsymbol{\partial}_{i^*} + k^u_{\mathbf{I}} \boldsymbol{\partial}_u \tag{3.7.25}$$

is a Killing vector of the block diagonal metric:

$$\mathfrak{g} = \begin{pmatrix} \widehat{g}_{ij^\star} & 0 \\ 0 & h_{uv} \end{pmatrix} \tag{3.7.26}$$

defined on the product manifold[7] $\widehat{\mathcal{SK}} \otimes \mathcal{QM}$.

Let us first focus on the manifold $\mathcal{QM}$. Triholomorphicity means that the Killing vector fields leave the HyperKähler structure invariant up to SU(2) rotations in the SU(2)-bundle defined by Eq. (3.6.2). Namely:

$$\mathcal{L}_{\mathbf{I}} K^x = \varepsilon^{xyz} K^y W^z_{\mathbf{I}} \; ; \; \mathcal{L}_{\mathbf{I}} \omega^x = \nabla W^x_{\mathbf{I}} \tag{3.7.27}$$

[7] Special Kähler geometry will be discussed in Chap. 4, yet we anticipate here that it is the geometrical structure imposed by $\mathcal{N} = 2$ supersymmetry on the scalars belonging to vector multiplets (the scalar partners of the gauge vectors). In our notations the Special Kähler manifold which describes the interaction of vector multiplets is denoted $\widehat{\mathcal{SK}}$ and all the Special Geometry Structures are endowed with a hat in order to distinguish this Special Kähler manifold from the other one which is incapsulated into the Quaternionic Kähler manifold $\mathcal{QM}$ describing the hypermultiplets when this latter happens to be in the image of the c-map. For all these concepts we refer the reader to Chap. 4. They are not necessary to understand the present constructions, yet they were essential part for their establishment in the original papers mentioned here above.

where $W^x_{\mathbf{I}}$ is an SU(2) compensator associated with the Killing vector $k^u_{\mathbf{I}}$. The compensator $W^x_{\mathbf{I}}$ necessarily fulfills the cocycle condition:

$$\mathscr{L}_{\mathbf{I}} W^x_{\mathbf{J}} - \mathscr{L}_{\mathbf{J}} W^x_{\mathbf{I}} + \varepsilon^{xyz} W^y_{\mathbf{I}} W^z_{\mathbf{J}} = f_{\mathbf{IJ}}^{\ \ \mathbf{L}} W^x_{\mathbf{L}} \tag{3.7.28}$$

In the HyperKähler case the SU(2)-bundle is flat and the compensator can be reabsorbed into the definition of the HyperKähler forms. In other words we can always find a map

$$\mathscr{Q}\mathscr{M} \longrightarrow L^x{}_y(q) \in \mathrm{SO}(3) \tag{3.7.29}$$

that trivializes the $\mathscr{S}\mathscr{U}$-bundle globally. Redefining:

$$K^{x\prime} = L^x{}_y(q)\, K^y \tag{3.7.30}$$

the new HyperKähler form obeys the stronger equation:

$$\mathscr{L}_{\mathbf{I}} K^{x\prime} = 0 \tag{3.7.31}$$

On the other hand, in the quaternionic case, the non-triviality of the $\mathscr{S}\mathscr{U}$-bundle forbids to eliminate the W-compensator completely. Due to the identification between HyperKähler forms and SU(2) curvatures Eq. (3.7.27) is rewritten as:

$$\mathscr{L}_{\mathbf{I}} \Omega^x = \varepsilon^{xyz} \Omega^y W^z_{\mathbf{I}} \; ; \; \mathscr{L}_{\mathbf{I}} \omega^x = \nabla W^x_{\mathbf{I}} \tag{3.7.32}$$

In both cases, anyhow, and in full analogy with the case of Kähler manifolds, to each Killing vector we can associate a triplet $\mathscr{P}^x_{\mathbf{I}}(q)$ of 0-form prepotentials. Indeed we can set:

$$\mathbf{i}_{\mathbf{I}} K^x = -\nabla \mathscr{P}^x_{\mathbf{I}} \equiv -(d\mathscr{P}^x_{\mathbf{I}} + \varepsilon^{xyz} \omega^y \mathscr{P}^z_{\mathbf{I}}) \tag{3.7.33}$$

where ∇ denotes the SU(2) covariant exterior derivative.

As in the Kähler case Eq. (3.7.33) defines a moment map:

$$\mathscr{P} : \mathscr{M} \longrightarrow \mathbb{R}^3 \otimes \mathbb{G}^* \tag{3.7.34}$$

where $\mathbb{G}^*$ denotes the dual of the Lie algebra $\mathbb{G}$ of the group $\mathscr{G}$. Indeed let $x \in \mathbb{G}$ be the Lie algebra element corresponding to the Killing vector $\vec{X}$; then, for a given $m \in \mathscr{M}$

$$\mu(m) : x \longrightarrow \mathscr{P}_{\vec{X}}(m) \in \mathbb{R}^3 \tag{3.7.35}$$

is a linear functional on $\mathscr{G}$. If we expand $\vec{X} = a^{\mathbf{I}} k_{\mathbf{I}}$ on a basis of Killing vectors $k_{\mathbf{I}}$ such that

$$[k_{\mathbf{I}}, k_{\mathbf{L}}] = f_{\mathbf{IL}}{}^{\mathbf{K}} k_{\mathbf{K}} \tag{3.7.36}$$

and we also choose a basis $\mathbf{i}_x$ $(x = 1, 2, 3)$ for $\mathbb{R}^3$ we get:

$$\mathscr{P}_{\vec{X}} = a^{\mathbf{I}} \mathscr{P}^x_{\mathbf{I}} \mathbf{i}_x \tag{3.7.37}$$

Furthermore we need a generalization of the equivariance defined by Eq. (3.7.10)

$$\vec{X} \circ \mathscr{P}_{\vec{Y}} = \mathscr{P}_{[\vec{X},\vec{Y}]} \tag{3.7.38}$$

In the HyperKähler case, the left-hand side of Eq. (3.7.38) is defined as the usual action of a vector field on a 0-form:

$$\vec{X} \circ \mathscr{P}_{\vec{Y}} = \mathbf{i}_{\vec{X}} \, d\mathscr{P}_{\vec{Y}} = X^u \frac{\partial}{\partial q^u} \mathscr{P}_{\vec{Y}} \tag{3.7.39}$$

The equivariance condition implies that we can introduce a triholomorphic Poisson bracket defined as follows:

$$\{\mathscr{P}_{\mathbf{I}}, \mathscr{P}_{\mathbf{J}}\}^x \equiv 2K^x(\mathbf{I}, \mathbf{J}) \tag{3.7.40}$$

leading to the triholomorphic Poissonian realization of the Lie algebra:

$$\{\mathscr{P}_{\mathbf{I}}, \mathscr{P}_{\mathbf{J}}\}^x = f^{\mathbf{K}}{}_{\mathbf{IJ}} \mathscr{P}^x_{\mathbf{K}} \tag{3.7.41}$$

which in components reads:

$$K^x_{uv} k^u_{\mathbf{I}} k^v_{\mathbf{J}} = \frac{1}{2} f^{\mathbf{K}}{}_{\mathbf{IJ}} \mathscr{P}^x_{\mathbf{K}} \tag{3.7.42}$$

In the quaternionic case, instead, the left-hand side of Eq. (3.7.38) is interpreted as follows:

$$\vec{X} \circ \mathscr{P}_{\vec{Y}} = \mathbf{i}_{\vec{X}} \nabla \mathscr{P}_{\vec{Y}} = X^u \nabla_u \mathscr{P}_{\vec{Y}} \tag{3.7.43}$$

where ∇ is the SU(2)-covariant differential. Correspondingly, the triholomorphic Poisson bracket is defined as follows:

$$\{\mathscr{P}_{\mathbf{I}}, \mathscr{P}_{\mathbf{J}}\}^x \equiv 2K^x(\mathbf{I}, \mathbf{J}) - \lambda \, \varepsilon^{xyz} \mathscr{P}^y_{\mathbf{I}} \mathscr{P}^z_{\mathbf{J}} \tag{3.7.44}$$

and leads to the Poissonian realization of the Lie algebra

$$\{\mathscr{P}_{\mathbf{I}}, \mathscr{P}_{\mathbf{J}}\}^x = f^{\mathbf{K}}{}_{\mathbf{IJ}} \mathscr{P}^x_{\mathbf{K}} \tag{3.7.45}$$

which in components reads:

$$K^x_{uv} k^u_{\mathbf{I}} k^v_{\mathbf{J}} - \frac{\lambda}{2} \varepsilon^{xyz} \mathscr{P}^y_{\mathbf{I}} \mathscr{P}^z_{\mathbf{J}} = \frac{1}{2} f^{\mathbf{K}}{}_{\mathbf{IJ}} \mathscr{P}^x_{\mathbf{K}} \tag{3.7.46}$$

Equation (3.7.46), which is the most convenient way of expressing equivariance in a coordinate basis was originally written in [4] and has played a fundamental role in the construction of supersymmetric actions for gauged $\mathcal{N} = 2$ supergravity both in $D = 4$ [4, 5] and in $D = 5$ [12].

3.8 Kähler Surfaces with One Continuous Isometry

As an illustration of the concepts introduced in the previous sections we consider here a class of very simple manifolds for which a lot of explicit calculations can be explicitly done, quite non trivial conceptual questions can be addressed and answered. These are 2-dimensional surfaces endowed with a one-dimensional continuous group of isometries $\mathcal{G}_{iso}$. As we advocate below the geometry of such manifolds is completely encoded in a single positive real function $V(\phi)$ of a single real coordinate ϕ. We name such a function the *potential*.[8] The main point is that any two-dimensional Euclidean manifold is actually complex and Kähler. This offers us the possibility of exemplifying all the structures we have discussed. We have to find the complex structure, the Kähler form and the Kähler potential. Furthermore since we have a Killing vector we can construct its moment map. Finally we can calculate the curvature. All these objects are functions of a single coordinate related with the initial potential $V(\phi)$ and its derivatives. Last but not least we have to decide the topological nature of the isometry group.

Within this class of manifolds we are able to construct several interesting examples that hopefully should clarify the non trivial aspects of the geometrical apparatus developed in previous sections. In particular, since we are dealing with 2-dimensional surfaces we can visualize them by means of their embedding in three-dimensional space.

With the above motivations let us consider Riemannian 2-dimensional manifolds Σ whose metric is the following one:

$$ds^2_\Sigma = p(U)\,dU^2 + q(U)\,dB^2 \tag{3.8.1}$$

$p(U), q(U)$ being two positive definite functions of their argument. The isometry group of the manifold Σ is generated by the Killing vector $\mathbf{k}_{[B]} = \partial_B$.

A fundamental geometrical question is whether $\mathbf{k}_{[B]}$ generates a *compact rotation symmetry*, or a *non compact symmetry either parabolic or hyperbolic*. We plan to discuss this issue in detail in the sequel.

Actually when $\Sigma = \Sigma_{max}$ is a constant curvature surface namely the coset manifold $\frac{\mathrm{SU}(1,1)}{\mathrm{U}(1)} \sim \frac{\mathrm{SL}(2,\mathbb{R})}{\mathrm{O}(2)}$, there is also a third possibility. In such a situation the Killing vector $\mathbf{k}_{[B]}$ can be the generator of a *dilatation*, namely it can correspond to

[8] This name is related with the use of this class of surfaces in supergravity inflationary models as described in [13–15], yet this is not relevant to us here. In this book our view point is just geometrical. Most of the material presented in this section was originally worked out in [13–15].

a non-compact but semi-simple element $\mathbf{d} = \begin{pmatrix} 1 & 0 \\ 0 & -1 \end{pmatrix}$ of the Lie algebra SL(2, $\mathbb{R}$) rather then to a nilpotent one $\mathbf{t} = \begin{pmatrix} 0 & 1 \\ 0 & 0 \end{pmatrix}$.

As all other two-dimensional surfaces, Σ has an underlying complex Kählerian structure that we can systematically uncover with the methods described in this chapter. The first step is to determine the complex structure with respect to which the metric (3.8.1) is hermitian. By definition an almost complex structure is a tensor $\mathfrak{J}_\alpha^\beta$ which squares to minus the identity:

$$\mathfrak{J}_\alpha^\beta \, \mathfrak{J}_\beta^\gamma = -\,\delta_\alpha^\gamma \tag{3.8.2}$$

The almost complex structure $\mathfrak{J}_\alpha^\beta$ becomes a true complex structure if its Nienhuis tensor vanishes:

$$N_{\alpha\beta}^\gamma \equiv \partial_{[\alpha} \, \mathfrak{J}_{\beta]}^\gamma - \mathfrak{J}_\alpha^\mu \, \mathfrak{J}_\beta^\nu \, \partial_{[\mu} \, \mathfrak{J}_{\nu]}^\gamma = 0 \tag{3.8.3}$$

Given a complex structure, a metric $g_{\alpha\beta}$ is hermitian with respect to it if the following identity is true:

$$g_{\alpha\beta} = \mathfrak{J}_\alpha^\gamma \, \mathfrak{J}_\beta^\delta \, g_{\gamma\delta} \tag{3.8.4}$$

Given the metric (3.8.1) there is a unique tensor $\mathfrak{J}_\alpha^\beta$, which simulatenously satisfies Eqs. (3.8.2), (3.8.3), (3.8.4) and it is the following:

$$\mathfrak{J} = \begin{pmatrix} 0 & \mathfrak{J}_B^U \\ \mathfrak{J}_U^B & 0 \end{pmatrix} = \begin{pmatrix} 0 & \sqrt{\frac{p(U)}{q(U)}} \\ -\sqrt{\frac{q(U)}{p(U)}} & 0 \end{pmatrix} \tag{3.8.5}$$

Next, according to theory, the Kähler 2-form is defined by:

$$\begin{aligned} \mathrm{K} = \mathrm{K}_{\alpha\beta} \, dx^\alpha \wedge dx^\beta &= g_{\alpha\gamma} \, \mathfrak{J}_\beta^\gamma \, dx^\alpha \wedge dx^\beta \\ &= -\sqrt{p(U)\,q(U)} \, dU \wedge dB \end{aligned} \tag{3.8.6}$$

and it is clearly closed. Hence the metric (3.8.1) is Kählerian and necessarily admits a representation in terms of a complex coordinate ζ and a Kähler potential $\mathscr{K}(\zeta\,,\,\overline{\zeta})$. In terms of the complex coordinate:

$$\zeta = \zeta(U, B) \tag{3.8.7}$$

the Kähler 2-form K in Eq. (3.8.6) should be rewritten as:

$$\mathrm{K} = \partial\,\overline{\partial}\,\mathscr{K} = \partial_\zeta \, \partial_{\overline{\zeta}} \, \mathscr{K} \, d\zeta \wedge d\overline{\zeta} \tag{3.8.8}$$

Next one aims at reproducing the Kählerian metric (3.8.1) in terms of a complex coordinate $\mathfrak{z} = \mathfrak{z}(U, B)$ and a Kähler potential $\mathscr{K}(\mathfrak{z}\,,\,\overline{\mathfrak{z}}) = \mathscr{K}^\star(\mathfrak{z}\,,\,\overline{\mathfrak{z}})$ such that:

$$\mathrm{K} = \mathrm{i}\,\partial\,\bar{\partial}\,\mathcal{K} = \mathrm{i}\partial_{\mathfrak{z}}\,\partial_{\bar{\mathfrak{z}}}\,\mathcal{K}\,d\mathfrak{z}\wedge d\bar{\mathfrak{z}} \;\; ; \;\; ds^2_{\Sigma} = \partial_{\mathfrak{z}}\,\partial_{\bar{\mathfrak{z}}}\,\mathcal{K}\,d\mathfrak{z}\otimes d\bar{\mathfrak{z}} \tag{3.8.9}$$

The complex coordinate $\mathfrak{z}$ is necessarily a solution of the complex structure equation:

$$\mathfrak{J}^{\beta}_{\alpha}\,\partial_\beta\,\mathfrak{z} = \mathrm{i}\partial_\alpha\,\mathfrak{z} \;\; \Rightarrow \;\; \sqrt{\frac{p(U)}{q(U)}}\,\partial_B\,\mathfrak{z}(U,B) = \mathrm{i}\,\partial_U\,\mathfrak{z}(U,B) \tag{3.8.10}$$

The general solution of such an equation is easily found. Define the linear combination[9]:

$$w \equiv \mathrm{i}\,C(U) - B \;\; ; \;\; C(U) = \int \sqrt{\frac{p(U)}{q(U)}}\,dU \tag{3.8.11}$$

and consider any holomorphic function $f(w)$. As one can immediately verify, the position $\mathfrak{z}(U,B) = f(w)$ solves Eq. (3.8.10). What is the appropriate choice of the holomorphic function $f(w)$? Locally (in an open neighborhood) this is an empty question, since the holomorphic function $f(w)$ simply corresponds to a change of coordinates and gives rise to the same Kähler metric in a different basis. Globally, however, there are significant restrictions that concern the range of the variables B and $C(U)$, namely the global topology of the manifold Σ. By definition B is the coordinate that, within Σ, parameterizes points along the $\mathcal{G}_\Sigma$-orbits, having denoted by $\mathcal{G}_\Sigma$ the isometry group. If $\mathcal{G}_\Sigma$ is compact, then B is a coordinate on the circle and it must be defined up to identifications $B \simeq B + 2\,n\,\pi$, where n is an integer. On the other hand if B is non compact its range extends on the full real line $\mathbb{R}$.

Furthermore, it is convenient to choose a canonical variable ϕ and codify the geometry of the surface in terms of a single positive potential function $V(\phi)$ rewriting it in the following way:

$$ds^2_g = d\phi^2 + \underbrace{\left(\frac{\mathrm{d}\sqrt{V(\phi)}}{\mathrm{d}\phi}\right)^2}_{f^2(\phi)} dB^2 \tag{3.8.12}$$

Hence we aim at a Kähler potential $\mathcal{K}(\mathfrak{z},\bar{\mathfrak{z}})$ that in terms of the variables $C(U)$ and B should actually depend only on C, being constant on the $\mathcal{G}$-orbits. Starting from the metric (3.8.1) we can always choose a canonical variable ϕ defined by the position:

$$\phi = \phi(U) = \int \sqrt{p(U)}\,dU \;\; ; \;\; d\phi = \sqrt{p(U)}\,dU \tag{3.8.13}$$

---

[9] As it follows from the present discussion the coordinate $C(U)$ has an intrinsic geometric characterization as that one which solves the differential equation of the complex structure. For the historical reasons explained in [13–15] we name C the Van Proeyen coordinate, abbreviated VP-coordinate.

and assuming that $\phi(U)$ can be inverted $U = U(\phi)$ we can rewrite (3.8.1) in the following canonical form:

$$ds^2_{can} = d\phi^2 + \left(\mathscr{P}'(\phi)\right)^2 dB^2 \quad ; \quad \mathscr{P}'(\phi) = \sqrt{q\left(U(\phi)\right)} \quad ; \quad \underbrace{\sqrt{p(U(\phi))}\,\frac{dU}{d\phi} = 1}_{\text{by construction}} \tag{3.8.14}$$

The reason to call the square root of $q\left(U(\phi)\right)$ with the name $\mathscr{P}'(\phi)$ is the interpretation of such a function as the derivative with respect to the canonical variable ϕ of the moment map of the Killing vector $\mathbf{k}_{[B]}$.

By using the canonical variable ϕ, the coordinate C defined in Eq. (3.8.11) becomes:

$$C(\phi) = C\left(U(\phi)\right) = \int \frac{d\phi}{\mathscr{P}'(\phi)} \tag{3.8.15}$$

and the metric $ds^2_\Sigma = ds^2_{can}$ of the Kähler surface Σ can be rewritten as:

$$ds^2_\Sigma = \frac{1}{2}\frac{\mathrm{d}^2 J}{\mathrm{d}C^2}\left(\mathrm{d}C^2 + \mathrm{d}B^2\right) \tag{3.8.16}$$

where the function $J(C)$ is defined as follows:

$$\mathscr{J}(\phi) \equiv 2\int \frac{\mathscr{P}(\phi)}{\mathscr{P}'(\phi)}\,d\phi \quad ; \quad J(C) \equiv \mathscr{J}\left(\phi(C)\right) \tag{3.8.17}$$

It appears from the above formula that the crucial step in working out the analytic form of the function $J(C)$ is the ability of inverting the relation between the coordinate C, defined by the integral (3.8.15), and the canonical one ϕ, a task which, in the general case, is quite hard in both directions. The indefinite integral (3.8.15) can be expressed in terms of special functions only in certain cases and even less frequently one has at his own disposal inverse functions. In any case the problem is reduced to quadratures and one can proceed further. Having already established in Eq. (3.8.11) the general solution of the complex structure equations, there are three possibilities that correspond, in the case of constant curvature manifolds Σ_{max}, to the three conjugacy classes of SL(2, $\mathbb{R}$) elements (elliptic, hyperbolic and parabolic). In the three cases $J(C)$ is identified with the Kähler potential $\mathscr{K}(\mathfrak{z}, \bar{\mathfrak{z}})$, but it remains to be decided whether the coordinate C is to be identified with the imaginary part of the complex coordinate, namely $C = \mathrm{Im}\,\mathfrak{z}$, with the logarithm of its modulus $C = \frac{1}{2}\log|\mathfrak{z}|^2$, or with a third combination of $\mathfrak{z}$ and $\bar{\mathfrak{z}}$, namely whether we choose the first the second or the third of the options listed below:

$$\mathfrak{z} = \left\{\begin{array}{lll} \zeta \equiv & \exp[-\mathrm{i}\,w] & = \underbrace{\exp[C(\phi)]}_{\rho(\phi)}\exp[\mathrm{i}B] \\ t \equiv & w & = \mathrm{i}\,C(\phi) - B \\ \hat{\zeta} \equiv & \mathrm{i}\tanh\left(-\frac{1}{2}w\right) & = \mathrm{i}\tanh\left(-\frac{1}{2}\left(\mathrm{i}\,C(\phi) - B\right)\right) \end{array}\right. \quad \Bigg| \quad C(\phi) \equiv \int \frac{1}{\mathscr{P}'(\phi)}\,d\phi \tag{3.8.18}$$

If we choose the first solution $\mathfrak{z} = \zeta$, that we name name of the *Disk-type*, we obtain that the basic isometry generated by the Killing vector $\mathbf{k}_{[B]}$ is a compact rotation symmetry. Choosing the second solution $\mathfrak{z} = t$, that we name of *Plane-type*, is appropriate instead to the case of a non compact shift symmetry. The third possibility mentioned above certainly occurs in the case of constant curvature surfaces Σ_{max} and leads to the interpretation of the B-shift as an SO(1, 1)-hyperbolic transformation.

In Sect. 3.8.5 we recall that the classification of a one dimensional isometry group as elliptic, parabolic or hyperbolic exists also for non maximally symmetric manifolds and it can be unambiguously formulated for *Hadamard manifolds* that are, by definition, simply connected, smooth Riemannian manifolds with a non positive definite curvature, *i.e.* $R(x) \leq 0,\ \forall x \in \Sigma$, having denoted by $R(x)$ the scalar curvature at the point x.

In the three cases mentioned in Eq. (3.8.18) the analytic form of the holomorphic Killing vector $\mathbf{k}_{[B]}$ is quite different:

$$\mathbf{k}_{[B]} = \begin{cases} \mathrm{i}\zeta\,\partial_\zeta & \equiv k^{\mathfrak{z}}\,\partial_{\mathfrak{z}} \Rightarrow k^{\mathfrak{z}} = \mathrm{i}\,\mathfrak{z} \quad ;\ \text{Disk-type, compact rotation} \\ \partial_t & \equiv k^{\mathfrak{z}}\,\partial_{\mathfrak{z}} \Rightarrow k^{\mathfrak{z}} = 1 \quad ;\ \text{Plane-type, non-compact shift} \\ \mathrm{i}\left(1+\hat{\zeta}^2\right)\partial_{\hat{\zeta}} & \equiv k^{\mathfrak{z}}\,\partial_{\mathfrak{z}} \Rightarrow k^{\mathfrak{z}} = \mathrm{i}\left(1 + \mathfrak{z}^2\right) \ ;\ \text{Disk-type, hyperbolic boost} \end{cases} \tag{3.8.19}$$

Choosing the complex structure amounts to the same as introducing one half of the missing information on the global structure of Σ, namely the range of the coordinate B. The other half is the range of the coordinate U or C.

Actually, by means of the constant curvature examples, a criterion able to discriminate the relevant topologies is encoded in the asymptotic behavior of the function $\partial_C^2 J(C)$ for large and small values of its argument, namely in the center of the bulk and on the boundary of the surface Σ. The main conclusions that we can reach by considering the case of constant curvature surfaces are those summarized below and are also encoded in Table 3.1:

(I) The global topology of the group $\mathcal{G}_\Sigma$ reflects into a different asymptotic behavior of the function $\partial_C^2 J(C)$ in the region that we can call the origin of the manifold. In the compact case the complex coordinate $\mathfrak{z}$ is charged with respect to U(1) and, for consistency, this symmetry should exist at all orders in an expansion of the line element ds_Σ^2 for small coordinates. Hence for $\mathfrak{z} \to 0$ the line element should approach the canonical one of a flat complex-manifold:

$$ds_\Sigma^2 \propto \mathrm{d}\mathfrak{z}\,\mathrm{d}\bar{\mathfrak{z}} \tag{3.8.20}$$

Assuming, as it is necessary for the U(1) interpretation of the B-shift symmetry, that $\mathfrak{z} = \zeta = \exp[\delta(C + \mathrm{i}\,B)]$, where δ is some real coefficient, Eq. (3.8.20) can be satisfied if and only if we have:

$$\lim_{C \to -\infty} \exp[-2\,\delta\,C]\,\partial_C^2 J(C) = \text{const.} \tag{3.8.21}$$

Table 3.1 Summary of the functions $V(\phi)$ defining the line element (3.8.12) which are obtained from constant curvature Kähler manifolds

Curv.	Isometry group	$V(\phi)$	$V(C)$	$V(\mathfrak{z})$	Comp. Struct.
$-\hat{\nu}^2$	U(1)	$\left(\cosh\left(\hat{\nu}\,\hat{\phi}\right) + \mu\right)^2$	$\left(\mu + \frac{2\,e^{4\,C\,\hat{\nu}^2}}{1-e^{4\,C\,\hat{\nu}^2}} + 1\right)^2$	$\frac{1}{\nu^4}\left(\frac{\mu+1-\mu\,\zeta\,\bar{\zeta}}{1-\zeta\,\bar{\zeta}}\right)^2$	$\zeta = e^{C - \mathrm{i}B}$
$-\hat{\nu}^2$	SO(1, 1)	$\left(\sinh\left(\hat{\nu}\,\hat{\phi}\right) + \mu\right)^2$	$\left(\mu + \tan\left(2\,C\,\hat{\nu}^2\right)\right)^2$	$\left(\frac{\bar{\zeta}(\zeta+\bar{\zeta})\zeta+\zeta+\bar{\zeta}+2\mu(\zeta\,\bar{\zeta}-1)}{4\zeta\,\bar{\zeta}-4}\right)^2$	$\zeta =$ $\mathrm{i}\tanh\left(\frac{1}{2}(B-\mathrm{i}C)\,\nu^2\right)$
$-\hat{\nu}^2$	parabolic	$\left(\exp\left(\hat{\nu}\,\hat{\phi}\right) + \mu\right)^2$	$\left(\mu + \frac{1}{2\,\hat{\nu}^2\,C}\right)^2$	$\left(\frac{1}{2}\,\mu + \frac{\mathrm{i}}{2\,\hat{\nu}^2}\,\left(t-\bar{t}\right)^{-1}\right)^2$	$t = \mathrm{i}\,C - B$
0	U(1)	$M^4\left[\left(\frac{\phi}{\phi_0}\right)^2 \pm 1\right]^2$	$M^4\left[\frac{e^{2a_2 C}}{\phi_0^2} \pm 1\right]^2$	$\frac{1}{4}\left(\mathfrak{z}\,\bar{\mathfrak{z}} - \frac{2\,a_0}{a_2}\right)^2$	$\mathfrak{z} = \exp\left[a_2\,(C - \mathrm{i}B)\right]$
0	parabolic	$(a_0 + a_1\,\phi)^2$	$(a_1\,C + \beta)^2$	$\frac{1}{2}\,(a_1 \mathrm{Im}\mathfrak{z} + \beta)^2$	$\mathfrak{z} = \mathrm{i}C - B$

or more precisely:

$$\begin{aligned} \partial_C^2 J(C) &\overset{C\to-\infty}{\approx} \text{const} \times \exp[2\,\delta\, C] + \text{subleading} \\ J(C) &\overset{C\to-\infty}{\approx} \text{const} \times \exp[2\,\delta\, C] + \text{subleading} \end{aligned} \tag{3.8.22}$$

The above stated is an intrinsic clue to establish the global topology of the Kähler surface Σ. In Sect. 3.8.5 we present some rigorous mathematical results that justify the above criterion to establish the compact nature of the gauged isometry. Indeed what, in heuristic jargon we call the origin of the manifold is, in rigorous mathematical language, the fixed point for all $\Gamma \in \mathscr{G}_\Sigma$, located in the interior of the manifold, whose existence is a necessary defining feature of an elliptic[10]isometry group $\mathscr{G}$.

(II) The above properties are general and apply to all surfaces of type (3.8.1)–(3.8.12). In the particular case of constant curvature Kähler surfaces there are five ways of writing the line-element (3.8.12), two associated with a flat Kähler manifold and three with the unique negative curvature two-dimensional symmetric space $\mathrm{SL}(2,\mathbb{R})/\mathrm{O}(2)$.

(III) Global topology amounts, at the end of the day, to giving the precise range of the coordinates C and B labeling the points of Σ. In the five constant curvature cases these ranges are as follows. In the elliptic and parabolic case C is in the range $[-\infty, 0]$, while it is in the range $[-\infty, +\infty]$ for the flat case and it is periodic in the hyperbolic case. The cooordinate B instead is periodic in the elliptic case, while it is unrestricted in the hyperbolic and parabolic cases. The manifold Σ in the flat case with B periodic is just a strip. It is instead the full plane in the flat parabolic case.

Our goal is to extend the above results to examples where the curvature of the Kähler surface Σ is not constant. In such examples we will verify the criterion that singles out the interpretation of the B-shift isometry as a parabolic shift-symmetry. In all such cases the range of the C coordinate is $[-\infty, 0]$[11] or $[-\infty, \infty]$. The limit $C \to 0$ always correspond to a boundary of the Kähler manifold Σ irrespectively whether the isometry group $\mathscr{G}_\Sigma$ is elliptic or parabolic. If the curvature is negative we always have:

$$\begin{aligned} \partial_C^2 J(C) &\overset{C\to 0}{\approx} \text{const} \times \frac{1}{C^2} + \text{subleading} \\ J(C) &\overset{C\to 0}{\approx} \text{const} \times \log[C] + \text{subleading} \end{aligned} \tag{3.8.23}$$

---

[10]Let us stress that this is true for Hadamard manifolds and possibly for CAT(k) manifolds, in any case for simple connected manifolds. In the presence of a non trivial fundamental group the presence of a fixed point is not necessary in order to establish the compact nature of the isometry group.

[11]Note that $[-\infty, 0]$ as range of the C-coordinate is conventional. Were it to be $[\infty, 0]$, we could just replace $C \to -C$ which is always possible since the Kähler metric is given by Eq. 3.8.16.

In case the curvature at $C = 0$ is zero, the gauge group is necessarily parabolic, since we cannot organize an exponential behavior of $J(C)$ for $C \to 0$. Such exponential behavior is instead requested by an elliptic isometry, so the only conclusion is that a limiting zero curvature at a boundary $C = 0$ can occur only in parabolic models and there we have:

$$\begin{aligned} \partial_C^2 J(C) &\overset{C \to 0}{\approx} \text{const} + \text{subleading} \\ J(C) &\overset{C \to 0}{\approx} \text{const} \times C^2 + \text{subleading} \end{aligned} \tag{3.8.24}$$

In the case of a parabolic structure of the isometry group $\mathscr{G}_\Sigma$, the locus $C = -\infty$ is always a boundary and not an interior fixed point which does not exist. Differently from Eq. (3.8.22) the asymptotic behavior of the metric and of the J-function is either:

$$\begin{aligned} \partial_C^2 J(C) &\overset{C \to -\infty}{\approx} \text{const} \times \frac{1}{C^2} + \text{subleading} \\ J(C) &\overset{C \to -\infty}{\approx} \frac{1}{R_\infty} \times \log[C] + \text{subleading} \end{aligned} \tag{3.8.25}$$

or

$$\begin{aligned} \partial_C^2 J(C) &\overset{C \to -\infty}{\approx} \text{const} + \text{subleading} \\ J(C) &\overset{C \to -\infty}{\approx} \text{const} \times C^2 + \text{subleading} \end{aligned} \tag{3.8.26}$$

The asymptotic behavior (3.8.25) obtains when the limit of the curvature for $C \to -\infty$ is $R_\infty < 0$. On the other hand, the exceptional asymptotic behavior (3.8.26) occurs when the limit of the curvature for $C \to -\infty$ is $R_\infty = 0$. As we did for the compact case, also for the parabolic case, in Sect. 3.8.5 we present rigorous mathematical arguments that sustain the heuristic criteria (3.8.25) and (3.8.26). Hence in the case where we deal with a parabolic isometry group, the Kähler potential has typically two logarithmic divergences one at $C = 0$, and one at $C = -\infty$, the two boundaries of the manifold Σ. One logarithm can be replaced by C^2 in case the limiting curvature on the corresponding boundary is zero. In other regions the behavior of J is different from logarithmic because of the non constant curvature.

Finally we can wonder what is the criterion to single out a hyperbolic characterization of the isometry group $\mathscr{G}_\Sigma$. A very simple answer arises from the example in the second line of Table 3.1. The hallmark of such isometries is a periodic coordinate C or anyhow a C that takes values in a finite range $[C_{min}, C_{max}]$. We will present an example of a non constant curvature Kähler surface with a hyperbolic isometry in Sect. 3.8.3.

There is still one subtle case of which we briefly discuss an example in Sect. 3.8.2. As we know there are two versions of flat manifolds, one where the selected isometry is a compact U(1) and one where it is a parabolic translation. In both cases the curvature is zero but in the former case the $J(C)$ function is:

$$J(C) \propto \exp[\delta\, C] \quad ; \quad \text{elliptic case} \tag{3.8.27}$$

while in the latter case we have

$$J(C) \propto C^2 \quad ; \quad \text{parabolic case} \tag{3.8.28}$$

Hence the following question arises. For Σ surfaces with a parabolic isometry group we foresaw the possibility, realized for instance in the example discussed in Sect. 3.8.4, that the limiting curvature might be zero on one of the boundaries so that the asymptotic behavior (3.8.25) is replaced by (3.8.26). In a similar way we might expect that there are elliptic models where the asymptotic behavior at $C \to \pm\infty$ is:

$$J(C) \stackrel{C\to\pm\infty}{\approx} \exp\left[\delta_\pm\, C\right] \tag{3.8.29}$$

one of the limits being interpreted as the symmetric fixed point in the interior of the manifold, the other being interpreted as the boundary on which the curvature should be zero. In Sect. 3.8.2 we will briefly sketch a model that realizes the above forseen situation. The corresponding manifold Σ has the topology of the disk. In the same section, as a counterexample, we consider a case where the same asymptotic (8.3.56) is realized in presence of an elliptic symmetry, yet $C \to -\infty$ no longer corresponds to an interior point, rather to a boundary. This is due to the non trivial homotopy group $\pi_1(\Sigma)$ of the surface which realizes such an asymptotic behavior. Being non-simply connected such Kähler surface is not a Hadamard manifold and presents new pathologies from the mathematical stand-point.

So let us turn to the analysis of the curvature.

3.8.1 *The Curvature and the Kähler Potential of the Surface Σ*

The curvature of a two-dimensional Kähler manifold with a one-dimensional isometry group can be written in two different ways in terms of the canonical coordinate ϕ or the coordinate C. In terms of the coordinate C we have the following formula:

$$\begin{aligned} R = R(C) &= -\tfrac{1}{2}\,\frac{J''''(C) - J'''(C)^2}{J''(C)^3} \\ &= -\tfrac{1}{2}\,\partial_C^2 \log\left[\partial_C^2 J(C)\right]\,\frac{1}{\partial_C^2 J(C)} \end{aligned} \tag{3.8.30}$$

which can be derived from the standard structural equations of the manifold [12]:

$$\begin{aligned}
0 &= \mathrm{d}E^1 + \omega \wedge E^2 \\
0 &= \mathrm{d}E^2 - \omega \wedge E^1 \\
\mathfrak{R} &\equiv \mathrm{d}\omega \equiv 2\,R\,E^1 \wedge E^2
\end{aligned} \tag{3.8.31}$$

by inserting into them the appropriate form of the zweibein:

$$E^1 = \sqrt{\frac{J''(C)}{2}}\,dC \quad ; \quad E^2 = \sqrt{\frac{J''(C)}{2}}\,dB \quad \Rightarrow \quad ds^2 = \tfrac{1}{2}\,J''(C)\left(dC^2 + dB^2\right) \tag{3.8.32}$$

Alternatively we can write the curvature in terms of the moment map $\mathscr{P}(\phi)$ or of the function $V(\phi) \propto \mathscr{P}^2(\phi)$ if we use the canonical coordinate ϕ and the corresponding appropriate zweibein:

$$E^1 = d\phi \quad ; \quad E^2 = \mathscr{P}'(\phi)\,dB \quad \Rightarrow \quad ds^2 = \left(d\phi^2 + \left(\mathscr{P}'(\phi)\right)^2 dB^2\right) \tag{3.8.33}$$

Upon insertion of Eq. (3.8.33) into (3.8.31) we get:

$$R(\phi) = -\tfrac{1}{2}\,\frac{\mathscr{P}'''(\phi)}{\mathscr{P}'(\phi)} = -\tfrac{1}{2}\left(\frac{V'''}{V'} - \tfrac{3}{2}\,\frac{V''}{V} - \tfrac{3}{4}\left(\frac{V'}{V}\right)^2\right) \tag{3.8.34}$$

The zero curvature and constant curvature cases can be easily analyzed. The general solution of the equation:

$$R(\phi) = -\tfrac{1}{2}\,\nu^2 \equiv -\,\hat{\nu}^2 \tag{3.8.35}$$

can be presented in terms of the moment map $\mathscr{P}(\phi)$ and of the canonical variable ϕ. We have:

$$\mathscr{P}(\phi) = a\,\exp(\nu\,\phi) + b\,\exp(-\,\nu\,\phi) + c \quad ; \quad a, b, c \in \mathbb{R} \tag{3.8.36}$$

In order to convert this solution in terms of the Jordan function $J(C)$ of the coordinate C, it is convenient to remark that, up to constant shift redefinitions and sign flips of the canonical variable $\phi \to \pm\phi + \kappa$, which leave the $d\phi^2$ part of the line-element invariant there are only three relevant cases:

(A) $a \neq 0$, $b \neq 0$ and $a/b > 0$. In this case, up to an overall constant, we can just set:

$$\mathscr{P}(\phi) = \cosh(\nu\,\phi) + \gamma \quad \Rightarrow \quad V(\phi) \propto (\cosh(\nu\,\phi) + \gamma)^2 \tag{3.8.37}$$

[12] The factor 2 introduced in this equation is chosen in order to have a normalization of what we name curvature that agrees with the normalization used in several papers of the physical literature.

(B) $a \neq 0,\ b \neq 0$ and $a/b < 0$. In this case we can just set:

$$\mathscr{P}(\phi) = \sinh(\nu\,\phi) + \gamma \quad \Rightarrow \quad V(\phi) \propto (\sinh(\nu\,\phi) + \gamma)^2 \tag{3.8.38}$$

(C) $a \neq 0,\ b = 0$. In this case we can just set:

$$\mathscr{P}(\phi) = \exp(\nu\,\phi) + \gamma \quad \Rightarrow \quad V(\phi) \propto (\exp(\nu\,\phi) + \gamma)^2 \tag{3.8.39}$$

Since our main goal is to understand the topology of the Kähler surface Σ and possibly to generalize the above three-fold classification of isometries to the non constant curvature case, it is very useful to recall how, in the above three cases, the corresponding (Euclidean) metric ds^2_ϕ is realized as the pull-back on the hyperboloid surface

$$X_1^2 + X_2^2 - X_3^2 = -1 \tag{3.8.40}$$

of the flat Lorentz metric in the three-dimensional Minkowski space of coordinates $\{X_1, X_2, X_3\}$. The manifold is always the same but the three different parameterizations single out different gaussian curves on the same surface. It is indeed an excellent exercise in differential geometry to see how the same space can be described in apparently very much different coordinate systems. Furthermore the gaussian curves being integral curves of different Killing vectors give visual appreciation of the different global character of elliptic, parabolic and hyperbolic isometries.

3.8.1.1 Embedding of Case (A)

Let us consider the case of the moment map of Eq. (3.8.37). The corresponding two-dimensional metric is:

$$ds^2_\phi = d\phi^2 + \sinh^2(\nu\,\phi)\,dB^2 \tag{3.8.41}$$

It is the pull-back of the (2, 1)-Lorentz metric onto the hyperboloid surface (3.8.40). Indeed setting:

$$\begin{aligned} X_1 &= \sinh(\nu\phi)\cos(B\nu) \\ X_2 &= \sinh(\nu\phi)\sin(B\nu) \\ X_3 &= \pm\cosh(\nu\phi) \end{aligned} \tag{3.8.42}$$

we obtain a parametric covering of the algebraic locus (3.8.40) and we can verify that:

$$\frac{1}{\nu^2}\left(dX_1^2 + dX_2^2 - dX_3^2\right) = d\phi^2 + \sinh^2(\nu\,\phi)\,dB^2 = ds^2_\phi \tag{3.8.43}$$

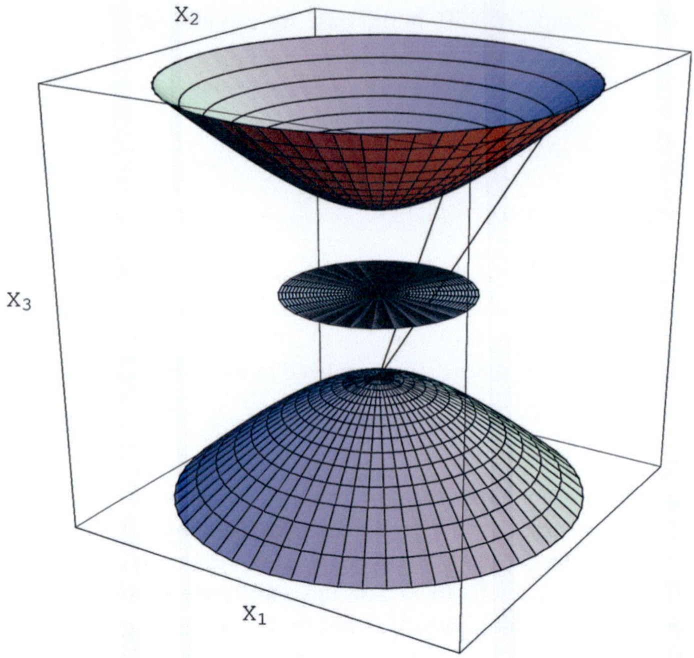

Fig. 3.1 In this figure we show the hyperboloid ruled by lines of constant ϕ that are circles and of constant B that are hyperbolae. In this figure we also show the stereographic projection of points of the hyperboloid onto points of the unit disk

A picture of the hyperboloid ruled by lines of constant ϕ and constant B according to the parametrization (3.8.42) is depicted in Fig. 3.1. In case of non constant curvature with a moment map which gives rise to a consistent U(1) interpretation of the isometry, the surface Σ is also a revolution surface but of a different curve than the hyperbola.

Setting:

$$f(\phi) = \mathscr{P}'(\phi) \tag{3.8.44}$$

we consider the parametric surface:

$$\begin{aligned} X_1 &= f(\phi) \cos B \\ X_2 &= f(\phi) \sin B \\ X_3 &= \pm g(\phi) \end{aligned} \tag{3.8.45}$$

where $g(\phi)$ is a function that satisfies the differential equation:

$$g'(\phi) = \sqrt{(f'(\phi))^2 - 1} \quad \Rightarrow \quad g(\phi) = \int \mathrm{d}\phi \sqrt{(f'(\phi))^2 - 1} \tag{3.8.46}$$

The pull back on the parametric surface (3.8.45) of the flat Minkowski metric:

$$ds_M^2 = dX_1^2 + dX_2^2 - dX_3^2 \tag{3.8.47}$$

reproduces the metric of the surface Σ under analysis:

$$ds_\Sigma^2 = \mathrm{d}\phi^2 + f^2(\phi)\,\mathrm{d}B^2 \tag{3.8.48}$$

Hence the revolution surface (3.8.45) is generically an explicit geometrical model of the Kähler manifolds Σ where the considered isometry is elliptic, namely a compact U(1). Note that the last integral in Eq. (3.8.46) can be performed and yields a real function only for those functions $f(\phi)$ that satisfy the condition $\left(f'(\phi)\right)^2 > 1$. Hence the condition:

$$\left(\mathscr{P}''(\phi)\right)^2 > 1 \tag{3.8.49}$$

is a necessary requirement for the U(1) interpretation of the gauged isometry which has to be true together with the asymptotic expansion criterion (3.8.22).

Applying to the present constant curvature case the general rule given in Eq. (3.8.15) that defines the coordinate C we get:

$$C(\phi) = \int \frac{d\phi}{\mathscr{P}'(\phi)} = \frac{\log\left(\tanh\left(\frac{\nu\phi}{2}\right)\right)}{\nu^2} \quad \Leftrightarrow \quad \phi = \frac{2\mathrm{Arctanh}\left(e^{C\nu^2}\right)}{\nu} \tag{3.8.50}$$

from which we deduce that the allowed range of the flat variable C, in which the canonical variable ϕ is real and goes from 0 to ∞, is the following one:

$$C \in [-\infty\,,\,0] \tag{3.8.51}$$

The Kähler potential function is easily calculated and we get:

$$J(C) = 2(\gamma+1)\,C - 2\,\frac{\log\left(1-e^{2C\nu^2}\right)}{\nu^2} + 2\,\frac{\log(2)}{\nu^2} \tag{3.8.52}$$

In this case the appropriate relation between ζ in the unit circle and the real variables C, B is the following:

$$\zeta = e^{\nu^2(\mathrm{i}B+C)} \tag{3.8.53}$$

3.8.1.2 Embedding of Case (B)

Consider the case of Eq. (3.8.38). The corresponding two-dimensional metric is:

$$ds_\phi^2 = \left(d\phi^2 + \cosh^2(\nu\,\phi)\,dB^2\right) \tag{3.8.54}$$

which can be shown to be another form of the pull-back of the Lorentz metric onto a hyperboloid surface. Indeed setting:

$$\begin{aligned} X_1 &= \cosh(\nu\phi)\sinh(B\nu) \\ X_2 &= \sinh(\nu\phi) \\ X_3 &= \pm\cosh(B\nu)\cosh(\nu\phi) \end{aligned} \tag{3.8.55}$$

we obtain a parametric covering of the algebraic locus (3.8.40) and we can verify that:

$$\frac{1}{\nu^2}\left(dX_1^2 + dX_2^2 - dX_3^2\right) = \left(d\phi^2 + \cosh^2(\nu\,\phi)\,dB^2\right) = ds_\phi^2 \tag{3.8.56}$$

A three-dimensional picture of the hyperboloid ruled by lines of constant ϕ and constant B is displayed in Fig. 3.2. For other surfaces Σ (if they exist and are regular) possessing a hyperbolic isometry we can realize their geometrical model considering the following parametric surface:

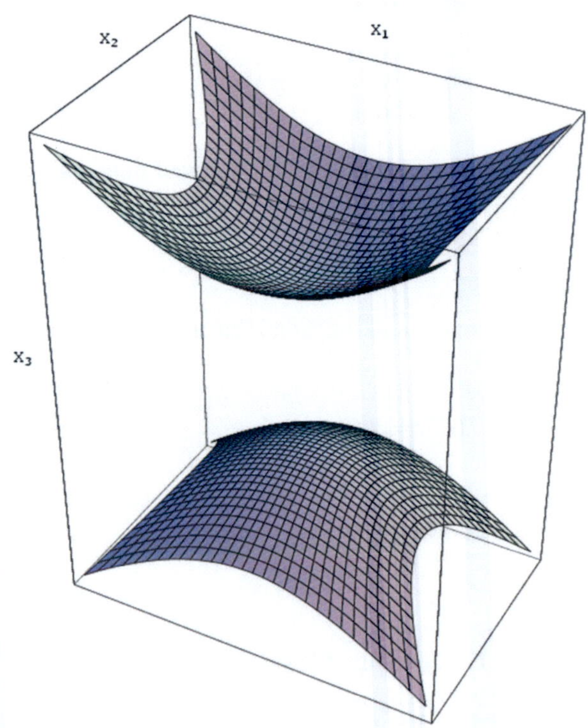

Fig. 3.2 The hyperboloid surface displayed in the parametrization (3.8.55). The lines drawn on the hyperboloid surface are those of constant B and constant ϕ respectively. Both of them are hyperbolae, in this case

$$
\begin{aligned}
X_1 &= f(\phi)\ \sinh B \\
X_2 &= g(\phi) \\
X_3 &= \pm\, f(\phi)\ \cosh B
\end{aligned}
\tag{3.8.57}
$$

where:

$$
f(\phi) \;=\; \mathscr{P}'(\phi) \tag{3.8.58}
$$

and where $g(\phi)$ is a function that satisfies the following differential equation:

$$
g'(\phi) \;=\; \sqrt{1\,+\,(f'(\phi))^2} \quad\Rightarrow\quad g(\phi) \;=\; \int \mathrm{d}\phi\,\sqrt{1\,+\,(f'(\phi))^2} \tag{3.8.59}
$$

Once again the pull-back of the flat Minkowski metric (3.8.47) on the parametric surface (3.8.57) reproduces the looked for metric of the Σ-surface:

$$
ds^2_\Sigma \;=\; \mathrm{d}\phi^2 \,+\, f^2(\phi)\,\mathrm{d}B^2 \tag{3.8.60}
$$

Which is the appropriate interpretation is dictated by the asymptotic behavior of the $J(C)$ function and of its second derivative, or alternatively by the equivalent mathematical criteria discussed in Sect. 3.8.5.

Applying to the present constant curvature case the general rule given in Eq. (3.8.15) that defines the coordinate C we get:

$$
C(\phi) \;=\; \int \frac{d\phi}{\mathscr{P}'(\phi)} \;=\; \frac{2\mathrm{Arctan}\left(\tanh\left(\frac{\nu\phi}{2}\right)\right)}{\nu^2} \quad\Leftrightarrow\quad \phi = \frac{2\mathrm{Arctanh}\left(\tan\left(\frac{C\nu^2}{2}\right)\right)}{\nu} \tag{3.8.61}
$$

from which we deduce that the allowed range of the flat variable C, in which the canonical variable ϕ is real and goes from $-\infty$ to ∞, is the following one:

$$
C \,\in\, \left[-\frac{\pi}{2\,\nu^2}\,,\,\frac{\pi}{2\,\nu^2}\right] \tag{3.8.62}
$$

The Kähler function $J(\phi)$ is easily calculated and we obtain:

$$
J(C) \;=\; 2\,\gamma\, C \;-\; \frac{2}{\nu^2}\,\log\left(\cos\left(C\nu^2\right)\right) \tag{3.8.63}
$$

In this case the appropriate relation between ζ in the unit circle and the real variables C, B is different, it is:

$$
\zeta \;=\; \mathrm{i}\tanh\left(\frac{1}{2}(B-\mathrm{i}\,C)\nu^2\right) \tag{3.8.64}
$$

3.8.1.3 Embedding of Case (C)

In the case the moment map is given by Eq. (3.8.39) the parameterization of the hyperboloid is the following one:

$$\begin{aligned} X_1 &= \frac{1}{2}\left(-e^{\nu\phi}B^2 + e^{\nu\phi} - \frac{e^{-\nu\phi}}{\nu^2}\right)\nu \\ X_2 &= Be^{\nu\phi}\nu \\ X_3 &= \frac{1}{2}\left(e^{\nu\phi}B^2 + e^{\nu\phi} + \frac{e^{-\nu\phi}}{\nu^2}\right)\nu \end{aligned} \tag{3.8.65}$$

Indeed upon insertion of Eq. (3.8.65) into (3.8.40) we see that for all values of B and ϕ the constraint defining the algebraic locus is satisfied. At the same time by means of an immediate calculation one finds:

$$\frac{1}{\nu^2}\left(dX_1^2 + dX_2^2 - dX_3^2\right) = d\phi^2 + e^{2\nu\phi}\,dB^2 = ds_\phi^2 \tag{3.8.66}$$

so that the considered metric is the pull-back of the three-dimensional Lorentz metric on the surface Σ parameterized as in Eq. (3.8.65). The integration of Eq. (3.8.15) is immediate and the coordinate $C(\phi)$ takes the following very simple invertible form:

$$C(\phi) = -\frac{e^{-\nu\phi}}{\nu^2} \Leftrightarrow \phi(C) = -\frac{\log\left(-C\nu^2\right)}{\nu} \tag{3.8.67}$$

The range of definition of C is:

$$C \in [-\infty, 0] \tag{3.8.68}$$

A three-dimensional picture of the hyperboloid ruled by lines of constant ϕ and constant B, according to Eq. (3.8.65) is displayed in Fig. 3.3.

The integration of Eq. (3.8.17) for the Kähler potential is equally immediate and using the inverse function $\phi(C)$ we obtain:

$$J(C) = 2\gamma\, C - \frac{2}{\nu^2}\log(-C) + \text{const} \tag{3.8.69}$$

From the form of Eq. (3.8.69) we conclude that in this case the appropriate solution of the complex structure equation is:

$$\mathfrak{z} = t = -\mathrm{i}C + B \tag{3.8.70}$$

so that the Kähler metric becomes proportional to the Poincaré metric in the upper complex plane (note that C is negative definite for the whole range of the canonical variable ϕ):

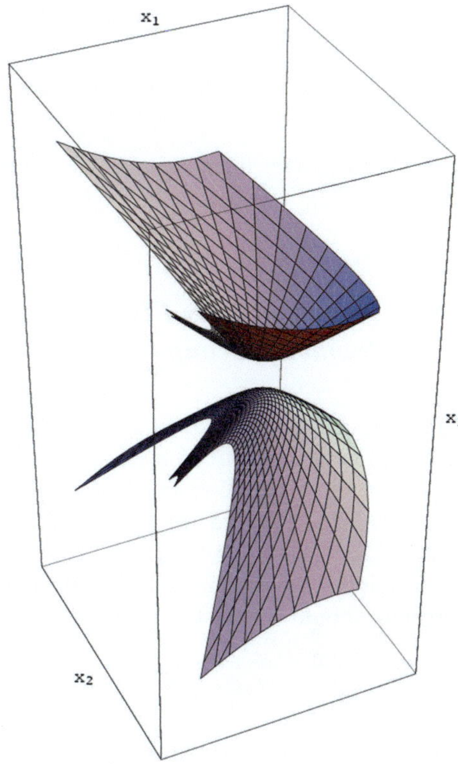

Fig. 3.3 The hyperboloid surface displayed in the parametrization (3.8.65). The lines drawn on the hyperboloid surface are those of constant B and constant ϕ respectively. The constant ϕ curves are parabolae and they are the orbits of the translation group

$$ds^2 = \tfrac{1}{2}\frac{\mathrm{d}^2 J}{\mathrm{d}C^2}\left(dC^2 + dB^2\right) = \frac{1}{4\,\nu^2}\,\frac{d\bar{t}\,dt}{(\mathrm{Im}t)^2} \tag{3.8.71}$$

As a consequence of Eq. (3.8.70), we see that the B-translation happens to be, in this case, a non-compact parabolic symmetry.

More generally for any surface Σ where the isometry of the metric:

$$\mathrm{d}s^2_\Sigma = \mathrm{d}\phi^2 + f^2(\phi)\,\mathrm{d}B^2 \tag{3.8.72}$$

is interpreted as a parabolic shift-symmetry we can construct a geometric model of Σ in three-dimensional Minkowski space by considering the following parametric surface:

$$\begin{aligned}
X_1 &= \frac{1}{2}\left(-f(\phi)B^2 + f(\phi) + g(\phi)\right)\\
X_2 &= Bf(\phi)\\
X_3 &= \frac{1}{2}\left(f(\phi)B^2 + f(\phi) - g(\phi)\right)
\end{aligned} \tag{3.8.73}$$

where $g(\phi)$ is a function that satisfies the differential equation:

$$f'(\phi)\, g'(\phi) = 1 \quad \Rightarrow \quad g(\phi) = \int \frac{1}{f'(\phi)}\, d\phi \tag{3.8.74}$$

The pull-back of the flat metric (3.8.47) onto the surface (3.8.73) is indeed the desired metric (3.8.72).

3.8.2 Asymptotically Flat Kähler Surfaces with an Elliptic Isometry Group

As announced above in this section we consider the problem of constructing a Kähler surface Σ with an elliptic isometry whose limiting curvature at the boundary vanishes $R_{\pm\infty} = 0$. In this case we can predict the asymptotic behavior of the function $J(C)$ for $C \to \pm\infty$. Indeed we know that for flat Kähler manifolds with an elliptic isometry, we have $J(C) \propto \exp[\delta\, C]$ for some value of $\delta \in \mathbb{R}$. Hence we expect that the function $J(C)$ for surfaces Σ with an elliptic isometry and a vanishing limiting curvature should behave has follows:

$$J(C) \overset{C\to\pm\infty}{\approx} \exp\left[\,\delta_{\pm}\, C\right] + \text{subleading terms} \tag{3.8.75}$$

There is however a fundamental subtlety that has to be immediately emphasized. If the topology of the surface Σ is the disk topology and Σ is simply connected $\pi_1(\Sigma) = 1$, then one of the two limits $C \to \infty$ has to be interpreted as the interior fixed point, required by Gromov criteria, for elliptic isometries in Hadamard manifolds (and possibly in CAT(k) manifolds). The other limit corresponds to the unique boundary of disk topology. On the other hand if $\pi_1(\Sigma) = \mathbb{Z}$ and the Kähler surface has the corona topology then there are two boundaries and the limiting curvature can be zero on both boundaries. We will illustrate this with two examples, respectively corresponding to the latter and to the former case.

3.8.2.1 The Catenoid Case with $\pi_1(\Sigma) = \mathbb{Z}$

We begin by considering explicit functions $J(C)$ that have the required asymptotic behavior and we try to work our way backward towards the canonical coordinate ϕ and the moment map $\mathscr{P}(\phi)$. In particular we want to make sure that the considered function $J(C)$ does indeed correspond to a compact isometry. This will certainly be the case if the corresponding metric is the pull-back of the flat three-dimensional Euclidean metric on a smooth surface of revolution.

To carry out such a program we consider the following one-parameter family of $J(C)$ functions:

$$J_{[\mu]}(C) = \frac{1}{8}\left(\mu\, C^2 + \cosh[2\, C]\right) \tag{3.8.76}$$

which fulfills condition (3.8.75), by construction. Many other examples can be obviously put forward, but this rather simple one is sufficient to single out the main subtlety that makes many asymptotically flat elliptic models pathological from the point of view of Gromov et al. classification of isometries. Using Eqs. (3.8.16) and (3.8.30) we write the metric and the curvature following from the $J(C)$ function of Eq. (3.8.76), obtaining

$$ds^2_{\Sigma} = \frac{1}{16}\,(2\mu + 4\cosh[2C])\,\left(dC^2 + dB^2\right) \tag{3.8.77}$$

$$R(C) = -\frac{4\mu\cosh(C) + 1}{(4\,\mu + \cosh[C])^3} \tag{3.8.78}$$

From these formulae we draw an important conclusion. In order for Σ to be a smooth manifold the curvature should not develop a pole neither in the interior nor on the boundary. This means that $4\mu + \cosh[C] > 0$ in the whole range of C. This is guaranteed if and only if $\mu > -\frac{1}{4}$. On the other hand, according to our previous discussions, in the case of an elliptic isometry, there should be, for a finite value of C, a zero of the metric coefficient. Such a zero is the fixed point that characterizes elliptic isometries of Hadamard manifolds. Looking at Eq. (3.8.77) we see that such a zero exists, if and only if $\mu < -\frac{1}{2}$. It follows that, at least in this family of models, there are no smooth manifolds that are asymptotically flat in the elliptic sense and fulfill the physical condition for U(1)-symmetry which corresponds to the Gromov et al. identification of elliptic isometries of Hadamard manifolds. At first sight one should draw the conclusion that, in the case of the $J(C)$ functions of Eq. (3.8.76), the isometry is not elliptic. Yet this is somehow strange, since at the boundary, where the curvature goes to zero, the form of $J(C)$ is precisely that which corresponds to elliptic isometries. Furthermore we will shortly show that for every value of μ the metric in Eq. (3.8.77) is just the metric of a smooth revolution surface. Actually for $\mu = 2$ such a revolution surface is the well-known **catenoid**, constructed by Bernoulli in 1744 as the first example of a minimal surface. Hence we arrive at a puzzle with Gromov et al. criteria, whose only resolution can be that the manifolds associated with the $J(C)$ functions of Eq. (3.8.76) are not Hadamard manifolds. From Eq. (3.8.78) we see that, provided $\mu > -\frac{1}{4}$, the curvature is negative definite and attains its maximal value $R = 0$ only on the boundary. Hence in relation with the curvature there is no violation of the properties defining a Hadamard manifold. The violation must be in another item of the definition. Considering the Definition 3.8.1 of Hadamard manifolds provided in Sect. 3.8.5 we realize that the only way out from the puzzle is that the surfaces corresponding to the $J(C)$ functions of Eq. (3.8.76) have to be **non simply connected**. That this is the case becomes visually obvious when we consider the plot of the surface in three-dimensional space-time (see Fig. 3.4), yet it is quite clear also analytically. For constant C the orbits of the isometry group spanned by $B \in [0, 2\,\pi]$ are circles of radius:

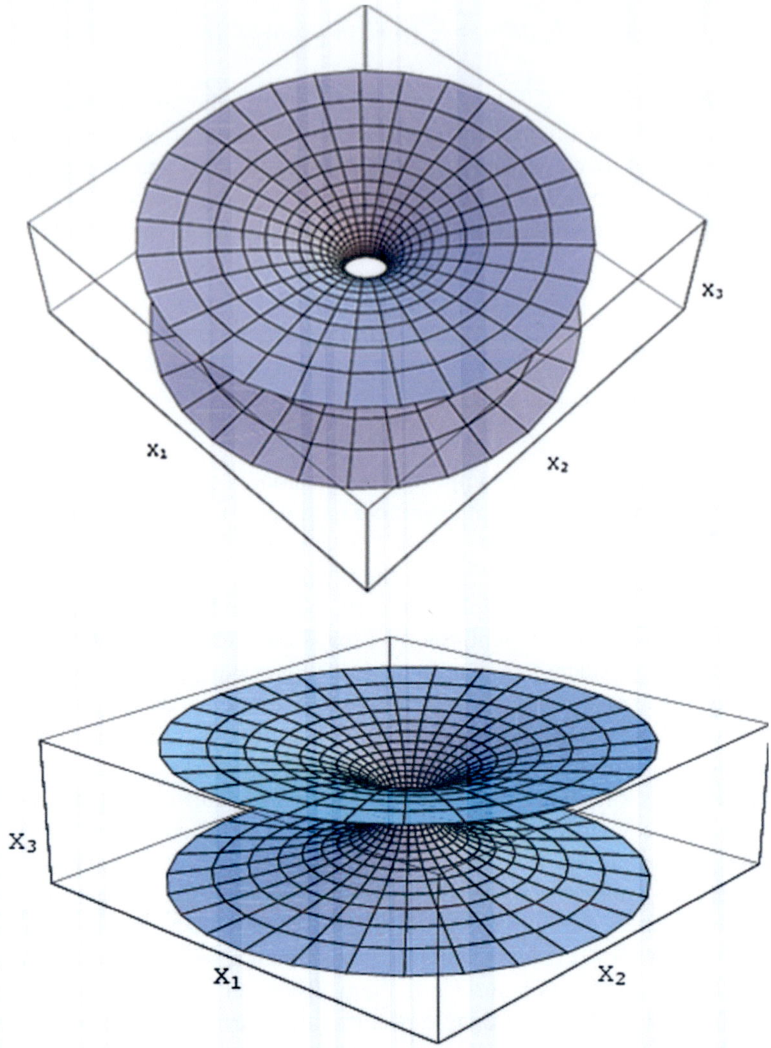

Fig. 3.4 In this picture we present two views of the **catenoid**, the revolution surface corresponding to $J_{[2]}(C) = \frac{1}{8}\left(2\,C^2 + \cosh[2\,C]\right)$. For large positive or negative values of C one is either in the superior or in inferior plane which is clearly flat with zero curvature. The center of the picture correspond instead to $C \to 0$ and is a sort of strongly negatively curved wormhole that connects the two asymptotic planes. Non simple connectedness is visually spotted. The circles on the surface winding around the throat cannot be contracted to zero and their homotopy class forms the non trivial element of the first homotopy group $\pi_1(\Sigma) = \mathbb{Z}$

$$r(C) = \frac{1}{4}\sqrt{2\mu + 4\cosh[2C]} \tag{3.8.79}$$

The fact that this radius has a minimum different from zero

$$r_{min} = \frac{1}{4}\sqrt{2\mu + 4} > 0 \tag{3.8.80}$$

is what spoils simple connectedness and prevents the existence of a fixed point for U(1). In this way the puzzle is resolved mathematically.

Having anticipated this conceptual discussion of their meaning let us work out the details of the models encoded in Eq. (3.8.76). Comparing Eqs. (3.8.16) and (3.8.14) we derive the relation between the canonical coordinate ϕ and C:

$$\phi = \sqrt{2}\int \sqrt{J''_{[\mu]}(C)}\, dC = \Phi_{[\mu]}(C) \equiv -\frac{1}{2}\,\mathrm{i}\,\sqrt{\mu+2}\, E\left(\mathrm{i}\, C \,\middle|\, \frac{4}{\mu+2}\right) \tag{3.8.81}$$

where $E\,(x\,|m)$ denotes the elliptic integral of its arguments. In the case $\mu = 2$ which turns out to be that of the **catenoid**, the function $\Phi_{[\mu]}(C)$ simplifies and it can be easily inverted in terms of elementary functions

$$\Phi_{[2]}(C) = \sinh(C) \Rightarrow \quad C(\phi) = \mathrm{ArcSinh}(C) \tag{3.8.82}$$

Substituting into the metric (3.8.77) one finds:

$$\mu = 2 \quad : \quad ds^2_\Sigma = \frac{\cosh^2(C)}{2}\left(dC^2 + dB^2\right) = \tfrac{1}{2}\left[d\phi^2 + \left(\phi^2 + 1\right) dB^2\right] \tag{3.8.83}$$

This implies that the derivative of the moment map is $\mathscr{P}'(\phi) = \sqrt{\phi^2 + 1}$ so that the moment map and the scalar potential are the following ones:

$$\begin{aligned} \mu = 2 : \mathscr{P}(\phi) &= \frac{1}{2}\left(\sqrt{\phi^2+1}\,\phi + \mathrm{ArcSinh}[\phi\,]\right) \quad \Rightarrow \\ V(\phi) &\propto \left(\sqrt{\phi^2+1}\,\phi + \mathrm{ArcSinh}[\phi\,]\right)^2 \end{aligned} \tag{3.8.84}$$

The metric (3.8.83) can be easily recognized to be the pull-back of the flat three-dimensional Euclidean metric:

$$ds^2_{\mathbb{E}^3} = dX_1^2 + dX_2^2 + dX_3^2 \tag{3.8.85}$$

on the following parametric surface:

$$
\begin{aligned}
X_1 &= \frac{\cos(B)\cosh(C)}{\sqrt{2}} \\
X_2 &= \frac{\cosh(C)\sin(B)}{\sqrt{2}} \\
X_3 &= \frac{C}{\sqrt{2}}
\end{aligned}
\tag{3.8.86}
$$

which is the classical catenoid. For other values of μ a similar parametric surface of revolution can be written in terms of appropriate functions of C. As we have already anticipated, although the catenoid is a rotation surface and its isometry is elliptic, its metric does not satisfy Gromov et al. criterion that requires the existence of a symmetric point. The reason for this pathology is the non trivial fundamental group $\pi_1(\Sigma)$.

Finally let us appreciate the nature of the same problem from the point of view of complex coordinates. If we introduce the complex coordinate:

$$
\zeta = \exp[C - \mathrm{i}\,B] \quad ; \quad \overline{\zeta} = \exp[C + \mathrm{i}\,B] \tag{3.8.87}
$$

and we insert it into the expression of (3.8.76) of the $J(C)$ function we easily obtain the Kähler potential:

$$
\mathcal{K}(\zeta,\overline{\zeta}) = 2\,J(C) = \frac{1}{16}\mu\log^2(\zeta\,\overline{\zeta}) + \frac{\zeta\,\overline{\zeta}}{8} + \frac{1}{8\,\zeta\,\overline{\zeta}} \tag{3.8.88}
$$

from which we obtain the metric:

$$
ds^2_{\Sigma} = \frac{d\zeta\,d\overline{\zeta}\,\left(\zeta\overline{\zeta}\,(\mu+\zeta\overline{\zeta})+1\right)}{8\,\left(\zeta\overline{\zeta}\right)^2} \overset{\mu\to 2}{\Longrightarrow} \frac{d\zeta\,d\overline{\zeta}\,\left(\zeta\overline{\zeta}+1\right)^2}{8\,\left(\zeta\overline{\zeta}\right)^2} \tag{3.8.89}
$$

Examining Eq. (3.8.89) we see that the metric diverges at the symmetry restoration point $\zeta = 0$ which now is the boundary of the manifold rather than its interior.

3.8.2.2 An Asymptotically Flat Kähler Surface with an Elliptic Isometry and $\pi_1(\Sigma) = 1$

Let us consider the following moment map written in terms of the canonical variable ϕ:

$$
\mathcal{P}(\phi) = \phi^2 - \frac{1}{2}\mathrm{ArcTan}\left(\phi^2\right) \tag{3.8.90}
$$

Using the standard formulae (3.8.15) for the calculation of the coordinate C we obtain:

$$C(\phi) = \log\left(\frac{\phi}{\sqrt[8]{2\phi^4+1}}\right) \quad \Leftrightarrow \quad \phi = \begin{cases} \pm\sqrt[4]{\sqrt{e^{8C}+e^{16C}}+e^{8C}} \\ \pm i\sqrt[4]{\sqrt{e^{8C}+e^{16C}}+e^{8C}} \\ \pm\sqrt[4]{\sqrt{e^{8C}-e^{16C}}+e^{8C}} \\ \pm i\sqrt[4]{\sqrt{e^{8C}-e^{16C}}+e^{8C}} \end{cases} \tag{3.8.91}$$

The eighth-root implies the existence of eight branches of the inverse function, that have to considered carefully. Indeed we can accept only those branches where ϕ turns out to be everywhere real. Six branches have to be rejected because of that reason and the only acceptable ones are the first two which are equivalent under the always possible sign revers of ϕ. In conclusion we have:

$$\phi = \sqrt[4]{\sqrt{e^{8C}+e^{16C}}+e^{8C}} \tag{3.8.92}$$

Using this branch the infinite interval $[-\infty\,,\,\infty]$ of the variable C is mapped into the semi-infinite interval $[0\,,\,\infty]$ of the variable ϕ. Indeed we have $C(0) = -\infty$, $C(\infty) = \infty$. In the canonical coordinate the form of the metric is:

$$ds^2_\Sigma = d\phi^2 + f^2(\phi)\,dB^2 \quad ; \quad f^2(\phi) = \left(\frac{\phi^5}{\phi^4+1}+\phi\right)^2 \tag{3.8.93}$$

and using Eq. (3.8.92) we can easily convert it to the C variable:

$$\begin{aligned} ds^2_\Sigma &= \tfrac{1}{2}\frac{d^2J}{dC^2}\left(dC^2+dB^2\right) \\ &= \frac{\sqrt{\sqrt{e^{8C}+e^{16C}}+e^{8C}}\left(2\sqrt{e^{8C}+e^{16C}}+2e^{8C}+1\right)^2}{\left(\sqrt{e^{8C}+e^{16C}}+e^{8C}+1\right)^2}\left(dC^2+dB^2\right) \end{aligned} \tag{3.8.94}$$

For $C \to -\infty$ the behavior of the metric coefficient is:

$$\tfrac{1}{2}\frac{d^2J}{dC^2} \overset{C\to-\infty}{\approx} e^{2C} + \frac{5e^{6C}}{2} + \mathcal{O}\left(e^{10C}\right) \;\Rightarrow\; J(C) \overset{C\to-\infty}{\approx} \tfrac{1}{2}e^{2C} \tag{3.8.95}$$

while for $C \to \infty$ it is the following:

$$\tfrac{1}{2}\frac{d^2J}{dC^2} \overset{C\to\infty}{\approx} 4\sqrt{2}e^{4C} - \frac{3e^{-4C}}{\sqrt{2}} + \mathcal{O}\left(e^{-12C}\right) \;\Rightarrow\; J(C) \overset{C\to\infty}{\approx} \tfrac{1}{2}\tfrac{1}{\sqrt{2}}e^{4C} \tag{3.8.96}$$

From previous considerations we see that $C \to -\infty$ corresponds to $\phi = 0$ and hence to the fixed point in the interior of the manifold, so that the exponential behavior of $J(C)$ is the expected one for an elliptic isometry. At the same time the exponential

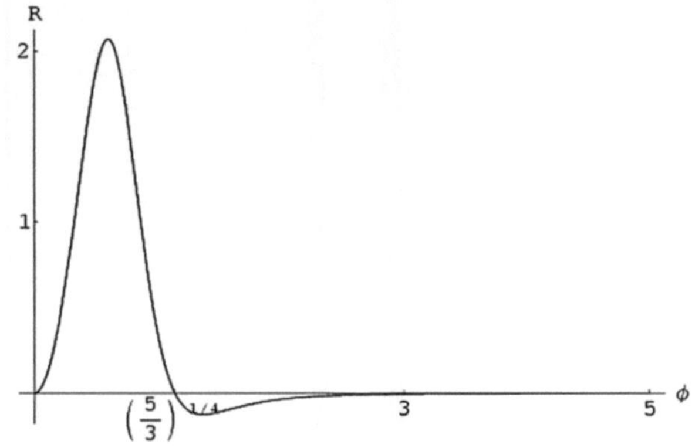

Fig. 3.5 In this picture we present the plot of the curvature for the elliptic model of Eq. (3.8.90). It is limited from above and has three zeros, one at the interior fixed point $\phi = 0$, a second one at $\phi = \left(\frac{5}{3}\right)^{1/4}$ and one on the boundary at $\phi = \infty$

behavior on the unique boundary implies that the limiting curvature on the boundary should be zero. Indeed from the standard formula (3.8.34) for the curvature we obtain:

$$R(\phi) = -\frac{2\phi^2\left(3\phi^4 - 5\right)}{\left(\phi^4+1\right)^2\left(2\phi^4+1\right)} \quad ; \quad R(0) = 0 \quad ; \quad R(\infty) = 0 \tag{3.8.97}$$

whose plot is displayed in Fig. 3.5. The vanishing of the limiting curvature is visually evident. Finally let us make sure that the isometry of this model is indeed elliptic. This we verify by showing that the metric (3.8.93) can be retrieved as the pull-back of the flat Lorentz metric in Minkowsian three-dimensional space (3.8.47) on the parametric revolution surface (3.8.45) defined by:

$$f(\phi) = \frac{\phi^5}{\phi^4+1} + \phi \quad ; \quad g(\phi) \equiv \int_0^\phi \sqrt{\frac{\sigma^4\left(\sigma^4+5\right)\left(3\sigma^8+9\sigma^4+2\right)}{\left(\sigma^4+1\right)^4}}\, d\sigma \tag{3.8.98}$$

Two views of this surface are presented in Fig. 3.6. It is evident from the picture that this surface is simply connected and that there is in the interior of the manifold a fixed point. It is given by $X_1 = X_2 = X_3 = 0$ which lies on the surface and where the radius of the U(1) orbit shrinks to zero.

3.8.3 An Example of a Non Maximally Symmetric Kähler Surface with an Isometry Group of the Hyperbolic Type

In order to exhibit an example of a surface with non constant curvature that has a hyperbolic isometry we consider the following moment map and potential:

$$V(\phi) = [\mathscr{P}(\phi)]^2 \quad ; \quad \mathscr{P}(\phi) = \phi + \sinh(\phi) \tag{3.8.99}$$

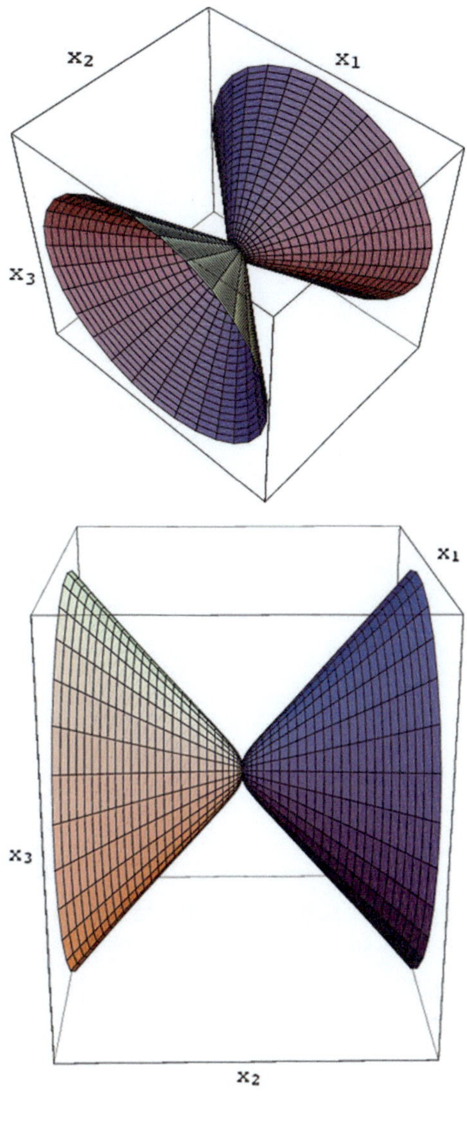

Fig. 3.6 In this picture we present two views of the revolution surface Σ associated with the elliptic model of Eq. (3.8.90). It is clearly regular and smooth everywhere

which yields:

$$\mathscr{P}'(\phi) = 1 + \cosh(\phi) \quad ; \quad ds^2_\Sigma = d\phi^2 + (1 + \cosh(\phi))^2 \, dB^2 \qquad (3.8.100)$$

According to the mathematical classification discussed in Sect. 3.8.5 the metric (3.8.100) has a hyperbolic type of isometry due to the two fixed points on the boundary of the manifold corresponding to the two singularities $\phi = \pm\infty$. The curvature of this manifold is finite but not constant. Indeed, applying Eq. (3.8.34) we obtain:

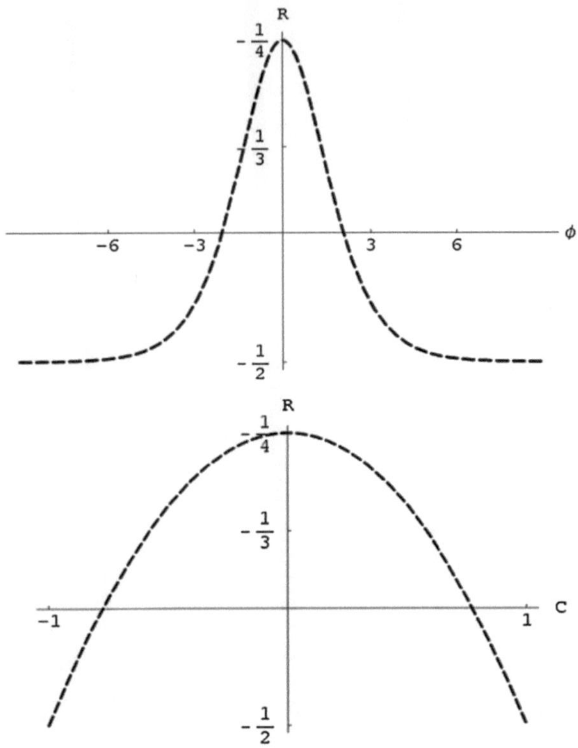

Fig. 3.7 In this figure we present the plot of the curvature of the surface Σ defined by Eq. (3.8.100) that has a hyperbolic isometry. The first picture displays the dependence of the curvature on the canonical coordinate ϕ, while the second picture displays its dependence on the coordinate C

$$R(\phi) = -\frac{\cosh(\phi)}{2(\cosh(\phi)+1)} \tag{3.8.101}$$

whose plot is presented in Fig. 3.7. In this case it is very simple to integrate the complex structure equation which defines the C-coordinate. We obtain:

$$C(\phi) = \tanh\left(\frac{\phi}{2}\right) \quad ; \quad \phi = 2\,\mathrm{ArcTanh}(C) \tag{3.8.102}$$

and we observe that in line with our general criteria for hyperbolic symmetry, the range of the C-coordinate is in this case finite:

$$C \in [-1\,,\,1] \tag{3.8.103}$$

From the integration of Eq. (3.8.17) that defines the J-function and the Kähler potential we obtain:

$$J(\phi) = 2\phi\tanh\left(\frac{\phi}{2}\right) = J(C) = 4C\,\mathrm{ArcTanh}(C) \tag{3.8.104}$$

Calculating the metric coefficient from (3.8.104) we get:

$$\frac{1}{2}\frac{d^2J}{dC^2} = \frac{4}{\left(C^2-1\right)^2} \quad ; \quad ds^2 = \frac{4}{\left(C^2-1\right)^2}\left(dC^2+dB^2\right) \tag{3.8.105}$$

displaying a polar singularity at both extrema of the C-range, namely at $C = \pm 1$.

In order to present a geometrical model of this Kähler manifold, we resort to the hyperbolic parametric surface encoded in formulae (3.8.57) and we calculate the relevant functions $f(\phi)$ and $g(\phi)$. In this case it is more convenient to express them in terms of the finite range coordinate C. We have:

$$f(\phi) = \cosh(\phi) + 1 = \frac{2}{1 - C^2} \tag{3.8.106}$$

and inserting the result into Eq. (3.8.59) we get:

$$g(C) = \frac{1}{8}\left(\frac{2C\left(C^2-3\right)}{\left(C^2-1\right)^2} + \log(C-1) - \log(C+1)\right) \tag{3.8.107}$$

The plots of these functions is presented in Fig. 3.8. In Fig. 3.9 we display the three dimensional shape of the parametric surface Σ realizing the desired Kähler manifold.

3.8.4 A Non Maximally Symmetric Kähler Manifold with Parabolic Isometry and Zero Curvature at One Boundary

As a final example we consider a parabolic model where the curvature at one of the two boundaries goes to zero so that the asymptotic behavior of the $J(C)$-function on that boundary becomes exceptional.

Let the moment map be the following one:

$$\mathscr{P}(\phi) = \exp[\nu\,\phi] + \mu\,\phi \tag{3.8.108}$$

The corresponding $f(\phi)$-function is:

$$f(\phi) = \mathscr{P}'(\phi) = \nu\,\exp[\nu\,\phi] + \mu \tag{3.8.109}$$

which has no zeros for finite ϕ if μ and ν have the same sign. If the two parameters have opposite signs there is such a zero and this creates a fixed point of the isometry $B \rightarrow B + c$ at finite ϕ which implies that the isometry is elliptic. Yet in case of opposite signs the curvature has a singularity so that any smooth Kähler manifold with a moment map of type (3.8.108) has a parabolic isometry group. Indeed using

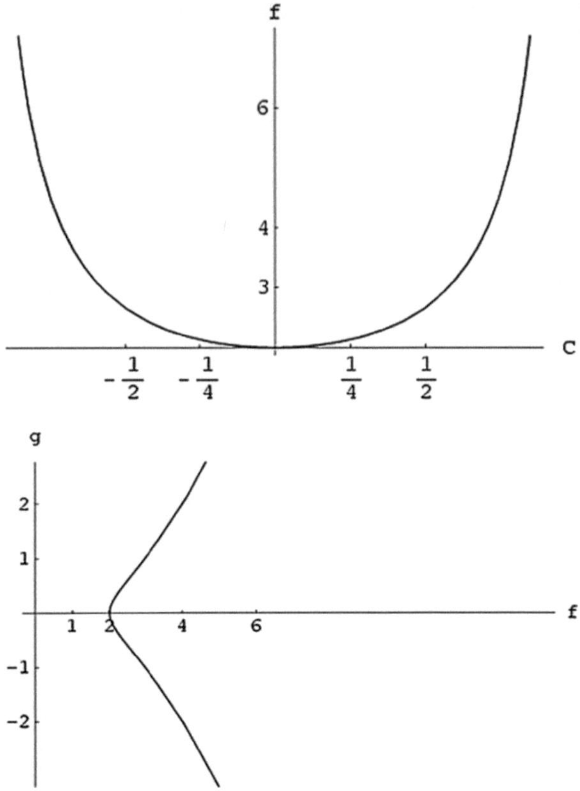

Fig. 3.8 In this picture we present the plots of the functions $f(C)$, $g(C)$ that define the realization of the Kähler manifold Σ associated with the potential (3.8.99) as a parametric surface in flat Minkowski three-dimensional space. The geometrical model is that appropriate to the hyperbolic character of the isometry $B \rightarrow B + c$. The first two pictures display the plot of g and f as functions of the VP coordinate C. The last plot is the parametric plot of the curve in the plane f, g. Geometrically this is the curve cut out by the surface Σ in any plane orthogonal to the axis X_2

Eq. (3.8.34) we can immediately calculate the curvature and we find:

$$R(\phi) = -\frac{e^{\nu\phi}\nu^3}{2\left(\mu + e^{\nu\phi}\nu\right)} \tag{3.8.110}$$

This shows what we just said. The manifold is smooth and singularity-free if and only if μ and ν have the same sign so that at no value of ϕ the denominator can develop a zero. Without loss of generality we can assume that $\nu > 0$ since the sign of ϕ can be flipped without changing its kinetic term. With this understanding it follows that also $\mu > 0$ for regularity.

Consider next the integral defining the VP coordinate C. We immediately obtain:

$$C(\phi) = \int \frac{1}{\mathscr{P}'(\phi)} d\phi = \frac{\phi}{\mu} - \frac{\log\left(\mu + e^{\nu\phi}\nu\right)}{\mu\nu} \tag{3.8.111}$$

The range of C is now easily determined considering the limits of the above function for $\phi = \pm\infty$. When $\mu > 0$, $\nu > 0$ we have:

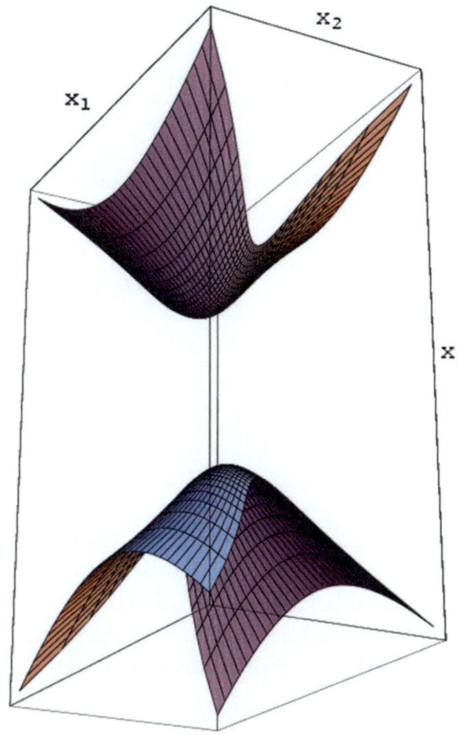

Fig. 3.9 In this figure we present the $3D$-plot of the surface Σ associated with the potential (3.8.99). The correct interpretation of the isometry in this case is that of a hyperbolic group. Indeed the hyperbolic embedding (3.8.57) in three-dimensional Minkowski space works beautifully and we have the smooth surface displayed here

$$C(-\infty) = -\infty \quad ; \quad C(\infty) = -\frac{\log[\nu]}{\mu\,\nu} \tag{3.8.112}$$

Hence $C \in \left[-\infty\,,\, -\frac{\log[\nu]}{\mu\,\nu}\right]$. The VP coordinate is always negative and it spans a semininfinite interval. Keeping this range in mind we can invert the relation (3.8.111) between ϕ and C obtaining:

$$\phi = -\frac{\log\left(\frac{e^{-C\mu\nu}}{\mu} - \frac{\nu}{\mu}\right)}{\nu} \tag{3.8.113}$$

The J-function is easily calculated from Eq. (3.8.17) and we find:

$$\mathscr{J}(\phi) = \frac{\nu^2\phi^2 + (2 - 2\nu\phi)\log\left(\frac{e^{\nu\phi}\nu}{\mu} + 1\right) - 2\text{Li}_2\left(-\frac{e^{\nu\phi}\nu}{\mu}\right)}{\nu^2} \tag{3.8.114}$$

where $\text{Li}_n(z)$ is the polylogarithmic function. Introducing in (3.8.114) the relation between ϕ and C, we get an explicit analytic expression for the $J(C)$ function, namely:

$$J(C) = \frac{1}{\nu^2}\left[\log^2\left(\frac{e^{-C\mu\nu} - \nu}{\mu}\right) + 2\left(\log\left(\frac{e^{-C\mu\nu} - \nu}{\mu}\right) + 1\right)\log\left(\frac{1}{1 - e^{C\mu\nu}\nu}\right)\right.$$
$$\left. -2\mathrm{Li}_2\left(1 + \frac{1}{e^{C\mu\nu}\nu - 1}\right)\right] \tag{3.8.115}$$

As for the metric, having the explicit expression (3.8.115), we easily calculate its second derivative and we find:

$$ds^2 = \frac{1}{2}\frac{\mathrm{d}^2 J}{\mathrm{d}C^2}\left(\mathrm{d}C^2 + \mathrm{d}B^2\right) = \frac{\mu^2}{\left(e^{C\mu\nu}\nu - 1\right)^2}\left(\mathrm{d}C^2 + \mathrm{d}B^2\right) \tag{3.8.116}$$

For $C \to -\infty$ the metric coefficient $\frac{1}{2}\frac{\mathrm{d}^2 J}{\mathrm{d}C^2}$ tends to a constant:

$$\frac{1}{2}\frac{\mathrm{d}^2 J}{\mathrm{d}C^2} \overset{C\to-\infty}{\approx} \mu^2 \quad \Rightarrow \quad J(C) \overset{C\to-\infty}{\approx} \frac{\mu^2}{2}C^2 \tag{3.8.117}$$

This asymptotic behavior differs from the usual logarithmic behavior of $J(C)$ at the boundary because at $C = -\infty$ and hence at $\phi = -\infty$ the curvature goes to zero.

In the other extremum of the C-range, namely for $C \to -\frac{\log[\nu]}{\mu\,\nu}$ the metric coefficient diverges and we have the standard logarithmic singularity. To see this, set $C = -\frac{\log[\nu]}{\mu\,\nu} - \xi$ and substitute it into the expression of the metric coefficient. We obtain:

$$\begin{aligned}\frac{1}{2}\frac{\mathrm{d}^2 J}{\mathrm{d}C^2} &= \frac{\mu^2}{\left(e^{\mu\nu\left(-\xi - \frac{\log(\nu)}{\mu\nu}\right)}\nu - 1\right)^2} \\ &\overset{\xi\to 0}{\approx} \frac{1}{\nu^2\xi^2} + \frac{\mu}{\nu\xi} + \frac{5\mu^2}{12} + \frac{1}{12}\mu^3\nu\xi + \mathcal{O}\left(\xi^2\right)\end{aligned} \tag{3.8.118}$$

and we conclude that, naming $C_0 = -\frac{\log[\nu]}{\mu\,\nu}$, we have:

$$J(C) \overset{C\to C_0}{\approx} \frac{2}{\nu^2}\log\left[C_0 - C\right] \tag{3.8.119}$$

This is the standard logarithmic singularity and the coefficient in front of the logarithm is indeed the inverse of the limiting curvature: $R_{C_0} = \frac{1}{2}\nu^2$.

This result confirms once again the relation between the asymptotic behavior of the $J(C)$ function and the character of the isometry group. For a parabolic isometry the asymtotic behavior is just that anticipated in Eqs. (3.8.25), (3.8.26). For a vanishing limiting curvature the correct asymptotic is (3.8.26).

The present example is very paedagical in order to avoid possible misconceptions. If we looked at the expression (3.8.116) and we forgot the precisely defined range of the variable C which is determined by the integration of the complex structure equation, we might be tempted to consider the same metric also for positive values of C. We would conclude that when $C \to \infty$ the metric coefficient goes to zero as $\exp[-\nu C]$. Then we would dispute that the last mentioned behavior indicates an elliptic interpretation of the isometry and advocate that there is a clash with our a priori knowledge that the isometry is instead parabolic. In fact there is no clash since the positive range of C is excluded and it is not to be considered. At the extrema of the C-interval, the function $J(C)$ displays the expected asymptotic behavior foreseen for the parabolic case.

3.8.5 On the Topology of Isometries

In this last subsection we provide a mathematically more rigorous illustration of the criteria discriminating among elliptic, parabolic and hyperbolic isometries of a two dimensional manifold whose metric is written in the standard form utilized throughout this section, namely:

$$ds^2 = d\phi^2 + f(\phi)^2 \, dB^2, \qquad (3.8.120)$$

In relation with the moment map issue, the function $f(\phi)$ is obviously the first derivative $\mathscr{P}'(\phi)$ with respect to the canonical coordinate ϕ of the moment map $\mathscr{P}(\phi)$. Considering the metric (3.8.120) as god-given, it obviously admits the one dimensional group of isometries $B \to B + c$ for any choice of the smooth function $f(\phi)$ parameterizing the metric coefficient and the question is what is the topology of such a group, is it compact or non-compact, and in the second case is it parabolic or hyperbolic. When we deal with a constant negative curvature manifold, namely with the coset $\mathrm{SL}(2,\mathbb{R})/\mathrm{O}(2)$ these questions have a precise answer within Lie algebra theory, since the considered one-dimensional group of isometries $\mathscr{G}_{iso}$ is necessarily a subgroup of $\mathrm{SL}(2,\mathbb{R})$ and as such its generator $\mathfrak{g} \in \mathfrak{sl}(2,\mathbb{R})$ can be of three types:

(a) $\mathfrak{g}$ is compact, which means that, as a matrix, in whatever representation of the Lie algebra $\mathfrak{sl}(2,\mathbb{R})$ it is diagonalizable and its eigenvalues are purely imaginary. In this case the one-dimensional subrgroup is topologically a circle $\mathbb{S}^1$ and isomorphic to $\mathrm{U}(1)$. We name *elliptic* the isometry group $\mathscr{G}_{iso}$ generated by such a $\mathfrak{g}$.

(b) $\mathfrak{g}$ is non-compact and semisimple, which means that, as a matrix, in whatever representation of the Lie algebra $\mathfrak{sl}(2,\mathbb{R})$, it is diagonalizable and its eigenvalues are real and non vanishing. In this case the one-dimensional subgroup is topologically a line $\mathbb{R}$ and it is isomorphic to $\mathrm{SO}(1,1)$. We name *hyperbolic* the isometry group $\mathscr{G}_{iso}$ generated by such a $\mathfrak{g}$.

(c) $\mathfrak{g}$ is non-compact and nilpotent, which means that, as a matrix, in whatever representation of the Lie algebra $\mathfrak{sl}(2,\mathbb{R})$, it is nilpotent and its eigenvalues are zero. In this case the one-dimensional subrgroup is topologically a line $\mathbb{R}$. We name *parabolic* the isometry group $\mathscr{G}_{iso}$ generated by such a $\mathfrak{g}$.

The interesting question is whether the characterization of an isometry as *elliptic,parabolic* or *hyperbolic* can be reformulated in pure geometrical terms and applied to cases where there is no ambient Lie algebra for the unique one-dimensional continuous isometry $\mathcal{G}_{iso}$. In this respect it is useful to remark that a metric of type (3.8.120) implies a fibre-bundle structure of the underlying two-dimensional manifold Σ:

$$\Sigma = \mathrm{P}(\mathbb{R}, \mathcal{F}, \mathcal{G}_{iso}) \rightarrow \mathbb{R} \tag{3.8.121}$$

where the base manifold is the real line $\mathbb{R}$ spanned by the coordinate $\phi \in [-\infty, +\infty]$, the structural group is the one-dimensional isometry group $\mathcal{G}_{iso}$ and the standard fibre $\mathcal{F}$ is a one dimensional space on which $\mathcal{G}_{iso}$ has a transitive action. In other words the manifold Σ is fibered into orbits of the isometry group. An explicit geometrical realization of this fibration in the three cases was already provided in the previous subsections by means of the three types of parametric surfaces encoded in:

1. Equation (3.8.45) which realize a surface in three-dimensional Minkowski space which is fibered in circles S^1 representating the orbits of an elliptic isometry group $\mathcal{G}_{iso}$.
2. Equation (3.8.57) which realize a surface in three-dimensional Minkowski space which is fibered in hyperbolae representating the orbits of a hyperbolic isometry group $\mathcal{G}_{iso}$.
3. Equation (3.8.73) which realize a surface in three-dimensional Minkowski space which is fibered in parabolae representating the orbits of a parabolic isometry group $\mathcal{G}_{iso}$.

As we argued in previous subsections, providing also some counterexamples, the subtle point is that the explicit geometric construction as a parametric surface fibered in circles, parabolae or hyperbolae, which a priori seems always possible, should lead to a smooth manifold having no singularity and being simply connected.

In more abstract terms the question was formulated by mathematicians for a single isometry Γ, even belonging to a discrete isometry group, not necessarily continuous and Lie, which can be characterized unambiguously as elliptic, parabolic, or hyperbolic, for Riemannian manifolds also of higher dimension than two, provided they are Hadamard manifolds.

Definition 3.8.1 A Hadamard manifold is a simply connected, geodesically complete Riemannian manifold $\mathcal{H} = (\mathcal{M}, g)$ whose scalar curvature $R(x)$ is **nonpositive definite and finite**, namely $-\infty < R(x) \leq 0, \quad \forall x \in \mathcal{M}$.

The virtue of Hadamard manifolds is that they allow for what is usually not available in generic Riemannian manifolds, namely the definition of a bilocal distance function $d(x, y)$ providing the absolute distance between any two points $x, y \in \mathcal{H}$. As we teach our students when introducing (pseudo)-Riemanian geometry and General Relativity, the concept of absolute space-(time) distance is lost in Differential Geometry and we can only define the length of any curve $\beta^\mu(t)$ ($t \in [0, 1]$), which starts at the point $x^\mu = \beta^\mu(0)$ and ends at the point $y^\mu = \beta^\mu(1)$. Given the metric $g_{\mu\nu}(x)$ we introduce the length functional which provides such a length:

$$\ell(\beta) = \int_0^1 \sqrt{g_{\mu\nu}\frac{d\beta^\mu}{dt}\frac{d\beta^\nu}{dt}}dt \tag{3.8.122}$$

The curves corresponding to extrema of the length functional are the geodesics, but in a generic Riemannian manifold there is no guarantee that for any two-points $x, y \in \mathscr{M}$ there is an arc of geodesic connecting them that is an absolute minimum of the length functional and that such minimum is unique and non-degenerate. Instead the hypotheses characterizing Hadamard manifolds guarantee precisely this (see, e.g. [16] and references therein) and one can define the distance function:

$$\forall x, y \in \mathscr{H} \quad : \quad d(x, y) = \text{infimum}\,[\ell(\beta)] \tag{3.8.123}$$

Hence restricting one's attention to Hadamard manifolds one can introduce a very useful geometrical concept that allows for a geometrical classification of isometries Γ:

$$\Gamma \quad : \quad \mathscr{M} \to \mathscr{M} \quad ; \quad \Gamma_\star\left[ds_g^2\right] = ds_g^2 \tag{3.8.124}$$

where $\Gamma_\star$ denotes the pull-back of Γ. The geometrical concept which provides the clue for such a classification is the displacement function defined below for any isometry Γ:

$$d_\Gamma(x) \equiv d(x, \Gamma x) \tag{3.8.125}$$

3.8.5.1 Classification of Isometries of Hadamard Manifolds $\mathscr{H} = (\mathscr{M}, g)$

The isometries of a Hadamerd manifold belong to the following types (see, e.g. [16] and references therein):

(a) **elliptic**, if $d_\Gamma(x)$ attains an absolute minimum of vanishing displacement $\min_{x\in\mathscr{M}} d_\Gamma(x) = 0$, or, to say it in other words, if and only if Γ has a fixed point $x_0 \in \mathscr{M}$ in the interior of the manifold for which $d\,(x_0, \Gamma x_0) = 0$.
(b) **hyperbolic**, if $d_\Gamma(x)$ attains an absolute minimum larger than zero $\min_{x\in\mathscr{M}} d_\Gamma(x) > 0$, or equivalently if Γ has two distinct fixed points on the boundary $\partial\mathscr{M}$ of $\mathscr{M}$
(c) **strictly parabolic**, if $d_\Gamma(x)$ never attains its infimum which is zero $\inf_{x\in\mathscr{M}} d_\Gamma(x) = 0$, or equivalently if Γ has just one fixed point on the boundary $\partial\mathscr{M}$ of $\mathscr{M}$;
(d) **mixed**, if $d_\Gamma(x)$ does not attain its the infimum which is larger than zero: $\inf_{x\in\mathscr{H}} d_\Gamma(x) > 0$.

The above classification of isometries is a generalisation to a nonconstant curvature case of the classification of isometries of the very particular constant curvature case, namely the Poincaré-Lobachevsky plane $\frac{\mathrm{SL}(2,\mathbb{R})}{\mathrm{O}(2)}$, where only the isometries (a), (b) and (c) are realized.

3.8.5.2 Application to the Kähler Surfaces considered in this Section

Not all Kähler surfaces Σ defined by Eq. (3.8.12) are Hadarmard since the curvature sometimes becomes positive in the interior of the manifold but most of them are such and moreover the limiting curvature of the boundary is non positive for all models. Therefore it makes sense to utilize the above geometric classification of isometries and verify that it just agrees with the criteria based on asymptotic expansions of the function $J(C)$ utilized in the previous subsections in order to discriminate among elliptic, parabolic and hyperbolic groups. Negative curvature guarantees the existence of a distance function, but probably in all considered examples such a distance function is well defined in spite of the existence of positive curvature domains in the deep interior of the manifold.

Hence with reference to the metric (3.8.120) let us consider the isometry Γ corresponding to B-shifts:

$$B \rightarrow \Gamma B = B + \delta , \tag{3.8.126}$$

where δ is a constant parameter, let us assume that the curvature

$$R = - \frac{\frac{d^2}{d\phi^2} f(\phi)}{f(\phi)} , \tag{3.8.127}$$

fulfills the Hadamard condition: $-\infty < R \leq 0$ and let us apply the classification scheme introduced above.

The first observation is the following. If the function $f(\phi)$ has neither a singularity nor a zero (i.e., if $f(\phi) \neq \pm\infty$ *and* $f(\phi) \neq 0$) both in the range of the coordinates $\{\phi, B\}$ corresponding to the interior of the manifold $\mathcal{M}$ and for those limiting values corresponding to the boundary $\{\phi, B\} \in \partial\mathcal{M}$ then the metric (3.8.120) has no coordinate singularity and the isometry (3.8.126) admits only one fixed point $B = \infty \in \partial\mathcal{M}$ on the boundary of the manifold. In this case the isometry Γ is strictly parabolic, according to item (c) of the above classification.

On the other hand, if the function $f(\phi)$ possesses a coordinate singularity at some value of $\phi = \phi_0 \in \mathcal{M}$ in the interior of $\mathcal{M}$, then in order to establish which is the type of the isometry Γ one has to introduce a new coordinate system $\{\phi, B\} \rightarrow \{\widetilde{\phi}, \widetilde{B}\}$ such that the metric expressed in terms of the new coordinates is non-singular in the vicinity of the former coordinate singularity. The existence of such a coordinate system is guaranteed by the non-singularity of the curvature and by the smoothness of the manifold. If in the newly constructed coordinate system the isometry has a fixed point corresponding to the former coordinate singularity then, according to item a) of the above classification, it is elliptic. Since this happens for all elements of the isometry group $\mathscr{G}_{iso}$, this latter is a compact U(1) and the appropriate complex structure is $\mathfrak{z} = \zeta = \exp[\delta (c - \mathrm{i}B)]$. Otherwise the isometry is certainly not elliptic and non-compact.

Summarizing, the necessary condition for the isometry Γ to be elliptic is that the function $f(\phi)$ has a zero or a pole in the interior of $\mathscr{M}$ at some $\phi = \phi_0 \equiv -\frac{a_1}{a_2}$, where a_1 and $a_2 > 0$ are arbitrary constant parameters. In case such a singularity is power-like, we conclude that in a neighborhood $\mathbb{U}_{\phi_0}$ of ϕ_0 we have:

$$f(\phi)|_{\phi \in \mathbb{U}_{\phi_0}} = (a_2 \phi + a_1)^n \tag{3.8.128}$$

where n is a positive or negative integer. Comparing Eq. (3.8.127) we see that the condition of a regular and finite curvature is fulfilled if and only if $n = 1$. In other words the function $f(\phi)$ has the following behavior at $\phi = \phi_0$:

$$f(\phi)|_{\phi \in \mathbb{U}_{\phi_0}} = a_2 \phi + a_1 + \mathscr{O}\left[(\phi - \phi_0)^3\right] \tag{3.8.129}$$

Correspondingly the curvature is zero at leading order:

$$R|_{\phi \in \mathbb{U}_{\phi_0}} = 0 + \mathscr{O}\left[(\phi - \phi_0)^3\right] \tag{3.8.130}$$

In the new coordinate system $\{x, y\}$, $\{\phi, B\} \to \{x, y\}$, defined by

$$x = \left(\phi + \frac{a_1}{a_2}\right) \cos(a_2 B), \quad y = \left(\phi + \frac{a_1}{a_2}\right) \sin(a_2 B), \tag{3.8.131}$$

the metric (3.8.120) becomes

$$\begin{aligned} ds^2|_{\phi \in \mathbb{U}_{\phi_0}} &\simeq d\phi^2 + (a_2 \phi + a_1)^2 dB^2 \\ &= dx^2 + dy^2, \end{aligned} \tag{3.8.132}$$

and the isometry transformations (3.8.126) takes the following form:

$$\{x, y\} \to \{x \cos\delta + y \sin\delta, -x \sin\delta + y \cos\delta\}, \tag{3.8.133}$$

The original coordinate singularity has disappeared, but in the new coordinates (3.8.131) the isometry (3.8.133) acquires the fixed point $\{x_0 = 0, y_0 = 0\}$, $\{0, 0\} \to \{0, 0\}$, in the interior of $\mathscr{M}$. Hence if the above situation is verified according to item a) of the above classification the isometry group is elliptic.

Consider next the behavior of the C-coordinate, defined by Eq. (3.8.15), in the neighborhood of ϕ_0. To leading order we have

$$\phi \to C \simeq \frac{1}{a_2} \ln(a_2 \phi + a_1) + \mathscr{O}\left[(\phi - \phi_0)^{-1}\right] \quad \Rightarrow \quad \phi_0 \Leftrightarrow C_0 = -\infty \tag{3.8.134}$$

so that the metric (3.8.120) becomes

$$ds^2|_{C \in \mathbb{U}_{C_0}} \simeq e^{2 a_2 C} \left(\mathrm{d}B^2 + \mathrm{d}C^2\right) \tag{3.8.135}$$

in the C_0-neighborhood $C \in \mathbb{U}_{C_0}$. Inspection of the latter formula shows that it reproduces the criterion to decide that the isometry is elliptic advocated in Eq. (3.8.22).

$$\frac{1}{2} \frac{d^2}{dC^2} J(C)|_{C \in \mathbb{U}_{C_0}} = e^{2 a_2 C}|_{C \in \mathbb{U}_{C_0}} \to 0 \tag{3.8.136}$$

Let us stress that the fixed point in the interior of the manifold required for an elliptic interpretation of the isometry group is just the origin of the manifold where the Kähler metric becomes approximately the flat one.

Let us now turn to the case where the singularity of the metric coefficient is of the exponential type, namely for $\phi_0 = \infty$ and for $\phi \in \mathbb{U}_{\phi_0}$, we have

$$f(\phi)|_{\phi \in \mathbb{U}_{\phi_0}} = a_1 e^{a_2 \phi}, \quad a_2 > 0 \tag{3.8.137}$$

this behavior is also consistent with the regularity of the curvature R (see Eq. (3.8.127)), which, in this case takes a finite negative value in the leading order approximation:

$$R|_{\phi \in \mathbb{U}_{\phi_0}} \simeq -a_2^2 + \text{subleading terms} \tag{3.8.138}$$

The metric (3.8.120) reproduces locally the metric of the hyperbolic (Poincaré - Lobachevsky) plane

$$ds^2|_{\phi \in \mathbb{U}_{\phi_0}} \approx d\phi^2 + a_1^2 e^{2a_2 \phi} dB^2 \tag{3.8.139}$$

for which it is well known that the value of $\phi_0 = \infty$ corresponds to the boundary $\partial\mathscr{M}$. If the function $f(\phi)$ does not have other singularities of the exponential type, but (3.8.137), then one can immediately conclude that the isometry (3.8.126) is strictly parabolic according to item c) of the above classification, since it possesses just a single fixed point $B = \infty$ on the boundary $\partial\mathscr{M}$.

If besides the singularity (3.8.137) the function $f(\phi)$ possesses a second exponential singularity at $\widetilde{\phi}_0 = -\infty$ for $\phi \in \mathbb{U}_{\widetilde{\phi}_0}$, namely

$$f(\phi)|_{\phi \in \mathbb{U}_{\widetilde{\phi}_0}} = \widetilde{a}_1 e^{-\widetilde{a}_2 \phi}, \quad \widetilde{a}_2 > 0, \tag{3.8.140}$$

then by the same token as above we come to the conclusion that the point $\widetilde{\phi}_0$ belongs to the boundary of another hyperbolic plane locally isomorphic to the neighborhood $U_{\widetilde{\phi}_0} \subset \mathscr{H}$ and that isometry (3.8.126) possesses a second fixed point on such a boundary. Hence the isometry is hyperbolic according to item (b) of the above classification and since this applies to all elements of the isometry group $\mathcal{G}_{iso}$ this latter is hyperbolic and isomorphic to SO(1, 1).

One can not exclude the existence of more sophisticated types of $f(\phi)$ singularities, besides the above described power-like and exponential one, that might be consistent with the regularity of the curvature R (3.8.127), yet in all examples considered in previous subsections no other singularities than these two are met.

Relying on these results we can summarize the geometric criteria for the classification of isometries in two-manifolds with a metric of type (3.8.120) which are of the Hadamard type

(a) **elliptic**, if the function $f(\phi)$ possesses a first order zero, i.e. $f(\phi)|_{\phi\in\mathbb{U}_{\phi_0}} = a_2\,(\phi - \phi_0)$;
(b) **hyperbolic**, if the function $f(\phi)$ possesses two different leading exponential singularities at $\phi_0^{(\pm)} = \pm\infty$, i.e. $f(\phi)|_{\phi\in\mathbb{U}_{\phi_0^{(\pm)}}} = a_1^{(\pm)}\,e^{\pm a_2^{(\pm)}\,\phi}$ and $a_2^{(\pm)} > 0$;
(c) **strictly parabolic**, if the function $f(\phi)$ possesses a single leading exponential singularity at either $\phi_0^{(+)} = +\infty$ or $\phi_0^{(-)} = -\infty$, i.e. $f(\phi)|_{\phi\in\mathbb{U}_{\phi_0^{(+)}}} = a_1^{(+)}\,e^{+a_2^{(+)}\,\phi}$ or $f(\phi)|_{\phi\in\mathbb{U}_{\phi_0^{(-)}}} = a_1^{(-)}\,e^{-a_2^{(-)}\,\phi}$ and $a_2^{(\pm)} > 0$.

The above characterization yields exactly the same result as the criteria based on the asymptotic behavior of $J(C)$ that have been utilized in the previous subsections and this happens also for such models that do not lead to exactly Hadamard manifolds, the curvature attaining somewhere also positive values. As an exemplification of the use of the above concepts we briefly reconsider from this point of view the flat models and the constant curvature models.

3.8.5.3 Flat Models

The flat metric

$$ds^2 = d\phi^2 + (a_2\,\phi + a_1)^2\,dB^2 \tag{3.8.141}$$

in case $a_2 \neq 0$ possesses a coordinate singularity at

$$\phi = -\frac{a_1}{a_2} \tag{3.8.142}$$

corresponding to a first order zero $f(\phi)$ at finite ϕ. According to the above classification this implies that the isometry $B \to B + \delta$ is elliptic.

In the case $a_2 = 0$ the metric (3.8.141) becomes

$$ds^2 = d\phi^2 + a_1^2\,dB^2 \tag{3.8.143}$$

and does not possess a coordinate singularity at all. This implies that the isometry $B \to B + \delta$ is strictly parabolic.

3.8.5.4 Constant Negative Curvature Models

Case (A)

$$ds^2 = d\phi^2 + \sinh^2(\nu\,\phi)\,dB^2 \tag{3.8.144}$$

This metric possesses a coordinate singularity at $\phi = 0$. In the neighborhood of $\phi = 0$ at leading order it behaves as follows

$$ds^2 \approx d\phi^2 + \nu^2\,\phi^2\,dB^2 \tag{3.8.145}$$

which modulo an inessential rescaling of the coordinate B and a shifting the coordinate ϕ reproduces the metric (3.8.141). Hence its isometry (3.8.126) is elliptic in this case.

Case (B)

$$ds^2 = d\phi^2 + \cosh^2(\nu\,\phi)\,dB^2 \tag{3.8.146}$$

This metric does not possess a coordinate singularity in the finite range of ϕ, but it has two exponential singularities of the type (3.8.137) and (3.8.140). Hence the isometry (3.8.126) is hyperbolic in this case.

Case (C)

$$ds^2 = d\phi^2 + e^{2\nu\phi}\,dB^2 \tag{3.8.147}$$

This metric does not possess a coordinate singularity in the finite range of ϕ, but it possesses a single exponential singularity either of the type (3.8.137) or of the type (3.8.140). Hence the isometry (3.8.126) is strictly parabolic in this case.

References

1. P.G. Fré, *A Conceptual History of Symmetry from Plato to the Superworld* (Springer, Berlin, 2018)
2. N.J. Hitchin, A. Karlhede, U. Lindström, M. Roček, Hyperkahler Metrics and Supersymmetry. Commun. Math. Phys. **108**, 535 (1987)
3. K. Galicki, A generalization of the momentum mapping construction for quaternionic Kahler manifolds. Commun. Math. Phys. **108**, 117 (1987)
4. R. D'Auria, S. Ferrara, P. Fré, Special and quaternionic isometries: general couplings in n = 2 supergravity and the scalar potential. Nucl. Phys. B **359**, 705 (1991)
5. L. Andrianopoli, M. Bertolini, A. Ceresole, R. D'Auria, S. Ferrara, P. Fre, T. Magri, N = 2 supergravity and N = 2 superYang-Mills theory on general scalar manifolds: symplectic covariance, gaugings and the momentum map. J. Geom. Phys. **23**, 111–189 (1997)
6. P.G. Fré, *Gravity, A Geometrical Course*, vol. 1, 2 (Springer Science and Business Media, Berlin, 2012)
7. J. Bagger, E. Witten, The gauge invariant supersymmetric nonlinear sigma model. Phys. Lett. **118B**, 103–106 (1982)
8. B. de Wit, F. Vanderseypen, A. Van Proeyen, Symmetry structure of special geometries. Nucl. Phys. B **400**, 463–524 (1993)

9. B. de Wit, A. Van Proeyen, Special geometry, cubic polynomials and homogeneous quaternionic spaces. Commun. Math. Phys. **149**, 307–334 (1992)
10. B. de Wit, A. Van Proeyen, Broken sigma model isometries in very special geometry. Phys. Lett. B **293**, 94–99 (1992)
11. P. Fre, Lectures on special Kahler geometry and electric - magnetic duality rotations. Nucl. Phys. Proc. Suppl. **45BC**, 59–114 (1996)
12. A. Ceresole, G. Dall'Agata, General matter coupled N = 2, D = 5 gauged supergravity. Nucl. Phys. B **585**, 143–170 (2000)
13. P. Fré, A.S. Sorin, Axial symmetric Kahler manifolds, the D-map of inflaton potentials and the Picard–Fuchs Equation. Fortsch. Phys. **62**, 26–77 (2014)
14. S. Ferrara, P. Fré, A.S. Sorin, On the topology of the inflaton field in minimal supergravity models. JHEP **04**, 095 (2014)
15. S. Ferrara, P. Fre, A.S. Sorin, On the gauged Khler isometry in minimal supergravity models of inflation. Fortsch. Phys. **62**, 277–349 (2014)
16. W. Ballmann, M. Gromov, V. Schroeder, *Manifolds of Nonpositive Curvature* (Birkhäuser, Boston, 1985)

Chapter 4
Special Geometries

> *La géométrie…est une science née à propos de l'expérience…nous avons créé l'espace qu'elle etudie, mais en l'adaptant au monde où nous vivons. Nous avons choisie l'espace le plus commode…*
>
> Henri Poincaré.

4.1 The Evolution of Geometry in the Second Half of the XXth Century

Relying for a complete historical account on the tale told in the twin book [1], let us summarize the steps that led, in the 1990's to Special Geometries.

4.1.1 Complex Geometry Rises to Prominence

On the purely mathematical front in the years from 1953 to 1955, Pierre Dolbeault introduced a new very important mathematical instrument: the $\overline{\partial}$-cohomology of the differential forms defined on complex analytic manifolds, namely the holomorphic analogue of de Rham cohomology defined on real manifolds. The essence of Dolbeault cohomology (described in Sect. 3.3) is the topic of Dolbeault's thesis, prepared by him under the direction of Henri Cartan, Élie's son and one of the closest friends of André Weil. The thesis was defended in Paris in 1955.

Complex Geometry and, within it Kähler Geometry, arose to high prominence in the three decades from 1950 to 1980. The language of fibre-bundles and characteristic classes was combined with the notion of holomorphicity and line-bundles, namely Principal Bundles whose structural group is the group of non vanishing complex numbers $\mathbb{C}^{\star}$, became ubiquitous in the discussion of complex manifolds.

A new innovative conception developed in this context, namely that of characterizing the geometry of base manifolds $\mathscr{M}$ by means of statements on the characteristic classes of bundles defined over them.

P. G. Frè, *Advances in Geometry and Lie Algebras from Supergravity*, Theoretical and Mathematical Physics,
https://doi.org/10.1007/978-3-319-74491-9_4

Fig. 4.1 On the left Eugenio Calabi (Milano, Italy 1923). On the right Shing–Tung Yau (Shantou, China 1949). Born Italian, Calabi is an American citizen. He graduated in 1946 from MIT and obtained his Ph.D from Princeton in 1950. He held temporary positions in Minnesota and in Princeton, then since 1967 to retirement he was Full Professor of Mathematics at the University of Pennsylvania, successor of Hans Rademacher. He came to the definition of Calabi–Yau n-folds while exploring the geometry of complex manifolds that support harmonic spinors. Born in China, Yau studied first at Hong Kong University, then he went to the USA where he got his Ph.D. in 1971 from Berkeley under the supervision of Chern. Post-doctoral fellow in Princeton and in Stony Brook, he became Professor in Stanford. Since 1987 he is Professor of Mathematics at Harvard University. Yau's proof of Calabi 1964 conjecture was published in 1977

The first example, which plays an important role in the sequel, is that of *Hodge–Kähler manifolds* that are Kähler manifolds $\mathscr{M}$ characterized by the existence of a line bundle $\mathscr{L} \to \mathscr{M}$, such that its first Chern Class coincides with the cohomology class of the Kähler 2-form: $c_1(\mathscr{L}) = [K]$.

Another important example is provided by Calabi–Yau n-folds. These latter were introduced by Eugenio Calabi (see Fig. 4.1) in 1964 with the definition of complex n-dimensional algebraic varieties $\mathscr{M}_n$, the first Chern class of whose tangent bundle vanishes: $c_1\,(T\mathscr{M}_n) = 0$. Later, the American-Chinese mathematician Shin–Tung Yau (see Fig. 4.1) proved the theorem that for Calabi–Yau n-folds, every $(1, 1)$ Dolbeault cohomology class contains a representative that can be identified with the Kähler 2-form of a Ricci flat Kähler metric: the Calabi–Yau metric.

4.1.2 On the Way to Special Geometries

Other notable examples of this way of thinking, applying both to complex and to real geometry are the *manifolds of restricted holonomy*. One considers Riemannian manifolds $\mathcal{M}_n$ in dimension n and their *spin bundles*, namely the principal bundles on which their spin connections ω^{ab} are defined as Ehresman connections. Generically such bundles have, as structural group, Spin(n), which is the double covering of SO(n), yet it may happen that ω^{ab} is Lie algebra–valued in a proper subalgebra $\mathbb{G} \subset \mathfrak{so}(n)$. Choosing algebras $\mathbb{G}$ for which this might happen and imposing that it should happen is a strong constraint on the geometry of the manifold $\mathcal{M}_n$.

Research on manifolds of restricted holonomy went on in the 1980s and 1990s in the mathematical community but, not too surprisingly, it was heavily stimulated by issues in theoretical physics and particularly in Superstring/Supergravity theory.

It is easy to understand why. The main input in Superstring/Supergravity is Supersymmetry, a generalization of Lie algebras where spinor representations and vector representations of groups SO(n) are transformed one into the other by new symmetry operators Q^α, dubbed the *supercharges*, that are themselves spinors. At the level of field theories we work with fibre-bundles and the fields we consider are sections of such bundles. Field theories can be supersymmetric if the supercharges Q^α find a field-theoretic realization which is a symmetry of the action, leaving the door open for its desired spontaneous breaking. It is quite intuitive that such a realization of the supercharges requires special restrictions on the bundles and this reflects into heavy constraints on the geometry of the base manifolds.

The above simple reasoning reveals what, in the opinion of this author, is the main conceptual contribution of Supergravity theories to the development of geometrical thought and, eventually, of physical thought, provisionally assuming that geometry and physics are, once properly interpreted, the same thing. Supersymmetry tackles with one of the most fundamental and so far unexplained pillars of physics, namely the separation of the physical world into bosons and fermions and the spin-statistics theorem. The distinction between vector and spinor representations is at the basis of all that and it is a distinctive property of the $\mathfrak{so}(n)$ Lie algebras, unexisting for the other simple Lie algebras. On the other hand the reduction of the tangent-bundle to an $\mathfrak{so}(n)$-bundle is the same thing as the existence of a metric and can be interpreted as gravity. Special Geometries arise because of supersymmetry, in order to allow the mixing of boson and fermions. It is the mathematical investigation of *Space* from this new viewpoint the new quality of geometrical studies inspired by supergravity. Before telling such a story we need to recall another mathematical conception, that was developed independently from Superstring/Supergravity yet found its most ample and fertile applications in the supersymmetric context.

4.1.3 The Geometry of Geometries

Let us recall Hermann Weyl's discussion of the ellipses, used by him to introduce his conception of mathematical thinking and reported by us in the twin book [1]. The coefficients a, b, c of the quadratic form quoted by Weyl are the first example of *moduli* and the portion of $\mathbb{R}^3$ where they are allowed to take values is the first example of a *moduli-space*. In complex algebraic geometry one considers loci of some projective space $\mathbb{P}_n(\mathbb{C})$ cut out by some homogeneous polynomial constraint of degree m:

$$0 = \mathscr{W}(a, X) = \sum_{i_1 \dots i_m} a_{i_1 \dots i_m} X^{i_1} \dots X^{i_m} \tag{4.1.1}$$

imposed on the $n+1$ homogeneous coordinates X^i $(i = 1, \dots, n+1)$. The complex coefficients $a_{i_1 \dots i_m}$ are also *moduli* and fill some complex manifold $\mathscr{M}$. If we consider the following constraint imposed on the metric tensor of some Riemannian manifold $\mathscr{M}_n$:

$$R_{\mu\nu}[g] = \lambda\, g_{\mu\nu} \tag{4.1.2}$$

where $R_{\mu\nu}[g]$ is the Ricci tensor and λ some constant, we actually write a set of differential equations for the metric tensor $g_{\mu\nu}$, which, on the manifold $\mathscr{M}_n$, generically admit a solution depending on a set of parameters $\{p_1, \dots p_r\}$, among which λ is included. Also these are moduli and they fill a space named *the moduli space of Einstein metrics* on $\mathscr{M}_n$.

Several other examples can be made of manifolds $\mathscr{M}_{mod}$ whose points correspond to the specification of a particular geometry within a class, for instance the moduli ρ^i of an instanton parameterize the solution of the self duality constraint[1]:

$$F^{\Lambda}_{\mu\nu}(\rho, x) = \tfrac{1}{2}\, \varepsilon_{\mu\nu\lambda\sigma}\, F^{\Lambda}_{\lambda\sigma}(\rho, x) \tag{4.1.3}$$

imposed on the field strength of a connection on a principal fibre bundle $P(\mathrm{G}, \mathscr{M}_4)$.

A new mathematical idea that is of outmost relevance both for physics and for mathematics is encoded in the following almost obvious argument. Being a manifold, the moduli space $\mathscr{M}_{mod}$ can support such geometrical structures like a metric, like a complex structure, or a fibration. We call this the *geometry of geometries*. There are several mathematical constructions, dictated by the mathematical nature of the objects of which we consider the moduli, that single out a canonical determination of the *geometry of geometries*, yet it is precisely at this level that the interaction between physics and mathematics becomes most profound and fertile. Indeed the geometry of geometries is typically what enters the supergravity lagrangians under the form of sigma-models for scalar fields that on one side are the spin zero members of supersymmetry multiplets,[2] while on the other side they are *moduli* of some

---

[1] *Clarification for readers with a mostly mathematical background*: in the physical literature *instantons* play a very important role. They are field configurations that in the Wick-rotated space-time with Euclidean signature satisfy first-order equations more restrictive than the second order Euler Lagrangian equations (the latter are implied by the former). In the *path integral* formulation of *quantum field theory*, instanton correspond to the absolute minimal of the action functional and provide the dominant contribution to quantum correlators. Depending on the type of considered fields instantons have different definitions. For gauge fields, instantons are the connections on the underlying principal fibre-bundle whose field strengths are self dual, namely satisfy Eq. (4.1.3).

[2] *Clarification for mathematicians:* the wording *supermultiplets* is universally used in the context of supersymmetric field theories to denote a finite set of standard fields of various spins that form a *unitary irreducible representation* of the supersymmetry algebra extending the Poincaré Lie algebra.

manifold, for a example a Calabi–Yau threefold, on which the superstring has been compactified.

This evenience produces a double check on the geometry of geometries. Its use in supersymmetric lagrangians, imposes strong constraints on the geometry of the scalar fields that, in many cases, have a recognizable solution in terms of known geometrical categories, in other cases it leads to the definition of new types of restricted geometries, generically dubbed *special geometries*. It is particularly rewarding that the *special geometries* selected by supersymmetry are just those apt to accomodate *the moduli spaces* of such mathematical structures as *the complex structures* or the *Kähler structures* of a compactification manifold like a Calabi–Yau threefold.

Altogether, a really new chapter has been written in the two decades from 1990 to 2010 in the history of geometry, where the distinction between physics and mathematics has become somewhat obsolete, ideas from one field compenetrating the other in an essential way.

4.1.4 The Advent of Special Geometries

The first instance of a special geometry was found by brute force, immediately after the discovery in 1976 by Sergio Ferrara, Daniel Freedman and Peter van Nieuwenhuizen of $\mathcal{N} = 1$, $d = 4$ supergravity (see Fig. 4.2). The next year, considering the coupling of a scalar multiplet to the newly found gravitational theory, the three supergravity founders, together with Breitenlohner, Gliozzi and Scherk, constructed a rather impressive and cumbersome lagrangian, depending on an arbitrary real function $G(A, B)$ of a scalar A and a pseudoscalar B and on all its derivatives up to the fourth one [2]. It was Bruno Zumino (see Fig. 4.3) who, in 1979, decoded the meaning of this monster, showing that $G(A, B)$ is just the Kähler potential of a Kähler metric, all of the introduced derivatives obtaining their adequate interpretation as metric, connection and curvature of the Kählerian manifold [3]. In this way the generalization to several scalar multiplets was singled out: it suffices to utilize an n-dimensional Kähler manifold.

Shortly after, the so named holomorphic superpotential introduced by physicists to describe fermion–scalar interactions and to produce a scalar potential consistent with supersymmetry, was also interpreted geometrically. The superpotential is just a holomorphic section of the Hodge line-bundle over the Kähler manifold.

In this way the firstly found special geometry was a known one, namely Hodge-Kähler geometry. This is not so for the next case.

At the beginning of the 1980's the next obvious case was the coupling of vector multiplets to $\mathcal{N} = 2$, $d = 4$ supergravity. Each multiplet contains a complex scalar field and the question was what is the geometry of the scalar manifold $\mathcal{M}_{scalar}$ in the case of several such multiplets. Certainly $\mathcal{M}_{scalar}$ had to be Kähler, since $\mathcal{N} = 2$ is in particular $\mathcal{N} = 1$. Yet the stronger supersymmetry imposes additional constraints so that $\mathcal{M}_{scalar}$ had to be a *special Kähler manifold*. A pioneering work on this problem was conducted in several different combinations by a group of French, Belgian,

Fig. 4.2 From left to right the three founders of Supergravity Theory, Daniel Freedman (1939), Sergio Ferrara (1945), Peter van Nieuwenhuizen (1938). Dan Freedman was born in the USA, graduated from Wisconsin University. He has been professor at Stony Brook University and he is currently full-professor at MIT. Sergio Ferrara born in Rome in 1945 graduated from la Sapienza University under the supervision of Raoul Gatto. Permanent Member of the CERN Theoretical Division for many years he is also professor of physics at UCLA. Peter van Nieuwenhuizen born in Holland in 1938, graduated in Utrecht under the supervision of Veltman, held various positions in the United States and since the middle 1980s he is full-professor of physics at Stony Brook University. The paper containing the lagrangian and the transformation rules of $\mathcal{N} = 1, d = 4$ supergravity was published by the three founders of the theory in 1976. Since then all the three have contributed extensively and in various different directions to the development of supergravity. Sergio Ferrara among the three has largely contributed to the development of special geometries

Fig. 4.3 Bruno Zumino (1923–2014). Born in 1923 in Rome, he graduated from the University La Sapienza in 1945. He died in 2014 in California, where he was emeritus professor of Berkeley University. For many years he was permanent member of the Theoretical Division at CERN. Zumino has given many important contributions to Theoretical Physics in several directions: supersymmetry, anomalies, conformal field theories, quantum groups

Fig. 4.4 On the left Antoine Van Proeyen (1953 Belgium), on the right Eugene Cremmer (Paris 1942). Antoine Van Proeyen graduated from KU Leuven and worked in several Laboratories and Universities, among which the École Normale of Paris, CERN Theoretical Division and Torino University, before becoming full-professor in Leuven. He is currently the Head of the Theoretical Physics Section at the K.U. Leuven. Since 1979, he has been involved in the construction of various supergravity theories, the resulting special geometries and their applications to phenomenology and cosmology. Cremmer is *directeur de recherche* of the CNRS working at the École Normale Supérieure of Paris. In 1978, together with Bernard Julia and Joël Scherk, he derived the space-time formulation of 11 dimensional supergravity theory, regarded today as the low energy limit of the so far mysterious M-theory. In the following few years, Cremmer, together with Bernard Julia, constructed the dimensional reductions of $d = 11$ supergravity, arriving in $d = 4$ at the maximal extended $\mathcal{N} = 8$ theory, whose structure is completely determined by the non-compact coset $\frac{E_{7(7)}}{SU(8)}$ accomodating the 70 scalars of the gravitational multiplet. Active research is going on at the present time to demonstrate that $\mathcal{N} = 8$ supergravity is a finite theory

Dutch, Swiss and Italian theoretical physicists in the papers mentioned in [4–6]. Using a special set of complex coordinates, the special Kähler manifolds that can accomodate the scalar fields of $\mathcal{N} = 2$ vector multiplets were described as those where the Kähler potential is obtained from a holomorphic prepotential according to a specific formula.

Once this was established, a natural question arose whether among so defined *special Kähler manifolds* there were symmetric spaces G/H. The answer to this question was given in Paris in 1985 by Eugene Cremmer and Antoine Van Proeyen (see Fig. 4.4) who, in a beautiful paper absolutely worth of Cartan's tradition [7], provided the exhaustive classification shown in the first column of Table 4.1. As one sees, exceptional Lie groups make their appearance in such a list through peculiar real forms. This was no longer a surprise for supergravity researchers since, four years before, the same Eugene Cremmer, in collaboration with Bernard Julia (see Fig. 4.5), had shown that the dimensional reduction of maximally extended supergravity from $D = 11$ down to $D = 10$, $D = 9, \ldots, D = 4$, $D = 3$ produces, as scalar manifolds, the following maximally split symmetric spaces:

$$M_D = \frac{E_{11-D(11-D)}}{H_c} \tag{4.1.4}$$

Table 4.1 List of special Kähler symmetric spaces with their Quaternionic Kähler c-map images. The number n denotes the complex dimension of the Special Kähler preimage. On the other hand $4n+4$ is the real dimension of the Quaternionic Kähler c-map image

$\mathcal{SK}_n$ Special Kähler manifold	$\mathcal{QM}_{4n+4}$ Quaternionic Kähler manifold	$\dim \mathcal{SK}_n = n$
$\frac{\mathrm{SU}(1,1)}{\mathrm{U}(1)}$	$\frac{\mathrm{G}_{2(2)}}{\mathrm{SU}(2)\times\mathrm{SU}(2)}$	$n=1$
$\frac{\mathrm{Sp}(6,\mathrm{R})}{\mathrm{SU}(3)\times\mathrm{U}(1)}$	$\frac{\mathrm{F}_{4(4)}}{\mathrm{USp}(6)\times\mathrm{SU}(2)}$	$n=6$
$\frac{\mathrm{SU}(3,3)}{\mathrm{SU}(3)\times\mathrm{SU}(3)\times\mathrm{U}(1)}$	$\frac{\mathrm{E}_{6(2)}}{\mathrm{SU}(6)\times\mathrm{SU}(2)}$	$n=9$
$\frac{\mathrm{SO}^\star(12)}{\mathrm{SU}(6)\times\mathrm{U}(1)}$	$\frac{\mathrm{E}_{7(-5)}}{\mathrm{SO}(12)\times\mathrm{SU}(2)}$	$n=15$
$\frac{\mathrm{E}_{7(-25)}}{\mathrm{E}_{6(-78)}\times\mathrm{U}(1)}$	$\frac{\mathrm{E}_{8(-24)}}{\mathrm{E}_{7(-133)}\times\mathrm{SU}(2)}$	$n=27$
$\frac{\mathrm{SL}(2,\mathbb{R})}{\mathrm{SO}(2)}\times\frac{\mathrm{SO}(2,2+p)}{\mathrm{SO}(2)\times\mathrm{SO}(2+p)}$	$\frac{\mathrm{SO}(4,4+p)}{\mathrm{SO}(4)\times\mathrm{SO}(4+p)}$	$n=3+p$
$\frac{\mathrm{SU}(p+1,1)}{\mathrm{SU}(p+1)\times\mathrm{U}(1)}$	$\frac{\mathrm{SU}(p+2,2)}{\mathrm{SU}(p+2)\times\mathrm{SU}(2)}$	$n=p+1$

Fig. 4.5 Bernard Julia (Paris 1952). He graduated from Université de Paris-Sud in 1978, and he is *directeur de recherche* of the CNRS working at the *École Normale Supŕieure*. In 1978, together with Eugne Cremmer and Joël Scherk, he constructed 11-dimensional supergravity. Shortly afterwards, Cremmer and Julia constructed the classical Lagrangian of four-dimensional $\mathcal{N}=8$ supergravity by dimensional reduction from the 11-dimensional theory

where:

$$\begin{aligned}
E_{5(5)} &\simeq D_{5(5)} \simeq \mathrm{SO}(5,5) \\
E_{4(4)} &\simeq A_{4(4)} \simeq \mathrm{SL}(5,\mathbb{R}) \\
E_{3(3)} &\simeq A_{1(1)} \times A_{2(2)} \simeq \mathrm{SL}(2,\mathbb{R}) \otimes \mathrm{SL}(3,\mathbb{R}) \\
E_{2(2)} &\simeq A_{1(1)} \times A_{1(1)} \simeq \mathrm{SL}(2,\mathbb{R}) \otimes \mathrm{SL}(2,\mathbb{R})
\end{aligned} \tag{4.1.5}$$

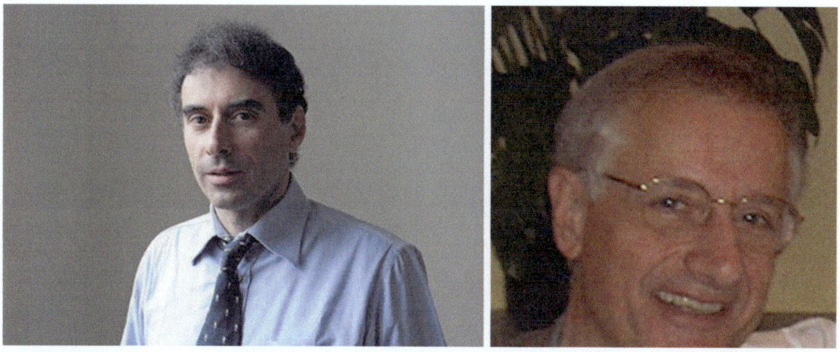

Fig. 4.6 On the left Leonardo Castellani (born 1953 in Freiburg, Switzerland). On the right Riccardo D'Auria (born 1940 in Rome). Leonardo Castellani studied physics at the University of Florence in Italy and obtained his Ph.D from Stony Brook University in the US, with a thesis written under the supervision of van Nieuwenhuizen. He had post-doctoral positions at Caltech and at CERN, then he became permanent Researcher in the Torino section of the National Institute of Nuclear Research (INFN) and in 1993 he was appointed full-professor of Theoretical Physics at the University of Eastern Piedmont, position that he holds at the present time. He is especially known for his contributions, together with D'Auria and Fré to the rheonomic formulation of supersymmetric theories, for his derivation together with Larry Romans of the list of G/H compactifications of d = 11 supergravity and more recently for developments in quantum group theories and, together with P.A. Grassi and R. Catenacci for the extension of Hodge theory to supermanifolds. Riccardo D'Auria studied at the University of Torino and graduated there with a thesis written under the supervision of Tullio Regge. He was for several years Associate Professor at the University of Torino, in 1987 he was appointed full-professor of Theoretical Physics at the University of Padua. Few years later he was offered a full professor chair at the Politecnico of Torino where he concluded his academic career becoming emeritus professor in 2011. D'Auria, together with Fré has been the founder of the rheonomic formulation of supergravity and also with Fré he introduced the notion of super Free Differential Algebras, that were singled out as the algebraic basis of all supergravity theories in dimension higher than four. In particular in 1982, D'Auria and Fré obtained the FDA formulation of d = 11 supergravity. D'Auria has given many more contributions to supergravity theory in particular in connection with special geometries, with the classification of black-hole solutions, with duality rotations, with the various formulations of the $d = 6$ theories and with several other aspects of the superworld

So exceptional Lie groups that had been regarded for long time as mathematical curiosities were brought to prominence by supergravity and in parallel also by superstring theory.

The fact that all such results were obtained in the *École Normale Supérieure de Paris* demonstrates the far reaching influence of Élie Cartan's tradition.

At the end of the eighties the intrinsic definition of *special Kähler geometry*, free from the use of special coordinates, was independently obtained with two different strategies by Andrew Strominger (see Fig. 4.7) and by Leonardo Castellani, Riccardo D'Auria and Sergio Ferrara (see Fig. 4.6).

While Strominger derived his definition from the properties of Calabi–Yau moduli spaces [8], Castellani, D'Auria and Ferrara [9, 10] (and later D'Auria Ferrara and Fré [11]) derived their own definition from the constraints imposed by supersymmetry on

Fig. 4.7 On the left Bernard Quirinus Petrus Joseph de Wit (born 1945 in the Netherlands). On the right Andrew Eben Strominger (born 1955 in the USA). Bernard de Wit studied theoretical physics at Utrecht University, where he got his PhD under the supervision of the Nobel Prize laureate Martinus Veltman in 1973. He held postdoc positions in Stony Brook, Utrecht and Leiden. He became a staff member at the National Institute for Nuclear and High Energy Physics (NIKHEF) in 1978, where he became head of the theory group in 1981. In 1984 he was appointed professor of theoretical physics at Utrecht University where he has stayed for the rest of his career. Bernard de Wit has given important contributions to the development of supergravity theory building, in collaboration mainly with Van Proeyen, the so named *conformal tensor calculus*. Together with Herman Nicolai he constructed the $\mathfrak{so}(8)$-gauged version of $\mathcal{N} = 8$ supergravity that has provided the paradigmatic example for all supergravity gaugings. Andrew Strominger completed his undergraduate studies at Harvard in 1977 before attending the University of California, Berkeley for his Master diploma. He received his PhD from MIT in 1982 under the supervision of Roman Jackiw. Prior to joining Harvard as a professor in 1997, he held a faculty position at the University of California, Santa Barbara. Strominger is especially known for introducing, together with Cumrun Vafa the string theory explanations of the microscopic origin of black hole entropy, originally calculated thermodynamically by Stephen Hawking and Jacob Bekenstein. Strominger, together with Philippe Candelas, Gary Horowitz and Edward Witten was the first proposer of Calabi–Yau threefolds as compactification manifolds for superstrings and supergravities in $\mathrm{d} = 10$

the curvature tensor of the Kählerian manifold. With some labour they also showed the full equivalence of the two definitions.

In the same years, Antoine Van Proeyen and Bernard de Wit (see Fig. 4.7), in some publications together with a younger collaborator, established a full classification of *homogeneous special geometries*, namely of special manifolds that admit a solvable transitive group of isometries [12–14]. They also explored the relation [12, 13] between *special Kähler geometries* and quaternionic geometries that can be obtained from them by means of a very interesting map, originally discovered by Cecotti [15] and further developed by Ferrara et al. in [16, 17]. So doing they came in touch with the classification of quaternionic manifolds with a transitive solvable group of motion that had been performed several years before by Alekseevsky [18, 19].

The map mentioned above is named the c-map and can be given a modern compact definition exhibited in [20]. Furthermore the c-map has a non Euclidean analogue, the $c^\star$-map that plays an important role in the discussion of supergravity based black-holes, another instance of geometry that will occupy us in later chapters.

4.1.5 A Survey of the Topics in This Chapter

In the sequel the special geometries motivated by supergravity will be thoroughly discussed and the properties of the c-map will be analyzed in detail. In that we closely follow the recent paper [20].[3] Indeed, coming to these topics our history of *Symmetry and Geometry* has reached the front of current research. Here physics and mathematics are fully entangled.

4.2 Special Kähler Geometry

In this section we present Special Kähler Geometry in a full-fledged rigorous mathematical form. Let us begin by summarizing some relevant concepts and definitions that are propaedeutical to the main definition.

4.2.1 Hodge–Kähler Manifolds

Consider a *line bundle* $\mathscr{L} \xrightarrow{\pi} \mathscr{M}$ over a Kähler manifold $\mathscr{M}$. By definition this is a holomorphic vector bundle of rank $r = 1$. For such bundles the only available Chern class is the first:

$$c_1(\mathscr{L}) = \frac{i}{2}\,\overline{\partial}\,\left(h^{-1}\,\partial\,h\right) = \frac{i}{2}\,\overline{\partial}\,\partial\log h \tag{4.2.1}$$

where the 1-component real function $h(z, \overline{z})$ is some hermitian fibre metric on $\mathscr{L}$. Let $\xi(z)$ be a holomorphic section of the line bundle $\mathscr{L}$: noting that under the action of the operator $\overline{\partial}\,\partial$ the term $\log\left(\overline{\xi}(\overline{z})\,\xi(z)\right)$ yields a vanishing contribution, we conclude that the formula in Eq. (4.2.1) for the first Chern class can be re-expressed as follows:

$$c_1(\mathscr{L}) = \frac{i}{2}\,\overline{\partial}\,\partial\log \parallel \xi(z) \parallel^2 \tag{4.2.2}$$

where $\parallel \xi(z) \parallel^2 = h(z, \overline{z})\,\overline{\xi}(\overline{z})\,\xi(z)$ denotes the norm of the holomorphic section $\xi(z)$.

Equation (4.2.2) is the starting point for the definition of Hodge–Kähler manifolds. A Kähler manifold $\mathscr{M}$ is a Hodge manifold if and only if there exists a line bundle

[3] An early review of Special Kähler Geometry was written by this author in 1996 in [21].

$\mathscr{L} \xrightarrow{\pi} \mathscr{M}$ such that its first Chern class equals the cohomology class of the Kähler two-form K:

$$c_1(\mathscr{L}) = [\,\mathrm{K}\,] \tag{4.2.3}$$

In local terms this means that there is a holomorphic section $\xi(z)$ such that we can write

$$\mathrm{K} = \frac{i}{2}\, g_{ij^\star}\, dz^i \wedge d\overline{z}^{j^\star} = \frac{i}{2}\, \overline{\partial}\, \partial \log \parallel \xi(z) \parallel^2 \tag{4.2.4}$$

Recalling the local expression of the Kähler metric in terms of the Kähler potential $g_{ij^\star} = \partial_i \partial_{j^\star} \mathscr{K}(z, \overline{z})$, it follows from Eq. (4.2.4) that if the manifold $\mathscr{M}$ is a Hodge manifold, then the exponential of the Kähler potential can be interpreted as the metric $h(z, \overline{z}) = \exp(\mathscr{K}(z, \overline{z}))$ on an appropriate line bundle $\mathscr{L}$.

4.2.2 Connection on the Line Bundle

On any complex line bundle $\mathscr{L}$ there is a canonical hermitian connection defined as:

$$\theta \equiv h^{-1}\, \partial\, h = \tfrac{1}{h}\, \partial_i h\, dz^i \;;\; \overline{\theta} \equiv h^{-1}\, \overline{\partial}\, h = \tfrac{1}{h}\, \partial_{i^\star} h\, d\overline{z}^{i^\star} \tag{4.2.5}$$

For the line-bundle advocated by the Hodge-Kähler structure we have

$$\left[\,\overline{\partial}\,\theta\,\right] = c_1(\mathscr{L}) = [\mathrm{K}] \tag{4.2.6}$$

and since the fibre metric h can be identified with the exponential of the Kähler potential we obtain:

$$\theta = \partial\,\mathscr{K} = \partial_i \mathscr{K}\, dz^i \;;\; \overline{\theta} = \overline{\partial}\,\mathscr{K} = \partial_{i^\star} \mathscr{K}\, d\overline{z}^{i^\star} \tag{4.2.7}$$

To define special Kähler geometry, in addition to the afore-mentioned line–bundle $\mathscr{L}$ we need a flat holomorphic vector bundle $\mathscr{SV} \longrightarrow \mathscr{M}$ whose sections play an important role in the construction of the supergravity Lagrangians. For reasons intrinsic to such constructions the rank of the vector bundle $\mathscr{SV}$ must be $2\,n_V$ where n_V is the total number of vector fields in the theory. If we have n-vector multiplets the total number of vectors is $n_V = n + 1$ since, in addition to the vectors of the vector multiplets, we always have the graviphoton sitting in the graviton multiplet. On the other hand the total number of scalars is $2n$. Suitably paired into n-complex fields z^i, these scalars span the n complex dimensions of the base manifold $\mathscr{M}$ to the rank $2n + 2$ bundle $\mathscr{SV} \longrightarrow \mathscr{M}$.

In the sequel we make extensive use of covariant derivatives with respect to the canonical connection of the line–bundle $\mathscr{L}$. Let us review its normalization. As it is well known there exists a correspondence between line–bundles and U(1)–bundles. If $\exp[f_{\alpha\beta}(z)]$ is the transition function between two local trivializations of the line–

bundle $\mathscr{L} \xrightarrow{\pi} \mathscr{M}$, the transition function in the corresponding principal U(1)–bundle $\mathscr{U} \longrightarrow \mathscr{M}$ is just $\exp[\mathrm{iIm} f_{\alpha\beta}(z)]$ and the Kähler potentials in two different charts are related by: $\mathscr{K}_\beta = \mathscr{K}_\alpha + f_{\alpha\beta} + \overline{f}_{\alpha\beta}$. At the level of connections this correspondence is formulated by setting: U(1)–connection $\equiv \mathscr{Q} = \mathrm{Im}\theta = -\frac{\mathrm{i}}{2}\left(\theta - \overline{\theta}\right)$. If we apply this formula to the case of the U(1)–bundle $\mathscr{U} \longrightarrow \mathscr{M}$ associated with the line–bundle $\mathscr{L}$ whose first Chern class equals the Kähler class, we get:

$$\mathscr{Q} = \frac{\mathrm{i}}{2}\left(\partial_i \mathscr{K} dz^i - \partial_{i^\star} \mathscr{K} d\overline{z}^{i^\star}\right) \tag{4.2.8}$$

Let now $\Phi(z, \overline{z})$ be a section of $\mathscr{U}^p$. By definition its covariant derivative is $\nabla \Phi = (d - ip\mathscr{Q})\Phi$ or, in components,

$$\nabla_i \Phi = (\partial_i + \tfrac{1}{2} p \partial_i \mathscr{K})\Phi \ ; \ \nabla_{i^*}\Phi = (\partial_{i^*} - \tfrac{1}{2} p \partial_{i^*} \mathscr{K})\Phi \tag{4.2.9}$$

A covariantly holomorphic section of $\mathscr{U}$ is defined by the equation: $\nabla_{i^*}\Phi = 0$. We can easily map each section $\Phi(z, \overline{z})$ of $\mathscr{U}^p$ into a section of the line–bundle $\mathscr{L}$ by setting:

$$\tilde{\Phi} = e^{-p\mathscr{K}/2}\Phi \, . \tag{4.2.10}$$

With this position we obtain:

$$\nabla_i \tilde{\Phi} = (\partial_i + p\partial_i \mathscr{K})\tilde{\Phi} \ ; \ \nabla_{i^*}\tilde{\Phi} = \partial_{i^*}\tilde{\Phi} \tag{4.2.11}$$

Under the map of Eq. (4.2.10) covariantly holomorphic sections of $\mathscr{U}$ flow into holomorphic sections of $\mathscr{L}$ and viceversa.

4.2.3 Special Kähler Manifolds

We are now ready to give the first of two equivalent definitions of special Kähler manifolds:

Definition 4.2.1 A Hodge Kähler manifold is **Special Kähler (of the local type)** if there exists a completely symmetric holomorphic 3-index section W_{ijk} of $(T^\star\mathscr{M})^3 \otimes \mathscr{L}^2$ (and its antiholomorphic conjugate $W_{i^*j^*k^*}$) such that the following identity is satisfied by the Riemann tensor of the Levi–Civita connection:

$$\begin{aligned}
\partial_{m^*} W_{ijk} &= 0 \quad \partial_m W_{i^*j^*k^*} = 0 \\
\nabla_{[m} W_{i]jk} &= 0 \quad \nabla_{[m} W_{i^*]j^*k^*} = 0 \\
\mathscr{R}_{i^*j\ell^*k} &= g_{\ell^*j} g_{ki^*} + g_{\ell^*k} g_{ji^*} - e^{2\mathscr{K}} W_{i^*\ell^*s^*} W_{tkj} g^{s^*t}
\end{aligned} \tag{4.2.12}$$

In the above equations ∇ denotes the covariant derivative with respect to both the Levi–Civita and the U(1) holomorphic connection of Eq. (4.2.8). In the case of W_{ijk}, the U(1) weight is $p = 2$.

Out of the W_{ijk} we can construct covariantly holomorphic sections of weight 2 and - 2 by setting:

$$C_{ijk} = W_{ijk}\, e^{\mathcal{K}} \quad ; \quad C_{i^\star j^\star k^\star} = W_{i^\star j^\star k^\star}\, e^{\mathcal{K}} \tag{4.2.13}$$

The flat bundle mentioned in the previous subsection apparently does not appear in this definition of special geometry. Yet it is there. It is indeed the essential ingredient in the second definition whose equivalence to the first we shall shortly provide.

Let $\mathscr{L} \xrightarrow{\pi} \mathscr{M}$ denote the complex line bundle whose first Chern class equals the cohomology class of the Kähler form K of an n-dimensional Hodge–Kähler manifold $\mathscr{M}$. Let $\mathscr{SV} \longrightarrow \mathscr{M}$ denote a holomorphic flat vector bundle of rank $2n+2$ with structural group $\mathrm{Sp}(2n+2, \mathbb{R})$. Consider tensor bundles of the type $\mathscr{H} = \mathscr{SV} \otimes \mathscr{L}$. A typical holomorphic section of such a bundle will be denoted by Ω and will have the following structure:

$$\Omega = \begin{pmatrix} X^\Lambda \\ F_\Sigma \end{pmatrix} \quad \Lambda, \Sigma = 0, 1, \ldots, n$$

By definition the transition functions between two local trivializations $U_i \subset \mathscr{M}$ and $U_j \subset \mathscr{M}$ of the bundle $\mathscr{H}$ have the following form:

$$\begin{pmatrix} X \\ F \end{pmatrix}_i = e^{f_{ij}} M_{ij} \begin{pmatrix} X \\ F \end{pmatrix}_j$$

where f_{ij} are holomorphic maps $U_i \cap U_j \to \mathbb{C}$ while M_{ij} is a constant $\mathrm{Sp}(2n+2, \mathbb{R})$ matrix. For a consistent definition of the bundle the transition functions are obviously subject to the cocycle condition on a triple overlap: $e^{f_{ij}+f_{jk}+f_{ki}} = 1$ and $M_{ij} M_{jk} M_{ki} = 1$.

Let $\mathrm{i}\langle\ |\ \rangle$ be the compatible hermitian metric on $\mathscr{H}$

$$\mathrm{i}\langle \Omega \,|\, \overline{\Omega} \rangle \equiv -\mathrm{i}\Omega^T \begin{pmatrix} 0 & \mathbb{1} \\ -\mathbb{1} & 0 \end{pmatrix} \overline{\Omega}$$

Definition 4.2.2 We say that a Hodge–Kähler manifold $\mathscr{M}$ is **special Kähler** if there exists a bundle $\mathscr{H}$ of the type described above such that for some section $\Omega \in \Gamma(\mathscr{H}, \mathscr{M})$ the Kähler two form is given by:

$$\mathrm{K} = \frac{\mathrm{i}}{2} \partial \overline{\partial} \log \left(\mathrm{i}\langle \Omega \,|\, \overline{\Omega} \rangle \right) = \frac{i}{2}\, g_{ij^\star}\, dz^i \wedge d\overline{z}^{j^\star} \tag{4.2.14}$$

From the point of view of local properties, Eq. (4.2.14) implies that we have an expression for the Kähler potential in terms of the holomorphic section Ω:

$$\mathscr{K} = -\log\left(\mathrm{i}\langle\Omega\,|\,\overline{\Omega}\rangle\right) = -\log\left[\mathrm{i}\left(\overline{X}^{\Lambda} F_{\Lambda} - \overline{F}_{\Sigma} X^{\Sigma}\right)\right] \tag{4.2.15}$$

The relation between the two definitions of special manifolds is obtained by introducing a non–holomorphic section of the bundle $\mathscr{H}$ according to:

$$V = \begin{pmatrix} L^{\Lambda} \\ M_{\Sigma} \end{pmatrix} \equiv e^{\mathscr{K}/2}\Omega = e^{\mathscr{K}/2}\begin{pmatrix} X^{\Lambda} \\ F_{\Sigma} \end{pmatrix} \tag{4.2.16}$$

so that Eq. (4.2.15) becomes:

$$1 = \mathrm{i}\langle V\,|\,\overline{V}\rangle = \mathrm{i}\left(\overline{L}^{\Lambda} M_{\Lambda} - \overline{M}_{\Sigma} L^{\Sigma}\right) \tag{4.2.17}$$

Since V is related to a holomorphic section by Eq. (4.2.16) it immediately follows that:

$$\nabla_{i^\star} V = \left(\partial_{i^\star} - \frac{1}{2}\partial_{i^\star}\mathscr{K}\right) V = 0 \tag{4.2.18}$$

On the other hand, from Eq. (4.2.16), defining:

$$\begin{aligned} U_i = \nabla_i V &= \left(\partial_i + \frac{1}{2}\partial_i\mathscr{K}\right) V \equiv \begin{pmatrix} f_i^{\Lambda} \\ h_{\Sigma|i} \end{pmatrix} \\ \overline{U}_{i^\star} = \nabla_{i^\star}\overline{V} &= \left(\partial_{i^\star} + \frac{1}{2}\partial_{i^\star}\mathscr{K}\right) \overline{V} \equiv \begin{pmatrix} \overline{f}^{\Lambda}_{i^\star} \\ \overline{h}_{\Sigma|i^\star} \end{pmatrix} \end{aligned}$$

it follows that:

$$\nabla_i U_j = \mathrm{i} C_{ijk}\, g^{k\ell^\star}\, \overline{U}_{\ell^\star} \tag{4.2.19}$$

where ∇_i denotes the covariant derivative containing both the Levi–Civita connection on the bundle $\mathscr{T}\mathscr{M}$ and the canonical connection θ on the line bundle $\mathscr{L}$. In Eq. (4.2.19) the symbol C_{ijk} denotes a covariantly holomorphic ($\nabla_{\ell^\star} C_{ijk} = 0$) section of the bundle $\mathscr{T}\mathscr{M}^3 \otimes \mathscr{L}^2$ that is totally symmetric in its indices. This tensor can be identified with the tensor of Eq. (4.2.13) appearing in Eq. (4.2.12). Alternatively, the set of differential equations:

$$\begin{aligned} &\nabla_i V = U_i \\ &\nabla_i U_j = \mathrm{i} C_{ijk} g^{k\ell^\star} U_{\ell^\star} \\ &\nabla_{i^\star} U_j = g_{i^\star j} V \\ &\nabla_{i^\star} V = 0 \end{aligned} \tag{4.2.20}$$

with V satisfying equation (4.2.17) give yet another definition of special geometry. In particular it is easy to find Eq. (4.2.12) as integrability conditions of (4.2.20).[4]

4.2.4 The Vector Kinetic Matrix $\mathcal{N}_{\Lambda\Sigma}$ in Special Geometry

In the construction of supergravity actions another essential item is the complex symmetric matrix $\mathcal{N}_{\Lambda\Sigma}$ whose real and imaginary parts are necessary in order to write the kinetic terms of the vector fields. From the physicist's viewpoint the matrix $\mathcal{N}_{\Lambda\Sigma}$ is an essential item since the Lagrangian cannot be written without it. From the mathematical viewpoint it is very much significant that the same $\mathcal{N}_{\Lambda\Sigma}$ constitutes an integral part of the Special Geometry set up. We provide its general definition in the following lines. Explicitly $\mathcal{N}_{\Lambda\Sigma}$ which, in relation to its interpretation in the case of Calabi–Yau threefolds, is named the *period matrix*, is defined by means of the following relations:

$$\overline{M}_\Lambda = \mathcal{N}_{\Lambda\Sigma}\overline{L}^\Sigma \quad ; \quad h_{\Sigma|i} = \mathcal{N}_{\Lambda\Sigma} f_i^\Sigma \tag{4.2.21}$$

which can be solved introducing the two $(n+1) \times (n+1)$ vectors

$$f_I^\Lambda = \begin{pmatrix} f_i^\Lambda \\ \overline{L}^\Lambda \end{pmatrix} \quad ; \quad h_{\Lambda|I} = \begin{pmatrix} h_{\Lambda|i} \\ \overline{M}_\Lambda \end{pmatrix}$$

and setting:

$$\mathcal{N}_{\Lambda\Sigma} = h_{\Lambda|I} \circ \left(f^{-1}\right)^I_\Sigma \tag{4.2.22}$$

Let us now consider the case where the Special Kähler manifold $\mathcal{SK}_n$ of complex dimension n has some isometry group $\mathrm{U}_{\mathcal{SK}}$. Compatibility with the Special Geometry structure requires the existence of a $2n+2$-dimensional symplectic representation of such a group that we name the **W** representation. In other words that there necessarily exists a symplectic embedding of the isometry group $\mathcal{SK}_n$

$$\mathrm{U}_{\mathcal{SK}} \mapsto \mathrm{Sp}(2\mathrm{n}+2, \mathbb{R}) \tag{4.2.23}$$

such that for each element $\xi \in \mathrm{U}_{\mathcal{SK}}$ we have its representation by means of a suitable real symplectic matrix:

$$\xi \mapsto \Lambda_\xi \equiv \begin{pmatrix} A_\xi & B_\xi \\ C_\xi & D_\xi \end{pmatrix} \tag{4.2.24}$$

[4]We omit the detailed proof that from Eq. (4.2.20) one obtains Eq. (4.2.12). The essential link between the two formulations resides in the second of Eq. (4.2.20) which identifies the tensor C_{ijk} with the expression of the derivative of U_i in terms of the same objects U_k.

satisfying the defining relation (in terms of the symplectic antisymmetric metric $\mathbb{C}$):

$$\underbrace{\Lambda_\xi^T \begin{pmatrix} \mathbf{0}_{n\times n} & \mathbf{1}_{n\times n} \\ -\mathbf{1}_{n\times n} & \mathbf{0}_{n\times n} \end{pmatrix} \Lambda_\xi}_{\equiv\mathbb{C}} = \underbrace{\begin{pmatrix} \mathbf{0}_{n\times n} & \mathbf{1}_{n\times n} \\ -\mathbf{1}_{n\times n} & \mathbf{0}_{n\times n} \end{pmatrix}}_{\mathbb{C}} \tag{4.2.25}$$

which implies the following relations on the $n \times n$ blocks:

$$\begin{aligned} A_\xi^T\, C_\xi - C_\xi^T\, A_\xi &= 0 \\ A_\xi^T\, D_\xi - C_\xi^T\, B_\xi &= \mathbf{1} \\ B_\xi^T\, C_\xi - D_\xi^T\, A_\xi &= -\mathbf{1} \\ B_\xi^T\, D_\xi - D_\xi^T\, B_\xi &= 0 \end{aligned} \tag{4.2.26}$$

Under an element of the isometry group the symplectic section Ω of Special Geometry transforms as follows:

$$\Omega\,(\xi \cdot z) = \Lambda_\xi\, \Omega\,(z) \tag{4.2.27}$$

As a consequence of its definition, under the same isometry the matrix $\mathcal{N}$ transforms by means of a generalized linear fractional transformation:

$$\mathcal{N}\,(\xi \cdot z, \xi \cdot \overline{z}) = \left(C_\xi + D_\xi\, \mathcal{N}(z,\overline{z})\right)\, \left(A_\xi + B_\xi\, \mathcal{N}(z,\overline{z})\right)^{-1} \tag{4.2.28}$$

4.3 The Quaternionic Kähler Geometry in the Image of the *c*-Map

The main object of study in the present section are those Quaternionic Kähler manifolds that are in the image of the c-map.[5] This latter

$$\text{c-map} \; : \; \mathcal{SK}_n \Longrightarrow \mathcal{QM}_{4n+4} \tag{4.3.1}$$

is a universal construction that starting from an arbitrary Special Kähler manifold $\mathcal{SK}_n$ of complex dimension n, irrespectively whether it is homogeneous or not, leads to a unique Quaternionic Kähler manifold $\mathcal{QM}_{4n+4}$ of real dimension $4n+4$ which contains $\mathcal{SK}_n$ as a submanifold. The precise modern definition of the c-map, originally introduced in [16, 17], is provided below.

[5]Not all non-compact, homogeneous Quaternionic Kähler manifolds which are relevant to supergravity (which are *normal*, i.e. exhibiting a solvable group of isometries having a free and transitive action on it) are in the image of the c-map, the only exception being the quaternionic projective spaces [14, 15].

Definition 4.3.1 Let $\mathcal{SK}_n$ be a special Kähler manifold whose complex coordinates we denote by z^i and whose Kähler metric we denote by $g_{ij^\star}$. Let moreover $\mathcal{N}_{\Lambda\Sigma}(z,\bar{z})$ be the symmetric period matrix defined by Eq. (4.2.22), introduce the following set of $4n+4$ coordinates:

$$\{q^u\} \equiv \underbrace{\{U,a\}}_{\text{2 real}} \bigcup \underbrace{\underbrace{\{z^i\}}_{\text{n complex}}}_{\text{2n real}} \bigcup \underbrace{\mathbf{Z} = \{Z^\Lambda\,,\,Z_\Sigma\}}_{\text{(2n+2) real}} \tag{4.3.2}$$

Let us further introduce the following $(2n+2)\times(2n+2)$ matrix $\mathcal{M}_4^{-1}$:

$$\mathcal{M}_4^{-1} = \left(\begin{array}{c|c} \mathrm{Im}\mathcal{N} + \mathrm{Re}\mathcal{N}\,\mathrm{Im}\mathcal{N}^{-1}\,\mathrm{Re}\mathcal{N} & -\mathrm{Re}\mathcal{N}\,\mathrm{Im}\,\mathcal{N}^{-1} \\ \hline -\,\mathrm{Im}\mathcal{N}^{-1}\,\mathrm{Re}\mathcal{N} & \mathrm{Im}\mathcal{N}^{-1} \end{array}\right) \tag{4.3.3}$$

which depends only on the coordinate of the Special Kähler manifold. The c-map image of $\mathcal{SK}_n$ is the unique Quaternionic Kähler manifold $\mathcal{QM}_{4n+4}$ whose coordinates are the q^u defined in (4.3.2) and whose metric is given by the following universal formula

$$ds^2_{\mathcal{QM}} = \frac{1}{4}\left(dU^2 + 4g_{ij^\star}\,dz^j\,d\bar{z}^{j^\star} + e^{-2U}\,(da + \mathbf{Z}^T\,\mathbb{C}d\mathbf{Z})^2 \;-\; 2\,e^{-U}\,d\mathbf{Z}^T\,\mathcal{M}_4^{-1}\,d\mathbf{Z}\right) \tag{4.3.4}$$

The metric (4.3.4) has the following positive definite signature

$$\mathrm{sign}\left[ds^2_{\mathcal{QM}}\right] = \left(\underbrace{+,\cdots,+}_{4+4n}\right) \tag{4.3.5}$$

since the matrix $\mathcal{M}_4^{-1}$ is negative definite.

In the case the Special Kähler pre-image is a symmetric space $\mathrm{U}_{\mathcal{SK}}/\mathrm{H}_{\mathcal{SK}}$, the manifold $\mathcal{QM}$ turns out to be symmetric spaces, $\mathrm{U}_Q/\mathrm{H}_Q$. We will come back to the issue of symmetric homogeneous Quaternionic Kähler manifolds in Sect. 4.3.4

4.3.1 The HyperKähler Two-Forms and the $\mathfrak{su}(2)$*-Connection*

The reason why we state that $\mathcal{QM}_{4n+4}$ is Quaternionic Kähler is that, by utilizing only the identities of Special Kähler Geometry we can construct the three complex structures $J_u^{x|v}$ satisfying the quaternionic algebra (3.6.6) the corresponding Hyper-

Kähler two-forms K^x and the $\mathfrak{su}(2)$ connection ω^x with respect to which they are covariantly constant.

The construction is extremely beautiful, it was found in [20] and it is the following one.

Consider the Kähler connection $\mathscr{Q}$ defined by Eq. (4.2.8) and furthermore introduce the following differential form:

$$\Phi = da + \mathbf{Z}^T \mathbb{C} \, d\mathbf{Z} \tag{4.3.6}$$

Next define the two dimensional representation of both the $\mathfrak{su}(2)$ connection and of the HyperKähler 2-forms as it follows:

$$\omega = \frac{\mathrm{i}}{\sqrt{2}} \sum_{x=1}^{3} \omega^x \gamma_x \tag{4.3.7}$$

$$\mathbf{K} = \frac{\mathrm{i}}{\sqrt{2}} \sum_{x=1}^{3} K^x \sigma_x \tag{4.3.8}$$

where γ_x denotes a basis of 2×2 Euclidean γ-matrices for which we utilize the following basis which is convenient in the explicit calculations we perform in later chapters[6]:

$$\gamma_1 = \begin{pmatrix} \frac{1}{\sqrt{2}} & 0 \\ 0 & -\frac{1}{\sqrt{2}} \end{pmatrix}$$

$$\gamma_2 = \begin{pmatrix} 0 & -\frac{i}{\sqrt{2}} \\ \frac{i}{\sqrt{2}} & 0 \end{pmatrix}$$

$$\gamma_3 = \begin{pmatrix} 0 & \frac{1}{\sqrt{2}} \\ \frac{1}{\sqrt{2}} & 0 \end{pmatrix} \tag{4.3.9}$$

These γ-matrices satisfy the following Clifford algebra:

$$\left\{ \gamma_x \,,\, \gamma_y \right\} = \delta^{xy} \, \mathbf{1}_{2\times 2} \tag{4.3.10}$$

and $\frac{\mathrm{i}}{2} \gamma_x$ provide a basis of generators of the $\mathfrak{su}(2)$ algebra.

Having fixed these conventions the expression of the quaternionic $\mathfrak{su}(2)$-connection in terms of Special Geometry structures is encoded in the following expression for the 2×2-matrix valued 1-form ω. Explicitly we have:

[6]The chosen γ-matrices are a permutation of the standard pauli matrices divided by $\sqrt{2}$ and multiplied by $\frac{\mathrm{i}}{2}$ can be used as a basis of anti-hermitian generators for the $\mathfrak{su}(2)$ algebra in the fundamental defining representation.

$$\omega = \begin{pmatrix} -\frac{\mathrm{i}}{2}\,\mathcal{Q} - \frac{\mathrm{i}}{4}\,e^{-U}\,\Phi & e^{-\frac{U}{2}}\,V^T\,\mathbb{C}\,\mathrm{d}\mathbf{Z} \\ -\,e^{-\frac{U}{2}}\,\overline{V}^T\,\mathbb{C}\,\mathrm{d}\mathbf{Z} & \frac{\mathrm{i}}{2}\,\mathcal{Q} + \frac{\mathrm{i}}{4}\,e^{-U}\,\Phi \end{pmatrix} \tag{4.3.11}$$

where V and $\overline{V}$ denote the covariantly holomorphic sections of Special geometry defined in Eq. (4.2.16). The curvature of this connection is obtained from a straightforward calculation:

$$\begin{aligned} \mathbf{K} &\equiv d\omega + \omega \wedge \omega \\ &= \begin{pmatrix} \mathfrak{u} & \mathfrak{v} \\ -\,\overline{\mathfrak{v}} & -\,\mathfrak{u} \end{pmatrix} \end{aligned} \tag{4.3.12}$$

the independent 2-form matrix elements being given by the following explicit formulae:

$$\begin{aligned} \mathfrak{u} &= -\mathrm{i}\frac{1}{2}\,K - \frac{1}{8} dS \wedge d\overline{S} - e^{-U}\,V^T\,\mathbb{C}\,\mathrm{d}\mathbf{Z} \wedge \overline{V}^T\,\mathbb{C}\,\mathrm{d}\mathbf{Z} - \frac{1}{4}\,e^{-U}\,\mathrm{d}\mathbf{Z}^T \wedge \mathbb{C}\,\mathrm{d}\mathbf{Z} \\ \mathfrak{v} &= e^{-\frac{U}{2}} \left(DV^T \wedge \mathbb{C}\,\mathrm{d}\mathbf{Z} - \frac{1}{2} dS \wedge V^T\,\mathbb{C}\,\mathrm{d}\mathbf{Z} \right) \\ \overline{\mathfrak{v}} &= e^{-\frac{U}{2}} \left(D\overline{V}^T \wedge \mathbb{C}\,\mathrm{d}\mathbf{Z} - \frac{1}{2} d\overline{S} \wedge \overline{V}^T\,\mathbb{C}\,\mathrm{d}\mathbf{Z} \right) \end{aligned} \tag{4.3.13}$$

where

$$K = \frac{\mathrm{i}}{2}\,g_{ij^\star}\,dz^i \wedge d\overline{z}^{j^\star} \tag{4.3.14}$$

is the Kähler 2-form of the Special Kähler submanifold and where we have used the following short hand notations:

$$dS = dU + \mathrm{i}\,e^{-U}\,\left(da + \mathbf{Z}^T\,\mathbb{C}\,\mathrm{d}\mathbf{Z}\right) \tag{4.3.15}$$

$$d\overline{S} = dU - \mathrm{i}\,e^{-U}\,\left(da + \mathbf{Z}^T\,\mathbb{C}\,\mathrm{d}\mathbf{Z}\right) \tag{4.3.16}$$

$$DV = dz^i\,\nabla_i V \tag{4.3.17}$$

$$D\overline{V} = d\overline{z}^{i^\star}\,\nabla_{i^\star} V \tag{4.3.18}$$

The three HyperKähler forms[7] K^x are easily extracted from Eqs. (4.3.12)–(4.3.13) by collecting the coefficients of the γ-matrix expansion and we need not to write their form which is immediately deduced. The relevant thing is that the components of K^x with an index raised through multiplication with the inverse of the quaternionic metric h^{uv} exactly satisfy the algebra of quaternionic complex structures (3.6.6). Explicitly we have:

[7]See Sect. 3.6 for notations.

$$\begin{aligned} K^x &= -\,\mathrm{i}\,4\sqrt{2}\,\mathrm{Tr}\left(\gamma^x\,\mathbf{K}\right) \equiv K^x_{uv}\,dq^u \wedge dq^v \\ J^{x|s}_u &= K^x_{uv}\,h^{vs} \\ J^{x|s}_u\,J^{y|v}_s &= -\delta^{xy}\,\delta^v_u + \varepsilon^{xyz}\,J^{z|v}_u \end{aligned} \tag{4.3.19}$$

The above formulae are not only the general proof that the Riemanniann manifold $\mathcal{QM}$ defined by the metric (4.3.4) is indeed a Quaternionic Kähler manifold, but, what is most relevant, they also provide an algorithm to write in terms of Special Geometry structures the tri-holomorphic moment map of the principal isometries possessed by $\mathcal{QM}$.

4.3.2 The Holomorphic Moment Map in Special Kähler Manifolds

In any Kähler manifold

$$\mathcal{P}_{\mathbf{I}}{}^x = -\frac{i}{2}\left(k^i_{\mathbf{I}}\partial_i\mathcal{K} - k^{\bar{\imath}}_{\mathbf{I}}\partial_{\bar{\imath}}\mathcal{K}\right) + \mathrm{Im}(f_{\mathbf{I}})\,, \tag{4.3.20}$$

where $f_{\mathbf{I}} = f_{\mathbf{I}}(z)$ is a holomorphic transformation on the line-bundle, defining a compensating Kähler transformation:

$$k^i_{\mathbf{I}}\partial_i\mathcal{K} + k^{\bar{\imath}}_{\mathbf{I}}\partial_{\bar{\imath}}\mathcal{K} = -f_{\mathbf{I}}(z) - \overline{f}_{\mathbf{I}}(\overline{z})\,. \tag{4.3.21}$$

We also have:

$$\mathfrak{T}_{\mathbf{I}}\cdot\Omega = \mathfrak{T}_{\mathbf{I}}\cdot\Omega + f_{\mathbf{I}}\,\Omega\,, \tag{4.3.22}$$

$$\mathfrak{T}_{\mathbf{I}}\cdot V + i\,\mathrm{Im}(f_{\mathbf{I}})\,V = k^i_{\mathbf{I}}\partial_i V + k^{\bar{\imath}}_{\mathbf{I}}\partial_{\bar{\imath}} V\,, \tag{4.3.23}$$

where $\mathfrak{T}_{\mathbf{I}}\cdot\Omega$ denotes the symplectic action of the isometry on the section V. If $\mathfrak{T}_{\mathbf{I}}$ is represented by the symplectic matrix $(\mathfrak{T}_{\mathbf{I}})_\alpha{}^\beta = -(\mathfrak{T}_{\mathbf{I}})^\beta{}_\alpha, \alpha,\ \beta = 1,\ldots,\ 2n+2$:

$$\mathfrak{T}^T_{\mathbf{I}}\,\mathbb{C} + \mathbb{C}\,\mathfrak{T}_{\mathbf{I}} = 0 \tag{4.3.24}$$

we have $(\mathfrak{T}_{\mathbf{I}}\cdot V)^\alpha = -\mathfrak{T}_{\mathbf{I}\beta}{}^\alpha\,V^\beta = \mathfrak{T}^\alpha_{\mathbf{I}\,\beta}\,V^\beta$. From (4.3.23) and (3.7.22) we derive the following useful symplectic-invariant expression for the moment maps:

$$\mathcal{P}_{\mathbf{I}}{}^x = -\overline{V}^\alpha\,\mathfrak{T}_{\mathbf{I}\alpha}{}^\beta\mathbb{C}_{\beta\gamma}\,V^\gamma\,. \tag{4.3.25}$$

Equations (3.7.22), (3.7.23), (4.3.23) generalize the corresponding formulae given in Sects. 7.1 and 7.2 of [22], where the condition $f_{\mathbf{I}} = 0$ was imposed, to gaugings of non-compact isometries which are associated with non-trivial compensating Kähler transformations and/or to gauged (non-compact) isometries whose symplectic action is not diagonal.

4.3.3 *Isometries of $\mathscr{QM}$ in the Image of the c-Map and Their Tri-Holomorphic Moment Maps*

Let us now consider the isometries of the metric (4.3.4). There are three type of isometries:

(a) The isometries of the $(2n + 3)$–dimensional Heisenberg algebra $\mathbb{H}$eis which is always present and is universal for any $(4n + 4)$–dimensional Quaternionic Kähler manifold in the image of the c-map. We describe it below.
(b) All the isometries of the pre-image Special Kähler manifold $\mathscr{SK}_n$ that are promoted to isometries of the image manifold in a way described below.
(c) The additional $2n + 4$ isometries that occur only when $\mathscr{SK}_n$ is a symmetric space and such, as a consequence, is also the c-map image $\mathscr{QM}_{4n+4}$. We will discuss these isometries in Sect. 4.3.4.

For the first two types of isometries (a) and (b) we are able to write general expressions for the tri-holomorphic moment maps that utilize only the structures of Special Geometry. In the case that the additional isometries (c) do exist we have another universal formula which can be used for all generators of the isometry algebra $\mathbb{U}_{\mathscr{Q}}$ and which relies on the identification of the generators of the $\mathfrak{su}(2) \subset \mathbb{H}$ subalgebra with the three complex structures. We will illustrate the details of such an identification while discussing the example of the S^3-model.

First of all let us fix the notation writing the general form of a Killing vector. This a tangent vector:

$$\begin{aligned}\mathbf{k} &= k^u(q)\,\partial_u \\ &= k^\diamond\,\frac{\partial}{\partial U} + k^i\,\frac{\partial}{\partial z^i} + k^{i^\star}\,\frac{\partial}{\partial \overline{z}^{i^\star}} + k^\bullet\,\frac{\partial}{\partial a} + k^\alpha\,\frac{\partial}{\partial \mathbf{Z}^\alpha} \\ &\equiv k^\diamond\,\partial_\diamond + k^i\,\partial_i + k^{i^\star}\,\partial_{i^\star} + k^\bullet\,\partial_\bullet + k^\alpha\,\partial_\alpha \end{aligned} \tag{4.3.26}$$

with respect to which the Lie derivative of the metric element (4.3.4) vanishes:

$$\ell_{\mathbf{k}}\,ds^2_{\mathscr{QM}} = 0 \tag{4.3.27}$$

4.3.3.1 Tri-Holomorphic Moment Maps for the Heisenberg Algebra Translations

First let us consider the isometries associated with the Heisenberg algebra. The transformation:

$$Z^\alpha \mapsto Z^\alpha + \Lambda^\alpha \quad ; \quad a \mapsto a - \Lambda^T \mathbb{C} \mathbf{Z} \tag{4.3.28}$$

where Λ^α is an arbitrary set of $2n+2$ real infinitesimal parameters is an infinitesimal isometry for the metric $ds^2_{\mathcal{QM}}$ in (4.3.4). It corresponds to the following Killing vector:

$$\begin{aligned} \vec{\mathbf{k}}_{[\Lambda]} &= \Lambda^\alpha \, \vec{\mathbf{k}}_\alpha \\ &= \Lambda^\alpha \, \partial_\alpha - \Lambda^T \, \mathbb{C} \mathbf{Z} \, \partial_\bullet \end{aligned} \tag{4.3.29}$$

whose components are immediately deduced by comparison of Eq. (4.3.29) with Eq. (4.3.26).

We are interested in determining the expression of the tri-holomorphic moment map $\mathfrak{P}_{[\Lambda]}$ which satisfies the defining equation:

$$\mathbf{i}_{[\Lambda]} \mathbf{K} \equiv \begin{pmatrix} \mathbf{i}_{[\Lambda]} \mathfrak{u} & \mathbf{i}_{[\Lambda]} \mathfrak{v} \\ -\mathbf{i}_{[\Lambda]} \bar{\mathfrak{v}} & -\mathbf{i}_{[\Lambda]} \mathfrak{u} \end{pmatrix} = \mathrm{d}\mathfrak{P}_{[\Lambda]} + \left[\omega \, , \, \mathfrak{P}_{[\Lambda]}\right] \tag{4.3.30}$$

The general solution to this problem is

$$\mathfrak{P}_{[\Lambda]} = \begin{pmatrix} -\frac{\mathrm{i}}{4} e^{-U} \, \Lambda^T \, \mathbb{C} \mathbf{Z} & \frac{1}{2} e^{-\frac{U}{2}} \, \Lambda^T \, C \, V \\ -\frac{1}{2} e^{-\frac{U}{2}} \, \Lambda^T \, C \, \overline{V} & \frac{\mathrm{i}}{4} e^{-U} \, \Lambda^T \, \mathbb{C} \mathbf{Z} \end{pmatrix} \tag{4.3.31}$$

4.3.3.2 Tri-Holomorphic Moment Map for the Heisenberg Algebra Central Charge

Consider next the isometry associated with the Heisenberg algebra central charge. The transformation:

$$a \mapsto a + \varepsilon \tag{4.3.32}$$

where ε is an arbitrary real small parameter is an infinitesimal isometry for the metric $ds^2_{\mathcal{QM}}$ in (4.3.4). It corresponds to the following Killing vector:

$$\varepsilon \, \vec{\mathbf{k}}_{[\bullet]} = \varepsilon \, \partial_\bullet \tag{4.3.33}$$

whose components are immediately deduced by comparison of Eq. (4.3.33) with Eq. (4.3.26).

We are interested in determining the expression of the tri-holomorphic moment map $\mathfrak{P}_{[\bullet]}$ which satisfies the defining equation analogous to Eq. (4.3.30):

$$\mathbf{i}_{[\bullet]} \mathbf{K} = \mathrm{d}\mathfrak{P}_{[\bullet]} + \left[\omega \, , \, \mathfrak{P}_{[\bullet]}\right] \tag{4.3.34}$$

The solution of this problem is even simpler than in the previous case. Explicitly we obtain:

$$\mathfrak{P}_{[\bullet]} = \begin{pmatrix} -\frac{\mathrm{i}}{8}\, e^{-U} & 0 \\ 0 & \frac{\mathrm{i}}{8}\, e^{-U} \end{pmatrix} \tag{4.3.35}$$

The explicit expression of the moment maps and Killing vectors associated with the Heisenberg isometries was used in the gauging of abelian subalgebras of the Heisenberg algebra, which is relevant to the description of compactifications of Type II superstring on a generalized Calabi–Yau manifold.

4.3.3.3 Tri-Holomorphic Moment Map for the Extension of $\mathscr{SK}_n$ Holomorphic Isometries

Next we consider the question how to write the moment map associated with those isometries that where already present in the original Special Kähler manifold $\mathscr{SK}_n$ which we c-mapped to a Quaternionic Kähler manifold.

Suppose that $\mathscr{SK}_n$ has a certain number of holomorphic Killing vectors $k^i_{\mathbf{I}}(z)$ satisfying equations (3.7.6), (3.7.7), (8.4.85) necessarily closing some Lie algebra $\mathfrak{g}_{\mathscr{SK}}$ among themselves.[8] Their holomorphic momentum-map is provided by Eq. (3.7.22). Necessarily every isometry of a special Kähler manifold has a linear symplectic $(2n+2)$-dimensional realization on the holomorphic section $\Omega(z)$ up to an overall holomorphic factor. This means that for each holomorphic Killing vector we have (see Eq. (4.3.22)):

$$k^i_{\mathbf{I}}(z)\,\partial_i\,\Omega(z) \;=\; \exp\left[f_{\mathbf{I}}(z)\right]\,\mathfrak{T}_{\mathbf{I}}\,\Omega(z)\,. \tag{4.3.36}$$

where $f_{\mathbf{I}}(z)$ the holomorphic Kähler compensator. Then it can be easily checked that the transformation:

$$z^i \;\mapsto\; z^i \,+\, k^i_{\mathbf{I}}(z) \quad ; \quad \mathbf{Z} \mapsto \mathbf{Z} + \mathfrak{T}_{\mathbf{I}}\,\mathbf{Z} \tag{4.3.37}$$

is an infinitesimal isometry of the metric (4.3.4) corresponding to the Killing vector:

$$\mathbf{k}_{\mathbf{I}} \;=\; k^i_{\mathbf{I}}(z)\,\partial_i \,+\, k^{i^\star}_{\mathbf{I}}(\bar{z})\,\partial_{i^\star} \,+\, (\mathfrak{T}_{\mathbf{I}})^{\alpha}{}_{\beta}\,\mathbf{Z}^{\beta}\,\partial_{\alpha} \tag{4.3.38}$$

Also in this case we are interested in determining the expression of the tri-holomorphic moment map $\mathfrak{P}_{[\mathbf{I}]}$ satisfying the defining equation:

$$\mathbf{i}_{\mathbf{k}_{\mathbf{I}}}\,\mathbf{K} = \mathrm{d}\mathfrak{P}_{[\mathbf{I}]} \,+\, \left[\omega\,,\,\mathfrak{P}_{[\mathbf{I}]}\right] \tag{4.3.39}$$

[8] *Clarification for mathematicians*: in the jargon ubiquitously utilized in the physical literature one says that a set of operators closes a Lie algebra when any of the commutators thereof belongs to the linear span of the same operators.

The solution is given by the expression below:

$$\mathfrak{P}_{[\mathbf{I}]} = \begin{pmatrix} \frac{\mathrm{i}}{4}\left(\mathscr{P}_{\mathbf{I}} + \frac{1}{2} e^{-U}\, \mathbf{Z}^T\, \mathbb{C}\, \mathfrak{T}_{\mathbf{I}}\, \mathbf{Z}\right) & -\frac{1}{2} e^{-U/2}\, V^T\, \mathbb{C}\, \mathfrak{T}_{\mathbf{I}}\, \mathbf{Z} \\ \frac{1}{2} e^{-U/2}\, \overline{V}^T\, \mathbb{C}\, \mathfrak{T}_{\mathbf{I}}\, \mathbf{Z} & -\frac{\mathrm{i}}{4}\left(\mathscr{P}_{\mathbf{I}} + \frac{1}{2} e^{-U}\, \mathbf{Z}^T\, \mathbb{C}\, \mathfrak{T}_{\mathbf{I}}\, \mathbf{Z}\right) \end{pmatrix} \tag{4.3.40}$$

where $\mathscr{P}_{\mathbf{I}}$ is the moment map of the same Killing vector in pure Special Geometry.

4.3.4 Homogeneous Symmetric Special Quaternionic Kähler Manifolds

When the Special Kähler manifold $\mathscr{SK}_n$ is a symmetric coset space, it turns out that the metric (4.3.4) is actually the symmetric metric on an enlarged symmetric coset manifold

$$\mathscr{QM}_{4n+4} = \frac{\mathrm{U}_Q}{\mathrm{H}_Q} \supset \frac{\mathrm{U}_{\mathscr{SK}}}{\mathrm{H}_{\mathscr{SK}}} \tag{4.3.41}$$

Naming $\Lambda[\mathfrak{g}]$ the **W**-representation of any finite element of the $\mathfrak{g} \in \mathrm{U}_{\mathscr{SK}}$ group, we have that the matrix $\mathscr{M}_4(z, \overline{z})$ transforms as follows:

$$\mathscr{M}_4\,(\mathfrak{g}\cdot z, \mathfrak{g}\cdot\overline{z}) = \Lambda[\mathfrak{g}]\,\mathscr{M}_4\,(z, \overline{z})]\,\Lambda^T[\mathfrak{g}] \tag{4.3.42}$$

where $\mathfrak{g}\cdot z$ denotes the non linear action of $\mathrm{U}_{\mathscr{SK}}$ on the scalar fields. Since the space $\frac{\mathrm{U}_{\mathscr{SK}}}{\mathrm{H}_{\mathscr{SK}}}$ is homogeneous, choosing any reference point z_0 all the others can be reached by a suitable group element $\mathfrak{g}_z$ such that $\mathfrak{g}_z \cdot z_0 = z$ and we can write:

$$\mathscr{M}_4^{-1}(z, \overline{z}) = \Lambda^T[\mathfrak{g}_z^{-1}]\,\mathscr{M}_4^{-1}(z_0, \overline{z}_0)]\,\Lambda[\mathfrak{g}_z^{-1}] \tag{4.3.43}$$

This allows to introduce a set of $4n+4$ vielbein defined in the following way:

$$E^I_{\mathscr{QM}} = \frac{1}{2}\left\{dU\,,\, \underbrace{e^i(z)}_{2n}\,,\, e^{-U}\,\left(da + \mathbf{Z}^T\mathbb{C}d\mathbf{Z}\right)\,,\, \underbrace{e^{-\frac{U}{2}}\,\Lambda[\mathfrak{g}_z^{-1}]\,\mathrm{d}\mathbf{Z}}_{2n+2}\right\} \tag{4.3.44}$$

and rewrite the metric (4.3.4) as it follows:

$$ds^2_{\mathscr{QM}} = E^I_{\mathscr{QM}}\,\mathfrak{q}_{IJ}\,E^J_{\mathscr{QM}} \tag{4.3.45}$$

where the quadratic symmetric constant tensor $\mathfrak{q}_{IJ}$ has the following form:

$$\mathfrak{q}_{IJ} = \left(\begin{array}{c|c|c|c} 1 & 0 & 0 & 0 \\ \hline 0 & \delta_{ij} & 0 & 0 \\ \hline 0 & 0 & 1 & 0 \\ \hline 0 & 0 & 0 & -2\,\mathscr{M}_4^{-1}(z_0,\bar{z}_0) \end{array}\right) \tag{4.3.46}$$

The above defined vielbein are endowed with a very special property namely they identically satisfy a set of Maurer Cartan equations:

$$dE^I_{\mathscr{Q}\mathscr{M}} - \frac{1}{2} f^I{}_{JK}\, E^J_{\mathscr{Q}\mathscr{M}} \wedge E^K_{\mathscr{Q}\mathscr{M}} = 0 \tag{4.3.47}$$

where f^I_{JK} are the structure constants of a solvable Lie algebra $\mathfrak{A}$ which can be identified as follows:

$$\mathfrak{A} = Solv\left(\frac{\mathrm{U}_{\mathscr{Q}}}{\mathrm{H}_{\mathscr{Q}}}\right) \tag{4.3.48}$$

In the above equation $Solv\left(\frac{\mathrm{U}_{\mathscr{Q}}}{\mathrm{H}_{\mathscr{Q}}}\right)$ denotes the Lie algebra of the solvable group manifold metrically equivalent to the non-comapact coset manifold $\frac{\mathrm{U}_{\mathscr{Q}}}{\mathrm{H}_{\mathscr{Q}}}$ according to what we explained in Sect. 2.5. In the case $\mathrm{U}_{\mathscr{S}\mathscr{K}}$ is a *maximally split* real form of a complex Lie algebra, then also $\mathrm{U}_{\mathscr{Q}}$ is maximally split and we have:

$$Solv\left(\frac{\mathrm{U}_{\mathscr{Q}}}{\mathrm{H}_{\mathscr{Q}}}\right) = \mathrm{Bor}\left(\mathbb{U}_{\mathscr{Q}}\right) \tag{4.3.49}$$

where $\mathrm{Bor}\left(\mathbb{U}_{\mathscr{Q}}\right)$ denotes the *Borel subalgebra* of the semi-simple Lie algebra $\mathbb{G}$, generated by its Cartan generators and by the step operators associated with all positive roots.

According to the mathematical theory summarized in Sect. 2.5 above, the very fact that the vielbein (4.3.44) satisfies the Maurer Cartan equations of the Lie algebra $Solv\left(\frac{\mathrm{U}_{\mathscr{Q}}}{\mathrm{H}_{\mathscr{Q}}}\right)$ implies that the metric (4.3.45) is the symmetric metric on the coset manifold $\frac{\mathrm{U}_{\mathscr{Q}}}{\mathrm{H}_{\mathscr{Q}}}$ which therefore admits continuous isometries associated with all the generators of the Lie algebra $\mathbb{U}_{\mathscr{Q}}$. For reader's convenience the list of Symmetric Special manifolds and of their Quaternionic Kähler counterparts in the image of the c-map is recalled in Table 4.1 which reproduces the results of [7], according to which there is a short list of Symmetric Homogeneous Special manifolds comprising five discrete cases and two infinite series.

Inspecting Eq. (1.7.19) we immediately realize that the Lie Algebra $\mathbb{U}_Q$ contains two universal Heisenberg subalgebras of dimension $(2n+3)$, namely:

$$\begin{aligned} \mathbb{U}_{\mathcal{Q}} \supset \mathbb{H}\mathrm{eis}_1 &= \mathrm{span}_{\mathbb{R}} \left\{ \mathbf{W}^{1\alpha} \, , \, \mathbb{Z}_1 \right\} \quad ; \quad \mathbb{Z}_1 \, = \, L_+ \, \equiv \, L^1 \, + \, L^2 \\ & \left[\mathbf{W}^{1\alpha} \, , \, \mathbf{W}^{1\beta} \right] = -\frac{1}{2} \mathbb{C}^{\alpha\beta} \, \mathbb{Z}_1 \quad ; \quad \left[\mathbb{Z}_1 \, , \, \mathbf{W}^{1\beta} \right] = 0 \end{aligned} \tag{4.3.50}$$

$$\begin{aligned} \mathbb{U}_{\mathcal{Q}} \supset \mathbb{H}\mathrm{eis}_2 &= \mathrm{span}_{\mathbb{R}} \left\{ \mathbf{W}^{2\alpha} \, , \, \mathbb{Z}_2 \right\} \quad ; \quad \mathbb{Z}_2 \, = \, L_- \, \equiv \, L^1 \, - \, L^2 \\ & \left[\mathbf{W}^{2\alpha} \, , \, \mathbf{W}^{2\beta} \right] = -\frac{1}{2} \mathbb{C}^{\alpha\beta} \, \mathbb{Z}_2 \quad ; \quad \left[\mathbb{Z}_2 \, , \, \mathbf{W}^{2\beta} \right] = 0 \end{aligned} \tag{4.3.51}$$

The first of these Heisenberg subalgebras of isometries is the universal one that exists for all Quaternionic Kähler manifolds $\mathcal{QM}_{4n+4}$ lying in the image of the c-map, irrespectively whether the pre-image Special Kähler manifold $\mathcal{SK}_n$ is a symmetric space or not. The tri-holomorphic moment map of these isometries was presented in Eqs. (4.3.31) and (4.3.35). The second Heisenberg algebra exists only in the case when the Quaternionic Kähler manifold $\mathcal{QM}_{4n+4}$ is a symmetric space.

From this discussion we also realize that the central charge $\mathbb{Z}_1$ is just the L_+ generator of a universal $\mathfrak{sl}(2,\mathbb{R})_E$ Lie algebra that exists only in the symmetric space case and which was named the Ehlers algebra in Sect. 1.7 where we presented the golden splitting (1.7.12). When $\mathfrak{sl}(2,\mathbb{R})_E$ does exist we can introduce the universal compact generator:

$$\mathfrak{S} \, \equiv \, L_+ \, - \, L_- \, = \, 2\,\lambda^2 \tag{4.3.52}$$

which rotates the two sets of Heisenberg translations one into the other:

$$\left[\mathfrak{S} \, , \, \mathbf{W}^{i\alpha} \right] \, = \, \varepsilon^{ij} \, \mathbf{W}^{j\alpha} \tag{4.3.53}$$

The gauging of this generator is a rather essential ingredient in the inclusion of one-field cosmological models into gauged $\mathcal{N} = 2$ supergravity as it was explained in [20].

4.3.4.1 The Tri-Holomorphic Moment Map in Homogeneous Symmetric Quaternionic Kähler Manifolds

In the case the Quaternionic Kähler manifold $\mathcal{QM}_{4n+4}$ is a homogeneous symmetric space $\frac{\mathrm{U}_{\mathcal{Q}}}{\mathrm{H}_{\mathcal{Q}}}$, the tri-holomorphic moment map associated with any generator of $\mathfrak{t} \in \mathbb{U}_{\mathcal{Q}}$ of the isometry Lie algebra can be easily constructed by means of the formula:

$$\mathcal{P}^x_{\mathfrak{t}} \, = \, \mathrm{Tr}_{[\mathbf{fun}]} \left(J^x \, \mathbb{L}^{-1}_{Solv} \, \mathfrak{t} \, \mathbb{L}_{Solv} \right) \tag{4.3.54}$$

where:

(a) J^x are the three generators of the $\mathfrak{su}(2)$ factor in the isotropy subalgebra $\mathbb{H} \, = \, \mathfrak{su}(2) \oplus \mathbb{H}'$, satisfying the quaternionic algebra (4.3.19). They should be normalized in such a way as to realize the following condition. Naming:

$$\Xi = \mathbb{L}_{Solv}^{-1}(q)\, \mathrm{d}\mathbb{L}_{Solv}(q) \tag{4.3.55}$$

the Maurer Cartan differential one-form, its projection on J^x should precisely yield the $\mathfrak{su}(2)$ one-form defined in Eq. (4.3.11):

$$\omega = -\frac{\mathrm{i}}{\sqrt{2}N_f}\sum_{x=1}^{3} \mathrm{Tr}_{[\mathbf{fun}]}\left(J^x\, \Xi\right)\gamma_x = \begin{pmatrix} -\frac{\mathrm{i}}{2}\mathscr{Q} - \frac{\mathrm{i}}{4}e^{-U}\Phi & e^{-\frac{U}{2}}\mathbf{V}^T\,\mathbb{C}\,\mathrm{d}\mathbf{Z} \\ -e^{-\frac{U}{2}}\overline{\mathbf{V}}^T\,\mathbb{C}\,\mathrm{d}\mathbf{Z} & \frac{\mathrm{i}}{2}\mathscr{Q} + \frac{\mathrm{i}}{4}e^{-U}\Phi \end{pmatrix} \tag{4.3.56}$$

In the above equation, which provides the precise link between the c-map description and the coset manifold description of the same geometry, $N_f = \dim \mathbf{fun}$ denotes the dimension of the fundamental representation of $\mathrm{U}_{\mathscr{Q}}$.

(b) The solvable coset representative $\mathbb{L}_{Solv}(q)$ is obtained by exponentiation of the Solvable Lie algebra:

$$\mathbb{L}_{Solv}(q) \simeq \exp\left[q \cdot Solv\left(\frac{\mathrm{U}_{\mathscr{Q}}}{\mathscr{H}_{\mathscr{Q}}}\right)\right] \tag{4.3.57}$$

but the detailed exponentiation rule has to be determined in such a way that projecting the same Maurer Cartan form (4.3.55) along an appropriate basis of generators $T_{I|Solv}$ of the solvable Lie algebra $Solv\left(\frac{\mathrm{U}_{\mathscr{Q}}}{\mathscr{H}_{\mathscr{Q}}}\right)$ we precisely obtain the vielbein E^I_{QM} defined in Eq. (4.3.44). This is summarized in the following general equations:

$$\begin{aligned} E^I_{\mathscr{Q}\mathscr{M}} &= \mathrm{Tr}_{[\mathbf{fun}]}\left(T^I_{Solv}\,\Xi\right) \\ \delta^I_J &= \mathrm{Tr}_{[\mathbf{fun}]}\left(T^I_{Solv}\,T_{I|Solv}\right) \\ \Xi &= E^I_{\mathscr{Q}\mathscr{M}}\,T_{I|Solv} \end{aligned} \tag{4.3.58}$$

In Eq. (4.3.58) by T^I_{Solv} we have denoted the conjugate (with respect to the trace) of the solvable Lie algebra generators.

A general comment is in order. The precise calibration of the basis of the solvable generators T^I_{Solv} and of their exponentiation outlined in Eq. (4.3.57) which allows the identification (4.3.58) is a necessary and quite laborious task in order to establish the bridge between the general c-map description of the quaternionic geometry and its actual realization in each symmetric coset model. This is also an unavoidable step in order to give a precise meaning to the very handy formula (4.3.54) for the tri-holomorphic map. It should also be noted that although (4.3.54) covers all the cases, the result of such a purely algebraic calculation is difficult to be guessed a priori. Hence educated guesses on the choice of generators whose gauging produces a priori determined features are difficult to be inferred from (4.3.54). The analytic structure of the tri-holomorphic moment map instead is much clearer in the c-map framework of formulae (4.3.31), (4.3.35), (4.3.40). The use of both languages and the construction of the precise bridge between them in each model is therefore an essential ingredient to understand the nature and the properties of candidate gaugings in whatever physical application.

References

1. P.G. Fré, *A Conceptual History of Symmetry from Plato to the Superworld* (Springer, Berlin, 2018)
2. S. Ferrara, D.Z. Freedman, P. van Nieuwenhuizen, P. Breitenlohner, F. Gliozzi, J. Scherk, Scalar multiplet coupled to supergravity. Phys. Rev. D **15**, 1013 (1977)
3. B. Zumino, Supersymmetry and Kahler manifolds. Phys. Lett. B **87**, 203 (1979)
4. E. Cremmer, C. Kounnas, A. Van Proeyen, J.P. Derendinger, S. Ferrara, B. De Wit, L. Girardello, Vector multiplets coupled to N = 2 supergravity: superHiggs effect, flat potentials and geometric structure. Nucl. Phys. B **250**, 385–426 (1985)
5. B. De Wit, P.G. Lauwers, R. Philippe, S.Q. Su, A. Van Proeyen, Gauge and matter fields coupled to N = 2 supergravity. Phys. Lett. B **134**, 37–43 (1984)
6. B. De Wit, A. Van Proeyen, Potentials and symmetries of general gauged N = 2 supergravity: Yang-Mills models. Nucl. Phys. B **245**, 89–117 (1984)
7. E. Cremmer, A. Van Proeyen, Classification of Kahler manifolds in N = 2 vector multiplet supergravity couplings. Class. Quantum Gravity **2**, 445 (1985)
8. A. Strominger, Special geometry. Commun. Math. Phys. **133**, 163–180 (1990)
9. L. Castellani, R. D'Auria, S. Ferrara, Special geometry without special coordinates. Class. Quantum Gravity **7**, 1767–1790 (1990)
10. L. Castellani, R. D'Auria, S. Ferrara, Special Kahler geometry: an intrinsic formulation from N = 2 space-time supersymmetry. Phys. Lett. B **241**, 57–62 (1990)
11. R. D'Auria, S. Ferrara, P. Fré, Special and quaternionic isometries: general couplings in n = 2 supergravity and the scalar potential, Nucl. Phys. B **359**, 705–740 (1991)
12. B. De Wit, A. Van Proeyen, Broken sigma model isometries in very special geometry. Phys. Lett. B **293**, 94–99 (1992)
13. B. De Wit, A. Van Proeyen, Special geometry, cubic polynomials and homogeneous quaternionic spaces. Commun. Math. Phys. **149**, 307–334 (1992)
14. B. De Wit, F. Vanderseypen, A. Van Proeyen, Symmetry structure of special geometries. Nucl. Phys. B **400**, 463–524 (1993)
15. S. Cecotti, Homogeneous Kahler manifolds and T algebras in N = 2 supergravity and superstrings. Commun. Math. Phys. **124**, 23–55 (1989)
16. R. D'Auria, S. Ferrara, M. Trigiante, C - map, very special quaternionic geometry and dual Kahler spaces. Phys. Lett. B **587**, 138–142 (2004)
17. S. Ferrara, S. Sabharwal, Quaternionic manifolds for type II superstring Vacua of Calabi-Yau spaces. Nucl. Phys. B **332**, 317–332 (1990)
18. D. Alekseevsky, Classification of quaternionic spaces with a transitive solvable group of motions. Math. USSR Izvestija **9**, 297–339 (1975)
19. V. Cortés, Alekseevskian spaces. Differ. Geom. Appl. **6**, 129–168 (1996)
20. P. Fré, A.S. Sorin, M. Trigiante, The c-map, Tits Satake subalgebras and the search for $\mathcal{N} = 2$ inflaton potentials. Fortsch. Phys. **63**, 198–258 (2015)
21. P. Fre, Lectures on special Kahler geometry and electric - magnetic duality rotations, Nucl. Phys. Proc. Suppl. **45**BC, 59–114 (1996)
22. L. Andrianopoli, M. Bertolini, A. Ceresole, R. D'Auria, S. Ferrara, P. Fre, T. Magri, N = 2 supergravity and N = 2 super Yang–Mills theory on general scalar manifolds: symplectic covariance, gaugings and the momentum map. J. Geom. Phys. **23**, 111–189 (1997)

Chapter 5
Solvable Algebras and the Tits Satake Projection

Quamlibet immani proiectu corporis extet, Lucretius, De Rerum Natura, 3, 987

5.1 Historical Introduction

In this chapter we are going to develop the details of a theory pertaining to Lie Algebras which, although it has its roots in mathematical work of the 1960s [1–3], contributed by two great algebrists, Jacques Tits and Ichiro Satake (see Fig. 5.1), yet fully revealed its profound significance for Geometry and Physics only much later, by the end of the XXth century, and within the context of supergravity.

The addressed topics is the Tits–Satake projection, a construction which, according to certain rules, from a class of homogeneous manifolds, extracts a single representative of the entire class. What is extremely surprising and inspiring is that such a projection, invented long before the advent of supergravity *special geometries*, has very nice properties with respect to *special structures*. Indeed it maps *special Kähler manifolds* into *special Kähler manifolds*, *quaternionic Kähler* into *quaternionic Kähler* and commutes with the c-map discussed in the previous section. Actually it also commutes with another map, the $c^{\star}$-map, which is relevant for the construction of supergravity black-hole solutions and will be illustrated in this chapter.

A conceptual procedure specially cheered by theoretical physicists is that of *Universality Classes*. Considering complex phenomena like, for instance, phase-transitions one looks for universal features that are the same for entire classes of such phenomena. After grouping the multitude of cases into *universality classes*, one tries to construct a theoretical model of the behavior shared by all elements of each class. A mathematical well founded projection is likely to provide a powerful weapon to this effect. Indeed one might expect that there are *universal features* shared by all cases that have the same projection and that the theoretical model of

P. G. Frè, *Advances in Geometry and Lie Algebras from Supergravity*, Theoretical and Mathematical Physics,
https://doi.org/10.1007/978-3-319-74491-9_5

Fig. 5.1 On the left J. Tits (1930 Uccle, Belgium). On the right Ichiro Satake (1927 Yamaguchi Japan - 2014 Tokyo Japan). Jacques Tits was born in Uccle, on the southern outskirts of Brussels. He graduated from the Free University of Brussels in 1950 with a dissertation *Généralisation des groupes projectifs basés sur la notion de transitivité*. From 1956 to 1962 Tits was an assistant at the University of Brussels. He became professor there in 1962 and remained in this role for two years before accepting a professorship at the University of Bonn in 1964. In 1973 he was offered the *Chair of Group Theory* at the *College de France* which he occupied until his retirement in 2000 being naturalised French citizen since 1974. Jacques Tits has given very prominent contributions to the advancement of Group Theory in many directions and he is especially known for the *Theory of Buildings*, which he founded, and for the *Tits alternative*, a theorem on the structure of finitely generated groups. After his retirement from the College the France, a special Vallée-Poussin Chair was created for him at the University of Louvain. Ichiro Satake was born in the Province of Yamaguchi in Japan and graduated from the University of Tokyo in 1959. He held various academic positions in the USA and since 1968 to his retirement in 1983 he was Full Professor of Mathematics at the University of California, Berkeley. He is specially known for his contributions to the theory of algebraic groups and for the Satake diagrams that classify the real forms of a complex Lie algebra

this shared behavior is encoded in the algebraic structure of the projection image. We will see that this is precisely what happens with the Tits–Satake projection that captures *universal geometrical features* of supergravity models.

Since the interplay between Mathematics and Theoretical Physics has been essential in the development of this new chapter of *homogeneous space geometry* we briefly recall the key facts of this short but intellectually intense history.

(1) In the early 1990s, as we have already reported, B. de Wit, A. Van Proeyen, F. Vanderseypen studied the classification of homogeneous special manifolds admitting a solvable transitive group of isometries [4–6]. This work extended and completed the results obtained several years before by Alekseevsky in

relation with the classification of quaternionic manifolds also admitting a transitive solvable group of isometries [7].

(2) In 1996–1998, L. Andrianopoli, R. D'Auria, S. Ferrara, P. Fré and M. Trigiante explored the general role of *solvable Lie algebras* in supergravity [8–10], pointing out that, since all homogenous scalar manifolds of all supergravity models are of the non-compact type, they all admit a description in terms of a solvable group manifold as we explained in Sect. 2.5. The solvable representation of the scalar geometry was shown to be particularly valuable in connection with the description of BPS black hole solutions of various supergravity models.

(3) In the years 1999–2005 Thibaut Damour, Marc Henneaux, Hermann Nicolai, Bernard Julia, F. Englert, P. Spindel and other collaborators, elaborating on old ideas of V.A. Belinsky, I.M. Khalatnikov, E.M. Lifshitz [11–13], introduced the conception of *rigid cosmic billiards* [14–27]. According to this conception the various dimensions of a higher dimensional gravitational theory are identified with the generators of the Cartan Subalgebra $\mathscr{H}$ of a supergravity motivated Lie algebra and cosmic evolution takes place in a Weyl chamber of $\mathscr{H}$. Considering the Cartan scalar fields as the coordinate of a fictitious ball, during cosmic evolution such a ball scatters on the walls of the Weyl chambers and this pictorial image of the phenomenon is at the origin of its denomination *cosmic billiard*. In this context the distinction between compact and non-compact directions of the Cartan subalgebra appeared essential and this brought the Tits Satake projection into the game.

(4) In 2003–2005 F. Gargiulo, K. Rulik, P. Fré, A.S. Sorin and M. Trigiante developed the conception of *soft cosmic billiards* [28–30], corresponding to exact, purely time dependent solutions of supergravity, including not only the Cartan fields but also those associated with roots which dynamically construct the Weyl chamber walls advocated by *rigid cosmic billiards*.

(5) In 2005, Fré, Gargiulo and Rulik constructed explicit examples of soft cosmic billiards in the case of a *non maximally split symmetric manifold*. In that context they analyzed the role of the Tits Satake projection and introduced the new mathematical concept of *Paint Group* [31].

(6) In 2007, P. Fré, F. Gargiulo, J. Rosseel, K. Rulik, M. Trigiante and A. Van Proeyen [32] axiomatized the Tits Satake projection for all homogeneous special geometries. They based their formulation of the projection on the intrinsic definition of the *Paint Group* as the group of outer automorphisms of the solvable transitive group of motion of the homogeneous manifold. This is the theory that will be explained in this chapter. Up to the knowledge of this author, this theory was never previously developed in the mathematical literature.

(7) In the years 2009–2011 the integration algorithm utilized in the framework of soft cosmic billiards was extended by P. Fré, A.S. Sorin and M. Trigiante to the case of spherical symmetric black-holes for manifolds in the image of the $c^\star$-map [33–35].

(8) In 2011, P. Fré, A.S. Sorin and M. Trigiante demonstrated that the classification of nilpotent orbits for a non maximally split Lie algebra depends only on its

Tits–Satake projection and it is a property of the Tits–Satake universality class (see Chap. 6).

Through the above sketched historical course, which unfolded in about a decade, the theory of the Tits–Satake projection has acquired a quite solid and ramified profile, intertwined with the c and $c^\star$ maps that opens new viewpoints and provides new classification tools in the geometry of homogeneous manifolds and symmetric spaces. Although the theory is distinctively algebraic and geometric, yet it is poorly known in the mathematical community due to its supergravity driven origins. Hopefully the present exposition will improve its status in the mathematical club.

We turn next to a systematic discussion of the $c^\star$-map environment where the Tits–Satake projection is best understood and most useful.

5.2 Physical-Mathematical Introduction

In the previous chapter we provided the definition of special Kähler geometry and of quaternionic Kähler geometry. In the context of $\mathcal{N} = 2$ supergravity, as we stressed there, the two types of geometries are respectively pertinent to the scalars included in the *vector multiplets* and to those pertinent to the *hypermultiplets*. The next main focus of attention was the c-map from Special Kähler Manifolds of complex dimension n to quaternionic Kähler manifolds of real dimension $4n + 4$:

$$\text{c-map} \quad : \quad \mathcal{SK}_n \rightarrow \mathcal{QM}_{(4n+4)} \tag{5.2.1}$$

What we did not emphasize in the previous chapter is that the c-map follows from the systematic procedure of dimensional reduction from a $D = 4$, $\mathcal{N} = 2$ supergravity theory to a $D = 3$ σ-model endowed with $\mathcal{N} = 4$ three-dimensional supersymmetry. We recall this point here since it helps understanding another very similar map that we are going to consider in this chapter and that we name the $c^\star$-map. Naming z^i the scalar fields that fill the special Kähler manifold $\mathcal{SK}_n$ and $g_{ij^\star}$ its metric, the $D = 3$ σ-model which encodes all the supergravity field equations after dimensional reduction on a space-like direction admits, as target manifold, a quaternionic manifold whose $4n + 4$ coordinates we name as follows:

$$\underbrace{\{U, a\}}_{2} \bigcup \underbrace{\{z^i\}}_{2n} \bigcup \underbrace{\mathbf{Z} = \{Z^\Lambda \, , \, Z_\Sigma\}}_{2n+2} \tag{5.2.2}$$

and whose quaternionic metric has the general form that we discussed at length in Chap. 4.

The $c^\star$-map arises in a similar way from dimensional reduction but along a time-like direction. Let us see in which context this takes place.

5.2.1 Black Holes and the Geometry of Geometries

In the last twenty years a lot of interest was devoted to study black-hole solutions of pure and matter coupled $\mathcal{N}$-extended supergravity theories, the case $\mathcal{N}=2$ being the most widely considered. Generally speaking a black-hole solution of matter coupled supergravity is an exact solution of the bosonic field equations where all the *items of geometry* that we have been so far studying are involved. Let us get an orientation on this exciting entanglement of several geometries.

The general form of a bosonic supergravity lagrangian in $D=4$ is the following one:

$$\mathcal{L}^{(4)} = \sqrt{|\det g|}\left[\frac{R[g]}{2} - \frac{1}{4}\partial_\mu\phi^a\partial^\mu\phi^b h_{ab}(\phi) + \mathrm{Im}\mathcal{N}_{\Lambda\Sigma}\, F^\Lambda_{\mu\nu}F^{\Sigma|\mu\nu}\right] + \frac{1}{2}\mathrm{Re}\mathcal{N}_{\Lambda\Sigma}\, F^\Lambda_{\mu\nu}F^\Sigma_{\rho\sigma}\varepsilon^{\mu\nu\rho\sigma}\,, \tag{5.2.3}$$

The fields included in the theory are the metric $g_{\mu\nu}(x)$, n_{v} abelian gauge fields A^Λ_ν, whose field strengths (or curvatures) we have denoted by $F^\Lambda_{\mu\nu} \equiv (\partial_\mu A^\Lambda_\nu - \partial_\nu A^\Lambda_\mu)/2$ and n_{s} scalar fields ϕ^a that parameterize a scalar manifold $\mathcal{M}^{D=4}_{scalar}$ that, for supersymmetry $\mathcal{N}>2$, is necessarily a coset manifold:

$$\mathcal{M}^{D=4}_{scalar} = \frac{\mathrm{U}_{\mathrm{D=4}}}{\mathrm{H}_c} \tag{5.2.4}$$

$\mathrm{U}_{\mathrm{D=4}}$ being a non-compact real form of a semi-simple Lie group, essentially fixed by supersymmetry and H_c its maximal compact subgroup. For $\mathcal{N}=2$ Eq. (5.2.4) is not obligatory yet it is possible: a well determined class of symmetric homogeneous manifolds that are special Kähler manifolds fall into the set up of the present general discussion.

Hence we see that we are dealing with geometries at three levels:

1. We deal with the geometry of space-time $\mathcal{M}^{st}_4$, encoded in its metric $g_{\mu\nu}$ which is dynamical, in the sense that we have to determine it through the solution of field equations, many possibilities being available, among which we have black-hole geometries with event horizons and all the rest.
2. We deal with connections on a fiber bundle $P\left(\mathcal{G}, \mathcal{M}^{st}_4\right)$, whose base manifold is the dynamically determined space-time $\mathcal{M}^{st}_4$ and whose structural group is an abelian group $\mathcal{G}$ of dimension equal to the number n_{v} of involved gauge fields. These connections are also dynamical in the sense that they have to be determined as solutions of the coupled field equations.
3. We deal with a fixed Riemannian geometry encoded in the target manifold (5.2.4) of which the scalar fields ϕ^a are local coordinates. Any solution of the coupled field equations defines a map

$$\phi \ : \ \mathscr{M}_4^{st} \ \to \ \mathscr{M}^{D=4}_{scalar} \tag{5.2.5}$$

of space-time into the scalar manifold.

There is still encoded into the lagrangian (5.2.3) another geometrical datum of utmost relevance. Let us describe it. Considering the n_v vector fields A^Λ_μ let

$$\mathscr{F}^{\pm|\Lambda}_{\mu\nu} \equiv \frac{1}{2}\left[F^\Lambda_{\mu\nu} \mp \mathrm{i}\,\frac{\sqrt{|\det g|}}{2}\varepsilon_{\mu\nu\rho\sigma}\,F^{\rho\sigma}\right] \tag{5.2.6}$$

denote the self-dual (respectively antiself-dual) parts of the field-strengths. As displayed in Eq. (5.2.3) they are non minimally coupled to the scalars via the symmetric complex matrix

$$\mathscr{N}_{\Lambda\Sigma}(\phi) = \mathrm{i}\,\mathrm{Im}\mathscr{N}_{\Lambda\Sigma} + \mathrm{Re}\mathscr{N}_{\Lambda\Sigma} \tag{5.2.7}$$

The key point is that the isometry group $\mathrm{U}_{\mathrm{D}=4}$ of the scalar manifold (5.2.4) is promoted to a symmetry of the entire lagrangian through the projective transformations of $\mathscr{N}_{\Lambda\Sigma}$ under the group action.

Indeed the field strengths $\mathscr{F}^{\pm|\Lambda}_{\mu\nu}$ plus their magnetic duals:

$$G_{\Lambda|\mu\nu} \equiv \frac{1}{2}\,\varepsilon_{\mu\nu}{}^{\rho\sigma}\,\frac{\delta\mathscr{L}^{(4)}}{\delta F^\Lambda_{\rho\sigma}} \tag{5.2.8}$$

fill up a $2\,n_v$-dimensional symplectic representation of $\mathbb{U}_{\mathrm{D}=4}$ which we call by the name of $\mathbf{W}$.

We rephrase the above statements by asserting that there is always a symplectic embedding of the duality group $\mathrm{U}_{D=4}$,

$$\mathrm{U}_{D=4} \mapsto \mathrm{Sp}(2\mathrm{n_v},\mathbb{R}) \quad ; \quad n_v \equiv \#\text{ of vector fields} \tag{5.2.9}$$

so that for each element $\xi \in \mathrm{U}_{D=4}$ we have its representation by means of a suitable real symplectic matrix:

$$\xi \mapsto \Lambda_\xi \equiv \begin{pmatrix} A_\xi & B_\xi \\ C_\xi & D_\xi \end{pmatrix} \tag{5.2.10}$$

satisfying the defining relation:

$$\Lambda^T_\xi \begin{pmatrix} \mathbf{0}_{n\times n} & \mathbf{1}_{n\times n} \\ -\mathbf{1}_{n\times n} & \mathbf{0}_{n\times n} \end{pmatrix} \Lambda_\xi = \begin{pmatrix} \mathbf{0}_{n\times n} & \mathbf{1}_{n\times n} \\ -\mathbf{1}_{n\times n} & \mathbf{0}_{n\times n} \end{pmatrix} \tag{5.2.11}$$

Under an element of the duality group the field strengths transform as follows:

$$\begin{pmatrix} \mathscr{F}^+ \\ \mathscr{G}^+ \end{pmatrix}' = \begin{pmatrix} A_\xi & B_\xi \\ C_\xi & D_\xi \end{pmatrix}\begin{pmatrix} \mathscr{F}^+ \\ \mathscr{G}^+ \end{pmatrix} \quad ; \quad \begin{pmatrix} \mathscr{F}^- \\ \mathscr{G}^- \end{pmatrix}' = \begin{pmatrix} A_\xi & B_\xi \\ C_\xi & D_\xi \end{pmatrix}\begin{pmatrix} \mathscr{F}^- \\ \mathscr{G}^- \end{pmatrix} \tag{5.2.12}$$

where, by their own definitions we get:

$$\mathscr{G}^{+} = \mathscr{N}\,\mathscr{F}^{+} \quad ; \quad \mathscr{G}^{-} = \overline{\mathscr{N}}\,\mathscr{F}^{-} \tag{5.2.13}$$

and the complex symmetric matrix $\mathscr{N}$ should transform as follows:

$$\mathscr{N}' = \left(C_\xi + D_\xi\,\mathscr{N}\right)\left(A_\xi + B_\xi\,\mathscr{N}\right)^{-1} \tag{5.2.14}$$

Choose a parametrization of the coset $\mathbb{L}(\phi) \in \mathrm{U}_{\mathrm{D}=4}$, which assigns a definite group element to every coset point identified by the scalar fields. Through the symplectic embedding (5.2.10) this produces a definite ϕ-dependent symplectic matrix

$$\begin{pmatrix} A(\phi) & B(\phi) \\ C(\phi) & D(\phi) \end{pmatrix} \tag{5.2.15}$$

in the $\mathbf{W}$-representation of $\mathrm{U}_{\mathrm{D}=4}$. In terms of its blocks the kinetic matrix $\mathscr{N}(\phi)$ is explicitly given by a formula that was found at the beginning of the 1980s by Gaillard-Zumino [36]:

$$\mathscr{N}(\phi) = [C(\phi) - i\,D(\phi)]\,[A(\phi) - i\,B(\phi)]^{-1}\ , \tag{5.2.16}$$

The matrix $\mathscr{N}$ is the same which appears in the definition of special Kähler geometry and it transforms according to Eq. (5.2.14).

Summarizing the geometrical structure of the bosonic supergravity lagrangian is essentially encoded in two data. The duality-isometry group $U_{D=3}$ and its symplectic representation $\mathbf{W}$ that corresponds to the embedding (5.2.9).

A brilliant discovery occurred in the first two decades of the XXIst century can be dubbed as the $D = 3$ approach to supergravity black-holes. Mainly originating from the contributions included in the following papers [37–43], it consists of the following.

The radial dependence of all the relevant functions parameterizing the supergravity solution can be viewed as the field equations of another one-dimensional σ-model where the evolution parameter τ is actually a monotonic function of the radial variable r and where the target manifold is a *pseudo-quaternionic manifold* $\mathscr{Q}^\star_{(4n+4)}$ related to the quaternionic manifold $\mathscr{Q}_{(4n+4)}$ in the following way. The coordinates of $\mathscr{Q}^\star_{(4n+4)}$ are the same as those of $\mathscr{Q}_{(4n+4)}$, while the two metrics differ only by a change of sign. Indeed we have

$$ds^2_{\mathscr{Q}} = \frac{1}{4}\left[dU^2 + 2\,g_{ij^\star}\,dz^i\,d\bar{z}^{j^\star} + \mathrm{e}^{-2U}\,(da + \mathbf{Z}^T\mathbb{C}d\mathbf{Z})^2 \;-\; 2\,e^{-U}\,d\mathbf{Z}^T\,\mathscr{M}_4(z,\bar{z})\,d\mathbf{Z}\right]$$

$$\Downarrow \text{ Wick rot.} \tag{5.2.17}$$

$$ds^2_{\mathscr{Q}^\star} = \frac{1}{4}\left[dU^2 + 2\,g_{ij^\star}\,dz^i\,d\bar{z}^{j^\star} + \mathrm{e}^{-2U}\,(da + \mathbf{Z}^T\mathbb{C}d\mathbf{Z})^2 \;+\; 2\,e^{-U}\,d\mathbf{Z}^T\,\mathscr{M}_4(z,\bar{z})\,d\mathbf{Z}\right] \tag{5.2.18}$$

In Eqs. (5.2.17) and (5.2.18), $\mathbb{C}$ denotes the $(2n+2) \times (2n+2)$ antisymmetric matrix defined over the fibers of the symplectic bundle characterizing special geometry, while the *negative definite*, $(2n+2) \times (2n+2)$ matrix $\mathcal{M}_4(z, \bar{z})$ is the one already introduced in Eq. (4.3.3). The pseudo-quaternionic metric is non-Euclidean and it has the following signature:

$$\text{sign}\left(ds^2_{\mathcal{Q}^\star}\right) = \left(\underbrace{+, \ldots, +}_{2n+2}, \underbrace{-, \ldots, -}_{2n+2}\right) \tag{5.2.19}$$

In this way we arrive at a *Geometry of the Geometries*. As solutions of the σ-model defined by the metric (5.2.18), all spherically symmetric black-holes correspond to geodesics and consequently a geodetic in the manifold $\mathcal{Q}^\star$ encodes all the geometrical structures listed below:

(a) A spherical black-hole metric,
(b) a spherical symmetric connection on the fiber bundle $P\left(\mathcal{G}, \mathcal{M}_4^{st}\right)$
(c) a spherical symmetric map from $\mathcal{M}_4^{st}$ into the manifold (5.2.4)

The indefinite signature (5.2.19) introduces a clear-cut distinction between non-extremal and extremal black-holes: the non-extremal ones correspond to time-like geodesics, while the extremal black-holes are associated with light-like ones. Space-like geodesics produce supergravity solutions with naked singularities [37].

In those cases where the Special Manifold $\mathcal{SK}_n$ is a symmetric space $\frac{\mathrm{U_{D=4}}}{\mathrm{H_{D=4}}}$ also the quaternionic manifold defined by the metric (5.2.17) is a symmetric coset manifold:

$$\frac{\mathrm{U_{D=3}}}{\mathrm{H_{D=3}}} \tag{5.2.20}$$

where $\mathrm{H_{D=3}} \subset \mathrm{U_{D=3}}$ is the *maximal compact subgroup* of the U-duality group, in three dimensions $\mathrm{U_{D=3}}$. The change of sign in the metric (5.2.19) simply turns the coset (5.2.20) into a new one:

$$\frac{\mathrm{U_{D=3}}}{\mathrm{H^\star_{D=3}}} \tag{5.2.21}$$

where $\mathrm{H^\star_{D=3}} \subset \mathrm{U_{D=3}}$ is another *non-compact maximal subgroup* of the U-duality group whose Lie algebra $\mathbb{H}^\star$ happens to be a different real form of the complexification of the Lie algebra $\mathbb{H}$ of $\mathrm{H_{D=3}}$. That such a different real form always exists within $\mathrm{U_{D=3}}$ is one of the group theoretical miracles of supergravity.

5.2.2 *The Lax Pair Description*

Once the problem of black-holes is reformulated in terms of geodesics within the coset manifold (5.2.21) a rich spectrum of additional mathematical techniques becomes available for its study and solution.

The most relevant of these techniques is the Lax pair representation of the supergravity field equations. According to a formalism reviewed in papers [34, 44], the fundamental evolution equation takes the following form:

$$\frac{d}{d\tau} L(\tau) + [W(\tau), L(\tau)] = 0 \tag{5.2.22}$$

where the so named Lax operator $L(\tau)$ and the connection $W(\tau)$ are Lie algebra elements of $\mathbb{U}$ respectively lying in the orthogonal subspace $\mathbb{K}$ and in the subalgebra $\mathbb{H}^\star$ in relation with the decomposition:

$$\mathbb{U} = \mathbb{H}^\star \oplus \mathbb{K} \tag{5.2.23}$$

As it was proven in [29, 33–35], both for the case of the coset (5.2.20) and the coset (5.2.21), the Lax pair representation (5.2.22) allows the construction of an explicit integration algorithm which provides the finite form of any supergravity solution in terms of two initial conditions, the Lax $L_0 = L(0)$ and the solvable coset representative $\mathbb{L}_0 = \mathbb{L}(0)$ at radial infinity $\tau = 0$.

The action of the global symmetry group $\mathrm{U}_{\mathrm{D}=3}$ on a geodesic can be described as follows: By means of a transformation $\mathrm{U}_{\mathrm{D}=3}/\mathrm{H}^\star$ we can move the "initial point" at $\tau = 0$ (described by $\mathbb{L}_0$) anywhere on the manifold, while for a fixed initial point we can act by means of $\mathrm{H}^\star$ on the "initial velocity vector", namely on L_0. Since the action of $\mathrm{U}_{\mathrm{D}=3}/\mathrm{H}^\star$ is transitive on the manifold, we can always bring the initial point to coincide with the origin (where all the scalar fields vanish) and classify the geodesics according to the $\mathrm{H}^\star$-orbit of the Lax matrix at radial infinity L_0. Since the evolution of the Lax operator occurs via a similarity transformation of L_0 by means of a time evolving element of the subgroup $\mathrm{H}^\star$, it will unfold within one $\mathrm{H}^\star$-orbit.

The main goal is then that of classifying all possible solutions by means of $\mathbb{H}^\star$-orbits within $\mathbb{K}$ which, in every supergravity based on homogeneous scalar geometries, is a well defined irreducible representation of $\mathbb{H}^\star$.

5.2.3 Nilpotent Orbits and Tits Satake Universality Classes

As it was discussed in [44] and in previous literature, regular extremal black-holes are associated with Lax operators $L(\tau)$ that are nilpotent at all times of their evolution. Hence the classification of extremal black-holes requires a classification of the orbits of nilpotent elements of the $\mathbb{K}$ space with respect to the stability subgroup $\mathbb{H}^\star \subset \mathrm{U}_{\mathrm{D}=3}$. This is a well posed, but difficult, mathematical problem. In [44] it was solved for the case of the special Käher manifold $\frac{\mathrm{SU}(1,1)}{\mathrm{U}(1)}$ which, upon time-like dimensional reduction to $D = 3$, yields the pseudo quaternionic manifold $\frac{\mathrm{G}_{(2,2)}}{\mathrm{SU}(1,1)\times\mathrm{SU}(1,1)}$. It would be desirable to extend the classification of such nilpotent orbits to supergravity models based on all the other special symmetric manifolds. Although these latter fall into a finite set of series, some of them are infinite and it might seem that we need to examine an infinite number of cases. This is not so because of a very important property of special geometries and of their quaternionic descendants.

This relates to the Tits–Satake (TS) projection of *special homogeneous (SH) manifolds*:

$$\mathscr{S}\mathscr{H} \stackrel{\text{Tits–Satake}}{\Longrightarrow} \mathscr{S}\mathscr{H}_{\text{TS}} \tag{5.2.24}$$

which was analysed in detail in [32], together with the allied concept of *Paint Group* that had been introduced previously in [31]. What it is meant by this wording is the following. It turns out that one can define an algorithm, the Tits–Satake projection π_{TS}, which works on the space of homogeneous manifolds with a solvable transitive group of motions $\mathscr{G}_M$, and with any such manifold associates another one of the same type. This map has a series of very strong distinctive features:

1. π_{TS} is a projection operator, so that several different manifolds $\mathscr{S}\mathscr{H}_i$ $(i = 1, \ldots, r)$ have the same image $\pi_{\text{TS}}\,(\mathscr{S}\mathscr{H}_i)$.
2. π_{TS} preserves the rank of $\mathscr{G}_M$ namely the dimension of the maximal Abelian semisimple subalgebra (Cartan subalgebra) of $\mathscr{G}_M$.
3. π_{TS} maps special homogeneous into special homogeneous manifolds. Not only. It preserves the two classes of manifolds discussed above, namely maps *special Kähler* into *special Kähler* and maps *Quaternionic* into *Quaternionic*
4. π_{TS} commutes with c-map, so that we obtain the following commutative diagram:

$$\begin{array}{ccc} \text{Special Kähler} & \stackrel{c\text{-map}}{\Longrightarrow} & \text{Quaternionic-Kähler} \\ \pi_{\text{TS}} \Downarrow & & \pi_{\text{TS}} \Downarrow \\ (\text{Special Kähler})_{\text{TS}} & \stackrel{c\text{-map}}{\Longrightarrow} & (\text{Quaternionic-Kähler})_{\text{TS}} \end{array} \tag{5.2.25}$$

The main consequence of the above features is that the whole set of special homogeneous manifolds and hence of associated supergravity models is distributed into a set of *universality classes* which turns out to be composed of extremely few elements.

If we confine ourselves to homogenous symmetric special geometries, which are those for which we can implement the integration algorithm based on the Lax pair representation, then the list of special symmetric manifolds contains only eight items among which two infinite series. They are displayed in the first column of Table 5.1. The c-map produces just as many quaternionic (Kähler) manifolds, that are displayed in the second column of the same table. Upon the Tits–Satake projection, this infinite set of models is organized into just five universality classes that are displayed on the third column of Table 5.1. The key-feature of the projection, relevant to our purposes is that all of its properties extend also to the *pseudo-quaternionic* manifolds produced by a time-like dimensional reduction. We can say that there exists a $c^\star$-map defined by this type of reduction, which associates a pseudo-quaternionic manifold with each special Kähler manifold. The Tits–Satake projection commutes also with the $c^\star$-map and we have another commutative diagram:

Table 5.1 The eight series of homogenous symmetric special Kähler manifolds (infinite and finite), their quaternionic counterparts and the grouping of the latter into five Tits Satake universality classes

Special Kähler $\mathcal{SK}_n$	Quaternionic $\mathcal{QM}_{4n+4}$	Tits Satake projection of quater $\mathcal{QM}_{\mathrm{TS}}$
$\frac{\mathrm{U(s+1,1)}}{\mathrm{U(s+1)\times U(1)}}$	$\frac{\mathrm{U(s+2,2)}}{\mathrm{U(s+2)\times U(2)}}$	$\frac{\mathrm{U(3,2)}}{\mathrm{U(3)\times U(2)}}$
$\frac{\mathrm{SU(1,1)}}{\mathrm{U(1)}}$	$\frac{\mathrm{G_{(2,2)}}}{\mathrm{SU(2)\times SU(2)}}$	$\frac{\mathrm{G_{(2,2)}}}{\mathrm{SU(2)\times SU(2)}}$
$\frac{\mathrm{SU(1,1)}}{\mathrm{U(1)}} \times \frac{\mathrm{SU(1,1)}}{\mathrm{U(1)}}$	$\frac{\mathrm{SO(3,4)}}{\mathrm{SO(3)\times SO(4)}}$	$\frac{\mathrm{SO(3,4)}}{\mathrm{SO(3)\times SO(4)}}$
$\frac{\mathrm{SU(1,1)}}{\mathrm{U(1)}} \times \frac{\mathrm{SO(p+2,2)}}{\mathrm{SO(p+2)\times SO(2)}}$	$\frac{\mathrm{SO(p+4,4)}}{\mathrm{SO(p+4)\times SO(4)}}$	$\frac{\mathrm{SO(5,4)}}{\mathrm{SO(5)\times SO(4)}}$
$\frac{\mathrm{Sp(6)}}{\mathrm{U(3)}}$ $\frac{\mathrm{SU(3,3)}}{\mathrm{SU(3)\times SU(3)\times U(1)}}$ $\frac{\mathrm{SO^\star(12)}}{\mathrm{SU(6)\times U(1)}}$ $\frac{\mathrm{E_{(7,-25)}}}{\mathrm{E_{(6,-78)}}\times U(1)}$	$\frac{\mathrm{F_{(4,4)}}}{\mathrm{Usp(6)\times SU(2)}}$ $\frac{\mathrm{E_{(6,-2)}}}{\mathrm{SU(6)\times SU(2)}}$ $\frac{\mathrm{E_{(7,-5)}}}{\mathrm{SO(12)\times SU(2)}}$ $\frac{\mathrm{E_{(8,-24)}}}{\mathrm{E_{(7,-133)}\times SU(2)}}$	$\frac{\mathrm{F_{(4,4)}}}{\mathrm{Usp(6)\times\ SU(2)}}$

$$
\begin{array}{ccc}
\text{Special Kähler} & \overset{c^\star\text{-map}}{\Longrightarrow} & \text{Pseudo-Quaternionic-Kähler} \\
\pi_{\mathrm{TS}} \Downarrow & & \pi_{\mathrm{TS}} \Downarrow \\
(\text{Special Kähler})_{\mathrm{TS}} & \overset{c^\star\text{-map}}{\Longrightarrow} & (\text{Pseudo-Quaternionic-Kähler})_{\mathrm{TS}}
\end{array}
\tag{5.2.26}
$$

By means of this token, we obtain Table 5.2, perfectly analogous to Table 5.1 where the Pseudo-Quaternionic manifolds associated which each symmetric special geometry are organized into five distinct Tits Satake universality classes.

Table 5.2 The eight series of homogenous symmetric special Kähler manifolds (infinite e finite), their Pseudo-Quaternionic counterparts and the grouping of the latter into five Tits Satake universality classes

Special Kähler $\mathcal{SK}_n$	Pseudo-quaternionic $\mathcal{QM}^\star_{4n+4}$	Tits Satake proj. of pseudo quater $\mathcal{QM}^\star_{\mathrm{TS}}$
$\frac{\mathrm{U(s+1,1)}}{\mathrm{U(s+1)\times U(1)}}$	$\frac{\mathrm{U(s+2,2)}}{\mathrm{U(s+1,1)\times U(1,1)}}$	$\frac{\mathrm{U(3,2)}}{\mathrm{U(2,1)\times U(1,1)}}$
$\frac{\mathrm{SU(1,1)}}{\mathrm{U(1)}}$	$\frac{\mathrm{G_{(2,2)}}}{\mathrm{SU(1,1)\times SU(1,1)}}$	$\frac{\mathrm{G_{(2,2)}}}{\mathrm{SU(1,1)\times SU(1,1)}}$
$\frac{\mathrm{SU(1,1)}}{\mathrm{U(1)}} \times \frac{\mathrm{SU(1,1)}}{\mathrm{U(1)}}$	$\frac{\mathrm{SO(3,4)}}{\mathrm{SO(2,1)\times SO(2,2)}}$	$\frac{\mathrm{SO(3,4)}}{\mathrm{SO(1,2)\times SO(2,2)}}$
$\frac{\mathrm{SU(1,1)}}{\mathrm{U(1)}} \times \frac{\mathrm{SO(p+2,2)}}{\mathrm{SO(p+2)\times SO(2)}}$	$\frac{\mathrm{SO(p+4,4)}}{\mathrm{SO(p+2,2)\times SO(2,2)}}$	$\frac{\mathrm{SO(5,4)}}{\mathrm{SO(3,2)\times SO(2,2)}}$
$\frac{\mathrm{Sp(6)}}{\mathrm{U(3)}}$ $\frac{\mathrm{SU(3,3)}}{\mathrm{SU(3)\times SU(3)\times U(1)}}$ $\frac{\mathrm{SO^\star(12)}}{\mathrm{SU(6)}\times U(1)}$ $\frac{\mathrm{E_{(7,-25)}}}{\mathrm{E_{(6,-78)}}\times U(1)}$	$\frac{\mathrm{F_{(4,4)}}}{\mathrm{Sp(6)\times SU(1,1)}}$ $\frac{\mathrm{E_{(6,-2)}}}{\mathrm{SU(3,3)\times SU(1,1)}}$ $\frac{\mathrm{E_{(7,-5)}}}{\mathrm{SO^\star(12)\times SU(1,1)}}$ $\frac{\mathrm{E_{(8,-24)}}}{\mathrm{E_{(7,-25)}\times SU(1,1)}}$	$\frac{\mathrm{F_{(4,4)}}}{\mathrm{Sp(6)\times SU(1,1)}}$

Hence we have the following:

Statement 5.2.1 *The number, structure and properties of* $\mathrm{H}^\star$ *orbits of* $\mathbb{K}$ *nilpotent elements depend only on the Tits Satake universality class and it is an intrinsic property of the class.*

So it suffices to determine the classification of nilpotent orbits for the five manifolds appearing in the third column of Table 5.2.

In Chap. 6 we will work out the details for the simplest case corresponding to the second line in Table 5.2. The details of the algorithm should be clear from such an illustration. In [45] the following case was studied in detail:

$$\mathscr{S}\mathscr{K}\mathscr{O}_{2s+2} \equiv \frac{\mathrm{SU}(1,1)}{\mathrm{U}(1)} \times \frac{\mathrm{SO}(2,2+2\mathrm{s})}{\mathrm{SO}(2)\times\mathrm{SO}(2+2\mathrm{s})} \tag{5.2.27}$$

which corresponds to one of the possible couplings of $2+2s$ vector multiplets.

Upon space-like dimensional reduction to $D=3$ and dualization of all the vector fields, a supergravity model of this type becomes a σ-model with the following quaternionic manifold as target space:

$$\mathscr{Q}\mathscr{M}_{(4,4+2s)} \equiv \frac{\mathrm{U_{D=3}}}{\mathrm{H}} = \frac{\mathrm{SO}(4,4+2\mathrm{s})}{\mathrm{SO}(4)\times\mathrm{SO}(4+2\mathrm{s})}\,. \tag{5.2.28}$$

as mentioned in Table 5.1. If we perform instead a time-like dimensional reduction, as it is relevant for the construction of black-hole solutions, we obtain an Euclidean σ-model where, as mentioned in Table 5.2 the target space is the following pseudo-quaternionic manifold:

$$\mathscr{Q}\mathscr{M}^\star_{(4,4+2s)} \equiv \frac{\mathrm{U_{D=3}}}{\mathrm{H}^\star} = \frac{\mathrm{SO}(4,4+2\mathrm{s})}{\mathrm{SO}(2,2)\times\mathrm{SO}(2,2+2\mathrm{s})}\,. \tag{5.2.29}$$

The Tits Satake projection of all such manifolds is:

$$\mathscr{Q}\mathscr{M}^\star_{\mathrm{TS}} = \frac{\mathrm{U}^{TS}_{\mathrm{D=3}}}{\mathrm{H}^\star_{TS}} = \frac{\mathrm{SO}(4,5)}{\mathrm{SO}(2,3)\times\mathrm{SO}(2,2)}\,. \tag{5.2.30}$$

We refer the reader to [45] for the explicit construction of nilpotent orbits pertaining to this example.

5.3 The Tits Satake Projection

The arguments exposed in the previous section should have convinced the reader of the high relevance of the Tits–Satake projection, both in the context of black-holes and in the context of other geometrical aspects of supergravity theory, a notable one being that of gauging. For this reason the remaining part of this chapter is devoted to the illustration of the rich mathematical theory underlying this projection.

In this section we explain the Tits–Satake projection of a metric solvable Lie algebra and how it is related to the notions of *paint* group $\mathrm{G}_{\mathrm{paint}}$ and *subpaint* group $\mathrm{G}_{\mathrm{subpaint}} \subset \mathrm{G}_{\mathrm{paint}}$. Although the Tits–Satake projection can be defined for general solvable Lie algebras, our main interest is in symmetric spaces and the just mentioned notions have been extracted precisely from the case of the Tits–Satake projections of solvable Lie algebras associated with symmetric spaces Solv(G/H). On these latter we focus.

5.3.1 The TS-Projection for Non Maximally Split Symmetric Spaces

Following the discussion of Sect. 2.4 let us recall that if the scalar manifold of supergravity is a *non maximally noncompact manifold* G/H the Lie algebra of the numerator group is some appropriate real form $\mathbb{G}_R$ of a complex Lie algebra $\mathbb{G}$. The Lie algebra $\mathbb{H}$ of the denominator H is the maximal compact subalgebra $\mathbb{H} \subset \mathbb{G}_R$. Denoting, as usual, by $\mathbb{K}$ the orthogonal complement of $\mathbb{H}$ in $\mathbb{G}_R$:

$$\mathbb{G}_R = \mathbb{H} \oplus \mathbb{K} \tag{5.3.1}$$

and defining as noncompact rank or rank of the coset G/H the dimension of the noncompact Cartan subalgebra (see Eq. (2.4.3), we obtain that $r_{\mathrm{nc}} \leq \mathrm{rank}(\mathbb{G})$, where the equality is the statement that the manifold is *maximally noncompact* (or '*maximally split*').

When the equality is strict, the manifold $\mathrm{G_R/H}$ is still metrically equivalent to a solvable group manifold but the form of the solvable Lie algebra $\mathrm{Solv(G_R/H)}$, whose structure constants define the Nomizu connection, is more complicated than in the maximally non-compact case. It was discussed and explained in Sect. 2.5.1. The Tits–Satake theory of non-compact cosets and split subalgebras is a classical topic in Differential Geometry and appears in some textbooks. Within such a mathematical framework there is a peculiar universal structure of the solvable algebra $\mathrm{Solv(G_R/H)}$ that had not been observed before [31] namely that of paint and subpaint groups which extends beyond symmetric spaces as it was demonstrated in [32].

Explicitly we have the following scheme. One can split the Cartan subalgebra into its compact and non-compact subalgebras as shown in Eq. (2.4.17) and these parts are orthogonal using the Cartan-Killing metric. Therefore, every vector in the dual of the full Cartan subalgebra, in particular every root α, can be decomposed into its transverse and parallel part to $\mathscr{H}^{\mathrm{nc}}$ as it was done in Eq. (2.4.19).

The Tits–Satake projection consists of two steps. First one sets all $\alpha_{\perp} = 0$, projecting the original root system $\Delta_{\mathbb{G}}$ onto a new system of vectors $\overline{\Delta}$ living in a Euclidean space of dimension equal to the non-compact rank r_{nc}. The set $\overline{\Delta}$ is called a restricted root system. It is not an ordinary root system in the sense that roots can occur with multiplicities different from one and $2\alpha_{||}$ can be a root if $\alpha_{||}$ is one. In the second step, one deletes the multiplicities of the restricted roots. Thus we have

$$\Pi_{\mathrm{TS}} \quad : \quad \Delta_{\mathbb{G}} \mapsto \Delta_{\mathrm{TS}},; \qquad \Delta_{\mathbb{G}} \stackrel{\alpha_\perp=0}{\longmapsto} \overline{\Delta} \underset{\text{multiplicities}}{\overset{\text{deleting}}{\longmapsto}} \Delta_{\mathrm{TS}}. \tag{5.3.2}$$

If $\overline{\Delta}$ contains no restricted root that is the double of another one, then Δ_{TS} is a root system of simple type. We will show later that this root subsystem defines a Lie algebra $\mathbb{G}_{\mathrm{TS}}$, *the Tits–Satake subalgebra* of $\mathbb{G}_R$:

$$\Delta_{\mathrm{TS}} = \text{root system of } \mathbb{G}_{\mathrm{TS}}, \qquad \mathbb{G}_{\mathrm{TS}} \subset \mathbb{G}_R. \tag{5.3.3}$$

The Tits–Satake subalgebra $\mathbb{G}_{\mathrm{TS}}$ is, as a consequence of its own definition, the maximally non-compact real section of its own complexification. For this reason, considering its maximal compact subalgebra $\mathbb{H}_{\mathrm{TS}} \subset \mathbb{G}_{\mathrm{TS}}$ we have a new smaller coset $\mathrm{G_{TS}/H_{TS}}$ which is maximally split and whose associated solvable algebra $\mathrm{Solv(G_{TS}/H_{TS})}$ has the standard structure utilized in [29] to prove complete integrability of supergravity compactified to 3 dimensions. This result demonstrates the relevance of the Tits–Satake projection.

In the case doubled restricted roots are present in $\overline{\Delta}$, the projection cannot be expressed in terms of a simple Lie algebra, but the concept remains the same. The root system is the so-called bc_r system, with $r = r_{\mathrm{nc}}$ the non-compact rank of the real form $\mathbb{G}$. It is the root system of a group $\mathrm{G_{TS}}$, which is now non-semi-simple. The manifold is similarly defined as $\mathrm{G_{TS}/H_{TS}}$, where $\mathrm{H_{TS}}$ is the maximal compact subgroup of $\mathrm{G_{TS}}$.

The next question is: what is the relation between the two solvable Lie algebras $\mathrm{Solv(G_R/H)}$ and $\mathrm{Solv(G_{TS}/H_{TS})}$? The answer can be formulated through the following statements A-E.

[A]

In a projection more than one higher dimensional vector can map to the same lower dimensional one. This means that in general there will be several roots of $\Delta_{\mathbb{G}}$ that have the same image in Δ_{TS}. The imaginary roots vanish under this projection, according to the definition of Sect. 2.5. Therefore, apart from these imaginary roots, there are two types of roots: those that have a distinct image in the projected root system and those that arrange into multiplets with the same projection. We can split the root spaces in subsets according to whether there is such a degeneracy or not. Calling $\Delta^+_{\mathbb{G}}$ and Δ^+_{TS} the sets of positive roots of the two root systems, we have the following scheme:

$$\begin{array}{llll} \Delta^+_{\mathbb{G}} & = \Delta^{\eta} & \bigcup \Delta^{\delta} & \bigcup \Delta_{\mathrm{comp}} \\ \downarrow \Pi_{\mathrm{TS}} & \downarrow \Pi_{\mathrm{TS}} & \downarrow \Pi_{\mathrm{TS}} & \\ \Delta^+_{\mathrm{TS}} & = \Delta^{\ell}_{\mathrm{TS}} & \bigcup \Delta^{s}_{\mathrm{TS}} & \end{array}$$

$$\forall \alpha^\ell \in \Delta^\ell_{\mathrm{TS}} \;:\; \dim \Pi^{-1}_{\mathrm{TS}}\left[\alpha^\ell\right] = 1, \qquad \forall \alpha^s \in \Delta^s_{\mathrm{TS}} \;:\; \dim \Pi^{-1}_{\mathrm{TS}}\left[\alpha^s\right] = m[\alpha^s] > 1. \tag{5.3.4}$$

The δ part thus contains all the roots that have multiplicities under the Tits–Satake projection while the roots in the η part have no multiplicities. These roots of type η are orthogonal to Δ_{comp}. Indeed, this follows from the fact that for any two root vectors α and β where there is no root of the form $\beta + m\alpha$ with m a non-zero integer, the inner product of β and α vanishes. It also follows from this definition that in maximally split symmetric spaces, in which case $\Delta_{\text{comp}} = \emptyset$, all root vectors are in Δ^η or Δ^ℓ (as the Tits–Satake projection is then trivialized).

These subsets moreover satisfy the following properties under addition of root vectors:

$\mathbb{G}$	$\mathbb{G}_{\text{TS}}$
$\Delta^\eta + \Delta^\eta \subset \Delta^\eta$	$\Delta^\ell_{\text{TS}} + \Delta^\ell_{\text{TS}} \subset \Delta^\ell_{\text{TS}}$
$\Delta^\eta + \Delta^\delta \subset \Delta^\delta$	$\Delta^\ell_{\text{TS}} + \Delta^s_{\text{TS}} \subset \Delta^s_{\text{TS}}$
$\Delta^\delta + \Delta^\delta \subset \Delta^\eta \bigcup \Delta^\delta$	$\Delta^s_{\text{TS}} + \Delta^s_{\text{TS}} \subset \Delta^\ell_{\text{TS}} \bigcup \Delta^s_{\text{TS}}$
$\Delta_{\text{comp}} + \Delta^\eta = \emptyset$	
$\Delta_{\text{comp}} + \Delta^\delta \subset \Delta^\delta$	

(5.3.5)

Because of this structure we can enumerate the generators of the solvable algebra $\text{Solv}(\text{G}_\text{R}/\text{H})$ in the following way:

$$\begin{aligned} \text{Solv}(\text{G}_\text{R}/\text{H}) &= \left\{H_i, \Phi_{\alpha^\ell}, \Omega_{\alpha^s|I}\right\} \\ H_i &\Rightarrow \text{Cartan generators} \\ \Phi_{\alpha^\ell} &\Rightarrow \eta - \text{roots} \\ \Omega_{\alpha^s|I} &\Rightarrow \delta - \text{roots} \quad ; \quad (I = 1, \ldots, m[\alpha^s]). \end{aligned} \tag{5.3.6}$$

The index I enumerating the m-roots of $\Delta_{\mathbb{G}_R}$ that have the same projection in Δ_{TS} is named the *paint index*.

[B]

There exists a *compact subalgebra* $\mathbb{G}_{\text{paint}} \subset \mathbb{G}_R$ which acts as an algebra of outer automorphisms (i.e. outer derivatives) of the solvable algebra $\text{Solv}_{\mathbb{G}_R} \equiv \text{Solv}(\text{G}_\text{R}/\text{H}) \subset \mathbb{G}_R$, namely:

$$\left[\mathbb{G}_{\text{paint}}\, , \, \text{Solv}_{\mathbb{G}_R}\right] \subset \text{Solv}_{\mathbb{G}_R}. \tag{5.3.7}$$

[C]

The Cartan generators H_i and the generators Φ_{α^ℓ} are singlets under the action of $\mathbb{G}_{\text{paint}}$, i.e. each of them commutes with the whole of $\mathbb{G}_{\text{paint}}$:

$$\left[H_i\, , \, \mathbb{G}_{\text{paint}}\right] = \left[\Phi_{\alpha^\ell}\, , \, \mathbb{G}_{\text{paint}}\right] = 0 \tag{5.3.8}$$

On the other hand, each of the multiplets of generators $\Omega_{\alpha^s|I}$ constitutes an orbit under the adjoint action of the paint group G_{paint}, i.e. a linear representation $\mathbf{D}[\alpha^s]$ which, for different roots α^s can be different:

$$\forall X \in \mathbb{G}_{\text{paint}} \quad : \quad \left[X \, , \, \Omega_{\alpha^s|I}\right] = \left(D^{[\alpha^s]}[X]\right)_I^{\ J} \, \Omega_{\alpha^s|J} \tag{5.3.9}$$

[D]

The *paint algebra* $\mathbb{G}_{\text{paint}}$ contains a subalgebra

$$\mathbb{G}^0_{\text{subpaint}} \subset \mathbb{G}_{\text{paint}} \tag{5.3.10}$$

such that with respect to $\mathbb{G}^0_{\text{subpaint}}$, each $m[\alpha^s]$-dimensional representation $\mathbf{D}[\alpha^s]$ branches as follows:

$$\mathbf{D}[\alpha^s] \stackrel{\mathbb{G}^0_{\text{subpaint}}}{\Longrightarrow} \underbrace{\mathbf{1}}_{\text{singlet}} \oplus \underbrace{\mathbf{J}}_{(m[\alpha^s]-1)-\text{dimensional}} \tag{5.3.11}$$

Accordingly we can split the range of the multiplicity index I as follows:

$$I = \{0, x\} \, , \qquad x = 1, \ldots, m[\alpha^s] - 1. \tag{5.3.12}$$

The index 0 corresponds to the singlet, while x ranges over the representation $\mathbf{J}$.

[E]

The tensor product $\mathbf{J} \otimes \mathbf{J}$ contains both the identity representation $\mathbf{1}$ and the representation $\mathbf{J}$ itself. Furthermore, there exists, in the representation $\bigwedge^3 \mathbf{J}$ a $\mathbb{G}^0_{\text{subpaint}}$-invariant tensor a^{xyz} such that the two solvable Lie algebras $\text{Solv}_{\mathbb{G}_R}$ and $\text{Solv}_{\mathbb{G}_{\text{TS}}}$ can be written as follows

$\text{Solv}_{\mathbb{G}_R}$	$\text{Solv}_{\mathbb{G}_{\text{TS}}}$
$\left[H_i \, , \, H_j\right] = 0$	$\left[H_i \, , \, H_j\right] = 0$
$\left[H_i \, , \, \Phi_{\alpha^\ell}\right] = \alpha_i^\ell \, \Phi_{\alpha^\ell}$	$\left[H_i \, , \, E^{\alpha^\ell}\right] = \alpha_i^\ell$
$\left[H_i \, , \, \Omega_{\alpha^s\|I}\right] = \alpha_i^s \, \Omega_{\alpha^s\|I}$	$\left[H_i \, , \, E^{\alpha^s}\right] = \alpha_i^s \, E^{\alpha^s}$
$\left[\Phi_{\alpha^\ell} \, , \, \Phi_{\beta^\ell}\right] = N_{\alpha^\ell \beta^\ell} \, \Phi_{\alpha^\ell + \beta^\ell}$	$\left[E^{\alpha^\ell} \, , \, E^{\beta^\ell}\right] = N_{\alpha^\ell \beta^\ell} \, E^{\alpha^\ell + \beta^\ell}$
$\left[\Phi_{\alpha^\ell} \, , \, \Omega_{\beta^s\|I}\right] = N_{\alpha^\ell \beta^s} \, \Omega_{\alpha^\ell + \beta^s\|I}$	$\left[E^{\alpha^\ell} \, , \, E^{\beta^s}\right] = N_{\alpha^\ell \beta^s} \, E^{\alpha^\ell + \beta^s}$
If $\alpha^s + \beta^s \in \Delta^\ell_{\text{TS}}$: $\left[\Omega_{\alpha^s\|I} \, , \, \Omega_{\beta^s\|J}\right] = \delta^{IJ} \, N_{\alpha^s \beta^s} \, \Phi_{\alpha^s + \beta^s}$	$\left[E^{\alpha^s} \, , \, E^{\beta^s}\right] = N_{\alpha^s \beta^s} \, E^{\alpha^s + \beta^s}$
If $\alpha^s + \beta^s \in \Delta^s_{\text{TS}}$: $\begin{cases} \left[\Omega_{\alpha^s\|0} \, , \, \Omega_{\beta^s\|0}\right] = N_{\alpha^s \beta^s} \, \Omega_{\alpha^s + \beta^s\|0} \\ \left[\Omega_{\alpha^s\|0} \, , \, \Omega_{\beta^s\|x}\right] = N_{\alpha^s \beta^s} \, \Omega_{\alpha^s + \beta^s\|x} \\ \left[\Omega_{\alpha^s\|x} \, , \, \Omega_{\beta^s\|y}\right] = N_{\alpha^s \beta^s} \, \left(\delta^{xy} \, \Omega_{\alpha^s + \beta^s\|0} + a^{xyz} \, \Omega_{\alpha^s + \beta^s\|z}\right) \end{cases}$	$\left[E^{\alpha^s} \, , \, E^{\beta^s}\right] = N_{\alpha^s \beta^s} \, E^{\alpha^s + \beta^s}$

(5.3.13)

where $N_{\alpha\beta} = 0$ if $\alpha + \beta \notin \Delta_{\text{TS}}$.

5.3.2 Paint and Subpaint Groups in an Example

We now want to illustrate the general structure described in the previous subsection through the analysis of a specific example of a non maximally split symmetric space. This will be both educational in order to clarify the notion of Tits–Satake projection and instrumental to extract a general systematics for the paint and subpaint groups, which we will later recognize in the entire classification of supergravity relevant symmetric spaces.

Hence let us consider the following quaternionic Kähler manifold:

$$\frac{\mathrm{G}_R}{\mathrm{H}} = \frac{\mathrm{E}_{8(-24)}}{\mathrm{E}_{7(-133)} \times \mathrm{SU}(2)} \tag{5.3.14}$$

which, according to Table 5.1 is the c-map image of the following special Kähler manifold

$$\frac{\mathrm{E}_{7(-25)}}{\mathrm{E}_{6(-78)} \times \mathrm{U}(1)} \tag{5.3.15}$$

The quaternionic nature of the chosen non maximally split symmetric space is signaled by the presence of the SU(2) factor in the denominator group and it is confirmed by the decomposition of the adjoint representation of the numerator group:

$$\mathbf{248} \overset{\mathrm{E}_{7(-133)}\times\mathrm{SU}(2)}{\Longrightarrow} (\mathbf{133}, \mathbf{1}) \oplus (\mathbf{1}, \mathbf{3}) \oplus (\mathbf{56}, \mathbf{2}) \tag{5.3.16}$$

Indeed the $4 \times 28 = 112$ coset generators being in the $(\mathbf{56}, \mathbf{2})$ of $\mathrm{E}_{7(-133)} \times \mathrm{SU}(2)$ are SU(2) doublets and transform symplectically under USp(56) transformations due to the symplectic embedding of the **56** representation of the compact E_7 group.

The quaternionic structure, however, is not relevant to our present discussion that focuses on the mechanisms of the Tits–Satake projection. By means of this latter we obtain the following result:

$$\Pi_{\mathrm{TS}} \quad : \quad \frac{\mathrm{E}_{8(-24)}}{\mathrm{E}_{7(-133)} \times \mathrm{SU}(2)} \longrightarrow \frac{\mathrm{F}_{4(4)}}{\mathrm{USp}(6) \times \mathrm{SU}(2)} \tag{5.3.17}$$

and we just note that the projected manifold is still quaternionic for similar reasons to those of (5.3.16). So the maximal non-compact Lie algebra $\mathrm{F}_{4(4)}$ is the Tits–Satake subalgebra of $\mathrm{E}_{8(-24)}$. Let us see how this happens, following step by step the scheme described in the previous section.

The rank of the complex E_8 algebra is 8 and, and in its real section $\mathrm{E}_{8(-24)}$ we can distinguish 4 compact and 4 non-compact Cartan generators. In a Euclidean orthonormal basis the complete E_8 root system is composed of the following 240 roots:

$$\Delta_{E_8} \equiv \left\{ \begin{array}{lr} \pm\varepsilon_i \pm \varepsilon_j & (i \neq j)\ \mathbf{112} \\ \underbrace{\pm\tfrac{1}{2}\varepsilon_1 \pm \tfrac{1}{2}\varepsilon_2 \pm \tfrac{1}{2}\varepsilon_3 \pm \tfrac{1}{2}\varepsilon_4 \pm \tfrac{1}{2}\varepsilon_5 \pm \tfrac{1}{2}\varepsilon_6 \pm \tfrac{1}{2}\varepsilon_7 \pm \tfrac{1}{2}\varepsilon_8}_{\text{even number of minus signs}} & \mathbf{128} \\ \hline & \mathbf{240} \end{array} \right\}, \tag{5.3.18}$$

and a convenient choice of the simple roots is provided by the following ones:

$$\begin{aligned} \alpha_1 &= \{0, 1, -1, 0, 0, 0, 0, 0\}, \\ \alpha_2 &= \{0, 0, 1, -1, 0, 0, 0, 0\}, \\ \alpha_3 &= \{0, 0, 0, 1, -1, 0, 0, 0\}, \\ \alpha_4 &= \{0, 0, 0, 0, 1, -1, 0, 0\}, \\ \alpha_5 &= \{0, 0, 0, 0, 0, 1, -1, 0\}, \\ \alpha_6 &= \{0, 0, 0, 0, 0, 1, 1, 0\}, \\ \alpha_7 &= \left\{-\frac{1}{2}, -\frac{1}{2}, -\frac{1}{2}, -\frac{1}{2}, -\frac{1}{2}, -\frac{1}{2}, -\frac{1}{2}, -\frac{1}{2}\right\}, \\ \alpha_8 &= \{1, -1, 0, 0, 0, 0, 0, 0\}. \end{aligned} \tag{5.3.19}$$

The corresponding Dynkin diagram is displayed in Fig. 5.2. where the roots α_3, α_4, α_5, α_6 have been marked in black. This indicates that these simple roots are imaginary, and Cartan generators as e.g. $\alpha_3^i \mathscr{H}_i$ belong to $\mathscr{H}^{\text{comp}}$. In this way these diagrams define both the real form $E_{8(-24)}$ and the corresponding Tits–Satake projection of the root system. The non-compact CSA $\mathscr{H}^{\text{nc}}$ is the orthogonal complement of $\mathscr{H}^{\text{comp}}$. Let us also note that the black roots form the Dynkin diagram of a D_4 algebra, *i.e* in its compact form the Lie algebra of SO(8). This is the origin of the paint group

$$\mathrm{G}_{\text{paint}} = \mathrm{SO}(8), \tag{5.3.20}$$

pertaining to this example. We shall identify it in a moment, but let us first perform the Tits–Satake projection on the root system. This case is particularly simple since the span of the simple imaginary roots $\boldsymbol{\alpha_3}, \boldsymbol{\alpha_4}, \boldsymbol{\alpha_5}, \boldsymbol{\alpha_6}$ is just given by the Euclidean space along the orthonormal axes ε_4, ε_5 ε_6, ε_7. The Euclidean space along the orthonormal axes ε_1, ε_2 ε_3, ε_8 is the non-compact CSA. Note that this is not the same as the span of $\boldsymbol{\alpha_1}, \boldsymbol{\alpha_2}, \boldsymbol{\alpha_7}, \boldsymbol{\alpha_8}$. Denoting the components of root vectors in the basis ε_i by α^i, the splitting (2.4.19) is very simple. We just have:

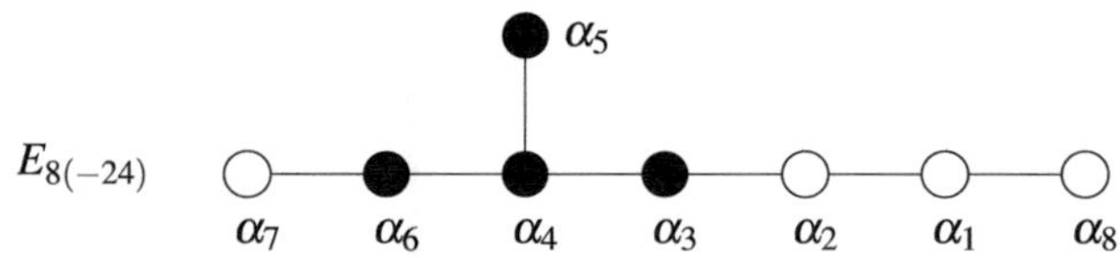

Fig. 5.2 The Tits–Satake diagram of $E_{8(-24)}$, rank = 8, split rank = 4, $\mathrm{G_{TS}} = \mathrm{F}_{4(4)}$

$$\alpha_\perp = \left\{\alpha^4\,,\,\alpha^5\,,\,\alpha^6\,,\,\alpha^3\right\} \quad ; \quad \alpha_\parallel = \left\{\alpha^1\,,\,\alpha^2\,,\,\alpha^7\,,\,\alpha^8\right\}, \tag{5.3.21}$$

and the projection (5.3.2) immediately yields the following restricted root system:

$$\Delta_{\mathrm{TS}} = \left\{\begin{array}{lr} \pm\varepsilon_i \pm \varepsilon_j \qquad (i \neq j \ ; \ i, j = 1, 2, 3, 8) & \mathbf{24} \\ \pm\varepsilon_i \qquad (i = 1, 2, 3, 8) & \mathbf{8} \\ \pm\frac{1}{2}\varepsilon_1 \pm \frac{1}{2}\varepsilon_2 \pm \frac{1}{2}\varepsilon_3 \pm \frac{1}{2}\varepsilon_8 & \mathbf{16} \\ \hline & \mathbf{48} \end{array}\right\}, \tag{5.3.22}$$

which can be recognized to be the root system of the simple complex algebra F_4.

With reference to the notations introduced in the previous section let us now identify the subsets Δ^η and Δ^δ in the positive root subsystem of $\Delta^+_{E_8}$ and their corresponding images in the projection, namely $\Delta^\ell_{\mathrm{TS}}$ and Δ^s_{TS}.

Altogether, performing the projection the following situation is observed:

- There are 24 roots that have null projection on the non-compact space, namely

$$\alpha_\parallel = 0 \Leftrightarrow \alpha = \pm\varepsilon_i \pm \varepsilon_j \quad ; \quad i, j = 4, 5, 6, 7. \tag{5.3.23}$$

  These roots, together with the four compact Cartan generators, form the root system of a D_4 algebra, whose dimension is exactly 28. In the chosen real form such a subalgebra of $\mathrm{E}_{8(-24)}$ is the compact algebra SO(8) and its exponential acts as the paint group, as already mentioned in (5.3.20). All the remaining roots have a non-vanishing projection on the compact space. In particular:
- There are 12 positive roots of E_8 that are exactly projected on the 12 positive long roots of F_4, namely the first line of (5.3.22), which we therefore identify with $\Delta^\ell_{\mathrm{TS}}$. For these roots we have $\alpha_\perp = 0$ and they constitute the Δ^η system mentioned above:

$$\Delta^+_{E_8} \supset \Delta^\eta_{\mathrm{TS}} = \left\{\varepsilon_i \pm \varepsilon_j\right\} = \Delta^\ell_{\mathrm{TS}} \quad ; \quad i < j \quad ; \quad i, j = 1, 2, 3, 8 \tag{5.3.24}$$

- There are 8 different positive roots of E_8 that have the same projection on each of the $12 = 4 \oplus 8$ positive short roots of F_4, i.e. the second and third line of (5.3.22). Namely the remaining $12 \times 8 = 96$ roots of E_8 are all projected on short roots of F_4. The set of F_4 positive short roots can be split as follows:

$$\begin{array}{|lr|c|} \hline \Delta^s_{\mathrm{TS}} = \Delta^s_{\mathrm{vec}} \bigcup \Delta^s_{\mathrm{spin}} \bigcup \Delta^s_{\overline{\mathrm{spin}}} & & \\ \Delta^s_{\mathrm{vec}} = \{\varepsilon_i\} & i = 1, 2, 3, 8 & \mathbf{4} \\ \Delta^s_{\mathrm{spin}} = \underbrace{\pm\frac{1}{2}\varepsilon_1 \pm \frac{1}{2}\varepsilon_2 \pm \frac{1}{2}\varepsilon_3 + \frac{1}{2}\varepsilon_8}_{\text{even number of minus signs}} & & \mathbf{4} \\ \Delta^s_{\overline{\mathrm{spin}}} = \underbrace{\pm\frac{1}{2}\varepsilon_1 \pm \frac{1}{2}\varepsilon_2 \pm \frac{1}{2}\varepsilon_3 + \frac{1}{2}\varepsilon_8}_{\text{odd number of minus signs}} & & \mathbf{4} \\ \hline & & \mathbf{12} \\ \hline \end{array} \tag{5.3.25}$$

Correspondingly the subset $\Delta^\delta \subset \Delta_{E_8}$ defined by its projection property $\Pi_{TS}\left(\Delta^\delta\right) = \Delta^s_{TS}$ is also split in three subsets as follows:

$$\begin{array}{|lcl|c|c|}
\hline
\Delta^\delta_+ & = & \Delta^\delta_{vec} \bigcup \Delta^\delta_{spin} & & \\
\Delta^\delta_{vec} & = & \left\{ \underbrace{\varepsilon_i}_{\alpha_\parallel} \oplus \underbrace{(\pm\varepsilon_j)}_{\alpha_\perp} \right\} \quad , \quad \begin{pmatrix} i = 1,2,3,8 \\ j = 4,5,6,7 \end{pmatrix} & 4\times\mathbf{8} & 32 \\
\Delta^\delta_{spin} & = & \left\{ \underbrace{\left(\pm\tfrac{1}{2}\varepsilon_1 \pm \tfrac{1}{2}\varepsilon_2 \pm \tfrac{1}{2}\varepsilon_3 + \tfrac{1}{2}\varepsilon_8\right)}_{\alpha_\parallel \quad \text{even \# of } - \text{signs}} \oplus \underbrace{\left(\pm\tfrac{1}{2}\varepsilon_4 \pm \tfrac{1}{2}\varepsilon_5 \pm \tfrac{1}{2}\varepsilon_6 \pm \tfrac{1}{2}\varepsilon_7\right)}_{\alpha_\perp \quad \text{even \# of} - \text{signs}} \right\} & 4\times\mathbf{8} & 32 \\
\Delta^\delta_{\overline{spin}} & = & \left\{ \underbrace{\left(\pm\tfrac{1}{2}\varepsilon_1 \pm \tfrac{1}{2}\varepsilon_2 \pm \tfrac{1}{2}\varepsilon_3 + \tfrac{1}{2}\varepsilon_8\right)}_{\alpha_\parallel \quad \text{odd \# of } - \text{ signs}} \oplus \underbrace{\left(\pm\tfrac{1}{2}\varepsilon_4 \pm \tfrac{1}{2}\varepsilon_5 \pm \tfrac{1}{2}\varepsilon_6 \pm \tfrac{1}{2}\varepsilon_7\right)}_{\alpha_\perp \quad \text{odd \# of } -} \right\} & 4\times\mathbf{8} & 32 \\
\hline
 & & & & \mathbf{96} \\
\hline
\end{array} \tag{5.3.26}$$

We can now verify the general statements made in the previous sections about the paint group representations to which the various roots are assigned. First of all we see that, as we claimed, the long roots of F_4, namely those 12 given in (5.3.24) are singlets under the paint group $G_{paint} = SO(8)$. All other roots fall into multiplets with the same Tits–Satake projection and each of these latter has always the same multiplicity, in our case $m = 8$ (compare with (5.3.9)). So the short roots of $F_{4(4)}$ fall into 8-dimensional representations of $G_{paint} = SO(8)$. But which ones? SO(8) has three kind of octets $\mathbf{8_v}$, $\mathbf{8_s}$ and $\mathbf{8_{\bar{s}}}$ and, as we stated, not every root α_s of the Tits–Satake algebra $\mathbb{G}_{TS}$ falls in the same representation $\mathbf{D}$ of the paint group although in this case all $\mathbf{D}[\alpha^s]$ have the same dimension. Looking back at our result we easily find the answer. The 4 positive roots in the subset Δ^δ_{vec} have as compact part $\alpha_\perp$ the weights of the vector representation of SO(8). Hence the roots of Δ^δ_{vec} are assigned to the $\mathbf{8_v}$ of the paint group. The 4 positive roots in Δ^δ_{spin} have instead as compact part the weights of the spinor representation of SO(8) and so they are assigned to the $\mathbf{8_s}$ irreducible representation. Finally, with a similar argument, we see that the 4 roots of $\Delta^\delta_{\overline{spin}}$ are in the conjugate spinor representation $\mathbf{8_{\bar{s}}}$. The last part of the general discussion of Sect. 5.3.1 is now easy to verify in the context of our example, namely that relevant to the subpaint group $G^0_{subpaint}$ (we will omit sometimes the 'subpaint' indication for convenience). According to (5.3.10)–(5.3.11) we have to find a subgroup $G^0 \subset SO(8)$ such that under reduction with respect to it, the three octet representations branch simultaneously as:

$$\begin{aligned}
\mathbf{8_v} &\xrightarrow{G^0} \mathbf{1} \oplus \mathbf{7}, \\
\mathbf{8_s} &\xrightarrow{G^0} \mathbf{1} \oplus \mathbf{7}, \\
\mathbf{8_{\bar{s}}} &\xrightarrow{G^0} \mathbf{1} \oplus \mathbf{7}.
\end{aligned} \tag{5.3.27}$$

Such group G^0 exists and it is uniquely identified as the 14 dimensional $\mathrm{G}_{2(-14)}$. Hence the subpaint group is $\mathrm{G}_{2(-14)}$. Considering now (5.3.13) we see that the commutation relations of the solvable Lie algebra $\mathrm{Solv}\left(\mathrm{E}_{8(-24)}/\mathrm{E}_{7(-133)}\times\mathrm{SU}(2)\right)$ precisely fall into the general form displayed in the first column of that table with the index $x = 1,\ldots,7$ spanning the fundamental 7-dimensional representation of $\mathrm{G}_{2(-14)}$ and the invariant antisymmetric tensor a^{xyz} being given by the $\mathrm{G}_{2(-14)}$-invariant octonionic structure constants. Indeed the representation $\mathbf{J}$ mentioned in Sect. 5.3.1 is the fundamental $\mathbf{7}$ and we have the decomposition:

$$\mathbf{7}\times\mathbf{7} = \underbrace{\mathbf{14}\oplus\mathbf{7}}_{\text{antisymmetric}}\ \oplus\underbrace{\mathbf{27}\oplus\mathbf{1}}_{\text{symmetric}}. \tag{5.3.28}$$

This shows that, as claimed in point [E] of the general discussion, the tensor product $\mathbf{J}\times\mathbf{J}$ contains both the singlet and $\mathbf{J}$.

In the example that is extensively discussed in [31], namely

$$\Pi_{\mathrm{TS}}\ :\ \frac{\mathrm{E}_{7(-5)}}{\mathrm{SO}(12)\times\mathrm{SU}(2)}\ \longrightarrow\ \frac{\mathrm{F}_{4(4)}}{\mathrm{USp}(6)\times\mathrm{SU}(2)} \tag{5.3.29}$$

the image of the Tits–Satake projection yields the same maximally split coset as in the case presently illustrated, although the original manifold is a different one. The only difference that distinguishes the two cases resides in the paint group. There we have $\mathrm{G}_{\mathrm{paint}} = \mathrm{SO}(3)\times\mathrm{SO}(3)\times\mathrm{SO}(3)$ and the subpaint group is identified as $\mathrm{G}^0_{\mathrm{subpaint}} = \mathrm{SO}(3)_{\mathrm{diag}}$. Correspondingly the index $x = 1, 2, 3$ spans the triplet representation of $\mathrm{SO}(3)$ which is the $\mathbf{J}$ appropriate to that case and the invariant tensor a^{xyz} is given by the Levi-Civita symbol ε^{xyz}.

Let us now consider the group theoretical meaning of the splitting of $\mathrm{F}_{4(4)}$ roots into the three subsets $\Delta^s_{\mathrm{vec}}, \Delta^s_{\mathrm{spin}}, \Delta^s_{\mathrm{TS},\overline{\mathrm{spin}}}$, which are assigned to different representations of the paint group $\mathrm{SO}(8)$. This is easily understood if we recall that there exists a subalgebra $\mathrm{SO}(4,4)\subset\mathrm{F}_{4(4)}$ with respect to which we have the following branching rule of the adjoint representation of $\mathrm{F}_{4(4)}$:

$$\mathbf{52}\ \overset{\mathrm{SO}(4,4)}{\rightarrow}\ \mathbf{28}^{\mathrm{nc}}\oplus\mathbf{8}^{\mathrm{nc}}_v\oplus\mathbf{8}^{\mathrm{nc}}_s\oplus\mathbf{8}^{\mathrm{nc}}_{\bar{s}} \tag{5.3.30}$$

The superscript nc is introduced just in order to recall that these are representations of the non-compact real form $\mathrm{SO}(4,4)$ of the D_4 Lie algebra. By $\mathbf{28}$, $\mathbf{8}_v$, $\mathbf{8}_s$ and $\mathbf{8}_{\bar{s}}$ we have already denoted and we continue to denote the homologous representations in the compact real form $\mathrm{SO}(8)$ of the same Lie algebra. The algebra $\mathrm{SO}(4,4)$ is regularly embedded and therefore its Cartan generators are the same as those of $\mathrm{F}_{4(4)}$. The 12 positive long roots of $\mathrm{F}_{4(4)}$ are the only positive roots of $\mathrm{SO}(4,4)$, while the three sets $\Delta^s_{\mathrm{vec}}, \Delta^s_{\mathrm{spin}}, \Delta^s_{\overline{\mathrm{spin}}}$ just correspond to the positive weights of the three representations $\mathbf{8}^{\mathrm{nc}}_v$, $\mathbf{8}^{\mathrm{nc}}_s$ and $\mathbf{8}^{\mathrm{nc}}_{\bar{s}}$, respectively. This is in agreement with the branching rule (5.3.30). So the conclusion is that the different paint group representation assignments of the various root subspaces correspond to the decomposition of the Tits–Satake algebra $F_{4(4)}$ with respect to what we can call the *sub Tits–Satake algebra* $\mathbb{G}_{subTS} = SO(4,4)$. We can just wonder how the concept of sub Tits–Satake algebra can be defined. This is very simple and obvious from our example. G_{subTS} is the normalizer of the paint group G_{paint} within the original group $G_{\mathbb{R}}$. Indeed there is a maximal subgroup:

$$SO(4,4) \times SO(8) \subset E_{8(-24)}, \tag{5.3.31}$$

with respect to which the adjoint of $E_{8(-24)}$ branches as follows:

$$\mathbf{248} \stackrel{SO(4,4)\times SO(8)}{\longrightarrow} (1, \mathbf{28}) \oplus (\mathbf{28}^{nc}, \mathbf{1}) \oplus (\mathbf{8}^{nc}_{\mathbf{v}}, \mathbf{8}_{\mathbf{v}}) \oplus (\mathbf{8}^{nc}_{\mathbf{s}}, \mathbf{8}_{\mathbf{s}}) \oplus (\mathbf{8}^{nc}_{\bar{\mathbf{s}}}, \mathbf{8}_{\bar{\mathbf{s}}}) \tag{5.3.32}$$

and the last three terms in this decomposition display the pairing between representations of the paint group and representations of the sub Tits–Satake group. Alternatively we can view the *subpaint group* $G^0_{subpaint} = G_{2(-14)}$ as the *normalizer* of the Tits–Satake subgroup $G_{TS} = F_{4(4)}$ within the original group $G_{\mathbb{R}} = E_{8(-24)}$. Indeed we have a subgroup

$$F_{4(4)} \times G_{2(-14)} \subset E_{8(-24)}, \tag{5.3.33}$$

such that the adjoint of $E_{8(-24)}$ branches as follows:

$$\mathbf{248} \stackrel{F_{4(4)}\times G_{2(-14)}}{\longrightarrow} (\mathbf{52}, \mathbf{1}) \oplus (\mathbf{1}, \mathbf{14}) \oplus (\mathbf{26}, \mathbf{7}) \tag{5.3.34}$$

The two decompositions (5.3.32) and (5.3.34) lead to the same decomposition with respect to the intersection group:

$$\begin{aligned} G_{intsec} &\equiv \left(G_{TS} \times G^0_{subpaint}\right) \bigcap \left(G_{subTS} \times G_{paint}\right) = G_{subTS} \times G^0_{subpaint} \\ &= \left(F_{4(4)} \times G_{2(-14)}\right) \bigcap (SO(4,4) \times SO(8)) = SO(4,4) \times G_{2(-14)}. \end{aligned} \tag{5.3.35}$$

We find

$$\begin{aligned} \mathbf{248} \rightarrow\ & (\mathbf{1}, \mathbf{14}) \oplus (\mathbf{1}, \mathbf{7}) \oplus (\mathbf{1}, \mathbf{7}) \oplus (\mathbf{8}^{nc}_{v}, \mathbf{7}) \oplus (\mathbf{8}^{nc}_{s}, \mathbf{7}) \oplus (\mathbf{8}^{nc}_{\bar{s}}, \mathbf{7}) \\ & \oplus (\mathbf{28}^{nc}, \mathbf{1}) \oplus (\mathbf{8}^{nc}_{v}, \mathbf{1}) \oplus (\mathbf{8}^{nc}_{s}, \mathbf{1}) \oplus (\mathbf{8}^{nc}_{\bar{s}}, \mathbf{1}). \end{aligned} \tag{5.3.36}$$

The adjoint of the Tits–Satake subalgebra $G_{TS} = F_{4(4)}$ is reconstructed by collecting together all the singlets with respect to the subpaint group $G^0_{subpaint}$. Alternatively the adjoint of the paint algebra $G_{paint} = SO(8)$ is reconstructed by collecting together all the singlets with respect to the *sub Tits–Satake algebra* $G_{subTS} = SO(4,4)$.

Finally, we can recognize the sub Tits–Satake algebra as the algebra generated by the CSA and roots Δ^{ℓ} (and their negatives) in the decomposition (5.3.4).

5.3.3 TS Projection for the Normed Solvable Algebras of Homogenous Special Manifolds

After our detailed discussion of the Tits–Satake projection in the above example of a specific symmetric space we can extract a general scheme that applies to all normal solvable Lie algebras. Let us discuss how the Tits–Satake projection can be reformulated relying on the paint and subpaint group structures. In Sect. 5.3.1 our starting point was the geometrical projection of the root system $\Delta_{\mathbb{G}}$ onto the non-compact Cartan subalgebra by setting, for each root $\alpha \in \Delta_{\mathbb{G}}$ its compact part $\alpha_{\perp}$ to zero. This is the operation that is no longer available in the general case of a solvable algebra. We now only have the solvable algebra, which corresponds to the non-compact part $\alpha_{\parallel}$. Indeed at the level of the solvable Lie algebra there is no notion of the compact Cartan generators. However, the structures that still persist and allow us to define the *Tits–Satake projection* are those of paint and subpaint groups. Indeed for all the solvable Lie algebras $\mathrm{Solv}\,(\mathscr{M})$ considered in the classification of homogeneous special geometries the following statements A-E are true:

[A1]

There exists a *compact algebra* $\mathbb{G}_{\text{paint}}$ which acts as an algebra of outer automorphisms (i.e. outer derivatives) of the solvable algebra $\mathrm{Solv}\,(\mathscr{M})$. The algebra $\mathbb{G}_{\text{paint}}$ is rigorously defined as follows. Given the solvable Lie algebra $\mathrm{Solv}\,(\mathscr{M})$ the corresponding Riemannian manifold $\mathscr{M} = \exp\,[\mathrm{Solv}\,(\mathscr{M})]$ has an algebra of isometries $\mathbb{G}^{\text{iso}}_{\mathscr{M}}$, which is normally larger than $\mathrm{Solv}\,(\mathscr{M})$, and for all special homogeneous manifolds $\mathscr{M}$ such algebras were studied and completely classified in [4, 5]. Obviously $\mathrm{Solv}\,(\mathscr{M}) \subset \mathbb{G}^{\text{iso}}_{\mathscr{M}}$. Let us define the subalgebra of automorphisms of the solvable Lie algebra in the standard way:

$$\begin{aligned} \mathbb{G}^{\text{iso}}_{\mathscr{M}} \supset \mathrm{Aut}\,[\mathrm{Solv}\,(\mathscr{M})] &= \\ \left\{ X \in \mathbb{G}^{\text{iso}}_{\mathscr{M}} \mid \forall \Psi \in \mathrm{Solv}\,(\mathscr{M}) \; : \; [X\,,\,\Psi] \in \mathrm{Solv}\,(\mathscr{M}) \right\} \end{aligned} \tag{5.3.37}$$

By its own definition the algebra $\mathrm{Aut}\,[\mathrm{Solv}\,(\mathscr{M})]$ contains $\mathrm{Solv}\,(\mathscr{M})$ as an ideal. Hence we can define the algebra of external automorphisms as the quotient:

$$\mathrm{Aut}_{\text{Ext}}\,[\mathrm{Solv}\,(\mathscr{M})] \equiv \frac{\mathrm{Aut}\,[\mathrm{Solv}\,(\mathscr{M})]}{\mathrm{Solv}\,(\mathscr{M})}, \tag{5.3.38}$$

and we identify $\mathbb{G}_{\text{paint}}$ as the maximal compact subalgebra of $\mathrm{Aut}_{\text{Ext}}\,[\mathrm{Solv}\,(\mathscr{M})]$. Actually we immediately see that

$$\mathbb{G}_{\text{paint}} = \mathrm{Aut}_{\text{Ext}}\,[\mathrm{Solv}\,(\mathscr{M})]\,. \tag{5.3.39}$$

Indeed, as a consequence of its own definition the algebra $\mathrm{Aut}_{\text{Ext}}\,[\mathrm{Solv}\,(\mathscr{M})]$ is composed of isometries which belong to the stabilizer subalgebra $\mathbb{H} \subset \mathbb{G}^{\text{iso}}_{\mathscr{M}}$ of any point of the manifold, since $\mathrm{Solv}\,(\mathscr{M})$ acts transitively. In virtue of the

Riemannian structure of $\mathcal{M}$ we have $\mathbb{H} \subset \mathfrak{so}(n)$ where $n = \dim(\text{Solv}(\mathcal{M}))$ and hence also $\text{Aut}_{\text{Ext}}[\text{Solv}(\mathcal{M})] \subset \mathfrak{so}(n)$ is a compact Lie algebra.

[A2]

We can now reformulate the notion of maximally non-compact or maximally split algebras in such a way that it applies to the case of all considered solvable algebras, independently whether they come from symmetric spaces or not. *The algebra* Solv $(\mathcal{M})$ *is maximally split if the paint algebra is trivial, namely:*

$$\text{Solv}(\mathcal{M}) = \text{maximally split} \Leftrightarrow \text{Aut}_{\text{Ext}}[\text{Solv}(\mathcal{M})] = \emptyset. \tag{5.3.40}$$

For maximally split algebras there is no Tits–Satake projection, namely the Tits–Satake subalgebra is the full algebra.

[B]

Let us now consider non maximally split algebras such that $\text{Aut}_{\text{Ext}}[\text{Solv}(\mathcal{M})] \neq \emptyset$. Let r be the rank of Solv $(\mathcal{M})$, namely the number of its Cartan generators H_i and n the number of its nilpotent generators $\mathcal{W}_\alpha$, namely the number of generalized roots $\boldsymbol{\alpha}$. The whole set of Cartan generators H_i, plus a subset of p nilpotent generators $\mathcal{W}_{\alpha^\ell}$ associated with roots $\boldsymbol{\alpha}^\ell$ that we name *long*, close a solvable subalgebra $\text{Solv}_{\text{subTS}} \subset$ Solv $(\mathcal{M})$ that is made of singlets under the action of the paint Lie algebra $\mathbb{G}_{\text{paint}}$, i.e.

$$\begin{aligned} \text{Solv}_{\text{subTS}} &= \text{span}\{H_i, \mathcal{W}_{\alpha^\ell}\}, \\ [\text{Solv}_{\text{subTS}}, \text{Solv}_{\text{subTS}}] &\subset \text{Solv}_{\text{subTS}}, \\ \forall X \in \mathbb{G}_{\text{paint}}, \forall \Psi \in \text{Solv}_{\text{subTS}} &: [X, \Psi] = 0. \end{aligned} \tag{5.3.41}$$

We name $\text{Solv}_{\text{subTS}}$ the *sub Tits–Satake algebra*. By definition $\text{Solv}_{\text{subTS}}$ has the same rank as the original solvable algebra Solv $(\mathcal{M})$. In all possible cases, it is the solvable Lie algebra of a symmetric maximally split coset $\mathbb{G}_{\text{subTS}}/\mathbb{H}_{\text{subTS}}$. In this way, eventually, we have the notion of a semisimple Lie algebra $\mathbb{G}_{\text{subTS}}$.

[C1]

Considering the orthogonal decomposition of the original solvable Lie algebra with respect to its *sub Tits–Satake algebra*:

$$\text{Solv}(\mathcal{M}) = \text{Solv}_{\text{subTS}} \oplus \mathbb{K}_{\text{short}}. \tag{5.3.42}$$

we find that the orthogonal subspace $\mathbb{K}_{\text{short}}$ necessarily decomposes into a sum of q subspaces:

$$\mathbb{K}_{\text{short}} = \bigoplus_{\wp=1}^{q} \mathbb{D}\left[\mathcal{P}_\wp^+, \mathbf{Q}_\wp\right], \tag{5.3.43}$$

where each $\mathbb{D}\left[\mathcal{P}_\wp^+, \mathbf{Q}_\wp\right]$ is the tensor product:

$$\mathbb{D}\left[\mathscr{P}^+_\wp, \mathbf{Q}_\wp\right] = \mathscr{P}^+_\wp \otimes \mathbf{Q}_\wp \tag{5.3.44}$$

of an irreducible module $\mathbf{Q}_\wp$ (i.e. representation) of the compact paint algebra $\mathbb{G}_{\text{paint}}$ with an irreducible module $\mathscr{P}^+_\wp$ of the solvable sub Tits–Satake algebra $\text{Solv}_{\text{subTS}}$. As we already noticed, $\text{Solv}_{\text{subTS}}$ is the maximal Borel subalgebra of the maximally split, semisimple, real Lie algebra $\mathbb{G}_{\text{subTS}}$. Hence an irreducible module $\mathscr{P}^+_\wp$ of $\text{Solv}_{\text{subTS}}$ necessarily decomposes in the following way:

$$\mathscr{P}^+_\wp = \bigoplus_{s=1}^{n_\wp} \mathbb{W}[\boldsymbol{\alpha}^{(\wp,s)}], \qquad n_\wp = \dim \mathscr{P}^+_\wp, \tag{5.3.45}$$

where each $\mathbb{W}[\boldsymbol{\alpha}^{(\wp,s)}]$ is an eigenspace of the CSA of $\mathbb{G}_{\text{subTS}}$, which coincides with that of $\text{Solv}_{\text{subTS}}$ and eventually with the CSA of the original Solv $(\mathscr{M})$. Explicitly this means:

$$\forall\, H_i \in \text{CSA}\,(\text{Solv}\,(\mathscr{M}))\ ,\ \forall \Psi \in \mathbb{W}[\boldsymbol{\alpha}^{(\wp,s)}] \otimes \mathbf{Q}_\wp \quad : \quad [H_i\ ,\ \Psi] = \alpha_i^{(\wp,s)}\,\Psi. \tag{5.3.46}$$

Furthermore the r-vectors of eigenvalues, which are roots of Solv $(\mathscr{M})$, are identified by (5.3.45) as the non negative weights of some irreducible module $\mathscr{P}_\wp$ of the simple Lie algebra $\mathbb{G}_{\text{subTS}}$:

$$\mathscr{P}_\wp = \mathscr{P}^+_\wp \oplus \mathscr{P}^-_\wp, \qquad \mathscr{P}^-_\wp = \bigoplus_{s=1}^{n_\wp} \mathbb{W}[-\boldsymbol{\alpha}^{(\wp,s)}]. \tag{5.3.47}$$

Indeed for the solvable Lie algebras Solv(G/H) of maximally split cosets the irreducible modules are easily constructed as *half-modules* of the full algebra $\mathbb{G}$, namely by taking the eigenspaces associated with non negative weights.

[C2]

The decomposition of $\mathbb{K}_{\text{short}}$ mentioned in (5.3.43) has actually a general form depending on the rank. We will discuss this here for the quaternionic-Kähler manifolds.

(r = 4) In this case there are just three modules of $\mathbb{G}_{\text{subTS}} = \text{SO}(4,4)$ involved in the sum of (5.3.43) namely $\mathscr{P}_{\mathbf{8_v}}$, $\mathscr{P}_{\mathbf{8_s}}$, $\mathscr{P}_{\mathbf{8_{\bar{s}}}}$, where $\mathbf{8_{v,s,\bar{s}}}$ denotes the vector, spinor and conjugate spinor representation, respectively. All these three modules are 8 dimensional, which means that for all of them there are 4 positive weights and 4 negative ones. Denoting these half spaces by $\mathbf{4}^+_{\mathbf{v,s,\bar{s}}}$, we can write:

$$\mathbb{K}_{\text{short}} = \left(\mathbf{4}^+_{\mathbf{v}}, \mathbf{Q}_{\mathbf{v}}\right) \oplus \left(\mathbf{4}^+_{\mathbf{s}}, \mathbf{Q}_{\mathbf{s}}\right) \oplus \left(\mathbf{4}^+_{\mathbf{\bar{s}}}, \mathbf{Q}_{\mathbf{\bar{s}}}\right), \tag{5.3.48}$$

where $\mathbf{Q}_{\mathbf{v,s,\bar{s}}}$ are three different irreducible modules of $\mathbb{G}_{\text{paint}}$ that we will discuss in later sections. The generic case is that where all three representations $\mathbf{Q}_{\mathbf{v,s,\bar{s}}}$ are non vanishing. Special cases where two of the three representations $\mathbb{G}_{\text{paint}}$ vanish do also exist. The limiting case is that where all three representations are deleted and the full algebra is just Solv $\left(\frac{\mathrm{SO}(4,4)}{\mathrm{SO}(4)\times\mathrm{SO}(4)}\right)$. Note that (5.3.48) is the generalization of the decomposition (5.3.32) applying to the case analyzed in detail above. There we have $\mathbb{G}_{\text{paint}} = \mathrm{SO}(8)$ and the aforementioned irreducible modules are:

$$\mathbf{Q_v} = \mathbf{8_v} \quad ; \quad \mathbf{Q_s} = \mathbf{8_s} \quad ; \quad \mathbf{Q_{\bar{s}}} = \mathbf{8_{\bar{s}}} \tag{5.3.49}$$

(r = 3) In this case there is only one module of $\mathbb{G}_{\text{subTS}} = \mathrm{SO}(3,4)$ involved in the sum of (5.3.43) namely $\mathscr{P}_{\mathbf{8_s}}$ where $\mathbf{8_s}$ denotes the 8 dimensional spinor representation of SO(3, 4). With a notation completely analogous to that employed above let 4_s^+ denote the space spanned by the eigenspaces pertaining to positive spinor weights. Then we can write:

$$\mathbb{K}_{\text{short}} = \left(\mathbf{4_s^+}, \mathbf{Q_s}\right), \tag{5.3.50}$$

(r = 2) In this case, there is one exceptional case, namely SG_5, where $G_R = G_{\text{subTS}} = G_{2(2)}$. In all other cases, there are two modules of SO(2, 2) involved in the sum of (5.3.43) and these are the spinor module $\mathscr{P}_{\mathbf{4_s}}$ and the vector module $\mathscr{P}_{\mathbf{4_v}}$. Both modules are 4-dimensional and in our adopted notations we can write:

$$\mathbb{K}_{\text{short}} = \left(\mathbf{2_s^+}, \mathbf{Q_s}\right) \oplus \left(\mathbf{2_v^+}, \mathbf{Q_v}\right). \tag{5.3.51}$$

(r = 1) In this case we have to distinguish between $G_{\text{subTS}} = \mathrm{SO}(1,1)$ or $G_{\text{subTS}} = \mathrm{SU}(1,1)$. When $G_{\text{subTS}} = \mathrm{SU}(1,1)$ we have:

$$\mathbb{K}_{\text{short}} = \left(\mathbf{1_s^+}, \mathbf{Q_s}\right), \tag{5.3.52}$$

where $\mathbf{1_s^+}$ denotes the positive weight subspace of the spinor representation of $\mathfrak{so}(1,2)$, i.e. the fundamental of $\mathfrak{su}(1,1)$, which is two-dimensional. The representation $\mathbf{Q_s}$ will be discussed later. When $G_{\text{subTS}} = \mathrm{SO}(1,1)$ on the other hand, we have:

$$\mathbb{K}_{\text{short}} = \left(\mathbf{1_s^+}, \mathbf{Q_s}\right) \oplus \left(\mathbf{1_v^+}, \mathbf{Q_v}\right). \tag{5.3.53}$$

In this case, $\mathbf{1_s^+}$ denotes a subspace of weight 1/2 with respect to $\mathbb{G}_{\text{subTS}} = \mathfrak{so}(1,1)$, while the subspace $\mathbf{1_v^+}$ has weight 1.

We can now note a regularity in the decomposition of $\mathbb{K}_{\text{short}}$. For all values of the rank we always have the space $(\mathscr{S}^+, \mathbf{Q_s})$ that associates a representation of the paint group to the half spinor representation of the sub Tits–Satake algebra. In the case of rank $r = 4$ in addition to this we also have the representations $\mathbf{Q_v}$ and $\mathbf{Q_{\bar{s}}}$, which we associate to what we can name the $\mathscr{V}^+$ and $\bar{\mathscr{S}}^+$ half modules. We have established a notation covering all the cases which enables us to proceed to the next point and give a general definition of the Tits–Satake projection.

[D]

The *paint algebra* $\mathbb{G}_{\text{paint}}$ contains a subalgebra

$$\mathbb{G}^0_{\text{subpaint}} \subset \mathbb{G}_{\text{paint}}, \tag{5.3.54}$$

such that with respect to $\mathbb{G}^0_{\text{subpaint}}$, each of the three irreducible representations $\mathbf{Q}_{\mathbf{v},\mathbf{s},\bar{\mathbf{s}}}$ branches as:

$$\mathbf{Q}_{\mathbf{v},\mathbf{s},\bar{\mathbf{s}}} \stackrel{\mathbb{G}^0_{\text{subpaint}}}{\Longrightarrow} \underbrace{\mathbf{1}}_{\text{singlet}} \oplus \mathbf{J}_{\mathbf{v},\mathbf{s},\bar{\mathbf{s}}}, \tag{5.3.55}$$

where the representation $\mathbf{J}_{\mathbf{v},\mathbf{s},\bar{\mathbf{s}}}$ is in general reducible.

[E]

The restriction to the singlets of $\mathbb{G}^0_{\text{subpaint}}$ defines a Lie subalgebra of Solv_{M}, namely, if we set:

$$\text{Solv}_{\text{TS}} \equiv \text{Solv}_{\text{subTS}} \oplus \left(\mathscr{V}^+, \mathbf{1}\right) \oplus \left(\mathscr{S}^+, \mathbf{1}\right) \oplus \left(\overline{\mathscr{S}}^+, \mathbf{1}\right), \tag{5.3.56}$$

we get:

$$[\text{Solv}_{\text{TS}} \,,\, \text{Solv}_{\text{TS}}] \subset \text{Solv}_{\text{TS}}. \tag{5.3.57}$$

Relying on all the above properties and structures described in points [A], [B], [C], [D] and [E], which turn out to hold true for every Solv $(\mathscr{M})$ considered in supergravity, irrespectively whether it is associated with a symmetric space or not, we can define the Tits–Satake projection at the level of solvable algebras by stating:

$$\begin{array}{l} \Pi_{\text{TS}} \;:\; \text{Solv}\,(\mathscr{M}) \longrightarrow \text{Solv}_{\text{TS}} \subset \text{Solv}\,(\mathscr{M}) \\ \Psi \in \text{Solv}_{\text{TS}} \;\; \text{if and only if} \;:\; \forall X \in \mathbb{G}^0_{\text{subpaint}} \;:\; [X, \Psi] = 0 \end{array} \tag{5.3.58}$$

In other words, we define the Tits–Satake solvable subalgebra Solv_{TS} as spanned by all the *singlets* under the *subpaint group* $\text{G}_{\text{subpaint}}$. By its very definition the Tits–Satake subalgebra contains the *sub Tits–Satake algebra* $\text{Solv}_{\text{subTS}} \subset \text{Solv}_{\text{TS}}$ which is made of singlets with respect to the full paint group G_{paint} The subtle points in the above definition of the Tits–Satake projection is given by point [D] and [E]. Namely it is a matter of fact, which is not obvious a priori, that the addition of the three modules (occasionally vanishing) $\mathscr{V}^+, \mathscr{S}^+, \overline{\mathscr{S}}^+$ to the sub Tits–Satake algebra $\text{Solv}_{\text{subTS}}$ always defines a new Lie algebra. Being true this implies that a subalgebra Solv_{TS} with the structure (5.3.56) exists in Solv_Q and $\mathbb{G}_{\text{subpaint}}$ is its stability subalgebra. Vice versa, the existence of a subpaint algebra such that the decomposition (5.3.55) is true, implies that the subspace (5.3.56) closes a subalgebra since the kernel of a subalgebra of automorphisms is necessarily a closed subalgebra.

5.4 The Systematics of Paint Groups

As we explained in Sect. 5.3.3, the Tits–Satake projection originally defined in terms of a geometrical projection of the root space, can be generalized to all solvable algebras of special geometries reformulating it in terms of the paint and subpaint group structures. The systematic procedure outlined there, started as step A] with the identification of the paint group. This is what we do now, unveiling a very elegant pattern of such paint groups.

As we claimed in the introduction, the specially fascinating property of the paint group is that it is invariant under both the **c**-map and the $\mathbf{c}^{\star}$-map, namely under dimensional reduction.

5.4.1 The Paint Group for Non-compact Symmetric Spaces

In Sect. 5.3.3, we defined the paint group as the group of external automorphisms of the solvable algebra associated with a certain homogeneous space (5.3.39). For non-compact symmetric spaces there exists another, more common, definition of the paint group. Referring to the presentation in the beginning of Sect. 5.3.1, the paint group is defined as a subgroup of $\mathbb{H}$, whose Cartan generators are those in $\mathscr{H}^{\text{comp}}$ and the roots are those in Δ_{comp} (and their negatives), i.e. those that have no component $\alpha_{||}$ in the decomposition (2.4.19).

As we mentioned already in the example in Sect. 5.3.2, a real form $\mathbb{G}_{\mathbb{R}}$ of the Lie algebra $\mathbb{G}$ is represented by the so-called Satake diagrams, which are Dynkin diagrams with the following extra decorations:

- Compact simple roots (those in Δ_{comp}) are denoted by filled circles.
- Simple roots that, upon setting $\alpha_{\perp} = 0$, project to the same restricted root are connected with a two-sided arrow. These are simple roots that necessarily belong to Δ^{δ}.

Given the Satake diagram the paint group can then be read from it in the following way. The black dots form a Dynkin diagram of the semi-simple type. The paint group then contains a factor corresponding to this painted subdiagram. This corresponds to the roots in Δ_{comp} and the elements of $\mathscr{H}^{\text{comp}}$ for which these roots have non-vanishing components. Furthermore, for every arrow, there is one additional SO(2)-factor that commutes with the rest of the paint group. These correspond to the additional generators in $\mathscr{H}^{\text{comp}}$. An example of this is given in Figs. 5.2 and 5.3. For the symmetric quaternionic spaces of rank 4, the paint groups are summarized in Table 5.3. The case 4 has already been extensively discussed. Here we can briefly explain the group theory of the case 2. It suffices to note that the $\mathrm{E}_{6(2)}$ Lie algebra contains $\mathrm{F}_{4(4)}$ as a maximal subalgebra and that the adjoint has the following branching rule:

$$\mathbf{78} \xrightarrow{\mathrm{F}_{4(4)}} \mathbf{52} \oplus \mathbf{26}. \tag{5.4.1}$$

Fig. 5.3 Satake diagram of $E_{6(2)}$. The paint group can be seen to be $\mathrm{SO}(2)^2$

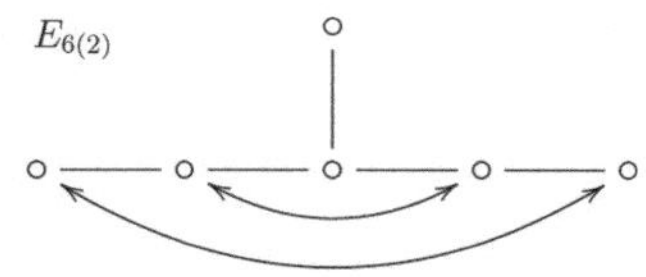

Table 5.3 Symmetric special Kähler manifolds and their corresponding quaternionic spaces. The last two columns indicate the paint and subpaint groups respectively. The spaces above the line are maximally non-compact and do not have any paint group

$C(h)$	Kähler	Quaternionic	$\mathrm{G}_{\mathrm{paint}}$	$\mathrm{G}^0_{\mathrm{subpaint}}$
1	$\frac{\mathrm{Sp}(6)}{\mathrm{U}(3)}$	$\frac{\mathrm{F}_{4(4)}}{\mathrm{USp}(6)\times\mathrm{SU}(2)}$	–	–
2	$\frac{\mathrm{SU}(3,3)}{\mathrm{SU}(3)\times\mathrm{SU}(3)\times\mathrm{U}(1)}$	$\frac{\mathrm{E}_{6(2)}}{\mathrm{SU}(2)\times\mathrm{SU}(6)}$	$\mathrm{SO}(2)^2$	$\mathbf{1}$
3	$\frac{\mathrm{SO}^*(12)}{\mathrm{SU}(6)\times\mathrm{U}(1)}$	$\frac{\mathrm{E}_{7(-5)}}{\mathrm{SO}(12)\times\mathrm{SU}(2)}$	$\mathrm{SO}(3)^3$	$\mathrm{SO}(3)_{\mathrm{diag}}$
4	$\frac{\mathrm{E}_{7(-25)}}{\mathrm{E}_{6(-78)}\times\mathrm{U}(1)}$	$\frac{\mathrm{E}_{8(-24)}}{\mathrm{E}_{7(-133)}\times\mathrm{SU}(2)}$	$\mathrm{SO}(8)$	$\mathrm{G}_{2(-14)}$

This shows that the subpaint group is empty since the normalizer of the Tits–Satake subalgebra $\mathrm{F}_{4(4)}$ is null. On the other hand, recalling the decomposition of the fundamental representation of $\mathrm{F}_{4(4)}$ with respect to the subalgebra SO(4, 4)

$$\mathbf{26} \overset{\mathrm{SO}(4,4)}{\longrightarrow} \mathbf{1} \oplus \mathbf{1} \oplus \mathbf{8}^{\mathrm{nc}}_{\mathbf{v}} \oplus \mathbf{8}^{\mathrm{nc}}_{\mathbf{s}} \oplus \mathbf{8}^{\mathrm{nc}}_{\bar{\mathbf{s}}}, \tag{5.4.2}$$

together with the branching rule of the adjoint given in (5.3.30), we conclude that under the subgroup $\mathrm{SO}(4,4) \times \mathrm{SO}(2)^2$ we have:

$$\begin{aligned} \mathbf{78} \overset{\mathrm{SO}(4,4)\times\mathrm{SO}(2)^2}{\longrightarrow} & (\mathbf{28}^{\mathrm{nc}}, \mathbf{1}, \mathbf{1}) \oplus \left(\mathbf{8}^{\mathrm{nc}}_{\mathbf{v}}, \mathbf{2}, \mathbf{1}\right) \oplus \left(\mathbf{8}^{\mathrm{nc}}_{\mathbf{s}}, \mathbf{1}, \mathbf{2}\right) \oplus (\mathbf{8}_{\bar{\mathbf{s}}^{\mathrm{nc}}}, \mathbf{1}, \mathbf{2}) \\ & \oplus (\mathbf{1}, \mathbf{1}, \mathbf{1}) \oplus (\mathbf{1}, \mathbf{1}, \mathbf{1}) \end{aligned} \tag{5.4.3}$$

which shows that the paint group is indeed $\mathrm{SO}(2)^2$ as claimed.

From (5.4.3) we also read off the representations $\mathbf{Q}_{\mathbf{v},\mathbf{s},\hat{\mathbf{s}}}$ defined by (5.3.48) that pertain to this case:

$$\mathbf{Q}_{\mathbf{v}} = (\mathbf{2}, \mathbf{1}) \quad ; \quad \mathbf{Q}_{\mathbf{s}} = (\mathbf{1}, \mathbf{2}) \quad ; \quad \mathbf{Q}_{\hat{\mathbf{s}}} = (\mathbf{1}, \mathbf{2}). \tag{5.4.4}$$

5.5 Classification of the Sugra-Relevant Symmetric Spaces and Their General Properties

Equipped with the powerful weapon of the Tits Satake projection which allows to organize them into universality classes, we can now make a complete survey of the symmetric spaces G/H that are relevant to supergravity theories and in particular to the construction of black-hole solutions. Indeed, as the reader cannot fail to appreciate there is a general group-theoretical framework underlying the construction of supergravity black holes which allows both for

(1) a classification of the relevant symmetric spaces,
(2) a general description of their structures which are relevant to the black hole solutions.

The presentation of both items in the above list is the goal of the present section. To achieve such a goal we need to emphasize a few general aspects of the decomposition (1.7.12) that relate to the underlying root systems and Dynkin diagrams. In the following we heavily rely on results presented several years ago in [46]. Indeed from the algebraic view-point a crucial property of the general decomposition in Eq. (1.7.12) is encoded into the following statements which are true for all the cases[1]:

1. The A_1 root-system associated with the $\mathfrak{sl}(2,\mathbb{R})_E$ algebra in the decomposition (1.7.12) is made of $\pm\,\psi$ where ψ is the highest root of $\mathbb{U}_{D=3}$.
2. Out of the r simple roots α_i of $\mathbb{U}_{D=3}$ there are $r-1$ that have grading zero with respect to ψ and just one α_W that has grading 1:

$$\begin{aligned} (\psi\,,\,\alpha_i) &= 0 \qquad i \neq W \\ (\psi\,,\,\alpha_W) &= 1 \end{aligned} \tag{5.5.1}$$

3. The only simple root α_W that has non vanishing grading with respect ψ is just the highest weight of the symplectic representation **W** of $\mathbb{U}_{D=4}$ to which the vector fields are assigned.
4. The Dynkin diagram of $\mathbb{U}_{D=4}$ is obtained from that of $\mathbb{U}_{D=3}$ by removing the dot corresponding to the special root α_W.
5. Hence we can arrange a basis for the simple roots of the rank r algebra $\mathbb{U}_{D=3}$ such that:

$$\begin{aligned} \alpha_i &= \{\overline{\alpha}_i, 0\} \quad ; \quad i \neq W \\ \alpha_W &= \left\{\overline{\mathbf{w}}_h, \frac{1}{\sqrt{2}}\right\} \\ \psi &= \left\{\mathbf{0}, \sqrt{2}\right\} \end{aligned} \tag{5.5.2}$$

where $\overline{\alpha}_i$ are $(r-1)$-component vectors representing a basis of simple roots for the Lie algebra $\mathbb{U}_{D=4}$, $\overline{\mathbf{w}}_h$ is also an $(r-1)$-vector representing the *highest weight* of the representation **W**.

[1] An apparent exception is given by the case of $\mathcal{N}=3$ supergravity. The extra complicacy, there, is that the duality algebra in $D=3$, namely $\mathbb{U}_{D=3}$ has rank $r+2$, rather than $r+1$ with respect to the rank of the algebra $\mathbb{U}_{D=4}$. Actually in this case there is an extra $\mathrm{U}(1)_Z$ factor that is active on the vectors, but not on the scalars and which is responsible for the additional complications. It happens in this case that there are two vector roots, one for the complex representation to which the vectors are assigned and one for its conjugate. They have opposite charges under $\mathrm{U}(1)_Z$. This case together with that of $\mathcal{N}=5$ supergravity and with one of the series of $\mathcal{N}=2$ theories completes the list of three *exotic models* which are anomalous also from the point of view of the Tits Satake projection (see below).

This means that the entire root system and the Cartan subalgebra of the $\mathbb{U}_{D=3}$ Lie algebra can be organized as follows:

$$
\begin{array}{lclcrcl}
\pm\psi & = & \pm\left(\mathbf{0}\,,\,\sqrt{2}\right) & ; & & & 2\\
\pm\hat{\alpha} & = & \pm\left(\alpha\,,\,\sqrt{2}\right) & ; & 2\times\#\text{ of roots} & = & 2\,n_r\\
\pm\hat{w} & = & \pm\left(w\,,\,\frac{\sqrt{2}}{2}\right) & ; & 2\times\#\text{ of weights} & = & 2\times\dim\mathbf{W}\\
\mathcal{H}^i & \in & \text{CSA}\subset\mathbb{U}_{D=4} & ; & \text{rank}\mathbb{U}_{D=4} & = & r\\
\mathcal{H}^\psi & & & & & & 1\\
\hline
 & & & & \dim\mathbb{U}_{D=4} & = & 3+\dim\mathbb{U}_{D=3}+2\times\dim\mathbf{W}
\end{array}
\tag{5.5.3}
$$

This organization of the Lie algebra is very important, as it was thoroughly discussed in [46], for the systematics of the Kač Moody extension which occurs when stepping down from D = 3 to D = 2 dimensions, but it is equally important in the present context to analyze the structure of the H*-subalgebra and the Tits Satake projection.

5.5.1 Tits Satake Projection

In most cases of lower supersymmetry, neither the algebra $\mathbb{U}_{D=4}$ nor the algebra $\mathbb{U}_{D=3}$ are **maximally split**. In short this means that the non-compact rank $r_{nc} < r$ is less than the rank of $\mathbb{U}$, namely not all the Cartan generators are non-compact. When this happens it means that the structure of black hole solutions is effectively determined by the *maximally split Tits Satake* subalgebra $\mathbb{U}^{TS} \subset \mathbb{U}$, whose rank is equal to r_{nc}. Effectively determined does not mean that solutions of the big system coincide with those of the smaller system rather it means that the former can be obtained from the latter by means of rotations of the *paint group*, $\mathrm{G}_{\text{paint}}$. As we have seen the Tits Satake algebra is obtained from the original algebra via a projection of the root system of $\mathbb{U}$ onto the subspace orthogonal to the compact part of the Cartan subalgebra of $\mathbb{U}^{TS}$:

$$
\Pi^{TS} \quad ; \quad \Delta_{\mathbb{U}} \mapsto \overline{\Delta}_{\mathbb{U}^{TS}} \tag{5.5.4}
$$

In Euclidean geometry $\overline{\Delta}_{\mathbb{U}^{TS}}$ is just a collection of vectors in r_{nc} dimensions; a priori there is no reason why it should be the root system of another Lie algebra. Yet as we illustrated, in most cases, $\overline{\Delta}_{\mathbb{U}^{TS}}$ turns out to be a Lie algebra root system and the maximal split Lie algebra corresponding to it, $\mathbb{U}^{TS}$, is, the Tits Satake subalgebra of the original non maximally split Lie algebra: $\mathbb{U}^{TS} \subset \mathbb{U}$. Such algebras $\mathbb{U}$ are called *non-exotic*. The *exotic* non compact algebras are those for which the system $\overline{\Delta}_{\mathbb{U}^{TS}}$ is not an admissible root system. In such cases there is no Tits Satake subalgebra $\mathbb{U}^{TS}$. Exotic algebras are very few and in supergravity they appear only in three instances that display additional pathologies relevant also for the black hole solutions. For the non exotic models we have that the decomposition (1.7.12) commutes with the projection, namely:

$$\begin{array}{rcl} \mathrm{adj}(\mathbb{U}_{D=3}) & = & \mathrm{adj}(\mathbb{U}_{D=4}) \oplus \mathrm{adj}(\mathfrak{sl}(2,\mathbb{R})_E) \oplus W_{(2,W)} \\ & \Downarrow & \\ \mathrm{adj}(\mathbb{U}^{TS}_{D=3}) & = & \mathrm{adj}(\mathbb{U}^{TS}_{D=4}) \oplus \mathrm{adj}(\mathfrak{sl}(2,\mathbb{R})_E) \oplus W_{(2,W^{TS})} \end{array} \tag{5.5.5}$$

In other words the projection leaves the A_1 Ehlers subalgebra untouched and has a non trivial effect only on the duality algebra $\mathbb{U}_{D=4}$. Furthermore the image under the projection of the highest root of $\mathbb{U}$ is the highest root of $\mathbb{U}^{TS}$:

$$\Pi^{TS} \quad : \quad \psi \to \psi^{TS} \tag{5.5.6}$$

The reason why the Tits Satake projection is relevant to us was first pointed out in [45] where the present author and his collaborators advocated that the classification of nilpotent orbits and hence of extremal black hole solutions depends only on the Tits Satake subalgebra and therefore is universal for all members of the same Tits Satake universality class. By this name we mean all algebras who share the same Tits Satake projection.

Having clarified these points we can proceed to present the classification of homogeneous symmetric spaces relevant to supergravity models and to black hole solutions.

5.5.2 *Classification of the Sugra-Relevant Symmetric Spaces*

The classification of the symmetric coset based supergravity models is exhaustive and it is presented in Tables 5.4 and 5.5. There are 16 universality classes of non-exotic models and 3 exceptional instances of exotic models which appear in the second table.

In the tables we have also listed the Paint groups and the subpaint groups. These latter are always compact and their different structures is what distinguishes the different elements belonging to the same class. As it was shown in [32] and extensively illustrated in the previous sections, these groups are dimensional reduction invariant, namely they are the same in $D = 4$ and in $D = 3$. Hence the representation $\mathbf{W}$, which in particular contains the electromagnetic charges of the hole, can be decomposed with respect to the Tits Satake subalgebra and the Paint group revealing a regularity structure inside each Tits Satake universality class which is at the heart of the classification of *charge orbits*. The same decomposition can be given also for the $\mathbb{K}^\star$ representation and this is at the heart of the classification of black holes according to nilpotent orbits.

Focusing on the non-exotic models, we note that the 16 classes have a quite different type of population. There are six one element classes whose single member is maximally split. They are the following ones and all have a distinguished standpoint within the panorama of supergravity theories:

Table 5.4 The 16 instances of *non-exotic* homogeneous symmetric scalar manifolds appearing in $D = 4$ supergravity. Non exotic means that the Tits Satake projection of the root system is a standard Lie Algebra root system. The 16 models are grouped according to their Tits Satake Universality classes. The time-like dimensional reduction is listed side by side. Within each class the models are distinguished by the different structure of the Paint group and of its subPaint subgroup. The Paint group is the same in D = 4 and in D = 3

#	TS D = 4	TS D = 3	Coset D = 4	Coset D = 3	Paint group	subP group	Susy
1	$\frac{E_{7(7)}}{SU(8)}$	$\frac{E_{8(8)}}{SO^\star(16)}$	$\frac{E_{7(7)}}{SU(8)}$	$\frac{E_{8(8)}}{SO^\star(16)}$	1	1	$\mathcal{N} = 8$
2	$\frac{SU(1,1)}{U(1)}$	$\frac{G_{2(2)}}{SL(2,R)\times SL(2,R)}$	$\frac{SU(1,1)}{U(1)}$	$\frac{G_{2(2)}}{SL(2,R)\times SL(2,R)}$	1	1	$\mathcal{N} = 2$ n = 1
3	$\frac{Sp(6,R)}{SU(3)\times U(1)}$	$\frac{F_{4(4)}}{Sp(6,R)\times SL(2,R)}$	$\frac{Sp(6,R)}{SU(3)\times U(1)}$	$\frac{F_{4(4)}}{Sp(6,R)\times SL(2,R)}$	1	1	$\mathcal{N} = 2$ $n = 6$
4			$\frac{SU(3,3)}{SU(3)\times SU(3)\times U(1)}$	$\frac{E_{6(2)}}{SU(3,3)\times SL(2,R)}$	SO(2) × SO(2)	1	$\mathcal{N} = 2$ $n = 9$
5			$\frac{SO^\star(12)}{SU(6)\times U(1)}$	$\frac{E_{7(-5)}}{SO^\star(12)\times SL(2,R)}$	SO(3) × SO(3) ×SO(3)	$SO(3)_d$	$\mathcal{N} = 6$ $\mathcal{N} = 2$ n = 16
6			$\frac{E_{7(-25)}}{E_{6(-78)}\times U(1)}$	$\frac{E_{8(-24)}}{E_{7(-25)}\times SL(2,R)}$	SO(8)	$G_{2(-14)}$	$\mathcal{N} = 2$ $n = 27$
7	$\frac{SL(2,\mathbb{R})}{O(2)} \times \frac{SO(2,1)}{SO(2)}$	$\frac{SO(4,3)}{SO(2,2)\times SO(2,1)}$	$\frac{SL(2,\mathbb{R})}{O(2)} \times \frac{SO(6,1)}{SO(6)}$	$\frac{SO(8,3)}{SO(6,2)\times SO(2,1)}$	SO(5)	SO(4)	$\mathcal{N} = 4$ n = 1
8	$\frac{SL(2,\mathbb{R})}{O(2)} \times \frac{SO(3,2)}{SO(3)\times SO(2)}$	$\frac{SO(5,4)}{SO(3,2)\times SO(2,2)}$	$\frac{SL(2,\mathbb{R})}{O(2)} \times \frac{SO(6,2)}{SO(6)\times SO(2)}$	$\frac{SO(8,4)}{SO(6,2)\times SO(2,2)}$	SO(4)	SO(3)	$\mathcal{N} = 4$ n = 2

(continued)

Table 5.4 (continued)

#	TS $D=4$	TS $D=3$	Coset $D=4$	Coset $D=3$	Paint group	subP group	Susy
9	$\frac{SL(2,\mathbb{R})}{O(2)} \times \frac{SO(4,3)}{SO(4)\times SO(3)}$	$\frac{SO(6,5)}{SO(4,2)\times SO(2,3)}$	$\frac{SL(2,\mathbb{R})}{O(2)} \times \frac{SO(6,3)}{SO(6)\times SO(3)}$	$\frac{SO(8,5)}{SO(6,2)\times SO(2,3)}$	SO(3)	SO(2)	$\mathcal{N}=4$ $n=3$
10	$\frac{SL(2,\mathbb{R})}{O(2)} \times \frac{SO(5,4)}{SO(5)\times SO(4)}$	$\frac{SO(7,6)}{SO(5,2)\times SO(2,4)}$	$\frac{SL(2,\mathbb{R})}{O(2)} \times \frac{SO(6,4)}{SO(6)\times SO(4)}$	$\frac{SO(8,6)}{SO(6,2)\times SO(2,4)}$	SO(2)	1	$\mathcal{N}=4$ $n=4$
11	$\frac{SL(2,\mathbb{R})}{O(2)} \times \frac{SO(6,5)}{SO(6)\times SO(5)}$	$\frac{SO(8,7)}{SO(6,2)\times SO(2,5)}$	$\frac{SL(2,\mathbb{R})}{O(2)} \times \frac{SO(6,5)}{SO(6)\times SO(5)}$	$\frac{SO(8,7)}{SO(6,2)\times SO(2,5)}$	1	1	$\mathcal{N}=4$ $n=5$
12	$\frac{SL(2,\mathbb{R})}{O(2)} \times \frac{SO(6,6)}{SO(6)\times SO(6))}$	$\frac{SO(8,8)}{SO(6,2)\times SO(2,6)}$	$\frac{SL(2,\mathbb{R})}{O(2)} \times \frac{SO(6,6)}{SO(6)\times SO(6))}$	$\frac{SO(8,8)}{SO(6,2)\times SO(2,6)}$	1	1	$\mathcal{N}=4$ $n=6$
13	$\frac{SL(2,\mathbb{R})}{O(2)} \times \frac{SO(6,7)}{SO(6)\times SO(7))}$	$\frac{SO(8,9)}{SO(6,2)\times SO(2,7)}$	$\frac{SL(2,\mathbb{R})}{O(2)} \times \frac{SO(6,6+p)}{SO(6)\times SO(6+p))}$	$\frac{SO(8,8+p)}{SO(6,2)\times SO(2,6+p)}$	SO(p)	SO(p − 1)	$\mathcal{N}=4$ $n=6+p$
14	$\frac{SL(2,\mathbb{R})}{O(2)} \times \frac{SO(2,1)}{SO(2)}$	$\frac{SO(4,3)}{SO(2,2)\times SO(2,1)}$	$\frac{SL(2,\mathbb{R})}{O(2)} \times \frac{SO(2,1)}{SO(2)}$	$\frac{SO(4,3)}{SO(2,2)\times SO(2,1)}$	1	1	$\mathcal{N}=2$ $n=2$
15	$\frac{SL(2,\mathbb{R})}{O(2)} \times \frac{SO(2,2)}{SO(2)\times SO(2)}$	$\frac{SO(4,4)}{SO(2,2)\times SO(2,2)}$	$\frac{SL(2,\mathbb{R})}{O(2)} \times \frac{SO(2,2)}{SO(2)\times SO(2)}$	$\frac{SO(4,4)}{SO(2,2)\times SO(2,2)}$	1	1	$\mathcal{N}=2$ $n=3$
16	$\frac{SL(2,\mathbb{R})}{O(2)} \times \frac{SO(2,3)}{SO(2)\times SO(3)}$	$\frac{SO(4,5)}{SO(2,2)\times SO(2,3)}$	$\frac{SL(2,\mathbb{R})}{O(2)} \times \frac{SO(2,2+p)}{SO(2)\times SO(2+p)}$	$\frac{SO(4,4+p)}{SO(2,2)\times SO(2,2+p)}$	SO(p)	SO(p − 1)	$\mathcal{N}=2$ $n=3+p$

Table 5.5 The 3 instances of *exotic* homogenous symmetric scalar manifolds appearing in $D = 4$ supergravity. Exotic means that the Tits Satake projection of the root system is not a standard Lie Algebra root system. Notwithstanding this anomaly the concept of Paint Group, according to its definition as group of external automorphisms of the solvable Lie algebra generating the non compact coset manifold still exists. The Paint group is the same in D = 4 and in D = 3

#	TS D = 4	TS D = 3	Coset D = 4	Coset D = 3	Paint group	subP group	Susy
1_e	bc_1	bc_2	$\frac{\mathrm{SU(p+1,1)}}{\mathrm{SU(p+1)\times U(1)}}$	$\frac{\mathrm{SU(p+2,2)}}{\mathrm{SU(p+1,1)\times SL(2,R)_{h^\star}}}$	$\mathrm{U(1)\times U(1)\times U(p)}$	$\mathrm{U(p-1)}$	$\mathcal{N}=2$ $\mathrm{n=p+1}$
2_e	bc_3	bc_4	$\frac{\mathrm{SU(p+1,3)}}{\mathrm{SU(p+1)\times SU(3)\times U(1)}}$	$\frac{\mathrm{SU(p+2,4)}}{\mathrm{SU(p+1,2)\times SU(1,2)\times U(1)}}$	$\mathrm{U(1)\times U(1)\times U(p)}$	$\mathrm{U(p-1)}$	$\mathcal{N}=3$ $\mathrm{n=p+1}$
3_e	bc_1	bc_2	$\frac{\mathrm{SU(5,1)}}{\mathrm{SU(5)\times U(1)}}$	$\frac{\mathrm{E_{6(-14)}}}{\mathrm{SO^\star(10)}\times SO(2)}$	$\mathrm{U(1)\times U(1)\times U(4)}$	$\mathrm{U(3)}$	$\mathcal{N}=5$

1. The $\mathcal{N} = 8$ supergravity theory, which is the maximal one in $D = 4$, (model 1).
2. The $\mathcal{N} = 2$ supergravity theory with a single vector multiplet and non-vanishing Yukawa coupling(model 2).
3. The $\mathcal{N} = 4$ supergravity theory with 5 vector multiplets (model 11).
4. The $\mathcal{N} = 4$ supergravity theory with 6 vector multiplets which is obtained compactifying a type II theory on a $T^6/\mathbb{Z}_2$ orbifold (model 12).
5. The $\mathcal{N} = 2$ theory with two vector multiplets and non vanishing Yukava couplings, usually called the st-model (model 14).
6. The $\mathcal{N} = 2$ theory with three vector multiplets and non vanishing Yukava couplings, usually called the stu-model (model 15).

Next we have two universality classes, each containing an infinite number of elements. They are

1. The $\mathcal{N} = 4$ supergravity theory with $n = 6 + p$ vector multiplets ($p \geq 1$), (model 13).
2. The $\mathcal{N} = 2$ supergravity theory with $n = 3 + p$ vector multiplets ($p \geq 1$) and non vanishing Yukawa couplings (model 16).

We still have the very interesting 4-element universality class whose maximally split representative corresponds to the maximally split special Kähler manifold $\frac{\mathrm{Sp}(6,\mathbb{R})}{\mathrm{SU}(3)\times\mathrm{U}(1)}$. This class contains the models 3, 4, 5, 6 distinguished by quite peculiar Paint groups. We will thoroughly analyze the structure of this class.

Finally we have the three exotic models whose common feature is that their group and subgroup all belong to the pseudo-unitary series $\mathrm{SU(p, q)}$. The general decomposition (1.7.12) still holds true, but the Tits Satake projection looses its significance.

5.5.3 Dynkin Diagram Analysis of the Principal Models

Next we analyze the form of the root systems of the $\mathbb{U}_{\mathrm{D}=3}$ algebras in relation with the decomposition (1.7.12).

$\mathcal{N} = 8$

This is the case of maximal supersymmetry and it is illustrated by Fig. 5.4.

In this case all the involved Lie algebras are maximally split and we have

$$\mathrm{adj}\, \mathrm{E}_{8(8)} = \mathrm{adj}\, \mathrm{E}_{7(7)} \oplus \mathrm{adj}\, \mathrm{SL}(2, \mathbb{R})_{\mathrm{E}} \oplus (\mathbf{2}, \mathbf{56}) \tag{5.5.7}$$

The highest root of $\mathrm{E}_{8(8)}$ is

$$\psi = 3\alpha_1 + 4\alpha_2 + 5\alpha_3 + 6\alpha_4 + 3\alpha_5 + 4\alpha_6 + 2\alpha_7 + 2\alpha_8 \tag{5.5.8}$$

and the unique simple root not orthogonal to ψ is $\alpha_8 = \alpha_W$, according to the labeling of roots as in Fig. 5.4. This root is the highest weight of the fundamental **56**-representation of $E_{7(7)}$.

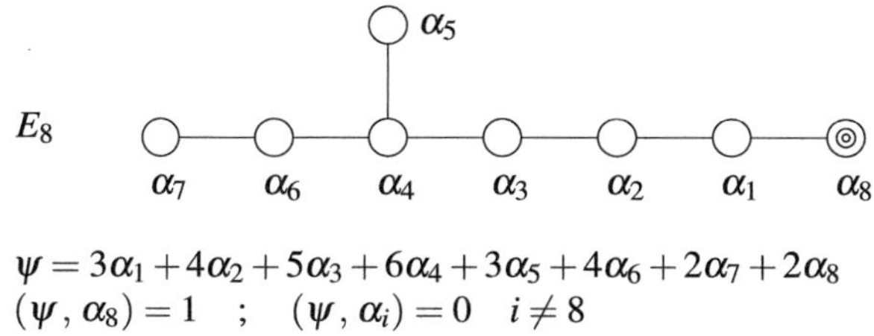

Fig. 5.4 The Dynkin diagram of $E_{8(8)}$. The only simple root which has grading one with respect to the highest root ψ is α_8 (painted with three circles). With respect to the algebra $\mathbb{U}_{D=4} = E_{7(7)}$ whose Dynkin diagram is obtained by removal of the multiple circle, α_8 is the highest weight of the symplectic representation of the vector fields, namely $\mathbf{W} = \mathbf{56}$

The well adapted basis of simple E_8 roots is constructed as follows:

$$\begin{aligned}
\alpha_1 &= \{1, -1, 0, 0, 0, 0, 0, 0\} &&= \{\overline{\alpha}_1, 0\} \\
\alpha_2 &= \{0, 1, -1, 0, 0, 0, 0, 0\} &&= \{\overline{\alpha}_2, 0\} \\
\alpha_3 &= \{0, 0, 1, -1, 0, 0, 0, 0\} &&= \{\overline{\alpha}_3, 0\} \\
\alpha_4 &= \{0, 0, 0, 1, -1, 0, 0, 0\} &&= \{\overline{\alpha}_4, 0\} \\
\alpha_5 &= \{0, 0, 0, 0, 1, -1, 0, 0\} &&= \{\overline{\alpha}_5, 0\} \\
\alpha_6 &= \{0, 0, 0, 0, 1, 1, 0, 0\} &&= \{\overline{\alpha}_6, 0\} \\
\alpha_7 &= \left\{-\tfrac{1}{2}, -\tfrac{1}{2}, -\tfrac{1}{2}, -\tfrac{1}{2}, -\tfrac{1}{2}, -\tfrac{1}{2}, \tfrac{1}{\sqrt{2}}, 0\right\} &&= \{\overline{\alpha}_7, 0\} \\
\alpha_8 &= \left\{-1, 0, 0, 0, 0, 0, -\tfrac{1}{\sqrt{2}}, \tfrac{1}{\sqrt{2}}\right\} &&= \left\{\mathbf{w}_h, \tfrac{1}{\sqrt{2}}\right\}
\end{aligned} \tag{5.5.9}$$

In this basis we recognize that the seven 7-vectors $\bar{\alpha}_i$ constitute a simple root basis for the E_7 root system, while:

$$\mathbf{w}_h = \left\{-1, 0, 0, 0, 0, 0, -\frac{1}{\sqrt{2}}\right\} \tag{5.5.10}$$

is the highest weight of the fundamental **56** dimensional representation. Finally in this basis the highest root ψ defined by Eq. (5.5.8) takes the expected form:

$$\psi = \{0, 0, 0, 0, 0, 0, 0, \sqrt{2}\} \tag{5.5.11}$$

$\mathcal{N} = 6$

In this case the $D = 4$ duality algebra is $\mathbb{U}_{D=4} = \mathrm{SO}^\star(12)$, whose maximal compact subgroup is $\mathrm{H} = \mathrm{SU}(6) \times \mathrm{U}(1)$. The scalar manifold (Fig. 5.5):

$$\mathcal{SK}_{N=6} \equiv \frac{\mathrm{SO}^\star(12)}{\mathrm{SU}(6) \times \mathrm{U}(1)} \tag{5.5.12}$$

is an instance of special Kähler manifold which can also be utilized in an $\mathcal{N} = 2$ supergravity context. The $D = 3$ algebra is $\mathbb{U}_{D=3} = E_{7(-5)}$. The 16 vector fields of

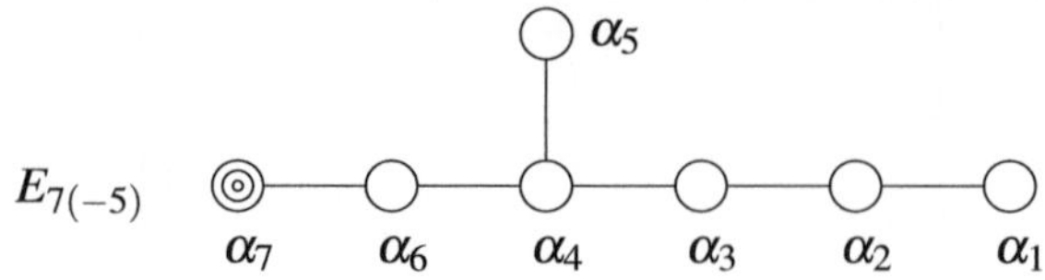

$$\psi = \alpha_1 + 2\alpha_2 + 3\alpha_3 + 4\alpha_4 + 2\alpha_5 + 3\alpha_6 + 2\alpha_7$$
$$(\psi, \alpha_7) = 1 \quad ; \quad (\psi, \alpha_i) = 0 \quad i \neq 7$$

Fig. 5.5 The Dynkin diagram of $E_{7(-5)}$. The only simple root which has grading one with respect to the highest root ψ is α_7 (painted with multiple circles). With respect to the algebra $\mathbb{U}_{D=4} = \mathrm{SO}^\star(12)$ whose Dynkin diagram is obtained by removal of the multiple circle, α_7 is the highest weight of the symplectic representation of the vector fields, namely the $\mathbf{W} = \mathbf{32}_s$

$D = 4$ $\mathcal{N} = 6$ supergravity with their electric and magnetic field strengths fill the spinor representation $\mathbf{32}_s$ of $\mathrm{SO}^\star(12)$, so that the decomposition (1.7.12), in this case becomes:

$$\mathrm{adj}\, \mathrm{E}_{7(-5)} = \mathrm{adj}\, \mathrm{SO}^\star(12) \oplus \mathrm{adj}\, \mathrm{SL}(2, \mathbb{R})_{\mathrm{E}} \oplus (\mathbf{2}, \mathbf{32}_s) \tag{5.5.13}$$

The simple root α_W is α_7 and the highest root is:

$$\psi = \alpha_1 + 2\alpha_2 + 3\alpha_3 + 4\alpha_4 + 2\alpha_5 + 3\alpha_6 + 2\alpha_7 \tag{5.5.14}$$

A well adapted basis of simple E_7 roots can be written as follows:

$$\begin{aligned}
\alpha_1 &= \{1, -1, 0, 0, 0, 0, 0\} && = \{\overline{\alpha}_1, 0\} \\
\alpha_2 &= \{0, 1, -1, 0, 0, 0, 0\} && = \{\overline{\alpha}_2, 0\} \\
\alpha_3 &= \{0, 0, 1, -1, 0, 0, 0\} && = \{\overline{\alpha}_3, 0\} \\
\alpha_4 &= \{0, 0, 0, 1, -1, 0, 0\} && = \{\overline{\alpha}_4, 0\} \\
\alpha_5 &= \{0, 0, 0, 0, 1, -1, 0\} && = \{\overline{\alpha}_5, 0\} \\
\alpha_6 &= \{0, 0, 0, 0, 1, 1, 0\} && = \{\overline{\alpha}_6, 0\} \\
\alpha_7 &= \left\{-\tfrac{1}{2}, -\tfrac{1}{2}, -\tfrac{1}{2}, -\tfrac{1}{2}, -\tfrac{1}{2}, -\tfrac{1}{2}, \tfrac{1}{\sqrt{2}}\right\} && = \{\mathbf{w}_h, \tfrac{1}{\sqrt{2}}\}
\end{aligned} \tag{5.5.15}$$

In this basis we recognize that the six 6-vectors $\bar{\alpha}_i$ $(i = 1, \ldots, 6)$ constitute a simple root basis for the $D_6 \simeq \mathrm{SO}^\star(12)$ root system, while:

$$\mathbf{w}_h = \left\{-\frac{1}{2}, -\frac{1}{2}, -\frac{1}{2}, -\frac{1}{2}, -\frac{1}{2}, -\frac{1}{2}, -\frac{1}{2}, -\frac{1}{2}\right\} \tag{5.5.16}$$

is the highest weight of the spinor **32**-dimensional representation of $\mathrm{SO}^\star(12)$. Finally in this basis the highest root ψ defined by Eq. (5.5.14) takes the expected form:

$$\psi = \{0, 0, 0, 0, 0, 0, \sqrt{2}\} \tag{5.5.17}$$

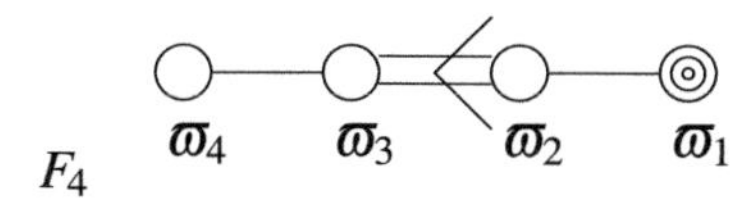

$$\psi = 2\varpi_1 + 3\varpi_2 + 4\varpi_3 + 2\varpi_4$$
$$(\psi\,,\varpi_1) = 2 \quad ; \quad (\psi\,,\varpi_i) = 0 \quad i \neq 1$$

Fig. 5.6 The Dynkin diagram of $F_{4(4)}$. The only root which is not orthogonal to the highest root is $\varpi_V = \varpi_1$. In the Tits Satake projection Π^{TS} the highest root ψ of $F_{4(4)}$ is the image of the highest root of $E_{7(-5)}$ and the root $\varpi_V = \varpi_1 = \Pi^{TS}(\alpha_7)$ is the image of the root associated with the vector fields

In this case, as in most cases of lower supersymmetry, neither the algebra $\mathbb{U}_{D=4}$ nor the algebra $\mathbb{U}_{D=3}$ are **maximally split**. The Tits Satake projection of $\mathrm{E}_{7(-5)}$ is $\mathrm{F}_{4(4)}$ and the explicit form of Eq. (5.5.5) is the following one:

$$\begin{array}{rcl} \mathrm{adj}(\mathrm{E}_{7(-5)}) &=& \mathrm{adj}(\mathrm{SO}^\star(12)) \oplus \mathrm{adj}(\mathrm{SL}(2,\mathbb{R})_\mathrm{E}) \oplus (\mathbf{2}, \mathbf{32}_s) \\ & \Downarrow & \\ \mathrm{adj}(\mathrm{F}_{4(4)}) &=& \mathrm{adj}(\mathrm{Sp}(6,\mathbb{R}) \oplus \mathrm{adj}(\mathrm{SL}(2,\mathbb{R})_\mathrm{E}) \oplus (\mathbf{2}, \mathbf{14}') \end{array} \tag{5.5.18}$$

The representation $\mathbf{14}'$ of $\mathrm{Sp}(6,\mathbb{R})$ is that of an antisymmetric symplectic traceless tensor:

$$\dim_{\mathrm{Sp}(6,\mathbb{R})} \widetilde{\begin{array}{|c|}\hline \\ \hline \\ \hline \\ \hline \end{array}} = \mathbf{14}' \tag{5.5.19}$$

The Dynkin diagram of the Tits Satake subalgebra $\mathfrak{f}_{4(4)}$ is discussed in Fig. 5.6.

$\mathcal{N} = 5$

The case of $\mathcal{N} = 5$ supergravity is described by Fig. 5.7 and it is one of the three exotic models whose Tits–Satake projection does not produce a Lie algebra root system.

In the $\mathcal{N} = 5$ theory the scalar manifold is a complex coset of rank $r = 1$,

$$\mathcal{M}_{\mathcal{N}=5,D=4} = \frac{\mathrm{SU}(1,5)}{\mathrm{SU}(5) \times \mathrm{U}(1)} \tag{5.5.20}$$

and there are **10** vector fields whose electric and magnetic field strengths are assigned to the **20**-dimensional representation of $\mathrm{SU}(1,5)$, which is that of an antisymmetric three-index tensor

$$\dim_{\mathrm{SU}(1,5)} \begin{array}{|c|}\hline \\ \hline \\ \hline \\ \hline \end{array} = 20 \tag{5.5.21}$$

The decomposition (1.7.12) takes the explicit form:

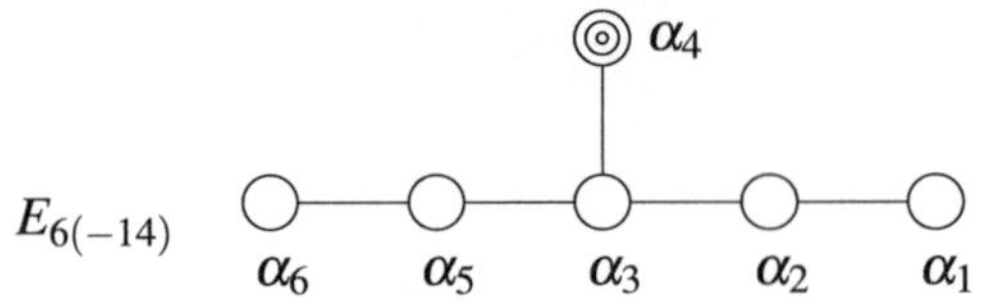

$$\psi = \alpha_1 + 2\alpha_2 + 3\alpha_3 + 2\alpha_4 + 2\alpha_5 + \alpha_6$$
$$(\psi\,,\,\alpha_4) = 1 \quad ; \quad (\psi\,,\,\alpha_i) = 0 \quad i \neq 4$$

Fig. 5.7 The Dynkin diagram of $E_{6(-14)}$. The only simple root which has grading one with respect to the highest root ψ is α_4 (painted with multiple circles). With respect to the algebra $\mathbb{U}_{D=4} = \mathrm{SU}(5,1)$) whose Dynkin diagram is obtained by removal of the black circle, α_4 is the highest weight of the symplectic representation of the vector fields, namely the $\mathbf{W} = \mathbf{20}$

$$\mathrm{adj}(\mathrm{E}_{6(-14)}) = \mathrm{adj}(\mathrm{SU}(1,5) \oplus \mathrm{adj}(\mathrm{SL}(2,\mathbb{R})_{\mathrm{E}}) \oplus (\mathbf{2}, \mathbf{20}) \tag{5.5.22}$$

and we have that the highest root of E_6, namely

$$\psi = \alpha_1 + 2\alpha_2 + 3\alpha_3 + 2\alpha_4 + 2\alpha_5 + \alpha_6 \tag{5.5.23}$$

has non vanishing scalar product only with the root α_4 in the form depicted in Fig. 5.7.

Writing a well adapted basis of E_6 roots is a little bit more laborious but it can be done. We find:

$$\begin{aligned}
\alpha_1 &= \left\{0, 0, -\tfrac{\sqrt{3}}{2}, \tfrac{1}{2\sqrt{5}}, \sqrt{\tfrac{6}{5}}, 0\right\} &&= \{\overline{\alpha}_1, 0\}\\
\alpha_2 &= \left\{-\tfrac{1}{\sqrt{2}}, \tfrac{1}{\sqrt{6}}, \tfrac{2}{\sqrt{3}}, 0, 0, 0\right\} &&= \{\overline{\alpha}_2, 0\}\\
\alpha_3 &= \left\{\sqrt{2}, 0, 0, 0, 0, 0\right\} &&= \{\overline{\alpha}_3, 0\}\\
\alpha_4 &= \left\{-\tfrac{1}{\sqrt{2}}, \tfrac{1}{\sqrt{6}}, -\tfrac{1}{\sqrt{3}}, \tfrac{1}{\sqrt{5}}, -\sqrt{\tfrac{3}{10}}, \tfrac{1}{\sqrt{2}}\right\} &&= \left\{\overline{\mathbf{w}}_h, \tfrac{1}{\sqrt{2}}\right\}\\
\alpha_5 &= \left\{-\tfrac{1}{\sqrt{2}}, -\sqrt{\tfrac{3}{2}}, 0, 0, 0, 0\right\} &&= \{\overline{\alpha}_4, 0\}\\
\alpha_6 &= \left\{0, \sqrt{\tfrac{2}{3}}, -\tfrac{1}{2\sqrt{3}}, -\tfrac{\sqrt{5}}{2}, 0, 0\right\} &&= \{\overline{\alpha}_5, 0\}
\end{aligned} \tag{5.5.24}$$

In this basis we can check that the five 5-vectors $\bar{\alpha}_i$ $(i = 1, \ldots, 5)$ constitute a simple root basis for the $A_5 \simeq \mathrm{SU}(1,5)$ root system, namely:

$$\langle \bar{\alpha}_i\,,\,\bar{\alpha}_j \rangle = \begin{pmatrix} 2 & -1 & 0 & 0 & 0\\ -1 & 2 & -1 & 0 & 0\\ 0 & -1 & 2 & -1 & 0\\ 0 & 0 & -1 & 2 & -1\\ 0 & 0 & 0 & -1 & 2 \end{pmatrix} = \text{Cartan matrix of } A_5 \tag{5.5.25}$$

while:

$$\mathbf{w}_h = \left\{ -\frac{1}{\sqrt{2}}, \frac{1}{\sqrt{6}}, -\frac{1}{\sqrt{3}}, \frac{1}{\sqrt{5}}, -\sqrt{\frac{3}{10}} \right\} \tag{5.5.26}$$

is the highest weight of the **20**-dimensional representation of SU(1, 5). Finally in this basis the highest root ψ defined by Eq. (5.5.23) takes the expected form:

$$\psi = \{0, 0, 0, 0, 0, 0, \sqrt{2}\} \tag{5.5.27}$$

$\mathcal{N} = 4$

The case of $\mathcal{N} = 4$ supergravity is the first where the scalar manifold is not completely fixed, since we can choose the number n_m of vector multiplets that we can couple to the graviton multiplet. In any case, once n_m is fixed the scalar manifold is also fixed and we have:

$$\mathcal{M}_{N=4,D=4} = \frac{\mathrm{SL}(2,\mathbb{R})_0}{\mathrm{O}(2)} \otimes \frac{\mathrm{SO}(6,\mathrm{n_m})}{\mathrm{SO}(6)\times\mathrm{SO}(\mathrm{n_m})} \tag{5.5.28}$$

The total number of vectors $n_v = 6 + n_m$ is also fixed and the symplectic representation **W** of the duality algebra

$$\mathbb{U}_{D=4} = \mathrm{SL}(2,\mathbb{R})_0 \times \mathrm{SO}(6,\mathrm{n_m}) \tag{5.5.29}$$

to which the vectors are assigned and which determines the embedding:

$$\mathrm{SL}(2,\mathbb{R})_0 \times \mathrm{SO}(6) \times \mathrm{SO}(\mathrm{n_m}) \mapsto \mathrm{Sp}(12 + 2\,\mathrm{n_m}, \mathbb{R}) \tag{5.5.30}$$

is also fixed, namely $\mathbf{W} = (\mathbf{2_0},\mathbf{6{+}n_m})$, $\mathbf{2_0}$ being the fundamental representation of $\mathrm{SL}(2,\mathbb{R})_0$ and $\mathbf{6{+}n_m}$ the fundamental vector representation of $\mathrm{SO}(6, \mathrm{n_m})$.

The $D = 3$ algebra is, $\mathbb{U}_{D=3} = \mathrm{SO}(8, \mathrm{n_m} + 2)$. Correspondingly the form taken by the general decomposition (1.7.12) is the following one:

$$\begin{aligned} \mathrm{adj}(\mathrm{SO}(8,\mathrm{n_m}+2)) &= \mathrm{adj}(\mathrm{SL}(2,\mathbb{R})_0) \oplus \mathrm{adj}(\mathrm{SO}(6,\mathrm{n_m})) \oplus \mathrm{adj}(\mathrm{SL}(2,\mathbb{R})_E) \\ &\oplus (\mathbf{2_E}, \mathbf{2_0},\mathbf{6{+}n_m}) \end{aligned} \tag{5.5.31}$$

where $\mathbf{2_{E,0}}$ are the fundamental representations respectively of $\mathrm{SL}(2,\mathbb{R})_E$ and of $\mathrm{SL}(2,\mathbb{R})_0$.

In order to give a Dynkin Weyl description of these algebras, we are forced to distinguish the case of an odd and even number of vector multiplets. In the first case both $\mathbb{U}_{D=3}$ and $\mathbb{U}_{D=4}$ are non simply laced algebras of the B-type, while in the second case they are both simply laced algebras of the D-type

$$n_m = \begin{cases} 2k & \to \quad \mathbb{U}_{D=4} \simeq D_{k+3} \\ 2k+1 & \to \quad \mathbb{U}_{D=4} \simeq B_{k+3} \end{cases} \tag{5.5.32}$$

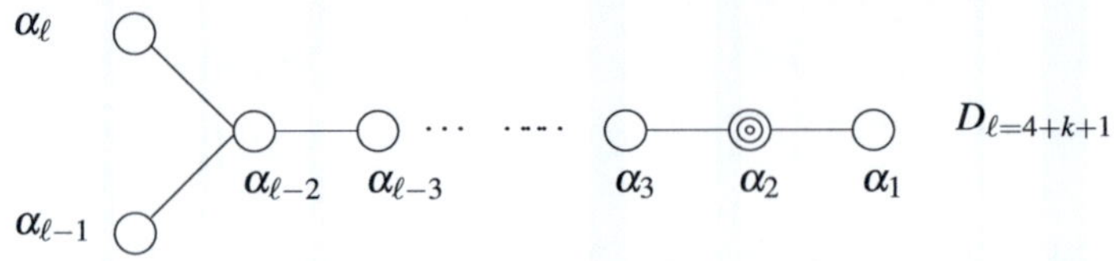

$$\psi = \alpha_1 + 2\alpha_2 + 2\alpha_3 + \cdots + 2\alpha_{\ell-2} + \alpha_{\ell-1} + \alpha_\ell$$
$$(\psi\,,\alpha_2) = 1 \quad ; \quad (\psi\,,\alpha_i) = 0 \quad i \neq 2$$

Fig. 5.8 The Dynkin diagram of D_{4+k+1}. The algebra D_{4+k+1} is that of the group SO(8, 2k + 2) corresponding to the σ-model reduction of $\mathcal{N} = 4$ supergravity coupled to $n_m = 2k$ vector multiplets. The only simple root which has non vanishing grading with respect to the highest one ψ is α_2. Removing it (black circle) we are left with the algebra $D_{4+k-1} \oplus A_1$ which is indeed the duality algebra in $D = 4$, namely SO(6, 2k) $\oplus$ SL(2, $\mathbb{R}$)$_0$. The root α_2 is the highest weight of the symplectic representation of the vector fields, namely the $\mathbf{W} = (\mathbf{2_0}, \mathbf{6} + \mathbf{2k})$

Just for simplicity and for shortness we choose to discuss only the even case $n_m = 2k$ which is described by Fig. 5.8.

In this case we consider the $\mathbb{U}_{D=3}$ = SO(8, 2k + 2) Lie algebra whose Dynkin diagram is that of D_{5+k}. Naming ε_i the unit vectors in an Euclidean ℓ-dimensional space where $\ell = 5 + k$, a well adapted basis of simple roots for the considered algebra is the following one:

$$\begin{aligned}
\alpha_1 &= \sqrt{2}\,\varepsilon_1 \\
\alpha_2 &= -\frac{1}{\sqrt{2}}\,\varepsilon_1 - \varepsilon_2 + \frac{1}{\sqrt{2}}\,\varepsilon_\ell \\
\alpha_3 &= \varepsilon_2 - \varepsilon_3 \\
\alpha_4 &= \varepsilon_3 - \varepsilon_4 \\
\ldots &= \ldots \\
\alpha_{\ell-1} &= \varepsilon_{\ell-2} - \varepsilon_{\ell-1} \\
\alpha_\ell &= \varepsilon_{\ell-2} + \varepsilon_{\ell-1}
\end{aligned} \tag{5.5.33}$$

which is quite different from the usual presentation but yields the correct Cartan matrix. In this basis the highest root of the algebra:

$$\psi = \alpha_1 + 2\alpha_2 + 2\alpha_3 + \cdots + 2\alpha_{\ell-2} + \alpha_{\ell-1} + \alpha_\ell \tag{5.5.34}$$

takes the desired form:

$$\psi = \sqrt{2}\,\varepsilon_\ell \tag{5.5.35}$$

In the same basis the $\alpha_W = \alpha_2$ root has also the expect form:

$$\alpha_W = \left(\mathbf{w}, \frac{1}{\sqrt{2}} \right) \tag{5.5.36}$$

where:

$$\mathbf{w} = -\frac{1}{\sqrt{2}} \varepsilon_1 - \varepsilon_2 \tag{5.5.37}$$

is the weight of the symplectic representation $\mathbf{W} = (\mathbf{2_0}, \mathbf{6} + \mathbf{2k})$. Indeed $-\frac{1}{\sqrt{2}} \varepsilon_1$ is the fundamental weight for the Lie algebra $\mathrm{SL}(2, \mathbb{R})_0$, whose root is $\alpha_1 = \sqrt{2}\, \varepsilon_1$, while $-\varepsilon_2$ is the highest weight for the vector representation of the algebra $\mathrm{SO}(6, 2\mathrm{k})$, whose roots are $\alpha_3, \alpha_4, \ldots, \alpha_\ell$.

Next we briefly comment on the Tits Satake projection. The algebra $\mathrm{SO}(8, \mathrm{n_m} + 2)$ is maximally split only for $n_m = 5, 6, 7$. The case $n_m = 6$, from the superstring view point, corresponds to the case of Neveu–Schwarz vector multiplets in a toroidal compactification. For a different number of vector multiplets, in particular for $n_m > 7$ the study of extremal black holes involves considering the Tits Satake projection, which just yields the universal algebra

$$\mathbb{U}^{TS}_{N=4, D=3} = \mathfrak{so}(8, 9) \tag{5.5.38}$$

5.6 Tits Satake Decompositions of the W Representations

One of the goals that we plan to pursue in Chap. 6 is the comparison of the classification of extremal black holes by means of *charge orbits* with their classification by means of $\mathrm{H}^\star$ *orbits*. Charge orbits means orbits of the $\mathrm{U}_{\mathrm{D}=4}$ group in the **W**-representation.

For this reason, in the present section we consider the decomposition of the **W**-representations with respect to Tits–Satake subalgebras and Paint groups for all the non-exotic models. The relevant **W**-representations are listed in Table 5.7. In Table 5.8 we listed the **W**-representations for the exotic models.

Given the paint algebra $\mathbb{G}_{\mathrm{paint}} \subset \mathbb{U}$ and the Tits Satake subalgebra $\mathbb{G}_{\mathrm{TS}} \subset \mathbb{U}$, one introduces, as we have seen, the *sub Tits Satake* and *sub paint* algebras as the centralizers of the paint algebra and of the Tits Satake algebra, respectively. In other words we have:

$$\mathfrak{s} \in \mathbb{G}_{\mathrm{subTS}} \subset \mathbb{G}_{\mathrm{TS}} \subset \mathbb{U} \quad \Leftrightarrow \quad \left[\mathfrak{s}\,,\, \mathbb{G}_{\mathrm{paint}}\right] = 0 \tag{5.6.1}$$

and

$$\mathfrak{t} \in \mathbb{G}_{\mathrm{subpaint}} \subset \mathbb{G}_{\mathrm{paint}} \subset \mathbb{U} \quad \Leftrightarrow \quad \left[\mathfrak{t}\,,\, \mathbb{G}_{\mathrm{TS}}\right] = 0 \tag{5.6.2}$$

As it was stressed repeatedly, a very important property of the paint and subpaint algebras is that they are conserved in the dimensional reduction, namely they are the same for $\mathbb{U}_{D=4}$ and $\mathbb{U}_{D=3}$.

In the next lines we analyze the decomposition of the **W**-representations with respect to these subalgebras for each Tits Satake universality class of non maximally split models. In the case of maximally split models there is no paint algebra and there is nothing with respect to which to decompose.

5.6.1 Universality Class $\mathfrak{sp}(6,\mathbb{R}) \Rightarrow \mathfrak{f}_{4(4)}$

In this case the sub Tits Satake Lie algebra is

$$\mathbb{G}_{\text{subTS}} = \mathfrak{sl}(2,\mathbb{R}) \oplus \mathfrak{sl}(2,\mathbb{R}) \oplus \mathfrak{sl}(2,\mathbb{R}) \subset \mathfrak{sp}(6,\mathbb{R}) = \mathbb{G}_{\text{TS}} \tag{5.6.3}$$

and the **W**-representation of the maximally split model decomposes as follows:

$$\mathbf{14}' \stackrel{\mathbb{G}_{\text{subTS}}}{\Longrightarrow} (\mathbf{2},\mathbf{1},\mathbf{1}) \oplus (\mathbf{1},\mathbf{2},\mathbf{1}) \oplus (\mathbf{1},\mathbf{1},\mathbf{2}) \oplus (\mathbf{2},\mathbf{2},\mathbf{2}) \tag{5.6.4}$$

This decomposition combines in the following way with the paint group representations in the various models belonging to the same universality class.

5.6.1.1 $\mathfrak{su}(3,3)$ Model

For this case the paint algebra is

$$\mathbb{G}_{\text{paint}} = \mathfrak{so}(2) \oplus \mathfrak{so}(2) \tag{5.6.5}$$

and the **W**-representation is the **20** dimensional of $\mathfrak{su}(3,3)$ corresponding to an antisymmetric tensor with a reality condition of the form:

$$t^{\star}_{\alpha\beta\gamma} = \frac{1}{3!}\,\varepsilon_{\alpha\beta\gamma\delta\eta\theta}\,t_{\delta\eta\theta} \tag{5.6.6}$$

The decomposition of this representation with respect to the Lie algebra $\mathbb{G}_{\text{paint}} \oplus \mathbb{G}_{\text{subTS}}$ is the following one:

$$\mathbf{20} \stackrel{\mathbb{G}_{\text{paint}}\oplus\mathbb{G}_{\text{subTS}}}{\Longrightarrow} (2,q_1|\mathbf{2},\mathbf{1},\mathbf{1}) \oplus (2,q_2|\mathbf{1},\mathbf{2},\mathbf{1}) \oplus (2,q_3|\mathbf{1},\mathbf{1},\mathbf{2}) \oplus (1,0|\mathbf{2},\mathbf{2},\mathbf{2}) \tag{5.6.7}$$

where $(2,q)$ means a doublet of $\mathfrak{so}(2) \oplus \mathfrak{so}(2)$ with a certain grading q with respect to the generators, while $(1,0)$ means the singlet that has 0 grading with respect to both generators. The subpaint algebra in this case is $\mathbb{G}_{\text{subpaint}} = 0$ and the decomposition of the same **W**-representation with respect to $\mathbb{G}_{\text{subpaint}} \oplus \mathbb{G}_{\text{TS}}$ is:

$$\mathbf{20} \stackrel{\mathbb{G}_{\text{subpaint}}\oplus\mathbb{G}_{\text{TS}}}{\Longrightarrow} \mathbf{6} \oplus \mathbf{14} \tag{5.6.8}$$

This follows from the decomposition of the **6** of $\mathfrak{sp}(6, \mathbf{R})$ with respect to the sub Tits Satake algebra (5.6.3):

$$\mathbf{6} \stackrel{\mathbb{G}_{\text{subTS}}}{\Longrightarrow} (\mathbf{2},\mathbf{1},\mathbf{1}) \oplus (\mathbf{1},\mathbf{2},\mathbf{1}) \oplus (\mathbf{1},\mathbf{1},\mathbf{2}) \tag{5.6.9}$$

5.6.1.2 $\mathfrak{so}^{\star}(12)$ Model

For this case the paint algebra is

$$\mathbb{G}_{\text{paint}} = \mathfrak{so}(3) \oplus \mathfrak{so}(3) \oplus \mathfrak{so}(3) \tag{5.6.10}$$

and the **W**-representation is the $\mathbf{32}_s$ dimensional spinorial representation of $\mathfrak{so}^{\star}(12)$. The decomposition of this representation with respect to the Lie algebra $\mathbb{G}_{\text{paint}} \oplus \mathbb{G}_{\text{subTS}}$ is the following one:

$$\mathbf{32}_s \stackrel{\mathbb{G}_{\text{paint}}\oplus\mathbb{G}_{\text{subTS}}}{\Longrightarrow} (\underline{2},\underline{2},\underline{1}|\mathbf{2},\mathbf{1},\mathbf{1}) \oplus (\underline{2},\underline{1},\underline{2}|\mathbf{1},\mathbf{2},\mathbf{1}) \oplus (\underline{1},\underline{1},\underline{2}|\mathbf{1},\mathbf{1},\mathbf{2}) \oplus (\underline{1},\underline{1},\underline{1}|\mathbf{2},\mathbf{2},\mathbf{2}) \tag{5.6.11}$$

where $\underline{2}$ means the doublet spinor representation of $\mathfrak{so}(3)$. The subpaint algebra in this case is $\mathbb{G}_{\text{paint}} = \mathfrak{so}(3)_{\text{diag}}$ and the decomposition of the same **W**-representation with respect to $\mathbb{G}_{\text{subpaint}} \oplus \mathbb{G}_{\text{TS}}$ is:

$$\mathbf{32}_s \stackrel{\mathbb{G}_{\text{TS}}\oplus\mathbb{G}_{\text{subpaint}}}{\Longrightarrow} (\mathbf{6}|\underline{3}) \oplus (\mathbf{14}'|\underline{1}) \tag{5.6.12}$$

This follows from the decomposition of the product $\underline{2} \times \underline{2}$ of $\mathfrak{so}(3)_{\text{diag}}$ times the Tits Satake algebra (5.6.3):

$$\underline{2} \times \underline{2} = \underline{3} \oplus \underline{1} \tag{5.6.13}$$

5.6.1.3 $\mathfrak{e}_{7(-25)}$ model

For this case the paint algebra is

$$\mathbb{G}_{\text{paint}} = \mathfrak{so}(8) \tag{5.6.14}$$

and the **W**-representation is the fundamental **56** dimensional representation of $\mathfrak{e}_{7(-25)}$ The decomposition of this representation with respect to the Lie algebra $\mathbb{G}_{\text{paint}} \oplus \mathbb{G}_{\text{subTS}}$ is the following one:

$$\mathbf{56} \stackrel{\mathbb{G}_{\text{paint}}\oplus\mathbb{G}_{\text{subTS}}}{\Longrightarrow} (\mathbf{8}_v|\mathbf{2},\mathbf{1},\mathbf{1}) \oplus (\mathbf{8}_s|\mathbf{1},\mathbf{2},\mathbf{1}) \oplus (\mathbf{8}_c|\mathbf{1},\mathbf{1},\mathbf{2}) \oplus (\mathbf{1}|\mathbf{2},\mathbf{2},\mathbf{2}) \tag{5.6.15}$$

where $\mathbf{8}_{v,s,c}$ are the three inequivalent eight-dimensional representations of $\mathfrak{so}(8)$, the vector, the spinor and the conjugate spinor. The subpaint algebra in this case is $\mathbb{G}_{\text{subpaint}} = \mathfrak{g}_{2(-14)}$ with respect to which all three 8-dimensional representations of $\mathfrak{so}(8)$ branch as follows:

$$\mathbf{8}_{v,s,c} \stackrel{\mathfrak{g}_{2(-14)}}{\Longrightarrow} \mathbf{7} \oplus \mathbf{1} \tag{5.6.16}$$

In view of this the decomposition of the same $\mathbf{W}$-representation with respect to $\mathbb{G}_{\text{subpaint}} \oplus \mathbb{G}_{\text{TS}}$ is:

$$\mathbf{56} \stackrel{\mathbb{G}_{\text{TS}} \oplus \mathbb{G}_{\text{subpaint}}}{\Longrightarrow} (\mathbf{6}|\mathbf{7}) \oplus (\mathbf{14}'|\mathbf{1}) \tag{5.6.17}$$

5.6.2 Universality Class $\mathfrak{sl}(2,\mathbb{R}) \oplus \mathfrak{so}(2,3) \Rightarrow \mathfrak{so}(4,5)$

This case corresponds to one of the possible infinite families of $\mathcal{N} = 2$ theories with a symmetric homogeneous special Kähler manifold and a number of vector multiplets larger than three ($n = 3 + p$). The other infinite family corresponds instead to one of the three exotic models.

The generic element of this infinite class corresponds to the following algebras:

$$\begin{aligned} \mathbb{U}_{\text{D}=4} &= \mathfrak{sl}(2,\mathbb{R}) \oplus \mathfrak{so}(2, 2+p) \\ \mathbb{U}_{\text{D}=3} &= \mathfrak{so}(4, 4+p) \end{aligned} \tag{5.6.18}$$

In this case the sub Tits Satake algebra is:

$$\begin{aligned} \mathbb{G}_{\text{subTS}} &= \mathfrak{sl}(2,\mathbb{R}) \oplus \mathfrak{sl}(2,\mathbb{R}) \oplus \mathfrak{sl}(2,\mathbb{R}) \\ &\simeq \mathfrak{sl}(2,\mathbb{R}) \oplus \mathfrak{so}(2,2) \ \subset \ \mathfrak{sl}(2,\mathbb{R}) \oplus \mathfrak{so}(2,3) \ = \ \mathbb{G}_{\text{TS}} \end{aligned} \tag{5.6.19}$$

an the paint and subpaint algebras are as follows:

$$\begin{aligned} \mathbb{G}_{\text{paint}} &= \mathfrak{so}(p) \\ \mathbb{G}_{\text{subpaint}} &= \mathfrak{so}(p-1) \end{aligned} \tag{5.6.20}$$

The symplectic $\mathbf{W}$ representation of $\mathbb{U}_{\text{D}=4}$ is the tensor product of the fundamental representation of $\mathfrak{sl}(2)$ with the fundamental vector representation of $\mathfrak{so}(2, 2+p)$, namely

$$\mathbf{W} \ = \ (\mathbf{2}|\mathbf{4}+p) \quad ; \quad \dim \mathbf{W} \ = \ 8 + 2p \tag{5.6.21}$$

The decomposition of this representation with respect to $\mathbb{G}_{\text{subTS}} \oplus \mathbb{G}_{\text{subpaint}}$ is the following one:

$$\mathbf{W} \stackrel{\mathbb{G}_{\text{subTS}} \oplus \mathbb{G}_{\textbf{subpaint}}}{\Longrightarrow} (\mathbf{2},\mathbf{2},\mathbf{2}|\mathbf{1}) \oplus (\mathbf{2},\mathbf{1},\mathbf{1}|\mathbf{1}) \oplus (\mathbf{2},\mathbf{1},\mathbf{1}|p-1) \tag{5.6.22}$$

where **2**, **2**, **2** denotes the tensor product of the three fundamental representations of $\mathfrak{sl}(2,\mathbb{R})^3$. Similarly **2**, **1**, **1** denotes the doublet of the first $\mathfrak{sl}(2,\mathbb{R})$ tensored with the singlets of the following two $\mathfrak{sl}(2,\mathbb{R})$ algebras. The representations appearing in (5.6.22) can be grouped in order to reconstruct full representations either of the complete Tits Satake or of the complete paint algebras. In this way one obtains:

$$\begin{aligned} \mathbf{W} &\stackrel{\mathbb{G}_{\text{subTS}} \oplus \mathbb{G}_{\textbf{paint}}}{\Longrightarrow} (\mathbf{2}, \mathbf{2}, \mathbf{2}|\mathbf{1}) \oplus (\mathbf{2}, \mathbf{1}, \mathbf{1}|p+1) \\ \mathbf{W} &\stackrel{\mathbb{G}_{\text{TS}} \oplus \mathbb{G}_{\textbf{subpaint}}}{\Longrightarrow} (\mathbf{2}, \mathbf{5}|\mathbf{1}) \oplus (\mathbf{2}, \mathbf{1}|p-1) \end{aligned} \tag{5.6.23}$$

5.6.3 Universality Class $\mathfrak{sl}(2,\mathbb{R}) \oplus \mathfrak{so}(6,7) \Rightarrow \mathfrak{so}(8,9)$

This case, which corresponds to an $\mathscr{N}=4$ theory with a number of vector multiplets larger than six ($n = 6 + p$) presents a very strong similarity with the previous $\mathscr{N}=2$ case.

The generic element of this infinite class corresponds to the following algebras:

$$\begin{aligned} \mathbb{U}_{\text{D}=4} &= \mathfrak{sl}(2,\mathbb{R}) \oplus \mathfrak{so}(6, 6+p) \\ \mathbb{U}_{\text{D}=3} &= \mathfrak{so}(8, 8+p) \end{aligned} \tag{5.6.24}$$

In this case the sub Tits Satake algebra is:

$$\mathbb{G}_{\text{subTS}} = \mathfrak{sl}(2,\mathbb{R}) \oplus \mathfrak{so}(6,6) \subset \mathfrak{sl}(2,\mathbb{R}) \oplus \mathfrak{so}(6,7) = \mathbb{G}_{\text{TS}} \tag{5.6.25}$$

an the paint and subpaint algebras are the same as in the previous $\mathscr{N}=2$ case, namely:

$$\begin{aligned} \mathbb{G}_{\text{paint}} &= \mathfrak{so}(p) \\ \mathbb{G}_{\text{subpaint}} &= \mathfrak{so}(p-1) \end{aligned} \tag{5.6.26}$$

The symplectic **W** representation of $\mathbb{U}_{\text{D}=4}$ is the tensor product of the fundamental representation of $\mathfrak{sl}(2)$ with the fundamental vector representation of $\mathfrak{so}(6, 6+p)$, namely

$$\mathbf{W} = (\mathbf{2}|\mathbf{12}+p) \quad ; \quad \dim \mathbf{W} = 24 + 2p \tag{5.6.27}$$

The decomposition of this representation with respect to $\mathbb{G}_{\text{subTS}} \oplus \mathbb{G}_{\text{subpaint}}$ is the following one:

$$\mathbf{W} \stackrel{\mathbb{G}_{\text{subTS}} \oplus \mathbb{G}_{\textbf{subpaint}}}{\Longrightarrow} (\mathbf{2}, \mathbf{12}|\mathbf{1}) \oplus (\mathbf{2}, \mathbf{1}|\mathbf{1}) \oplus (\mathbf{2}, \mathbf{1}|p) \tag{5.6.28}$$

Just as above the three representations appearing in (5.6.28) can be grouped in order to obtain either representation of the complete Tits Satake or of the complete paint algebras. This yields

$$\begin{aligned} \mathbf{W} &\stackrel{\mathbb{G}_{\text{subTS}} \oplus \mathbb{G}_{\textbf{paint}}}{\Longrightarrow} (\mathbf{2}, \mathbf{12}|\mathbf{1}) \oplus (\mathbf{2}, \mathbf{1}|p+1) \\ \mathbf{W} &\stackrel{\mathbb{G}_{\text{TS}} \oplus \mathbb{G}_{\textbf{subpaint}}}{\Longrightarrow} (\mathbf{2}, \mathbf{13}|\mathbf{1}) \oplus (\mathbf{2}, \mathbf{1}|p) \end{aligned} \quad (5.6.29)$$

5.6.4 *The Universality Classes* $\mathfrak{sl}(2,\mathbb{R}) \oplus \mathfrak{so}(6,n) \Rightarrow \mathfrak{so}(8,n+2)$ *with* $n \le 5$

These classes correspond to the $\mathcal{N} = 4$ theories with a number $n = 1, 2, 3, 4, 5$ of vector multiplets. In each case we have the following algebras:

$$\begin{aligned} \mathbb{U}_{\text{D}=4} &= \mathfrak{sl}(2,\mathbb{R}) \oplus \mathfrak{so}(6,n) \\ \mathbb{U}_{\text{D}=3} &= \mathfrak{so}(8,n+2) \end{aligned} \quad (5.6.30)$$

In all these cases the Tits Satake and sub Tits Satake algebras are:

$$\begin{aligned} \mathbb{G}_{\text{TS}} &= \mathfrak{sl}(2,\mathbb{R}) \oplus \mathfrak{so}(n+1,n) \\ \mathbb{G}_{\text{subTS}} &= \mathfrak{sl}(2,\mathbb{R}) \oplus \mathfrak{so}(n,n) \end{aligned} \quad (5.6.31)$$

and the paint and subpaint algebras are:

$$\begin{aligned} \mathbb{G}_{\text{paint}} &= \mathfrak{so}(6-n) \\ \mathbb{G}_{\text{subpaint}} &= \mathfrak{so}(5-n) \end{aligned} \quad (5.6.32)$$

The symplectic **W** representation is the tensor product of the doublet representation of $\mathfrak{sl}(2)$ with the fundamental representation of $\mathfrak{so}(6,n)$, namely

$$\mathbf{W} = (\mathbf{2}, \mathbf{6}+\mathbf{n}) \quad (5.6.33)$$

and its decomposition with respect to the $\mathbb{G}_{\text{subTS}} \oplus \mathbb{G}_{\text{subpaint}}$ algebra is as follows

$$\mathbf{W} \stackrel{\mathbb{G}_{\text{subTS}} \oplus \mathbb{G}_{\textbf{subpaint}}}{\Longrightarrow} (\mathbf{2}, \mathbf{2n}|\mathbf{1}) \oplus (\mathbf{2}, \mathbf{1}|\mathbf{1}) \oplus (\mathbf{2}, \mathbf{1}|\mathbf{5}-\mathbf{n}) \quad (5.6.34)$$

which, with the same procedure as above leads to:

$$\begin{aligned} \mathbf{W} &\stackrel{\mathbb{G}_{\text{subTS}} \oplus \mathbb{G}_{\textbf{paint}}}{\Longrightarrow} (\mathbf{2}, \mathbf{2n}|\mathbf{1}) \oplus (\mathbf{2}, \mathbf{1}|\mathbf{6}-\mathbf{n}) \\ \mathbf{W} &\stackrel{\mathbb{G}_{\text{TS}} \oplus \mathbb{G}_{\textbf{subpaint}}}{\Longrightarrow} (\mathbf{2}, \mathbf{2n}+\mathbf{1}|\mathbf{1}) \oplus (\mathbf{2}, \mathbf{1}|\mathbf{5}-\mathbf{n}) \end{aligned} \quad (5.6.35)$$

5.6.5 W-Representations of the Maximally Split Non Exotic Models

In the previous subsections we have analysed the Tits–Satake decomposition of the **W**-representation for all those models that are non maximally split. The remaining models are the maximally split ones for which there is no paint algebra and the Tits Satake projection is the identity map. For reader's convenience we have extracted the list of such models and presented it in Table 5.6. As we see from the table we have essentially five type of models:

1. The $E_{7(7)}$ model corresponding to $\mathscr{N}=8$ supergravity where the **W**-representation is the fundamental **56**.
2. The SU(1, 1) non exotic model where the **W**-representation is the $j=\frac{3}{2}$ of $\mathfrak{so}(1,2)\sim\mathfrak{su}(1,1)$
3. The Sp(6, ℝ) model where the **W**-representation is the **14**′ (antisymmetric symplectic traceless three-tensor).
4. The models $\mathfrak{sl}(2,\mathbb{R})\oplus\mathfrak{so}(q,q)$ where the **W**-representation is the (2, 2q), namely the tensor product of the two fundamentals.
5. The models $\mathfrak{sl}(2,\mathbb{R})\oplus\mathfrak{so}(q,q+1)$ where the **W**-representation is the (2, 2q + 1), namely the tensor product of the two fundamentals.

Therefore, for the above maximally split models, the *charge classification* of black holes reduces to the classification of $U_{D=4}$ orbits in the mentioned **W**-representations. Actually such orbits are sufficient also for the non maximally split models. Indeed each of the above 5-models correspond to one Tits Satake universality class and, within each universality class, the only relevant part of the **W**-representation is the subpaint group singlet which is universal for all members of the class. This is precisely what we verified in the previous subsections.

For instance for all members of the universality class of Sp(6, ℝ), the **W**-representation splits as follows with respect to the subalgebra $\mathfrak{sp}(6,\mathbb{R})\oplus\mathbb{G}_{\text{subpaint}}$:

$$\mathbf{W} \overset{\mathfrak{sp}(6,\mathbb{R})\oplus\mathbb{G}_{\text{subpaint}}}{\Longrightarrow} \left(\mathbf{6}\,|\,\mathscr{D}_{\text{subpaint}}\right) + \left(\mathbf{14}'\,|\,\mathbf{1}_{\text{subpaint}}\right) \tag{5.6.36}$$

where the representation $\mathscr{D}_{\text{subpaint}}$ is the following one for the three non-maximally split members of the class:

$$\mathscr{D}_{\text{subpaint}} = \begin{cases} \mathbf{1} \text{ of } \;\; \mathbf{1} & \text{for the } \mathfrak{su}(3,3) - \text{model} \\ \mathbf{3} \text{ of } \mathfrak{so}(3) & \text{for the } \mathfrak{so}^\star(12) - \text{model} \\ \mathbf{7} \text{ of } \mathfrak{g}_{2(-14)} & \text{for the } \mathfrak{e}_{7(-25)} - \text{model} \end{cases} \tag{5.6.37}$$

Clearly the condition:

$$\left(\mathbf{6}\,|\,\mathscr{D}_{\text{subpaint}}\right) = 0 \tag{5.6.38}$$

imposed on a vector in the **W**-representation breaks the group $U_{D=4}$ to its Tits Satake subgroup. The key point is that each **W**-orbit of the big group $U_{D=4}$ crosses the locus

Table 5.6 The list of *non-exotic* homogenous symmetric scalar manifolds appearing in $D = 4$ supergravity which are also maximally split. For these models the paint group is the identity group

#	TS D = 4	TS D = 3	Coset D = 4	Coset D = 3	Paint group	subP group	Susy
1	$\frac{E_{7(7)}}{SU(8)}$	$\frac{E_{8(8)}}{SO^\star(16)}$	$\frac{E_{7(7)}}{SU(8)}$	$\frac{E_{8(8)}}{SO^\star(16)}$	1	1	$\mathcal{N} = 8$
2	$\frac{SU(1,1)}{U(1)}$	$\frac{G_{2(2)}}{SL(2,R)\times SL(2,R)}$	$\frac{SU(1,1)}{U(1)}$	$\frac{G_{2(2)}}{SL(2,R)\times SL(2,R)}$	1	1	$\mathcal{N} = 2$ n = 1
3	$\frac{Sp(6,R)}{SU(3)\times U(1)}$	$\frac{F_{4(4)}}{Sp(6,R)\times SL(2,R)}$	$\frac{Sp(6,R)}{SU(3)\times U(1)}$	$\frac{F_{4(4)}}{Sp(6,R)\times SL(2,R)}$	1	1	$\mathcal{N} = 2$ $n = 6$
11	$\frac{SL(2,\mathbb{R})}{O(2)} \times \frac{SO(6,5)}{SO(6)\times SO(5)}$	$\frac{SO(8,7)}{SO(6,2)\times SO(2,5)}$	$\frac{SL(2,\mathbb{R})}{O(2)} \times \frac{SO(6,5)}{SO(6)\times SO(5)}$	$\frac{SO(8,7)}{SO(6,2)\times SO(2,5)}$	1	1	$\mathcal{N} = 4$ n = 5
12	$\frac{SL(2,\mathbb{R})}{O(2)} \times \frac{SO(6,6)}{SO(6)\times SO(6))}$	$\frac{SO(8,8)}{SO(6,2)\times SO(2,6)}$	$\frac{SL(2,\mathbb{R})}{O(2)} \times \frac{SO(6,6)}{SO(6)\times SO(6))}$	$\frac{SO(8,8)}{SO(6,2)\times SO(2,6)}$	1	1	$\mathcal{N} = 4$ n = 6
13	$\frac{SL(2,\mathbb{R})}{O(2)} \times \frac{SO(6,7)}{SO(6)\times SO(7))}$	$\frac{SO(8,9)}{SO(6,2)\times SO(2,7)}$	$\frac{SL(2,\mathbb{R})}{O(2)} \times \frac{SO(6,7)}{SO(6)\times SO(7))}$	$\frac{SO(8,9)}{SO(6,2)\times SO(2,7)}$	1	1	$\mathcal{N} = 4$ n = 7
14	$\frac{SL(2,\mathbb{R})}{O(2)} \times \frac{SO(2,1)}{SO(2)}$	$\frac{SO(4,3)}{SO(2,2)\times SO(2,1)}$	$\frac{SL(2,\mathbb{R})}{O(2)} \times \frac{SO(2,1)}{SO(2)}$	$\frac{SO(4,3)}{SO(2,2)\times SO(2,1)}$	1	1	$\mathcal{N} = 2$ n = 2
15	$\frac{SL(2,\mathbb{R})}{O(2)} \times \frac{SO(2,2)}{SO(2)\times SO(2)}$	$\frac{SO(4,4)}{SO(2,2)\times SO(2,2)}$	$\frac{SL(2,\mathbb{R})}{O(2)} \times \frac{SO(2,2)}{SO(2)\times SO(2)}$	$\frac{SO(4,4)}{SO(2,2)\times SO(2,2)}$	1	1	$\mathcal{N} = 2$ n = 3
16	$\frac{SL(2,\mathbb{R})}{O(2)} \times \frac{SO(2,3)}{SO(2)\times SO(3)}$	$\frac{SO(4,5)}{SO(2,2)\times SO(2,3)}$	$\frac{SL(2,\mathbb{R})}{O(2)} \times \frac{SO(2,3)}{SO(2)\times SO(3)}$	$\frac{SO(4,5)}{SO(2,2)\times SO(2,3)}$	1	1	$\mathcal{N} = 2$ n = 4

(5.6.38) so that the classification of Sp(6, $\mathbb{R}$) orbits in the $\mathbf{14}'$-representation exhausts the classification of **W**-orbits for all members of the universality class.

In order to prove that the gauge (5.6.38) is always reachable it suffices to show that the representation $\left(\mathbf{6}\,|\,\mathscr{D}_{\text{subpaint}}\right)$ always appears at least once in the decomposition of the Lie algebra $\mathbb{U}_{D=4}$ with respect to the subalgebra $\mathfrak{sp}(6,\mathbb{R})\oplus\mathbb{G}_{\text{subpaint}}$. The corresponding parameters of the big group can be used to set to zero the projection of the **W**-vector onto $\left(\mathbf{6}\,|\,\mathscr{D}_{\text{subpaint}}\right)$.

The required condition is easily verified since we have:

$$\begin{array}{lll}
\underbrace{\text{adj}\,\mathfrak{su}(3,3)}_{\mathbf{35}} & \stackrel{\mathfrak{sp}(6,\mathbb{R})}{\Longrightarrow} & \underbrace{\text{adj}\,\mathfrak{sp}(6,\mathbb{R})}_{\mathbf{21}}\oplus\mathbf{6}\oplus\mathbf{6}\oplus\mathbf{1}\oplus\mathbf{1} \\
\underbrace{\text{adj}\,\mathfrak{so}^{\star}(12)}_{\mathbf{66}} & \stackrel{\mathfrak{sp}(6,\mathbb{R})\oplus\mathfrak{so}(3)}{\Longrightarrow} & \underbrace{\text{adj}\,\mathfrak{sp}(6,\mathbb{R})}_{\mathbf{21}}\oplus\underbrace{\text{adj}\,\mathfrak{so}(3)}_{\mathbf{3}}\oplus(\mathbf{6},\mathbf{3})\oplus(\mathbf{6},\mathbf{3})\oplus(\mathbf{1},\mathbf{3})\oplus(\mathbf{1},\mathbf{3}) \\
\underbrace{\text{adj}\,\mathfrak{e}_{7(-25)}}_{\mathbf{133}} & \stackrel{\mathfrak{sp}(6,\mathbb{R})\oplus\mathfrak{g}_{2(-14)}}{\Longrightarrow} & \underbrace{\text{adj}\,\mathfrak{sp}(6,\mathbb{R})}_{\mathbf{21}}\oplus\underbrace{\text{adj}\,\mathfrak{g}_{2(-14)}}_{\mathbf{14}}\oplus(\mathbf{6},\mathbf{7})\oplus(\mathbf{6},\mathbf{7})\oplus(\mathbf{1},\mathbf{7})\oplus(\mathbf{1},\mathbf{7})
\end{array} \tag{5.6.39}$$

The reader cannot avoid being impressed by the striking similarity of the above decompositions which encode the very essence of Tits Satake universality. Indeed the representations of the common Tits Satake subalgebra appearing in the decomposition of the adjoint are the same for all members of the class. They are simply uniformly assigned to the fundamental representation of the subpaint algebra which is different in the three cases. The representation $\left(\mathbf{6}\,|\,\mathscr{D}_{\text{subpaint}}\right)$ appears twice in these decompositions and can be used to reach the gauge (5.6.38) as we claimed above.

For the models of type $\mathfrak{sl}(2,\mathbb{R})\oplus\mathfrak{so}(q,q+p)$ having $\mathfrak{sl}(2,\mathbb{R})\oplus\mathfrak{so}(q,q+1)$ as Tits Satake subalgebra and $\mathfrak{so}(p-1)$ as subpaint algebra the decomposition of the **W**-representation is the following one:

$$\mathbf{W}\;=\;(\mathbf{2},\mathbf{2q}+\mathbf{p})\;\stackrel{\mathfrak{sl}(2,\mathbb{R})\oplus\mathfrak{so}(q,q+1)\oplus\mathfrak{so}(p-1)}{\Longrightarrow}\;(\mathbf{2},\mathbf{2q}+\mathbf{1}|\mathbf{1})\;\oplus\;(\mathbf{2},\mathbf{1}|\mathbf{p}-\mathbf{1}) \tag{5.6.40}$$

and the question is whether each $\mathfrak{sl}(2,\mathbb{R})\oplus\mathfrak{so}(q,q+p)$ orbit in the $(\mathbf{2},\mathbf{2q}+\mathbf{p})$ representation intersects the $\mathfrak{sl}(2,\mathbb{R})\oplus\mathfrak{so}(q,q+1)\oplus\mathfrak{so}(p-1)$-invariant locus:

$$(\mathbf{2},\mathbf{1}|\mathbf{p}-\mathbf{1})\;=\;0 \tag{5.6.41}$$

The answer is yes since we always have enough parameters in the coset

$$\frac{\mathrm{SL}(2,\mathbb{R})\times\mathrm{SO}(\mathrm{q},\mathrm{q}+\mathrm{p})}{\mathrm{SL}(2,\mathbb{R})\times\mathrm{SO}(\mathrm{q},\mathrm{q}+1)\times\mathrm{SO}(\mathrm{p}-1)} \tag{5.6.42}$$

to reach the desired gauge (5.6.41). Indeed let us observe the decomposition:

$$
\begin{aligned}
\text{adj}\ [\mathfrak{sl}(2,\mathbb{R}) \oplus \mathfrak{so}(q,q+p)] = &\ \text{adj}\ [\mathfrak{sl}(2,\mathbb{R})] \oplus \text{adj}\ [\mathfrak{so}(q,q+1)] \\
&\oplus \text{adj}\ [\mathfrak{so}(p-1)] \oplus (\mathbf{1}, \mathbf{2q+1}|\mathbf{p-1})
\end{aligned}
\tag{5.6.43}
$$

The $2q+1$ vectors of $\mathfrak{so}(p-1)$ appearing in (5.6.43) are certainly sufficient to set to zero the 2 vectors of $\mathfrak{so}(p-1)$ appearing in $\mathbf{W}$.

The conclusion therefore is that the classification of charge-orbits for all supergravity models can be performed by restriction to the Tits Satake sub-model. The same we show, in the next section, to be true at the level of the classification based on H* orbits of the Lax operators, so that the final comparison of the two classifications can be performed by restriction to the Tits Satake subalgebras.

5.7 Tits Satake Reduction of the $\mathbb{H}^\star$ Subalgebra and of Its Representation $\mathbb{K}^\star$

As we show in Chap. 6, in the σ-model approach to black hole solutions one arrives at the new coset manifold (4.3.41). The structure of the enlarged group $\mathrm{U}_{D=3}$ and of its Lie algebra $\mathbb{U}_{D=3}$ was discussed in Eq. (1.7.12). The subgroups $\mathbb{H}^\star$ are listed in Table 5.7 for the non exotic models and in Table 5.8 for the exotic ones. The coset generators fall into a representation of $\mathbb{H}^\star$ that we name $\mathbb{K}^\star$. The Lax operator L_0 which determines the spherically symmetric black hole solution up to boundary conditions of the scalar fields at infinity is just an element of such a representation:

$$
L_0 \in \mathbb{K}^\star \tag{5.7.1}
$$

so that the classification of spherical black holes is reduced to the classification of $\mathbb{H}^\star$ orbits in the $\mathbb{K}^\star$ representation. On the other hand, in Chap. 6, we demonstrate how nilpotent orbits can be associated to multicenter solutions.

We focus on non-exotic models that admit a regular Tits Satake projection.

A first general remark concerns the structure of $\mathbb{H}^\star$ in all those models that correspond to $\mathcal{N}=2$ supersymmetry. In these cases the $\mathbb{H}^\star$ subalgebra is isomorphic to $\mathfrak{sl}(2,\mathbb{R}) \oplus \mathbb{U}_{D=4}$ so that we have a decomposition of the $\mathbb{U}_{D=3}$ Lie algebra with respect to $\mathbb{H}^\star$ completely analogous to that in Eq. (1.7.12), namely:

$$
\text{adj}(\mathbb{U}_{D=3}) = \underbrace{\text{adj}(\widehat{\mathbb{U}_{D=4}}) \oplus \text{adj}(\mathfrak{sl}(2,\mathbb{R})_{\mathrm{h}^\star})}_{\mathbb{H}^\star} \oplus \underbrace{(2_{\mathrm{h}^\star}, \widehat{\mathbf{W}})}_{\mathbb{K}^\star} \tag{5.7.2}
$$

Hence the representation $\mathbb{K}^\star$ which contains the Lax operators has a structure analogous to the representation which contains the generators of $\mathbb{U}_{D=4}$ that originate from the vector fields, namely: $(2_{\mathrm{h}^\star}, \widehat{\mathbf{W}})$. This means that in all these models, by means of exactly the same argument as utilized above, we can always reach the gauge where

Table 5.7 Table of $\mathbb{H}^\star$ subalgebras of $\mathbb{U}_{D=3}$, $\mathbb{K}^\star$-representations and **W** representations of $\mathbb{U}_{D=4}$ for the supergravity models based on *non-exotic* scalar symmetric spaces

#	$\mathbb{U}_{D=3}$	$\mathbb{H}^\star$	$\mathbb{K}^\star$	$\mathbb{U}_{D=4}$	Rep. W	$\mathbb{H}_c$
1	$\mathfrak{e}_{8(8)}$	$\mathfrak{so}^\star(16)$	$\mathbf{128}_s$	$\mathfrak{e}_{7(7)}$	$\mathbf{56}$	$\mathfrak{su}(8)$
2	$\mathfrak{g}_{2(2)}$	$\widehat{\mathfrak{sl}(2,\mathrm{R})} \oplus \mathfrak{sl}(2,\mathrm{R})_{\mathrm{h}^\star}$	$(\mathbf{4}_{3/2}, \mathbf{2}_{h^\star})$	$\mathfrak{sl}(2,\mathrm{R})$	$\mathbf{4}_{3/2}$	$\mathfrak{so}(2)$
3	$\mathfrak{f}_{4(4)}$	$\widehat{\mathfrak{sp}(6,\mathrm{R})} \oplus \mathfrak{sl}(2,\mathrm{R})_{\mathrm{h}^\star}$	$(\widehat{\mathbf{14}'}, \mathbf{2}_{h^\star})$	$\mathfrak{sp}(6,\mathrm{R})$	$\mathbf{14}'$	$\mathfrak{u}(3)$
4	$\mathfrak{e}_{6(2)}$	$\widehat{\mathfrak{su}(3,3)} \oplus \mathfrak{sl}(2,\mathrm{R})_{\mathrm{h}^\star}$	$(\widehat{\mathbf{20}}, \mathbf{2}_{h^\star})$	$\mathfrak{su}(3,3)$	$\mathbf{20}$	$\mathfrak{su}(3) \oplus \mathfrak{su}(3) \oplus \mathfrak{u}(1)$
5	$\mathfrak{e}_{7(-5)}$	$\widehat{\mathfrak{so}^\star(12)} \oplus \mathfrak{sl}(2,\mathrm{R})_{\mathrm{h}^\star}$	$(\widehat{\mathbf{32}_{spin}}, \mathbf{2}_{h^\star})$	$\mathfrak{so}^\star(12)$	$\mathbf{32}_{spin}$	$\mathfrak{u}(6)$
6	$\mathfrak{e}_{8(-24)}$	$\widehat{\mathfrak{e}_{7(-25)}} \oplus \mathfrak{sl}(2,\mathrm{R})_{\mathrm{h}^\star}$	$(\widehat{\mathbf{56}}, \mathbf{2}_{h^\star})$	$\mathfrak{e}_{7(-25)}$	$\mathbf{56}$	$\mathfrak{u}(6)$
7	$\mathfrak{so}(8,3)$	$\mathfrak{so}(6,2) \oplus \mathfrak{so}(2,1)$	$(\mathbf{8}, \mathbf{3})$	$\mathfrak{so}(6,1) \oplus \mathfrak{sl}(2,\mathrm{R})$	$(\mathbf{7}, \mathbf{2})$	$\mathfrak{so}(6) \oplus \mathfrak{u}(1)$
8	$\mathfrak{so}(8,4)$	$\mathfrak{so}(6,2) \oplus \mathfrak{so}(2,2)$	$(\mathbf{8}, \mathbf{4})$	$\mathfrak{so}(6,2) \oplus \mathfrak{sl}(2,\mathrm{R})$	$(\mathbf{8}, \mathbf{2})$	$\mathfrak{so}(6) \oplus \mathfrak{so}(2) \oplus \mathfrak{u}(1)$
9	$\mathfrak{so}(8,5)$	$\mathfrak{so}(6,2) \oplus \mathfrak{so}(2,3)$	$(\mathbf{8}, \mathbf{5})$	$\mathfrak{so}(6,3) \oplus \mathfrak{sl}(2,\mathrm{R})$	$(\mathbf{9}, \mathbf{2})$	$\mathfrak{so}(6) \oplus \mathfrak{so}(3) \oplus \mathfrak{u}(1)$
10	$\mathfrak{so}(8,6)$	$\mathfrak{so}(6,2) \oplus \mathfrak{so}(2,4)$	$(\mathbf{8}, \mathbf{6})$	$\mathfrak{so}(6,4) \oplus \mathfrak{sl}(2,\mathrm{R})$	$(\mathbf{10}, \mathbf{2})$	$\mathfrak{so}(6) \oplus \mathfrak{so}(4) \oplus \mathfrak{u}(1)$
11	$\mathfrak{so}(8,7)$	$\mathfrak{so}(6,2) \oplus \mathfrak{so}(2,5)$	$(\mathbf{8}, \mathbf{7})$	$\mathfrak{so}(6,5) \oplus \mathfrak{sl}(2,\mathrm{R})$	$(\mathbf{11}, \mathbf{2})$	$\mathfrak{so}(6) \oplus \mathfrak{so}(5) \oplus \mathfrak{u}(1)$
12	$\mathfrak{so}(8,8)$	$\mathfrak{so}(6,2) \oplus \mathfrak{so}(2,6)$	$(\mathbf{8}, \mathbf{8})$	$\mathfrak{so}(6,6) \oplus \mathfrak{sl}(2,\mathrm{R})$	$(\mathbf{12}, \mathbf{2})$	$\mathfrak{so}(6) \oplus \mathfrak{so}(6) \oplus \mathfrak{u}(1)$
13	$\mathfrak{so}(8,8+\mathrm{p})$	$\mathfrak{so}(6,2) \oplus \mathfrak{so}(2,6+p)$	$(\mathbf{8}, \mathbf{8+p})$	$\mathfrak{so}(6,6+p) \oplus \mathfrak{sl}(2,\mathrm{R})$	$(\mathbf{12+p}, \mathbf{2})$	$\mathfrak{so}(6) \oplus \mathfrak{so}(6+p) \oplus \mathfrak{u}(1)$
14	$\mathfrak{so}(4,3)$	$\widehat{\mathfrak{sl}(2,\mathrm{R})} \oplus \widehat{\mathfrak{so}(2,1)} \oplus \mathfrak{sl}(2,\mathrm{R})_{\mathrm{h}^\star}$	$(\widehat{\mathbf{2}}, \widehat{\mathbf{3}}, \mathbf{2}_{h^\star})$	$\mathfrak{sl}(2,\mathrm{R}) \oplus \mathfrak{so}(2,1)$	$(\mathbf{2}, \mathbf{3})$	$\mathfrak{so}(2) \oplus \mathfrak{u}(1)$
15	$\mathfrak{so}(4,4)$	$\widehat{\mathfrak{sl}(2,\mathrm{R})} \oplus \widehat{\mathfrak{so}(2,2)} \oplus \mathfrak{sl}(2,\mathrm{R})_{\mathrm{h}^\star}$	$(\widehat{\mathbf{2}}, \widehat{\mathbf{4}}, \mathbf{2}_{h^\star})$	$\mathfrak{sl}(2,\mathrm{R}) \oplus \mathfrak{so}(2,2)$	$(\mathbf{2}, \mathbf{4})$	$\mathfrak{so}(2) \oplus \mathfrak{so}(2) \oplus \mathfrak{u}(1)$
16	$\mathfrak{so}(4,4+\mathrm{p})$	$\widehat{\mathfrak{sl}(2,\mathrm{R})} \oplus \widehat{\mathfrak{so}(2,2+p)} \oplus \mathfrak{sl}(2,\mathrm{R})_{\mathrm{h}^\star}$	$(\widehat{\mathbf{2}}, \widehat{\mathbf{4+p}}, \mathbf{2}_{h^\star})$	$\mathfrak{sl}(2,\mathrm{R}) \oplus \mathfrak{so}(2,2)$	$(\mathbf{2}, \mathbf{4+p})$	$\mathfrak{so}(2) \oplus \mathfrak{so}(2+p) \oplus \mathfrak{u}(1)$

Table 5.8 Table of $\mathbb{H}^\star$ subalgebras of $\mathbb{U}_{D=3}$, $\mathbb{K}^\star$-representations and **W** representations of $\mathbb{U}_{D=4}$ for the supergravity models based on *exotic* scalar symmetric spaces

#	$\mathbb{U}_{\mathrm{D=3}}$	$\mathbb{H}^\star$	$\mathbb{K}^\star$	$\mathbb{U}_{\mathrm{D=4}}$	Symp. rep. W	$\mathbb{H}_c$
1_e	$\mathfrak{su}(\mathrm{p}+2,2) \oplus \mathfrak{sl}(2,\mathrm{R})_{\mathrm{h}^\star}$	$\widehat{\mathfrak{su}(\mathrm{p}+1,1)} \oplus \widehat{\mathfrak{u}(1)}$	$(\mathbf{p}+\mathbf{2}, \mathbf{2}_{h^\star})$	$\mathfrak{su}(\mathrm{p}+1,1) \oplus \mathfrak{u}(1)$	$\mathbf{p}+\mathbf{2}$	$\mathfrak{su}(\mathrm{p}+1) \oplus \mathfrak{u}(1)$
2_e	$\mathfrak{su}(\mathrm{p}+2,4) \oplus \mathfrak{su}(1,2)$	$\mathfrak{su}(\mathrm{p}+1,2) \oplus \mathfrak{u}(1)$	$(\mathbf{p}+\mathbf{3}, \mathbf{3})$	$\mathfrak{su}(\mathrm{p}+1,1) \oplus \mathfrak{u}(1)$	$\mathbf{p}+\mathbf{4}$	$\mathfrak{su}(\mathrm{p}+1) \oplus \mathfrak{su}(3) \oplus \mathfrak{u}(1)$
3_e	$\mathfrak{e}_{6(-14)}$	$\mathfrak{so}^\star(10) \oplus \mathfrak{so}(2)$	$(\mathbf{16}_s, \mathbf{2})$	$\mathfrak{su}(5,1)$	$\mathbf{10}$	$\mathfrak{u}(5)$

the $\mathbb{K}^\star$ representation is localized on the image of the Tits Satake projection $\mathbb{K}^\star_{\text{TS}}$. For instance, for the models in the $\mathfrak{f}_{4(4)}$ universality class we have:

$$\mathbb{H}^\star_{\text{TS}} = \mathfrak{sl}(2,\mathbb{R})_{\text{h}^\star} \oplus \widehat{\mathfrak{sp}(6,\text{R})} \tag{5.7.3}$$

and:

$$\begin{aligned}
\mathbb{H}^\star &\overset{\mathbb{H}^\star_{\text{TS}}\oplus\mathbb{G}_{\text{subpaint}}}{\Longrightarrow} && \text{adj}\,\mathfrak{sl}(2,\mathbb{R})_{\text{h}^\star} \oplus \text{adj}\,\widehat{\mathfrak{sp}(6,\text{R})} \\
& && \oplus \left(\mathbf{6}\,|\,\mathscr{D}_{\text{subpaint}}\right) \oplus \left(\mathbf{6}\,|\,\mathscr{D}_{\text{subpaint}}\right) \\
& && \oplus \left(\mathbf{1}\,|\,\mathscr{D}_{\text{subpaint}}\right) \oplus \left(\mathbf{1}\,|\,\mathscr{D}_{\text{subpaint}}\right) \\
\mathbb{K}^\star &\overset{\mathbb{H}^\star_{\text{TS}}\oplus\mathbb{G}_{\text{subpaint}}}{\Longrightarrow} && \left(2_{\text{h}^\star}\,,\,\mathbf{14}'\,|\,\mathbf{1}_{\text{subpaint}}\right) \oplus \left(2_{\text{h}^\star}\,,\,\mathbf{6}\,|\,\mathscr{D}_{\text{subpaint}}\right)
\end{aligned} \tag{5.7.4}$$

and the two representations $\left(\mathbf{6}\,|\,\mathscr{D}_{\text{subpaint}}\right)$ appearing in the adjoint representation of $\mathbb{H}^\star$ can be utilized to get rid of $\left(2_{\text{h}^\star}\,,\,\mathbf{6}\,|\,\mathscr{D}_{\text{subpaint}}\right)$ appearing in the decomposition of $\mathbb{K}^\star$.

What is important to stress is that, although isomorphic $\mathbb{H}^\star$ and $\mathfrak{sl}(2,\mathbb{R}) \oplus \mathbb{U}_{D=4}$ are different subalgebras of $\mathbb{U}_{D=3}$:

$$\mathbb{U}_{D=3} \supset \mathfrak{sl}(2,\mathbb{R})_{\text{h}^\star} \neq \mathfrak{sl}(2,\mathbb{R})_E \subset \mathbb{U}_{D=3} \quad ; \quad \mathbb{U}_{D=3} \supset \widehat{\mathbb{U}_{D=4}} \neq \mathbb{U}_{D=4} \subset \mathbb{U}_{D=3} \tag{5.7.5}$$

Moreover, while the decomposition (1.7.12) is universal and holds true for all supergravity models, the structure (5.7.3) of the $\mathbb{H}^\star$ subalgebra is peculiar to the $\mathscr{N}=2$ models. In other cases the structure of $\mathbb{H}^\star$ is different.

The reduction to the Tits Satake projection however is universal and applies to all non maximally split cases.

Indeed the remaining cases are of the form:

$$\frac{\text{U}_{D=3}}{\text{H}^\star} = \frac{\text{SO}(2+\text{q},\text{q}+2+\text{p})}{\text{SO}(\text{q},2)\times\text{SO}(2,\text{q}+\text{p})} \tag{5.7.6}$$

leading to

$$\mathbb{K}^\star = (\mathbf{q+2},\mathbf{q+p+2}) \overset{\mathfrak{so}(q,2)\oplus\mathfrak{so}(2,q+1)\oplus\mathfrak{so}(p-1)}{\Longrightarrow} (\mathbf{q+2},\mathbf{q+1},\mathbf{1}) \oplus (\mathbf{q+2},\mathbf{1},\mathbf{p-1}) \tag{5.7.7}$$

where:

$$\mathfrak{so}(q,2) \oplus \mathfrak{so}(2,q+1) = \mathbb{H}^\star_{\text{TS}} \tag{5.7.8}$$

$$\mathfrak{so}(p-1) = \mathbb{G}_{\text{subpaint}} \tag{5.7.9}$$

Considering the coset:

$$\frac{\text{H}^\star}{\text{H}^\star_{\text{TS}}\times\text{G}_{\text{subpaint}}} = \frac{\text{SO}(2,\text{q}+\text{p})}{\text{SO}(\text{q}+1,2)\times\text{SO}(\text{p}-1)} \tag{5.7.10}$$

we see that its $(q+3)\times(p-1)$ parameters are arranged into the

$$(\mathbf{q+3|p-1}) \tag{5.7.11}$$

representation of $\mathfrak{so}(q+1,2)\oplus\mathfrak{so}(p-1)$ and can be used to put to zero the component $(\mathbf{q+2,1,p-1})$ in the decomposition (5.7.7). Note that the $\mathcal{N}=4$ cases with more than 6 vector multiplets are covered by the above formulae by setting:

$$q = 6 \quad ; \quad p > 1 \tag{5.7.12}$$

Similarly the $\mathcal{N}=2$ cases with more than 3 vector multiplets are covered by the above formulae by setting:

$$q = 2 \quad ; \quad p > 1 \tag{5.7.13}$$

Finally the $\mathcal{N}=4$ cases with less than 6 vector multiplets are covered by the above formulae by setting:

$$q = n \quad ; \quad p = 6-n \quad ; \quad n = 1,2,3,4,5 \tag{5.7.14}$$

5.8 The General Structure of the $\mathbb{H}^\star \oplus \mathbb{K}^\star$ Decomposition in the Maximally Split Models

In the previous section we have shown that all $\mathrm{H}^\star$ orbits in the $\mathbb{K}^\star$ representation cross the locus defined by:

$$\Pi_{\mathrm{TS}}\left(\mathbb{K}^\star\right) = \mathbb{K}^\star \tag{5.8.1}$$

where Π_{TS} is the Tits–Satake projection.

In other words just as for the $\mathbf{W}$-representation of $\mathrm{U}_{D=4}$, it suffices to classify the orbits $\mathrm{H}^\star_{\mathrm{TS}}$ in the $\mathbb{K}^\star_{\mathrm{TS}}$ representation. In view of this result, in the present section we study the general structure of the $\mathbb{H}^\star\oplus\mathbb{K}^\star$ decomposition for maximally split algebras $\mathbb{U}_{D=3}$.

A key point in our following discussion is provided by the structure of the root system of $\mathbb{U}_{D=3}$ as described in Sect. 5.5.3. The entire set of positive roots can be written as follows:

$$0 < \mathfrak{a} = \begin{cases} \alpha = \{\overline{\alpha}, 0\} \\ \mathfrak{w} = \left\{\overline{\mathbf{w}}, \frac{1}{\sqrt{2}}\right\} \\ \psi = \left\{0, \sqrt{2}\right\} \end{cases} \tag{5.8.2}$$

where $\overline{\alpha} > 0$ denotes the set of all positive roots of $\mathbb{U}_{D=4}$, while $\overline{\mathbf{w}}$ denotes the complete set of weights (positive, negative and null) of the $\mathbf{W}$ representation of

$\mathbb{U}_{D=4}$. The root ψ is the highest root of the $\mathbb{U}_{D=3}$ root system and is also the root of the Ehlers subalgebra $\mathfrak{sl}(2,\mathbb{R})_E$. Accordingly, a basis of the Cartan subalgebra of $\mathbb{U}_{D=3}$ is constructed as follows:

$$\underbrace{\text{CSA}}_{\text{of } \mathbb{U}_{D=3}} = \text{span of} \left\{ \underbrace{\mathscr{H}_1 \, , \, \mathscr{H}_2 \, , \, \dots \, , \, \mathscr{H}_r}_{\text{CSA generators of } \mathbb{U}_{D=4}} \, , \quad \underbrace{\mathscr{H}_\psi}_{\text{CSA generator of } \mathfrak{sl}(2,\mathbb{R})_E} \right\} \tag{5.8.3}$$

For all maximally split Lie algebras $\mathbb{U}$ of rank $r+1$, the maximal compact subalgebra $\mathbb{H} \subset \mathbb{U}$ is generated by:

$$T^{\mathfrak{a}} = E^{\mathfrak{a}} - E^{-\mathfrak{a}} \tag{5.8.4}$$

while the complementary orthogonal space $\mathbb{K}$ is generated by

$$K^{\mathfrak{a}} = E^{\mathfrak{a}} + E^{-\mathfrak{a}} \tag{5.8.5}$$

$$K^I = \mathscr{H}^I \quad ; \quad I = 1, \dots, r+1 \tag{5.8.6}$$

The splitting $\mathbb{H}^\star \oplus \mathbb{K}^\star$ is obtained by means of just one change of sign which, thanks to the structure (5.8.2) of the root system is consistent, namely still singles out a subalgebra.

The generators of the $\mathbb{H}^\star$ subalgebra are as follows:

$$\begin{aligned} T^{\alpha}_{\star} &= E^{\alpha} - E^{-\alpha} \\ T^{\mathfrak{w}}_{\star} &= E^{\mathfrak{w}} + E^{-\mathfrak{w}} \\ T^{\psi}_{\star} &= E^{\psi} - E^{-\psi} \end{aligned} \tag{5.8.7}$$

while the generators of the $\mathbb{K}^\star$ complementary subspace are as follows:

$$\begin{aligned} K^{\alpha}_{\star} &= E^{\alpha} + E^{-\alpha} \\ K^{\mathfrak{w}}_{\star} &= E^{\mathfrak{w}} - E^{-\mathfrak{w}} \\ K^{\psi}_{\star} &= E^{\psi} + E^{-\psi} \\ K^I &= \mathscr{H}^I \quad ; \quad I = 1, \dots, r+1 \end{aligned} \tag{5.8.8}$$

From Eq. (5.8.7) we see that $\mathbb{H}^\star$ contains the maximal compact subalgebra of the original $\mathbb{U}_{D=4}$ and the maximal compact subalgebra $\mathfrak{so}(2) \subset \mathfrak{sl}(2,R)_E$ of the Ehlers group. Using this structure we can now compare the classification of $\mathbb{K}^\star$ orbits with the classification of **W**-orbits.

5.9 $\mathbb{K}^\star$ Orbits Versus W-Orbits

In the σ-model approach the complete black hole spherically symmetric supergravity solution is obtained from two data,[2] namely the Lax operator L_0 evaluated at spatial infinity (see Eq. (5.7.1)) and the coset representative $\mathbb{L}_0$ also evaluated at spatial infinity. In terms of these data one defines the matrix of conserved Noether charges:

$$Q^{Noether} = \mathbb{L}_0 \, L_0 \, \mathbb{L}_0^{-1} = \mathbb{L}(\tau) \, L(\tau) \, \mathbb{L}^{-1}(\tau) \tag{5.9.1}$$

from which the electromagnetic charges of the black hole, belonging to the **W**-representation of $\mathrm{U}_{D=4}$, can be obtained by means of the following trace:

$$\mathcal{Q}^{\mathbf{w}} = \mathrm{Tr}\left(Q^{Noether} \, \mathcal{T}^{\mathbf{w}}\right) \tag{5.9.2}$$

where

$$\mathcal{T}^{\mathbf{w}} \propto E^{\mathfrak{w}} \tag{5.9.3}$$

are the generators of the solvable Lie algebra corresponding to the **W**-representation.

It is important to stress that, because of physical boundary conditions, the coset representative at spatial infinity $\mathbb{L}_0$ belongs to the subgroup $\mathrm{U}_{D=4} \subset \mathrm{U}_{D=3}$. Indeed it simply encodes the boundary values at infinity of the $D = 4$ scalar fields:

$$\mathrm{U}_{D=3} \supset \mathrm{U}_{D=4} \ni \mathbb{L}_0 = \exp\left[\phi_0^\alpha \, E^\alpha + \sum_{i=1}^{r} \phi_0^i \, \mathcal{H}_i\right] \tag{5.9.4}$$

Using this information in Eq. (5.9.2) we obtain

$$\mathcal{Q}^{\mathbf{w}} = \mathrm{Tr}\left(L_0 \, \mathbb{L}_0^{-1}(\phi) \mathcal{T}^{\mathbf{w}} \mathbb{L}_0(\phi)\right) = R(\phi)^{\mathbf{w}}{}_{\mathbf{w}'} \, \mathcal{Q}^{\mathbf{w}'} \tag{5.9.5}$$

where:

$$\mathcal{Q}^{\mathbf{w}'} = \mathrm{Tr}\left(L_0 \, \mathcal{T}^{\mathbf{w}'}\right) \tag{5.9.6}$$

are the electromagnetic charges obtained with no scalar field dressing at infinity and

$$R(\phi)^{\mathbf{w}}{}_{\mathbf{w}'} \in \mathrm{U}_{D=4} \tag{5.9.7}$$

is the matrix representing the group element $\mathbb{L}_0(\phi)$ in the **W**-representation.

This result has a very significant consequence. The scalar field dressing at infinity simply rotates the charge vector along the same **W**-orbit and is therefore irrelevant.

Hence we conclude that for each Lax operator, the **W**-orbit of charges is completely determined and unique. The next question is whether the charge-orbit **W** is

[2]See papers [34, 44, 45] for detailed explanations.

the same for all Lax operators belonging to the same H*-orbit. As already anticipated, the answer is no and it is quite easy to produce counter examples.

Yet if we impose the condition that the Taub-NUT charge should be zero:

$$\text{Tr}\left(L_0\, L_-^E\right) = 0 \qquad (5.9.8)$$

then for all Lax operators in the same H*, satisfying the additional constraint (5.9.8), the corresponding charges $Q^w = \text{Tr}\,(L_0\, T^w)$ fall into the same **W**-orbit.

We were not able to prove this statement, but we assert it as a *conjecture*, since we analyzed many cases and it was always true, no counter example being ever found.

In the case of multicenter non spherically symmetric solutions our conjecture appears to be true as long as we impose the condition of vanishing of the Taub-NUT current:

$$j^{TN} = 0 \qquad (5.9.9)$$

So doing, at every pole of the involved harmonic functions, we obtain a black hole that always falls into the same **W**-orbit.

What happens instead when the Taub-NUT current is turned on cannot be predicted in general terms at the present status of our knowledge and more study is certainly in order.

The reader will understand the meaning of the last two paragraphs by carefully reading Chap. 6. In the present one we outlined the entire beautiful group-theoretical machinery that sustains the construction and classification of black-hole geometries addressed there.

References

1. I. Satake, On representations and compactifications of symmetric riemannian spaces. Ann. Math. **71**, 77–110 (1960). Second Series
2. J. Tits, Classification of algebraic semisimple groups, in Algebraic Groups and Discontinuous Subgroups (Proceedings of Symposia in Pure Mathematics, Boulder, Colorado, 1965), p. 33–62
3. A. Borel, J. Tits, Groupes réductifs. Publications Mathémathiques de l'IHES **27**, 55–151 (1965); Compléments à l'article **41**, 253–276 (1972)
4. B. de Wit, A. Van Proeyen, Broken sigma model isometries in very special geometry. Phys. Lett. B **293**, 94–99 (1992)
5. B. de Wit, F. Vanderseypen, A. Van Proeyen, Symmetry structure of special geometries. Nucl. Phys. B **400**, 463–524 (1993)
6. B. de Wit, A. Van Proeyen, Isometries of special manifolds (1995)
7. D.V. Alekseevsky, V. Cortes, C. Devchand, A. Van Proeyen, Polyvector super Poincare algebras. Commun. Math. Phys. **253**, 385–422 (2004)
8. L. Andrianopoli, R. D'Auria, S. Ferrara, P. Fré, M. Trigiante, R-R scalars, U-duality and solvable Lie algebras. Nucl. Phys. B **496**, 617–629 (1997), arXiv:hep-th/9611014
9. L. Andrianopoli, R. D'Auria, S. Ferrara, P. Fré, R. Minasian, M. Trigiante, Solvable Lie algebras in type IIA, type IIB and M-theories. Nucl. Phys. B **493**, 249–277 (1997), arXiv:hep-th/9612202

10. M. Trigiante, Dualities in supergravity and solvable Lie algebras, Ph.D. thesis, Swansea University, 1998
11. V. Belinsky, I. Khalatnikov, E. Lifshitz, Oscillatory approach to a singular point in the relativistic cosmology. Adv. Phys. **19**, 525–573 (1970)
12. V. Belinsky, I. Khalatnikov, E. Lifshitz, A general solution of the einstein equations with a time singularity. Adv. Phys. **31**, 639–667 (1982)
13. J.K. Erickson, D.H. Wesley, P.J. Steinhardt, N. Turok, Kasner and mixmaster behavior in universes with equation of state $w \geq 1$. Phys. Rev. D **69**, 063514 (2004)
14. J. Demaret, M. Henneaux, P. Spindel, Nonoscillatory behavior in vacuum Kaluza–Klein cosmologies. Phys. Lett. **164B**, 27–30 (1985)
15. J. Demaret, J.L. Hanquin, M. Henneaux, P. Spindel, A. Taormina, The fate of the mixmaster behavior in vacuum inhomogeneous Kaluza–Klein cosmological models. Phys. Lett. B **175**, 129–132 (1986)
16. J. Demaret, Y. De Rop, M. Henneaux, Chaos in nondiagonal spatially homogeneous cosmological models in space-time dimensions ≤ 10. Phys. Lett. B **211**, 37–41 (1988)
17. T. Damour, M. Henneaux, B. Julia, H. Nicolai, Hyperbolic Kac–Moody algebras and chaos in Kaluza–Klein models. Phys. Lett. B **509**, 323–330 (2001)
18. T. Damour, S. de Buyl, M. Henneaux, C. Schomblond, Einstein billiards and overextensions of finite dimensional simple Lie algebras. JHEP **08**, 030 (2002)
19. T. Damour, M. Henneaux, H. Nicolai, Cosmological billiards. Class. Quantum Gravity **20**, R145–R200 (2003)
20. S. de Buyl, M. Henneaux, B. Julia, L. Paulot, Cosmological billiards and oxidation. Fortsch. Phys. **52**, 548–554 (2004). [PoSjhw2003, 015 (2003)]
21. J. Brown, O.J. Ganor, C. Helfgott, M theory and E(10): billiards, branes, and imaginary roots. JHEP **08**, 063 (2004)
22. F. Englert, M. Henneaux, L. Houart, From very-extended to overextended gravity and M-theories. JHEP **02**, 070 (2005)
23. T. Damour, Cosmological singularities, Einstein billiards and Lorentzian Kac–Moody algebras, in *Miami Waves 2004: Conference on Geometric Analysis, Nonlinear Wave Equations and General Relativity Coral Gables, Florida, 4-10 January 2004*, (2005). [J. Hyperbol. Diff. Equat. (2005)]
24. T. Damour, Poincare, relativity, billiards and symmetry, in *Proceedings, Symposium Henri Poincare, Brussels, Belgium, 8-9 October 2004*, (2005), p. 149
25. M. Henneaux, B. Julia, Hyperbolic billiards of pure D = 4 supergravities. JHEP **05**, 047 (2003)
26. A. Keurentjes, The group theory of oxidation. Nucl. Phys. B **658**, 303–347 (2003)
27. A. Keurentjes, The group theory of oxidation 2: cosets of nonsplit groups. Nucl. Phys. B **658**, 348–372 (2003)
28. P. Fré, V. Gili, F. Gargiulo, A.S. Sorin, K. Rulik, M. Trigiante, Cosmological backgrounds of superstring theory and solvable algebras: oxidation and branes. Nucl. Phys. B **685**, 3–64 (2004)
29. P. Fré, A.S. Sorin, Integrability of supergravity billiards and the generalized Toda lattice equation. Nucl. Phys. B **733**, 334–355 (2006)
30. P. Fre, A.S. Sorin, The Weyl group and asymptotics: all supergravity billiards have a closed form general integral. Nucl. Phys. B **815**, 430–494 (2009)
31. P. Fré, F. Gargiulo, K. Rulik, Cosmic billiards with painted walls in non-maximal supergravities: a worked out example. Nucl. Phys. B **737**, 1–48 (2006). https://doi.org/10.1016/j.nuclphysb.2005.10.023, arXiv:hep-th/0507256
32. P. Fre, F. Gargiulo, J. Rosseel, K. Rulik, M. Trigiante, A. Van Proeyen, Tits–Satake projections of homogeneous special geometries. Class. Quantum Gravity **24**, 27–78 (2007)
33. P. Fré, A. Sorin, Supergravity black holes and billiards and liouville integrable structure of dual borel algebras. JHEP **03**, 066 (2010)
34. P. Fré, A. Sorin, The integration algorithm for nilpotent orbits of $g/h^{\star}$ lax systems: for extremal black holes (2009)

35. W. Chemissany, J. Rosseel, M. Trigiante, T. Van Riet, The full integration of black hole solutions to symmetric supergravity theories. Nucl. Phys. B **830**, 391 (2010)
36. M. Gaillard, Z. Bruno, Duality rotations for interacting fields. Nucl. Phys. B **193**, 221 (1981)
37. P. Breitenlohner, D. Maison, G.W. Gibbons, Four-dimensional black holes from Kaluza–Klein theories. Commun. Math. Phys. **120**, 295 (1988)
38. M. Gunaydin, A. Neitzke, B. Pioline, A. Waldron, Bps black holes, quantum attractor flows and automorphic forms. Phys. Rev. D **73**, 084019 (2006)
39. B. Pioline, Lectures on black holes, topological strings and quantum attractors. Class. Quantum Gravity **23**, S981 (2006)
40. M. Gunaydin, A. Neitzke, B. Pioline, A. Waldron, Quantum attractor flows. JHEP **0709**, 056 (2007)
41. D. Gaiotto, W. Li, M. Padi, Non-supersymmetric attractor flow in symmetric spaces. JHEP **12**, 093 (2007)
42. E. Bergshoeff, W. Chemissany, A. Ploegh, M. Trigiante, T. Van Riet, Generating geodesic flows and supergravity solutions. Nucl. Phys. B **812**, 343 (2009)
43. G. Bossard, H. Nicolai, K. Stelle, Universal bps structure of stationary supergravity solutions. JHEP **0907**, 003 (2009)
44. P. Frè, A. Sorin, M. Trigiante, Integrability of supergravity black holes and new tensor classifiers of regular and nilpotent orbits. JHEP **1204**, 015 (2012)
45. P. Frè, A. Sorin, M. Trigiante, Black hole nilpotent orbits and tits satake universality classes (2011)
46. P. Fre', F. Gargiulo, K. Rulik, M. Trigiante, The general pattern of Kac Moody extensions in supergravity and the issue of cosmic billiards. Nucl. Phys. B **741**, 42–82 (2006)

Chapter 6
Black Holes and Nilpotent Orbits

Deep into that darkness peering, long I stood there,
wondering, fearing, doubting,
dreaming dreams no mortal
ever dared to dream before.

Edgar Allan Poe

6.1 Historical Introduction

When on September 14th 2015 the gravitational wave signal emitted 1.5 billion year ago by two coalescing black stars was detected at LIGO I and LIGO II, we not only obtained a new spectacular confirmation of General Relativity but we actually saw the dynamical process of formation of the most intriguing objects populating the Universe, namely black holes (Fig. 6.1).

Black Holes are on one side physical objects capable of interacting with the emission of enormous quantities of energy, on the other side they are just pure geometries. Indeed a classical black-hole is nothing else but a solution of Einstein equations which are just geometrical statements on the curvature tensor.

6.1.1 Black Holes in Supergravity and Superstrings

A new season of research in Black Hole theory started in the middle nineties of the XXth century with the contributions of Sergio Ferrara, Renata Kallosh, Andrew Strominger and Cumrun Vafa, that are described in the following short summary:

1. In 1995 R. Kallosh, S. Ferrara and A. Strominger considered black holes in the context of $\mathcal{N} = 2$ supergravity and introduced the notion of attractors [1, 2].
2. In 1996 S. Ferrara (see Fig. 4.2) and R. Kallosh (see Fig. 6.2) formalized the attractor mechanism for supergravity black holes [1, 2].

P. G. Frè, *Advances in Geometry and Lie Algebras from Supergravity*, Theoretical and Mathematical Physics,
https://doi.org/10.1007/978-3-319-74491-9_6

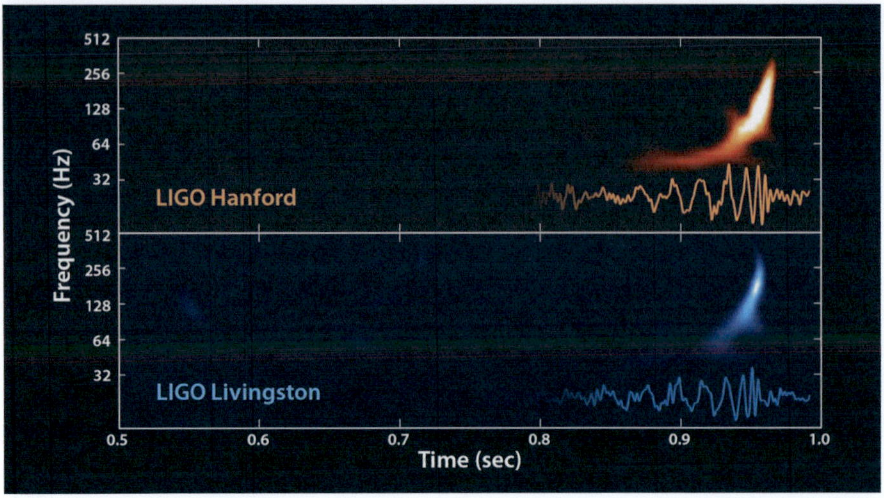

Fig. 6.1 The gravitational wave signal emitted in the coalescence of two black holes which occurred 1.5 billion of years ago was simultaneously detected September 14th 2015 by the two interferometers LIGO I and LIGO II

3. In 1996 A. Strominger (see Fig. 4.7) and C. Vafa (see Fig. 6.3) showed that an extremal BPS black hole in $d = 5$ has a horizon area that exactly counts the number of string microstates it corresponds to [3].[1]
4. In the years 1997–2000 the horizon area of BPS supergravity black holes was interpreted in terms of a symplectic invariant constructed with the black hole electromagnetic charges (for a review containing also an extensive bibliography see [11]).
5. In the years 2006–2009 new insights extended the attractor mechanism to non BPS black-holes [12–25].
6. Since 2010 new exact integration techniques for Sugra Black Holes were found by A. Sorin, P. Fré, M. Trigiante and their younger collaborators [26–33].

6.1.2 *Black Holes in This Chapter*

The intriguing relation between Geometry and Physics arises at several levels, the most profound and challenging being provided by the identification of the *horizon area* with the *statistical entropy* of the mysterious dynamical system which is encoded in a *classical black solution*.

[1]There followed a vast literature some items of which are are quoted in [4–10].

Fig. 6.2 Renata Kallosh (on the left) born in Moscow in 1943 completed her Bachelor's from Moscow State University in 1966 and obtained her Ph.D. from Lebedev Physical Institute, Moscow in 1968. She then held a position, as professor, at the same institute, before moving to CERN for a year in 1989. Kallosh joined Stanford University in 1990 and continues to work there. She is married with the famous cosmologist Andrei Linde. Renata Kallosh is renowned for her pioneering contributions with Ferrara to the attractor mechanism in supergravity black holes, for her studies in supergravity cosmology and for her early work with A. Van Proeyen on the AdS/CFT correspondence. Indeed Kallosh and Van Proeyen were the first to propose the interpretation of the anti de Sitter group as the conformal group on a brane boundary. Anna Ceresole (on the right), born 1961 in Torino, graduated from Torino University in 1984 with a thesis on Kaluza Klein supergravity written under the supervision of Hermann Nicolai and the author of this book. In 1989 she obtained her Ph.D. from Stony Brook University under the supervision of Peter van Nieuwenhuizen. Post doctoral fellow at Caltech for two years she was Assistant Professor at the Politecnico di Torino for several years. Then she became Senior Research Scientist of INFN and joined the Torino University String Group. Anna Ceresole has given many important contributions to the development of supergravity, in particular in relation with special Kähler Geometry and black hole charges, duality transformations, gaugings and inflaton potentials. She has worked both with younger students and post-doc and, in different combinations, with all the main actors in the development of supergravity theory

We are not going to touch upon the physics of black holes and on the exciting question of their interpretation in terms of *microstates*, yet we cannot avoid discussing their several nested geometrical aspects, glimpses of which were already provided in Chap. 5.

We emphasized there that in the context of supergravity a black hole solution of Einstein equations comes equipped with other associated geometrical data, namely those encoded in a set of electromagnetic fields that are connections on suitable bundles and those encoded in scalar fields that describe a map from 4-dimensional space-time $\mathscr{M}_4$ to *special manifolds* $\mathscr{S}\mathscr{K}_n$. We also stressed the remarkable picture of a black-hole solution as a map from a three-dimensional Euclidean manifold $\mathscr{M}_3$ to a Lorentzian pseudo-quaternionic manifold $\mathscr{Q}_r$ lying in the image of the $c^\star$-map.

Fig. 6.3 Cumrun Vafa (on the left) was born in Tehran, Iran in 1960. He graduated from Alborz High School and went to the US in 1977. He got his undergraduate degree from the Massachusetts Institute of Technology with a double major in physics and mathematics. He received his Ph.D. from Princeton University in 1985 under the supervision of Edward Witten. He then became a junior fellow at Harvard, where he later got a junior faculty position. In 1989 he was offered a senior faculty position, and he has been there ever since. Currently, he is the Donner Professor of Science at Harvard University. Vafa's most relevant achievement is, together with Strominger, the first example of interpretation of the Bekenstein Hawking black hole entropy in terms of superstring microstates. He has also given pioneering contributions to topological strings, F-theory and to the general vision named *geometric engineering of quantum field theories*, which is a programme aimed at decoding quantum field theories in terms of algebraic geometry constructions. Dieter Luest (on the right) born 1956 in Chicago, graduated from the Ludvig Maximillian University in Muenchen in 1985. He was postdoctoral fellow in Caltech, Pasadena, in the Max Planck Institute in Muenchen and at CERN in Geneva. From 1993 to 2004 he was full professor of Quantum Field Theory at the von Humboldt University in Berlin. Since 2004 he made return to Muenchen where he is both full professor at the Ludwig Maximilan University and Research Director at the Max Planck Institute. Dieter Luest has given very important contributions in a large variety of topics connected with String Theory and Supergravity, in particular in relation with Black Hole solutions, D-brane engineering, Calabi Yau compactifications, double geometries, flux compactifications and string cosmology

This last viewpoint corresponds to the σ-model approach to black-hole solutions and it was developed in the last two decades.

If the special manifold $\mathscr{SK}_n = \frac{\mathrm{U_{D=4}}}{\mathrm{H_{D=4}}}$ is a symmetric coset manifold, then also the pseudo-quaternionic manifold $\mathscr{Q}_r = \frac{\mathrm{U_{D=3}}}{\mathrm{H_{D=3}}}$ is such and the classification of possible extremal black-hole solutions is turned into an algebraic problem that is the contemporary frontier of research in Lie algebra theory: *the classification of nilpotent orbits*.

In this chapter we analyze in detail the new very rich *geometric lore* which emerges from the issue of black–hole constructions within the σ-model approach. Here all the issues discussed in previous chapters enter the game in an essential way:

1. Special Kähler Geometry,
2. Lie Algebra invariants,
3. $c^\star$ map,
4. Tits Satake projection and its universality classes,

5. Weyl Group and its extensions,
6. Classification of nilpotent orbits.

In view of the deep relation between quantum physics and geometry encapsulated into black-holes it is to be expected that all the intriguing geometrical relations listed above are the tip of an iceberg of theoretical knowledge yet to be uncovered.

Hence let us resume the σ-model approach to black-holes.

6.2 The σ-Model Approach to Black-Hole Resumed

We start from Eq. (5.2.21) and from the golden splitting (1.7.12) which we rewrite as follows:

$$\mathrm{adj}(\mathbb{U}_{D=3}) = \mathrm{adj}(\mathbb{U}_{D=4}) \oplus \mathrm{adj}(\mathfrak{sl}(2,\mathbb{R})_E) \oplus W_{(2,\mathbf{W})} \tag{6.2.1}$$

where $\mathbf{W}$ is the **symplectic** representation of $\mathbb{U}_{D=4}$ to which the electric and magnetic field strengths are assigned.

Next we consider a gravity coupled three-dimensional Euclidean σ-model, whose fields

$$\Phi^A(x) \equiv \{U(x), a(x), \phi(x), Z(x)\}$$

describe mappings:

$$\Phi : \mathscr{M}_3 \to \mathscr{Q} \tag{6.2.2}$$

from a three-dimensional manifold $\mathscr{M}_3$, whose metric we denote by $\gamma_{ij}(x)$, to the target space $\mathscr{Q}$. The action of this σ-model is the following:

$$\mathscr{A}^{[3]} = \int \sqrt{\det\gamma}\, \mathfrak{R}[\gamma]\, d^3x + \int \sqrt{\det\gamma}\, \mathscr{L}^{(3)}\, d^3x \tag{6.2.3}$$

$$\begin{aligned}\mathscr{L}^{(3)} = \Big(&\partial_i U\, \partial_j U + h_{rs}\, \partial_i\phi^r\, \partial_j\phi^s \\ &+ e^{-2U}\left(\partial_i a + \mathbf{Z}^T \mathbb{C}\partial_i \mathbf{Z}\right)\left(\partial_j a + \mathbf{Z}^T \mathbb{C}\partial_j \mathbf{Z}\right) + 2\, e^{-U}\, \partial_i \mathbf{Z}^T\, \mathscr{M}_4\, \partial_j \mathbf{Z}\Big)\, \gamma^{ij}\end{aligned} \tag{6.2.4}$$

where $\mathfrak{R}[\gamma]$ denotes the scalar curvature of the metric γ_{ij}.

The field equations of the σ-model are obtained by varying the action both in the metric γ_{ij} and in the fields $\Phi^A(x)$. The Einstein equation reads as usual:

$$\mathfrak{R}_{ij} - \tfrac{1}{2}\gamma_{ij}\, \mathfrak{R} = \mathfrak{T}_{ij} \tag{6.2.5}$$

where:

$$\mathfrak{T}_{ij} = \frac{\delta \mathscr{L}^{(3)}}{\delta \gamma^{ij}} - \gamma_{ij}\, \mathscr{L}^{(3)} \tag{6.2.6}$$

is the stress energy tensor, while the matter field equations assume the standard form:

$$\frac{1}{\sqrt{\det\gamma}}\,\gamma^{ij}\,\partial_i\left[\sqrt{\det\gamma}\,\frac{\delta\mathscr{L}^{(3)}}{\delta\,\partial^j\Phi^A}\right]-\frac{\delta\mathscr{L}^{(3)}}{\delta\Phi^A}=0 \tag{6.2.7}$$

As it is well known, in $D=3$ there is no propagating graviton and the Riemann tensor is completely determined by the Ricci tensor, namely, via Einstein equations, by the stress-energy tensor of the matter fields.[2]

Extremal solutions of the σ-model are those for which the three-dimensional metric can be consistently chosen flat:

$$\gamma_{ij}=\delta_{ij} \tag{6.2.8}$$

corresponding to a vanishing stress-energy tensor:

$$\begin{aligned}&\partial_i U\,\partial_j U+h_{rs}\,\partial_i\phi^r\,\partial_j\phi^s+e^{-2U}\left(\partial_i a+\mathbf{Z}^T\mathbb{C}\partial_i\mathbf{Z}\right)\left(\partial_j a+\mathbf{Z}^T\mathbb{C}\partial_j\mathbf{Z}\right)\\&+2\,e^{-U}\,\partial_i\mathbf{Z}^T\,\mathscr{M}_4\,\partial_j\mathbf{Z}=0\end{aligned} \tag{6.2.9}$$

We will see in the sequel how the nilpotent orbits of the group $\mathrm{H}^\star$ in the $\mathbb{K}^\star$ representation can be systematically associated with general extremal solutions of the field equations.

6.2.1 *Oxidation Rules for Extremal Multicenter Black Holes*

Let us now describe the oxidation rules, namely the procedure by means of which to every configuration of the three-dimensional fields $\Phi(x)=\{U(x),a(x),\phi(x),Z(x)\}$, satisfying the field equations (6.2.7) and also the extremality condition (6.2.9), we can associate a well defined configuration of the four-dimensional fields satisfying the field equations of supergravity that follow from the lagrangian (5.2.3). We might write such oxidation rules for general solutions of the σ-model, also non extremal, yet given our present goal we confine ourselves to spell out such rule in the extremal case, which is somewhat simpler since it avoids the extra complications related with the three-dimensional metric γ_{ij}.

In order to write the $D=4$ fields, the first necessary item we have to determine is the Kaluza–Klein vector field $\mathbf{A}^{[KK]}=A_i^{[KK]}\,dx^i$. This latter is worked out through the following dualization procedure:

[2]*Clarification for mathematicians*: General Relativity in $D=3=1\oplus 2$ dimensions is a rather empty field theory. Einstein equations do not describe the propagation of any particle since there are no solutions of the wave-type and the only degree of freedom is the analogue of the Newton potential. Mathematically this follows from the fact that the Riemann tensor is fully determined by the Ricci tensor and the latter is identified by Einstein equations with the stress-energy tensor of matter fields.

$$\begin{aligned}
\mathbf{F}^{[KK]} &= d\mathbf{A}^{[KK]} \\
\mathbf{F}^{[KK]} &= -\varepsilon_{ijk}\, dx^i \wedge dx^j \left[\exp[-2U]\left(\partial^k a + Z\,\mathbb{C}\,\partial^k Z\right)\right]
\end{aligned} \tag{6.2.10}$$

Given the Kaluza–Klein vector we can write the four-dimensional metric which is the following:

$$ds^2 = -\exp[U]\left(dt + \mathbf{A}^{[KK]}\right)^2 + \exp[-U]\, dx^i \otimes dx^j\, \delta_{ij} \tag{6.2.11}$$

The vielbein description of the same metric is immediate. We just write:

$$\begin{aligned}
ds^2 &= -E^0 \otimes E^0 + E^i \otimes E^i \\
E^0 &= \exp[\tfrac{U}{2}]\left(dt + \mathbf{A}^{[KK]}\right) \\
E^i &= \exp[-\tfrac{U}{2}]\, dx^i
\end{aligned} \tag{6.2.12}$$

Next we can present the form of the electromagnetic field strengths:

$$\begin{aligned}
\mathbf{F}^{\Lambda} &= \mathbb{C}^{\Lambda M}\partial_i Z_M\, dx^i \wedge \left(dt + \mathbf{A}^{[KK]}\right) \\
&\quad + \varepsilon_{ijk} dx^i \wedge dx^j \left[\exp[-U]\left(\mathrm{Im}\mathcal{N}^{-1}\right)^{\Lambda\Sigma}\left(\partial^k Z_\Sigma + \mathrm{Re}\mathcal{N}_{\Sigma\Gamma}\partial^k Z^\Gamma\right)\right]
\end{aligned} \tag{6.2.13}$$

Next we define the electromagnetic charges and the Taub-NUT charges for multicenter solutions. Considering the metric (6.2.11) the black hole centers are defined by the zeros of the warp-factor $\exp[U(\mathbf{x})]$. In a composite m-black hole solution there are m three-vectors $\mathbf{r}_\alpha$ ($\alpha = 1, \ldots, m$), such that:

$$\lim_{\mathbf{x}\to\mathbf{r}_\alpha} \exp[U(\mathbf{x})] = 0 \tag{6.2.14}$$

Each of these zeros defines a non trivial homology two-cycle $\mathbb{S}^2_\alpha$ of the 4-dimensional space-time which surrounds the singularity $\mathbf{r}_\alpha$. The electromagnetic charges of the individual holes are obtained by integrating the field strengths and their duals on such homology cycles.

$$\begin{pmatrix} p^\Lambda \\ q_\Sigma \end{pmatrix}_\alpha = \frac{1}{4\pi\sqrt{2}}\begin{pmatrix} \int_{\mathbb{S}^2_\alpha} \mathbf{F}^\Lambda \\ \int_{\mathbb{S}^2_\alpha} \mathbf{G}_\Sigma \end{pmatrix} \equiv \frac{1}{4\pi}\int_{\mathbb{S}^2_\alpha} j^{EM} \tag{6.2.15}$$

Utilizing the form of the field strengths we obtain the explicit formula:

$$\begin{aligned}
\mathcal{Q}_\alpha \equiv \begin{pmatrix} p^\Lambda \\ q_\Sigma \end{pmatrix}_\alpha &= \frac{1}{4\pi\sqrt{2}}\int_{\mathbb{S}^2_\alpha} \varepsilon_{ijk} dx^i \wedge dx^j \left[\exp[-U]\,\mathcal{M}_4\,\partial^k Z\right. \\
&\quad \left. + \exp[-2U]\left(\partial^k a + Z\,\mathbb{C}\,\partial^k Z\right)\mathbb{C}\,Z\right]
\end{aligned} \tag{6.2.16}$$

which provides m-sets of electromagnetic charges associated with the solution. Similarly we have m Taub-NUT charges defined by:

$$\mathbf{n}_\alpha = -\frac{1}{4\pi}\int_{\mathbb{S}^2_\alpha} \varepsilon_{ijk} dx^i \wedge dx^j \exp[-2U]\left(\partial^k a + Z\,\mathbb{C}\,\partial^k Z\right) \equiv \frac{1}{4\pi}\int_{\mathbb{S}^2_\alpha} j^{TN} \tag{6.2.17}$$

6.2.1.1 Reduction to the Spherical Case

The spherical symmetric one-center solutions are retrieved from the general case by assuming that all the three-dimensional fields depend only on one radial coordinate:

$$\tau = -\frac{1}{r} \quad ; \quad r = \sqrt{x_1^2 + x_2^2 + x_3^2} \tag{6.2.18}$$

On functions only of τ we have the identity:

$$\partial_i f(\tau) = -x^i \tau^3 \frac{d}{d\tau} f(\tau) \tag{6.2.19}$$

and introducing polar coordinates:

$$\begin{aligned} x_1 &= \frac{1}{\tau} \cos\theta \\ x_2 &= \frac{1}{\tau} \sin\theta \sin\varphi \\ x_3 &= \frac{1}{\tau} \sin\theta \cos\varphi \end{aligned} \tag{6.2.20}$$

we obtain:

$$\tau^3 \varepsilon_{ijk} x^i dx^j \wedge dx^k = -2 \sin\theta \, d\theta \wedge d\varphi \tag{6.2.21}$$

By using these identities and restricting one's attention to the extremal case, the action of the σ-model (6.2.3) reduces to:

$$\begin{aligned} \mathscr{A} &= \int d\tau \, \mathscr{L} \\ \mathscr{L} &= \dot{U}^2 + h_{rs} \dot{\varphi}^r \dot{\varphi}^s + e^{-2U} (\dot{a} + \mathbf{Z}^T \mathbb{C} \dot{\mathbf{Z}})^2 + 2 e^{-U} \dot{\mathbf{Z}}^T \mathscr{M}_4 \dot{\mathbf{Z}} \end{aligned} \tag{6.2.22}$$

where the dot denotes derivatives with respect to the τ variable. The σ-model field equations take the standard form of the Euler Lagrangian equations:

$$\frac{\mathrm{d}}{\mathrm{d}\tau}\frac{\mathrm{d}\mathscr{L}}{\mathrm{d}\dot{\Phi}} = \frac{\mathrm{d}\mathscr{L}}{\mathrm{d}\Phi} \tag{6.2.23}$$

and the extremality conditions (6.2.9) reduces to:

$$\mathscr{L} = \dot{U}^2 + h_{rs}\,\dot{\varphi}^r\,\dot{\varphi}^s + e^{-2U}\,(\dot{a} + \mathbf{Z}^T\mathbb{C}\dot{\mathbf{Z}})^2 + 2\,e^{-U}\,\dot{\mathbf{Z}}^T\,\mathscr{M}_4\,\dot{\mathbf{Z}} = 0 \tag{6.2.24}$$

It appears from this that spherical extremal black holes are in one-to-one correspondence with light-like geodesics of the manifold $\mathscr{Q}$.

The Reduced Oxidation Rules

In the spherical case the above discussed oxidation rules reduce as follows. For the metric we have

$$ds^2_{(4)} = -\,e^{U(\tau)}\,(dt + 2\,\mathbf{n}\,\cos\theta\,d\varphi)^2 + e^{-\,U(\tau)}\left[\frac{1}{\tau^4}\,d\tau^2 + \frac{1}{\tau^2}\left(d\theta^2 + \sin^2\theta\,d\phi^2\right)\right] \tag{6.2.25}$$

where $\mathbf{n}$ denotes the Taub-NUT charge obtained from the form of the Kaluza–Klein field strength:

$$\begin{aligned}\mathbf{F}^{KK} &= -2\,\mathbf{n}\,\sin\theta\,d\theta\,\wedge\,d\varphi\\ \mathbf{n} &= \left(\dot{a} + Z\,\mathbb{C}\,\dot{Z}\right)\end{aligned} \tag{6.2.26}$$

The electromagnetic field-strengths are instead the following ones:

$$F^\Lambda = 2\,p^\Lambda\,\sin\theta\,d\theta\wedge d\varphi + \dot{Z}_\Lambda d\tau\wedge(dt + 2\mathbf{n}\,\cos\theta\,d\varphi) \tag{6.2.27}$$

where the magnetic charges p^Λ are extracted from the reduction of the general formula (6.2.16), namely:

$$\mathscr{Q}^M = \begin{pmatrix} p^\Lambda \\ q_\Sigma \end{pmatrix} = \sqrt{2}\left[e^{-U}\,\mathscr{M}_4\,\dot{Z} - \mathbf{n}\,\mathbb{C}\,Z\right]^M \tag{6.2.28}$$

6.3 The $\mathfrak{g}_{2(2)}$ Lie Algebra and the S^3 Model

In Sect. 1.6 we discussed the structure of the smallest exceptional Lie algebra $\mathfrak{g}_2$ and we anticipated that it plays an important role in relation with the simplest example of special Kähler geometry and of its quaternionic images under the c and the $c^\star$ maps. Indeed the simplest example of special Kähler geometry occurs when we have only one complex scalar coordinate z which parameterizes the complex lower half-plane endowed with the standard Poincaré metric. In other words[3]:

$$g_{z\bar{z}} dz\, d\bar{z} = \frac{3}{4} \frac{1}{(\mathrm{Im} z)^2} dz\, d\bar{z} \tag{6.3.1}$$

From the point of view of geometry the lower half-plane is the symmetric coset manifold $\frac{\mathrm{SL(2,R)}}{\mathrm{SO(2)}} \sim \frac{\mathrm{SU(1,1)}}{\mathrm{U(1)}}$.

According to the presented theory and to Table 5.2 the c-map and $c^\star$-map images of this special Kähler manifold are:

$$\begin{aligned} c\left[\frac{\mathrm{SU}(1,1)}{\mathrm{U}(1)}\right] &= \frac{\mathrm{G}_{2(2)}}{\mathrm{SU}(2) \times \mathrm{SU}(2)} \\ c^\star\left[\frac{\mathrm{SU}(1,1)}{\mathrm{U}(1)}\right] &= \frac{\mathrm{G}_{2(2)}}{\mathrm{SU}(1,1) \times \mathrm{SU}(1,1)} \end{aligned} \tag{6.3.2}$$

and the architecture of the (pseudo)-quaternionic manifold is algebraically governed by the golden splitting (1.7.21) and analytically determined by the explicit form of the $\mathcal{N}$-matrix of special geometry appearing in Eqs. (5.2.17) and (5.2.18).

In our discussion of supergravity black-holes from the point of view of the $D = 3$ σ-model and of nilpotent orbits, the master model we will constantly utilize is the simplest one based on the above mentioned one dimensional special Kähler manifold traditionally dubbed the S^3 model. Hence we are interested in the explicit derivation of its special geometry items.

The manifold $\frac{\mathrm{SU(1,1)}}{\mathrm{U(1)}}$ admits a standard solvable parametrization constructed as it follows. Let:

$$L_0 = \frac{1}{2}\begin{pmatrix} 1 & 0 \\ 0 & -1 \end{pmatrix} \quad ; \quad L_+ = \frac{1}{2}\begin{pmatrix} 0 & 1 \\ 0 & 0 \end{pmatrix} \quad ; \quad L_- = \frac{1}{2}\begin{pmatrix} 0 & 0 \\ 1 & 0 \end{pmatrix} \tag{6.3.3}$$

be the standard three generators of the $\mathfrak{sl}(2,\mathbb{R})$ Lie algebra satisfying the commutation relations $\left[L_0, L_\pm\right] = \pm L_\pm$ and $\left[L_+, L_-\right] = 2L_0$. The coset manifold $\frac{\mathrm{SU(1,1)}}{\mathrm{U(1)}}$ is metrically equivalent with the solvable group manifold generated by L_0 and L_+. Correspondingly we can introduce the coset representative:

$$\mathbb{L}_4(\phi, y) = \exp[y\, L_1] \exp[\varphi\, L_0] = \begin{pmatrix} e^{\varphi/2} & e^{-\varphi/2} y \\ 0 & e^{-\varphi/2} \end{pmatrix} \tag{6.3.4}$$

Generic group elements of $\mathrm{SL}(2,\mathbb{R})$ are just 2×2 real matrices with determinant one:

$$\mathrm{SL}(2,\mathbb{R}) \ni \mathfrak{A} = \begin{pmatrix} a & b \\ c & d \end{pmatrix} \quad ; \quad ad - bc = 1 \tag{6.3.5}$$

[3]The special overall normalization of the Poincaré metric is chosen in order to match the general definitions of special geometry applied to the present case.

and their action on the lower half-plane is defined by usual fractional linear transformations:

$$\mathfrak{A} \quad : \quad z \to \frac{a\,z\,+\,b}{c\,z\,+d} \tag{6.3.6}$$

The correspondence between the lower complex half-plane $\mathbb{C}_-$ and the solvable -parameterized coset (6.3.4) is easily established observing that the entire set of $\text{Im}z < 0$ complex numbers is just the orbit of the number i under the action of $\mathbb{L}(\phi, y)$:

$$\mathbb{L}_4(\phi, y) \quad : \quad \text{i} \to \frac{-e^{\varphi/2}\,\text{i}\,+\,e^{-\varphi/2}\,y}{e^{-\varphi/2}} \;=\; y \,-\, \text{i}e^{\varphi} \tag{6.3.7}$$

This simple argument shows that we can rewrite the coset representative $\mathbb{L}(\phi, y)$ in terms of the complex scalar field z as follows:

$$\mathbb{L}_4(z) \;=\; \begin{pmatrix} \sqrt{|\text{Im}z\,|} & \frac{\text{Re}z}{\sqrt{|\text{Im}z\,|}} \\ 0 & \frac{1}{\sqrt{|\text{Im}z\,|}} \end{pmatrix} \tag{6.3.8}$$

The issue of special Kähler geometry becomes clear at this stage. If we did not consider the symplectic vector bundle, the choice of the coset metric would be sufficient and nothing more would have to be said. The point is that we still have to define the $\mathcal{N}$–matrix associated with the flat symplectic bundle which enters the definition of special Kähler geometry. On the same base manifold $\text{SL}(2,\mathbb{R})/\text{SO}(2)$ we have different special structures which lead to different physical models and to different duality groups $\text{U}_{\text{D}=3}$ upon reduction to $D = 3$. The special structure is determined by the choice of the symplectic embedding $\text{SL}(2,\mathbb{R}) \to \text{Sp}(4,\mathbb{R})$. The symplectic embedding that defines our master model and which eventually leads to the duality group $\text{U}_{\text{D}=3} = \text{G}_{2(2)}$ is cubic and it was already described in Sect. 1.7.1.1. It is explicitly given by Eq. (1.7.28).

The 2×2 blocks A, B, C, D of the 4×4 symplectic matrix $\Lambda\,(\mathfrak{A})$ are easily readable from Eq. (1.7.28) so that, assuming that the matrix $\mathfrak{A}(z)$ is the coset representative of the manifold $\text{SU}(1,1)/\text{U}(1)$, we can apply the Gaillard-Zumino formula (5.2.16) and obtain the explicit form of the kinetic matrix $\mathcal{N}_{\Lambda\Sigma}$:

$$\overline{\mathcal{N}} \;=\; \begin{pmatrix} -\frac{2ac-ibc+iad+2bd}{a^2+b^2} & -\frac{\sqrt{3}(c+id)(ac+bd)}{(a-ib)(a+ib)^2} \\ -\frac{\sqrt{3}(c+id)(ac+bd)}{(a-ib)(a+ib)^2} & -\frac{(c+id)^2(2ac+ibc-iad+2bd)}{(a-ib)(a+ib)^3} \end{pmatrix} \tag{6.3.9}$$

Inserting the specific values of the entries a, b, c, d corresponding to the coset representative (6.3.8), we get the explicit dependence of the $\mathcal{N}$-matrix on the complex coordinate z:

$$\overline{\mathcal{N}}_{\Lambda\Sigma}(z) \;=\; \begin{pmatrix} -\frac{3z+\bar{z}}{2z\bar{z}} & -\frac{\sqrt{3}(z+\bar{z})}{2z\bar{z}^2} \\ -\frac{\sqrt{3}(z+\bar{z})}{2z\bar{z}^2} & -\frac{z+3\bar{z}}{2z\bar{z}^3} \end{pmatrix} \tag{6.3.10}$$

This might conclude the determination of the quaternionic or pseudo-quaternionic metric of our master example, yet we have not yet seen the special Kähler structure induced by the cubic embedding. Let us present it.

The key point is the construction of the required holomorphic symplectic section $\Omega(z)$. As usual the transformation properties of a geometrical object indicate the way to build it explicitly. For consistency we should have that:

$$\Omega\left(\frac{a\,z+b}{c\,z+d}\right) = f(z)\,\Lambda(\mathfrak{A})\,\Omega(z) \tag{6.3.11}$$

where $\Lambda(\mathfrak{A})$ is the symplectic representation (1.7.28) of the considered $\mathrm{SL}(2,\mathbb{R})$ matrix $\begin{pmatrix} a & b \\ c & d \end{pmatrix}$ and $f(z)$ is the associated transition function for that line-bundle whose Chern-class is the Kähler class of the base-manifold. The identification of the symplectic fibres with the cubic symmetric representation provide the construction mechanism of Ω. Consider a vector $\begin{pmatrix} v_1 \\ v_2 \end{pmatrix}$ that transforms in the fundamental doublet representation of $\mathrm{SL}(2,\mathbb{R})$. On one hand we can identify the complex coordinate z on the lower half-plane as $z = v_1/v_2$, on the other we can construct a symmetric three-index tensor taking the tensor products of three v_i, namely: $t_{ijk} = v_i\,v_j\,v_k$. Dividing the resulting tensor by v_2^3 we obtain a four vector:

$$\widehat{\Omega}(z) = \frac{1}{v_2^3}\begin{pmatrix} v_1^3 \\ v_1^2\,v_2 \\ v_1\,v_2^2 \\ v_2^3 \end{pmatrix} = \begin{pmatrix} z^3 \\ z^2 \\ z \\ 1 \end{pmatrix} \tag{6.3.12}$$

Next, recalling the change of basis (1.7.25), (1.7.26) required to put the cubic representation into a standard symplectic form we set:

$$\Omega(z) = S\,\widehat{\Omega}(z) = \begin{pmatrix} -\sqrt{3}z^2 \\ z^3 \\ \sqrt{3}z \\ 1 \end{pmatrix} \tag{6.3.13}$$

and we can easily verify that this object transforms in the appropriate way. Indeed we obtain:

$$\Omega\left(\frac{a\,z+b}{c\,z+d}\right) = (c\,z+d)^{-3}\,\Lambda(\mathfrak{A})\,\Omega(z) \tag{6.3.14}$$

The pre-factor $(c\,z+d)^{-3}$ is the correct one for the prescribed line-bundle. To see this let us first calculate the Kähler potential and the Kähler form. Inserting (6.3.13) into Eq. (4.2.15) we get:

$$\mathscr{K} = -\log\left(\mathrm{i}\langle\Omega \mid \bar{\Omega}\rangle\right) = -\log\left(-\mathrm{i}(z-\bar{z})^3\right)$$
$$\mathrm{K} = \frac{\mathrm{i}}{2\pi}\,\partial\,\bar{\partial}\,\mathscr{K} = \frac{\mathrm{i}}{2\pi}\,\frac{3}{(\mathrm{Im}z)^2}dz\wedge d\bar{z} \tag{6.3.15}$$

This shows that the constructed symplectic bundle leads indeed to the standard Poincaré metric and the exponential of the Kähler potential transforms with the prefactor $(c\,z + d)^3$ whose inverse appears in Eq. (6.3.14).

To conclude let us show that the special geometry definition of the period matrix $\mathscr{N}$ agrees with the Gaillard-Zumino definition holding true for all symplectically embedded cosets. To this effect we calculate the necessary ingredients:

$$\nabla_z V(z) = \exp\left[\frac{\mathscr{K}}{2}\right](\partial_z\Omega(z) + \partial_z\mathscr{K}\,\Omega(z)) = \begin{pmatrix} \frac{\sqrt{3}z(z+2\bar{z})}{(z-\bar{z})\sqrt{-i(z-\bar{z})^3}} \\ -\frac{3z^2\bar{z}}{(z-\bar{z})\sqrt{-i(z-\bar{z})^3}} \\ -\frac{\sqrt{3}(2z+\bar{z})}{(z-\bar{z})\sqrt{-i(z-\bar{z})^3}} \\ -\frac{3}{(z-\bar{z})\sqrt{-i(z-\bar{z})^3}} \end{pmatrix} \equiv \begin{pmatrix} f_z^\Lambda \\ h_{\Sigma z} \end{pmatrix} \tag{6.3.16}$$

Then according to Eq. (4.2.21) we obtain:

$$f_I^\Lambda = \begin{pmatrix} \frac{\sqrt{3}z(z+2\bar{z})}{(z-\bar{z})\sqrt{-i(z-\bar{z})^3}} & -\frac{2\sqrt{6}\bar{z}^2}{(-i(z-\bar{z}))^{3/2}} \\ -\frac{3z^2\bar{z}}{(z-\bar{z})\sqrt{-i(z-\bar{z})^3}} & \frac{2\sqrt{2}\bar{z}^3}{(-i(z-\bar{z}))^{3/2}} \end{pmatrix}$$
$$h_{\Lambda|I} = \begin{pmatrix} -\frac{\sqrt{3}(2z+\bar{z})}{(z-\bar{z})\sqrt{-i(z-\bar{z})^3}} & \frac{2\sqrt{6}\bar{z}}{(-i(z-\bar{z}))^{3/2}} \\ -\frac{3}{(z-\bar{z})\sqrt{-i(z-\bar{z})^3}} & \frac{2\sqrt{2}}{(-i(z-\bar{z}))^{3/2}} \end{pmatrix} \tag{6.3.17}$$

and applying definition (4.2.21) we exactly retrieve the same form of $\mathscr{N}_{\Lambda\Sigma}$ as given in Eq. (6.3.10).

For completeness and also for later use we calculate the remaining items pertaining to special geometry, in particular the symmetric C-tensor. From the general definition (4.2.18) applied to the present one-dimensional case we get:

$$\nabla_z U_z = \mathrm{i}\,C_{zzz}\,h^{zz^\star}\,\bar{U}_{z^\star} \quad\Rightarrow\quad C_{zzz} = -\frac{6\mathrm{i}}{(z-z^\star)^3} \tag{6.3.18}$$

As for the standard Levi-Civita connection we have:

$$\Gamma^z_{zz} = \frac{2}{z-z^\star} \quad ; \quad \Gamma^{z^\star}_{z^\star z^\star} = -\frac{2}{z-z^\star} \quad ; \quad \text{all other components vanish} \tag{6.3.19}$$

This concludes our illustration of the cubic special Kähler structure on $\frac{\mathrm{SL}(2,\mathbb{R})}{\mathrm{SO}(2)}$.

6.3.1 *The Quartic Invariant*

In the cubic spin $j = \frac{3}{2}$ representation of $\mathrm{SL}(2, \mathbb{R})$ there is a quartic invariant which plays an important role in the discussion of black-holes. As it happens for all the other supergravity models, the quartic invariant of the symplectic vector of magnetic and electric charges:

$$\mathscr{Q} = \begin{pmatrix} p^\Lambda \\ q_\Sigma \end{pmatrix} \tag{6.3.20}$$

is related to the entropy of the extremal black-holes, the latter being its square root. The origin of the quartic invariant is easily understood in terms of the symmetric tensor t_{ijk}. Using the $\mathrm{SL}(2, \mathbb{R})$-invariant antisymmetric symbol ε^{ij} we can construct an invariant order four polynomial in the tensor t_{ijk} by writing:

$$\mathfrak{I}_4 \propto \varepsilon^{ai}\, \varepsilon^{bj}\, \varepsilon^{pl}\, \varepsilon^{qm}\, \varepsilon^{kr}\, \varepsilon^{cn}\, t_{abc}\, t_{ijk}\, t_{pqr}\, t_{lmn} \tag{6.3.21}$$

If we use the standard basis $t_{111}, t_{112}, t_{122}, t_{222}$, we rotate it with the matrix (1.7.25) and we identify the components of the resultant vector with those of the charge vector $\mathscr{Q}$ the explicit form of the invariant quartic polynomial is the following one:

$$\mathfrak{I}_4 = \frac{1}{3\sqrt{3}} q_2 p_1^3 + \frac{1}{12} q_1^2 p_1^2 - \frac{1}{2} p_2 q_1 q_2 p_1 - \frac{1}{3\sqrt{3}} p_2 q_1^3 - \frac{1}{4} p_2^2 q_2^2 \tag{6.3.22}$$

where we have also chosen a specific overall normalization which turns out to be convenient in the sequel.

6.4 Attractor Mechanism, the Entropy and Other Special Geometry Invariants

One of the most important features of supergravity black-holes is the attractor mechanism discovered in the nineties by Ferrara and Kallosh for the case of BPS solutions[4] [1, 2] and in recent time extended to non-BPS cases [12–14, 21–25]. According to this mechanism, if we focus on spherical symmetric configurations, the evolving

[4] *Clarification for mathematicians:* the acronym BPS stands for Bogomolny, Prasad and Sommerfeld. It is a notion occuring in the theory of monopoles where one always derives a bound according to which the energy (or mass) of a quasi-particle corresponding to a localized solution of non linear propagation equations is always larger or equal than some kind of *charge* carried by the quasi-particle. BPS states are those that saturate the bound and typically correspond to shortened representations of the space-time group. In the case of supergravity black–holes the BPS bound relates the mass of the hole with the modulus of the central charge of the supersymmetry algebra. Because of the scope of this book we omit the original definition of the central charge in terms of superalgebras and we confine to give its expression in terms of special Kähler geometrical items (see Eq. (6.4.4)).

scalar fields $z^i(\tau)$ flow to fixed values at the horizon of the black-hole ($\tau = -\infty$), which do not depend from their initial values at infinite radius ($\tau = 0$) but only on the electromagnetic charges p, q.

In order to establish the relation of the quartic invariant $\mathfrak{I}_4$ defined in Eq. (6.3.22) with the black-hole entropy and review the attractor mechanism, we must briefly recall the essential items of black hole field equations in the *geodesic potential approach* [10]. In this framework we do not consider all the fields listed in Eq. (5.2.2). We introduce only the warp factor $U(\tau)$ and the original scalar fields of $D = 4$ supergravity. The information about vector gauge fields is encoded solely in the set of electric and magnetic charges $\mathcal{Q}$ defined by Eq. (6.3.20) which is retrieved in Eq. (6.2.28). Under these conditions the correct field equations for an $\mathcal{N} = 2$ black-hole are derived from the geodesic one dimensional field-theory described by the following lagrangian:

$$
\begin{aligned}
S_{eff} &\equiv \int \mathcal{L}_{eff}(\tau)\, d\tau \quad ; \quad \tau = -\frac{1}{r} \\
\mathcal{L}_{eff}(\tau) &= \tfrac{1}{4}\left(\frac{dU}{d\tau}\right)^2 + g_{ij^\star}\,\frac{dz^i}{d\tau}\,\frac{dz^{j^\star}}{d\tau} + e^U\, V_{BH}(z, \bar{z}, \mathcal{Q})
\end{aligned} \tag{6.4.1}
$$

where, by definition, the *geodesic potential* $V(z, \bar{z}, \mathcal{Q})$ is given by the following formula in terms of the matrix $\mathcal{M}_4$ introduced in Eq. (4.3.4):

$$
V_{BH}(z, \bar{z}, \mathcal{Q}) \;=\; \tfrac{1}{4}\, \mathcal{Q}^t\, \mathcal{M}_4^{-1}\,(\mathcal{N})\; \mathcal{Q} \tag{6.4.2}
$$

The effective lagrangian (6.4.1) is derived from the σ-model lagrangian (6.2.24) upon substitution of the first integrals of motion corresponding to the electromagnetic charges (6.2.28) under the condition that the Taub-NUT charge, defined in (6.2.17), vanishes[5] ($\mathbf{n} = 0$). Indeed, when the Taub-NUT charge $\mathbf{n}$ vanishes, which will be our systematic choice, we can invert the above mentioned relations, expressing the derivatives of the Z^M fields in terms of the charge vector $\mathcal{Q}^M$ and the inverse of the matrix $\mathcal{M}_4$. Upon substitution in the $D = 3$ sigma model lagrangian (4.3.4) we obtain the effective lagrangian for the $D = 4$ scalar fields z^i and the warping factor U given by Eqs. (6.4.1)–(6.4.3).

The important thing is that, thanks to various identities of special geometry, the effective geodesic potential admits the following alternative representation:

$$
V_{BH}(z, \bar{z}, \mathcal{Q}) = -\tfrac{1}{2}\left(|Z|^2 + |Z_i|^2\right) \equiv -\tfrac{1}{2}\left(Z\,\bar{Z} + Z_i g^{ij^\star}\,\bar{Z}_{j^\star}\right) \tag{6.4.3}
$$

[5] As we are going to see later, each orbit of Lax operators always contains representatives such that the Taub-NUT charge is zero. Alternatively from a dynamical system point of view the Taub-NUT charge can be annihilated by setting a constraint which is consistent with the hamiltonian and which reduces the dimension of the system by one unit. The problem of black hole physics is therefore equivalent to the sigma model based on an appropriate codimension one hypersurface in the coset manifold $G/H^\star$.

where the symbol Z denotes the complex scalar field valued central charge of the supersymmetry algebra:

$$Z \equiv V^T \mathbb{C} \mathcal{Q} = M_\Sigma p^\Sigma - L^\Lambda q_\Lambda \tag{6.4.4}$$

and Z_i denote its covariant derivatives:

$$\begin{aligned} Z_i &= \nabla_i Z = U_i \mathbb{C} \mathcal{Q} \quad ; \quad Z^{j^\star} = g^{j^\star i} Z_i \\ \bar{Z}_{j^\star} &= \nabla_{j^\star} Z = \bar{U}_{j^\star} \mathbb{C} \mathcal{Q} \quad ; \quad \bar{Z}^i = g^{ij^\star} \bar{Z}_{j^\star} \end{aligned} \tag{6.4.5}$$

Equation (6.4.3) is a result in special geometry whose proof can be found in several articles and reviews of the late nineties.[6]

6.4.1 *Critical Points of the Geodesic Potential and Attractors*

The structure of the geodesic potential illustrated above allows for a detailed discussion of its critical points, which are relevant for the asymptotic behavior of the scalar fields.

By definition, critical points correspond to those values of z^i for which the first derivative of the potential vanishes: $\partial_i V_{BH} = 0$. Utilizing the fundamental identities of special geometry and Eq. (6.4.3), the vanishing derivative condition of the potential can be reformulated as follows:

$$0 = 2\, Z_i\, \bar{Z} + \mathrm{i} C_{ijk}\, \bar{Z}^j\, \bar{Z}^k \tag{6.4.6}$$

From this equation it follows that there are three possible types of critical points:

$$\begin{array}{lll} Z_i = 0 \; ; \; Z \neq 0 \; ; & & \text{BPS attractor} \\ Z_i \neq 0 \; ; \; Z = 0 \; ; \; \mathrm{i}\, C_{ijk}\, \bar{Z}^j\, \bar{Z}^k = & 0 & \text{non BPS attractor I} \\ Z_i \neq 0 \; ; \; Z \neq 0 \; ; \; \mathrm{i}\, C_{ijk}\, \bar{Z}^j\, \bar{Z}^k = & -2\, Z_i\, \bar{Z} & \text{non BPS attractor II} \end{array} \tag{6.4.7}$$

It should be noted that in the case of one-dimensional special geometries, like the S^3-model, only BPS attractors and non BPS attractors of type II are possible. Indeed non BPS attractors of type I are forbidden unless C_{zzz} vanishes identically.

In order to characterize the various type of attractors, the authors of [20] and [34] introduced a certain number of special geometry invariants that obey different and characterizing relations at attractor points of different type. They are defined as follows. Let us introduce the symbols:

$$N_3 \equiv C_{ijk}\, \bar{Z}^i\, \bar{Z}^j\, \bar{Z}^k \quad ; \quad \bar{N}_3 \equiv C_{i^\star j^\star k^\star}\, Z^{i^\star}\, Z^{j^\star}\, Z^{k^\star} \tag{6.4.8}$$

[6] See for instance the lecture notes [11].

and let us set:

$$\begin{array}{ll} i_1 = Z\bar{Z} & ; \; i_2 = Z_i \bar{Z}_{j^\star} g^{ij^\star} \\ i_3 = \frac{1}{6}\left(Z N_3 + \bar{Z}\bar{N}_3\right) & ; \; i_4 = \mathrm{i}\,\frac{1}{6}\left(Z N_3 - \bar{Z}\bar{N}_3\right) \\ i_5 = C_{ijk}\, C_{\bar{\ell}\bar{m}\bar{n}}\, \bar{Z}^j\, \bar{Z}^k\, Z^{\bar{m}}\, Z^{\bar{n}}\, g^{i\bar{\ell}}\; ; & \end{array} \tag{6.4.9}$$

An important identity satisfied by the above invariants, that depend both on the scalar fields z^i and the charges (p, q), is the following one:

$$\mathfrak{I}_4(p,q) = \frac{1}{4}(i_1 - i_2)^2 + i_4 - \frac{1}{4}\, i_5 \tag{6.4.10}$$

where $\mathfrak{I}_4(p,q)$ is the quartic symplectic invariant that depends only on the charges (see Eq. (6.3.22)). This means that in the above combination the dependence on the fields z^i cancels identically.

In the case of the one-dimensional S^3 model there are two additional identities [34] that read as follows:

$$i_2^2 = \frac{3}{4}\, i_5\, ; \; i_3^2 + i_4^2 = 4 i_1 \left(\frac{i_2}{3}\right)^3 \quad ; \quad \text{for the } S^3 \text{ model} \tag{6.4.11}$$

In [20] it was proposed that the three types of critical points can be characterized by the following relations among the above invariants holding at the attractor point:

At BPS Attractor Points

we have:

$$i_1 \neq 0 \quad ; \quad i_2 = i_3 = i_4 = i_5 = 0 \quad ; \tag{6.4.12}$$

At Non BPS Attractor Points of Type I

we have:

$$i_2 \neq 0 \quad ; \quad i_1 = i_3 = i_4 = i_5 = 0 \tag{6.4.13}$$

At Non BPS Attractor Points of Type II

we have:

$$i_2 = 3 i_1 \quad ; \quad i_3 = 0 \quad ; \quad i_4 = -2\, i_1^2 \quad ; \quad i_5 = 12\, i_1^2 \tag{6.4.14}$$

These relations follow from the definition of the critical point with the use of standard special geometry manipulations. Their values resides in that they inform us in a simple way about the nature of the black-hole solution we are considering. Indeed they provide a partial classification of solution orbits since, given a configuration of charges (p, q), whose structure depends, as we are going to see, from the choice of an $\mathrm{H}^\star$ orbit for the Lax operator, we can calculate the possible critical points of the corresponding geodesic potential and find out to which type they belong. We might expect several different critical points for each (p, q)-choice, yet it turns out that there is only one and it always belongs to the same type for all elements of the same H⋆ orbit. This fact, whose *a priori proof* has still to be given, implies that a classification of attractor points is also a partial classification of Lax operator orbits. We shall come back on this crucial issue later on. Yet it is appropriate to emphasize the word *partial classification*. Although the type of fixed point is the same for each element of the same orbit we should by no means assume that fixed point types select orbits. Indeed there are Lax operators belonging to different H⋆ orbits that have the same electromagnetic charges and therefore define the same fixed point. Furthermore the fact that a Lax operator defines certain charges and hence an associated fixed point does not imply that the solution generated by such Lax will necessarily reach that fixed point. The solution can break up at a finite value of τ, stopping before the fixed point is attained. Hence the classification of fixed points is not a classification of H⋆ orbits although the two classifications have partial relations to each other.

6.4.2 Fixed Scalars at BPS Attractor Points

In the case of BPS attractors we can find the explicit expression in terms of the (p,q)-charges for the scalar field fixed values at the critical point.

By means of standard special geometry manipulations the BPS critical point equation

$$\nabla_j Z = 0 \quad ; \quad \nabla_{j^\star} \bar{Z} = 0 \tag{6.4.15}$$

can be rewritten in the following celebrated form which, in the late nineties, appeared in numerous research and review papers (see for instance [11]):

$$p^\Lambda = \mathrm{i}\left(Z_{fix}\,\bar{L}^\Lambda_{fix} - \bar{Z}_{fix}\,L^\Lambda_{fix}\right) \tag{6.4.16}$$

$$q_\Sigma = \mathrm{i}\left(Z_{fix}\,\bar{M}^{fix}_\Sigma - \bar{Z}_{fix}\,M^{fix}_\Sigma\right) \tag{6.4.17}$$

Using the explicit form of the symplectic section $\Omega(z)$ given in Eq. (6.3.13), we can easily solve Eq. (6.4.17) for the S^3 model and obtain the following fixed scalars:

$$z_{fixed} = -\frac{p_1 q_1 + 3p_2 q_2 + \mathrm{i}\,6\sqrt{\mathfrak{I}_4(p,q)}}{2\left(q_1^2 + \sqrt{3}p_1 q_2\right)} \tag{6.4.18}$$

where $\mathfrak{I}_4(p,q)$ is the quartic invariant defined in Eq. (6.3.22). In fact, one can give the BPS solution in a closed form by replacing in the expression (6.4.18) z_{fixed} the quantized charges with harmonic functions

$$q_\Lambda \to H_\Lambda \equiv h_\Lambda - \sqrt{2}\,q_\Lambda\,\tau \;;\; p^\Lambda \to H^\Lambda \equiv h^\Lambda - \sqrt{2}\,p^\Lambda\,\tau \tag{6.4.19}$$

The same substitution allows to describe the radial evolution of the warp factor:

$$e^{-U} = \frac{1}{2}\sqrt{\mathfrak{I}_4(H^\Lambda, H_\Lambda)} \tag{6.4.20}$$

The constants h^Λ, h_Λ in the harmonic functions are subject to two conditions: one originates from the requirement of asymptotic flatness ($\lim_{\tau\to 0^-} e^U = 1$), while the other reads $h^\Lambda q_\Lambda - h_\Lambda p^\Lambda = 0$. The remaining two free parameters are fixed by the choice of the value of z at radial infinity.

By replacing the fixed values (6.4.18) into the expression (6.4.3) for the potential we find:

$$V_{BH}\left(z_{fixed}, \bar{z}_{fixed}\,,\,\mathcal{Q}\right) = -\sqrt{\mathfrak{I}_4(p,q)} \tag{6.4.21}$$

The above result implies that the horizon area in the case of an extremal BPS black-hole is proportional to the square root of $\mathfrak{I}_4(p,q)$ and, as such, depends only on the charges[7] The argument goes as follows.

Consider the behavior of the warp factor $\exp[-U]$ in the vicinity of the horizon, when $\tau \to -\infty$. For regular black-holes the near horizon metric must factorize as follows:

$$ds^2_{\text{near hor.}} \approx \underbrace{-\frac{1}{r_H^2\,\tau^2}\,dt^2 + r_H^2\left(\frac{d\tau}{\tau}\right)^2}_{\text{AdS}_2\text{ metric}} + \underbrace{r_H^2\left(d\theta^2\,\sin^2\theta\,d\phi^2\right)}_{\text{S}^2\text{ metric}} \tag{6.4.22}$$

where r_H is the Schwarzschild radius defining the horizon. This implies that the asymptotic behavior of the warp factor, for $\tau \to -\infty$ is the following one:

$$\exp[-U] \sim r_H^2\,\tau^2 \tag{6.4.23}$$

In the same limit the scalar fields go to their fixed values and their derivatives become essentially zero. Hence near the horizon we have:

$$\begin{aligned}\left(\dot{U}\right)^2 \approx \frac{4}{\tau^2} \quad &; \quad g_{ij^\star}\frac{dz^i}{d\tau}\frac{dz^{j^\star}}{d\tau} \approx 0\\ e^U\,V_{BH}(z,\bar{z},\mathcal{Q}) &\approx \frac{1}{r_H^2\,\tau^2}\,V\left(z_{fixed}, \bar{z}_{fixed}\,,\,\mathcal{Q}\right)\end{aligned} \tag{6.4.24}$$

Since for extremal black-holes the sum of the above three terms vanishes (see Eq. (6.2.3)), we conclude that:

$$r_H^2 = -V_{BH}\left(z_{fixed}, \bar{z}_{fixed}\,,\,\mathcal{Q}\right) \tag{6.4.25}$$

which yields

$$\text{Area}_H = 4\pi\,r_H^2 = 4\pi\sqrt{\mathfrak{I}_4(p,q)} \tag{6.4.26}$$

---

[7] *Clarification for mathematicians*: for a short but comprehensive introduction to the theory of Black Holes we refer the interested reader to Chaps. 2 and 3 of Volume II of [35] by the present author.

6.5 A Counter Example: The Extremal Kerr Metric

In this section, in order to better clarify the notion of extremality provided by conditions (6.2.8)–(6.2.9) we consider the physically relevant counter-example of the extremal Kerr metric. Such static solution of Einstein equations is certainly encoded in the σ-model approach yet it is not extremal in the sense of Eqs. (6.2.8)–(6.2.9) and therefore it is not related to any nilpotent orbit. Indeed the extremal Kerr metric is a solution of pure gravity and as such its σ-model representation lies in the Euclidean submanifold:

$$\frac{\mathrm{SL}(2,\mathbb{R})}{\mathrm{O}(2)} \tag{6.5.1}$$

for which the coset tangent space $\mathbb{K}$ contains no nilpotent elements.

Instead the so named BPS Kerr–Newman metric, which is not extremal in the sense of General Relativity and actually displays a naked singularity, is extremal in the sense of Eqs. (6.2.8)–(6.2.9) and can be retrieved in one of the nilpotent orbits of the S^3-model. We will show that explicitly in Sect. 6.11.4.

As a preparation to such discussions let us recall the general form of the Kerr–Newman metric which we represent in polar coordinates as it follows:

$$\begin{aligned}
ds^2_{KN} &= -V^0 \otimes V^0 + \sum_{i=1}^{3} V^i \otimes V^i && (6.5.2)\\
V^0 &= \frac{\delta(r)}{\sigma(r,\theta)} \left(dt \;-\; \alpha \sin^2\theta \, d\phi\right) && (6.5.3)\\
V^1 &= \frac{\sigma(r,\theta)}{\delta(r)}\, dr && (6.5.4)\\
V^2 &= \sigma(r,\theta)\, d\theta && (6.5.5)\\
V^3 &= \frac{\sin(\theta)}{\sigma(r,\theta)} \left(\left(r^2+\alpha^2\right) d\phi \;-\; \alpha\, dt\right) && (6.5.6)\\
\delta(r) &= \sqrt{q^2+r^2+\alpha^2-2mr} && (6.5.7)\\
\sigma(r,\theta) &= \sqrt{r^2+\alpha^2\cos^2(\theta)} && (6.5.8)
\end{aligned}$$

Parameters of the Kerr–Newman solution are the mass m, the electric charge q and the angular momentum $J = m\,\alpha$ of the Black Hole. The two particular cases we shall consider in this paper correspond to:

(a) The extremal Kerr solution: $q = 0$ and $m = \alpha$.
(b) The BPS Kerr–Newman solution $q = m$, arbitrary α.

Let us then focus now on the extremal Kerr solution. With the choice $m = \alpha, q = 0$, the metric (6.5.2) can be rewritten in the following form:

$$ds^2_{EK} = -\exp[U]\left(dt + \mathbf{A}^{[KK]}\right)^2 + \exp[-U]\,\gamma_{ij}\, dy^i \otimes dy^j \tag{6.5.9}$$

where $y^i = \{r, \theta, \phi\}$ are the polar coordinates, the three dimensional metric γ_{ij} is the following one:

$$\gamma_{ij} = \begin{pmatrix} \frac{2r^2-\alpha^2+\alpha^2\cos(2\theta)}{2r^2} & 0 & 0 \\ 0 & r^2 - \frac{\alpha^2}{2} + \frac{1}{2}\alpha^2\cos(2\theta) & 0 \\ 0 & 0 & r^2\sin^2(\theta) \end{pmatrix} \tag{6.5.10}$$

the warp factor is:

$$U = \log\left[\frac{r^2 - \alpha^2\sin^2(\theta)}{(r+\alpha)^2 + \alpha^2\cos^2(\theta)}\right] \tag{6.5.11}$$

and the Kaluza Klein vector has the following appearance:

$$\mathbf{A}^{[KK]} = \frac{2\alpha^2(r+\alpha)\sin^2(\theta)}{r^2 - \alpha^2\sin^2(\theta)}\, d\phi \tag{6.5.12}$$

In presence of the metric γ_{ij} the duality relation between the Kaluza Klein vector field and the σ-model scalar field a reads as follows:

$$\mathbf{F}^{[KK]}_{ij} \equiv \partial_{[i}\mathbf{A}^{[KK]}_{j]} = \exp[-2U]\sqrt{\det\gamma}\,\varepsilon_{ijk}\,\gamma^{k\ell}\,\partial_\ell\, a \tag{6.5.13}$$

and it is solved by:

$$a = -\frac{2\alpha^2\cos(\theta)}{2r^2 + 4\alpha r + 3\alpha^2 + \alpha^2\cos(2\theta)} \tag{6.5.14}$$

In this way, by means of inverse engineering we have showed how the extremal Kerr metric is retrieved in the σ-model approach. The crucial point is that the metric γ_{ij} is not flat and hence such a configuration of the U, a fields does not correspond to an extremal solution of the σ-model field equations. Indeed calculating the curvature two-form of the three-dimensional metric (6.5.10) we find

$$\mathfrak{R}^{12} = \frac{4\alpha^2\left(2r^2 + \alpha^2 - \alpha^2\cos(2\theta)\right)}{\left(2r^2 - \alpha^2 + \alpha^2\cos(2\theta)\right)^3}\, e^1 \wedge e^2 \tag{6.5.15}$$

$$\mathfrak{R}^{13} = \frac{4\alpha^2}{\left(2r^2 - \alpha^2 + \alpha^2\cos(2\theta)\right)^2}\, e^1 \wedge e^3 \tag{6.5.16}$$

$$\mathfrak{R}^{23} = -\frac{4\alpha^2}{\left(2r^2 - \alpha^2 + \alpha^2\cos(2\theta)\right)^2}\, e^2 \wedge e^3 \tag{6.5.17}$$

where

$$e^1 = \frac{dr\sqrt{\frac{\cos(2\theta)\alpha^2}{r^2} - \frac{\alpha^2}{r^2} + 2}}{\sqrt{2}} \tag{6.5.18}$$

$$e^2 = d\theta\sqrt{r^2 - \frac{\alpha^2}{2} + \frac{1}{2}\alpha^2\cos(2\theta)} \tag{6.5.19}$$

$$e^3 = d\phi\, r\sin(\theta) \tag{6.5.20}$$

is the *dreibein* corresponding to (6.5.10).

Hopefully this explicit calculation should have convinced the reader that the extremal Kerr solution and, by the same token, also the extremal Kerr–Newman solution are not extremal in the σ-model sense and are retrieved in regular rather than in nilpotent orbits[8] of $U/H^\star$.

6.6 The Standard Triple Classification of Nilpotent Orbits

The construction and classification of nilpotent orbits in semi-simple Lie algebras is a relatively new field of mathematics which has already generated a vast literature. Notwithstanding this, a well established set of results ready to use by physicists is not yet available mainly because existing classifications are concerned with orbits with respect to the full complex group $G_{\mathbb{C}}$ or of one of its real forms $G_{\mathbb{R}}$ [36], which is not exactly what the problem of supergravity black-holes requires (i.e. the classification of the nilpotent $H^\star$-orbits in $\mathbb{K}$). Furthermore the complexity of the existing mathematical papers and books is rather formidable and their reading not too easy. Yet the main mathematical idea underlying all classification schemes is very simple and intuitive and can be rephrased in a language very familiar to physicists, namely that of angular momentum. Such rephrasing allows for what we named a *practitioner's approach* to the method of triples. In other words after decoding this method in terms of angular momentum we can derive case by case the needed results by using a relatively elementary algorithm supplemented with some hints borrowed from the mathematical literature.

[8] *Clarification for mathematicians*: Extremal in the GR sense means something different than extremal in the σ-model sense. As we mentioned above the extremal Kerr solution, according to General Relativity is the solution where $m = \alpha$. In the σ-model sense any extremal solution corresponds to a light-like geodesic of the of the $U/H^\star$ manifold. Light-like geodesics, on their turn are associated with $H^\star$ orbits of nilpotent U Lie algebra elements. As shown above the extremal Kerr solution is obtained from a $U/H^\star$ geodesic that is not light-like so it is not extremal in the σ-model sense.

6.6.1 Presentation of the Method

In this section we shall denote the isometry group $\mathrm{U}_{D=3}$ by $\mathrm{G}_{\mathbb{R}}$ to emphasize that it is a real form of some complex semisimple Lie group.

We will present the practitioner's argument in the form of an ordered list.

1. The basic theorem proved by mathematicians (the Jacobson–Morozov theorem [36]) is that any nilpotent element of a Lie algebra $X \in \mathfrak{g}$ can be regarded as belonging $(X = x)$ to a triple of elements $\{x, y, h\}$ that satisfy the standard commutation relations of the $\mathfrak{sl}(2)$ Lie algebra, namely:

$$[h\,,\,x] = x \quad ; \quad [h\,,\,y] = -\,y \quad ; \quad [x\,,\,y] = 2\,h \tag{6.6.1}$$

   Hence the classification of nilpotent orbits is just the classification of embeddings of an $\mathfrak{sl}(2)$ Lie algebra in the ambient one, modulo conjugation by the full group $\mathrm{G}_{\mathbb{R}}$ or by one of its subgroups. In our case the relevant subgroup is $\mathrm{H}^{\star} \subset \mathrm{G}_{\mathbb{R}}$.
2. The second relevant point in our decoding is that embeddings of subalgebras $\mathfrak{h} \subset \mathfrak{g}$ are characterized by the branching law of any representation of $\mathfrak{g}$ into irreducible representations of $\mathfrak{h}$. Clearly two embeddings might be conjugate only if their branching laws are identical. Embeddings with different branching laws necessarily belong to different orbits. In the case of the $\mathfrak{sl}(2) \sim \mathfrak{so}(1,2)$ Lie algebra, irreducible representations are uniquely identified by their spin j, so that the branching law is expressed by listing the angular momenta $\{j_1, j_2, \ldots j_n\}$ of the irreducible blocks into which any representation of the original algebra, for instance the fundamental, decomposes with respect to the embedded subalgebra. The dimensions of each irreducible module is $2j+1$ so that an a priori constraint on the labels $\{j_1, j_2, \ldots j_n\}$ characterizing an orbit is the summation rule:

$$\sum_{i=1}^{n}(2j_i + 1) \;=\; N \;=\; \text{dimension of the fundamental representation} \tag{6.6.2}$$

   Taking into account that j_i are integer or half integer numbers, the sum rule (6.6.2) is actually a partition of N into integers and this explains why mathematicians classify nilpotent orbits starting from partitions of N and use Young tableaux in the process.
3. The next observation is that the central element h of any triple is by definition a diagonalizable (semisimple) non-compact element of the Lie algebra and as such it can always be rotated into the Cartan subalgebra by means of a $\mathrm{G}_{\mathbb{R}}$ transformation. In the case of interest to us, the Cartan subalgebra $\mathscr{C}$ can be chosen, as we will do, inside the subalgebra $\mathbb{H}^{\star}$ and consequently we can argue that for any standard triple $\{x, y, h\}$ the central element is inside that subalgebra:

$$h \,\in\, \mathbb{H}^{\star} \tag{6.6.3}$$

Since we shall work with real representations of $\mathrm{G}_{\mathbb{R}}$, we choose a basis in which h is a symmetric matrix. Indeed there are two possibilities: either $x \in \mathbb{H}^\star$ or $x \in \mathbb{K}$. In the first case we have $y \in \mathbb{H}^\star$, while in the second we have $y \in \mathbb{K}$. This follows from matrix transposition. Given x, the element y is just its transposed $y = x^T$ and transposition maps $\mathbb{H}^\star$ into $\mathbb{H}^\star$ and $\mathbb{K}$ into $\mathbb{K}$. Since it is already in $\mathbb{H}^\star$, in order to rotate the central element h into the Cartan subalgebra it suffices an $\mathrm{H}^\star$ transformation. Therefore to classify $\mathrm{H}^\star$ orbits of nilpotent $\mathbb{K}$ elements we can start by considering central elements h belonging to the Cartan subalgebra $\mathscr{C}$ chosen inside $\mathbb{H}^\star$.

4. The central element h of the standard triple, chosen inside the Cartan subalgebra, is identified by its eigenvalues and by their ordering with respect to a standard basis. Since h is the third component of the angular momentum, *i.e.* the operator J_3, its eigenvalues in a representation of spin j are $-j, -j+1, \dots, j-1, j$. Hence if we choose a branching law $\{j_1, j_2, \dots j_n\}$, we also decide the eigenvalues of h and consequently its components along a standard basis of simple roots. The only indeterminacy which remains to be resolved is the order of the available eigenvalues.
5. The question which remains to be answered is how much we can order the eigenvalues of Cartan elements by means of $\mathrm{H}^\star$ group rotations. The answer is given in terms of the generalized Weyl group $\mathscr{GW}$ and the Weyl group $\mathscr{W}$.
6. The generalized Weyl group is the discrete group generated by all matrices of the form:

$$\mathscr{O}_\alpha = \exp\left[\theta_\alpha \left(E^\alpha - E^{-\alpha}\right)\right] \tag{6.6.4}$$

where $E^{\pm\alpha}$ are the step operators associated with the roots $\pm\alpha$ and the angle θ_α is chosen in such a way that it realizes the α-reflection on a Cartan subalgebra element $\beta \cdot \mathscr{H}$ associated with a vector β:

$$\begin{aligned} O_\alpha\, \beta \cdot \mathscr{H}\, O_\alpha^{-1} &= \sigma_\alpha(\beta) \cdot \mathscr{H} \\ \sigma_\alpha(\beta) &\equiv \beta - 2\,\frac{(\alpha\,,\,\beta)}{(\alpha\,,\,\alpha)}\,\alpha \end{aligned} \tag{6.6.5}$$

The generalized Weyl group has the property that for each of its elements $\gamma \in \mathscr{GW}$ and for each element $h \in \mathscr{C}$ of the Cartan subalgebra $\mathscr{C}$, we have:

$$\gamma\, h\, \gamma^{-1} = h' \in \mathscr{C} \tag{6.6.6}$$

7. The generalized Weyl group contains a normal subgroup $\mathscr{HW} \subset \mathscr{GW}$, named the Weyl stability group and defined by the property that for each element $\xi \in \mathscr{HW}$ and for each Cartan subalgebra element $h \in \mathscr{HW}$ we have:

$$\gamma\, h\, \gamma^{-1} = h \tag{6.6.7}$$

8. The proper Weyl group is defined as the quotient of the generalized Weyl group with respect to the Weyl stability subgroup:

$$\mathscr{W} \equiv \frac{\mathscr{G}\mathscr{W}}{\mathscr{H}\mathscr{W}} \tag{6.6.8}$$

9. The above definition of the Weyl group shows that we can distinguish among its elements those that can be realized by H* transformations, namely those whose corresponding generalized Weyl group elements satisfy the condition $O^T \eta O = \eta$ and those that are outside of H*.
10. If we were to consider nilpotent orbits with respect to the whole group G we would just have to mod out all Weyl transformations. In the case of H* orbits this is too much since the entire Weyl group is not contained in H* as we just said. The rotations that have to be modded out are those of the intersection of the generalized Weyl group $\mathscr{G}\mathscr{W}_H$ with H*, namely:

$$\mathscr{G}\mathscr{W}_H \equiv \mathscr{G}\mathscr{W} \bigcap \mathrm{H}^\star \tag{6.6.9}$$

It should be noted that the Weyl stability subgroup is always contained in H* so that, by definition, it is also a subgroup of $\mathscr{G}\mathscr{W}_H$:

$$\mathscr{H}\mathscr{W} \subset \mathscr{G}\mathscr{W}_H \tag{6.6.10}$$

which happens to be normal. Hence we can define the ratio

$$\mathscr{W}_H \equiv \frac{\mathscr{G}\mathscr{W}_H}{\mathscr{H}\mathscr{W}} \tag{6.6.11}$$

which is a subgroup of the Weyl group.
11. There is a simple method to find directly $\mathscr{W}_H$. The Weyl group is the symmetry group of the root system Δ. When we choose the Cartan subalgebra inside H* the root system splits into two disjoint subsets:

$$\Delta = \Delta_H \bigoplus \Delta_K \tag{6.6.12}$$

respectively containing the roots represented in $\mathbb{H}^\star$ and those represented in $\mathbb{K}$. Clearly the looked for subgroup $\mathscr{W}_H \subset \mathscr{W}$ is composed by those Weyl elements which do not mix Δ_H with Δ_K and thus respect the splitting (6.6.12). According to this viewpoint, given a Cartan element h corresponding to a partition $\{j_1, j_2, \dots j_n\}$, we consider its Weyl orbit and we split this Weyl orbit into m suborbits corresponding to the m cosets:

$$\frac{\mathscr{W}}{\mathscr{W}_H} \quad ; \quad m \equiv \frac{|\mathscr{W}|}{|\mathscr{W}_H|} \tag{6.6.13}$$

Each Weyl suborbit corresponds to an H*-orbit of the neutral elements h in the standard triples. We just have to separate those triples whose x and y elements lie in $\mathbb{K}$ from those whose x and y elements lie in $\mathbb{H}^\star$. By construction if the x and y elements of one triple lie in $\mathbb{K}$, the same is true for all the other triples in the same $\mathscr{W}_H$ orbit. Weyl transformations outside $\mathscr{W}_H$ mix instead $\mathbb{K}$-triples with $\mathbb{H}^\star$ ones.

12. The construction described in the above points fixes completely the choice of the central element h in a standard triple providing a standard representative of an H* orbit. The work would be finished if the choice of h uniquely fixed also x and $y = x^T$ that are our main target. This is not so. Given h one can impose the commutation relations:

$$[h\,,\,x] = x \tag{6.6.14}$$
$$\left[x\,,\,x^T\right] = 2\,h \tag{6.6.15}$$

as a set of algebraic equations for x. Typically these equations admit more than one solution.[9] The next task is that of arranging such solutions in orbits with respect to the stability subgroup $\mathscr{S}_h \subset$ H* of the central element. Typically such a group is the product, direct or semidirect, of the discrete group $\mathscr{HW}$, which stabilizes any Cartan Lie algebra element, with a continuous subgroup of H* which stabilizes only the considered central element h. The presence of such a continuous part of the stabilizer $\mathscr{S}_h$ manifests itself in the presence of continuous parameters in the solution of the second equation (6.6.15) at fixed h.

13. When there are no continuous parameters in the solution of Eq. (6.6.15) what we have to do is quite simple. We just need to verify which solutions are related to which by means of $\mathscr{HW}$ transformations and we immediately construct the $\mathscr{HW}$-orbits. Each $\mathscr{HW}$ orbit of x solutions corresponds to an independent H* orbit of nilpotent operators.

14. When continuous parameters are left over in the solutions space, signaling the existence of a continuous part in the $\mathscr{S}_h$ stabilizer, the direct construction of $\mathscr{S}_h$ orbits is more involved and time consuming. An alternative method, however, is available to distribute the obtained solutions into distinct orbits which is based on invariants. Let us define the non-compact operator:

$$X_c \equiv \mathrm{i}\left(x \; - \; x^T\right) \tag{6.6.16}$$

and consider its adjoint action on the maximal compact subalgebra $\mathbb{H} \subset \mathbb{U}$ which, by construction, has the same dimension as $\mathbb{H}^\star$. We name β-labels the spectrum of eigenvalues of that adjoint matrix[10]:

[9] Such solutions actually correspond to different $\mathrm{G}_{\mathbb{R}}$-orbits [36].

[10] In the literature, see [36], β-labels are defined as the value of the simple roots β^i of the complexification $\mathbb{H}_{\mathbb{C}}$ of $\mathbb{H}^\star$ on the non-compact element X_c, viewed as a Cartan element of $\mathbb{H}_{\mathbb{C}}$ in the Weyl chamber of (β^i). We find it more practical to work with the equivalent characterization (6.6.17).

$$\beta - \text{label} = \text{Spectrum}\left[\text{adj}_{\mathbb{H}}\left(X_c\right)\right] \tag{6.6.17}$$

Since the spectrum is an invariant property with respect to conjugation, x-solutions that have different β-labels belong to different $\mathrm{H}^\star$ orbits necessarily. Actually they even belong to different orbits with respect to the full group U. In fact there exists a one-to-one correspondence between nilpotent U orbits in $\mathbb{U}$ and β-labels, which directly follows from the celebrated Kostant-Sekiguchi theorem [36]. So we arrange the different solutions of Eq. (6.6.15) into orbits by grouping them according to their β-labels.

15. The set of possible β-labels at fixed choice of the partition $\{j_1, j_2, \dots j_n\}$ is predetermined since it corresponds to the set of γ-labels [37]. Let us define these latter. Given the central element h of the triple, we consider its adjoint action on the subalgebra $\mathbb{H}^\star$ and we set:

$$\gamma - \text{label} = \text{Spectrum}\left[\text{adj}_{\mathbb{H}^\star}\left(h\right)\right] \tag{6.6.18}$$

Obviously all h-operators in the same $\mathscr{W}_H$-orbit have the same γ-label. Hence the set of possible γ-labels corresponding to the same partition $\{j_1, j_2, \dots j_n\}$ contains at most as many elements as the order of lateral classes $\frac{\mathscr{W}}{\mathscr{W}_H}$. The actual number can be less when some $\mathscr{W}_H$-orbits of h-elements coincide.[11] Given the set of γ-labels pertaining to one $\{j_1, j_2, \dots j_n\}$-partition the set of possible β-labels pertaining to the same partition is the same. We know a priori that the solutions to Eq. (6.6.15) will distribute in groups corresponding to the available β-labels. Typically all available β-labels will be populated, yet for some partition $\{j_1, j_2, \dots j_n\}$ and for some chosen γ-label one or more β-labels might be empty.

16. The above discussion shows that by naming α-label the partition $\{j_1, j_2, \dots j_n\}$ (branching rule of the fundamental representation of $\mathbb{U}$ with respect to the embedded $\mathfrak{sl}(2)$) the orbits can be classified and named with a triple of indices:

$$\mathscr{O}^{\alpha}_{\gamma\beta} \tag{6.6.19}$$

the set of $\gamma\beta$-labels available for each α-label being determined by means of the action of the Weyl group as we have thoroughly explained.

What we have described in the above list is a concrete algorithm to single out standard triple representatives of nilpotent $\mathrm{H}^\star$ orbits of $\mathbb{K}$ operators. In the next section we apply it to the example of the $\mathfrak{g}_{(2,2)}$ model in order to show how it works.

[11] Note that the action of certain Weyl group elements $g \in \mathscr{W}$ on specific h.s can be the identity: $g \cdot h = h$. When such stabilizing group elements g are inside $\mathscr{W}_H$ the number of different h.s inside each lateral classes is accordingly reduced. If there are stabilizing elements g that are not inside $\mathscr{W}_H$ than two or more $\mathscr{W}_H$ orbits coincide.

6.7 The Nilpotent Orbits of the $\mathfrak{g}_{(2,2)}$ Model

In the present section we consider the classification of nilpotent H*-orbits in $\mathfrak{g}_{(2,2)}$ by using the algorithm described in the previous section.

6.7.1 *The Weyl and the Generalized Weyl Groups for* $\mathfrak{g}_{(2,2)}$

According to our general discussion the most important tools for the orbit classification are the generalized Weyl groups and its subgroups.

We begin with the structure of the Weyl group for the $\mathfrak{g}_{(2,2)}$ root system Δ_{g2}. By definition this is the group of rotations in a two-dimensional plane generated by the reflections along all the roots contained in Δ_{g2}. Abstractly the structure of the group is given by the semidirect product of the permutation group of three object S_3 with a $\mathbb{Z}_2$ factor:

$$\mathscr{W} = S_3 \ltimes \mathbb{Z}_2 \tag{6.7.1}$$

Correspondingly the order of the group is:

$$|\mathscr{W}| = 12 \tag{6.7.2}$$

An explicit realization by means of 2×2 orthogonal matrices is the following one:

$$\begin{aligned}
\text{Id} &= \begin{pmatrix} 1 & 0 \\ 0 & 1 \end{pmatrix} \; ; \alpha_1 = \begin{pmatrix} -1 & 0 \\ 0 & 1 \end{pmatrix} \; ; \alpha_2 = \begin{pmatrix} -\frac{1}{2} & \frac{\sqrt{3}}{2} \\ \frac{\sqrt{3}}{2} & \frac{1}{2} \end{pmatrix} \\
\alpha_3 &= \begin{pmatrix} \frac{1}{2} & \frac{\sqrt{3}}{2} \\ \frac{\sqrt{3}}{2} & -\frac{1}{2} \end{pmatrix} \; ; \alpha_4 = \begin{pmatrix} \frac{1}{2} & -\frac{\sqrt{3}}{2} \\ -\frac{\sqrt{3}}{2} & -\frac{1}{2} \end{pmatrix} \; ; \alpha_5 = \begin{pmatrix} -\frac{1}{2} & -\frac{\sqrt{3}}{2} \\ -\frac{\sqrt{3}}{2} & \frac{1}{2} \end{pmatrix} \\
\alpha_6 &= \begin{pmatrix} 1 & 0 \\ 0 & -1 \end{pmatrix} \; ; \xi_1 = \begin{pmatrix} -1 & 0 \\ 0 & -1 \end{pmatrix} \; ; \xi_2 = \begin{pmatrix} -\frac{1}{2} & -\frac{\sqrt{3}}{2} \\ \frac{\sqrt{3}}{2} & -\frac{1}{2} \end{pmatrix} \\
\xi_3 &= \begin{pmatrix} -\frac{1}{2} & \frac{\sqrt{3}}{2} \\ -\frac{\sqrt{3}}{2} & -\frac{1}{2} \end{pmatrix} \; ; \xi_4 = \begin{pmatrix} \frac{1}{2} & -\frac{\sqrt{3}}{2} \\ \frac{\sqrt{3}}{2} & \frac{1}{2} \end{pmatrix} \; ; \xi_5 = \begin{pmatrix} \frac{1}{2} & \frac{\sqrt{3}}{2} \\ -\frac{\sqrt{3}}{2} & \frac{1}{2} \end{pmatrix}
\end{aligned} \tag{6.7.3}$$

where Id is the identity element, α_i $(i = 1, \ldots, 6)$ denote the reflections along the corresponding roots and ξ_i $(i = 1, \ldots, 5)$ are the additional elements created by products of reflections. The multiplication table of this group is displayed below:

0	Id	α_1	α_2	α_3	α_4	α_5	α_6	ξ_1	ξ_2	ξ_3	ξ_4	ξ_5
Id	Id	α_1	α_2	α_3	α_4	α_5	α_6	ξ_1	ξ_2	ξ_3	ξ_4	ξ_5
α_1	α_1	Id	ξ_4	ξ_2	ξ_3	ξ_5	ξ_1	α_6	α_3	α_4	α_2	α_5
α_2	α_2	ξ_5	Id	ξ_4	ξ_1	ξ_3	ξ_2	α_4	α_6	α_5	α_3	α_1
α_3	α_3	ξ_3	ξ_5	Id	ξ_2	ξ_1	ξ_4	α_5	α_4	α_1	α_6	α_2
α_4	α_4	ξ_2	ξ_1	ξ_3	Id	ξ_4	ξ_5	α_2	α_1	α_3	α_5	α_6
α_5	α_5	ξ_4	ξ_2	ξ_1	ξ_5	Id	ξ_3	α_3	α_2	α_6	α_1	α_4
α_6	α_6	ξ_1	ξ_3	ξ_5	ξ_4	ξ_2	Id	α_1	α_5	α_2	α_4	α_3
ξ_1	ξ_1	α_6	α_4	α_5	α_2	α_3	α_1	Id	ξ_5	ξ_4	ξ_3	ξ_2
ξ_2	ξ_2	α_4	α_5	α_1	α_3	α_6	α_2	ξ_5	ξ_3	Id	ξ_1	ξ_4
ξ_3	ξ_3	α_3	α_6	α_4	α_1	α_2	α_5	ξ_4	Id	ξ_2	ξ_5	ξ_1
ξ_4	ξ_4	α_5	α_1	α_2	α_6	α_4	α_3	ξ_3	ξ_1	ξ_5	ξ_2	Id
ξ_5	ξ_5	α_2	α_3	α_6	α_5	α_1	α_4	ξ_2	ξ_4	ξ_1	Id	ξ_3

(6.7.4)

Next let us discuss the structure of the generalized Weyl group. In this case $\mathscr{GW}$ is composed by 48 elements and its stability subgroup $\mathscr{HW} \sim \mathbb{Z}_2 \times \mathbb{Z}_2$ is made by the following four 7×7 matrices belonging to the $\mathrm{G}_{(2,2)}$ group:

$$hw_1 = \begin{pmatrix} -1 & 0 & 0 & 0 & 0 & 0 & 0 \\ 0 & 1 & 0 & 0 & 0 & 0 & 0 \\ 0 & 0 & -1 & 0 & 0 & 0 & 0 \\ 0 & 0 & 0 & 1 & 0 & 0 & 0 \\ 0 & 0 & 0 & 0 & -1 & 0 & 0 \\ 0 & 0 & 0 & 0 & 0 & 1 & 0 \\ 0 & 0 & 0 & 0 & 0 & 0 & -1 \end{pmatrix} ; \; hw_2 = \begin{pmatrix} 0 & 0 & 0 & 0 & 0 & 0 & -1 \\ 0 & 0 & 0 & 0 & 0 & -1 & 0 \\ 0 & 0 & 0 & 0 & -1 & 0 & 0 \\ 0 & 0 & 0 & -1 & 0 & 0 & 0 \\ 0 & 0 & -1 & 0 & 0 & 0 & 0 \\ 0 & -1 & 0 & 0 & 0 & 0 & 0 \\ -1 & 0 & 0 & 0 & 0 & 0 & 0 \end{pmatrix}$$

$$hw_3 = \begin{pmatrix} 0 & 0 & 0 & 0 & 0 & 0 & 1 \\ 0 & 0 & 0 & 0 & 0 & -1 & 0 \\ 0 & 0 & 0 & 0 & 1 & 0 & 0 \\ 0 & 0 & 0 & -1 & 0 & 0 & 0 \\ 0 & 0 & 1 & 0 & 0 & 0 & 0 \\ 0 & -1 & 0 & 0 & 0 & 0 & 0 \\ 1 & 0 & 0 & 0 & 0 & 0 & 0 \end{pmatrix} ; \; \mathrm{Id} = \begin{pmatrix} 1 & 0 & 0 & 0 & 0 & 0 & 0 \\ 0 & 1 & 0 & 0 & 0 & 0 & 0 \\ 0 & 0 & 1 & 0 & 0 & 0 & 0 \\ 0 & 0 & 0 & 1 & 0 & 0 & 0 \\ 0 & 0 & 0 & 0 & 1 & 0 & 0 \\ 0 & 0 & 0 & 0 & 0 & 1 & 0 \\ 0 & 0 & 0 & 0 & 0 & 0 & 1 \end{pmatrix} \tag{6.7.5}$$

In order to complete the description of the generalized Weyl group it is now sufficient to write one representative for each equivalence class of the quotient:

$$\frac{\mathscr{GW}}{\mathscr{HW}} \simeq \mathscr{W} \tag{6.7.6}$$

We have:

$$
\alpha_1 \sim \begin{pmatrix} 0 & 0 & -1 & 0 & 0 & 0 & 0 \\ 0 & 0 & 0 & 0 & 0 & 1 & 0 \\ 1 & 0 & 0 & 0 & 0 & 0 & 0 \\ 0 & 0 & 0 & -1 & 0 & 0 & 0 \\ 0 & 0 & 0 & 0 & 0 & 0 & -1 \\ 0 & 1 & 0 & 0 & 0 & 0 & 0 \\ 0 & 0 & 0 & 0 & 1 & 0 & 0 \end{pmatrix} ; \ \alpha_2 \sim \begin{pmatrix} -\frac{1}{2} & 0 & 0 & -\frac{1}{\sqrt{2}} & 0 & 0 & \frac{1}{2} \\ 0 & \frac{1}{2} & -\frac{1}{2} & 0 & -\frac{1}{2} & -\frac{1}{2} & 0 \\ 0 & -\frac{1}{2} & -\frac{1}{2} & 0 & \frac{1}{2} & -\frac{1}{2} & 0 \\ -\frac{1}{\sqrt{2}} & 0 & 0 & 0 & 0 & 0 & -\frac{1}{\sqrt{2}} \\ 0 & -\frac{1}{2} & \frac{1}{2} & 0 & -\frac{1}{2} & -\frac{1}{2} & 0 \\ 0 & -\frac{1}{2} & -\frac{1}{2} & 0 & -\frac{1}{2} & \frac{1}{2} & 0 \\ \frac{1}{2} & 0 & 0 & -\frac{1}{\sqrt{2}} & 0 & 0 & -\frac{1}{2} \end{pmatrix}
$$

$$
\alpha_3 \sim \begin{pmatrix} -\frac{1}{2} & -\frac{1}{2} & 0 & 0 & 0 & -\frac{1}{2} & -\frac{1}{2} \\ \frac{1}{2} & -\frac{1}{2} & 0 & 0 & 0 & \frac{1}{2} & -\frac{1}{2} \\ 0 & 0 & \frac{1}{2} & \frac{1}{\sqrt{2}} & \frac{1}{2} & 0 & 0 \\ 0 & 0 & -\frac{1}{\sqrt{2}} & 0 & \frac{1}{\sqrt{2}} & 0 & 0 \\ 0 & 0 & \frac{1}{2} & -\frac{1}{\sqrt{2}} & \frac{1}{2} & 0 & 0 \\ \frac{1}{2} & \frac{1}{2} & 0 & 0 & 0 & -\frac{1}{2} & -\frac{1}{2} \\ -\frac{1}{2} & \frac{1}{2} & 0 & 0 & 0 & \frac{1}{2} & -\frac{1}{2} \end{pmatrix} ; \ \alpha_4 \sim \begin{pmatrix} -\frac{1}{2} & 0 & 0 & -\frac{1}{\sqrt{2}} & 0 & 0 & \frac{1}{2} \\ 0 & -\frac{1}{2} & \frac{1}{2} & 0 & \frac{1}{2} & \frac{1}{2} & 0 \\ 0 & \frac{1}{2} & \frac{1}{2} & 0 & -\frac{1}{2} & \frac{1}{2} & 0 \\ -\frac{1}{\sqrt{2}} & 0 & 0 & 0 & 0 & 0 & -\frac{1}{\sqrt{2}} \\ 0 & \frac{1}{2} & -\frac{1}{2} & 0 & \frac{1}{2} & \frac{1}{2} & 0 \\ 0 & \frac{1}{2} & \frac{1}{2} & 0 & \frac{1}{2} & -\frac{1}{2} & 0 \\ \frac{1}{2} & 0 & 0 & -\frac{1}{\sqrt{2}} & 0 & 0 & -\frac{1}{2} \end{pmatrix}
$$

$$
\alpha_5 \sim \begin{pmatrix} -\frac{1}{2} & -\frac{1}{2} & 0 & 0 & 0 & -\frac{1}{2} & -\frac{1}{2} \\ -\frac{1}{2} & \frac{1}{2} & 0 & 0 & 0 & -\frac{1}{2} & \frac{1}{2} \\ 0 & 0 & -\frac{1}{2} & -\frac{1}{\sqrt{2}} & -\frac{1}{2} & 0 & 0 \\ 0 & 0 & -\frac{1}{\sqrt{2}} & 0 & \frac{1}{\sqrt{2}} & 0 & 0 \\ 0 & 0 & -\frac{1}{2} & \frac{1}{\sqrt{2}} & -\frac{1}{2} & 0 & 0 \\ -\frac{1}{2} & -\frac{1}{2} & 0 & 0 & 0 & \frac{1}{2} & \frac{1}{2} \\ -\frac{1}{2} & \frac{1}{2} & 0 & 0 & 0 & \frac{1}{2} & -\frac{1}{2} \end{pmatrix} ; \ \alpha_6 = \begin{pmatrix} 0 & 0 & -1 & 0 & 0 & 0 & 0 \\ 0 & 0 & 0 & 0 & 0 & -1 & 0 \\ -1 & 0 & 0 & 0 & 0 & 0 & 0 \\ 0 & 0 & 0 & -1 & 0 & 0 & 0 \\ 0 & 0 & 0 & 0 & 0 & 0 & 1 \\ 0 & -1 & 0 & 0 & 0 & 0 & 0 \\ 0 & 0 & 0 & 0 & 1 & 0 & 0 \end{pmatrix} \tag{6.7.7}
$$

and

$$
\xi_1 \sim \begin{pmatrix} -1 & 0 & 0 & 0 & 0 & 0 & 0 \\ 0 & -1 & 0 & 0 & 0 & 0 & 0 \\ 0 & 0 & 1 & 0 & 0 & 0 & 0 \\ 0 & 0 & 0 & 1 & 0 & 0 & 0 \\ 0 & 0 & 0 & 0 & 1 & 0 & 0 \\ 0 & 0 & 0 & 0 & 0 & -1 & 0 \\ 0 & 0 & 0 & 0 & 0 & 0 & -1 \end{pmatrix} ; \ \xi_2 \sim \begin{pmatrix} 0 & 0 & -\frac{1}{2} & -\frac{1}{\sqrt{2}} & -\frac{1}{2} & 0 & 0 \\ \frac{1}{2} & \frac{1}{2} & 0 & 0 & 0 & -\frac{1}{2} & -\frac{1}{2} \\ -\frac{1}{2} & -\frac{1}{2} & 0 & 0 & 0 & -\frac{1}{2} & -\frac{1}{2} \\ 0 & 0 & \frac{1}{\sqrt{2}} & 0 & -\frac{1}{\sqrt{2}} & 0 & 0 \\ \frac{1}{2} & -\frac{1}{2} & 0 & 0 & 0 & -\frac{1}{2} & \frac{1}{2} \\ \frac{1}{2} & -\frac{1}{2} & 0 & 0 & 0 & \frac{1}{2} & -\frac{1}{2} \\ 0 & 0 & \frac{1}{2} & -\frac{1}{\sqrt{2}} & \frac{1}{2} & 0 & 0 \end{pmatrix}
$$

$$
\xi_3 \sim \begin{pmatrix} 0 & -\frac{1}{2} & -\frac{1}{2} & 0 & \frac{1}{2} & -\frac{1}{2} & 0 \\ 0 & \frac{1}{2} & \frac{1}{2} & 0 & \frac{1}{2} & -\frac{1}{2} & 0 \\ -\frac{1}{2} & 0 & 0 & -\frac{1}{\sqrt{2}} & 0 & 0 & \frac{1}{2} \\ \frac{1}{\sqrt{2}} & 0 & 0 & 0 & 0 & 0 & \frac{1}{\sqrt{2}} \\ -\frac{1}{2} & 0 & 0 & \frac{1}{\sqrt{2}} & 0 & 0 & \frac{1}{2} \\ 0 & -\frac{1}{2} & \frac{1}{2} & 0 & \frac{1}{2} & \frac{1}{2} & 0 \\ 0 & \frac{1}{2} & -\frac{1}{2} & 0 & \frac{1}{2} & \frac{1}{2} & 0 \end{pmatrix} ; \ \xi_4 \sim \begin{pmatrix} 0 & -\frac{1}{2} & -\frac{1}{2} & 0 & \frac{1}{2} & -\frac{1}{2} & 0 \\ 0 & -\frac{1}{2} & -\frac{1}{2} & 0 & -\frac{1}{2} & \frac{1}{2} & 0 \\ \frac{1}{2} & 0 & 0 & \frac{1}{\sqrt{2}} & 0 & 0 & -\frac{1}{2} \\ \frac{1}{\sqrt{2}} & 0 & 0 & 0 & 0 & 0 & \frac{1}{\sqrt{2}} \\ \frac{1}{2} & 0 & 0 & -\frac{1}{\sqrt{2}} & 0 & 0 & -\frac{1}{2} \\ 0 & \frac{1}{2} & -\frac{1}{2} & 0 & -\frac{1}{2} & -\frac{1}{2} & 0 \\ 0 & \frac{1}{2} & -\frac{1}{2} & 0 & \frac{1}{2} & \frac{1}{2} & 0 \end{pmatrix}
$$

$$
\xi_5 \sim \begin{pmatrix} 0 & 0 & -\frac{1}{2} & -\frac{1}{\sqrt{2}} & -\frac{1}{2} & 0 & 0 \\ -\frac{1}{2} & -\frac{1}{2} & 0 & 0 & 0 & \frac{1}{2} & \frac{1}{2} \\ \frac{1}{2} & \frac{1}{2} & 0 & 0 & 0 & \frac{1}{2} & \frac{1}{2} \\ 0 & 0 & \frac{1}{\sqrt{2}} & 0 & -\frac{1}{\sqrt{2}} & 0 & 0 \\ -\frac{1}{2} & \frac{1}{2} & 0 & 0 & 0 & \frac{1}{2} & -\frac{1}{2} \\ -\frac{1}{2} & \frac{1}{2} & 0 & 0 & 0 & -\frac{1}{2} & \frac{1}{2} \\ 0 & 0 & \frac{1}{2} & -\frac{1}{\sqrt{2}} & \frac{1}{2} & 0 & 0 \end{pmatrix} ; \tag{6.7.8}
$$

We can explicitly verify that all the elements of the $\mathscr{H}\mathscr{W}$ subgroup are in $\mathrm{H}^\star = \mathfrak{su}(1,1) \times \mathfrak{su}(1,1)$ since they satisfy the condition:

$$hw_i^T\, \eta\, hw_i \;=\; \eta \tag{6.7.9}$$

where

$$\eta \;=\; \begin{pmatrix} -1 & 0 & 0 & 0 & 0 & 0 & 0 \\ 0 & -1 & 0 & 0 & 0 & 0 & 0 \\ 0 & 0 & 1 & 0 & 0 & 0 & 0 \\ 0 & 0 & 0 & 1 & 0 & 0 & 0 \\ 0 & 0 & 0 & 0 & 1 & 0 & 0 \\ 0 & 0 & 0 & 0 & 0 & -1 & 0 \\ 0 & 0 & 0 & 0 & 0 & 0 & -1 \end{pmatrix} \tag{6.7.10}$$

is the invariant metric which defines the $\mathrm{H}^\star$ subgroup. Note that here we use all the conventions and the definitions introduced in [32].

The next required ingredient of our construction is the subgroup $\mathscr{W}_H$. As it was shown in [32], when we diagonalize the adjoint action of a Cartan Subalgebra contained in the $\mathbb{H}^\star$ subalgebra, the root system of the $\mathfrak{g}_2$ Lie algebra (see Fig. 6.4), decomposes in two subsystems Δ_H and Δ_K such that the step operators corresponding to roots in Δ_H belong to $\mathbb{H}^\star$ while the step operators corresponding to roots in Δ_K belong to $\mathbb{K}$. The subsystem Δ_H is composed by the roots $\pm\alpha_3$, $\pm\alpha_5$, while Δ_K is made by the remaining ones. The subgroup $\mathscr{W}_H \subset \mathscr{W}$ can be easily derived. It is made by all those elements of the Weyl group which map Δ_H into itself and Δ_K into itself, as well. Referring to the previously introduced notation, we easily see that (Fig. 6.5):

$$\mathscr{W}_H \;=\; \{\mathrm{Id}, \alpha_3, \alpha_5, \xi_1\} \tag{6.7.11}$$

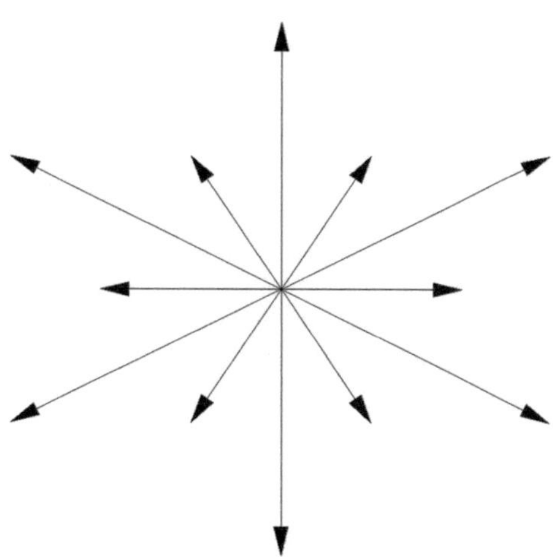

Fig. 6.4 The $\mathfrak{g}_2$ root system Δ_{g2} is made of six positive roots and of their negatives

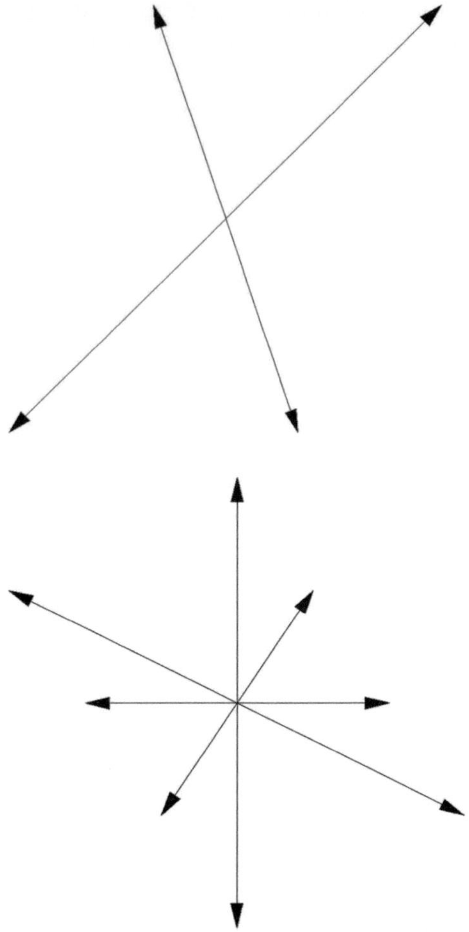

Fig. 6.5 The root system Δ_{g2} splits in two subsystems, the system Δ_H on the left, the system Δ_K on the right

Abstractly the structure of $\mathscr{W}_H$ is the following:

$$\mathscr{W}_H \sim \mathbb{Z}_2 \times \mathbb{Z}_2 \tag{6.7.12}$$

since all of its elements square to the identity.

There are three lateral classes in $\mathscr{W}/\mathscr{W}_H$, respectively associated with the identity element and with the reflection along the two simple roots.

$$[\text{Id}] = \{\text{Id}, \alpha_3, \alpha_5, \xi_1\} \tag{6.7.13}$$

$$[\alpha_1] = \{\alpha_1, \alpha_6, \xi_3, \xi_4\} \tag{6.7.14}$$

$$[\alpha_2] = \{\alpha_2, \alpha_4, \xi_2, \xi_5\} \tag{6.7.15}$$

It follows that for each partition $\{j_1, j_2, \ldots j_n\}$ (α-label) there are three possible γ-labels and three possible β-labels. It remains to be seen for which combinations of these γ and β-labels there exist an x-operator purely contained in $\mathbb{K}$ which completes the standard triple.

6.7.2 The Table of $\frac{\mathrm{G}_{(2,2)}}{\mathrm{SU}(1,1)\times\mathrm{SU}(1,1)}$ Nilpotent Orbits

In order to derive the desired table of nilpotent orbits we begin from the first step namely from partitions or, said differently, from α-labels.

6.7.2.1 α-Labels

Taking into account the restriction (see [36]) that every half-integer spin j should appear an even number of times we easily conclude that the possible branching laws of the 7-dimensional fundamental representation of $\mathfrak{g}_{(2,2)}$ into irreducible representations of $\mathfrak{sl}(2)$ are the following ones:

$$\alpha_1 - \text{label} = [\text{j=3}] \tag{6.7.16}$$

$$\alpha_2 - \text{label} = [\text{j=1}] \times 2[j = 1/2] \tag{6.7.17}$$

$$\alpha_3 - \text{label} = 2[\text{j=1}] \times [j = 0] \tag{6.7.18}$$

$$\alpha_4 - \text{label} = 2[\text{j=1/2}] \times 3[j = 0] \tag{6.7.19}$$

6.7.2.2 γ-Labels

Analyzing the two Eqs. (6.6.14), (6.6.15) for the x-triple element at fixed h we find the following result:

α_1 In this sector there are x operators in $\mathbb{K}$ only for the second lateral class (6.7.14). This means that there is only one γ-label which has the following form:

$$\gamma_1 = \{\pm 8, \pm 4, 0, 0\} \equiv \{8_1, 4_1, 0_1\} \tag{6.7.20}$$

The notation introduced in Eq. (6.7.20) is based on the following observation. The dimension of $\mathbb{H}$ or $\mathbb{H}^\star$ is six and every eigenvalue appears together with its negative. Hence it suffices to mention the non-negative eigenvalues (including the zero) with their multiplicity (all zeros appear in pairs as well). It follows that the β-label is also unique so that in this sector there is only one nilpotent orbit.

α_2 For this partition the $\mathscr{W}_H$ orbits (6.7.13) and (6.7.14) coincide: within them we find x operators in $\mathbb{K}$. In the third $\mathscr{W}_H$ orbit there are no solutions for x in $\mathbb{K}$. So we have only one γ-label:

Table 6.1 Classification of nilpotent orbits of $\frac{\mathrm{G}_{(2,2)}}{\mathrm{SU}(1,1)\times\mathrm{SU}(1,1)}$

N	d_n	α − label	$\gamma\beta$ − labels	Orbits	$\mathscr{W}_H$ − classes
1	7	[j=3]	$\gamma\beta_1 = \{8_1 4_1 0_1\}$	$\mathscr{O}_1^1$	$(\times, \gamma_1, \times)$
2	3	[j=1] × 2[$j = 1/2$]	$\gamma\beta_1 = \{3_1 1_1 0_1\}$	$\mathscr{O}_1^2$	$(\gamma_1, \gamma_1, \times)$
7	3	2[j=1] × [$j = 0$]	$\gamma\beta_1 = \{4_1 0_2\}$ $\gamma\beta_2 = \{2_2 0_1\}$	β_1 β_2 γ_1: $\mathscr{O}_{1,1}^3$ $\mathscr{O}_{1,2}^3$ γ_2: $\mathscr{O}_{2,1}^3$ $\mathscr{O}_{2,2}^3$	$(\gamma_1, \gamma_2, \gamma_2)$
4	2	2[j=1/2] × 3[$j = 0$]	$\gamma\beta_1 = \{1_2 0_1\}$	$\mathscr{O}_1^4$	$(0, \gamma_1, \gamma_1)$

$$\gamma_1 = \{3_1, 1_1, 0_1\} \tag{6.7.21}$$

and consequently only one nilpotent orbit.

α_3 For this partition the $\mathscr{W}_H$ orbits (6.7.14) and (6.7.15) coincide while the first is distinct. We find solutions for x in $\mathbb{K}$ both for the first $\mathscr{W}_H$-orbit (6.7.13) and for the coinciding subsequent two. That means that we have two γ-labels

$$\gamma_1 = \{4_1, 0_2\} \tag{6.7.22}$$

$$\gamma_2 = \{2_2, 0_1\} \tag{6.7.23}$$

Considering the solutions for x both in the case of γ_1 and γ_2 they group in two non empty classes corresponding to β-labels β_1 and β_2. This means that we have a total of 4 nilpotent orbits from this sector.

α_4 For this partition the situation is similar to that of partition one and two. There are no $\mathbb{K}$ solutions for x in the first $\mathscr{W}_H$ orbit while there are such solutions in the second and third $\mathscr{W}_H$-orbits, which coincide. Hence there is only one γ-label:

$$\gamma_1 = \{1_2, 0_1\} \tag{6.7.24}$$

and one nilpotent orbit.

In Table 6.1 the results we have described are summarized.

6.8 Construction of Multicenter Solutions Associated with Nilpotent Orbits

In this section we summarize in purely mathematical terms the algorithm that associates extremal black hole solutions of supergravity to nilpotent orbits of the Lie algebra $\mathbb{U}$. As the reader will appreciate the algorithm is completely sequential and constructive so that it can be easily implemented by means of computer codes.

For spherically symmetric black holes the construction of solutions is associated with nilpotent orbits in the following way. A representative of the $\mathrm{H}^\star$ orbit is a standard triple $\{h, X, Y\}$ and hence an embedding of an $\mathfrak{sl}(2, \mathbb{R})$ Lie algebra:

$$[h, X] = 2X \quad ; \quad [h, Y] = -2Y \quad ; \quad [X, Y] = 2h \tag{6.8.1}$$

into $\mathbb{U}_{\mathrm{D}=3}$ in such a way that $h \in \mathbb{H}^\star$ and $X, Y \in \mathbb{K}^\star$. The nilpotent operator X is identified with the Lax operator L_0 at Euclidean time $\tau = 0$ and the corresponding solution depending on τ is constructed by using the algorithm described in [27, 29, 32].

In the multicenter approach of [15–19, 38] one utilizes the standard triple to single out a nilpotent subalgebra $\mathbb{N}$, as follows. One diagonalizes the adjoint action of the central element h of the triple on the Lie Algebra $\mathbb{U}_{\mathrm{D}=3}$:

$$\left[h\,,\, C_\mu\right] = \mu\, C_\mu \tag{6.8.2}$$

The set of all eigen-operators C_μ corresponding to positive gradings $\mu > 0$ spans a subalgebra $\mathbb{N} \subset \mathbb{U}_{\mathrm{D}=3}$ which is necessarily nilpotent

$$\mathbb{N} = \mathrm{span}\,[C_2\,,\, C_3\,,\, \ldots\,,\, C_{max}] \tag{6.8.3}$$

Such a nilpotent subalgebra has an intersection $\mathbb{N} \bigcap \mathbb{K}^\star$ with the space $\mathbb{K}^\star$ which is not empty since at least the operator $C_2 = X$ is present by definition of a standard triple. The next steps of the construction are as follows.

6.8.1 The Coset Representative in the Symmetric Gauge

Given a basis A^i of the space $\mathbb{N}_{\mathbb{K}} \equiv \mathbb{N} \bigcap \mathbb{K}^\star$, whose dimension we denote:

$$\ell \equiv \dim \mathbb{N}_{\mathbb{K}} \tag{6.8.4}$$

and a basis B^α of the subalgebra $\mathbb{N}_{\mathbb{H}} \equiv \mathbb{N} \bigcap \mathbb{H}^\star$, whose dimension we denote

$$\mathfrak{m} \equiv \dim \mathbb{N}_{\mathbb{H}} \tag{6.8.5}$$

we can construct a map:

$$\mathfrak{H} \;:\; \mathbb{R}^3 \to \mathbb{N}_{\mathbb{K}} \tag{6.8.6}$$

by writing:

$$\mathbb{N}_{\mathbb{K}} \ni \mathfrak{H}(\mathbf{x}) = \sum_{i=1}^{\ell} \mathfrak{h}_i(\mathbf{x})\, A^i \tag{6.8.7}$$

By construction, the point dependent Lie algebra element $\mathfrak{H}(\mathbf{x})$ is nilpotent of a certain maximal degree $\mathrm{d_n}$, so that its exponential map to the nilpotent group $\mathrm{N} \subset \mathrm{U_{D=3}}$ truncates to a finite sum:

$$\mathscr{Y}(x) \; = \; \exp\left[\mathfrak{H}(\mathbf{x})\right] \; = \; \mathbf{1} \, + \, \sum_{a=1}^{\mathrm{d_n}} \frac{1}{a!} \, \mathfrak{H}^a(\mathbf{x}) \tag{6.8.8}$$

The above constructed object realizes an explicit $\mathbf{x}$-dependent coset representative from which we can construct the Maurer Cartan left-invariant one form:

$$\Sigma \; = \; \mathscr{Y}^{-1} \partial_i \mathscr{Y} \, dx^i \tag{6.8.9}$$

Next let us decompose Σ along the $\mathbb{K}^\star$ subspace and the $\mathbb{H}^\star$ subalgebra, respectively. This is done by setting:

$$\mathbf{P} \; = \; \mathrm{Tr}(\Sigma \, K^A) K_A \quad ; \quad \Omega \; = \; \mathrm{Tr}(\Sigma \, H^m) H_m \tag{6.8.10}$$

where K_A and H_m denote a basis of generators for the two considered subspaces, K^A and H^m being their duals:

$$\mathrm{Tr}(K^A \, K_B) \; = \; \delta^A_B \quad ; \quad \mathrm{Tr}(H^m \, H_n) \; = \; \delta^m_n \quad ; \quad \mathrm{Tr}(K^A \, H_n) \; = \; 0 \tag{6.8.11}$$

Denoting:

$$^\star\mathbf{P} \; \equiv \; \tfrac{1}{2} \, \varepsilon_{ijk} \, \delta^{im} \, \mathbf{P}_m \, dx^j \wedge dx^k \tag{6.8.12}$$

the Hodge-dual of the coset vielbein

$$\mathbf{P} \; = \; \mathbf{P}_m \, dx^m \tag{6.8.13}$$

the field equations of the three dimensional σ-model reduce to the following one:

$$d^\star\mathbf{P} \; = \; \Omega \wedge {}^\star\mathbf{P} \; - \; {}^\star\mathbf{P} \wedge \Omega \tag{6.8.14}$$

Actually, since $\mathbb{N} \subset \mathbb{U}_{D=3}$ forms a nilpotent subalgebra the constructed object $\mathscr{Y}$ realizes a map from the three-dimensional space to the much smaller coset manifold:

$$\mathscr{Y} \;\; : \;\; \mathbb{R}^3 \to \frac{\mathrm{N}}{\mathrm{N_H}} \tag{6.8.15}$$

and due to the polynomial form of the coset representative the final equations of motion obtain a triangular solvable form that we describe here below. Since the algebra $\mathbb{N}$ is nilpotent, its derivative series terminates, namely we have:

$$\mathbb{N} \supset \mathscr{D}\mathbb{N} \supset \ldots \supset \mathscr{D}^n\mathbb{N} \supset \mathscr{D}^{n+1}\mathbb{N} = \mathbf{0} \tag{6.8.16}$$

where at each step $\mathscr{D}^i\mathbb{N}$ is a proper subspace of $\mathscr{D}^{i-1}\mathbb{N}$. Correspondingly let us define:

$$\mathscr{D}^i\mathbb{N}_{\mathbb{K}} = \mathscr{D}^i\mathbb{N}\bigcap\mathbb{K}^\star \tag{6.8.17}$$

the intersections of the derivative subalgebras with the $\mathbb{K}^\star$ subspace and let us introduce the complementary orthogonal subspaces:

$$\mathscr{D}^i\mathbb{N}_{\mathbb{K}} = \mathbb{N}_K^{(i)} \oplus \mathscr{D}^{i+1}\mathbb{N}_{\mathbb{K}} \tag{6.8.18}$$

This yields an orthogonal graded decomposition of the space $\mathbb{N}_{\mathbb{K}}$ of the following form:

$$\mathbb{N}_{\mathbb{K}} = \bigoplus_{a=0}^{n} \mathbb{N}_{\mathbb{K}}^{(a)} \tag{6.8.19}$$

The space $\mathbb{N}_{\mathbb{K}}^{(0)}$ contains those generators that cannot be produced by any commutator within the algebra, $\mathbb{N}_{\mathbb{K}}^{(1)}$ contains those generators that are produced in simple commutators, $\mathbb{N}_{\mathbb{K}}^{(2)}$ contains those that are produced in double commutators and so on. Let us name

$$\ell_a = \dim\mathbb{N}_{\mathbb{K}}^{(a)} \quad ; \quad \sum_{a}^{n}\ell_a = \ell \tag{6.8.20}$$

Correspondingly we can arrange the ℓ functions $\mathfrak{h}_i(\mathbf{x})$ according to the graded decomposition (6.8.19), by writing:

$$\mathfrak{H}(\mathbf{x}) = \sum_{\alpha=0}^{n}\underbrace{\sum_{i=1}^{\ell_\alpha}\mathfrak{h}_i^{(\alpha)}(\mathbf{x})\,A_\alpha^i}_{\in\mathbb{N}_{\mathbb{K}}^{(\alpha)}} \tag{6.8.21}$$

and Eq. (6.8.14) take the following triangular form:

$$\begin{aligned}
\nabla^2\mathfrak{h}_i^{(0)} &= 0\\
\nabla^2\mathfrak{h}_i^{(1)} &= \mathfrak{F}_i^{(1)}\left(\mathfrak{h}^{(0)}, \nabla\mathfrak{h}^{(0)}\right)\\
\nabla^2\mathfrak{h}_i^{(2)} &= \mathfrak{F}_i^{(2)}\left(\mathfrak{h}^{(0)}, \nabla\mathfrak{h}^{(0)}, \mathfrak{h}^{(1)}, \nabla\mathfrak{h}^{(1)}\right)\\
\ldots &= \ldots\\
\nabla^2\mathfrak{h}_i^{(n)} &= \mathfrak{F}_i^{(n)}\left(\mathfrak{h}^{(0)}, \nabla\mathfrak{h}^{(0)}, \mathfrak{h}^{(1)}, \nabla\mathfrak{h}^{(1)}, \ldots, \mathfrak{h}^{(n-1)}, \nabla\mathfrak{h}^{(n-1)}\right),
\end{aligned} \tag{6.8.22}$$

where ∇^2 denotes the three-dimensional Laplacian and at each level α, by $\mathfrak{F}_i^{(\alpha)}(\ldots)$ we denote an $\mathfrak{so}(3)$ invariant polynomial of all the functions h^β up to level $\alpha - 1$ and of their derivatives.

Therefore the first ℓ_0 functions $\mathfrak{h}_i^{(0)}$ are just harmonic functions, while the higher ones satisfy Laplace equation with a source that is provided by the previously determined functions.

6.8.2 Transformation to the Solvable Gauge

Given the symmetric coset representative $\mathscr{Y}(\mathbf{x})$, parameterized by functions $\mathfrak{h}_i^{(\alpha)}(\mathbf{x})$ which satisfy the field equations (6.8.22), in order to retrieve the corresponding supergravity fields satisfying supergravity field equations, we need to solve a technical, yet quite crucial problem. We need to construct a new upper triangular coset representative:

$$\mathbb{L}(\mathscr{Y}) = \begin{pmatrix} L_{1,1}(\mathscr{Y}) & L_{1,2}(\mathscr{Y}) & \cdots & L_{1,n-1}(\mathscr{Y}) & L_{1,n}(\mathscr{Y}) \\ 0 & L_{2,2}(\mathscr{Y}) & \cdots & L_{2,n-1}(\mathscr{Y}) & L_{2,n}(\mathscr{Y}) \\ 0 & 0 & L_{3,3}(\mathscr{Y}) & \cdots & L_{3,n}(\mathscr{Y}) \\ \vdots & \cdots & 0 & \cdots & \vdots \\ 0 & 0 & \cdots & 0 & L_{3,n}(\mathscr{Y}) \end{pmatrix} \tag{6.8.23}$$

which depends algebraically on the matrix entries of $\mathscr{Y}$ and satisfies the following equivalence condition

$$\mathbb{L}(\mathscr{Y})\,\mathscr{Q}(\mathscr{Y}) = \mathscr{Y} \quad ; \quad \mathscr{Q}(\mathscr{Y}) \in \mathrm{H}^\star \tag{6.8.24}$$

where, as specified above, $\mathscr{Q}(\mathscr{Y})$ is a suitable element of the subgroup $\mathrm{H}^\star$. It should be stressed that in the existing literature, this transition from the symmetric to the solvable gauge, which is compulsory in order to make the construction of the black hole solutions explicit, has been advocated, yet it has been left to *ad hoc* procedures to be invented case by case.

Actually a universal and very elegant solution of such a problem exists and was found, from a different perspective, by the author of the present book in collaboration with A. Sorin. It was presented in [27–30, 32]. Defining the following determinants:

$$\mathfrak{D}_i(\mathscr{Y}) := \mathrm{Det}\begin{pmatrix} \mathscr{Y}_{1,1} & \cdots & \mathscr{Y}_{1,i} \\ \vdots & \vdots & \vdots \\ \mathscr{Y}_{i,1} & \cdots & \mathscr{Y}_{i,i} \end{pmatrix}, \quad \mathfrak{D}_0(\mathscr{Y}) := 1 \tag{6.8.25}$$

the matrix elements of the inverse of the upper triangular coset representative satisfying both Eqs. (6.8.23) and (6.8.24) are given by the following expressions:

$$\left(\mathbb{L}(\mathscr{Y})^{-1}\right)_{ij} \equiv \frac{1}{\sqrt{\mathfrak{D}_i(\mathscr{Y})\mathfrak{D}_{i-1}(\mathscr{Y})}}\mathrm{Det}\begin{pmatrix} \mathscr{Y}_{1,1} & \dots & \mathscr{Y}_{1,i-1} & \mathscr{Y}_{1,j} \\ \vdots & \vdots & \vdots & \vdots \\ \mathscr{Y}_{i,1} & \dots & \mathscr{Y}_{i,i-1} & \mathscr{Y}_{i,j} \end{pmatrix} \tag{6.8.26}$$

Equation (6.8.26) provides a universal non-trivial and very elegant solution to the gauge-change problem and makes the entire construction based on harmonic functions truly algorithmic from the start to the very end.

6.8.3 Extraction of the Three Dimensional Scalar Fields

The result of the procedure described in the previous section is a triangular coset representative $\mathbb{L}(\mathfrak{h}_i^{(\alpha)})$ whose entries are polynomial and square root of polynomials in the functions $\mathfrak{h}_i^{(\alpha)}(x)$. The extraction of the scalar fields $\{U(x), a(x), Z(x), \phi(x)\}$ can now be performed according to the rules already presented in [32], which we recall here in full.

The general form of the solvable coset representative in terms of the fields is the following one:

$$\mathbb{L}(\Phi) = \exp\left[-a\,L_+^E\right]\exp\left[\sqrt{2}\,Z^M\,\mathscr{W}_M\right]\mathbb{L}_4(\phi)\exp\left[U\,L_0^E\right] \tag{6.8.27}$$

where $L_0^E, L_\pm^E$ are the generators of the Ehlers group and $\mathscr{W}^M \equiv W^{1M}$ are the generators in the W-representation, according to the general structure (1.7.13) of the $\mathbb{U}_{D=3}$ Lie algebra; furthermore $\mathbb{L}_4(\phi)$ is the coset representative of the $D=4$ scalar coset manifold immersed in the $\mathrm{U}_{\mathrm{D}=3}$ group. From this structure, identifying $\mathbb{L}(\Phi) = \mathbb{L}(\mathfrak{h}_i^{(\alpha)})$ we deduce the following iterative procedure for the extraction of the relevant fields:

First of all we can determine the warp factor U by means of the following simple formula:

$$U(\mathfrak{h}) = \log\left[\tfrac{1}{2}\,\mathrm{Tr}\left(\mathbb{L}(\mathfrak{h})\,L_+^E\,\mathbb{L}^{-1}(\mathfrak{h})\,L_-^E\right)\right] \tag{6.8.28}$$

Secondly we obtain the fields ϕ_i as follows. Defining the functionals

$$\Xi_i(\mathfrak{h}) = \mathrm{Tr}\left(\mathbb{L}^{-1}(\mathfrak{h})\,T_i\,\mathbb{L}(\tau)\right) \tag{6.8.29}$$

from the form of the coset representative (6.8.27) it follows that Ξ_i depend only on the $D=4$ scalar fields and, according to the explicit form of the $D=4$ coset, one can work out the scalar fields ϕ_i.

The knowledge of U, ϕ_i allows to define:

$$\Omega(\mathfrak{h}) = \mathbb{L}(\mathfrak{h})\exp\left[-U\,L_0^E\right]\mathbb{L}_4(\phi)^{-1} \tag{6.8.30}$$

from which we extract the Z^M fields by means of the following formula:

$$Z^M(\mathfrak{h}) = \frac{1}{2\sqrt{2}} \mathrm{Tr}\left[\Omega(\mathfrak{h})\, \mathscr{W}_M^T\right] \tag{6.8.31}$$

where T means transposed. Finally the knowledge of $Z^M(\mathfrak{h})$ allows to extract the a field by means of the following trace:

$$a(\mathfrak{h}) = -\tfrac{1}{2}\mathrm{Tr}\left[\Omega(\mathfrak{h})\, \exp\left[-\sqrt{2}\, Z^M(\mathfrak{h})\, \mathscr{W}_M\right] L_+^E\right] \tag{6.8.32}$$

6.9 General Properties of the Black Hole Solutions and Structure of Their Poles

Having discussed the structure of supergravity solutions in terms of black-boxes that are a set of harmonic functions and of their descendants generated through the solution of the hierarchical equations (6.8.22), it is appropriate to study the general form of the geometries one obtains in this way and the properties of the available harmonic functions.

First of all, naming:

$$\mathfrak{W} = \exp[U(x)] \tag{6.9.1}$$

the warp factor that defines the 4-dimensional metric (6.2.11), we would like to investigate the general properties of the corresponding geometries. For the case where the Kaluza–Klein monopole is zero $\mathbf{A}^{[KK]} = 0$ we can write the general form of the curvature two-form of such spaces and therefore the intrinsic form of the Riemann tensor. Using the vielbein formalism introduced in Eq. (6.2.12) we obtain:

$$\begin{aligned}
\mathfrak{R}^{0i} &= -\,\mathfrak{W}\,\nabla^i\nabla_k\mathfrak{W}\; E^0 \wedge E^k - 2\,\nabla^i\mathfrak{W}\nabla_k\mathfrak{W}\; E^0 \wedge E^k \\
\mathfrak{R}^{ij} &= -\,2\,\mathfrak{W}\,\nabla^{[i}\nabla_k\mathfrak{W}\; E^{j]} \wedge E^k + (\nabla\mathfrak{W}\cdot\nabla\mathfrak{W})\,\nabla_k\mathfrak{W}\; E^i \wedge E^j
\end{aligned} \tag{6.9.2}$$

where the derivatives used in the above equations are defined as follows. Let the flat metric in three dimension be described by a Euclidean *dreibein* e^i such that:

$$\begin{aligned}
ds^2_{flat} &= \sum_{i=1}^{3} e^i \otimes e^i \\
E^i &= \frac{1}{\mathfrak{W}}\, e^i
\end{aligned} \tag{6.9.3}$$

then the total differential of the warp factor expanded along e^i yields the derivatives $\nabla_k\mathfrak{W}$, namely:

$$d\,\mathfrak{W} = \nabla_k\mathfrak{W}\, e^k \tag{6.9.4}$$

Next let us consider the general form of harmonic functions. These latter form a linear space since any linear combination of harmonic functions is still harmonic. There are three types of building blocks that we can use:

a Real center pole:

$$\mathscr{H}_\alpha(\mathbf{x}) = \frac{1}{|\mathbf{x} - \mathbf{x}_\alpha|} \tag{6.9.5}$$

b Real part of an imaginary center pole:

$$\mathscr{R}_\alpha(\mathbf{x}) = \mathrm{Re}\left[\frac{1}{|\mathbf{x} - \mathrm{i}\,\mathbf{x}_\alpha|}\right] \tag{6.9.6}$$

c Imaginary part of an imaginary center pole:

$$\mathscr{I}_\alpha(\mathbf{x}) = \mathrm{Im}\left[\frac{1}{|\mathbf{x} - \mathrm{i}\,\mathbf{x}_\alpha|}\right] \tag{6.9.7}$$

Hence the most general harmonic function can be written as the following sum:

$$\mathrm{Harm}(\mathbf{x}) = h_\infty + \sum_\alpha \frac{p_\alpha}{|\mathbf{x} - \mathbf{x}_\alpha|} + \sum_\beta q_\beta \,\mathrm{Re}\left[\frac{1}{|\mathbf{x} - \mathrm{i}\,\mathbf{x}_\beta|}\right] + \sum_\gamma k_\gamma \,\mathrm{Im}\left[\frac{1}{|\mathbf{x} - \mathrm{i}\,\mathbf{x}_\gamma|}\right] \tag{6.9.8}$$

where the constant h_∞ is the boundary value of the harmonic function at infinity far from all the poles. In order to study the behavior of $\mathrm{Harm}(\mathbf{x})$ in the vicinity of a real pole ($|\mathbf{x} - \mathbf{x}_\alpha| << 1$) it is convenient to adopt local polar coordinates:

$$\begin{aligned} x^1 - x^1_\alpha &= r \cos\theta \\ x^2 - x^2_\alpha &= r \sin\theta \sin\phi \\ x^3 - x^3_\alpha &= r \sin\theta \cos\phi \end{aligned} \tag{6.9.9}$$

In this coordinates the harmonic function is approximated by:

$$\mathrm{Harm}(\mathbf{x}) \simeq h_\alpha + \frac{p_\alpha}{r} \tag{6.9.10}$$

where the effective constant h_α encodes the finite part of the function contributed by all the other poles. In polar coordinates the Laplacian operator on functions of r becomes:

$$\Delta = \frac{d^2}{dr^2} + \frac{2}{r}\frac{d}{dr} \tag{6.9.11}$$

The general outcome of the construction procedure outlined in the previous section is that the warp factor is the square root of a rational function of n harmonic functions, where $n = \dim \mathbb{N}_{\mathbb{K}}$

$$\mathfrak{W}(\mathbf{x}) = \sqrt{\frac{\mathbb{P}\left(\widehat{\text{Harm}}_1(\mathbf{x}), \ldots, \widehat{\text{Harm}}_n(\mathbf{x})\right)}{\mathbb{Q}\left(\widehat{\text{Harm}}_1(\mathbf{x}), \ldots, \widehat{\text{Harm}}_n(\mathbf{x})\right)}} \tag{6.9.12}$$

where $\mathbb{P}$ and $\mathbb{Q}$ are two polynomials. By $\widehat{\text{Harm}}_1(\mathbf{x})$ we denote both harmonic functions and their descendants generated by the hierarchical system (6.8.22). For a given multicenter solution it is convenient to enumerate all the poles displayed by one or the other of the harmonic functions and in the vicinity of each of those poles we will have:

$$\widehat{\text{Harm}}_i(\mathbf{x}) \simeq \frac{p_i}{r^{m_i}} \tag{6.9.13}$$

where $p_i \neq 0$ if the considered pole belongs to the considered function and it is zero otherwise. Furthermore if $\text{Harm}_i(\mathbf{x})$ is one of the level one harmonic function the exponent $m_i = 1$. Otherwise it is bigger, but in any case $m_i \geq 1$. Taking this into account the effective behavior of the warp factor will always be of the following form:

$$\mathfrak{W}(\mathbf{x}) \simeq r^{\ell_\alpha} \sqrt{c_\alpha} \tag{6.9.14}$$

where ℓ is some integer or half integer power (positive or negative) and c_α is a constant. In order for the pole to be a regular point of the solution, two conditions have to be satisfied:

1. The constant $c_\alpha > 0$ must be positive so that the warp factor is real.
2. The power $\ell_\alpha \geq 1$ so that the Riemann tensor does not diverge at the pole.

The second condition follows from the form (6.9.2) of the Riemann tensor which implies that all of its components behave as:

$$\mathfrak{R}^{ab}_{cd} \simeq r^{2\ell_\alpha - 2} \times \text{const} \tag{6.9.15}$$

Near the pole the metric behaves as follows:

$$ds^2 \simeq -\sqrt{c_\alpha}\, r^{\ell_\alpha} dt^2 + \frac{1}{\sqrt{c_\alpha}} \frac{1}{r^{\ell_\alpha}} \left[dr^2 + r^2 \left(d\theta^2 + \sin^2\theta\, d\phi^2\right)\right] \tag{6.9.16}$$

In order for the pole to be an event horizon of finite or of vanishing area, we must have $2 - \ell_\alpha > 0$, so that the volume of the two-sphere described by $\left(d\theta^2 + \sin^2\theta\, d\phi^2\right)$ does not diverge. Hence for regular black holes we have only three possibilities:

$$\underbrace{\ell_\alpha = 2}_{\text{Large Black Holes}} \quad ; \quad \underbrace{\ell_\alpha = \tfrac{3}{2}}_{\text{Small Black Holes}} \quad ; \quad \underbrace{\ell_\alpha = 1}_{\text{Very Small Black Holes}} \tag{6.9.17}$$

When we are in the case of Large Black Holes, the near horizon geometry is approximated by that:

$$\text{AdS}_2 \times \mathbb{S}^2 \tag{6.9.18}$$

The case of the harmonic functions with an imaginary center requires a different treatment. Their near singularity behavior is best analyzed by using spheroidal coordinates.

These are easily introduced by setting:

$$\begin{aligned} x^1 &= \sqrt{r^2+\alpha^2}\,\sin\theta\,\sin\phi \\ x^2 &= \sqrt{r^2+\alpha^2}\,\sin\theta\,\cos\phi \\ x^3 &= r\cos\theta \end{aligned} \tag{6.9.19}$$

where r, θ, ϕ are the new coordinates and α is a deformation parameter which represents the position of the center in the complex plane. In terms of these coordinates the flat Euclidean three-dimensional metric takes the following form:

$$\begin{aligned} ds^2_{\mathbb{E}^3} = d\Omega^2_{spheroidal} \equiv & \frac{\left(r^2+\alpha^2\cos^2\theta\right)dr^2}{r^2+\alpha^2} + \left(r^2+\alpha^2\right)\sin^2\theta\,d\phi^2 \\ & + \left(r^2+\alpha^2\cos^2\theta\right)\,d\theta^2 \end{aligned} \tag{6.9.20}$$

and the two harmonic functions that correspond to the real and imaginary part of a complex harmonic function with center on the imaginary z-axis at α-distance from zero are:

$$\mathscr{P}_\alpha(r,\theta) = \frac{r}{r^2+\alpha^2\cos^2\theta} \tag{6.9.21}$$

$$\mathscr{R}_\alpha(r,\theta) = \frac{\alpha\cos\theta}{r^2+\alpha^2\cos^2\theta} \tag{6.9.22}$$

and the Hodge duals of their gradients, in spheroidal coordinates have the following form:

$$\begin{aligned} \star\,\nabla\mathscr{P}_\alpha = & \frac{\sin\theta}{\left(r^2+\alpha^2\cos^2\theta\right)^2}\left[2\alpha^2 r\cos\theta\sin\theta\,dr\wedge d\phi\right. \\ & \left. + \left(r^2+\alpha^2\right)\left(r^2-\alpha^2\cos^2\theta\right)d\theta\wedge d\phi\right] \end{aligned} \tag{6.9.23}$$

$$\begin{aligned} \star\nabla\mathscr{R}_\alpha = & \frac{\alpha\sin\theta}{\left(r^2+\alpha^2\cos^2\theta\right)^2}\left[\left(\alpha^2\cos^2\theta-r^2\right)\sin\theta dr\wedge d\phi\right. \\ & \left. +2r\left(r^2+\alpha^2\right)\cos\theta d\theta\wedge d\phi\right] \end{aligned} \tag{6.9.24}$$

These are the building blocks we can use to construct Kerr–Newman like solutions and we shall outline a pair of examples in the sequel.

6.10 The Example of the S^3 Model: Classification of the Nilpotent Orbits

As an illustration of the general procedure we explore the case of the S^3 model, leading to the $\mathrm{G}_{2,2}$ group in $D = 3$. The detailed classification of the nilpotent orbits pertaining to this case was derived in Sect. 6.7. According to it, for the case of the coset manifold[12]:

$$\frac{\mathrm{U}_{\mathrm{D=3}}}{\mathrm{H}^\star} = \frac{\mathrm{G}_{(2,2)}}{\widehat{\mathrm{SL}(2,\mathbb{R})} \times \mathrm{SL}(2,\mathbb{R})_{\mathrm{h}^\star}} \tag{6.10.1}$$

there just seven distinct nilpotent orbits of the $\mathrm{H}^\star = \widehat{\mathrm{SL}(2,\mathbb{R})} \times \mathrm{SL}(2,\mathbb{R})_{\mathrm{h}^\star}$ subgroup in the $\mathbb{K}^\star$ representation $\left(2, \frac{3}{2}\right)$, which are enumerated by the three set of labels $\alpha\beta\gamma$ and are denoted $\mathscr{O}^{\alpha}_{\beta\gamma}$ as described in Table 6.1. An explicit choice of a representative for each of the seven orbits is provided below.

$$\mathscr{O}^1_{11} = \begin{pmatrix} \sqrt{\frac{3}{2}} & \frac{\sqrt{\frac{5}{2}}}{2} & \sqrt{\frac{3}{2}} & \frac{\sqrt{5}}{2} & 0 & \frac{\sqrt{\frac{5}{2}}}{2} & 0 \\ \frac{\sqrt{\frac{5}{2}}}{2} & \sqrt{6} & -\frac{\sqrt{\frac{5}{2}}}{2} & -\sqrt{3} & -\frac{\sqrt{\frac{5}{2}}}{2} & 0 & \frac{\sqrt{\frac{5}{2}}}{2} \\ -\sqrt{\frac{3}{2}} & \frac{\sqrt{\frac{5}{2}}}{2} & -\sqrt{\frac{3}{2}} & \frac{\sqrt{5}}{2} & 0 & \frac{\sqrt{\frac{5}{2}}}{2} & 0 \\ -\frac{\sqrt{5}}{2} & \sqrt{3} & \frac{\sqrt{5}}{2} & 0 & \frac{\sqrt{5}}{2} & -\sqrt{3} & -\frac{\sqrt{5}}{2} \\ 0 & \frac{\sqrt{\frac{5}{2}}}{2} & 0 & \frac{\sqrt{5}}{2} & \sqrt{\frac{3}{2}} & \frac{\sqrt{\frac{5}{2}}}{2} & \sqrt{\frac{3}{2}} \\ \frac{\sqrt{\frac{5}{2}}}{2} & 0 & -\frac{\sqrt{\frac{5}{2}}}{2} & \sqrt{3} & -\frac{\sqrt{\frac{5}{2}}}{2} & -\sqrt{6} & \frac{\sqrt{\frac{5}{2}}}{2} \\ 0 & \frac{\sqrt{\frac{5}{2}}}{2} & 0 & \frac{\sqrt{5}}{2} & -\sqrt{\frac{3}{2}} & \frac{\sqrt{\frac{5}{2}}}{2} & -\sqrt{\frac{3}{2}} \end{pmatrix} \tag{6.10.2}$$

$$\mathscr{O}^4_{11} = \begin{pmatrix} \frac{1}{2} & 0 & 0 & 0 & \frac{1}{2} & 0 & 0 \\ 0 & 0 & 0 & 0 & 0 & 0 & 0 \\ 0 & 0 & \frac{1}{2} & 0 & 0 & 0 & \frac{1}{2} \\ 0 & 0 & 0 & 0 & 0 & 0 & 0 \\ -\frac{1}{2} & 0 & 0 & 0 & -\frac{1}{2} & 0 & 0 \\ 0 & 0 & 0 & 0 & 0 & 0 & 0 \\ 0 & 0 & -\frac{1}{2} & 0 & 0 & 0 & -\frac{1}{2} \end{pmatrix} \tag{6.10.3}$$

$$\mathscr{O}^2_{11} = \begin{pmatrix} \frac{1}{2} & 0 & -\frac{1}{2} & 0 & 0 & 0 & 0 \\ 0 & 1 & 0 & \frac{1}{\sqrt{2}} & 0 & 0 & 0 \\ \frac{1}{2} & 0 & -\frac{1}{2} & 0 & 0 & 0 & 0 \\ 0 & -\frac{1}{\sqrt{2}} & 0 & 0 & 0 & \frac{1}{\sqrt{2}} & 0 \\ 0 & 0 & 0 & 0 & \frac{1}{2} & 0 & -\frac{1}{2} \\ 0 & 0 & 0 & -\frac{1}{\sqrt{2}} & 0 & -1 & 0 \\ 0 & 0 & 0 & 0 & \frac{1}{2} & 0 & -\frac{1}{2} \end{pmatrix} \tag{6.10.4}$$

[12]For the rationale of our notation we refer the reader to previous Sect. 5.8.

$$\mathscr{O}^3_{11} = \begin{pmatrix} 1 & 0 & 0 & -\frac{1}{\sqrt{2}} & 0 & 0 & 0 \\ 0 & 1 & -\frac{1}{2} & 0 & \frac{1}{2} & 0 & 0 \\ 0 & \frac{1}{2} & 0 & 0 & 0 & -\frac{1}{2} & 0 \\ \frac{1}{\sqrt{2}} & 0 & 0 & 0 & 0 & 0 & \frac{1}{\sqrt{2}} \\ 0 & -\frac{1}{2} & 0 & 0 & 0 & \frac{1}{2} & 0 \\ 0 & 0 & \frac{1}{2} & 0 & -\frac{1}{2} & -1 & 0 \\ 0 & 0 & 0 & -\frac{1}{\sqrt{2}} & 0 & 0 & -1 \end{pmatrix} \tag{6.10.5}$$

$$\mathscr{O}^3_{22} = \begin{pmatrix} 1 & 0 & 0 & \frac{1}{\sqrt{2}} & 0 & 0 & 0 \\ 0 & 1 & -\frac{1}{2} & 0 & -\frac{1}{2} & 0 & 0 \\ 0 & \frac{1}{2} & 0 & 0 & 0 & \frac{1}{2} & 0 \\ -\frac{1}{\sqrt{2}} & 0 & 0 & 0 & 0 & 0 & -\frac{1}{\sqrt{2}} \\ 0 & \frac{1}{2} & 0 & 0 & 0 & \frac{1}{2} & 0 \\ 0 & 0 & -\frac{1}{2} & 0 & -\frac{1}{2} & -1 & 0 \\ 0 & 0 & 0 & \frac{1}{\sqrt{2}} & 0 & 0 & -1 \end{pmatrix} \tag{6.10.6}$$

$$\mathscr{O}^3_{21} = \begin{pmatrix} -1 & 0 & 0 & \frac{1}{\sqrt{2}} & 0 & 0 & 0 \\ 0 & 0 & -\frac{1}{2} & 0 & -\frac{1}{2} & 0 & 0 \\ 0 & \frac{1}{2} & -1 & 0 & 0 & \frac{1}{2} & 0 \\ -\frac{1}{\sqrt{2}} & 0 & 0 & 0 & 0 & 0 & -\frac{1}{\sqrt{2}} \\ 0 & \frac{1}{2} & 0 & 0 & 1 & \frac{1}{2} & 0 \\ 0 & 0 & -\frac{1}{2} & 0 & -\frac{1}{2} & 0 & 0 \\ 0 & 0 & 0 & \frac{1}{\sqrt{2}} & 0 & 0 & 1 \end{pmatrix} \tag{6.10.7}$$

$$\mathscr{O}^3_{12} = \begin{pmatrix} -1 & 0 & 0 & -\frac{1}{\sqrt{2}} & 0 & 0 & 0 \\ 0 & 0 & -\frac{1}{2} & 0 & \frac{1}{2} & 0 & 0 \\ 0 & \frac{1}{2} & -1 & 0 & 0 & -\frac{1}{2} & 0 \\ \frac{1}{\sqrt{2}} & 0 & 0 & 0 & 0 & 0 & \frac{1}{\sqrt{2}} \\ 0 & -\frac{1}{2} & 0 & 0 & 1 & \frac{1}{2} & 0 \\ 0 & 0 & \frac{1}{2} & 0 & -\frac{1}{2} & 0 & 0 \\ 0 & 0 & 0 & -\frac{1}{\sqrt{2}} & 0 & 0 & 1 \end{pmatrix} \tag{6.10.8}$$

Each orbit representative $\mathscr{O}^\alpha_{\beta\gamma}$ identifies a standard triple $\{h, X, Y\}$ and hence an embedding of an $\mathfrak{sl}(2, \mathbb{R})$ Lie algebra:

$$[h, X] = 2X \quad ; \quad [h, Y] = -2Y \quad ; \quad [X, Y] = 2h \tag{6.10.9}$$

into $\mathfrak{g}_{(2,2)}$ in such a way that $h \in \mathbb{H}^\star$ and $X, Y \in \mathbb{K}^\star$. The triple is obtained by setting:

$$X_{\alpha|\beta\gamma} \equiv \mathscr{O}^{\alpha}_{\beta\gamma} \quad ; \quad Y_{\alpha|\beta\gamma} \equiv X^{T}_{\alpha|\beta\gamma} \quad ; \quad h_{\alpha|\beta\gamma} \equiv \left[X_{\alpha|\beta\gamma}, Y_{\alpha|\beta\gamma}\right] \tag{6.10.10}$$

The relevant item in the construction of solutions based on the integration of equations in the symmetric gauge is provided by the central element of the triple $h_{\alpha|\beta\gamma}$ which defines the gradings. In the present example of the S^3 model, it turns out the orbits having the same α and γ labels but different β-labels have the same central element, namely:

$$h_{\alpha|\beta\gamma} = h_{\alpha|\beta'\gamma} \tag{6.10.11}$$

so that the solutions pertaining both to orbit $\mathscr{O}^{\alpha}_{\beta\gamma}$ and to orbit $\mathscr{O}^{\alpha}_{\beta'\gamma}$ are obtained from the same construction and are distinguished only by different choices in the space of the available harmonic functions parameterizing the general solution.

The explicit form of the central elements are the following ones:

Large Orbit $\mathscr{O}^{1}_{11}$: Central Element

$$h_{1|11} = \begin{pmatrix} 0 & 0 & -1 & 0 & 5 & 0 & 0 \\ 0 & 0 & 0 & \sqrt{2} & 0 & 0 & 0 \\ -1 & 0 & 0 & 0 & 0 & 0 & 5 \\ 0 & \sqrt{2} & 0 & 0 & 0 & \sqrt{2} & 0 \\ 5 & 0 & 0 & 0 & 0 & 0 & -1 \\ 0 & 0 & 0 & \sqrt{2} & 0 & 0 & 0 \\ 0 & 0 & 5 & 0 & -1 & 0 & 0 \end{pmatrix}$$

$$\text{Eigenvalues}\left[\tfrac{1}{2}\, h_{1|11}\right] = \{-3, 3, -2, 2, -1, 1, 0\} \tag{6.10.12}$$

Very Small Orbit $\mathscr{O}^{4}_{11}$: Central Element

$$h_{4|11} = \begin{pmatrix} 0 & 0 & 0 & 0 & -1 & 0 & 0 \\ 0 & 0 & 0 & 0 & 0 & 0 & 0 \\ 0 & 0 & 0 & 0 & 0 & 0 & -1 \\ 0 & 0 & 0 & 0 & 0 & 0 & 0 \\ -1 & 0 & 0 & 0 & 0 & 0 & 0 \\ 0 & 0 & 0 & 0 & 0 & 0 & 0 \\ 0 & 0 & -1 & 0 & 0 & 0 & 0 \end{pmatrix}$$

$$\text{Eigenvalues}\left[\tfrac{1}{2}\, h_{4|11}\right] = \left\{-\frac{1}{2}, -\frac{1}{2}, \frac{1}{2}, \frac{1}{2}, 0, 0, 0\right\} \tag{6.10.13}$$

Small Orbit $\mathscr{O}^2_{11}$: Central Element

$$h_{2|11} = \begin{pmatrix} 0 & 0 & 1 & 0 & 0 & 0 & 0 \\ 0 & 0 & 0 & -\sqrt{2} & 0 & 0 & 0 \\ 1 & 0 & 0 & 0 & 0 & 0 & 0 \\ 0 & -\sqrt{2} & 0 & 0 & 0 & -\sqrt{2} & 0 \\ 0 & 0 & 0 & 0 & 0 & 0 & 1 \\ 0 & 0 & 0 & -\sqrt{2} & 0 & 0 & 0 \\ 0 & 0 & 0 & 0 & 1 & 0 & 0 \end{pmatrix}$$

$$\text{Eigenvalues}\left[\tfrac{1}{2}\, h_{2|11}\right] = \left\{-1, 1, -\frac{1}{2}, -\frac{1}{2}, \frac{1}{2}, \frac{1}{2}, 0\right\} \tag{6.10.14}$$

Large BPS Orbit $\mathscr{O}^3_{11}$: Central Element

$$h_{3|11} = h_{3|21} = \begin{pmatrix} 0 & 0 & 0 & \sqrt{2} & 0 & 0 & 0 \\ 0 & 0 & 1 & 0 & -1 & 0 & 0 \\ 0 & 1 & 0 & 0 & 0 & 1 & 0 \\ \sqrt{2} & 0 & 0 & 0 & 0 & 0 & -\sqrt{2} \\ 0 & -1 & 0 & 0 & 0 & -1 & 0 \\ 0 & 0 & 1 & 0 & -1 & 0 & 0 \\ 0 & 0 & 0 & -\sqrt{2} & 0 & 0 & 0 \end{pmatrix}$$

$$\text{Eigenvalues}\left[\tfrac{1}{2}\, h_{3|11}\right] = \{-1, -1, 1, 1, 0, 0, 0\} \tag{6.10.15}$$

Large Non BPS Orbit $\mathscr{O}^3_{22}$: Central Element

$$h_{3|12} = h_{3|22} = \begin{pmatrix} 0 & 0 & 0 & -\sqrt{2} & 0 & 0 & 0 \\ 0 & 0 & 1 & 0 & 1 & 0 & 0 \\ 0 & 1 & 0 & 0 & 0 & -1 & 0 \\ -\sqrt{2} & 0 & 0 & 0 & 0 & 0 & \sqrt{2} \\ 0 & 1 & 0 & 0 & 0 & -1 & 0 \\ 0 & 0 & -1 & 0 & -1 & 0 & 0 \\ 0 & 0 & 0 & \sqrt{2} & 0 & 0 & 0 \end{pmatrix}$$

$$\text{Eigenvalues}\left[\tfrac{1}{2}\, h_{3|22}\right] = \{-1, -1, 1, 1, 0, 0, 0\} \tag{6.10.16}$$

6.11 Explicit Construction of the Multicenter Black Holes Solutions of the S^3 Model

Having enumerated the central elements for the independent orbits we proceed to the construction and discussion of the corresponding black hole solutions, whose properties are summarized in Table 6.2.

Table 6.2 Properties of the $\mathfrak{g}_{(2,2)}$ orbits in the S^3 model. The structure of the electromagnetic charge vector is that obtained for solutions with vanishing Taub-NUT current. The symbol $\triangleright$ is meant to denote semidirect product. $\mathscr{S}_{\mathbf{W}}$ denotes the subgroup of the $D = 4$ duality group which leaves the charge vector invariant, while $\mathfrak{S}_{\mathrm{H}^\star}$ denotes the subgroup of the $\mathrm{H}^\star$ isotropy group of the $D = 3$ sigma-model which leaves invariant the X element of the standard triple. This latter is the Lax operator in the one-dimensional spherical symmetric approach

Name of orbit	pq charges	Quart. Inv. $\mathfrak{I}_4$	$\mathbf{W}$ – stab. group $\mathscr{S}_{\mathbf{W}} \subset \mathfrak{sl}(2,\mathbb{R})$	$\mathrm{H}^\star$ – stab. group $\mathfrak{S}_{\mathrm{H}^\star} \subset \widehat{\mathfrak{sl}(2,\mathbb{R}) \oplus \mathfrak{sl}(2,\mathbb{R})_{\mathrm{h}^\star}}$	dim $\mathbb{N}$	dim $\mathbb{N} \bigcap \mathbb{K}^\star$
$\mathscr{O}^4_{11}$	$\begin{pmatrix} 0 \\ 0 \\ 0 \\ q \end{pmatrix}$	0	$\begin{pmatrix} 1 & 0 \\ c & 1 \end{pmatrix}$	$\underbrace{\mathrm{ISO}(1,1)}_{3\text{ gen.}}$	3	3
$\mathscr{O}^2_{11}$	$\begin{pmatrix} \sqrt{3}\,p \\ 0 \\ 0 \\ 0 \end{pmatrix}$	0	$\mathbf{1}$	$\underbrace{\mathrm{SO}(1,1) \triangleright \mathbb{R}}_{2\text{ gen.}}$	4	3
$\mathscr{O}^3_{11}$	$\begin{pmatrix} 0 \\ p \\ -\sqrt{3}q \\ 0 \end{pmatrix}$	$9\,p\,q^3 > 0$	$\mathbb{Z}_3$	$\underbrace{\mathbb{R}}_{1\text{ gen.}}$ $A^2 = 0$	5	4
$\mathscr{O}^3_{22}$	$\begin{pmatrix} 0 \\ p \\ \sqrt{3}q \\ 0 \end{pmatrix}$	$-9\,p\,q^3 < 0$	$\mathbf{1}$	$\underbrace{\mathbb{R}}_{1\text{ gen.}}$ $A^3 = 0$	3	3
$\mathscr{O}^1_{11}$	$\begin{pmatrix} \frac{1}{2}\sqrt{\frac{3}{2}}p \\ 0 \\ \frac{7}{6}p \\ \sqrt{2}q \end{pmatrix}$	$\frac{1}{128}p^3 \times (49p + 72q)$	$\mathbf{1}$	$\mathbf{1}$	6	4

6.11.1 The Very Small Black Holes of $\mathscr{O}^4_{11}$

We begin with the smallest orbits which, in a sense that will become clear further on, represent the elementary blocks in terms of which bigger black holes are constructed.

Focusing on any orbit $\mathscr{O}^\alpha_{\beta\gamma}$ and considering the nilpotent element of the corresponding triple $X_{\alpha|\beta\gamma} \in \mathbb{K}^\star$ as a Lax operator L_0, we easily workout the electromagnetic charges by calculating the traces displayed below (see Sect. 5.9, for more explanations)

$$\mathscr{Q}^{\mathbf{w}} = \mathrm{Tr}(X_{\alpha|\beta\gamma}\mathscr{T}^{\mathbf{w}}) \tag{6.11.1}$$

W-Representation

In the case of the orbit $\mathcal{O}^4_{11}$ we obtain:

$$\mathcal{Q}^{\mathbf{W}}_{4|11} = (0, 0, 0, 1) \tag{6.11.2}$$

Substituting such a result in the expression for the quartic symplectic invariant (see [32]):

$$\mathfrak{I}_4 = \frac{1}{4}\left(4\sqrt{3}Q_4Q_1^3 + 3Q_3^2Q_1^2 - 18Q_2Q_3Q_4Q_1 - Q_2\left(4\sqrt{3}Q_3^3 + 9Q_2Q_4^2\right)\right) \tag{6.11.3}$$

of the **W** representation which happens to be the spin $\frac{3}{2}$ of $\mathfrak{sl}(2, \mathbb{R})$ we find:

$$\mathfrak{I}_4 = 0 \tag{6.11.4}$$

The result is meaningful since, by calculating the trace $\mathrm{Tr}(X_{4|11}L^E_+) = 0$, we can also check that the Taub-NUT charge vanishes. We can also address the question whether there are subgroups of the original duality group in four-dimensions SL(2, R) that leave the charge vector (6.11.2) invariant. Using the explicit form of the $j = \frac{3}{2}$ representation displayed in Eq. (3.13) of [32], we realize that indeed such group exists and it is the parabolic subgroup described below:

$$\forall c \in \mathbb{R} \quad : \quad \begin{pmatrix} 1 & 0 \\ c & 1 \end{pmatrix} \in \mathcal{S}_{4|11} \subset \mathrm{SL}(2, \mathbb{R}) \tag{6.11.5}$$

This stability subgroup together with the vanishing of the quartic invariant are the intrinsic definition of the **W**-orbit pertaining to very small black holes.

H*-Stability Subgroup

In a parallel way we can pose the question what is the stability subgroup of the nilpotent element $X_{4|11}$ in $\mathrm{H}^\star = \widehat{\mathfrak{sl}(2, \mathbb{R})} \oplus \mathfrak{sl}(2, \mathbb{R})_{\mathrm{h}^\star}$ (For further explanations on H* and its structure see Sect. 5.8). The answer is the following:

$$\mathfrak{S}_{4|11} = \mathrm{ISO}(1, 1) \tag{6.11.6}$$

A generic element of the corresponding Lie algebra is a linear combination of three generators J, T_1, T_2, satisfying the commutation relations:

$$\begin{aligned} [J\,,\,T_1] &= \frac{1}{\sqrt{2}}T_1 + \frac{3}{2\sqrt{6}}T_2 \\ [J\,,\,T_2] &= \frac{3}{2\sqrt{2}}T_1 \quad ; \quad [T_1\,,\,T_2] = 0 \end{aligned} \tag{6.11.7}$$

It is explicitly given by the following matrix:

$$\omega J + x T_1 + y T_2 = \begin{pmatrix} 0 & -\frac{x}{2\sqrt{2}} & \frac{\omega}{2\sqrt{2}} & -\frac{x}{2} & 0 & -\frac{1}{2}\sqrt{\frac{3}{2}}y & 0 \\ \frac{x}{2\sqrt{2}} & 0 & -\frac{1}{2}\sqrt{\frac{3}{2}}y & -\frac{\omega}{2} & \frac{x}{2\sqrt{2}} & 0 & -\frac{1}{2}\sqrt{\frac{3}{2}}y \\ \frac{\omega}{2\sqrt{2}} & -\frac{1}{2}\sqrt{\frac{3}{2}}y & 0 & -\frac{x}{2} & 0 & -\frac{x}{2\sqrt{2}} & 0 \\ -\frac{x}{2} & -\frac{\omega}{2} & \frac{x}{2} & 0 & -\frac{x}{2} & -\frac{\omega}{2} & \frac{x}{2} \\ 0 & \frac{x}{2\sqrt{2}} & 0 & \frac{x}{2} & 0 & \frac{1}{2}\sqrt{\frac{3}{2}}y & \frac{\omega}{2\sqrt{2}} \\ \frac{1}{2}\sqrt{\frac{3}{2}}y & 0 & -\frac{x}{2\sqrt{2}} & -\frac{\omega}{2} & \frac{1}{2}\sqrt{\frac{3}{2}}y & 0 & -\frac{x}{2\sqrt{2}} \\ 0 & \frac{1}{2}\sqrt{\frac{3}{2}}y & 0 & \frac{x}{2} & \frac{\omega}{2\sqrt{2}} & \frac{x}{2\sqrt{2}} & 0 \end{pmatrix} \tag{6.11.8}$$

Nilpotent Algebra $\mathbb{N}_{4|11}$

Considering next the adjoint action of the central element $h_{4|11}$ on the subspace $\mathbb{K}^\star$ we find that its eigenvalues are the following ones:

$$\text{Eigenvalues}_{4|11}^{\mathbb{K}^\star} = \{-2, 2, -1, -1, 1, 1, 0, 0\} \tag{6.11.9}$$

Therefore the three eigenoperators A_1, A_2, A_3 corresponding to the positive eigenvalues 2, 1, 1, respectively, form the restriction to $\mathbb{K}^\star$ of a nilpotent algebra $\mathbb{N}_{4|11}$. In this case A_i commute among themselves so that $\mathbb{N}_{4|11} = \mathbb{N}_{4|11} \bigcap \mathbb{K}^\star$ and it is abelian. This structure of the nilpotent algebra implies that for the orbit $\mathscr{O}_{11}^4$ we have only three functions $\mathfrak{h}_i^0$ which will be harmonic and independent.

Explicitly we set:

$$\mathfrak{H}(\mathfrak{h}_1, \mathfrak{h}_2, \mathfrak{h}_3) = \sum_{i=1}^{3} \mathfrak{h}_i A_i = \begin{pmatrix} -\mathfrak{h}_1 & \mathfrak{h}_3 & 0 & -\sqrt{2}\mathfrak{h}_3 & -\mathfrak{h}_1 & -\mathfrak{h}_2 & 0 \\ \mathfrak{h}_3 & 0 & -\mathfrak{h}_2 & 0 & \mathfrak{h}_3 & 0 & -\mathfrak{h}_2 \\ 0 & \mathfrak{h}_2 & -\mathfrak{h}_1 & \sqrt{2}\mathfrak{h}_3 & 0 & -\mathfrak{h}_3 & -\mathfrak{h}_1 \\ \sqrt{2}\mathfrak{h}_3 & 0 & \sqrt{2}\mathfrak{h}_3 & 0 & \sqrt{2}\mathfrak{h}_3 & 0 & \sqrt{2}\mathfrak{h}_3 \\ \mathfrak{h}_1 & -\mathfrak{h}_3 & 0 & \sqrt{2}\mathfrak{h}_3 & \mathfrak{h}_1 & \mathfrak{h}_2 & 0 \\ -\mathfrak{h}_2 & 0 & \mathfrak{h}_3 & 0 & -\mathfrak{h}_2 & 0 & \mathfrak{h}_3 \\ 0 & -\mathfrak{h}_2 & \mathfrak{h}_1 & -\sqrt{2}\mathfrak{h}_3 & 0 & \mathfrak{h}_3 & \mathfrak{h}_1 \end{pmatrix} \tag{6.11.10}$$

Considering $\mathfrak{H}(\mathfrak{h}_1, \mathfrak{h}_2, \mathfrak{h}_3)$ as a Lax operator and calculating its Taub-NUT charge and electromagnetic charges we find:

$$\mathbf{n}_{TN} = -2\,\mathfrak{h}_2 \quad ; \quad \mathscr{Q} = \left(0\,,\, 2\mathfrak{h}_2\,,\, -2\sqrt{3}\mathfrak{h}_3\,,\, -2\mathfrak{h}_1\right) \tag{6.11.11}$$

This implies that constructing the multi-centre solution with harmonic functions the condition $\mathfrak{h}_2 = 0$ should be sufficient to annihilate the Taub-NUT current.

For later convenience let us change the normalization in the basis of harmonic functions as follows:

$$\mathfrak{h}_1^{(0)} = \frac{1}{\sqrt{2}}\mathscr{H}_1 \quad ; \quad \mathfrak{h}_2^{(0)} = \frac{1}{2}(1 - \mathscr{H}_2) \quad ; \quad \mathfrak{h}_3^{(0)} = \frac{1}{\sqrt{2}}\mathscr{H}_3 \tag{6.11.12}$$

Implementing the symmetric coset construction with:

$$\mathcal{Y}\left(\mathcal{H}_1, \mathcal{H}_2, \mathcal{H}_3\right) \equiv \exp\left[\mathfrak{H}\left(\frac{1}{\sqrt{2}}\mathcal{H}_1, \frac{1}{2}\left(1 - \mathcal{H}_2\right), \frac{1}{\sqrt{2}}\mathcal{H}_3\right)\right] \tag{6.11.13}$$

and calculating the upper triangular coset representative $\mathbb{L}(\mathcal{Y})$ according to Eq. (6.8.26) we find a relatively simple expression which, however, is still too large to be displayed. Yet the extraction of the σ-model scalar fields produces a quite compact answer which we list below:

$$\exp[-U] = \sqrt{\mathcal{H}_2^2 - 3\mathcal{H}_3^2 + \mathcal{H}_1} \tag{6.11.14}$$

$$\operatorname{Im} z = \frac{\sqrt{\mathcal{H}_2^2 - 3\mathcal{H}_3^2 + \mathcal{H}_1}}{\mathcal{H}_2^2 - \mathcal{H}_3^2 + \mathcal{H}_1} \tag{6.11.15}$$

$$\operatorname{Re} z = -\frac{\sqrt{2}\mathcal{H}_3}{\mathcal{H}_2^2 - \mathcal{H}_3^2 + \mathcal{H}_1} \tag{6.11.16}$$

$$Z^M = \begin{pmatrix} \frac{\sqrt{6}\mathcal{H}_3^2}{\mathcal{H}_2^2-3\mathcal{H}_3^2+\mathcal{H}_1} \\ \frac{(\mathcal{H}_2-2\mathcal{H}_3)(\mathcal{H}_2+\mathcal{H}_3)^2+\mathcal{H}_1\mathcal{H}_2}{\sqrt{(\mathcal{H}_2^2-3\mathcal{H}_3^2+\mathcal{H}_1)^2}} \\ -\frac{\sqrt{3}\mathcal{H}_3}{\mathcal{H}_2^2-3\mathcal{H}_3^2+\mathcal{H}_1} \\ \frac{\mathcal{H}_2^2-3\mathcal{H}_3^2+\mathcal{H}_1-1}{\sqrt{2}(\mathcal{H}_2^2-3\mathcal{H}_3^2+\mathcal{H}_1)} \end{pmatrix} \tag{6.11.17}$$

$$a = \frac{\mathcal{H}_2^3 + \left(-3\mathcal{H}_3^2 + \mathcal{H}_1 + 1\right)\mathcal{H}_2 - 2\mathcal{H}_3^3}{\sqrt{2}\left(\mathcal{H}_2^2 - 3\mathcal{H}_3^2 + \mathcal{H}_1\right)} \tag{6.11.18}$$

The Taub-NUT Current

Given this explicit result we can turn to the explicit oxidation formulae described in Sect. 6.2.1 and calculate the Taub-NUT current which is the integrand of Eq. (6.2.17). We find:

$$j^{TN} = \sqrt{2}\,{}^\star\nabla\,\mathcal{H}_2 \tag{6.11.19}$$

Hence the vanishing of the Taub-NUT current is guaranteed by the very simple condition:

$$\mathcal{H}_2 = \alpha \quad ; \quad \nabla\mathcal{H}_2 = 0 \tag{6.11.20}$$

where α is just a constant. This confirms the preliminary analysis obtained from the Lax operator which requires a vanishing component of the Lax along the second generator A_2 of the nilpotent algebra.

General Form of the Solution

Imposing this condition we arrive at the following form of the solution depending on two harmonic functions $\mathscr{H}_1$, $\mathscr{H}_3$:

$$\exp[-U] = \sqrt{\alpha^2 - 3\mathscr{H}_3^2 + \mathscr{H}_1} \tag{6.11.21}$$

$$z = \mathrm{i}\,\frac{1}{\sqrt{\alpha^2 - 3\mathscr{H}_3^2 + \mathscr{H}_1}} - \frac{\sqrt{2}\,\mathscr{H}_3}{\alpha^2 - 3\mathscr{H}_3^2 + \mathscr{H}_1} \tag{6.11.22}$$

$$j^{TN} = 0 \tag{6.11.23}$$

$$j^{EM} = \star\nabla \begin{pmatrix} 0 \\ 0 \\ \sqrt{3}\,\mathscr{H}_3 \\ -\frac{1}{\sqrt{2}}\,\mathscr{H}_1 \end{pmatrix} \tag{6.11.24}$$

Obviously the physical range of the solution is determined by the condition $(\alpha^2 - 3\mathscr{H}_3^2 + \mathscr{H}_1) > 0$ which can always be arranged, by tuning the parameters contained in the harmonic functions.

To this effect let us discuss the nature of the black holes encompassed by this solution, that, by definition, are located at the poles of the harmonic functions $\mathscr{H}_1$, $\mathscr{H}_3$.

According to the argument developed in Sect. 6.9, in the vicinity of each pole $|\mathbf{x} - \mathbf{x}_I| = r < \varepsilon$ we can choose polar coordinates centered at $\mathbf{x}_\alpha$ and the behavior of the harmonic functions, for $\varepsilon \to 0$ is the following one:

$$\mathscr{H}_1 \sim a_1 + \frac{b_1}{r} \tag{6.11.25}$$

$$\mathscr{H}_3 \sim a_3 + \frac{b_3}{r} \tag{6.11.26}$$

which corresponds to the following behavior of the warp factor:

$$\exp[-U] \sim \sqrt{\alpha^2 - 3a_3^2 - \frac{3b_3^2}{r^2} + a_1 + \frac{b_1}{r} - \frac{6a_3b_3}{r}} \tag{6.11.27}$$

In order for the warp factor to be real for all values of $r \to 0$ we necessarily find

$$\begin{aligned} b_3 &= 0 \\ b_1 &> 0 \\ \alpha^2 - 3a_3^2 + a_1 &> 0 \end{aligned} \tag{6.11.28}$$

Since conditions (6.11.28) hold true for each available pole, it means the harmonic function $\mathscr{H}_3$ has actually no pole and is therefore equal to some constant. The boundary condition of asymptotic flatness fixes the value of such a constant:

$$\lim_{r\to\infty} \exp[-U] = 1 \quad \Leftrightarrow \quad \mathscr{H}_3 = \frac{\sqrt{\alpha^2 + \mathscr{H}_1(\infty) - 1}}{\sqrt{3}} \tag{6.11.29}$$

Under such conditions in the vicinity of each pole $\mathbf{x}_\alpha$, the warp factor has the following behavior:

$$|\mathbf{x} - \mathbf{x}_\alpha|^2 \exp[-U] \overset{\mathbf{x}\to\mathbf{x}_\alpha}{\sim} \sqrt{b_1}\, |\mathbf{x} - \mathbf{x}_\alpha|^{3/2} + \mathscr{O}\left(|\mathbf{x} - \mathbf{x}_\alpha|^{5/2}\right) \tag{6.11.30}$$

leading to a vanishing horizon area:

$$\text{Area}_{\,\mathrm{H}_\alpha} = \lim_{\mathbf{x}\to\mathbf{x}_\alpha} |\mathbf{x} - \mathbf{x}_\alpha|^2 \exp[-U] = 0 \tag{6.11.31}$$

At the same time using the form of the electromagnetic current in Eq. (6.11.24) and the behavior of the harmonic function in the vicinity of the poles we obtain the charge vector of each black hole encompassed by the solution:

$$\mathscr{Q}_\alpha = \int_{\mathbb{S}^2_\alpha} j^{EM} = \begin{pmatrix} 0 \\ 0 \\ 0 \\ -\frac{1}{\sqrt{2}}\, q_\alpha \end{pmatrix} \quad ; \quad \text{where } q_\alpha = b_1 \text{ for pole } \mathbf{x}_\alpha \tag{6.11.32}$$

Summarizing

For the regular multicenter solutions associated with the orbit 4|11 all blacks holes localized at each pole are of the same type, namely they are very small black holes with vanishing horizon area and a charge vector $\mathscr{Q}$ belonging to **W**-orbit which is characterized by both a vanishing quartic invariant and the existence of a continuous parabolic stability subgroup of $\mathrm{SL}(2,\mathbb{R})$. Every black hole is a repetition in a different place of the spherical symmetric black hole which gives its name to the orbit.

6.11.2 *The Small Black Holes of* $\mathscr{O}^2_{11}$

Next let us consider the orbit $\mathscr{O}^2_{11}$.

W-Representation

Applying the same strategy as in the previous case, from the general formula we obtain

$$\mathscr{Q}^{\mathbf{W}}_{2|11} = \mathrm{Tr}(X_{2|11}\mathscr{T}^{\mathbf{W}}) = \left(\sqrt{3}, 0, 0, 0\right) \tag{6.11.33}$$

Substituting such a result in the expression for the quartic symplectic invariant (see Eq. (6.11.3) we find:

$$\mathfrak{I}_4 = 0 \tag{6.11.34}$$

Just as before we stress that this result is meaningful since, by calculating the trace $\mathrm{Tr}(X_{2|11} L_{+}^{E}) = 0$, we can also check that the Taub-NUT charge vanishes. Addressing the question whether there are subgroups of the original duality group in four-dimensions SL(2, R) that leave the charge vector (6.11.33) invariant we realize that such a group contains only the identity

$$\mathrm{SL}(2, \mathbb{R}) \supset \mathscr{S}_{2|11} = \mathbf{1} \tag{6.11.35}$$

Hence we clearly establish the intrinsic difference between the two type of small black holes at the level of the **W**-representation. Both have vanishing quartic invariant, yet only the orbit 4|11 has a residual symmetry.

H*-Stability Subgroup

Considering next the stability subgroup of the nilpotent element $X_{2|11}$ in $\mathrm{H}^{\star} = \widehat{\mathfrak{sl}(2, \mathbb{R})} \oplus \mathfrak{sl}(2, \mathbb{R})_{\mathrm{h}^{\star}}$ we obtain:

$$\mathfrak{S}_{2|11} = \mathrm{SO}(1, 1) \triangleright \mathbb{R} \tag{6.11.36}$$

A generic element of the corresponding Lie algebra is a linear combination of two generators J, T, satisfying the commutation relations:

$$[J\,,\,T] = \frac{3}{2\sqrt{6}}\,T \tag{6.11.37}$$

We do not give its explicit form which we do not use in the sequel.

Nilpotent Algebra $\mathbb{N}_{4|11}$

Considering next the adjoint action of the central element $h_{2|11}$ on the subspace $\mathbb{K}^{\star}$ we find that its eigenvalues are the following ones:

$$\mathrm{Eigenvalues}_{4|11}^{\mathbb{K}^{\star}} = \{-3, 3, -2, 2, -1, 1, 0, 0\} \tag{6.11.38}$$

Therefore the three eigenoperators A_3, A_2, A_1 corresponding to the positive eigenvalues 3, 2, 1, respectively, form the restriction to $\mathbb{K}^{\star}$ of a nilpotent algebra $\mathbb{N}_{2|11}$. In this case A_i do not all commute among themselves so that, differently from the previous case we have $\mathbb{N}_{4|11} \neq \mathbb{N}_{4|11} \bigcap \mathbb{K}^{\star}$. In particular we find a new generator:

$$B \in \mathbb{H}^{\star} \tag{6.11.39}$$

which completes a four-dimensional algebra with the following commutation relations:

$$
\begin{aligned}
0 &= [A_3\,,\,A_2] \;=\; [A_1\,,\,A_3] \\
B &= [A_2\,,\,A_1] \\
0 &= [B\,,\,A_1] \\
0 &= [B\,,\,A_2] \\
0 &= [B\,,\,A_3]
\end{aligned}
\tag{6.11.41}
$$

(6.11.40)

As in the previous case, the structure of the nilpotent algebra implies that for the orbit $\mathscr{O}^2_{11}$ we have only three functions $\mathfrak{h}^0_i$ which will be harmonic and independent. This is so because $\mathscr{D}^2\mathbb{N}_{2|11} = 0$ and $\mathscr{D}\mathbb{N}_{2|11} \bigcap \mathbb{K}^\star = 0$.

Explicitly we set:

$$
\mathfrak{H}(\mathfrak{h}_1, \mathfrak{h}_2, \mathfrak{h}_3) = \textstyle\sum_{i=1}^3 \mathfrak{h}_i\, A_i =
$$

$$
\begin{pmatrix}
-\mathfrak{h}_2 & \mathfrak{h}_1-\mathfrak{h}_3 & \mathfrak{h}_2 & -\sqrt{2}\mathfrak{h}_1-\sqrt{2}\mathfrak{h}_3 & 0 & -3\mathfrak{h}_1-\mathfrak{h}_3 & 0 \\
\mathfrak{h}_1-\mathfrak{h}_3 & -2\mathfrak{h}_2 & \mathfrak{h}_3-3\mathfrak{h}_1 & -\sqrt{2}\mathfrak{h}_2 & \mathfrak{h}_1+\mathfrak{h}_3 & 0 & -3\mathfrak{h}_1-\mathfrak{h}_3 \\
-\mathfrak{h}_2 & 3\mathfrak{h}_1-\mathfrak{h}_3 & \mathfrak{h}_2 & \sqrt{2}\mathfrak{h}_1-\sqrt{2}\mathfrak{h}_3 & 0 & -\mathfrak{h}_1-\mathfrak{h}_3 & 0 \\
\sqrt{2}\mathfrak{h}_1+\sqrt{2}\mathfrak{h}_3 & \sqrt{2}\mathfrak{h}_2 & \sqrt{2}\mathfrak{h}_1-\sqrt{2}\mathfrak{h}_3 & 0 & \sqrt{2}\mathfrak{h}_1-\sqrt{2}\mathfrak{h}_3 & -\sqrt{2}\mathfrak{h}_2 & \sqrt{2}\mathfrak{h}_1+\sqrt{2}\mathfrak{h}_3 \\
0 & -\mathfrak{h}_1-\mathfrak{h}_3 & 0 & \sqrt{2}\mathfrak{h}_1-\sqrt{2}\mathfrak{h}_3 & -\mathfrak{h}_2 & 3\mathfrak{h}_1-\mathfrak{h}_3 & \mathfrak{h}_2 \\
-3\mathfrak{h}_1-\mathfrak{h}_3 & 0 & \mathfrak{h}_1+\mathfrak{h}_3 & \sqrt{2}\mathfrak{h}_2 & \mathfrak{h}_3-3\mathfrak{h}_1 & 2\mathfrak{h}_2 & \mathfrak{h}_1-\mathfrak{h}_3 \\
0 & -3\mathfrak{h}_1-\mathfrak{h}_3 & 0 & -\sqrt{2}\mathfrak{h}_1-\sqrt{2}\mathfrak{h}_3 & -\mathfrak{h}_2 & \mathfrak{h}_1-\mathfrak{h}_3 & \mathfrak{h}_2
\end{pmatrix}
\tag{6.11.42}
$$

Considering $\mathfrak{H}(\mathfrak{h}_1, \mathfrak{h}_2, \mathfrak{h}_3)$ as a Lax operator and calculating its Taub-NUT charge and electromagnetic charges we find:

$$
\mathbf{n}_{TN} = -2\,(3\mathfrak{h}_1 + \mathfrak{h}_3) \quad ; \quad \mathscr{Q} = \left\{-2\sqrt{3}\mathfrak{h}_2, 6\mathfrak{h}_1 - 2\mathfrak{h}_3, -2\sqrt{3}\,(\mathfrak{h}_1 + \mathfrak{h}_3)\,, 0\right\}
\tag{6.11.43}
$$

This implies that constructing the multi-centre solution with harmonic functions the condition $\mathfrak{h}_3 = -3\,\mathfrak{h}_1$ might be sufficient to annihilate the Taub-NUT current. We shall demonstrate that in this case the condition is slightly more complicated.

For later convenience let us change the normalization in the basis of harmonic functions as follows:

$$
\mathfrak{h}^{(0)}_1 = \tfrac{1}{4}\mathscr{H}_3 \;\; ; \;\; \mathfrak{h}^{(0)}_2 = \tfrac{1}{2}\,(1 - \mathscr{H}_2) \;\; ; \;\; \mathfrak{h}^{(0)}_3 = \tfrac{1}{4}\mathscr{H}_1
\tag{6.11.44}
$$

Implementing the symmetric coset construction with:

$$
\mathscr{Y}\,(\mathscr{H}_3, \mathscr{H}_2, \mathscr{H}_1) \equiv \exp\left[\mathfrak{H}\left(\tfrac{1}{4}\mathscr{H}_3, \tfrac{1}{2}\,(1 - \mathscr{H}_2)\,, \tfrac{1}{4}\mathscr{H}_1\right)\right]
\tag{6.11.45}
$$

calculating the upper triangular coset representative $\mathbb{L}(\mathscr{Y})$ according to equations (6.8.26) and extracting the σ-model scalar fields we obtain the answer which we list below:

$$\exp[-U] = \frac{1}{2}\sqrt{-\mathscr{H}_3^2 + \left(4\mathscr{H}_1^3 + 6\mathscr{H}_2\mathscr{H}_1\right)\mathscr{H}_3 + \mathscr{H}_2^2\left(3\mathscr{H}_1^2 + 4\mathscr{H}_2\right)} \tag{6.11.46}$$

$$\operatorname{Im} z = \frac{\sqrt{-\mathscr{H}_3^2 + \left(4\mathscr{H}_1^3 + 6\mathscr{H}_2\mathscr{H}_1\right)\mathscr{H}_3 + \mathscr{H}_2^2\left(3\mathscr{H}_1^2 + 4\mathscr{H}_2\right)}}{2\left(\mathscr{H}_1^2 + \mathscr{H}_2\right)} \tag{6.11.47}$$

$$\operatorname{Re} z = \frac{\mathscr{H}_3 - \mathscr{H}_2\mathscr{H}_1}{2\left(\mathscr{H}_1^2 + \mathscr{H}_2\right)} \tag{6.11.48}$$

$$Z^M = \begin{pmatrix} \frac{\sqrt{\frac{3}{2}}\left(\mathscr{H}_3^2 - 2\mathscr{H}_1\left(2\mathscr{H}_1^2 + 3\mathscr{H}_2 - 1\right)\mathscr{H}_3 + \mathscr{H}_2\left(-4\mathscr{H}_2^2 + \left(4 - 3\mathscr{H}_1^2\right)\mathscr{H}_2 + 2\mathscr{H}_1^2\right)\right)}{\mathscr{H}_3^2 - 2\left(2\mathscr{H}_1^3 + 3\mathscr{H}_2\mathscr{H}_1\right)\mathscr{H}_3 - \mathscr{H}_2^2\left(3\mathscr{H}_1^2 + 4\mathscr{H}_2\right)} \\ \frac{\sqrt{2}\left(2\mathscr{H}_1^3 + 3\mathscr{H}_2\mathscr{H}_1 - \mathscr{H}_3\right)}{-\mathscr{H}_3^2 + \left(4\mathscr{H}_1^3 + 6\mathscr{H}_2\mathscr{H}_1\right)\mathscr{H}_3 + \mathscr{H}_2^2\left(3\mathscr{H}_1^2 + 4\mathscr{H}_2\right)} \\ \frac{\sqrt{6}\left(\mathscr{H}_1\mathscr{H}_2^2 + \mathscr{H}_3\left(2\mathscr{H}_1^2 + \mathscr{H}_2\right)\right)}{\mathscr{H}_3^2 - 2\left(2\mathscr{H}_1^3 + 3\mathscr{H}_2\mathscr{H}_1\right)\mathscr{H}_3 - \mathscr{H}_2^2\left(3\mathscr{H}_1^2 + 4\mathscr{H}_2\right)} \\ \frac{4\mathscr{H}_3\mathscr{H}_1^3 + 3\mathscr{H}_2^2\mathscr{H}_1^2 + \mathscr{H}_3^2}{\sqrt{2}\left(-\mathscr{H}_3^2 + \left(4\mathscr{H}_1^3 + 6\mathscr{H}_2\mathscr{H}_1\right)\mathscr{H}_3 + \mathscr{H}_2^2\left(3\mathscr{H}_1^2 + 4\mathscr{H}_2\right)\right)} \end{pmatrix} \tag{6.11.49}$$

$$a = \frac{\mathscr{H}_3\left(-6\mathscr{H}_1^2 - 3\mathscr{H}_2 + 1\right) - \mathscr{H}_1\left(3\mathscr{H}_2^2 + 3\mathscr{H}_2 + 2\mathscr{H}_1^2\right)}{\mathscr{H}_3^2 - 2\left(2\mathscr{H}_1^3 + 3\mathscr{H}_2\mathscr{H}_1\right)\mathscr{H}_3 - \mathscr{H}_2^2\left(3\mathscr{H}_1^2 + 4\mathscr{H}_2\right)} \tag{6.11.50}$$

The Taub-NUT Current

Given this explicit result we can turn to the explicit oxidation formulae described in Sect. 6.2.1 and calculate the Taub-NUT current which is the integrand of Eq. (6.2.17). We find:

$$j^{TN} = \tfrac{1}{2}\left(^\star\nabla\mathscr{H}_3 + 3\left(\mathscr{H}_2{}^\star\nabla\mathscr{H}_1 - \mathscr{H}_1{}^\star\nabla\mathscr{H}_2\right)\right) \tag{6.11.51}$$

Analyzing Eq. (6.11.51) we see that there are just two possible solutions to the condition $j^{TN} = 0$:

(case a) $\mathscr{H}_3 = \beta = \text{const}$; $\mathscr{H}_1 = 0$. With this condition we obtain:

$$\exp[-U] = \frac{1}{2}\sqrt{4\mathscr{H}_2^3 - \beta^2} \tag{6.11.52}$$

$$z = \frac{\beta + \mathrm{i}\sqrt{4\mathscr{H}_2^3 - \beta^2}}{2\mathscr{H}_2} \tag{6.11.53}$$

$$j^{EM} = \star\nabla \begin{pmatrix} -\sqrt{\frac{3}{2}}\mathscr{H}_2 \\ 0 \\ 0 \\ 0 \end{pmatrix} \tag{6.11.54}$$

(case b) $\mathscr{H}_3 = \beta = \text{const}$; $\mathscr{H}_2 = 0$

$$\exp[-U] = \frac{1}{2}\sqrt{\beta\left(4\mathscr{H}_1^3 - \beta\right)} \tag{6.11.55}$$

$$z = \frac{\beta + \mathrm{i}\sqrt{\beta\left(4\mathscr{H}_1^3 - \beta\right)}}{2\mathscr{H}_3^2} \tag{6.11.56}$$

$$j^{EM} = \begin{pmatrix} 0 \\ 0 \\ -\sqrt{\frac{3}{2}}\mathscr{H}_1 \\ 0 \end{pmatrix} \tag{6.11.57}$$

It might seem that these two solutions correspond to different types of black holes but this is not the case, as we now show. From the asymptotic flatness boundary condition we find that the value of β is fixed in terms of the value at infinity of the corresponding harmonic function $\mathscr{H}_{1,2}$, which of course must satisfy the necessary condition for reality of the solution $\mathscr{H}_{1,2}(\infty) \geq 1$:

$$\begin{cases} \beta = \quad 2\sqrt{[\mathscr{H}_2(\infty)]^3 - 1} & \text{case a} \\ \beta = 2\left([\mathscr{H}_1(\infty)]^3 + \sqrt{[\mathscr{H}_1(\infty)]^6 - 1}\right) & \text{case b} \end{cases} \tag{6.11.58}$$

In the vicinity of a pole by means of the usual argument we obtain the following behavior of the warp factor:

$$|\mathbf{x} - \mathbf{x}_\alpha|^2 \exp[-U] \overset{\mathbf{x}\to\mathbf{x}_\alpha}{\sim} \begin{cases} \sqrt{b_2^3}\sqrt{|\mathbf{x} - \mathbf{x}_\alpha|} + \mathscr{O}\left(|\mathbf{x} - \mathbf{x}_\alpha|^{3/2}\right) & : \text{case a} \\ \sqrt{\beta\, b_1^3}\sqrt{|\mathbf{x} - \mathbf{x}_\alpha|} + \mathscr{O}\left(|\mathbf{x} - \mathbf{x}_\alpha|^{3/2}\right) & : \text{case b} \end{cases} \tag{6.11.59}$$

Hence in both cases the horizon area vanishes at all poles $\mathbf{x}_\alpha$ and the reality conditions are satisfied choosing the appropriate sign of $b_{1,2}$. The charge vector has the same structure for all black holes encompassed in the first or in the second solution, namely:

$$\mathscr{Q}_\alpha = \begin{cases} \left\{-\sqrt{\frac{3}{2}}\,p_\alpha\,,\,0\,,\,0\,,\,0\right\} : p_\alpha = b_2 & \text{for pole } \alpha \\ \left\{0\,,\,0\,,\,-\sqrt{\frac{3}{2}}\,q_\alpha\,,\,0\right\} : q_\alpha = b_1 & \text{for pole } \alpha \end{cases} \tag{6.11.60}$$

In both cases the quartic invariant $\mathfrak{I}_4$ is zero for all black holes in the solutions, yet one might still doubt whether the **W**-orbit for the two cases might be different. It is not so, since a direct calculation shows that the image in the $j = \frac{3}{2}$ representation $\Lambda[\mathfrak{A}]$[13] of the following $\mathrm{SL}(2,\mathbb{R})$ element:

$$\mathfrak{A} = \begin{pmatrix} 0 & \frac{p}{q} \\ -\frac{q}{p} & 0 \end{pmatrix} \tag{6.11.61}$$

[13] See [32] for details, in particular Eq. (3.13) of that reference for the explicit form of the spin $\frac{3}{2}$ matrices.

maps the charge vector $\mathcal{Q}_{[q]} = \{0\,,\,0\,,\,-q\,,\,0\}$, into the charge vector $\mathcal{Q}_{[p]} = \{p\,,\,0\,,\,0\,,\,0\}$, namely we have $\Lambda[\mathfrak{A}]\,\mathcal{Q}_{[q]} = \mathcal{Q}_{[p]}$. Hence the two solutions we have here discussed simply give different representatives of the same **W**-orbit.

SUMMARY

Just as in the previous case for a multicenter solution associated with the $\mathcal{O}^2_{11}$ orbit all the black holes included in one solution are of the same type, namely small black holes with the same identical properties.

6.11.3 The Large BPS Black Holes of $\mathcal{O}^3_{11}$

Next let us consider the orbit $\mathcal{O}^3_{11}$, which in the spherical symmetric case leads to BPS Black holes with a finite horizon area.

W-Representation

In order to better appreciate the structure of these solutions, let us slightly generalize our orbit representative, writing the following nilpotent matrix that depends on two parameters (p, q) to be interpreted later as the magnetic and the electric charge of the hole:

$$X_{3|11}(p,q) = \begin{pmatrix} q & 0 & 0 & -\frac{q}{\sqrt{2}} & 0 & 0 & 0 \\ 0 & \frac{p+q}{2} & -\frac{p}{2} & 0 & \frac{q}{2} & 0 & 0 \\ 0 & \frac{p}{2} & \frac{q-p}{2} & 0 & 0 & -\frac{q}{2} & 0 \\ \frac{q}{\sqrt{2}} & 0 & 0 & 0 & 0 & 0 & \frac{q}{\sqrt{2}} \\ 0 & -\frac{q}{2} & 0 & 0 & \frac{p-q}{2} & \frac{p}{2} & 0 \\ 0 & 0 & \frac{q}{2} & 0 & -\frac{p}{2} & \frac{1}{2}(-p-q) & 0 \\ 0 & 0 & 0 & -\frac{q}{\sqrt{2}} & 0 & 0 & -q \end{pmatrix} \tag{6.11.62}$$

The standard triple representative mentioned in Eq. (6.10.5) is just the particular case $X_{3|11}(1,1)$. Applying the same strategy as in the previous case, from the general formula we obtain

$$\mathcal{Q}^{\mathbf{W}}_{3|11} = \mathrm{Tr}(X_{3|11}(p,q)\mathcal{T}^{\mathbf{W}}) = \left(0,\, p,\, -\sqrt{3}q,\, 0\right) \tag{6.11.63}$$

Substituting such a result in the expression for the quartic symplectic invariant (see Eq. (6.11.3)) we find:

$$\mathfrak{I}_4 = 9\,p\,q^3 > 0 \quad \text{if } p \text{ and } q \text{ have the same sign} \tag{6.11.64}$$

Just as before we stress that this result is meaningful since, by calculating the trace $\mathrm{Tr}(X_{3|11}L^E_+) = 0$, we can also check that the Taub-NUT charge vanishes. Furthermore we note that the condition that p and q have the same sign was singled out in [32] as the defining condition of the orbit O^3_{11} which, in the spherical symmetry approach leads to regular BPS solutions. The choice of opposite signs was proved in [32] to correspond to a different H⋆ orbit, the non diagonal O^3_{21} which instead contains only singular solutions. Here we will show another important and intrinsically four dimensional reason to separate the two cases.

Addressing the question whether there are subgroups of the original duality group in four-dimensions SL(2, R) that leave the charge vector (6.11.63) invariant we realize that such a subgroup exists and is the finite cyclic group of order three:

$$\mathrm{SL}(2,\mathbb{R}) \supset \mathscr{S}_{3|11} = \mathbb{Z}_3 \tag{6.11.65}$$

$\mathscr{S}_{3|11}$ is made by the following three elements:

$$\mathbf{1} = \begin{pmatrix} 1 & 0 \\ 0 & 1 \end{pmatrix} \tag{6.11.66}$$

$$\mathfrak{B} = \begin{pmatrix} -\frac{1}{2} & -\frac{\sqrt{3}}{2}\sqrt{\frac{p}{q}} \\ \frac{\sqrt{3}}{2}\sqrt{\frac{q}{p}} & -\frac{1}{2} \end{pmatrix} \tag{6.11.67}$$

$$\mathfrak{B}^2 = \begin{pmatrix} -\frac{1}{2} & \frac{\sqrt{3}}{2}\sqrt{\frac{p}{q}} \\ -\frac{\sqrt{3}}{2}\sqrt{\frac{q}{p}} & -\frac{1}{2} \end{pmatrix} \quad ; \quad \mathfrak{B}^3 = \mathbf{1} \tag{6.11.68}$$

It is evident that such a $\mathbb{Z}_3$ subgroup exists if and only if the two charges p, q have the same sign. Otherwise the corresponding matrices develop imaginary elements and migrate to SL(2, $\mathbb{C}$). The existence of this isotropy group $\mathbb{Z}_3$ can be considered the very definition of the **W**-orbit corresponding to BPS black holes. Indeed let us name $\lambda = \sqrt{\frac{p}{q}}$ and consider the algebraic condition imposed on a generic charge vector: $\mathscr{Q} = \{Q_1, Q_2, Q_3, Q_4\}$ by the request that it should admit the above described $\mathbb{Z}_3$ stability group:

$$\Lambda[\mathfrak{B}]\mathscr{Q} = \mathscr{Q} \quad \Leftrightarrow \quad \mathscr{Q} = \left(\sqrt{3}\lambda^2 Q_4, -\frac{\lambda^2 Q_3}{\sqrt{3}}, Q_3, Q_4\right) \tag{6.11.69}$$

It is evident from the above explicit result that the charge vectors having this symmetry depend only on three parameters (λ^2, Q_3, Q_4). The very relevant fact is that substituting this restricted charge vector in the general formula (6.11.3) for the quartic invariant we obtain:

$$\mathfrak{I}_4 = \lambda^2\left(Q_3^2 + 3\lambda^2 Q_4^2\right)^2 > 0 \tag{6.11.70}$$

Hence the $\mathbb{Z}_3$ guarantees that the quartic invariant is a perfect square and hence positive. It is an intrinsic restriction characterizing the $\mathbb{W}$-orbit.

H*-Stability Subgroup

Considering next the stability subgroup of the nilpotent element $X_{3|11}(1,1)$ in $\mathrm{H}^\star = \widehat{\mathfrak{sl}(2,\mathbb{R})} \oplus \mathfrak{sl}(2,\mathbb{R})_{\mathrm{h}^\star}$ we obtain:

$$\mathfrak{S}_{3|11} = \mathbb{R} \qquad (6.11.71)$$

the group being generated by a matrix $\mathbb{A}_{3|11}$ of nilpotency degree 2:

$$\mathbb{A}^2_{3|11} = \mathbf{0} \qquad (6.11.72)$$

We do not give its explicit form which we do not use in the sequel.

Nilpotent Algebra $\mathbb{N}_{3|11}$

Considering next the adjoint action of the central element $h_{3|11}$ on the subspace $\mathbb{K}^\star$ we find that its eigenvalues are the following ones:

$$\text{Eigenvalues}^{\mathbb{K}^\star}_{3|11} = \{-2, -2, -2, -2, 2, 2, 2, 2\} \qquad (6.11.73)$$

Therefore the four eigenoperators A_1, A_2, A_3, A_4 corresponding to the four positive eigenvalues 2, respectively, form the restriction to $\mathbb{K}^\star$ of a nilpotent algebra $\mathbb{N}_{3|11}$. Also in this case the A_i do not all commute among themselves so that, we have $\mathbb{N}_{3|11} \neq \mathbb{N}_{3|11} \bigcap \mathbb{K}^\star$. In particular we find a new generator:

$$B \in \mathbb{H}^\star \qquad (6.11.74)$$

which completes a five-dimensional algebra with the following commutation relations:

$$\begin{aligned} \left[A_i\,,\,A_j\right] &= \Omega_{ij}\,B \\ [B\,,\,A_i] &= 0 \\ B &= \begin{pmatrix} 0 & 0 & -1 & 1 \\ 0 & 0 & -1 & -1 \\ 1 & 1 & 0 & 0 \\ -1 & 1 & 0 & 0 \end{pmatrix} \end{aligned} \qquad (6.11.75)$$

The structure of the nilpotent algebra implies that for the orbit $\mathscr{O}^3_{11}$ we have only four functions $\mathfrak{h}^0_i$ which will be harmonic and independent. This is so because $\mathscr{D}^2\mathbb{N}_{3|11} = 0$ and $\mathscr{D}\mathbb{N}_{3|11} \bigcap \mathbb{K}^\star = 0$.

Explicitly we set:

$$\mathfrak{H}(\mathfrak{h}_1, \mathfrak{h}_2, \mathfrak{h}_3, \mathfrak{h}_4) = \sum_{i=1}^4 \mathfrak{h}_i\, A_i =$$

$$\begin{pmatrix} 2\mathfrak{h}_3 & \mathfrak{h}_1-2\mathfrak{h}_2 & 2\mathfrak{h}_1-\mathfrak{h}_2 & -\sqrt{2}\mathfrak{h}_3 & -3\mathfrak{h}_2 & -3\mathfrak{h}_1 & 0 \\ \mathfrak{h}_1-2\mathfrak{h}_2 & \mathfrak{h}_3-\mathfrak{h}_4 & \mathfrak{h}_4 & \sqrt{2}\mathfrak{h}_2-2\sqrt{2}\mathfrak{h}_1 & \mathfrak{h}_3 & 0 & -3\mathfrak{h}_1 \\ \mathfrak{h}_2-2\mathfrak{h}_1 & -\mathfrak{h}_4 & \mathfrak{h}_3+\mathfrak{h}_4 & \sqrt{2}\mathfrak{h}_1-2\sqrt{2}\mathfrak{h}_2 & 0 & -\mathfrak{h}_3 & -3\mathfrak{h}_2 \\ \sqrt{2}\mathfrak{h}_3 & 2\sqrt{2}\mathfrak{h}_1-\sqrt{2}\mathfrak{h}_2 & \sqrt{2}\mathfrak{h}_1-2\sqrt{2}\mathfrak{h}_2 & 0 & \sqrt{2}\mathfrak{h}_1-2\sqrt{2}\mathfrak{h}_2 & \sqrt{2}\mathfrak{h}_2-2\sqrt{2}\mathfrak{h}_1 & \sqrt{2}\mathfrak{h}_3 \\ 3\mathfrak{h}_2 & -\mathfrak{h}_3 & 0 & \sqrt{2}\mathfrak{h}_1-2\sqrt{2}\mathfrak{h}_2 & -\mathfrak{h}_3-\mathfrak{h}_4 & -\mathfrak{h}_4 & 2\mathfrak{h}_1-\mathfrak{h}_2 \\ -3\mathfrak{h}_1 & 0 & \mathfrak{h}_3 & 2\sqrt{2}\mathfrak{h}_1-\sqrt{2}\mathfrak{h}_2 & \mathfrak{h}_4 & \mathfrak{h}_4-\mathfrak{h}_3 & \mathfrak{h}_1-2\mathfrak{h}_2 \\ 0 & -3\mathfrak{h}_1 & 3\mathfrak{h}_2 & -\sqrt{2}\mathfrak{h}_3 & \mathfrak{h}_2-2\mathfrak{h}_1 & \mathfrak{h}_1-2\mathfrak{h}_2 & -2\mathfrak{h}_3 \end{pmatrix} \tag{6.11.76}$$

Considering $\mathfrak{H}(\mathfrak{h}_1, \mathfrak{h}_2, \mathfrak{h}_3, \mathfrak{h}_4)$ as a Lax operator and calculating its Taub-NUT charge and electromagnetic charges we find:

$$\mathbf{n}_{TN} = -6\mathfrak{h}_1 \quad ; \quad \mathscr{Q} = \left\{2\sqrt{3}\left(\mathfrak{h}_2 - 2\mathfrak{h}_1\right), -2\mathfrak{h}_4, -2\sqrt{3}\mathfrak{h}_3, -6\mathfrak{h}_2\right\} \tag{6.11.77}$$

This implies that constructing the multi-centre solution with harmonic functions the condition $\mathfrak{h}_1 = 0$ might be sufficient to annihilate the Taub-NUT current. We shall demonstrate that also in this case the condition is slightly more complicated. This emphasizes the difference between the Lax operator one-dimensional approach and the multicenter construction based on harmonic functions.

For later convenience let us change the normalization in the basis of harmonic functions as follows:

$$\mathfrak{h}_1^{(0)} = \frac{1}{\sqrt{12}}\mathscr{H}_1 \quad ; \quad \mathfrak{h}_2^{(0)} = \frac{1}{\sqrt{12}}\mathscr{H}_2 \quad ; \quad \mathfrak{h}_3^{(0)} = \frac{1}{2}\left(\mathscr{H}_3 - 1\right) \quad ; \quad \mathfrak{h}_4^{(0)} = \frac{1}{2}\left(\mathscr{H}_4 + 1\right) \tag{6.11.78}$$

Implementing the symmetric coset construction with:

$$\mathscr{Y}\left(\mathscr{H}_1, \mathscr{H}_2, \mathscr{H}_3, \mathscr{H}_4\right) \equiv \exp\left[\mathfrak{H}\left(\frac{1}{\sqrt{12}}\mathscr{H}_1, \frac{1}{\sqrt{12}}\mathscr{H}_2, \frac{1}{2}\left(\mathscr{H}_3 - 1\right), \frac{1}{2}\left(\mathscr{H}_4 + 1\right)\right)\right] \tag{6.11.79}$$

calculating the upper triangular coset representative $\mathbb{L}(\mathscr{Y})$ according to Eq. (6.8.26) and extracting the σ-model scalar fields we obtain an explicit but rather messy answer which we omit. In particular we obtain the Taub-NUT current in the following form:

$$j^{TN} = \sum_{i=1}^{4} \mathfrak{R}_i\left(\mathscr{H}\right) \nabla \mathscr{H}_i \tag{6.11.80}$$

where $\mathfrak{R}_i\left(\mathscr{H}\right)$ are rational functions of the four harmonic functions, the maximal degree of involved polynomials being 16. A priori, imposing the vanishing of the Taub-NUT current is a problem without guaranteed solutions. In the 4-dimensional linear space of the harmonic functions we can introduce r-linear relations of the form:

$$0 = V_\alpha^i \mathscr{H}_i \quad ; \quad \alpha = 1, \ldots, r \tag{6.11.81}$$

Let U_a^i be a set of $4 - r$ linear independent 4-vectors orthogonal to the vectors V_α^i. Then it must happen that on the locus defined by Eqs. (6.11.81), the following rational functions should also vanish

$$0 = \mathfrak{P}_a(\mathscr{H}) \equiv U_a^i \, \mathfrak{R}_i\,(\mathscr{H}) \quad ; \quad (a = 1, \ldots, r - 4) \tag{6.11.82}$$

For generic rational functions this will never happen, yet we know that for our system such solutions should exist and in want of a clear cut algorithm it is a matter of ingenuity to find them. We do not find any solution with $r = 1$ but we find two nice solutions with $r = 2$. They are the following ones:

(a) $\mathscr{H}_1 = \mathscr{H}_2 = 0$. The complete form of the supergravity solution corresponding to this choice is:

$$\exp[-U] = \sqrt{-\mathscr{H}_3^3\,\mathscr{H}_4} \tag{6.11.83}$$

$$z = \mathrm{i}\,\frac{\sqrt{-\mathscr{H}_3^3\,\mathscr{H}_4}}{\mathscr{H}_3^2} \tag{6.11.84}$$

$$j^{TN} = 0 \tag{6.11.85}$$

$$j^{EM} = \star\nabla \begin{pmatrix} 0 \\ \frac{\mathscr{H}_4}{\sqrt{2}} \\ \sqrt{\frac{3}{2}}\mathscr{H}_3 \\ 0 \end{pmatrix} \tag{6.11.86}$$

(b) $\mathscr{H}_1 = 0$, $\mathscr{H}_3 = -\mathscr{H}_4$. The complete form of the supergravity solution corresponding to this choice is:

$$\exp[-U] = \sqrt{-\frac{\mathscr{H}_2^4}{3} - 2\mathscr{H}_4^2\mathscr{H}_2^2 + \mathscr{H}_4^4} \tag{6.11.87}$$

$$z = \frac{2\mathscr{H}_2\mathscr{H}_4 - \mathrm{i}\sqrt{-\mathscr{H}_2^4 - 6\mathscr{H}_4^2\mathscr{H}_2^2 + 3\mathscr{H}_4^4}}{\sqrt{3}\left(\mathscr{H}_2^2 - \mathscr{H}_4^2\right)} \tag{6.11.88}$$

$$j^{TN} = 0 \tag{6.11.89}$$

$$j^{EM} = \star\nabla \begin{pmatrix} -\frac{\mathscr{H}_2}{\sqrt{2}} \\ \frac{\mathscr{H}_4}{\sqrt{2}} \\ -\sqrt{\frac{3}{2}}\mathscr{H}_4 \\ \sqrt{\frac{3}{2}}\mathscr{H}_2 \end{pmatrix} \tag{6.11.90}$$

We can now make some comments about the two solutions. First of all both in case a) and in case b) we have to fix the asymptotic value of the harmonic functions at spatial infinity $r = \infty$, in such a way as to obtain asymptotic flatness. This is quite easy and we do not dwell on it. Secondly we have to fix the parameters of the harmonic functions in such a way that the warp factor is always real on the whole physical range. These conditions are also easily spelled out:

$$\begin{array}{ll} a) & -\mathscr{H}_3\mathscr{H}_4 > 0 \\ b) & -\frac{\mathscr{H}_2^4}{3} - 2\mathscr{H}_4^2\mathscr{H}_2^2 + \mathscr{H}_4^4 > 0 \end{array} \tag{6.11.91}$$

and in a multicenter solution can be easily arranged adjusting the coefficients of each pole. Thirdly we can comment about the structure of the charge vector that we obtain at each pole:

$$\mathscr{H}_i \sim a_i + \frac{Q_i}{|x - x_\alpha|} \tag{6.11.92}$$

In case (a) and (b) we respectively obtain:

$$\mathscr{Q}_\alpha = \begin{pmatrix} 0 \\ \frac{Q_4}{\sqrt{2}} \\ \sqrt{\frac{3}{2}}Q_3 \\ 0 \end{pmatrix} \tag{6.11.93}$$

$$\mathscr{Q}_\alpha = \begin{pmatrix} -\frac{Q_2}{\sqrt{2}} \\ \frac{Q_4}{\sqrt{2}} \\ -\sqrt{\frac{3}{2}}Q_4 \\ \sqrt{\frac{3}{2}}Q_2 \end{pmatrix} \tag{6.11.94}$$

Comparing with Eqs. (6.11.69), (6.11.70) we see that in both cases the structure of these charges is that imposed by the $\mathbb{Z}_3$ invariance which characterizes BPS black holes. The necessary choice of signs in the case (a)

$$\frac{Q_4}{Q_3} < 0 \tag{6.11.95}$$

is the same which is required by the reality of the warp factor. Hence in case (b) all the black holes encompassed by the solution at each pole are finite area BPS black holes. In case (a) the same is true for all the poles common to the harmonic function $\mathscr{H}_3$ and $\mathscr{H}_4$: they are finite area BPS black holes. Yet we can envisage the situation where some poles of $\mathscr{H}_3$ are not shared by $\mathscr{H}_4$ and viceversa. In this case the pole of $\mathscr{H}_4$ defines a very small black hole, while the pole of $\mathscr{H}_3$ defines a small black hole. This is confirmed by the fact that a charge vector of type $\{0, p, 0, 0\}$ is mapped into $\{0, 0, 0, p\}$ by $\Lambda\left[\begin{pmatrix} 0 & -1 \\ 1 & 0 \end{pmatrix}\right]$ and as such admits a parabolic subgroup of stability $\Lambda\left[\begin{pmatrix} 1 & b \\ 0 & 1 \end{pmatrix}\right]$.

Summary

For a multicenter solution associated with the $\mathscr{O}^3_{11}$ orbit there are two possibilities either all the black holes included in one solution are regular, finite area, BPS black holes, either we have a mixture of very small and small black holes. A finite area BPS black hole emerges when the center of a very small black hole coincides with the center of a small one. This provides the challenging suggestion that a BPS black hole can be considered quantum mechanically as a composite object where the "quarks" are small and very small black holes.

6.11.4 BPS Kerr–Newman Solution

Next we want to show how this orbit encompasses also the BPS Kerr–Newman solution that was found by Luest et al. in [39].

To this effect we go back to the general formulae for the scalar fields in this orbit and we make the following reduction from four to two independent harmonic functions:

$$\mathscr{H}_2 = 0 \quad ; \quad \mathscr{H}_4 = -\tfrac{1}{3}\,\mathscr{H}_3 \tag{6.11.96}$$

With such a choice the expressions for all the scalar fields dramatically simplify and we obtain:

$$\mathfrak{W} = \frac{\sqrt{3}}{\mathscr{H}_1^2 + \mathscr{H}_3^2} \tag{6.11.97}$$

$$z = \mathrm{i}\,\frac{1}{\sqrt{3}} \tag{6.11.98}$$

$$Z = \begin{pmatrix} -\frac{3\mathscr{H}_1}{\sqrt{2}(\mathscr{H}_1^2+\mathscr{H}_3^2)} \\ \frac{\mathscr{H}_1^2+(\mathscr{H}_3-3)\mathscr{H}_3}{\sqrt{2}(\mathscr{H}_1^2+\mathscr{H}_3^2)} \\ -\frac{\sqrt{\frac{3}{2}}(\mathscr{H}_1^2+(\mathscr{H}_3-1)\mathscr{H}_3)}{\mathscr{H}_1^2+\mathscr{H}_3^2} \\ -\frac{\mathscr{H}_1}{\sqrt{6}(\mathscr{H}_1^2+\mathscr{H}_3^2)} \end{pmatrix} \tag{6.11.99}$$

$$a = \frac{5\mathscr{H}_1}{\sqrt{3}\left(\mathscr{H}_1^2 + \mathscr{H}_3^2\right)} \tag{6.11.100}$$

Utilizing the above expressions in the final oxidation formulae we obtain the following result for the Taub-Nut current and for the electromagnetic currents:

$$j^{TN} = \frac{2\left(\star\nabla\mathscr{H}_1\mathscr{H}_3 \;-\; \star\nabla\mathscr{H}_3\mathscr{H}_1\right)}{\sqrt{3}} \tag{6.11.101}$$

$$
j^{EM} = \begin{pmatrix} \frac{2\star\nabla\mathscr{H}_3\mathscr{H}_1\left(\mathscr{H}_1^2+(\mathscr{H}_3-2)\mathscr{H}_3\right)-\star\nabla\mathscr{H}_1\left((2\mathscr{H}_3+1)\mathscr{H}_1^2+\mathscr{H}_3^2(2\mathscr{H}_3-3)\right)}{\sqrt{2}\left(\mathscr{H}_1^2+\mathscr{H}_3^2\right)} \\ \frac{\star\nabla\mathscr{H}_3\left(3\mathscr{H}_1^2-\mathscr{H}_3^2\right)-4\star\nabla\mathscr{H}_1\mathscr{H}_1\mathscr{H}_3}{3\sqrt{2}\left(\mathscr{H}_1^2+\mathscr{H}_3^2\right)} \\ \frac{\sqrt{\frac{3}{2}}\left(4\star\nabla\mathscr{H}_1\mathscr{H}_1\mathscr{H}_3+\star\nabla\mathscr{H}_3\left(\mathscr{H}_3^2-3\mathscr{H}_1^2\right)\right)}{\mathscr{H}_1^2+\mathscr{H}_3^2} \\ \frac{2\star\nabla\mathscr{H}_3\mathscr{H}_1\left(\mathscr{H}_1^2+(\mathscr{H}_3-6)\mathscr{H}_3\right)-\star\nabla\mathscr{H}_1\left((2\mathscr{H}_3+3)\mathscr{H}_1^2+\mathscr{H}_3^2(2\mathscr{H}_3-9)\right)}{\sqrt{6}\left(\mathscr{H}_1^2+\mathscr{H}_3^2\right)} \end{pmatrix} \tag{6.11.102}
$$

Next identifying the two harmonic functions with those introduced in Eqs. (6.9.21)–(6.9.24), according to:

$$
\mathscr{H}_1 = 3^{\frac{1}{4}}\,(1 + m\,\mathscr{P}) \quad ; \quad \mathscr{H}_3 = 3^{\frac{1}{4}}\,m\,\mathscr{R} \tag{6.11.103}
$$

we obtain the following result for the warp-factor:

$$
\exp[U] = \frac{(m+r)^2 + \alpha^2\cos^2(\theta)}{r^2 + \alpha^2\cos^2(\theta)} \tag{6.11.104}
$$

and for the Kaluza–Klein vector:

$$
\mathbf{A}^{[KK]} = \omega \equiv \frac{m(m+2r)\alpha\sin^2(\theta)}{r^2+\alpha^2\cos^2(\theta)}\,d\phi \tag{6.11.105}
$$

Indeed one can easily check that, in the spheroidal coordinates (6.9.19) with flat metric Eq. (6.9.20) we have:

$$
2m\,(\star\nabla\mathscr{P}\,\mathscr{R} - \mathscr{P}\star\nabla\mathscr{R}) = \mathrm{d}\omega \tag{6.11.106}
$$

where $\star\nabla$ denotes the Hodge dual of the exterior derivative d. Writing the corresponding final form of the metric:

$$
ds^2_{BPSKN} = -\exp[U]\,(dt + \omega)^2 + \exp[-U]\,d\Omega^2_{spheroidal} \tag{6.11.107}
$$

we can easily check that it is just the Kerr–Newman metric (6.5.2) with $q = m$. The only necessary step, in order to verify such an identity is a redefinition of the coordinate r. If in the metric (6.5.2) one replaces $r \to r + m$, then (6.5.2) becomes identical to (6.11.107).

It is interesting to consider the expressions for the vector field strengths that solve the Maxwell-Einstein system together with the BPS Kerr–Newman metric. For the first two field strengths (magnetic), from Eq. (6.11.102) we find:

$$\begin{aligned}
F^1 = & -\frac{1}{\sqrt{2}\left(r^2+\alpha^2\cos^2\theta\right)^2\left((m+r)^2+\alpha^2\cos^2\theta\right)}\left(\sqrt[4]{3}m\alpha\sin\theta\,(((-3\right.\\
& +2\sqrt[4]{3}\big)\alpha^4\cos^4\theta\\
& +m\left(2\sqrt[4]{3}m+m+2\left(1+\sqrt[4]{3}\right)r\right)\alpha^2\cos^2\theta\\
& -r(m+r)^2\left(2\sqrt[4]{3}m+\left(-3+2\sqrt[4]{3}\right)r\right)\Big)\sin\theta dr\wedge d\phi\\
& +2\left(r^2+\alpha^2\right)\cos\theta\left(\left(\left(-2+\sqrt[4]{3}\right)m\right.\right.\\
& +\left(-3+2\sqrt[4]{3}\right)r\Big)\alpha^2\cos^2\theta+(m+r)\left(\sqrt[4]{3}m^2\right.\\
& \left.\left.\left.+\left(-1+3\sqrt[4]{3}\right)rm+\left(-3+2\sqrt[4]{3}\right)r^2\right)\right)d\theta\wedge d\phi\right)\Big)
\end{aligned} \tag{6.11.108}$$

$$\begin{aligned}
F^2 = & \frac{1}{\sqrt{2}3^{3/4}\left(r^2+\alpha^2\cos^2\theta\right)^2\left((m+r)^2+\alpha^2\cos^2\theta\right)}\left(m\sin\theta\left(\alpha^2\left(-2\cos\theta\sin\theta r^3\right.\right.\right.\\
& \left.+m^2\sin 2\theta r-2(2m+r)\alpha^2\cos^3\theta\sin\theta\right)dr\wedge d\phi\\
& -\frac{1}{8}\left(r^2+\alpha^2\right)\left(8r^4+16mr^3+8m^2r^2+\alpha^4\right.\\
& \left.\left.\left.-8\alpha^2\left(-3m^2-6rm+\alpha^2\right)\cos^2\theta-\alpha^4\cos(4\theta)\right)d\theta\wedge d\phi\right)\right)
\end{aligned} \tag{6.11.109}$$

while for the second two we get:

$$\begin{aligned}
G^3 = & \frac{1}{\sqrt{2}\left(r^2+\alpha^2\cos^2\theta\right)^2\left((m+r)^2+\alpha^2\cos^2\theta\right)}\left(3^{3/4}m\sin\theta\left(\left(\sin 2\theta r^3\right.\right.\right.\\
& -2m^2\cos\theta\sin\theta r\\
& \left.+2(2m+r)\alpha^2\cos^3\theta\sin\theta\right)dr\wedge d\phi\alpha^2\\
& +\frac{1}{8}\left(r^2+\alpha^2\right)\left(8r^4+16mr^3+8m^2r^2+\alpha^4\right.\\
& \left.\left.\left.-8\alpha^2\left(-3m^2-6rm+\alpha^2\right)\cos^2\theta-\alpha^4\cos(4\theta)\right)d\theta\wedge d\phi\right)\right)
\end{aligned} \tag{6.11.110}$$

$$\begin{aligned}
G^4 = & -\frac{1}{\sqrt{2}\left(r^2+\alpha^2\cos^2\theta\right)^2\left((m+r)^2+\alpha^2\cos^2\theta\right)}\left(m\alpha\sin\theta\,((\right.\\
& -\left(-2+33^{3/4}\right)\alpha^4\cos^4\theta\\
& +m\left(\left(2+3^{3/4}\right)m+2\left(1+3^{3/4}\right)r\right)\alpha^2\cos^2\theta\\
& +r(m+r)^2\left(\left(-2+33^{3/4}\right)r-2m\right)\big)\sin\theta dr\wedge d\phi\\
& -2\left(r^2+\alpha^2\right)\cos\theta\left(-m^3+\left(-4+3^{3/4}\right)rm^2+\left(-5+43^{3/4}\right)r^2m\right.\\
& +\left(-2+33^{3/4}\right)r^3\\
& \left.\left.+\left(\left(-1+23^{3/4}\right)m+\left(-2+33^{3/4}\right)r\right)\alpha^2\cos^2\theta\right)d\theta\wedge d\phi\right)\big)
\end{aligned} \tag{6.11.111}$$

The above expressions are rather formidable, yet considering them in some limit their meaning can be decoded. First of all we recall that in the limit $\alpha \to 0$ the metric (6.11.107) becomes the Reissner–Nordstrom metric. Correspondingly in the same limit the above four-vector of field strengths degenerates into:

$$\begin{pmatrix} F^1 \\ F^2 \\ G^3 \\ G^4 \end{pmatrix} \overset{\alpha \to 0}{\Longrightarrow} \begin{pmatrix} 0 \\ -\frac{m \sin(\theta) d\theta \wedge d\phi}{\sqrt{2} 3^{3/4}} \\ \frac{3^{3/4} m \sin(\theta) d\theta \wedge d\phi}{\sqrt{2}} \\ 0 \end{pmatrix} \tag{6.11.112}$$

showing that the black hole charges $\left(0, -\frac{m}{\sqrt{2}\, 3^{1/4}}, \frac{m\, 3^{1/4}}{\sqrt{2}}, 0\right)$ have the correct form for a BPS black hole and are endowed with the characteristic $\mathbb{Z}_3$ symmetry.

Also in the $\alpha \neq 0$ we can easily determine the black hole charges by integrating the field strengths on a two-sphere of very large radius $r \to \infty$. For this purpose it is important to evaluate the asymptotic expansion of the field strengths for large radius. We find:

$$\begin{pmatrix} F^1 \\ F^2 \\ G^3 \\ G^4 \end{pmatrix} \overset{r \to \infty}{\simeq} \begin{pmatrix} -\frac{\sqrt{2}\sqrt[4]{3}\left(-3+2\sqrt[4]{3}\right) m\alpha \cos\theta \sin\theta d\theta \wedge d\phi}{r} + \mathcal{O}\left(\frac{1}{r^2}\right) \\ -\frac{m \sin\theta d\theta \wedge d\phi}{\sqrt{2} 3^{3/4}} + \mathcal{O}\left(\frac{1}{r^2}\right) \\ \frac{3^{3/4} m \sin\theta d\theta \wedge d\phi}{\sqrt{2}} + \mathcal{O}\left(\frac{1}{r^2}\right) \\ \frac{\sqrt{2}\left(-2+3 3^{3/4}\right) m\alpha \cos\theta \sin\theta d\theta \wedge d\phi}{r} + \mathcal{O}\left(\frac{1}{r^2}\right) \end{pmatrix} \tag{6.11.113}$$

and the integration on the angular variables produces the same result as for the corresponding Reissner–Nordstrom black hole:

$$\mathcal{Q}_{BPSKN} = \left(0, -\frac{m}{\sqrt{2}\, 3^{1/4}}, \frac{m\, 3^{1/4}}{\sqrt{2}}, 0\right) \tag{6.11.114}$$

In conclusion the BPS Kerr–Newman solution is a deformation of the Reissner–Nordstrom BPS black hole. It is extremal in the σ-model sense and for this reason could be retrieved from the nilpotent orbit construction. However it is not extremal in the sense of General Relativity since the mass is less than $\sqrt{q^2 + \alpha^2}$ being equal to m. For this reason we are below the limit of the cosmic censorship, there is no horizon and we have instead a naked singularity.

The important message is that, notwithstanding the deformation and the presence of a Kaluza–Klein vector, the structure of the charges is that pertaining to the orbit where the solution has been constructed, namely the BPS orbit $\mathcal{O}^3_{11}$.

6.11.5 The Large Non BPS Black Holes of $\mathscr{O}^3_{22}$

Next let us consider the orbit $\mathscr{O}^3_{22}$, which in the spherical symmetric case leads to non BPS Black holes with a finite horizon area.

W-Representation

As in the previous case, in order to better appreciate the structure of these solutions, let us slightly generalize our orbit representative, writing the following nilpotent matrix that depends on two parameters (p, q)

$$X_{3|22}(p,q) = \begin{pmatrix} q & 0 & 0 & \frac{q}{\sqrt{2}} & 0 & 0 & 0 \\ 0 & \frac{p+q}{2} & -\frac{p}{2} & 0 & -\frac{q}{2} & 0 & 0 \\ 0 & \frac{p}{2} & \frac{q-p}{2} & 0 & 0 & \frac{q}{2} & 0 \\ -\frac{q}{\sqrt{2}} & 0 & 0 & 0 & 0 & 0 & -\frac{q}{\sqrt{2}} \\ 0 & \frac{q}{2} & 0 & 0 & \frac{p-q}{2} & \frac{p}{2} & 0 \\ 0 & 0 & -\frac{q}{2} & 0 & -\frac{p}{2} & \frac{1}{2}(-p-q) & 0 \\ 0 & 0 & 0 & \frac{q}{\sqrt{2}} & 0 & 0 & -q \end{pmatrix} \quad (6.11.115)$$

The standard triple representative mentioned in Eq. (6.10.6) is just the particular case $X_{3|22}(1,1)$. Applying the usual strategy from the general formula we obtain

$$\mathscr{Q}^{\mathbf{W}}_{3|22} = \mathrm{Tr}(X_{3|22}(p,q)\mathscr{T}^{\mathbf{W}}) = \left(0, p, \sqrt{3}q, 0\right) \quad (6.11.116)$$

Substituting such a result in the expression for the quartic symplectic invariant (see Eq. (6.11.3) we find:

$$\mathfrak{I}_4 = -9\,p\,q^3 < 0 \quad \text{if } p \text{ and } q \text{ have the same sign} \quad (6.11.117)$$

This result is meaningful since, by calculating the trace $\mathrm{Tr}(X_{3|22}L^E_{+}) = 0$, we find that the Taub-NUT charge vanishes. Furthermore we note that the condition that p and q have the same sign was singled out in [32] as the defining condition of the orbit O^3_{22} which, in the spherical symmetry approach leads to regular non BPS solutions. The choice of opposite signs was proved in [32] to correspond to a different $\mathrm{H}^\star$ orbit, the non diagonal O^3_{12} which instead contains only singular solutions.

Addressing the question of stability subgroups of the original duality group in four-dimensions SL(2, R), we realize that for the charge vector (6.11.116) this subgroup is just trivial:

$$\mathrm{SL}(2,\mathbb{R}) \supset \mathscr{S}_{3|22} = \mathbf{1} \quad (6.11.118)$$

$\mathrm{H}^\star$-Stability Subgroup

Considering next the stability subgroup of the nilpotent element $X_{3|22}(1,1)$ in $\mathrm{H}^\star = \widehat{\mathfrak{sl}(2,\mathbb{R})} \oplus \mathfrak{sl}(2,\mathbb{R})_{\mathrm{h}^\star}$ we obtain:

$$\mathfrak{S}_{3|22} = \mathbb{R} \quad (6.11.119)$$

the group being generated by a matrix $\mathbb{A}_{3|22}$ of nilpotency degree 2:

$$\mathbb{A}_{3|22}^3 = \mathbf{0} \tag{6.11.120}$$

We do not give its explicit form which we do not use in the sequel.

Nilpotent Algebra $\mathbb{N}_{3|22}$

Considering next the adjoint action of the central element $h_{3|22}$ on the subspace $\mathbb{K}^\star$ we find that its eigenvalues are the following ones:

$$\text{Eigenvalues}_{3|22}^{\mathbb{K}^\star} = \{-4, 4, -2, -2, 2, 2, 0, 0\} \tag{6.11.121}$$

Therefore the three eigenoperators A_1, A_2, A_3 corresponding to the three positive eigenvalues $4, 2, 2$, respectively, form the restriction to $\mathbb{K}^\star$ of a nilpotent algebra $\mathbb{N}_{3|22}$. In this case the A_i do all commute among themselves so that we have $\mathbb{N}_{3|22} = \mathbb{N}_{3|22} \bigcap \mathbb{K}^\star$ and it is abelian. The abelian structure of the nilpotent algebra implies that for the orbit $\mathscr{O}^3_{22}$ we have only three functions $\mathfrak{h}_i^0$ which will be harmonic and independent. This is so because $\mathscr{D}\mathbb{N}_{3|22} = 0$

Explicitly we set:

$$\mathfrak{H}(\mathfrak{h}_1, \mathfrak{h}_2, \mathfrak{h}_3) = \sum_{i=1}^3 \mathfrak{h}_i \, A_i =$$

$$\begin{pmatrix}
2\mathfrak{h}_3 & \mathfrak{h}_1 - 2\mathfrak{h}_2 & 2\mathfrak{h}_1 - \mathfrak{h}_2 & -\sqrt{2}\mathfrak{h}_3 & -3\mathfrak{h}_2 & -3\mathfrak{h}_1 & 0 \\
\mathfrak{h}_1 - 2\mathfrak{h}_2 & \mathfrak{h}_3 & 0 & \sqrt{2}\mathfrak{h}_2 - 2\sqrt{2}\mathfrak{h}_1 & \mathfrak{h}_3 & 0 & -3\mathfrak{h}_1 \\
\mathfrak{h}_2 - 2\mathfrak{h}_1 & 0 & \mathfrak{h}_3 & \sqrt{2}\mathfrak{h}_1 - 2\sqrt{2}\mathfrak{h}_2 & 0 & -\mathfrak{h}_3 & -3\mathfrak{h}_2 \\
\sqrt{2}\mathfrak{h}_3 & 2\sqrt{2}\mathfrak{h}_1 - \sqrt{2}\mathfrak{h}_2 & \sqrt{2}\mathfrak{h}_1 - 2\sqrt{2}\mathfrak{h}_2 & 0 & \sqrt{2}\mathfrak{h}_1 - 2\sqrt{2}\mathfrak{h}_2 & \sqrt{2}\mathfrak{h}_2 - 2\sqrt{2}\mathfrak{h}_1 & \sqrt{2}\mathfrak{h}_3 \\
3\mathfrak{h}_2 & -\mathfrak{h}_3 & 0 & \sqrt{2}\mathfrak{h}_1 - 2\sqrt{2}\mathfrak{h}_2 & -\mathfrak{h}_3 & 0 & 2\mathfrak{h}_1 - \mathfrak{h}_2 \\
-3\mathfrak{h}_1 & 0 & \mathfrak{h}_3 & 2\sqrt{2}\mathfrak{h}_1 - \sqrt{2}\mathfrak{h}_2 & 0 & -\mathfrak{h}_3 & \mathfrak{h}_1 - 2\mathfrak{h}_2 \\
0 & -3\mathfrak{h}_1 & 3\mathfrak{h}_2 & -\sqrt{2}\mathfrak{h}_3 & \mathfrak{h}_2 - 2\mathfrak{h}_1 & \mathfrak{h}_1 - 2\mathfrak{h}_2 & -2\mathfrak{h}_3
\end{pmatrix} \tag{6.11.122}$$

Considering $\mathfrak{H}(\mathfrak{h}_1, \mathfrak{h}_2, \mathfrak{h}_3)$ as a Lax operator and calculating its Taub-NUT charge and electromagnetic charges we find:

$$\mathbf{n}_{TN} = -6\mathfrak{h}_1 \quad ; \quad \mathscr{Q} = \left\{2\sqrt{3}\left(\mathfrak{h}_2 - 2\mathfrak{h}_1\right), 0, -2\sqrt{3}\mathfrak{h}_3, -6\mathfrak{h}_2\right\} \tag{6.11.123}$$

This implies that constructing the multi-centre solution with harmonic functions the condition $\mathfrak{h}_1 = 0$ might be sufficient to annihilate the Taub-NUT current. In this case we will be lucky and such a condition suffices.

For later convenience let us change the normalization in the basis of harmonic functions as follows:

$$\mathfrak{h}_1^{(0)} = \mathscr{H}_1 \quad ; \quad \mathfrak{h}_2^{(0)} = \tfrac{1}{2}\left(1 - \mathscr{H}_2\right) \quad ; \quad \mathfrak{h}_3^{(0)} = \tfrac{1}{2}\left(1 - \mathscr{H}_3\right) \tag{6.11.124}$$

Implementing the symmetric coset construction with:

$$\mathscr{Y}\left(\mathscr{H}_1, \mathscr{H}_2, \mathscr{H}_3\right) \equiv \exp\left[\mathfrak{H}\left(\mathscr{H}_1, \tfrac{1}{2}\left(1-\mathscr{H}_2\right), \tfrac{1}{2}\left(1-\mathscr{H}_3\right)\right)\right] \tag{6.11.125}$$

calculating the upper triangular coset representative $\mathbb{L}(\mathscr{Y})$ according to Eq. (6.8.26) and extracting the σ-model scalar fields we obtain an explicit expression which is sufficiently simple to be displayed:

$$\exp[-U] = \sqrt{\mathscr{H}_2\mathscr{H}_3^3 - 4\mathscr{H}_1^2} \tag{6.11.126}$$

$$\operatorname{Im} z = \frac{\sqrt{\mathscr{H}_2\mathscr{H}_3^3 - 4\mathscr{H}_1^2}}{\mathscr{H}_3^2} \tag{6.11.127}$$

$$\operatorname{Re} z = -\frac{2\mathscr{H}_1}{\mathscr{H}_3^2} \tag{6.11.128}$$

$$Z^M = \begin{pmatrix} -\frac{\sqrt{6}\mathscr{H}_1\mathscr{H}_3}{4\mathscr{H}_1^2-\mathscr{H}_2\mathscr{H}_3^3} \\ \frac{4\mathscr{H}_1^2-(\mathscr{H}_2-1)\mathscr{H}_3^3}{\sqrt{2}\left(4\mathscr{H}_1^2-\mathscr{H}_2\mathscr{H}_3^3\right)} \\ \frac{\sqrt{\frac{3}{2}}\left(4\mathscr{H}_1^2-\mathscr{H}_2(\mathscr{H}_3-1)\mathscr{H}_3^2\right)}{4\mathscr{H}_1^2-\mathscr{H}_2\mathscr{H}_3^3} \\ \frac{\sqrt{2}\mathscr{H}_1\mathscr{H}_2}{4\mathscr{H}_1^2-\mathscr{H}_2\mathscr{H}_3^3} \end{pmatrix} \tag{6.11.129}$$

$$a = -\frac{\mathscr{H}_1\left(\mathscr{H}_2 + 3\mathscr{H}_3 - 2\right)}{4\mathscr{H}_1^2 - \mathscr{H}_2\mathscr{H}_3^3} \tag{6.11.130}$$

Using these results we easily obtain the Taub-NUT current in the following form:

$$j^{TN} = 2 \star \nabla\mathscr{H}_1 \tag{6.11.131}$$

In this case the predicted condition $\mathscr{H}_1 = 0$ is sufficient to annihilate the Taub-NUT current and we obtain an extremely simple result.[14] The complete form of the supergravity solution corresponding to this choice is:

$$\exp[-U] = \sqrt{\mathscr{H}_3^3\mathscr{H}_2} \tag{6.11.132}$$

$$z = \mathrm{i}\,\frac{\sqrt{\mathscr{H}_3^3\mathscr{H}_2}}{\mathscr{H}_3^2} \tag{6.11.133}$$

$$j^{TN} = 0 \tag{6.11.134}$$

[14] Actually even the condition $\mathscr{H}_1 = \text{const}$ suffices to annihilate the Taub-NUT charge allowing for a non trivial real part of the z-field. However in this section we analyze the case $\mathscr{H}_1 = 0$ for its remarkable simplicity.

$$j^{EM} = \star\nabla \begin{pmatrix} 0 \\ -\frac{\mathscr{H}_2}{\sqrt{2}} \\ -\sqrt{\frac{3}{2}}\mathscr{H}_3 \\ 0 \end{pmatrix} \tag{6.11.135}$$

Comparing with the case of the large BPS orbit we see that the only difference is the relative sign of the harmonic functions in the electromagnetic current. What we said for the BPS black holes extends to the non BPS ones in the same way.

Summary

For a multicenter solution associated with the $\mathscr{O}^3_{22}$ orbit we have a mixture of very small and small black holes as in the case of the orbit $\mathscr{O}^3_{22}$. Also here a finite area non BPS black hole emerges when the center of a very small black comes to coincides with the center of a small one. The only difference is the relative sign of the two charges. With equal signs we construct a non BPS state, while with opposite charges we construct a BPS one. This reinforces the conjecture that at the quantum level finite black holes can be interpreted as composite states.

This conjecture is also supported by an angular momentum analysis. Looking at the representations in Table 6.1, we see that the representation $2(j = 1) + (j = 0)$ that corresponds to BPS and non BPS large black holes can be obtained by summing the representation $(j = 1) + 2(j = \frac{1}{2})$ that corresponds to small black holes with the representation $3(j = 0) + 2(j = \frac{1}{2})$ that corresponds to very small black holes. Consider the following table:

1	$\frac{1}{2}$	$\frac{1}{2}$	0	$-\frac{1}{2}$	$-\frac{1}{2}$	-1
0	$\frac{1}{2}$	$-\frac{1}{2}$	0	$\frac{1}{2}$	$-\frac{1}{2}$	0
1	1	0	0	0	-1	-1

the numbers in the first line are the eigenvalues of the central element h in the triplet (h, X, Y) characterizing the orbit $\mathscr{O}^4_{11}$. The second line contains the eigenvalues for the central element of the triplet of the orbit $\mathscr{O}^4_{11}$. In the last line we have the eigenvalues for the h in the triplet characterizing the orbit $\mathscr{O}^3_{i,j}$. We realize that the coincidence of centres correspond to the identification of a new SL(2, R) subgroup which is the direct sum of the original two associated with the two small black holes.

6.11.6 The Largest Orbit $\mathscr{O}^1_{11}$

Next let us consider the orbit $\mathscr{O}^1_{11}$, which in the spherical symmetric case leads only to singular solutions.

W-Representation

Applying the usual strategy from the general formula we obtain a charge vector

$$\mathscr{Q}^{\mathbf{w}}_{1|11} = \mathrm{Tr}(X_{1|11}(p,q)\mathscr{T}^{\mathbf{w}}) \tag{6.11.136}$$

which has no invariance:

$$\mathrm{SL}(2,\mathbb{R}) \supset \mathscr{S}_{1|11} = \mathbf{1} \tag{6.11.137}$$

and yields a quartic invariant generically different from zero:

$$\mathfrak{I}_4 \neq 0 \tag{6.11.138}$$

Because of our simplified choice of the representative the Taub-NUT charge is not zero and only later we will enforce the vanishing of the Taub-NUT current on the harmonic function parameterized solution.

H*-Stability Subgroup

Considering next the stability subgroup of the nilpotent element $X_{1|11}$ in $\mathrm{H}^\star = \widehat{\mathfrak{sl}(2,\mathbb{R})} \oplus \mathfrak{sl}(2,\mathbb{R})_{\mathrm{h}^\star}$ we obtain that it is trivial:

$$\mathfrak{S}_{1|11} = \mathbf{1} \tag{6.11.139}$$

Nilpotent Algebra $\mathbb{N}_{1|11}$

Considering next the adjoint action of the central element $h_{1|11}$ on the subspace $\mathbb{K}^\star$ we find that its eigenvalues are the following ones:

$$\mathrm{Eigenvalues}^{\mathbb{K}^\star}_{3|22} = \{-5, 5, -3, 3, -1, -1, 1, 1\} \tag{6.11.140}$$

Therefore the four eigenoperators A_1, A_2, A_3, A_4 corresponding to the four positive eigenvalues 5, 3, 1, 1, respectively, form the restriction to $\mathbb{K}^\star$ of a nilpotent algebra $\mathbb{N}_{1|11}$. In this case the A_i do not all commute among themselves so that we have $\mathbb{N}_{1|11} \neq \mathbb{N}_{1|11} \bigcap \mathbb{K}^\star$. The full algebra involves also two operators $B_1, B_2 \in \mathbb{H}^\star$ and the full set of commutation relations is the following one:

$$\begin{aligned}
0 &= [A_1\,,\,A_2] = [A_1\,,\,A_3] = [A_1\,,\,A_4]\\
0 &= [A_2\,,\,A_3]\\
0 &= [B_1\,,\,B_2] = [B_1\,,\,A_1] = [B_1\,,\,A_2]\\
0 &= [B_1\,,\,A_4] = [B_2\,,\,A_1] = [B_2\,,\,A_3]\\
B_1 &= [A_2\,,\,A_4]\\
B_2 &= [A_3\,,\,A_4]\\
-16\,A_1 &= [B_1\,,\,A_3]\\
-16\,A_1 &= [B_2\,,\,A_1]
\end{aligned}$$

$$24\,A_2 = [B_2\,,\,A_4] \tag{6.11.141}$$

By inspection of Eq. (6.11.141) we easily see that:

$$\mathscr{D}\mathbb{N}_{1|11} = \text{span}\,\{B_1, B_2, A_1, A_2\} \quad ; \quad \mathscr{D}\mathbb{N}_{1|11} \bigcap \mathbb{K}^\star = \text{span}\,\{A_1, A_2\} \tag{6.11.142}$$

$$\mathscr{D}^2\mathbb{N}_{1|11} = \text{span}\,\{A_1\} = \mathscr{D}^2\mathbb{N}_{1|11} \bigcap \mathbb{K}^\star \tag{6.11.143}$$

This structure of the nilpotent algebra implies that for the orbit $\mathscr{O}^1_{11}$ we have only two functions $\mathfrak{h}^0_3$, $\mathfrak{h}^0_4$ which are harmonic and independent. The other two functions $\mathfrak{h}^2_1$, $\mathfrak{h}^1_2$, obey instead equations in which the previous two play the role of sources. Not surprisingly $\mathfrak{h}^2_1$, $\mathfrak{h}^1_2$ correspond to the higher gradings 5 and 3, while $\mathfrak{h}^0_3$, $\mathfrak{h}^0_4$ correspond to the gradings 1, 1. More precisely $\mathfrak{h}^1_2$ receives source contributions only from $\mathfrak{h}^0_3$, $\mathfrak{h}^0_4$, while $\mathfrak{h}^2_1$ receives source contributions from $\mathfrak{h}^1_2$, $\mathfrak{h}^0_3$, $\mathfrak{h}^0_4$

Explicitly we set:

$$\mathfrak{H}(\mathfrak{h}_1, \ldots, \mathfrak{h}_4) = \textstyle\sum_{i=1}^4 \mathfrak{h}_i\, A_i =$$

$$\begin{pmatrix}
\mathfrak{h}_1+\mathfrak{h}_4 & \frac{\mathfrak{h}_2}{3}-\mathfrak{h}_3 & \mathfrak{h}_4 & \frac{\sqrt{2}\mathfrak{h}_2}{3}-\sqrt{2}\mathfrak{h}_3 & -\mathfrak{h}_1 & -\mathfrak{h}_2-\mathfrak{h}_3 & 0\\
\frac{\mathfrak{h}_2}{3}-\mathfrak{h}_3 & 2\mathfrak{h}_4 & \mathfrak{h}_2+\mathfrak{h}_3 & -\sqrt{2}\mathfrak{h}_4 & \mathfrak{h}_3-\frac{\mathfrak{h}_2}{3} & 0 & -\mathfrak{h}_2-\mathfrak{h}_3\\
-\mathfrak{h}_4 & -\mathfrak{h}_2-\mathfrak{h}_3 & \mathfrak{h}_1-\mathfrak{h}_4 & \frac{\sqrt{2}\mathfrak{h}_2}{3}-\sqrt{2}\mathfrak{h}_3 & 0 & \frac{\mathfrak{h}_2}{3}-\mathfrak{h}_3 & -\mathfrak{h}_1\\
\sqrt{2}\mathfrak{h}_3-\frac{\sqrt{2}\mathfrak{h}_2}{3} & \sqrt{2}\mathfrak{h}_4 & \frac{\sqrt{2}\mathfrak{h}_2}{3}-\sqrt{2}\mathfrak{h}_3 & 0 & \frac{\sqrt{2}\mathfrak{h}_2}{3}-\sqrt{2}\mathfrak{h}_3 & -\sqrt{2}\mathfrak{h}_4 & \sqrt{2}\mathfrak{h}_3-\frac{\sqrt{2}\mathfrak{h}_2}{3}\\
\mathfrak{h}_1 & \frac{\mathfrak{h}_2}{3}-\mathfrak{h}_3 & 0 & \frac{\sqrt{2}\mathfrak{h}_2}{3}-\sqrt{2}\mathfrak{h}_3 & \mathfrak{h}_4-\mathfrak{h}_1 & -\mathfrak{h}_2-\mathfrak{h}_3 & \mathfrak{h}_4\\
-\mathfrak{h}_2-\mathfrak{h}_3 & 0 & \mathfrak{h}_3-\frac{\mathfrak{h}_2}{3} & \sqrt{2}\mathfrak{h}_4 & \mathfrak{h}_2+\mathfrak{h}_3 & -2\mathfrak{h}_4 & \frac{\mathfrak{h}_2}{3}-\mathfrak{h}_3\\
0 & -\mathfrak{h}_2-\mathfrak{h}_3 & \mathfrak{h}_1 & \frac{\sqrt{2}\mathfrak{h}_2}{3}-\sqrt{2}\mathfrak{h}_3 & -\mathfrak{h}_4 & \frac{\mathfrak{h}_2}{3}-\mathfrak{h}_3 & -\mathfrak{h}_1-\mathfrak{h}_4
\end{pmatrix} \tag{6.11.144}$$

Considering $\mathfrak{H}(\mathfrak{h}_1, \ldots, \mathfrak{h}_4)$ as a Lax operator and calculating its Taub-NUT charge and electromagnetic charges we find:

$$\mathbf{n}_{TN} = -2(\mathfrak{h}_2+\mathfrak{h}_3) \quad ; \quad \mathscr{Q} = \left\{-2\sqrt{3}\mathfrak{h}_4, -2\,(\mathfrak{h}_2+\mathfrak{h}_3)\,, \frac{2\,(\mathfrak{h}_2-3\mathfrak{h}_3)}{\sqrt{3}}, -2\mathfrak{h}_1\right\} \tag{6.11.145}$$

This implies that constructing the multi-centre solution with harmonic functions the condition $\mathfrak{h}_2 = -\mathfrak{h}_3$ might be sufficient to annihilate the Taub-NUT current.

Implementing the symmetric coset construction with:

$$\mathscr{Y}\,(\mathfrak{h}_1, \ldots, \mathfrak{h}_4) \equiv \exp\left[\mathfrak{H}\,(\mathfrak{h}_1, \ldots, \mathfrak{h}_4)\right] \tag{6.11.146}$$

and imposing the field equations (6.8.14) we obtain the following conditions:

$$
\begin{aligned}
0 = &\frac{224}{5}\nabla\mathfrak{h}_3\circ\nabla\mathfrak{h}_3\,\mathfrak{h}_4^3 - \frac{16}{5}\,\mathfrak{h}_3\Delta\,\mathfrak{h}_3\,\mathfrak{h}_4^3 - \frac{416}{5}\nabla\,\mathfrak{h}_3\circ\nabla\mathfrak{h}_4\,\mathfrak{h}_3\,\mathfrak{h}_4^2 + \frac{16}{5}\,\mathfrak{h}_3^2\Delta\,\mathfrak{h}_4\,\mathfrak{h}_4^2\\
&+\frac{192}{5}\nabla\,\mathfrak{h}_4\circ\nabla\mathfrak{h}_4\,\mathfrak{h}_3^2\,\mathfrak{h}_4 + \frac{32}{3}\nabla\mathfrak{h}_2\circ\nabla\mathfrak{h}_3\,\mathfrak{h}_4 - \frac{8}{3}\,\mathfrak{h}_3\Delta\mathfrak{h}_2\,\mathfrak{h}_4 - \frac{8}{3}\,\mathfrak{h}_2\Delta\,\mathfrak{h}_3\,\mathfrak{h}_4\\
&-\frac{16}{3}\nabla\,\mathfrak{h}_3\circ\nabla\mathfrak{h}_4\,\mathfrak{h}_2 - \frac{16}{3}\nabla\mathfrak{h}_2\circ\nabla\mathfrak{h}_4\,\mathfrak{h}_3 + \Delta\,\mathfrak{h}_1 + \frac{16}{3}\,\mathfrak{h}_2\,\mathfrak{h}_3\Delta\mathfrak{h}_4\\
0 = &\,4\Delta\mathfrak{h}_3\,\mathfrak{h}_4^2 - 8\nabla\,\mathfrak{h}_3\circ\nabla\mathfrak{h}_4\,\mathfrak{h}_4 - 4\,\mathfrak{h}_3\Delta\mathfrak{h}_4\,\mathfrak{h}_4 + 8\nabla\mathfrak{h}_4\circ\nabla\mathfrak{h}_4\,\mathfrak{h}_3 + \Delta\mathfrak{h}_2\\
0 = &\,\Delta\mathfrak{h}_3\\
0 = &\,\Delta\mathfrak{h}_4
\end{aligned}
\tag{6.11.147}
$$

Solutions of the above system can be quite complicated and can encompass many different types of behaviors, yet what is generically true is that the contributions from the source term introduces in $\mathfrak{h}_1$ and $\mathfrak{h}_2$ poles $1/r^p$ stronger than $p = 1$, while $\mathfrak{h}_3$ and $\mathfrak{h}_4$ have only simple poles. Hence if the structure of the polynomials in the functions $\mathfrak{h}_{1,2,3,4}$ is such that at simple poles the divergence of the inverse warp factor is already too strong or the coefficient already becomes imaginary, introducing stronger poles can only make the situation worse. For this reason we confine ourselves to analyze solutions encompassed in this orbit in which the source terms vanish identically upon the implementation of some identifications.

There are few different reductions with such a property and we choose just one that has also the additional feature of annihilating the Taub-NUT current. It is the following one:

$$\mathfrak{h}_3 \;=\; \mathfrak{h}_4 \;=\; -\,\mathfrak{h}_2 \;\equiv\; \mathfrak{h} \tag{6.11.148}$$

The reader can easily check that with the choice (6.11.148) the system of equations (6.11.147) reduces to:

$$\Delta\mathfrak{h} \;=\; \Delta\mathfrak{h}_1 \;=\; 0 \tag{6.11.149}$$

For later convenience let us change the normalization in the basis of harmonic functions as follows:

$$\mathfrak{h}_4 = \tfrac{1}{4}\mathscr{H} \quad ; \quad \mathfrak{h}_3 = \tfrac{1}{4}\mathscr{H} \quad ; \quad \mathfrak{h}_2 = -\tfrac{1}{4}\mathscr{H} \quad ; \quad \mathfrak{h}_1 = -\tfrac{1}{4} + \mathscr{W} \tag{6.11.150}$$

calculating the upper triangular coset representative $\mathbb{L}(\mathscr{Y})$ according to Eq. (6.8.26) and extracting the σ-model scalar fields we obtain explicit expressions which are sufficiently simple to be displayed:

$$\exp[U] = \frac{8\sqrt{15}}{\sqrt{-(\mathscr{H}+2)^3\left(\mathscr{H}^5+10\mathscr{H}^4+40\mathscr{H}^3+80\mathscr{H}^2-60(4\mathscr{W}+1)\right)}} \tag{6.11.151}$$

$$\mathrm{Im}\, z = \frac{3\sqrt{15}(\mathscr{H}+2)}{\sqrt{-\frac{\mathscr{H}+2}{\mathscr{H}^2(\mathscr{H}(\mathscr{H}(\mathscr{H}+10)+40)+80)-60(4\mathscr{W}+1)}}\left((\mathscr{H}(\mathscr{H}(\mathscr{H}+10)+20)-40)\mathscr{H}^2+90(4\mathscr{W}+1)\right)} \tag{6.11.152}$$

$$\mathrm{Re}\, z = \frac{15\mathscr{H}(\mathscr{H}+2)(\mathscr{H}+4)}{\mathscr{H}^5+10\mathscr{H}^4+20\mathscr{H}^3-40\mathscr{H}^2+360\mathscr{W}+90} \tag{6.11.153}$$

We skip the form of the Z fields and of a but we mention their consequences, namely the Taub-NUT current

$$j^{TN} = 0 \tag{6.11.154}$$

and the electromagnetic currents

$$j^{EM} = \star\nabla \left\{ \frac{1}{2}\sqrt{\frac{3}{2}}\mathscr{H}, 0, \frac{7\mathscr{H}}{6}, \sqrt{2}\mathscr{W} \right\} \tag{6.11.155}$$

This shows that a black hole belonging to this orbit has a charge vector $\mathscr{Q} = \left\{ \frac{1}{2}\sqrt{\frac{3}{2}}p, 0, \frac{7p}{6}, \sqrt{2}q \right\}$, whose quartic invariant is:

$$\mathfrak{I}_4 = \frac{1}{128}p^3(49p + 72q) \tag{6.11.156}$$

This latter can be positive or negative depending on the choices for p and q. The problem, however, is that this solution is always singular around all poles of $\mathscr{H}$. Indeed setting:

$$\mathscr{H} \sim \frac{p}{r} \quad ; \quad \mathscr{W} \sim \frac{q}{r} \tag{6.11.157}$$

we find that for $r \to 0$ the inverse warp factor behaves as follows:

$$\exp[-U] \sim \frac{\sqrt{-p^8}}{8\sqrt{15}r^4} + \frac{\sqrt{-p^8}}{\sqrt{15}pr^3} + \frac{\sqrt{\frac{3}{5}}\sqrt{-p^8}}{p^2r^2} + \frac{4\sqrt{-p^8}}{\sqrt{15}p^3r} + \frac{\sqrt{\frac{3}{5}}p^3(p+5q)}{\sqrt{-p^8}} + \mathscr{O}(r) \tag{6.11.158}$$

The coefficient $\sqrt{-p^8}$ indicates that approaching the pole the warp factor becomes imaginary at a finite distance from it and the would be horizon $r = 0$ is never reached. If it were reached, the divergence $\frac{1}{r^4}$ would imply an infinite area of the horizon. As we know from our general discussion the Riemann tensor diverges if the warp factor goes to zero faster than r^2 so that the would be horizon would actually be a singularity. Yet since the warp factor becomes imaginary at a finite distance from the pole it remains open the question if solutions of this type can be prolonged by suitably changing the coordinate system. In that case they might acquire a physical meaning. So far such a question has not been tackled but it deserves to be.

6.12 Conclusions on the Episteme Contained in This Chapter

In this very long chapter we have tackled quite advanced issues of current or of quite recent research. Although all the inspiring motivations come from *Supergravity*, the material here presented is of genuine algebraic and geometrical character; indeed it might be understood and treated within the scope of pure Mathematics. As usual, the role of supersymmetry was just that of directing our choices, leading us to focus on *special manifolds* endowed with *special geometries*.

Actually the methods and the constructions considered in this chapter are general and might be dealt with no knowledge of supermultiplets and supercharges. Additional inspiration coming from *Supergravity* is encoded in the strategic attention paid to the *Tits–Satake projection* and to *Tits–Satake universality classes*, which, however, are purely mathematical phenomena, self-contained in Lie algebra theory.

Even the very final physical motivation of constructing *extremal black-hole solutions* might be forgotten once, in the spirit of *the geometry of geometries*, a physical–geometrical problem has been mapped into another purely geometrical one.

Thus let us summarize into a list of points the mathematical logic of what we have been discussing in the present chapter.

(A) The problem of constructing extremal black-hole solutions is reduced to the construction and classification of mappings:

$$\Phi \quad : \quad \mathbb{R}^3 \Longrightarrow \mathcal{M}_s \tag{6.12.1}$$

where $(\mathcal{M}_s, g)$ is a pseudo-Riemmannian manifold and the map Φ satisfies both the σ-model equations of motion and the stress-tensor vanishing condition:

$$\partial_i \left(\frac{\partial \Phi^\mu}{\partial x^i} \nabla_\mu \Phi^\nu \right) = 0 \quad ; \quad g_{\mu\nu}(\Phi)\, \partial_i \Phi^\mu \, \partial_j \Phi^\nu = 0 \tag{6.12.2}$$

(B) The geometrical problem posed in (A) can be considered for any Lorentzian-manifold $\mathcal{M}_s$ but, instructed by supersymmetry, we localize it on the homogeneous manifolds:

$$\mathcal{M}_s = \frac{\mathrm{U}_{\mathrm{D}=3}}{\mathrm{H}^\star} \tag{6.12.3}$$

listed in Table 5.4 that are in the image of the $c^\star$-map and have a structure fitting the golden splitting (1.7.12)

(C) For the reasons discussed at length in previous sections and chapters we are actually interested only in those maps of the type (6.12.1) where:

$$\Phi\left[\mathbb{R}^3\right] \subset \frac{\mathrm{U}^{\mathrm{TS}}_{\mathrm{D}=3}}{\mathrm{H}^\star_{\mathrm{TS}}} \subset \frac{\mathrm{U}_{\mathrm{D}=3}}{\mathrm{H}^\star} \tag{6.12.4}$$

namely where the image of the three-dimensional space $\mathbb{R}^3$ lies entirely inside the Tits-Satake submanifold.

(D) These $\mathrm{H}^\star$–orbits of solutions can be classified and explicitly constructed thanks to an algorithm, thoroughly explained in Sect. 6.8, that associates such solutions to each $\mathrm{H}^\star$–orbit of nilpotent operators $X \in \mathbb{K}$, where $\mathbb{K}$ is the orthogonal complement of the subalgebra $\mathbb{H}^\star \subset \mathbb{U}$. The classification of U-nilpotent orbits is a frontier topic in Mathematics and, further specialized to $\mathrm{H}^\star \subset \mathrm{U}$ orbits, involves items and techniques generically not yet available in the mathematical supermarket, like the generalized Weyl group $\mathscr{GW}$ and the H-Weyl subgroup $\mathscr{W}_{\mathrm{H}}$.

(E) Within the class of manifolds in the image of the $c^{\star}$-map, the problem of $\mathrm{H}^{\star}$ nilpotent orbits acquires very special features because of the special nature of the subgroup $\mathrm{H}^{\star}$. These special features are ultimately related with the golden splitting structure (1.7.12) which is on its turn a land-mark of special geometries. The complicated mechanisms here at work relate the classification of $\mathrm{H}^{\star}$–orbits with the classification of $\mathrm{U}_{\mathrm{D}=4}$–orbits in the **W**-representation.

(F) The association of the considered mathematical problem with extremal black-holes provides the features pointed out in (E) with physical interpretations in terms of electromagnetic charges, horizon areas and fixed scalars. Yet we might complete ignore such interpretations and ask ourself the question of what is the abstract, purely mathematical meaning of such relations as that between $\mathrm{U}_{\mathrm{D}=4}$–orbits in the **W**-representation and $\mathrm{H}^{\star}$ nilpotent orbits. Such a study has not yet been performed but might be the source of new precious insights.

Generally speaking the problem considered in this chapter unveils new very profound aspects of Special Geometries pertaining both to the scope of Geometry and of Lie Algebra Theory. As we tried to emphasize in point (F) of the above list a mathematical reformulation of all the mechanisms spotted in this context might be of great moment. We might find clues to some generalization of the golden splitting that goes beyond both supersymmetry and even homogeneous spaces and opens some new direction in differential and algebraic geometry. Inspiring clues come probably from a careful analysis of Weyl subgroups and the characterization among them of those that can be regarded as H-subgroups.

In this context an inspiring observation appears to be the one highlighted in previous pages that regular finite horizon black-holes can be regarded as bound-states of small or very small black-holes. An in depth investigation of the proper mathematics lurking behind this feature is potentially capable of revealing new exciting perspectives both in geometry and physics.

References

1. S. Ferrara, R. Kallosh, A. Strominger, N = 2 extremal black holes. Phys. Rev. D **52**, 5412–5416 (1995)
2. S. Ferrara, R. Kallosh, Supersymmetry and attractors. Phys. Rev. D **54**, 1514–1524 (1996)
3. A. Strominger, C. Vafa, Microscopic origin of the Bekenstein-Hawking entropy. Phys. Lett. B **379**, 99 (1996)
4. M. Bertolini, M. Trigiante, Microscopic entropy of the most general four-dimensional BPS black hole. JHEP **10**, 002 (2000)
5. A. Dabholkar, Microstates of nonsupersymmetric black holes. Phys. Lett. B **402**, 53–58 (1997)
6. A. Ghosh, P. Mitra, Counting of black hole microstates. Indian J. Phys. **80**, 867 (2006)
7. F. Larsen, A String model of black hole microstates. Phys. Rev. D **56**, 1005–1008 (1997)
8. J.M. Maldacena, A. Strominger, E. Witten, Black hole entropy in M theory. JHEP **12**, 002 (1997)
9. A. Strominger, Black hole entropy from near horizon microstates. JHEP **02**, 009 (1998)

10. G. Gibbons, R. Kallosh, B. Kol, Moduli, scalar charges, and the first law of black hole thermodynamics. Phys. Rev. Lett. **77**, 4992–4995 (1996)
11. R. D'Auria, P. Fré, BPS black-holes in supergravity: duality groups, p-branes, central charges and entropy, in *Classical and Quantum Black Holes*, ed. by P. Fré, V. Gorini, G. Magli, U. Moschella (IOP Publishing Ltd, 1999), pp. 137–272
12. L. Andrianopoli, R. D'Auria, S. Ferrara, M. Trigiante, Extremal black holes in supergravity. Lect. Notes Phys. **737**, 661–727 (2008)
13. S. Bellucci, S. Ferrara, M. Gunaydin, A. Marrani, Charge orbits of symmetric special geometries and attractors. Int. J. Mod. Phys. A **21**, 5043–5098 (2006)
14. S. Bellucci, S. Ferrara, A. Marrani, On some properties of the attractor equations. Phys. Lett. B **635**, 172 (2006)
15. G. Bossard, Extrenal black holes and nilpotent orbits, in *Proceedings, 16th International Congress on Mathematical Physics (ICMP09)* (Prague, Czech Republic, 2009)
16. G. Bossard, The extremal black holes of N = 4 supergravity from so(8,2+n) nilpotent orbits. Gen. Rel. Grav. **42**, 539–565 (2010)
17. G. Bossard, H. Nicolai, Multi-black holes from nilpotent lie algebra orbits. Gen. Rel. Grav. **42**, 509 (2010)
18. G. Bossard, H. Nicolai, K. Stelle, Universal bps structure of stationary supergravity solutions. JHEP **0907**, 003 (2009)
19. G. Bossard, C. Ruef, Interacting non-BPS black holes. Gen. Rel. Grav. **44**, 21 (2012)
20. A. Ceresole, S. Ferrara, A. Marrani, Small N = 2 extremal black holes in special geometry. Phys. Lett. B **693**, 366–372 (2010)
21. A. Giryavets, New attractors and area codes. JHEP **0603**, 020 (2006)
22. K. Goldstein, N. Iizuka, R.P. Jena, S.P. Trivedi, Non-supersymmetric attractors. Phys. Rev. D **72**, 124021 (2005)
23. R. Kallosh, JHEP. New attractors **0512**, 022 (2005)
24. R. Kallosh, N. Sivanandam, M. Soroush, The non-BPS black hole attractor equation. JHEP **0603**, 060 (2006)
25. P. Tripathy, S. Trivedi, Non-supersymmetric attractors in string theory. JHEP **0603**, 022 (2006)
26. W. Chemissany, J. Rosseel, M. Trigiante, T. Van Riet, The full integration of black hole solutions to symmetric supergravity theories. Nucl. Phys. B **830**, 391 (2010)
27. W. Chemissany, P. Fre, J. Rosseel, A. Sorin, M. Trigiante, T. Van Riet, Black holes in supergravity and integrability. JHEP **2010**, 080 (1009)
28. W. Chemissany, P. Fré, A. Sorin, The integration algorithm of lax equation for both generic lax matrices and generic initial conditions. Nucl. Phys. B **833**, 220 (2010)
29. P. Fré, A. Sorin, *The Integration Algorithm for Nilpotent Orbits of* $g/h^\star$ *Lax Systems: for Extrenal Black Holes* (2009)
30. P. Fré, A. Sorin, Supergravity black holes and billiards and liouville integrable structure of dual borel algebras. JHEP **03**, 066 (2010)
31. P. Frè, A. Sorin, M. Trigiante, *Black Hole Nilpotent Orbits and Tits Satake Universality Classes* (2011)
32. P. Frè, A. Sorin, M. Trigiante, Integrability of supergravity black holes and new tensor classifiers of regular and nilpotent orbits. JHEP **1204**, 015 (2012)
33. P. Fré, A.S. Sorin, Integrability of supergravity billiards and the generalized Toda lattice equation. Nucl. Phys. B **733**, 334–355 (2006)
34. A. Ceresole, G. Dall'Agata, S. Ferrara, A. Yeranyan, First order flows for N = 2 extremal black holes and duality invariants. Nucl. Phys. B **824**, 239–253 (2010)
35. P. G. Fré, *Gravity, a Geometrical Course*, vols. 1, 2 (Springer Science & Business Media, Berlin, 2012)
36. D.H. Collingwood, W.McGovern, *Nilpotent Orbits in Semisimple Lie Algebras* (Van Nostrand Reinhold, 1993)

37. S. Kim, J. Lindman, J. Hörnlund, J. Palmkvist, A. Virmani, Extremal solutions of the s^3 model and nilpotent $g_{2(2)}$ orbits. JHEP **2010**, 072 (1008)
38. G. Bossard, Octonionic black holes. JHEP **05**, 113 (2012)
39. K. Behrndt, D. Lust, W. Sabra, Stationary solutions of n = 2 supergravity. Nucl. Phys. B **510**, 264–288 (1998)

Chapter 7
E_7, F_4 and Supergravity Scalar Potentials

To him who looks upon the world rationally, the world in its turn presents a rational aspect. The relation is mutual.

Georg Wilhelm Friedrich Hegel

7.1 Historical Introduction

As we mentioned in previous chapters, the exceptional Lie algebras, for long time regarded as mathematical curiosities, came to the forefront of research with the advent of supergravity.

Their first sensational appearance took place with the work of Cremmer and Julia (see Figs. 4.4 and 4.5) who constructed ungauged $\mathcal{N} = 8$ supergravity in $D = 4$ by means of dimensional reduction of the $D = 11$ theory and demonstrated that the entire structure of the lagrangian is governed by a duality symmetry $\mathrm{E}_{7(7)}$ [1]. Indeed the 70 scalar fields comprised in the unique graviton multiplet are the coordinates of the symmetric space:

$$\mathcal{M}_{scalar} = \frac{\mathrm{E}_{7(7)}}{\mathrm{SU}(8)} \tag{7.1.1}$$

and the field strengths of the **28** vector fields, together with their magnetic duals span the fundamental **56** representation of $E_{7(7)}$, which Élie Cartan had constructed in his doctoral thesis. Actually Cremmer and Julia proved that in the dimensional reduction of the $D = 11$ theory compactified on a torus T^d one obtains, for $d = 2, 3, \ldots, 8$, all the Lie algebras of the series $\mathrm{E}_{d(d)}$ according to the scheme displayed in Table 7.1.

In 1982, Bernard de Wit and Hermann Nicolai (see Fig. 7.1) constructed the first example of a gauged supergravity, namely $\mathcal{N} = 8$ with the compact gauge group $\mathrm{SO}(8) \subset \mathrm{SU}(8) \subset E_{7(7)}$ [2, 3]. Their seminal paper was extremely influential because of two separate reasons:

P. G. Frè, *Advances in Geometry and Lie Algebras from Supergravity*, Theoretical and Mathematical Physics,
https://doi.org/10.1007/978-3-319-74491-9_7

Table 7.1 Scalar geometries in maximal supergravities

$D = 9$	$E_{2(2)} \equiv SL(2, \mathbb{R}) \otimes O(1, 1)$	$\mathscr{H} = O(2)$	$\dim_{\mathbb{R}} (\mathscr{G}/\mathscr{H}) = 3$
$D = 8$	$E_{3(3)} \equiv SL(3, \mathbb{R}) \otimes SL(2, \mathbb{R})$	$\mathscr{H} = O(2) \otimes O(3)$	$\dim_{\mathbb{R}} (\mathscr{G}/\mathscr{H}) = 7$
$D = 7$	$E_{4(4)} \equiv SL(5, \mathbb{R})$	$\mathscr{H} = O(5)$	$\dim_{\mathbb{R}} (\mathscr{G}/\mathscr{H}) = 14$
$D = 6$	$E_{5(5)} \equiv O(5, 5)$	$\mathscr{H} = O(5) \otimes O(5)$	$\dim_{\mathbb{R}} (\mathscr{G}/\mathscr{H}) = 25$
$D = 5$	$E_{6(6)}$	$\mathscr{H} = Usp(8)$	$\dim_{\mathbb{R}} (\mathscr{G}/\mathscr{H}) = 42$
$D = 4$	$E_{7(7)}$	$\mathscr{H} = SU(8)$	$\dim_{\mathbb{R}} (\mathscr{G}/\mathscr{H}) = 70$
$D = 3$	$E_{8(8)}$	$\mathscr{H} = O(16)$	$\dim_{\mathbb{R}} (\mathscr{G}/\mathscr{H}) = 128$

Fig. 7.1 Herman Nicolai (1952) graduated in 1978 from Karlsruhe University with a thesis written under the supervision of Julius Wess. He obtained his habilitation from Heidelberg University in 1983. He was postdoctoral fellow and staff member at CERN from 1979 to 1986. Full Professor of Theoretical Physics at Hamburg University from 1986 to 1996, since 1997 he is Research Director at the Max Planck Institute for Gravitational Physics in Potsdam (Albert Einstein Institute). In 1980, together with Bernard de Wit (see Fig. 4.7) Nicolai constructed the first example of gauged $\mathscr{N} = 8$ supergravity, the *de Wit–Nicolai theory* with SO(8) gauge group. Nicolai has given outstanding contribution to Kaluza Klein supergravity and to the algebraic foundations of M-theory introducing the study of hyperbolic Lie algebras like E_{10} and E_{11}. He also contributed to the first development of cosmic billiards

(a) On one side it provided the explicit interacting form of $D = 11$ supergravity compactified on anti de Sitter space times the round seven sphere:

$$\mathscr{M}_{11} \simeq \mathrm{AdS}_4 \times \mathbb{S}^7 \tag{7.1.2}$$

and consistently truncated to the massless modes.

(b) On the other side it provided a paradigm for the gauging of supergravity theories via the introduction of a gauge algebra, of fermion shifts in the supersymmetry transformation rules of the fermions and of a scalar potential that is always given by a quadratic form in such fermion shifts.

The same years 1982–1986 witnessed the first golden season of Kaluza–Klein supergravity. After the seminal paper by Freund and Rubin [4], in a quick succession of papers, the various solutions of $\mathrm{D} = 11$ supergravity of the form:

$$\mathscr{M}_{11} \simeq \mathrm{AdS}_4 \times \mathrm{M}^7 \tag{7.1.3}$$

where $\mathrm{M}^7 = \frac{\mathrm{G}}{\mathrm{H}}$ is a 7-dimensional compact coset manifold were found:

1. In [5] the case where M^7 is one the $\mathrm{M}^{\mathrm{pqr}}$ manifolds with isometry $\mathrm{G} = \mathrm{SU}(3) \times \mathrm{SU}(2) \times \mathrm{U}(1)$.
2. In [6] the case where M^7 is the seven sphere but there is an additional internal flux.
3. In [7] the case where M^7 is the given by the squashed seven sphere whose isometry is $\mathrm{G} = \mathrm{SO}(5) \times \mathrm{SO}(3)$.
4. In [8] the case where M^7 is given by the $\mathrm{Q}^{\mathrm{pqr}}$ manifolds with isometry $G = \mathrm{SU}(2) \times \mathrm{SU}(2) \times \mathrm{SU}(2)$.
5. In [9] where M^7 is given by the $\mathrm{N}^{\mathrm{pqr}}$ spaces with isometry $\mathrm{SU}(3)$.
6. In [10] where all the remaining cases were classified.

Relying on general techniques of harmonic analysis on homogeneous spaces, the number of preserved supersymmetry and the Kaluza Klein spectra of all such compactifications were also determined in [11–15]. There followed an intense research activity where the two approaches top-bottom and bottom-up were compared. It soon became evident that compactifications on $\mathrm{AdS}_4 \times \mathrm{M}^7$ and $\mathrm{AdS}_4 \times \widehat{\mathrm{M}}^7$ can be connected at the level of the $D = 4$ theory if the two manifolds M^7 and $\widehat{\mathrm{M}^7}$ are diffeomorphic as topological manifolds, although with different Einstein metrics. In that case the second solution corresponds to a different extremum of the potential in the gauged supergravity obtained from the first compactification. The paradigmatic example is that of the round and squashed seven sphere that have the same topology but two different Einstein metrics. Both compactification are encoded in the de Wit–Nicolai theory. The round seven sphere corresponds to an $\mathrm{SO}(8)$-symmetric extremum of the potential and preserves $\mathscr{N} = 8$, supersymmetries, while the second compactification corresponds to an extremum with $\mathrm{SO}(5) \times \mathrm{SO}(3)$ symmetry and preserves only $\mathscr{N} = 1$ supersymmetry. When the compact manifolds M^7 and $\widehat{\mathrm{M}}^7$ have different topology the low energy supergravities corresponding to the two compactifications are different theories, either different gaugings of the same ungauged supergravity or even different gaugings of different ungauged supergravities with a different content of matter multiplets.

This state of affairs promoted the search for new supergravity gaugings and for their interpretation in terms of compactification either of $D = 11$ supergravity or of $D = 10$ supergravities of type IIA or IIB.

In a series of papers dating 1985 [16–22] Christopher Hull and Nick Warner determined an entire list of gaugings of maximal $\mathcal{N} = 8$ supergravity based on gauge Lie algebras that are not semi-simple. We recall in the present chapter how that list can be systematically retrieved and exhausts the classification of so named *electric gaugings* of maximal supergravity.

There followed an intermission of about ten years when the research interests in the field of superstring and supergravity were mostly focused on Calabi–Yau compactifications, non perturbative dualities and D-branes.

After 1997 a second golden season for Kaluza–Klein supergravity opened up with the advent of the AdS/CFT correspondence introduced by Maldacena [23] and by Kallosh and Van Proeyen [24]. In the new perspective promoted by this conceptual scheme, gauged supergravities corresponding to AdS_5 or AdS_4 backgrounds were viewed as a tool to calculate quantum correlators of Conformal Field Theories on the boundary (see also the discussion of these viewpoints in Chap. 8, in particular, Sect. 8.1.2). In this new vision the Kaluza–Klein spectra were revisited as describing towers of superconformal BPS multiplets on the boundary $\partial\mathrm{AdS}_5$ [25–28] or on the boundary $\partial\mathrm{AdS}_4$ [29–32]. The geometry of the metric cone on the compact manifold G/H:

$$\mathscr{C}(\mathrm{G/H}) \Rightarrow dr^2 + r^2\, d\Omega^2_{\mathrm{G/H}} \tag{7.1.4}$$

raised to prominence as the key to the geometrical engineering of the dual superconformal theories.

Obviously the interest in classifying, constructing and interpreting supergravity gaugings came once again to prominence.

In 1998 the present author together with Mario Trigiante (see Fig. 7.2) and younger collaborators addressed the classification of electric gaugings of the $\mathcal{N} = 8$ theory in a paper [33] that turned out to be seminal. Analyzing the so named T-identities introduced by de Wit and Nicolai, which are field dependent relations between the fermion shifts necessary for the consistency and supersymmetry of the action, it was possible to reduce them to a purely algebraic constraint imposed on the *embedding matrix*. The very concept of the embedding matrix, which will be reviewed in the present chapter was formulated for the first time in [33]. The algebraic t-identity was solved in [33], yielding an exhaustive classification of the electric gaugings of $\mathcal{N} = 8$ supergravity.

In the next few years Mario Trigiante and Henning Samtleben, also in collaboration with Bernard de Wit, generalized the setup of the embedding matrix, now known in the literature under the name of *embedding tensor*, and developed a general scheme for the construction and classification of gaugings in all type of supergravities [34–36]. A very much detailed review and classification of all gauged supergravities is provided in [37]. Very good lectures on the relation between gauged supergravity and flux compactifications are presented in [38].

The interest in the scalar potentials that can be obtained from supergravity gaugings grew more and more in the last decade in connection with the following issues:

Fig. 7.2 Mario Trigiante (on the left) was born in Bari in 1970. He graduated from Pisa University in 1994 and he obtained his Ph.D. in Theoretical Physics from SISSA in 1998 writing a thesis under the supervision of this book's author. Postdoctoral Fellow first at Wales University in Swansea, then in Utrecht, since 2004 he is Associate Professor at the Politecnico of Torino. Mario Trigiante together with this book's author started a project aimed at the strategical use of solvable Lie algebras in supergravity. He has given many important contributions in supergravity black hole theory, in the systematic approach to supergravity gaugings, in the inclusion of cosmological models into supergravity and in the classification of nilpotent orbits. Hennig Santleben, graduated from Hamburg University and in 1998 he obtained his PhD writing a thesis under the supervision of Hermann Nicolai. He was postdoctoral fellow in Utrecht and currently he is Professor of Theoretical Physics at the École Normale Superiéure of Lyon. Together with Mario Trigiante, Santleben set the foundations of the embedding tensor formalism for supergravity gaugings. The very first idea of the embedding matrix was put forward in [33] by Cordaro et al.

1. Search of stable and unstable de Sitter vacua.
2. Patterns of spontaneous symmetry breaking.
3. Inflaton potentials and the inclusion of realistic inflationary models into supergravity.

The first example of a non-compact gauging of $\mathcal{N} = 2$ supergravity leading to a stable de Sitter vacuum was found by the present author together with Mario Trigiante and Antoine Van Proeyen in [39]. Such vacua are quite rare and their search justifies a systematic scanning of all gauging constructions.

Summarizing we can say that one more deep geometrical aspect of supergravity is encoded in the supergravity scalar potentials. From a mathematical point of view the scalar potential is a map from the scalar manifold to the real line:

$$\mathscr{V}(\phi) \quad : \quad \mathscr{M}_{scalar} \rightarrow \mathbb{R} \tag{7.1.5}$$

Differently from the case of non supersymmetric field theories, where the potential $\mathscr{V}(\phi)$ is just any function, in supergravity, the determination of the scalar potential $\mathscr{V}(\phi)$ is a complicated algebro-geometric issue that involves all the structures of special geometries and the algebraic subtleties of the embedding tensor. In this chapter we plan to illustrate these construction considering two different paradigmatic cases. The first concerns maximal $\mathscr{N} = 8$ supergravity where we illustrate the $\mathfrak{e}_{7(7)}$ algebraic machinery underlying the classification of gaugings. The second case-study relates instead with the inclusion of Starobinsky-like inflaton potentials in $\mathscr{N} = 2$ supergravity models whose origin is traced back to the c-map and to the universal structure of the sub Tits Satake algebra.

Having spelled out our goals we turn to a more technical description of the items we shall be working with.

7.1.1 Gaugings and Vacua

The conventional lore is that a vacuum of gravity or supergravity is a configuration with maximal symmetry, namely with Lorentz invariance $\mathrm{SO}(1, \mathrm{D}-1)$ in D-dimensions. Adding translation invariance one ends up with either Poincaré, or de Sitter, or anti de Sitter symmetry, which forces the vacuum expectation values of all scalar fields to be constant. Conventional vacua are also effectively characterized by their properties with respect to supersymmetry breaking or preservation. Hence we begin our analysis of supergravity gaugings by recalling the general properties of conventional vacua and of the possible supersymmetry breaking patterns, that, as it will immediately appear, encode fundamental information about the basic new ingredients produced by the gaugings, namely the fermion shifts.

7.1.2 General Aspects of Supergravity Gaugings and Susy Breaking

Let us begin by recalling some very general aspects of the super-Higgs mechanism in extended supergravity that were codified in the literature of the early and middle eighties [40–43] and were further analyzed and extended in the middle nineties [44–47]. Because of the fundamental property of extended supergravity that the scalar potential is generated by the gauging procedure, the discussion of spontaneous supersymmetry breaking goes hand in hand with the discussion of possible gaugings.

7.1.2.1 Supersymmetry Breaking in Conventional Vacua

A conventional vacuum of $p+2$-dimensional supergravity corresponds to a space-time geometry with a maximally extended group of isometries, namely with $\frac{1}{2}(p+2)(p+3)$ Killing vectors. This means that the metric $ds^2 = g_{\mu\nu}dx^\mu \otimes dx^\nu$ necessarily has constant curvature in $p+2$-dimensions and is one of the following three:

$$\mathcal{M}_{space-time} = \begin{cases} AdS_{p+2} & ; \text{ negative curvature} \\ \text{Minkowski}_{\text{p+2}} & ; \text{ zero curvature} \\ \text{dS}_{\text{p+2}} & ; \text{ positive curvature} \end{cases} \tag{7.1.1}$$

At the same time, in order to be consistent with this maximal symmetry, the *v.e.v.*s of the scalar fields, $<\phi^i>=\phi^i_0$ must be constant and be extrema of the scalar potential:

$$\left.\frac{\partial \mathcal{V}}{\partial \phi^i}\right|_{\phi=\phi_0} = 0\,, \tag{7.1.2}$$

Minkowski space occurs when $\mathcal{V}(\phi_0)=0$, anti de Sitter space AdS_{p+2} occurs when $\mathcal{V}(\phi_0)<0$ and finally de Sitter space dS_{p+2} is generated by $\mathcal{V}(\phi_0)>0$. To be definite we focus on the 4-dimensional case, but all the mechanisms and properties we describe below have straightforward counterparts in higher dimensions. So let us state that in relation with the super-Higgs mechanism, there are just three relevant items of the entire $D=4$ supergravity construction that have to be considered.

1. The *gravitino mass matrix* $S_{AB}(\phi)$, namely the non-derivative scalar field dependent term that appears in the gravitino supersymmetry transformation rule:

$$\delta\psi_{A|\mu} = \mathcal{D}_\mu\,\varepsilon_A + S_{AB}\,(\phi)\;\gamma_\mu\,\varepsilon^B + \ldots, \tag{7.1.3}$$

   and reappears as a mass term in the Lagrangian:

$$\mathcal{L}^{\text{SUGRA}} = \ldots + \text{const}\left(S_{AB}(\phi)\,\overline{\psi}^A_\mu\,\gamma^{\mu\nu}\,\psi^B_\nu + S^{AB}(\phi)\,\overline{\psi}_{A|\mu}\,\gamma^{\mu\nu}\,\psi_{B|\nu}\right) \tag{7.1.4}$$

2. The *fermion shifts*, namely the non-derivative scalar field dependent terms in the supersymmetry transformation rule of the spin $\frac{1}{2}$ fields:

$$\begin{aligned} \delta\,\lambda^i_R &= \text{derivative terms} + \Sigma_A{}^i\,(\phi)\;\varepsilon^A\,, \\ \delta\,\lambda^i_L &= \text{derivative terms} + \Sigma^{A|i}\,(\phi)\;\varepsilon_A\,. \end{aligned} \tag{7.1.5}$$

3. The scalar potential itself, $\mathcal{V}(\phi)$.

These three items are related by a general supersymmetry Ward identity, firstly discovered in the context of gauged $\mathcal{N}=8$ supergravity [2, 3] and later extended to all supergravities [40, 42, 43, 48, 49], that, in the conventions of [41, 50–52] reads as follows:

$$24\, S_{AC}\, S^{CB} - 4\, K_{i,j}\, \Sigma_A{}^i\, \Sigma^{B|j} = -\delta_A^B\, \mathscr{V}\,, \tag{7.1.6}$$

where $K_{i,j}$ is the kinetic matrix of the spin-1/2 fermions. The numerical coefficients appearing in (7.1.6) depend on the normalization of the kinetic terms of the fermions, while $A, B, \dots = 1, \dots, \mathscr{N}$ are $SU(\mathscr{N})$ indices that enumerate the supersymmetry charges. We also follow the standard convention that the upper or lower position of such indices denotes definite chiral projections of Majorana spinors, right or left, depending on the species of fermions considered.[1] The position denotes also the way of transforming of the fermion with respect to $SU(\mathscr{N})$, with lower indices in the fundamental and upper indices in the fundamental bar. In this way we have $S^{AB} = (S_{AB})^\star$ and $\Sigma_A{}^i = \left(\Sigma^{B|i}\right)^\star$. Finally, the index i is a collective index that enumerates all spin-1/2 fermions λ^i present in the theory.[2]

The corresponding fermion shifts are defined by

$$\delta\, \lambda^i = \text{derivative terms} + \Sigma_A{}^i\, (\phi)\, \varepsilon^A\,. \tag{7.1.7}$$

A vacuum configuration ϕ_0 that preserves $\mathscr{N}_0$ supersymmetries is characterized by the existence of $\mathscr{N}_0$ vectors $\rho^A_{(\ell)}$ ($\ell = 1, \dots, \mathscr{N}_0$) of $SU(\mathscr{N})$, such that

$$\begin{aligned} S_{AB}\,(\phi_0)\; \rho^A_{(\ell)} &= e^{i\theta}\, \sqrt{\tfrac{-\mathscr{V}\,(\phi_0)}{24}}\; \rho_{A(\ell)}\,, \\ \Sigma_A{}^i\,(\phi_0)\; \rho^A_{(\ell)} &= 0\,, \end{aligned} \tag{7.1.8}$$

where θ is an irrelevant phase. Indeed, consider the spinor

$$\varepsilon^A(x) = \sum_{\ell=1}^{\mathscr{N}_0} \rho^A_{(\ell)} \varepsilon^{(\ell)}(x)\,, \tag{7.1.9}$$

where $\varepsilon^{(\ell)}(x)$ are $\mathscr{N}_0$ independent solutions of the equation for covariantly constant spinors in AdS_4 (or Minkowski space) with $2\,e = \sqrt{-\mathscr{V}(\phi_0)/24}$:

$$D^{(\mathrm{AdS})}_a \varepsilon(x) \equiv \left(\partial_a - \frac{1}{4}\omega^{bc}{}_a \gamma_{bc} - 2\, e\, \gamma_5 \gamma_a\right) \varepsilon(x) = 0\,, \tag{7.1.10}$$

The integrability of Eq. (7.1.10) is guaranteed by the expression of the AdS_4 curvature, $R^{ab}{}_{cd} = -16\, e^2\, \delta^{ab}_{cd}$, that corresponds to the Ricci tensor:

$$\mathscr{R}_{ab} = -24\, e^2\, \eta_{ab} = \frac{1}{4}\, \mathscr{V}(\phi_0)\, \eta_{ab}, \tag{7.1.11}$$

[1] For instance, we have $\gamma_5\, \varepsilon_A = \varepsilon_A$ and $\gamma_5\, \varepsilon^A = -\varepsilon^A$.

[2] We denote by λ^i the right handed chiral projection while λ_i are the left handed ones.

Then it follows that under supersymmetry transformations of parameter (7.1.9) the chosen vacuum configuration $\phi = \phi_0$ is invariant.[3] That such a configuration is a true vacuum follows from another property proved, for instance, in [43]: all vacua that admit at least one vector ρ^A satisfying Eq. (7.1.8) are automatically extrema of the potential, namely they satisfy Eq. (7.1.2). Furthermore, for constant scalar field configurations the Ricci tensor must be $\mathscr{R}_{\mu\nu} = \frac{1}{4}\mathscr{V}(\phi_0)\, g_{\mu\nu}$ as in Eq. (7.1.11).

The above integrability argument can be easily generalized to all dimensions and to all numbers of supersymmetries $\mathscr{N}$. Consider a supergravity action in D dimensions that, once reduced to the gravitational plus scalar field sector, has the following normalization:

$$A^{[D]}_{grav+scal} = \int d^D x\, \sqrt{-g}\left[2\, R[g] + \alpha\, \frac{1}{2} g_{ij}(\phi)\, \partial^\mu\, \phi^i\, \partial_\mu \phi^j - \mathscr{V}(\phi)\right] \quad (7.1.12)$$

where α is a normalization constant that can vary from case to case since it can always be reabsorbed into the definition of the scalar metric but the scalar potential $\mathscr{V}$ has an unambiguous and unique normalization with respect to the Einstein term. For constant field configurations ϕ_0 the Einstein equations derived from (7.1.12) imply that:

$$R_{\mu\nu} = \frac{1}{2(D-2)}\, \mathscr{V}(\phi_0)\, g_{\mu\nu} \quad (7.1.13)$$

Then the Riemann tensor of an anti de Sitter space AdS_D consistent with Eq. (7.1.13) is necessarily the following:

$$R^{\rho\sigma}_{\mu\nu} = \frac{1}{(D-1)(D-2)}\, \mathscr{V}(\phi_0)\, \delta^{[\rho}_{[\mu}\, \delta^{\sigma]}_{\nu]} \quad (7.1.14)$$

Consider next the equation for a covariantly constant spinor in AdS_D. Its general form is:

$$D^{(\mathrm{AdS})}_\mu \varepsilon \equiv \mathscr{D}_\mu \varepsilon(x) - \mu\, \gamma_\mu\, \varepsilon = \left(\partial_\mu - \frac{1}{4}\omega^{bc}{}_\mu \gamma_{bc} - \mu\, \gamma_\mu\right) = 0\,, \quad (7.1.15)$$

where the parameter μ is fixed by integrability in terms of the vacuum value of potential $\mathscr{V}(\phi_0)$. Indeed from the condition $D^{(\mathrm{AdS})}_{[\mu} D^{(\mathrm{AdS})}_{\nu]} = 0$ we immediately get:

$$|\mu|^2 = \frac{1}{4}\, \frac{|\mathscr{V}(\phi_0)|}{(D-1)(D-2)} \quad (7.1.16)$$

On the other hand the general form of the gravitino transformation rule is, independently from the number of space-time dimensions, that given in Eq. (7.1.3), so that, in a conventional vacuum with an unbroken supersymmetry μ is to be inter-

[3] As already stressed, the v.e.v.s of all the fermions are zero and Eq. (7.1.8) guarantees that they remain zero under supersymmetry transformations of parameters (7.1.9).

preted as an **eigenvalue** of the gravitino mass-matrix. So the general conditions for the preservation of $\mathcal{N}_0$ supersymmetries in D dimensions are fully analogous to those in Eq. (7.1.8) and correspond to the existence of $\mathcal{N}_0$ independent vectors $\rho^A_{(\ell)}$ ($\ell = 1, \dots, \mathcal{N}_0$), such that:

$$\begin{aligned} S_{AB}\,(\phi_0)\,\rho^A_{(\ell)} &= e^{i\theta}\,\sqrt{\frac{|\mathcal{V}(\phi_0)|}{4(D-2)(D-1)}}\,\rho_{A(\ell)}\,, \\ \Sigma_A{}^i\,(\phi_0)\,\rho^A_{(\ell)} &= 0\,, \end{aligned} \tag{7.1.17}$$

By extension of language the vectors $\rho^A_{(\ell)}$ are named **Killing spinors**.

7.2 Electric Gaugings of $\mathcal{N} = 8$ Supergravity in $D = 4$

To illustrate the general ideas in a case of maximal supersymmetry, we consider the possible *gaugings* of the $\mathcal{N} = 8$ theory in four dimensions. The complete classification that can be reached in this case constitutes an inspiring paradigm.

Considering the $\mathcal{N} = 8$, $D = 4$ case we recall that here there is no other multiplet besides the graviton multiplet which contains the graviton $g_{\mu\nu}$, 8 gravitinos $\psi_{A|\mu}\,dx^\mu$, 28 one-form gauge fields $A^{\Lambda\Sigma}_\mu\,dx^\mu = -A^{\Sigma\Lambda}_\mu\,dx^\mu$ transforming in the 28 antisymmetric representation of the electric subgroup $\mathrm{SL}(8,\mathbb{R}) \subset \mathrm{E}_{7(7)}$, 56 spin $1/2$ dilatinos $\chi_{ABC} = \gamma_5\chi_{ABC}$ (anti-symmetric in ABC) and 70 scalars parametrizing the $\mathrm{E}_{7(7)}/\mathrm{SU}(8)$ coset manifold. We have labeled the vector fields with a pair of anti-symmetric indices, each of them ranging on 8 values $\Lambda, \Sigma, \Delta, \Pi, = 1, \dots, 8$ and transforming in the fundamental representation of $\mathrm{SL}(8,\mathbb{R})$. The capital latin indices carried by the fermionic fields range also on eight values $A, B, C, = 1, \dots, 8$ but they are covariant under the maximal compact subgroup $\mathrm{SU}(8) \subset \mathrm{E}_{7(7)}$ rather than the non compact $\mathrm{SL}(8,\mathbb{R}) \subset \mathrm{E}_{7(7)}$. As in previous sections, also here we use the convention that upper and lower SU(8) indices denote different chirality projections of the fermion fields: $\psi^A = -\gamma_5\psi^A$ and $\chi^{ABC} = -\gamma_5\chi^{ABC}$.

7.2.1 *The Bosonic Action*

In order to proceed further we need to fix our conventions for $\mathcal{N} = 8$ supergravity and for its gaugings. We adopt those utilized in papers [33, 53]. We introduce the coset representative $\mathbb{L}$ of $\frac{E_{(7)7}}{SU(8)}$ in the **56** representation of $E_{(7)7}$:

$$\mathbb{L} = \frac{1}{\sqrt{2}}\left(\begin{array}{c|c} f + ih & \overline{f} + i\overline{h} \\ \hline f - ih & \overline{f} - i\overline{h} \end{array}\right) \tag{7.2.1}$$

where the submatrices (h, f) are 28×28 matrices indexed by antisymmetric pairs Λ, Σ, A, B, in which $\Lambda, \Sigma = 1, \ldots, 8$, $A, B = 1, \ldots, 8$. The first pair transforms under $E_{7(7)}$ while the second one transforms under $SU(8)$:

$$(h, f) = \left(h_{\Lambda\Sigma|AB}, f^{\Lambda\Sigma}_{\ \ AB}\right) \tag{7.2.2}$$

Note that $\mathbb{L} \in \mathrm{Usp}(28, 28)$. The vielbein P_{ABCD} and the $SU(8)$ connection $\Omega_A^{\ B}$ of $\frac{E_{7(7)}}{SU(8)}$ are computed from the left invariant 1-form $\mathbb{L}^{-1}d\mathbb{L}$:

$$\mathbb{L}^{-1}d\mathbb{L} = \left(\begin{array}{c|c} \delta^{[A}_{\ \ [C}\Omega^{B]}_{\ \ D]} & \overline{P}^{ABCD} \\ \hline P_{ABCD} & \delta_{[A}^{\ \ [C}\overline{\Omega}^{\ \ D]}_{B]} \end{array}\right) \tag{7.2.3}$$

where $P_{ABCD} \equiv P_{ABCD,\alpha} d\Phi^\alpha \quad (\alpha = 1, \ldots, 70)$ is completely antisymmetric and satisfies the reality condition

$$P_{ABCD} = \frac{1}{24}\varepsilon_{ABCDEFGH}\overline{P}^{EFGH} \tag{7.2.4}$$

The bosonic lagrangian of gauged $\mathcal{N} = 8$ supergravity is the following

$$\begin{aligned}\mathcal{L} = \int \sqrt{-g}\, d^4x \Big(2R + \mathrm{Im}\mathcal{N}_{\Lambda\Sigma|\Gamma\Delta}\mathcal{F}^{\Lambda\Sigma}_{\mu\nu}\mathcal{F}^{\Gamma\Delta|\mu\nu} + \frac{1}{6}P_{ABCD,i}\overline{P}^{ABCD}_j\partial_\mu\Phi^i\partial^\mu\Phi^j + \\ + \frac{1}{2}\mathrm{Re}\mathcal{N}_{\Lambda\Sigma|\Gamma\Delta}\frac{\varepsilon^{\mu\nu\rho\sigma}}{\sqrt{-g}}\mathcal{F}_{\mu\nu}^{\ \ \Lambda\Sigma}\mathcal{F}^{\Gamma\Delta}_{\ \ \rho\sigma} - \mathcal{V}(\phi)\Big)\end{aligned} \tag{7.2.5}$$

where:

$$\mathcal{F}^{\Lambda\Sigma} = dA^{\Lambda\Sigma} + C^{\Lambda\Sigma}_{\ \ \ \Gamma\Delta,\Phi\Xi}\, A^{\Gamma\Delta} \wedge A^{\Phi\Xi} \tag{7.2.6}$$

having denoted $C^{\Lambda\Sigma}_{\ \ \ \Gamma\Delta,\Phi\Xi}$ the structure constant of the gauge Lie algebra. The 28 one-forms $A^{\Lambda\Sigma}_\mu\, dx^\mu = -A^{\Sigma\Lambda}_\mu\, dx^\mu$ transform in the 28 antisymmetric representation of the electric subgroup $SL(8, \mathbb{R}) \subset E_{7(7)}$. We have labeled these vector fields with a pair of antisymmetric indices. The curvature two-form is defined as

$$R^{ab} = d\omega^{ab} - \omega^a_{\ c} \wedge \omega^{cb}. \tag{7.2.7}$$

and the kinetic matrix $\mathcal{N}_{\Lambda\Sigma|\Gamma\Delta}$ is given by:

$$\mathcal{N} = hf^{-1} \quad \rightarrow \quad \mathcal{N}_{\Lambda\Sigma|\Gamma\Delta} = h_{\Lambda\Sigma|AB}\, f^{-1\ AB}_{\ \ \ \ \Gamma\Delta}. \tag{7.2.8}$$

The same matrix relates the (anti)self-dual electric and magnetic 2-form field strengths, namely, setting

$$\mathscr{F}^{\pm\,\Lambda\Sigma} = \frac{1}{2}\left(\mathscr{F} \pm \mathrm{i} \star \mathscr{F}\right)^{\Lambda\Sigma} \tag{7.2.9}$$

one has

$$\begin{aligned}
\mathscr{G}^-_{\Lambda\Sigma} &= \overline{\mathscr{N}}_{\Lambda\Sigma|\Gamma\Delta}\mathscr{F}^{-\,\Gamma\Delta} \\
\mathscr{G}^+_{\Lambda\Sigma} &= \mathscr{N}_{\Lambda\Sigma|\Gamma\Delta}\mathscr{F}^{+\,\Gamma\Delta}
\end{aligned} \tag{7.2.10}$$

where the "dual" field strengths $\mathscr{G}^{\pm}_{\Lambda\Sigma}$ are defined as $\mathscr{G}^{\pm}_{\Lambda\Sigma} = \frac{i}{2}\frac{\delta\mathscr{L}}{\delta\mathscr{F}^{\pm\,\Lambda\Sigma}}$. Note that the 56 dimensional (anti)self-dual vector $\left(\mathscr{F}^{\pm\,\Lambda\Sigma}, \mathscr{G}^{\pm}_{\Lambda\Sigma}\right)$ transforms covariantly under $U \in \mathrm{Sp}(56, \mathbb{R})$

$$\begin{aligned}
&U\begin{pmatrix}\mathscr{F}\\ \mathscr{G}\end{pmatrix} = \begin{pmatrix}\mathscr{F}'\\ \mathscr{G}'\end{pmatrix} \quad ; \quad U = \begin{pmatrix}A & B\\ C & D\end{pmatrix} \\
&A^t C - C^t A = 0 \\
&B^t D - D^t B = 0 \\
&A^t D - C^t B = \mathbf{1}
\end{aligned} \tag{7.2.11}$$

The matrix transforming the coset representative $\mathbb{L}$ from the Usp (28, 28) basis, Eq. (7.2.1), to the real $\mathrm{Sp}(56, \mathbb{R})$ basis is the Cayley matrix:

$$\mathbb{L}_{\mathrm{Usp}} = \mathscr{C}\mathbb{L}_{\mathrm{Sp}}\mathscr{C}^{-1} \qquad \mathscr{C} = \begin{pmatrix}\mathbb{1} & \mathrm{i}\mathbb{1}\\ \mathbb{1} & -\mathrm{i}\mathbb{1}\end{pmatrix} \tag{7.2.12}$$

implying

$$\begin{aligned}
f &= \frac{1}{\sqrt{2}}\left(A - \mathrm{i}B\right) \\
h &= \frac{1}{\sqrt{2}}\left(C - \mathrm{i}D\right)
\end{aligned} \tag{7.2.13}$$

The only object which we need to manipulate to get command of the theory is the coset representative $\mathbb{L}(\phi)$ parametrizing the equivalence classes of $\mathrm{E}_{7(7)}/\mathrm{SU}(8)$. Just to fix ideas and avoiding the subtleties of the solvable decomposition we can think of $\mathbb{L}(\phi)$ as the exponential of the 70-dimensional coset $\mathbb{K}$ in the orthogonal decomposition of the Lie algebra:

$$\mathfrak{e}_{7(7)} = \mathfrak{su}(8) \oplus \mathbb{K} \tag{7.2.14}$$

In practice this means that we can write:

$$\mathbb{L}(\phi) = \exp\begin{pmatrix}0 & \phi^{EFGH}\\ \phi_{ABCD} & 0\end{pmatrix} = \begin{pmatrix}u^{\Lambda\Xi}{}_{AB} & v^{\Lambda\Xi CD}\\ v_{\Delta\Gamma AB} & u_{\Delta\Gamma}{}^{CD}\end{pmatrix} \tag{7.2.15}$$

where the 70 parameters ϕ_{ABCD} satisfy the self-duality condition[4]:

$$\phi_{ABCD} = \frac{1}{4!}\varepsilon_{ABCDEFGH}\phi^{EFGH} \tag{7.2.16}$$

The interaction structure of the theory is fully encoded in the following geometrical data:

1. The symplectic geometry of the scalar coset manifold $\mathrm{E}_{7(7)}/\mathrm{SU}(8)$
2. The choice of the gauge group $\mathcal{G}_{gauge} \subset \mathrm{SL}(8, \mathbb{R}) \subset \mathrm{E}_{7(7)}$.

In this chapter we mainly need the second item of the this list, yet we need to recollect some information on the other items.

Let us first recall that

$$g_{ij} = \frac{1}{6}P_{ABCD,i}\overline{P}_j^{ABCD} \tag{7.2.17}$$

appearing in the scalar field kinetic term of the lagrangian (7.2.5) is the unique $E_{7(7)}$ invariant metric on the scalar coset manifold.

The coset representative $\mathbb{L}$ as defined by (7.2.15) is in the Usp(28, 28) representation. There are actually four bases where the 56×56 matrix $\mathbb{L}(\phi)$ can be written:

1. The SpD(56)-basis
2. The UspD(28, 28)-basis
3. The SpY(56)-basis
4. The UspY(28, 28)-basis

corresponding to two cases where $\mathbb{L}$ is symplectic real (SpD(56),SpY(56)) and two cases where it is pseudo-unitary symplectic (UspD(56),UspY(56)). This further distinction in a pair of subcases corresponds to choosing either a basis composed of weights or of Young tableaux. By relying on (7.2.15) we have chosen to utilize the UspY(28, 28)-basis which is directly related to the SU(8) indices carried by the fundamental fields of supergravity. However, for the description of the gauge generators the Dynkin basis is more convenient. We can optimize the advantages of both bases introducing a mixed one where the coset representative $\mathbb{L}$ is multiplied on the left by the constant matrix $\mathcal{S}$ performing the transition from the pseudo-unitary Young basis to the real symplectic Dynkin basis. Explicitly we have:

$$\begin{pmatrix} u^{AB} \\ v_{AB} \end{pmatrix} = \mathcal{S}\begin{pmatrix} W^i \\ W^{i+28} \end{pmatrix} (i, 1, \ldots 28) \tag{7.2.18}$$

where

$$\mathcal{S} = \begin{pmatrix} \mathbf{S} & \mathbf{0} \\ \mathbf{0} & \mathbf{S}^\star \end{pmatrix} \mathcal{C} = \frac{1}{\sqrt{2}}\begin{pmatrix} \mathbf{S} & \mathrm{i}\,\mathbf{S} \\ \mathbf{S}^\star & -\mathrm{i}\,\mathbf{S}^\star \end{pmatrix} \tag{7.2.19}$$

the 28×28 matrix $\mathbf{S}$ being unitary:

[4]Here we have used the notation, $\phi^{ABCD} \equiv (\phi_{ABCD})^*$.

$$\mathbf{S}^\dagger \mathbf{S} = \mathbb{1} \tag{7.2.20}$$

The explicit form of the U(28) matrix **S** was given in Sect. 5.4 of [54] while the weights of the $E_{7(7)}$ **56** representation are listed in Table 7.2. In the **Dynkin basis** the basis vectors of the real symplectic representation are eigenstates of the Cartan generators with eigenvalue one of the 56 weight vectors ($\pm\boldsymbol{\Lambda} = \{\Lambda_1, \ldots, \Lambda_7\}$ pertaining to the representation:

$$(W = 1, \ldots 56) : |\,W\,\rangle = \begin{cases} |\boldsymbol{\Lambda}\rangle \quad : \quad H_i|\,\boldsymbol{\Lambda}\,\rangle \quad = \quad \Lambda_i\,|\,\boldsymbol{\Lambda}\,\rangle \quad (\Lambda = 1, \ldots 28) \\ |-\boldsymbol{\Lambda}\rangle : \quad H_i|-\boldsymbol{\Lambda}\,\rangle = -\Lambda_i\,|-\boldsymbol{\Lambda}\,\rangle \quad (\Lambda = 1, \ldots 28) \end{cases}$$

$$|\,V\,\rangle = f^\Lambda\,|\boldsymbol{\Lambda}\rangle \,\oplus\, g_\Lambda\,|-\boldsymbol{\Lambda}\rangle$$

or in matrix notation

$$\mathbf{V}_{SpD} = \begin{pmatrix} f^\Lambda \\ g_\Sigma \end{pmatrix} \tag{7.2.21}$$

In the **Young basis**, instead, the basis vectors of the complex pseudounitary representation correspond to the natural basis of the $\mathbf{28} + \overline{\mathbf{28}}$ antisymmetric representation of the maximal compact subgroup SU(8). In other words, in this realization of the fundamental $\mathrm{E}_{7(7)}$ representation a generic vector is of the following form:

$$|V\rangle = u^{AB}\,\boxed{\begin{matrix} A \\ B \end{matrix}} \,\oplus\, v_{AB}\,\boxed{\begin{matrix} \overline{A} \\ \overline{B} \end{matrix}} \quad ; \quad (A, B = 1, \ldots, 8)$$

or in matrix notation

$$\mathbf{V}_{UspY} = \begin{pmatrix} u^{AB} \\ v_{AB} \end{pmatrix} \tag{7.2.22}$$

To complete the illustration of the bosonic lagrangian we need to discuss the scalar potential $\mathscr{V}(\phi)$. This cannot be done without referring to the supersymmetry transformation rules since, as we have explained in Sect. 7.1.2, the potential is determined by the fundamental relation (7.1.6) that gives it as a quadratic form in terms of the *fermion shifts*. These latter appear in the supersymmetry transformation rules of the fermionic fields and are the primary objects determined by the choice of the gauge algebra.

7.2.2 *Supersymmetry Transformation Rules of the Fermi Fields*

Since the $\mathscr{N} = 8$ theory has no matter multiplets the fermions are just, as already pointed out, the **8** spin 3/2 gravitinos and the **56** spin 1/2 dilatinos. The two numbers **8** and **56** have been written boldfaced since they also single out the dimensions of the two irreducible SU(8) representations to which the two kind of fermions are respectively assigned, namely the fundamental and the three times antisymmetric:

$$\psi_{\mu|A} \leftrightarrow \boxed{A} \equiv \mathbf{8} \quad ; \quad \chi_{ABC} \leftrightarrow \begin{array}{|c|}\hline A \\ \hline B \\ \hline C \\ \hline \end{array} \equiv \mathbf{56} \tag{7.2.23}$$

Following the conventions of [53] the fermionic supersymmetry transformation rules are written as follows:

$$\begin{aligned} \delta\psi_{A\mu} &= \nabla_\mu \varepsilon_A \; - \frac{1}{4} T^{(-)}_{AB|\rho\sigma} \gamma^{\rho\sigma} \gamma_\mu \varepsilon^B + S_{AB} \, \gamma_\mu \, \varepsilon^B \; + \cdots \\ \delta\chi_{ABC} &= \; 4\mathrm{i} \; P_{ABCD|i} \partial_\mu \Phi^i \gamma^\mu \varepsilon^D - 3 T^{(-)}_{[AB|\rho\sigma} \gamma^{\rho\sigma} \varepsilon_{C]} + \Sigma^D_{ABC} \, \varepsilon_D \cdots \end{aligned} \tag{7.2.24}$$

where $T^{-}_{AB|\mu\nu}$ is the antiselfdual part of the graviphoton field strength, $P_{ABCD|i}$ is the already mentioned vielbein of the scalar coset manifold completely antisymmetric in $ABCD$ and satisfying the same pseudoreality condition as our choice of the scalars ϕ_{ABCD}:

$$P_{ABCD} = \frac{1}{4!} \varepsilon_{ABCDEFGH} \overline{P}^{EFGH} \, . \tag{7.2.25}$$

By comparison with Eqs. (7.1.3) and (7.1.5) we see that S_{AB}, Σ^D_{ABC} are the appropriate *gravitino mass matrix* and *fermion shifts*. Recalling also the normalization of the fermion kinetic terms:

$$\mathcal{L}^{kin}_{fermion} = \int d^4x \left[2 \left(\overline{\psi}^A_\mu \, \gamma_\nu \nabla_\rho \psi_{A|\mu} + \text{h.c} \right) - \mathrm{i} \sqrt{-g} \, \frac{1}{24} \left(\overline{\chi}^{ABC} \gamma^\mu \, \nabla_\mu \chi_{ABC} - \text{h.c.} \right) \right] \tag{7.2.26}$$

the general Ward identity (7.1.6) takes, in this theory, the following explicit form:

$$- \mathcal{V} \, \delta^A_B = 24 \, S_{AM} \, S^{BM} \; - \frac{1}{6} \Sigma^{PQR}_A \, \Sigma^B_{PQR} \tag{7.2.27}$$

What we need is the explicit expression of the two items appearing in the supersymmetry transformations (7.2.24) in terms of the coset representatives. For the graviphoton such an expression is *independent of the gauging* and coincides with that appearing in the case of ungauged supergravity. On the contrary, the expression of the scalar vielbein and of the fermion shifts, involves the choice of the gauge group and can be given only upon introducing the *gauged Maurer Cartan equations*. Hence we first recall the structure of the graviphoton and then we turn our attention to the second kind of items entering the transformation rules that are the most relevant ones in our discussion.

7.2.2.1 The Graviphoton Field Strength

We introduce the multiplet of electric and magnetic field strengths:

$$\mathbf{V}_{\mu\nu} \equiv \begin{pmatrix} F^{\Lambda\Sigma}_{\mu\nu} \\ G_{\Delta\Pi|\mu\nu} \end{pmatrix} \tag{7.2.28}$$

where

$$\begin{aligned} G_{\Delta\Pi|\mu\nu} &= -\mathrm{Im}\mathcal{N}_{\Delta\Pi,\Lambda\Sigma}\, \widetilde{F}^{\Lambda\Sigma}_{\mu\nu} - \mathrm{Re}\mathcal{N}_{\Delta\Pi,\Lambda\Sigma}\, F^{\Lambda\Sigma}_{\mu\nu} \\ \widetilde{F}^{\Lambda\Sigma}_{\mu\nu} &= \frac{1}{2}\,\varepsilon_{\mu\nu\rho\sigma}\, F^{\Lambda\Sigma|\rho\sigma} \end{aligned} \tag{7.2.29}$$

The 56-component field strength vector $\mathbf{V}_{\mu\nu}$ transforms in the real symplectic representation of the U-duality group $E_{7(7)}$. We can also write a column vector containing the 28 components of the graviphoton field strengths and their complex conjugate:

$$\mathbf{T}_{\mu\nu} \equiv \begin{pmatrix} T_{\mu\nu}{}^{|AB} \\ T_{\mu\nu|AB} \end{pmatrix} \quad T^{|AB}_{\mu\nu} = \left(T_{\mu\nu|AB}\right)^{\star} \tag{7.2.30}$$

in which the upper and lower components transform in the canonical *Young basis* of SU(8) for the $\overline{\mathbf{28}}$ and **28** representation respectively.

The relation between the graviphoton field strength vectors and the electric magnetic field strength vectors involves the coset representative in the SpY(56) representation and it is the following one:

$$\mathbf{T}_{\mu\nu} = -\mathscr{C}\,\mathbb{C}\,\mathbb{L}^{-1}_{SpY}(\phi)\,\mathbf{V}_{\mu\nu} \tag{7.2.31}$$

The matrix $\mathbb{C}$ being the symplectic invariant matrix. Equation (7.2.31) reveals that the graviphotons transform under the SU(8) compensators associated with the $E_{7(7)}$ rotations. It is appropriate to express the upper and lower components of **T** in terms of the self-dual and antiself-dual parts of the graviphoton field strengths, since only the latter enters the transformation rules (7.2.24).

These components are defined as follows:

$$\begin{aligned} T^{+|AB}_{\mu\nu} &= \frac{1}{2}\left(T^{|AB}_{\mu\nu} + \frac{\mathrm{i}}{2}\,\varepsilon_{\mu\nu\rho\sigma}\, g^{\rho\lambda} g^{\sigma\pi}\, T^{|AB}_{\lambda\pi}\right) \\ T^{-}_{AB|\mu\nu} &= \frac{1}{2}\left(T_{AB|\mu\nu} - \frac{\mathrm{i}}{2}\,\varepsilon_{\mu\nu\rho\sigma}\, g^{\rho\lambda} g^{\sigma\pi}\, T_{AB|\lambda\pi}\right) \end{aligned} \tag{7.2.32}$$

As shown in [54] the following equalities hold true:

$$T_{\mu\nu}{}^{|AB} = T^{+|AB}_{\mu\nu} \quad ; \quad T_{\mu\nu|AB} = T^{-}_{\mu\nu|AB} \tag{7.2.33}$$

and we can simply write:

$$\mathbf{T}_{\mu\nu} \equiv \begin{pmatrix} T^{+|AB}_{\mu\nu} \\ T^{-}_{\mu\nu|AB} \end{pmatrix} \tag{7.2.34}$$

7.2.3 The Embedding Matrix

The main item in the construction of the electrically gauged supergravity is the *embedding matrix* whose conception was introduced in [33] and later extended to more general (not necessarily electric) gaugings under the name of *embedding tensor* [34–36]. In [33] the general form of the embedding matrix was derived by solving so named *t-identities*. Because of its central role in the present discussion let us recall the setup of the embedding matrix following [33].

The generators of the electric subalgebra $\mathrm{SL}(8, \mathbb{R}) \subset \mathrm{E}_{7(7)}$ have the following form

$$G_\alpha = \begin{pmatrix} q^{\Lambda\Sigma}{}_{\Pi\Delta}(\alpha) & p^{\Lambda\Sigma\Psi\Xi}(\alpha) \\ p_{\Delta\Gamma\Pi\Delta}(\alpha) & q_{\Lambda\Sigma}{}^{\Psi\Xi}(\alpha) \end{pmatrix} \tag{7.2.35}$$

where the matrices q and p are real and have the following form

$$\begin{aligned} q^{\Lambda\Sigma}{}_{\Pi\Delta} &= 2\delta^{[\Lambda}{}_{[\Pi} q^{\Sigma]}{}_{\Delta]} = \frac{2}{3}\delta^{[\Lambda}{}_{[\Pi} q^{\Sigma]\Gamma}{}_{\Delta]\Gamma} \,, \\ p_{\Delta\Gamma\Pi\Omega} &= \frac{1}{24}\varepsilon_{\Delta\Gamma\Pi\Omega\Lambda\Sigma\Psi\Xi}\, p^{\Lambda\Sigma\Psi\Xi} \,. \end{aligned} \tag{7.2.36}$$

The index $\alpha = 1, \ldots 63$ in (7.2.35) spans the adjoint representation of $\mathrm{SL}(8, \mathbb{R})$ according to some chosen basis and we can freely raise and lower the greek indices $\Lambda, \Sigma, \ldots$ because of the reality of the representation.

The fundamental item in the gauging construction is the 28×63 constant embedding matrix:

$$\mathscr{E} \equiv e^{\alpha}_{\Lambda\Sigma} \tag{7.2.37}$$

transforming under $\mathrm{SL}(8, \mathbb{R})$ as its indices specify, namely in the tensor product of the adjoint with the antisymmetric **28**. This matrix specifies which generators of $\mathrm{SL}(8, \mathbb{R})$ are gauged and by means of which vector fields in the 28-dimensional stock. In particular, using this matrix $\mathscr{E}$, one writes the gauge 1-form as:

$$A \equiv A^{\Lambda\Sigma} e^{\alpha}_{\Lambda\Sigma} G_\alpha \tag{7.2.38}$$

As already stressed, the main result of [33] was the determination of the most general form and the analysis of the embedding matrix $e^{\alpha}_{\Lambda\Sigma}$. This general form was obtained by solving the algebraic t-identity which is a linear equation imposed on the embedding matrix $\mathscr{E}$ by the request that the lagrangian should be supersymmetric. These identities were solved by means of a computer program and a 36-parameter solution was found which we do not display explicitly since it is encoded in too large formulae.

Let us summarize the derivation of this result. In terms of the gauge 1-form A and of the coset representative $\mathbb{L}(\phi)$ we can write the *gauged left-invariant 1-form*:

$$\Omega = \mathbb{L}^{-1}d\mathbb{L} + g\mathbb{L}^{-1}A\mathbb{L} \tag{7.2.39}$$

which belongs to the $E_{7(7)}$ Lie algebra in the UspY(28, 28) representation and defines the *gauged* scalar vielbein P^{ABEF} and the SU(8) connection $Q_D{}^B$:

$$\Omega = \begin{pmatrix} 2\delta^{[A}_{[C}\, Q_{D]}{}^{B]} & P^{ABEF} \\ P_{CDGH} & -2\delta^{[E}_{[G}\, Q^{F]}{}_{H]} \end{pmatrix} \tag{7.2.40}$$

Because of its definition the 1-form Ω satisfies *gauged Maurer Cartan equations*:

$$d\Omega + \Omega \wedge \Omega = g\big[F^{\Lambda\Sigma} - (\sqrt{2}(u^{\Lambda\Sigma}{}_{AB} + v_{\Lambda\Sigma AB})\overline{\psi}^A \wedge \psi^B + \text{h.c.})\big]e^{\alpha}_{\Lambda\Sigma}\mathbb{L}^{-1}G_\alpha\mathbb{L}\,, \tag{7.2.41}$$

with $F^{\Lambda\Sigma}$ the supercovariant field strength of the vectors $A^{\Lambda\Sigma}$. Let us focus on the last factor in Eq. (7.2.41):

$$\mathbf{U}_\alpha \equiv \mathbb{L}^{-1}G_\alpha\mathbb{L} = \begin{pmatrix} \mathscr{A}(\alpha) & \mathscr{B}(\alpha) \\ \overline{\mathscr{B}(\alpha)} & \overline{\mathscr{A}(\alpha)} \end{pmatrix} \tag{7.2.42}$$

Since $\mathbf{U}_\alpha$ is an $E_{7(7)}$ Lie algebra element, for each gauge generator G_α we necessarily have:

$$\begin{aligned} \mathscr{A}^{AB}{}_{CD}(\alpha) &= \frac{2}{3}\,\delta^{[A}_{[C}\,\mathscr{A}^{B]M}{}_{D]M} \\ \mathscr{B}^{ABFG}(\alpha) &= \mathscr{B}^{[ABFG]}(\alpha) \end{aligned} \tag{7.2.43}$$

Comparing with Eq. (7.2.41) we see that the scalar field dependent SU(8) tensors multiplying the gravitino bilinear terms are the following ones:

$$\begin{aligned} T^A{}_{BCD} &\equiv (u^{\Omega\Sigma}{}_{CD} + v_{\Omega\Sigma CD})\, e^{\alpha}_{\Omega\Sigma}\, \mathscr{A}^{AM}{}_{BM}(\alpha) \\ Z^{ABEF}_{CD} &\equiv (u^{\Omega\Sigma}{}_{CD} + v_{\Omega\Sigma CD})\, e^{\alpha}_{\Omega\Sigma}\, \mathscr{B}^{ABEF}(\alpha) \end{aligned} \tag{7.2.44}$$

As shown in the original papers by de Wit and Nicolai [2, 3] (or Hull [16–22]) and reviewed in [55], closure of the supersymmetry algebra[5] and hence existence of the corresponding gauged supergravity models is obtained *if and only if* the following T-identities are satisfied:

[5]In the rheonomy approach closure of the Bianchi identities.

$$T^A{}_{BCD} = T^A{}_{[BCD]} + \frac{2}{7}\delta^A_{[C} T^M{}_{D]MB} \tag{7.2.45}$$

$$Z^{CD}_{ABEF} = \frac{4}{3}\delta^{[C}{}_{[A} T^{D]}{}_{BEF]} \tag{7.2.46}$$

Equations (7.2.45) and (7.2.46) have a clear group theoretical meaning. Namely, they state that both the $T^A{}_{BCD}$ tensor and the Z^{ABEF}_{CD} tensor can be expressed in a basis spanned by two irreducible SU(8) tensors corresponding to the **420** and **36** representations respectively:

$$\overset{\circ}{T}{}^A{}_{BCD} \equiv \varepsilon^{AI_1\ldots I_7} \begin{array}{|c|c|} \hline I_1 & B \\ \hline I_2 & C \\ \hline I_3 & D \\ \hline I_4 \\ \cline{1-1} I_5 \\ \cline{1-1} I_6 \\ \cline{1-1} I_7 \\ \cline{1-1} \end{array} \equiv \mathbf{420} \quad ; \quad \overset{\circ}{T}_{DB} \equiv \begin{array}{|c|c|} \hline D & B \\ \hline \end{array} \equiv \mathbf{36}$$

To see this let us consider first Eq. (7.2.45). In general a tensor of type $T^A{}_{B[CD]}$ would have $8 \times 8 \times 28$ components and contain several irreducible representations of SU(8). However, as a consequence of Eq. (7.2.45) only the representations **420**, **28** and **36** can appear. (see Fig. 7.3). In addition, since the $\mathscr{A}$ tensor, being in the adjoint of SU(8), is traceless also the T-tensor appearing in (7.2.45) is traceless: $T^A{}_{ABC} = 0$. Combining this information with Eq. (7.2.45) we obtain

$$T^M{}_{[AB]M} = 0, \tag{7.2.47}$$

Equation (7.2.47) is the statement that the **28** representation appearing in Fig. 7.3 vanishes so that the $T^A{}_{B[CD]}$ tensor is indeed expressed solely in terms of the irreducible tensors (7.2.47).

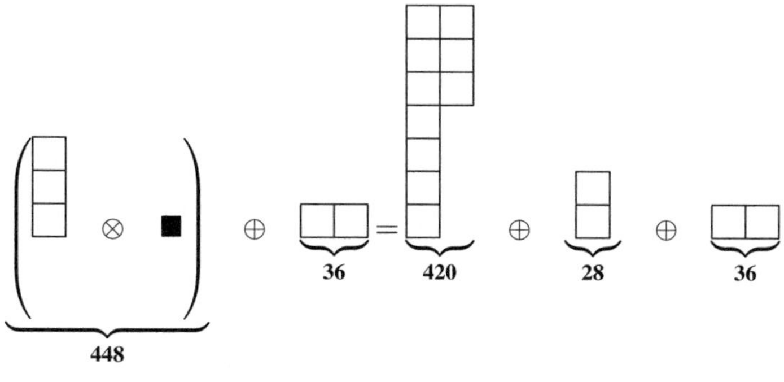

Fig. 7.3 Decomposition of a tensor of type $T^A{}_{BCD}$ into irreducible representations

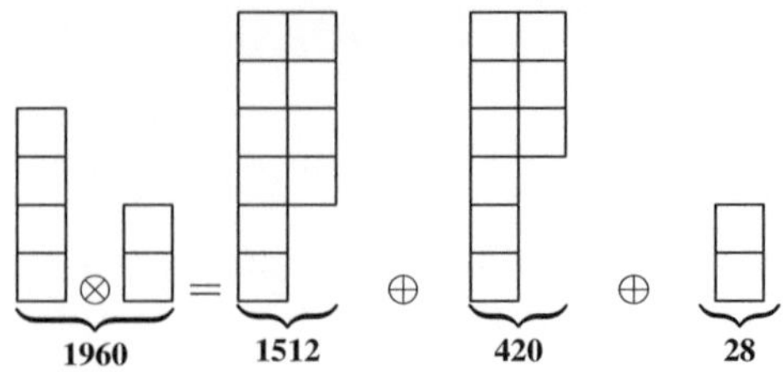

Fig. 7.4 Decomposition of a tensor of type Z^{CD}_{ABEF} into irreducible representations

A similar argument can be given to interpret the second T-identity (7.2.46). A tensor of type $Z^{[ABEF]}_{[CD]}$ contains, a priori, 70×28 components and contains the irreducible representations **1512**, **420** and **28** (see Fig. 7.4). Using Eq. (7.2.46) one immediately sees that the representations **1512** and **28** must vanish and that the surviving **420** is proportional through a fixed coefficient to the **420** representations appearing in the decomposition of the $T^A{}_{B[CD]}$ tensor.

In view of this discussion, the T-identities can be rewritten as follows in the basis of the independent irreducible tensors

$$\overset{\circ}{T}{}^A{}_{BCD} = T^A{}_{[BCD]} \quad ; \quad \overset{\circ}{T}_{AB} = T^M{}_{AMB} \tag{7.2.48}$$

The irreducible tensors **420** and **36** can be identified, through a suitable coefficient fixed by Bianchi identities, with the fermion shifts appearing in the supersymmetry transformation rules (7.2.24):

$$\Sigma^A_{BCD} = \sigma \; \overset{\circ}{T}{}^A{}_{BCD} \quad ; \quad S_{DB} = s \, \overset{\circ}{T}_{DB} \tag{7.2.49}$$

Finally, as shown by de Wit and Nicolai [2, 3] the crucial Ward identity (7.2.27) is satisfied if and only if the ratio between the two constants in Eq. (7.2.49) is:

$$\frac{s^2}{\sigma^2} = \frac{1}{392} \tag{7.2.50}$$

7.2.4 *Algebraic Characterization of the Gauge Group Embedding $G_{gauge} \longrightarrow \mathrm{SL}(8, \mathbb{R})$*

As we have seen in the previous section the existence of gauged supergravity models relies on a peculiar pair of identities to be satisfied by the T-tensors. Therefore a classification of all possible electric gaugings involves a parametrization of all $\mathrm{SL}(8, \mathbb{R})$ subalgebras that lead to satisfied T-identities. Since the T-tensors are scalar field dependent objects it is not immediately obvious how such a program can be carried through. On the other hand since the problem is algebraic in nature (one looks for all Lie subalgebras of $\mathrm{SL}(8, \mathbb{R})$ fulfilling a certain property) it is clear that it should admit a completely algebraic formulation. It turns out that such an algebraic formulation is possible and actually very simple. Indeed the T-identities imposed on the T-tensors are nothing else but a single algebraic equation imposed on the embedding matrix $\mathcal{E}$ introduced in Eq. (7.2.37). This is what we outline next.

To begin with we recall a general and obvious constraint to be satisfied by $\mathcal{E}$ which embeds a subalgebra of the SL(8, $\mathbb{R}$) Lie algebra into its **28** irreducible representation: the **vectors** should be in the **coadjoint representation** of the gauge group. Hence under the reduction to $G_{gauge} \subset \mathrm{SL}(8, \mathbb{R})$ we must obtain the following decomposition of the entire set of the electric vectors:

$$\mathbf{28} \overset{G_{gauge}}{\rightarrow} \mathbf{coadj}G_{gauge} \oplus \mathcal{R} \tag{7.2.51}$$

where $\mathcal{R}$ denotes the subspace of vectors not entering the adjoint representation of G_{gauge} which is not necessarily a representation of G_{gauge} itself.

Next in order to reduce the field dependent T-identities to an algebraic equation on $\mathcal{E}$ we introduce the following constant tensors[6]:

$$t^{(1)}_{\Omega\Sigma}{}^{\Pi\Gamma}{}_{\Delta\Lambda} \equiv \sum_\alpha e^\alpha_{\Omega\Sigma}\, q^{\Pi\Gamma}{}_{\Delta\Lambda}(\alpha) \quad , \quad t^{(2)}_{\Omega\Sigma}{}^{\Pi\Gamma\Delta\Lambda} \equiv \sum_\alpha e^\alpha_{\Omega\Sigma}\, p^{\Pi\Gamma\Delta\Lambda}(\alpha)\,. \tag{7.2.52}$$

In terms of $t^{(1)}$ and $t^{(2)}$ the field dependent T-tensor is rewritten as

$$\begin{aligned} T^A{}_{BCD} = (u^{\Omega\Sigma}{}_{CD} + v_{\Omega\Sigma CD})\big[\, & t^{(1)}_{\Omega\Sigma}{}^{\Pi\Gamma}{}_{\Delta\Lambda}\,(u^{AM}{}_{\Pi\Gamma}\, u^{\Delta\Lambda}{}_{BM} - v^{AM\Phi\Gamma}\, v_{\Delta\Lambda BM}) \\ & + t^{(2)}_{\Omega\Sigma}{}^{\Pi\Gamma\Delta\Lambda}\,(u^{AM}{}_{\Pi\Gamma}\, v^{\Delta\Lambda}{}_{BM} - v^{AM\Phi\Gamma}\, u_{\Delta\Lambda BM})\big]\,. \end{aligned} \tag{7.2.53}$$

By means of lengthy algebraic manipulations in [33] the following statement was shown to be true

Theorem 7.2.1 *The field dependent T-identities are fully equivalent to the following algebraic equation:*

$$t^{(1)}_{\Omega\Sigma}{}^{\Pi\Gamma}{}_{\Delta\Lambda} + t^{(1)}_{\Delta\Lambda}{}^{\Pi\Gamma}{}_{\Omega\Sigma} + t^{(2)}_{\Pi\Gamma}{}^{\Delta\Lambda\Omega\Sigma} = 0 \tag{7.2.54}$$

Here we omit the proof but we stress the relevance of the result. All possible gauged supergravities have been put into one-to-one correspondence with the inequivalent solutions of an algebraic equation to be satisfied by the embedding matrix.

The algebraic t-identity (7.2.54) is a linear equation imposed on the embedding matrix $\mathcal{E}$. In [33] it was solved by means of a computer program yielding a 36-parameter solution. It was then shown that all the 36 parameters could be absorbed by means of conjugations with elements of the electric subgroup leaving only a finite

[6]For example, in the de Wit–Nicolai theory, where one gauges $G_{gauge} = \mathrm{SO}(8)$ we have:

$$t^{(1)}_{\Omega\Sigma}{}^{\Pi\Gamma}{}_{\Delta\Lambda} = \delta^{[\Pi}_{[\Delta}\delta_{\Lambda][\Omega}\delta^{\Gamma]}_{\Sigma]}\,, \qquad t^{(2)}_{\Omega\Sigma}{}^{\Pi\Gamma\Delta\Lambda} = 0\,.$$

discrete set of inequivalent solutions corresponding to as many inequivalent compact and non compact subalgebras of SL(8, $\mathbb{R}$). In order to describe this result more explicitly we need to discuss the embedding of the electric group in some detail.

7.2.5 *Embedding of the Electric Group*

The first information we need to specify is the explicit embedding of the electric subalgebra SL(8, $\mathbb{R}$) into the U-duality algebra $E_{7(7)}$. For this latter we adopt the conventions and notations of [54].

7.2.5.1 The $E_{7(7)}$ Algebra: Roots and Weights

We consider the standard E_7 Dynkin diagram (see Fig. 7.5) and we name α_i ($i = 1, \ldots, 7$) the corresponding simple roots. An explicit representation of the simple roots in Euclidean 7-dimensional space is the following one:

$$
\begin{aligned}
\alpha_1 &= \{1, -1, 0, 0, 0, 0, 0\} \\
\alpha_2 &= \{0, 1, -1, 0, 0, 0, 0\} \\
\alpha_3 &= \{0, 0, 1, -1, 0, 0, 0\} \\
\alpha_4 &= \{0, 0, 0, 1, -1, 0, 0\} \\
\alpha_5 &= \{0, 0, 0, 0, 1, -1, 0\} \\
\alpha_6 &= \{0, 0, 0, 0, 1, 1, 0\} \\
\alpha_7 &= \left\{-\frac{1}{2}, -\frac{1}{2}, -\frac{1}{2}, -\frac{1}{2}, -\frac{1}{2}, -\frac{1}{2}, \frac{1}{\sqrt{2}}\right\}
\end{aligned}
\tag{7.2.55}
$$

Having fixed this basis, each $E_{7(7)}$ root is intrinsically identified by its Dynkin labels, namely by its integer valued components in the simple root basis (Fig. 7.5).

Having identified the roots, the next step we need is the construction of the real fundamental representation SpD(56) of our U-duality Lie algebra $E_{7(7)}$. For this we need the corresponding weight vectors $\mathbf{W}$.

A particularly relevant property of the maximally non-compact real sections of a simple complex Lie algebra is that all its irreducible representations are real. $E_{7(7)}$ is the maximally non compact real section of the complex Lie algebra E_7, hence all its irreducible representations Γ are real. This implies that if an element of the weight

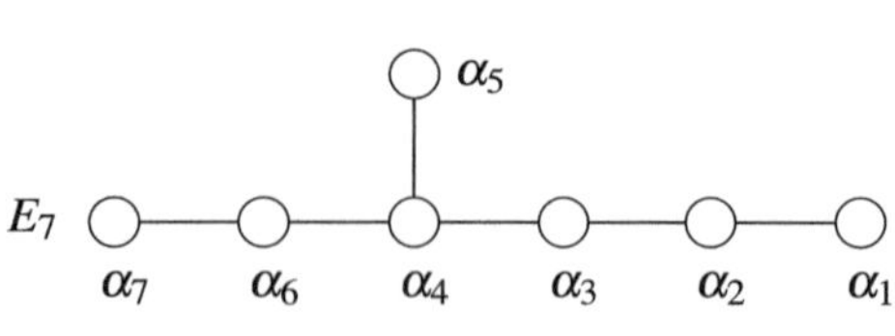

Fig. 7.5 E_7 Dynkin diagram and root labeling

lattice $\mathbf{W} \in \Lambda_w$ is a weight of a given irreducible representation $\mathbf{W} \in \Gamma$ then also its negative is a weight of the same representation: $-\mathbf{W} \in \Gamma$. Indeed changing sign to the weights corresponds to complex conjugation.

According to standard Lie algebra lore every irreducible representation of a simple Lie algebra $\mathbb{G}$ is identified by a unique *highest weight* $\mathbf{W}_{max}$. Furthermore all weights can be expressed as integral non-negative linear combinations of the *simple* weights $\mathbf{W}_\ell$ ($\ell = 1, \ldots, r = \text{rank}(\mathbb{G})$), whose components are named the Dynkin labels of the weight. The simple weights $\mathbf{W}_i$ of $\mathbb{G}$ are the generators of the dual lattice to the root lattice and are defined by the condition:

$$\frac{2(\mathbf{W}_i\,,\,\boldsymbol{\alpha}_j)}{(\boldsymbol{\alpha}_j\,,\,\boldsymbol{\alpha}_j)} = \delta_{ij} \tag{7.2.56}$$

In the simply laced $E_{7(7)}$ case, the previous equation simplifies as follows

$$(\mathbf{W}_i\,,\,\boldsymbol{\alpha}_j) = \delta_{ij} \tag{7.2.57}$$

where $\boldsymbol{\alpha}_j$ are the simple roots. Using the Dynkin diagram of $E_{7(7)}$ (see Fig. 7.5) from Eq. (7.2.57) we can easily obtain the explicit expression of the simple weights that are listed in Table 7.2.

The Dynkin labels of the highest weight of an irreducible representation Γ give the Dynkin labels of the representation. Therefore the representation is usually denoted by $\Gamma[n_1, \ldots, n_r]$. All the weights $\mathbf{W}$ belonging to the representation Γ can be described by r integer non-negative numbers q^ℓ defined by the following equation:

$$\mathbf{W}_{max} - \mathbf{W} = \sum_{\ell=1}^{r} q^\ell \boldsymbol{\alpha}_\ell \tag{7.2.58}$$

where α_ℓ are the simple roots. According to this standard formalism the fundamental real representation SpD(56) of $E_{7(7)}$ is $\Gamma[1, 0, 0, 0, 0, 0, 0]$ and the expression of its weights in terms of q^ℓ is given in Table 7.2, the highest weight being $\mathbf{W}^{(51)}$.

We can now explain the specific ordering of the weights we have adopted.

First of all we have separated the 56 weights in two groups of 28 elements so that the first group:

$$\boldsymbol{\Lambda}^{(n)} = \mathbf{W}^{(n)} \quad n = 1, \ldots, 28 \tag{7.2.59}$$

are the weights for the irreducible **28** dimensional representation of the *electric* subgroup $SL(8, \mathbb{R}) \subset E_{7(7)}$. The remaining group of 28 weight vectors are the weights for the transposed representation of the same group that we name $\overline{\mathbf{28}}$.

Secondly the 28 weights $\boldsymbol{\Lambda}$ have been arranged according to the decomposition with respect to the T-*duality* subalgebra $SO(6, 6) \subset E_7(7)$. From a superstring point of view the first 16 correspond to Ramond–Ramond vectors and are the weights of the spinor representation of SO(6, 6) while the last 12 are associated with Neveu–Schwarz fields and correspond to the weights of the vector representation of SO(6, 6).

Table 7.2 Weights of the **56** representation of $E_{7(7)}$

Weight q^ℓ		Weight q^ℓ	
Name	Vector	Name	Vector
$\mathbf{W}^{(1)}$	$= \{2, 3, 4, 5, 3, 3, 1\}$	$\mathbf{W}^{(2)}$	$= \{2, 2, 2, 2, 1, 1, 1\}$
$\mathbf{W}^{(3)}$	$= \{1, 2, 2, 2, 1, 1, 1\}$	$\mathbf{W}^{(4)}$	$= \{1, 1, 2, 2, 1, 1, 1\}$
$\mathbf{W}^{(5)}$	$= \{1, 1, 1, 2, 1, 1, 1\}$	$\mathbf{W}^{(6)}$	$= \{1, 1, 1, 1, 1, 1, 1\}$
$\mathbf{W}^{(7)}$	$= \{2, 3, 3, 3, 1, 2, 1\}$	$\mathbf{W}^{(8)}$	$= \{2, 2, 3, 3, 1, 2, 1\}$
$\mathbf{W}^{(9)}$	$= \{2, 2, 2, 3, 1, 2, 1\}$	$\mathbf{W}^{(10)}$	$= \{2, 2, 2, 2, 1, 2, 1\}$
$\mathbf{W}^{(11)}$	$= \{1, 2, 2, 2, 1, 2, 1\}$	$\mathbf{W}^{(12)}$	$= \{1, 1, 2, 2, 1, 2, 1\}$
$\mathbf{W}^{(13)}$	$= \{1, 1, 1, 2, 1, 2, 1\}$	$\mathbf{W}^{(14)}$	$= \{1, 2, 2, 3, 1, 2, 1\}$
$\mathbf{W}^{(15)}$	$= \{1, 2, 3, 3, 1, 2, 1\}$	$\mathbf{W}^{(16)}$	$= \{1, 1, 2, 3, 1, 2, 1\}$
$\mathbf{W}^{(17)}$	$= \{2, 2, 2, 2, 1, 1, 0\}$	$\mathbf{W}^{(18)}$	$= \{1, 2, 2, 2, 1, 1, 0\}$
$\mathbf{W}^{(19)}$	$= \{1, 1, 2, 2, 1, 1, 0\}$	$\mathbf{W}^{(20)}$	$= \{1, 1, 1, 2, 1, 1, 0\}$
$\mathbf{W}^{(21)}$	$= \{1, 1, 1, 1, 1, 1, 0\}$	$\mathbf{W}^{(22)}$	$= \{1, 1, 1, 1, 1, 0, 0\}$
$\mathbf{W}^{(23)}$	$= \{3, 4, 5, 6, 3, 4, 2\}$	$\mathbf{W}^{(24)}$	$= \{2, 4, 5, 6, 3, 4, 2\}$
$\mathbf{W}^{(25)}$	$= \{2, 3, 5, 6, 3, 4, 2\}$	$\mathbf{W}^{(26)}$	$= \{2, 3, 4, 6, 3, 4, 2\}$
$\mathbf{W}^{(27)}$	$= \{2, 3, 4, 5, 3, 4, 2\}$	$\mathbf{W}^{(28)}$	$= \{2, 3, 4, 5, 3, 3, 2\}$
$\mathbf{W}^{(29)}$	$= \{1, 1, 1, 1, 0, 1, 1\}$	$\mathbf{W}^{(30)}$	$= \{1, 2, 3, 4, 2, 3, 1\}$
$\mathbf{W}^{(31)}$	$= \{2, 2, 3, 4, 2, 3, 1\}$	$\mathbf{W}^{(32)}$	$= \{2, 3, 3, 4, 2, 3, 1\}$
$\mathbf{W}^{(33)}$	$= \{2, 3, 4, 4, 2, 3, 1\}$	$\mathbf{W}^{(34)}$	$= \{2, 3, 4, 5, 2, 3, 1\}$
$\mathbf{W}^{(35)}$	$= \{1, 1, 2, 3, 2, 2, 1\}$	$\mathbf{W}^{(36)}$	$= \{1, 2, 2, 3, 2, 2, 1\}$
$\mathbf{W}^{(37)}$	$= \{1, 2, 3, 3, 2, 2, 1\}$	$\mathbf{W}^{(38)}$	$= \{1, 2, 3, 4, 2, 2, 1\}$
$\mathbf{W}^{(39)}$	$= \{2, 2, 3, 4, 2, 2, 1\}$	$\mathbf{W}^{(40)}$	$= \{2, 3, 3, 4, 2, 2, 1\}$
$\mathbf{W}^{(41)}$	$= \{2, 3, 4, 4, 2, 2, 1\}$	$\mathbf{W}^{(42)}$	$= \{2, 2, 3, 3, 2, 2, 1\}$
$\mathbf{W}^{(43)}$	$= \{2, 2, 2, 3, 2, 2, 1\}$	$\mathbf{W}^{(44)}$	$= \{2, 3, 3, 3, 2, 2, 1\}$
$\mathbf{W}^{(45)}$	$= \{1, 2, 3, 4, 2, 3, 2\}$	$\mathbf{W}^{(46)}$	$= \{2, 2, 3, 4, 2, 3, 2\}$
$\mathbf{W}^{(47)}$	$= \{2, 3, 3, 4, 2, 3, 2\}$	$\mathbf{W}^{(48)}$	$= \{2, 3, 4, 4, 2, 3, 2\}$
$\mathbf{W}^{(49)}$	$= \{2, 3, 4, 5, 2, 3, 2\}$	$\mathbf{W}^{(50)}$	$= \{2, 3, 4, 5, 2, 4, 2\}$
$\mathbf{W}^{(51)}$	$= \{0, 0, 0, 0, 0, 0, 0\}$	$\mathbf{W}^{(52)}$	$= \{1, 0, 0, 0, 0, 0, 0\}$
$\mathbf{W}^{(53)}$	$= \{1, 1, 0, 0, 0, 0, 0\}$	$\mathbf{W}^{(54)}$	$= \{1, 1, 1, 0, 0, 0, 0\}$
$\mathbf{W}^{(55)}$	$= \{1, 1, 1, 1, 0, 0, 0\}$	$\mathbf{W}^{(56)}$	$= \{1, 1, 1, 1, 0, 1, 0\}$

7.2.5.2 The Matrices of the Fundamental 56 Representation

Equipped with the weight vectors we can now proceed to the explicit construction of the **SpD(56)** representation of $E_{7(7)}$. In the construction of [33], the basis vectors are the 56 weights, according to the enumeration of Table 7.2. What we need are the 56×56 matrices associated with the 7 Cartan generators H_{α_i} $(i = 1, \ldots, 7)$ and with the 126 step operators E^α that are defined by:

$$\begin{aligned}\left[SpD_{56}\left(H_{\boldsymbol{\alpha}_i}\right)\right]_{nm} &\equiv \langle \mathbf{W}^{(n)} | \, H_{\boldsymbol{\alpha}_i} \, | \mathbf{W}^{(m)} \rangle \\ \left[SpD_{56}\left(E^{\boldsymbol{\alpha}}\right)\right]_{nm} &\equiv \langle \mathbf{W}^{(n)} | \, E^{\boldsymbol{\alpha}} \, | \mathbf{W}^{(m)} \rangle\end{aligned} \tag{7.2.60}$$

Following [33] let us begin with the Cartan generators. As a basis of the Cartan subalgebra we use the generators $H_{\boldsymbol{\alpha}_i}$ defined by the commutators:

$$\left[E^{\boldsymbol{\alpha}_i}, E^{-\boldsymbol{\alpha}_i}\right] \equiv H_{\boldsymbol{\alpha}_i} \tag{7.2.61}$$

In the SpD(56) representation the corresponding matrices are diagonal and of the form:

$$\langle \mathbf{W}^{(p)} | \, H_{\boldsymbol{\alpha}_i} \, | \mathbf{W}^{(q)} \rangle = \left(\mathbf{W}^{(p)}, \boldsymbol{\alpha}_i\right) \delta_{pq} \quad ; \quad (p, q = 1, \ldots, 56) \tag{7.2.62}$$

The scalar products

$$\left(\boldsymbol{\Lambda}^{(n)} \cdot \mathbf{h}, -\boldsymbol{\Lambda}^{(m)} \cdot \mathbf{h}\right) = \left(\mathbf{W}^{(p)} \cdot \mathbf{h}\right) \quad ; \quad (n, m = 1, \ldots, 28 \, ; \; p = 1, \ldots, 56) \tag{7.2.63}$$

are to be understood in the following way:

$$\mathbf{W}^{(p)} \cdot \mathbf{h} = \sum_{i=1}^{7} \left(\mathbf{W}^{(p)}, \boldsymbol{\alpha}_i\right) h^i \tag{7.2.64}$$

Next we construct the matrices associated with the step operators. Here the first observation is that it suffices to consider the positive roots. Because of the reality of the representation, the matrix associated with the negative of a root is just the transposed of that associated with the root itself:

$$E^{-\alpha} = \left[E^{\alpha}\right]^T \leftrightarrow \langle \mathbf{W}^{(n)} | \, E^{-\boldsymbol{\alpha}} \, | \mathbf{W}^{(m)} \rangle = \langle \mathbf{W}^{(m)} | \, E^{\boldsymbol{\alpha}} \, | \mathbf{W}^{(n)} \rangle \tag{7.2.65}$$

The method followed in [33] to obtain the matrices for all the positive roots is that of constructing first the 56×56 matrices for the step operators $E^{\boldsymbol{\alpha}_\ell}$ $(\ell = 1, \ldots, 7)$ associated with the simple roots and then generating all the others through their commutators. The construction rules for the SpD(56) representation of the six operators $E^{\boldsymbol{\alpha}_\ell}$ $(\ell \neq 5)$ are:

$$\ell \neq 5 \quad \begin{cases} \langle \mathbf{W}^{(n)} | \, E^{\boldsymbol{\alpha}_\ell} \, | \mathbf{W}^{(m)} \rangle = \delta_{\mathbf{W}^{(n)}, \mathbf{W}^{(m)} + \boldsymbol{\alpha}_\ell} & ; \; n, m = 1, \ldots, 28 \\ \langle \mathbf{W}^{(n+28)} | \, E^{\boldsymbol{\alpha}_\ell} \, | \mathbf{W}^{(m+28)} \rangle = -\delta_{\mathbf{W}^{(n+28)}, \mathbf{W}^{(m+28)} + \boldsymbol{\alpha}_\ell} & ; \; n, m = 1, \ldots, 28 \end{cases} \tag{7.2.66}$$

The six simple roots $\boldsymbol{\alpha}_\ell$ with $\ell \neq 5$ belong also to the Dynkin diagram of the electric subgroup SL(8, $\mathbb{R}$). Thus their shift operators have a block diagonal action on the **28** and $\overline{\mathbf{28}}$ subspaces of the $SpD(56)$ representation that are irreducible under the electric subgroup. From Eq. (7.2.66) we conclude that:

$$\ell \neq 5 \quad \mathrm{SpD}_{56}\left(E^{\alpha_\ell}\right) = \begin{pmatrix} A[\boldsymbol{\alpha}_\ell] & \mathbf{0} \\ \mathbf{0} & -A^T[\boldsymbol{\alpha}_\ell] \end{pmatrix} \tag{7.2.67}$$

the 28×28 block $A[\boldsymbol{\alpha}_\ell]$ being defined by the first line of Eq. (7.2.66).

On the contrary the operator E^{α_5}, corresponding to the only root of the E_7 Dynkin diagram that is not also part of the A_7 diagram is represented by a matrix whose non-vanishing 28×28 blocks are off-diagonal. We have

$$\mathrm{SpD}_{56}\left(E^{\alpha_5}\right) = \begin{pmatrix} \mathbf{0} & B[\boldsymbol{\alpha}_5] \\ C[\boldsymbol{\alpha}_5] & \mathbf{0} \end{pmatrix} \tag{7.2.68}$$

where both $B[\boldsymbol{\alpha}_5] = B^T[\boldsymbol{\alpha}_5]$ and $C[\boldsymbol{\alpha}_5] = C^T[\boldsymbol{\alpha}_5]$ are symmetric 28×28 matrices. More explicitly the matrix $\mathrm{SpD}_{56}\left(E^{\alpha_5}\right)$ is given by:

$$\begin{aligned} \langle \mathbf{W}^{(n)} | \, E^{\alpha_5} \, | \mathbf{W}^{(m+28)} \rangle &= \langle \mathbf{W}^{(m)} | \, E^{\alpha_5} \, | \mathbf{W}^{(n+28)} \rangle \\ \langle \mathbf{W}^{(n+28)} | \, E^{\alpha_5} \, | \mathbf{W}^{(m)} \rangle &= \langle \mathbf{W}^{(m+28)} | \, E^{\alpha_5} \, | \mathbf{W}^{(n)} \rangle \end{aligned} \tag{7.2.69}$$

with

$$\begin{array}{lll} \langle \mathbf{W}^{(7)} | \, E^{\alpha_5} \, | \mathbf{W}^{(44)} \rangle = -1 & \langle \mathbf{W}^{(8)} | \, E^{\alpha_5} \, | \mathbf{W}^{(42)} \rangle = \ \ 1 & \langle \mathbf{W}^{(9)} | \, E^{\alpha_5} \, | \mathbf{W}^{(43)} \rangle = -1 \\ \langle \mathbf{W}^{(14)} | \, E^{\alpha_5} \, | \mathbf{W}^{(36)} \rangle = \ \ 1 & \langle \mathbf{W}^{(15)} | \, E^{\alpha_5} \, | \mathbf{W}^{(37)} \rangle = -1 & \langle \mathbf{W}^{(16)} | \, E^{\alpha_5} \, | \mathbf{W}^{(35)} \rangle = -1 \\ \langle \mathbf{W}^{(29)} | \, E^{\alpha_5} \, | \mathbf{W}^{(6)} \rangle = -1 & \langle \mathbf{W}^{(34)} | \, E^{\alpha_5} \, | \mathbf{W}^{(1)} \rangle = -1 & \langle \mathbf{W}^{(49)} | \, E^{\alpha_5} \, | \mathbf{W}^{(28)} \rangle = \ \ 1 \\ \langle \mathbf{W}^{(50)} | \, E^{\alpha_5} \, | \mathbf{W}^{(27)} \rangle = -1 & \langle \mathbf{W}^{(55)} | \, E^{\alpha_5} \, | \mathbf{W}^{(22)} \rangle = -1 & \langle \mathbf{W}^{(56)} | \, E^{\alpha_5} \, | \mathbf{W}^{(21)} \rangle = \ \ 1 \end{array} \tag{7.2.70}$$

In this way we have completed the construction of the $E^{\boldsymbol{\alpha}_\ell}$ operators associated with simple roots. For the matrices associated with higher roots we just proceed iteratively in the following way. As usual we organize the roots by height:

$$\alpha = n^\ell \, \alpha_\ell \quad \rightarrow \quad \mathrm{ht}\, \alpha = \sum_{\ell=1}^{7} n^\ell \tag{7.2.71}$$

and for the roots $\alpha_i + \alpha_j$ of height ht $= 2$ we set:

$$\mathrm{SpD}_{56}\left(E^{\alpha_i+\alpha_j}\right) \equiv \left[\mathrm{SpD}_{56}\left(E^{\alpha_i}\right) \, , \, \mathrm{SpD}_{56}\left(E^{\alpha_i}\right)\right] \quad ; \quad i < j \tag{7.2.72}$$

Next for the roots of ht $= 3$ that can be written as $\alpha_i + \beta$ where α_i is simple and ht $\beta \ = 2$ we write:

$$\mathrm{SpD}_{56}\left(E^{\alpha_i+\beta}\right) \equiv \left[\mathrm{SpD}_{56}\left(E^{\alpha_i}\right) \, , \, \mathrm{SpD}_{56}\left(E^{\beta}\right)\right] \tag{7.2.73}$$

Obtained the matrices for the roots of ht $= 3$ one proceeds in a similar way for those of the next height and so on up to exhaustion of all the 63 positive roots.

This concludes our description of the algorithm by means of which our computer program constructed all the 133 matrices spanning the $E_{7(7)}$ Lie algebra in the SpD(56) representation. A fortiori, if we specify the embedding we have the matrices generating the electric subgroup $SL(8, \mathbb{R})$.

7.2.5.3 The SL(8, ℝ) Subalgebra

The Electric $\mathfrak{sl}(8, \mathbb{R})$ subalgebra is identified in $E_{7(7)}$ by specifying its simple roots β_i spanning the standard A_7 Dynkin diagram. The Cartan generators are the same for the $E_{7(7)}$ Lie algebra as for the $SL(8, \mathbb{R})$ subalgebra and if we give β_i every other generator is defined. The basis we have chosen is the following one:

$$\begin{array}{lcl} \beta_1 = & \alpha_2 + 2\alpha_3 + 3\alpha_4 + 2\alpha_5 + 2\alpha_6 + \alpha_7 & ; \ \beta_2 = \alpha_1 \\ \beta_3 = & \alpha_2 & ; \ \beta_4 = \alpha_3 \\ \beta_5 = & \alpha_4 & ; \ \beta_6 = \alpha_6 \\ \beta_7 = & \alpha_7 & \end{array} \tag{7.2.74}$$

The complete set of positive roots of $SL(8\mathbb{R})$ is then composed of 28 elements that we name ρ_i $(i = 1, \ldots, 28)$ and that are enumerated according to our chosen order in Table 7.3.

Hence the 63 generators of the $SL(8, \mathbb{R})$ Lie algebra are:

$$\begin{array}{ccc} \text{The 7 Cartan generators} & C_i = H_{\alpha_i} & i = 1, \ldots, 7 \\ \text{The 28 positive root generators} & E^{\rho_i} & i = 1, \ldots, 28 \\ \text{The 28 negative root generators} & E^{-\rho_i} & i = 1, \ldots, 28 \end{array} \tag{7.2.75}$$

and since the 56×56 matrix representation of each $E_{7(7)}$ Cartan generator or step operator was constructed in the previous subsection it is obvious that it is in particular given for the subset of those that belong to the $SL(8, \mathbb{R})$ subalgebra. The basis of this matrix representation is provided by the weights enumerated in Table 7.2.

In this way we have concluded our illustration of the basis in which the algebraic t-identity was solved in [33]. The result is the 28×63 matrix:

$$\mathcal{E}(h, \ell) \longrightarrow e^{\alpha}_{W}(h, \ell) \tag{7.2.76}$$

where the index W runs on the 28 negative weights of Table 7.2, while the index α runs on all the $SL(8, \mathbb{R})$ generators according to Eq. (7.2.75). The matrix $\mathcal{E}(h, p)$ depends on 36 parameters that we have named:

$$\begin{array}{l} h_i \ \ i = 1, \ldots, 8 \\ \ell_i \ \ i = 1, \ldots, 28 \end{array} \tag{7.2.77}$$

and its entries are explicitly displayed in tables given in [33]. The distinction between the h_i parameters and the ℓ_i parameters has been drawn in the following way:

Table 7.3 The choice of the order of the SL(8, ℝ) roots

$\rho_1 \equiv \beta_2$
$\rho_2 \equiv \beta_2 + \beta_3$
$\rho_3 \equiv \beta_2 + \beta_3 + \beta_4$
$\rho_4 \equiv \beta_2 + \beta_3 + \beta_4 + \beta_5$
$\rho_5 \equiv \beta_2 + \beta_3 + \beta_4 + \beta_5 + \beta_6$
$\rho_6 \equiv \beta_3$
$\rho_7 \equiv \beta_3 + \beta_4$
$\rho_8 \equiv \beta_3 + \beta_4 + \beta_5$
$\rho_9 \equiv \beta_3 + \beta_4 + \beta_5 + \beta_6$
$\rho_{10} \equiv \beta_4$
$\rho_{11} \equiv \beta_4 + \beta_5$
$\rho_{12} \equiv \beta_4 + \beta_5 + \beta_6$
$\rho_{13} \equiv \beta_5$
$\rho_{14} \equiv \beta_5 + \beta_6$
$\rho_{15} \equiv \beta_6$
$\rho_{16} \equiv \beta_1 + \beta_2 + \beta_3 + \beta_4 + \beta_5 + \beta_6 + \beta_7$
$\rho_{17} \equiv \beta_2 + \beta_3 + \beta_4 + \beta_5 + \beta_6 + \beta_7$
$\rho_{18} \equiv \beta_3 + \beta_4 + \beta_5 + \beta_6 + \beta_7$
$\rho_{19} \equiv \beta_4 + \beta_5 + \beta_6 + \beta_7$
$\rho_{20} \equiv \beta_5 + \beta_6 + \beta_7$
$\rho_{21} \equiv \beta_6 + \beta_7$
$\rho_{22} \equiv \beta_1$
$\rho_{23} \equiv \beta_1 + \beta_2$
$\rho_{24} \equiv \beta_1 + \beta_2 + \beta_3$
$\rho_{25} \equiv \beta_1 + \beta_2 + \beta_3 + \beta_4$
$\rho_{26} \equiv \beta_1 + \beta_2 + \beta_3 + \beta_4 + \beta_5$
$\rho_{27} \equiv \beta_1 + \beta_2 + \beta_3 + \beta_4 + \beta_5 + \beta_6$
$\rho_{28} \equiv \beta_7$

- The 8 parameters h_i are those that never multiply a Cartan generator
- The 28 parameters ℓ_i are those that multiply at least one Cartan generator.

In other words if we set all the $\ell_i = 0$ the gauge subalgebra $G_{gauge} \subset \mathrm{SL}(8, \mathbb{R})$ is composed solely of step operators while if you switch on also the ℓ_i.s then some Cartan generators appear in the Lie algebra. This distinction is very useful in classifying the independent solutions.

7.2.6 *Classification of Electrically Gauged $\mathcal{N} = 8$ Supergravities*

Equipped with the explicit solution of the t-identity encoded in the embedding matrix $\mathcal{E}$ one can address the complete classification of the *electrically gauged supergravity* models.

The complete set of possible theories found in [33] coincides with the gaugings found by Hull [16–22] (together with the ones simply outlined by Hull [16–22]) in the middle of the eighties and correspond to all possible non-compact real forms of the SO(8) Lie algebra plus a number of Inonu–Wigner contractions thereof. The method of [33] is algorithmic and allows to construct any model in this class. It is particularly suited for our present purposes.

We have to begin our discussion with two observations:

1. The solution of t-identities encoded in the matrix $\mathcal{E}(h, \ell)$ is certainly overcomplete since we are still free to conjugate any gauge algebra G_{gauge} with an arbitrary finite element of the electric group $g \in \mathrm{SL}(8, \mathbb{R})$: $G'_{gauge} = g\, G_{gauge}\, g^{-1}$ yields a completely physically equivalent gauging as G_{gauge}. This means that we need to consider the $\mathrm{SL}(8, \mathbb{R})$ transformations of the matrix $\mathcal{E}(h, \ell)$ defined as:

$$\forall g \in \mathrm{SL}(8, \mathbb{R}) \; : \; g \cdot \mathcal{E}(h, \ell) \equiv D_{28}(g^{-1})\, \mathcal{E}(h, \ell)\, D_{63}(g) \tag{7.2.78}$$

   where $D_{28}(g)$ and $D_{63}(g)$ denote the matrices of the **28** and the **63** representation respectively. If two set of parameters $\{h, \ell\}$ and $\{h', \ell'\}$ are related by an $\mathrm{SL}(8, \mathbb{R})$ conjugation, in the sense that:

$$\exists g \in \mathrm{SL}(8, \mathbb{R}) \; : \; \mathcal{E}(h', \ell') = g \cdot \mathcal{E}(h, \ell) \tag{7.2.79}$$

   then the theories described by $\{h, \ell\}$ and $\{h', \ell'\}$ are the same theory. In other words what we need is the space of orbits of $\mathrm{SL}(8, \mathbb{R})$ inequivalent embedding matrices.
2. Possible theories obtained by choosing a set of $\{h, \ell\}$ parameters are further restricted by the constraints that

   - The selected generators of $\mathrm{SL}(8, \mathbb{R})$ should close a Lie subalgebra G_{gauge}
   - The selected vectors (=weights, see Table 7.2) should transform in the coadjoint representation $Coadj\left(G_{gauge}\right)$

In view of these observations a natural question we should pose is the following one: *is there a natural way to understand why the number of parameters on which the embedding matrix depends is, a part from an immaterial overall constant, precisely 35?* The answer is immediate and inspiring. Because of point (2) in the above list of properties the 28 linear combinations of $\mathrm{SL}(8, \mathbb{R})$ generators:

$$T_W \equiv e_W{}^{\alpha}\, (h, \ell)\; G_\alpha \tag{7.2.80}$$

must span the adjoint representation of a 28-dimensional subalgebra $G_{gauge}(h, \ell)$ of the $SL(8, \mathbb{R})$ algebra.

Naming $\mathscr{G}_{gauge}(h, \ell)$ the corresponding Lie subgroup, because of its very definition we have that the matrix $\mathscr{E}(h, \ell)$ is invariant under transformations of $\mathscr{G}_{gauge}(h, \ell)$[7]:

$$\forall \gamma \in \mathscr{G}_{gauge}(h, \ell \subset \mathrm{SL}(8, \mathbb{R}) : \quad \gamma \cdot \mathscr{E}(h, \ell) = \mathscr{E}(h, \ell) \tag{7.2.81}$$

Comparing Eq. (7.2.81) with (7.2.78) we see that having fixed a matrix $\mathscr{E}(h, \ell)$ and hence an algebra $\mathscr{G}_{gauge}(h, \ell)$, according to point (1) of the above discussion we can obtain a 35-dimensional orbit of equivalent embedding matrices:

$$\mathscr{E}\left(h'(\mu), \ell'(\mu)\right) \equiv g(\mu) \cdot \mathscr{E}(h, \ell) \quad \text{where} \quad g(\mu) \in \frac{\mathrm{SL}(8, \mathbb{R})}{\mathscr{G}_{gauge}(h, \ell)} \tag{7.2.82}$$

Hence, $35 = 63 - 28$ is the dimension of the coset manifold $\mathrm{SL}(8, \mathbb{R})/\mathscr{G}_{gauge}(h, \ell)$ and $\mathscr{E}\left(h'(\mu), \ell'(\mu)\right)$ is the embedding matrix for the family of conjugated, isomorphic, Lie algebras:

$$\mathscr{G}_{gauge}\left(h'(\mu), \ell'(\mu)\right) = g^{-1}(\mu)\, \mathscr{G}_{gauge}(h, \ell)\, g(\mu) \tag{7.2.83}$$

An essential and a priori unexpected conclusion was drown in [33] from this discussion.

Proposition 7.2.1 *The gauged $\mathscr{N} = 8$ supergravity models cannot depend on more than a single continuous parameter (=coupling constant), even if they correspond to gauging a multidimensional abelian algebra.*

Since the explicit solution of the algebraic t-identities has produced an embedding matrix $\mathscr{E}(h, \ell)$ depending on no more than 36-parameters, then the only continuous parameter which is physically relevant is the overall proportionality constant. The remaining 35-parameters can be reabsorbed by $\mathrm{SL}(8, \mathbb{R})$ conjugations according to Eq. (7.2.83)

In other words what it was found is that the space of orbits one was looking for is a discrete space. The classifications of electrically gauged supergravity models is just a classification of gauge algebras a single coupling constant being assigned to each case. This is considerably different from other supergravities with less supersymmetries, like the $\mathscr{N} = 2$ case. There gauging a group G_{gauge} involves as many coupling constants as there are simple factors in G_{gauge}. So in those cases not only we have a much wider variety of possible gauge algebras but also we have lagrangians depending on several continuous parameters. In the $\mathscr{N} = 8$ case supersymmetry constraints the theory in a much stronger way. This is an yield of supersymmetry

[7] Note that some of the 28 generators of $\mathscr{G}_{gauge}(h, \ell) \subset \mathrm{SL}(8, \mathbb{R})$ may be represented trivially in the adjoint representation, but in this case also the corresponding group transformations leave the embedding matrix invariant.

and not of Lie algebra theory. It is the algebraic t-identity, enforced by the closure of Bianchi identities, that admits a general solution depending only on 36-parameters. If the solution depended on $35 + m$ parameters then we might have introduced m relevant continuous parameters into the Lagrangian.

Relying on these observations one is left with the problem of classifying the orbit space already knowing that it is composed of a finite number of discrete elements. Orbits are characterized in terms of invariants, so we have to ask ourselves what is the natural invariant associated with the embedding matrix $\mathscr{E}(h, \ell)$. The answer is once again very simple. It is the *signature* of the *Killing–Cartan 2-form* for the resulting gauge algebra $\mathscr{G}_{gauge}(h, \ell)$. Consider the 28 generators (7.2.80) and define:

$$\begin{aligned} \eta_{W_1 W_2}(h, \ell) &\equiv \text{Tr}\left(T_{W_1} T_{W_1}\right) \\ &= e^{\alpha}_{W_1}(h, \ell)\, e^{\beta}_{W_2}(h, \ell)\, \text{Tr}\left(G_\alpha G_\beta\right) \\ &= e^{\alpha}_{W_1}(h, \ell)\, e^{\beta}_{W_2}(h, \ell)\, B_{\alpha\beta} \end{aligned} \tag{7.2.84}$$

where the trace Tr is taken over any representation and the constant matrix $B_{\alpha\beta} \equiv \text{Tr}\left(G_\alpha G_\beta\right)$ is the Killing–Cartan 2-form of the SL(8, $\mathbb{R}$) Lie algebra. The 28×28 matrix is the Killing–Cartan 2-form of the gauge algebra G_{gauge}. As it is well known from general Lie algebra theory, by means of suitable changes of bases inside the same Lie algebra the matrix $\eta_{W_1 W_2}(h, \ell)$ can be diagonalized and its eigenvalues can be reduced to be either of modulus one or zero. What cannot be done since it corresponds to an intrinsic characterization of the Lie algebra is to change the signature of $\eta_{W_1 W_2}(h, \ell)$, namely the ordered set of 28 signs (or zeros) appearing on the principal diagonal when $\eta_{W_1 W_2}(h, \ell)$ is reduced to diagonal form. Hence what is constant throughout an SL(8, $\mathbb{R}$) orbit is the signature. Let us name Σ (orbit) the 28 dimensional vector characterizing the signature of an orbit:

$$\Sigma\,(\text{orbit}) \equiv \text{signature}\left[\eta_{W_1 W_2}\left(h'(\mu), \ell'(\mu)\right)\right] \tag{7.2.85}$$

From our discussion we conclude that

Proposition 7.2.2 *The classification of gauged $\mathcal{N} = 8$ models has been reduced to the classification of the signature vectors Σ (orbit) of Eq.* (7.2.85).

The procedure to calculate Σ (orbit) associated with an orbit $\eta_{W_1 W_2}\left(h'(\mu), \ell'(\mu)\right)$ is that of choosing the representative $\left(h'(\mu_0), \ell'(\mu_0)\right)$ for which the corresponding matrix $\eta_{W_1 W_2}\left(h'(\mu_0), \ell'(\mu_0)\right)$ is diagonal and then to evaluate the signs of the diagonal elements. In principle finding the appropriate $h'(\mu_0)$, $\ell'(\mu_0)$ could be a difficult task since we are supposed to diagonalize a 28×28 matrix. However our choice of coordinates on the parameter space is such that our task becomes very simple. Using the results for the embedding matrix we can calculate the matrix $\eta_{W_1 W_2}(h, \ell)$ and for generic values of h_i and ℓ_i we find that all of its 28×28 entries are non vanishing; yet setting $\ell_i = 0$ the matrix becomes automatically diagonal and we get:

Table 7.4 Electric gauge algebras

Algebra	n_+	n_-	n_0	$\{h_1, h_2, h_3, h_4, h_5, h_6, h_7\}$	Dimension
SO(8)	28	0	0	$\{1, 1, -1, 1, -1, 1, 1, -1\}$	28
SO(1, 7)	21	7	0	$\{1, 1, -1, 1, -1, 1, 1, 1\}$	28
SO(2, 6)	16	12	0	$\{-1, 1, -1, 1, -1, 1, 1, 1\}$	28
SO(3, 5)	13	15	0	$\{-1, -1, -1, 1, -1, 1, 1, 1\}$	28
SO(4, 4)	12	16	0	$\{-1, -1, 1, 1, -1, 1, 1, 1\}$	28
SO(5, 3)	13	15	0	$\{-1, -1, 1, -1, -1, 1, 1, 1\}$	28
SO(6, 2)	16	12	0	$\{-1, -1, 1, -1, 1, 1, 1, 1\}$	28
SO(7, 1)	21	7	0	$\{-1, -1, 1, -1, 1, -1, 1, 1\}$	28
CSO(1, 7)	0	0	28	$\{0, 0, 0, 0, 0, 0, 0, 1\}$	7
CSO(2, 6)	1	0	27	$\{-1, 0, 0, 0, 0, 0, 0, 1\}$	13
CSO(3, 5)	3	0	25	$\{-1, -1, 0, 0, 0, 0, 0, 1\}$	18
CSO(4, 4)	6	0	22	$\{-1, -1, 1, 0, 0, 0, 0, 1\}$	22
CSO(5, 3)	10	0	18	$\{-1, -1, 1, -1, 0, 0, 0, 1\}$	25
CSO(6, 2)	15	0	13	$\{-1, -1, 1, -1, 1, 0, 0, 1\}$	27
CSO(7, 1)	21	0	7	$\{-1, -1, 1, -1, 1, -1, 0, 1\}$	28
CSO(1, 1, 6)	0	1	27	$\{1, 0, 0, 0, 0, 0, 0, 1\}$	13
CSO(1, 2, 5)	1	2	25	$\{1, -1, 0, 0, 0, 0, 0, 1\}$	18
CSO(2, 1, 5)	1	2	25	$\{1, 1, 0, 0, 0, 0, 0, 1\}$	18
CSO(1, 3, 4)	3	3	22	$\{1, -1, 1, 0, 0, 0, 0, 1\}$	22
CSO(2, 2, 4)	2	4	22	$\{1, 1, 1, 0, 0, 0, 0, 1\}$	22
CSO(3, 1, 4)	3	3	22	$\{1, 1, -1, 0, 0, 0, 0, 1\}$	22
CSO(1, 4, 3)	6	4	18	$\{1, -1, 1, -1, 0, 0, 0, 1\}$	25
CSO(2, 3, 3)	4	6	18	$\{1, 1, 1, -1, 0, 0, 0, 1\}$	25
CSO(3, 2, 3)	4	6	18	$\{1, 1, -1, -1, 0, 0, 0, 1\}$	25
CSO(4, 1, 3)	6	4	18	$\{1, 1, -1, 1, 0, 0, 0, 1\}$	25
CSO(1, 5, 2)	10	5	13	$\{1, -1, 1, -1, 1, 0, 0, 1\}$	27
CSO(2, 4, 2)	7	8	13	$\{1, 1, 1, -1, 1, 0, 0, 1\}$	27
CSO(3, 3, 2)	6	9	13	$\{1, 1, -1, -1, 1, 0, 0, 1\}$	27
CSO(4, 2, 2)	7	8	13	$\{1, 1, -1, 1, 1, 0, 0, 1\}$	27
CSO(5, 1, 2)	10	5	13	$\{1, 1, -1, 1, -1, 0, 0, 1\}$	27
CSO(1, 6, 1)	15	6	7	$\{1, -1, 1, -1, 1, -1, 0, 1\}$	28
CSO(2, 5, 1)	11	10	7	$\{1, 1, 1, -1, 1, -1, 0, 1\}$	28
CSO(3, 4, 1)	9	12	7	$\{1, 1, -1, -1, 1, -1, 0, 1\}$	28
CSO(4, 3, 1)	9	12	7	$\{1, 1, -1, 1, 1, -1, 0, 1\}$	28
CSO(5, 2, 1)	11	10	7	$\{1, 1, -1, 1, -1, -1, 0, 1\}$	28
CSO(6, 1, 1)	15	6	7	$\{1, 1, -1, 1, -1, 1, 0, 1\}$	28

Table 7.5 Generators of electric gauge algebras in the $\ell_i = 0$ frame

Electric vector	Gauge generator
$\mathbf{W}^{(35)}$	$\leftrightarrow h_2 E_{-\beta_2} - h_1 E_{\beta_2}$
$\mathbf{W}^{(36)}$	$\leftrightarrow h_3 E_{-\beta_2-\beta_3} + h_1 E_{\beta_2+\beta_3}$
$\mathbf{W}^{(37)}$	$\leftrightarrow h_4 E_{-\beta_2-\beta_3-\beta_4} - h_1 E_{\beta_2+\beta_3+\beta_4}$
$\mathbf{W}^{(38)}$	$\leftrightarrow h_5 E_{-\beta_2-\beta_3-\beta_4-\beta_5} + h_1 E_{\beta_2+\beta_3+\beta_4+\beta_5}$
$\mathbf{W}^{(30)}$	$\leftrightarrow h_6 E_{-\beta_2-\beta_3-\beta_4-\beta_5-\beta_6} - h_1 E_{\beta_2+\beta_3+\beta_4+\beta_5+\beta_6}$
$\mathbf{W}^{(45)}$	$\leftrightarrow -h_7 E_{-\beta_2-\beta_3-\beta_4-\beta_5-\beta_6-\beta_7} + h_1 E_{\beta_2+\beta_3+\beta_4+\beta_5+\beta_6+\beta_7}$
$\mathbf{W}^{(51)}$	$\leftrightarrow h_1 E_{-\beta_1} + h_8 E_{\beta_1}$
$\mathbf{W}^{(52)}$	$\leftrightarrow h_2 E_{-\beta_1-\beta_2} + h_8 E_{\beta_1+\beta_2}$
$\mathbf{W}^{(53)}$	$\leftrightarrow h_3 E_{-\beta_1-\beta_2-\beta_3} - h_8 E_{\beta_1+\beta_2+\beta_3}$
$\mathbf{W}^{(54)}$	$\leftrightarrow h_4 E_{-\beta_1-\beta_2-\beta_3-\beta_4} + h_8 E_{\beta_1+\beta_2+\beta_3+\beta_4}$
$\mathbf{W}^{(55)}$	$\leftrightarrow h_5 E_{-\beta_1-\beta_2-\beta_3-\beta_4-\beta_5} - h_8 E_{\beta_1+\beta_2+\beta_3+\beta_4+\beta_5}$
$\mathbf{W}^{(56)}$	$\leftrightarrow h_6 E_{-\beta_1-\beta_2-\beta_3-\beta_4-\beta_5-\beta_6} + h_8 E_{\beta_1+\beta_2+\beta_3+\beta_4+\beta_5+\beta_6}$
$\mathbf{W}^{(29)}$	$\leftrightarrow -h_7 E_{-\beta_1-\beta_2-\beta_3-\beta_4-\beta_5-\beta_6-\beta_7} - h_8 E_{\beta_1+\beta_2+\beta_3+\beta_4+\beta_5+\beta_6+\beta_7}$
$\mathbf{W}^{(43)}$	$\leftrightarrow -h_3 E_{-\beta_3} - h_2 E_{\beta_3}$
$\mathbf{W}^{(42)}$	$\leftrightarrow -h_4 E_{-\beta_3-\beta_4} + h_2 E_{\beta_3+\beta_4}$
$\mathbf{W}^{(39)}$	$\leftrightarrow -h_5 E_{-\beta_3-\beta_4-\beta_5} - h_2 E_{\beta_3+\beta_4+\beta_5}$
$\mathbf{W}^{(31)}$	$\leftrightarrow -h_6 E_{-\beta_3-\beta_4-\beta_5-\beta_6} + h_2 E_{\beta_3+\beta_4+\beta_5+\beta_6}$
$\mathbf{W}^{(46)}$	$\leftrightarrow h_7 E_{-\beta_3-\beta_4-\beta_5-\beta_6-\beta_7} - h_2 E_{\beta_3+\beta_4+\beta_5+\beta_6+\beta_7}$
$\mathbf{W}^{(44)}$	$\leftrightarrow h_4 E_{-\beta_4} + h_3 E_{\beta_4}$
$\mathbf{W}^{(40)}$	$\leftrightarrow h_5 E_{-\beta_4-\beta_5} - h_3 E_{\beta_4+\beta_5}$
$\mathbf{W}^{(32)}$	$\leftrightarrow h_6 E_{-\beta_4-\beta_5-\beta_6} + h_3 E_{\beta_4+\beta_5+\beta_6}$
$\mathbf{W}^{(47)}$	$\leftrightarrow -h_7 E_{-\beta_4-\beta_5-\beta_6-\beta_7} - h_3 E_{\beta_4+\beta_5+\beta_6+\beta_7}$
$\mathbf{W}^{(41)}$	$\leftrightarrow -h_5 E_{-\beta_5} - h_4 E_{\beta_5}$
$\mathbf{W}^{(33)}$	$\leftrightarrow -h_6 E_{-\beta_5-\beta_6} + h_4 E_{\beta_5+\beta_6}$
$\mathbf{W}^{(48)}$	$\leftrightarrow h_7 E_{-\beta_5-\beta_6-\beta_7} - h_4 E_{\beta_5+\beta_6\vert\beta_7}$
$\mathbf{W}^{(34)}$	$\leftrightarrow h_6 E_{-\beta_6} + h_5 E_{\beta_6}$
$\mathbf{W}^{(49)}$	$\leftrightarrow -h_7 E_{-\beta_6-\beta_7} - h_5 E_{\beta_6+\beta_7}$
$\mathbf{W}^{(50)}$	$\leftrightarrow h_7 E_{-\beta_7} - h_6 E_{\beta_7}$

$$\begin{aligned}\eta\,(h, \ell = 0) = \mathrm{diag}\Big\{ & -h_7 h_8,\ h_1 h_6,\ h_2 h_6,\ -h_3 h_6,\ h_4 h_6,\ -h_5 h_6,\ h_1 h_2, \\ & -h_1 h_3,\ h_1 h_4,\ -h_1 h_5,\ -h_2 h_5,\ h_3 h_5,\ -h_4 h_5,\ h_2 h_4, \\ & -h_2 h_3,\ -h_3 h_4,\ h_1 h_7,\ h_2 h_7,\ -h_3 h_7,\ h_4 h_7,\ -h_5 h_7, \\ & h_6 h_7,\ -h_1 h_8,\ -h_2 h_8,\ h_3 h_8,\ -h_4 h_8,\ h_5 h_8,\ -h_6 h_8 \Big\}\end{aligned} \tag{7.2.86}$$

Hence all possible signatures Σ $(orbit)$ are obtained by assigning to the parameters h_i the values $1, -1, 0$ in all possible ways. Given an h vector constructed in this way we have then to check that the corresponding 28 generators (7.2.80) close a Lie subalgebra and accept only those for which this happens. Clearly such an algorithm can be easily implemented by means of a computer program. The result is provided by a table of SL(8, $\mathbb{R}$) Lie subalgebras identified by a corresponding acceptable h-vector. This result is displayed in Table 7.4. In this table in addition to the h-vector that identifies it we have displayed the signature of the Killing–Cartan form by writing the numbers n_+,n_-,n_0 of its positive, negative and zero eigenvalues. In addition we have also written the actual dimension of the gauge algebra namely the number of generators that have a non-vanishing representations or correspondingly the number of gauged vectors that are gauged (=paired with a non vanishing generator).

By restricting the matrix ${e_W}^\alpha$ to the parameters h_i we can immediately write the correspondence between the vectors $\mathbf{W}^{(28+i)}$ and the generators of the gauge algebra that applies to all the gaugings we have classified above. For the reader's convenience this correspondence is summarized in Table 7.5, where it suffices to substitute the corresponding values of h_i to obtain the generators of each gauge algebra expressed as linear combinations of the 56 positive and negative root step operators of SL(8, $\mathbb{R}$).

7.3 Embedding of the Group L_{168} into $E_{7(7)}$

In Sect. 1.3 we considered the simple group $L_{168} \sim$ PSL(2, 7) and we shew that it acts crystallographically on $\Lambda_{\mathrm{root}[A_7]}$ and, consequently, also on the dual weight lattice $\Lambda_{\mathrm{root}[A_7]}$.

This clearly raises the question of the embedding of L_{168} into SL(8, $\mathbb{R}$) $\subset E_{7(7)}$ suggesting that we might consider gaugings where the embedding matrix is invariant under L_{168}. As we are going to show, this condition uniquely determines the gauged supergravity model, once the embedding is stated. As examples we will consider two embeddings and derive the corresponding gauged supergravity models.

7.3.1 *Embedding of* L_{168} *into the Weyl Group*

Let us consider the Weyl group Weyl $[\mathfrak{e}_7]$ of the $\mathfrak{e}_7$ Lie algebra. It is necessarily a finite subgroup of the maximal compact subgroup of $E_{7(7)}$ namely of SU(8):

$$\mathrm{Weyl}\,[\mathfrak{e}_7] \subset \mathrm{SU}(8) \subset E_{7(7)} \tag{7.3.1}$$

Since the electric subgroup SL(8, $\mathbb{R}$) is regularly embedded into $E_{7(7)}$, namely shares with it the same Cartan subalgebra, it follows that the Weyl group of the smaller algebra is a subgroup of the Weyl group of the larger one:

$$\mathrm{Weyl}\,[\mathfrak{a}_7] \subset \mathrm{Weyl}\,[\mathfrak{e}_7] \tag{7.3.2}$$

As it is well known the Weyl group of any $\mathfrak{a}_\ell$ Lie algebra is isomorphic to the symmetric group $S_{\ell+1}$ and can be realized by means of integer valued orthogonal matrices in dimension $\ell + 1$. Applied to our case this means that:

$$\mathrm{Weyl}\,[\mathfrak{a}_7] \sim S_8 \subset \mathrm{SO}(8,\mathbb{Z}) \tag{7.3.3}$$

It is known that the group L_{168} is a subgroup of S_8, hence it follows that we have a chain of embeddings:

$$L_{168} \subset \mathrm{Weyl}\,[\mathfrak{a}_7] \subset \mathrm{SO}(8,\mathbb{Z}) \subset \mathrm{SL}(8,\mathbb{R}) \subset E_{7(7)} \tag{7.3.4}$$

Following this chain we can determine the action of the group L_{168} on the embedding matrix and work out the gaugings that are L_{168}-invariant. What we need is just the form of the generators of L_{168} inside $\mathrm{SO}(8,\mathbb{Z})$.

Let us recall that the simple group under consideration is abstractly defined in terms of the following generators and relations:

$$L_{168} = \left(R, S, T \parallel R^2 = S^3 = T^7 = RST = (TSR)^4 = \mathbf{e}\right) \tag{7.3.5}$$

In view of the above we just need to specify three orthogonal integer valued matrices 8×8 that satisfy relations (7.3.5). This is done below:

$$T = \begin{pmatrix} 0&1&0&0&0&0&0&0\\ 0&0&1&0&0&0&0&0\\ 0&0&0&1&0&0&0&0\\ 0&0&0&0&1&0&0&0\\ 0&0&0&0&0&1&0&0\\ 0&0&0&0&0&0&1&0\\ 1&0&0&0&0&0&0&0\\ 0&0&0&0&0&0&0&1 \end{pmatrix} \; ; \quad S = \begin{pmatrix} 0&1&0&0&0&0&0&0\\ 0&0&0&0&0&0&0&1\\ 0&0&0&0&0&0&1&0\\ 0&0&0&1&0&0&0&0\\ 0&0&1&0&0&0&0&0\\ 0&0&0&0&0&1&0&0\\ 0&0&0&0&1&0&0&0\\ 1&0&0&0&0&0&0&0 \end{pmatrix} \tag{7.3.6}$$

$$R = \begin{pmatrix} 0&0&0&0&0&0&0&1\\ 0&0&0&0&0&0&1&0\\ 0&0&0&1&0&0&0&0\\ 0&0&1&0&0&0&0&0\\ 0&0&0&0&0&1&0&0\\ 0&0&0&0&1&0&0&0\\ 0&1&0&0&0&0&0&0\\ 1&0&0&0&0&0&0&0 \end{pmatrix} \tag{7.3.7}$$

Given these generators we immediately construct the immersion of the group L_{168} into the 28 representation of $\mathrm{SL}(8,\mathbb{R})$ by means of an antisymmetrized tensor product:

$$\begin{aligned} \mathscr{T}^{[\Lambda\Sigma]}{}_{[\Delta\Gamma]} &= T^{[\Lambda}{}_{[\Delta}\, T^{\Sigma]}{}_{\Gamma]} \\ \mathscr{S}^{[\Lambda\Sigma]}{}_{[\Delta\Gamma]} &= S^{[\Lambda}{}_{[\Delta}\, S^{\Sigma]}{}_{\Gamma]} \\ \mathscr{R}^{[\Lambda\Sigma]}{}_{[\Delta\Gamma]} &= R^{[\Lambda}{}_{[\Delta}\, R^{\Sigma]}{}_{\Gamma]} \end{aligned} \tag{7.3.8}$$

This provides the immersion of the group L_{168} into the fundamental 56 representation of $\mathrm{E}_{7(7)}$ by means of the following definitions:

$$\begin{aligned}
\mathfrak{T} &= \left(\begin{array}{c|c} \mathscr{T} & 0 \\ \hline 0 & \mathscr{T} \end{array}\right) \in \mathrm{E}_{7(7)} \\
\mathfrak{S} &= \left(\begin{array}{c|c} \mathscr{S} & 0 \\ \hline 0 & \mathscr{S} \end{array}\right) \in \mathrm{E}_{7(7)} \\
\mathfrak{R} &= \left(\begin{array}{c|c} \mathscr{R} & 0 \\ \hline 0 & \mathscr{R} \end{array}\right) \in \mathrm{E}_{7(7)}
\end{aligned} \tag{7.3.9}$$

Next we can easily derive the action of the subgroup $\mathrm{L}_{168} \subset \mathrm{Weyl}[\mathrm{e}_7]$ on the $\mathfrak{e}_7$ simple roots and consequently on the $\mathfrak{e}_7$ root lattice. It suffice to define:

$$\begin{aligned}
\mathfrak{T}^{-1}\, H_{\alpha_i}\, \mathfrak{T} &= H_{\alpha_j}\, \mathbf{T}_{ji} \\
\mathfrak{S}^{-1}\, H_{\alpha_i}\, \mathfrak{S} &= H_{\alpha_j}\, \mathbf{S}_{ji} \\
\mathfrak{R}^{-1}\, H_{\alpha_i}\, \mathfrak{R} &= H_{\alpha_j}\, \mathbf{R}_{ji}
\end{aligned} \tag{7.3.10}$$

where H_{α_i} are the Cartan generators of $\mathrm{E}_{7(7)}$ in the 56-dimensional representation dual to the simple roots:

$$\left[E^{\alpha_i}\,,\, E^{-\alpha_i}\right] = H_{\alpha_i} \tag{7.3.11}$$

The explicit form of the 7×7 matrices $\mathbf{R}, \mathbf{S}, \mathbf{T}$ is given below:

$$\mathbf{T} = \begin{pmatrix} 0&0&0&0&1&-1&1 \\ 1&0&0&0&1&-2&2 \\ 0&1&0&0&1&-3&3 \\ 0&0&1&0&1&-4&4 \\ 0&0&0&0&1&-2&2 \\ 0&0&0&1&0&-3&3 \\ 0&0&0&0&0&-1&2 \end{pmatrix} \; ; \quad \mathbf{S} = \begin{pmatrix} 0&0&0&0&1&0&-1 \\ 0&0&-1&1&1&0&-2 \\ 0&-1&0&1&1&0&-3 \\ 0&-1&0&1&2&-1&-3 \\ 0&0&0&0&1&0&-2 \\ 0&-1&0&0&2&0&-2 \\ -1&0&0&0&1&0&-1 \end{pmatrix} \tag{7.3.12}$$

$$\mathbf{R} = \begin{pmatrix} 0&0&0&0&1&-1&0 \\ 0&-1&1&0&1&-1&-1 \\ -1&0&1&0&1&-1&-2 \\ -1&0&1&-1&2&0&-3 \\ 0&0&0&0&1&0&-2 \\ -1&0&0&0&1&0&-2 \\ 0&0&0&0&0&0&-1 \end{pmatrix} \tag{7.3.13}$$

We can now easily verify that L_{168} is crystallographic with respect to the $\mathfrak{e}_7$-root lattice as indeed it should be since it is a subgroup of $\mathrm{Weyl}[\mathfrak{e}_7]$. It suffices to check that the integer valued matrices $\mathbf{R}, \mathbf{S}, \mathbf{T}$ satisfy:

$$\mathbf{T}^T\, \mathscr{C}_{\mathfrak{e}_7}\, \mathbf{T} = \mathbf{S}^T\, \mathscr{C}_{\mathfrak{e}_7}\, \mathbf{S} = \mathbf{R}^T\, \mathscr{C}_{\mathfrak{e}_7}\, \mathbf{R} = \mathscr{C}_{\mathfrak{e}_7} \tag{7.3.14}$$

where:

$$\mathscr{C}_{\mathfrak{e}_7} = \begin{pmatrix} 2 & -1 & 0 & 0 & 0 & 0 & 0 \\ -1 & 2 & -1 & 0 & 0 & 0 & 0 \\ 0 & -1 & 2 & -1 & 0 & 0 & 0 \\ 0 & 0 & -1 & 2 & -1 & -1 & 0 \\ 0 & 0 & 0 & -1 & 2 & 0 & 0 \\ 0 & 0 & 0 & -1 & 0 & 2 & -1 \\ 0 & 0 & 0 & 0 & 0 & -1 & 2 \end{pmatrix} \tag{7.3.15}$$

is the Cartan matrix of $\mathfrak{e}_7$.

Equation (7.3.13) imply that L_{168} is embedded into SO(7) as we already know. Given the matrix which transforms the e_7 simple root basis into the a_7 simple root basis:

$$\beta_i = \Pi_{ij}\,\alpha_j$$
$$\Pi = \begin{pmatrix} 0&1&2&3&2&2&1 \\ 1&0&0&0&0&0&0 \\ 0&1&0&0&0&0&0 \\ 0&0&1&0&0&0&0 \\ 0&0&0&1&0&0&0 \\ 0&0&0&0&0&1&0 \\ 0&0&0&0&0&0&1 \end{pmatrix} \tag{7.3.16}$$

We can easily convert the generators of the discrete group L_{168} from the $\mathfrak{e}_7$ root lattice to the $\mathfrak{a}_7$ root lattice by setting:

$$\mathbf{T}_{\mathfrak{a}} = \left(\Pi^T\right)^{-1} \mathbf{T}\,\Pi^T \quad ; \quad \mathbf{S}_{\mathfrak{a}} = \left(\Pi^T\right)^{-1} \mathbf{T}\,\Pi^T \quad ; \quad \mathbf{R}_{\mathfrak{a}} = \left(\Pi^T\right)^{-1} \mathbf{R}\,\Pi^T \tag{7.3.17}$$

and we get:

$$\mathbf{T}_{\mathfrak{a}} = \begin{pmatrix} 0&0&0&0&0&-1&1 \\ 1&0&0&0&0&-1&1 \\ 0&1&0&0&0&-1&1 \\ 0&0&1&0&0&-1&1 \\ 0&0&0&1&0&-1&1 \\ 0&0&0&0&1&-1&1 \\ 0&0&0&0&0&0&1 \end{pmatrix} \quad ; \quad \mathbf{S}_{\mathfrak{a}} = \begin{pmatrix} 0&0&0&0&0&0&-1 \\ 1&0&0&0&0&0&-1 \\ 1&0&0&-1&1&0&-1 \\ 1&0&-1&0&1&0&-1 \\ 1&0&-1&0&1&-1&0 \\ 1&0&-1&0&0&0&0 \\ 1&-1&0&0&0&0&0 \end{pmatrix}$$

$$\mathbf{R}_{\mathfrak{a}} = \begin{pmatrix} 0&0&0&0&0&0&-1 \\ 0&0&0&0&0&-1&0 \\ 0&0&-1&1&0&-1&0 \\ 0&-1&0&1&0&-1&0 \\ 0&-1&0&1&-1&0&0 \\ 0&-1&0&0&0&0&0 \\ -1&0&0&0&0&0&0 \end{pmatrix} \tag{7.3.18}$$

The integer valued matrices $\mathbf{R}_\mathfrak{a}, \mathbf{S}_\mathfrak{a}, \mathbf{T}_\mathfrak{a}$ satisfy:

$$\mathbf{T}_\mathfrak{a}^T \mathscr{C}_{\mathfrak{a}_7} \mathbf{T}_\mathfrak{a} = \mathbf{S}_\mathfrak{a}^T \mathscr{C}_{\mathfrak{a}_7} \mathbf{S}_\mathfrak{a} = \mathbf{R}_\mathfrak{a}^T \mathscr{C}_{\mathfrak{a}_7} \mathbf{R}_\mathfrak{a} = \mathscr{C}_{\mathfrak{a}_7} \tag{7.3.19}$$

where:

$$\mathscr{C}_{\mathfrak{a}_7} = \begin{pmatrix} 2 & -1 & 0 & 0 & 0 & 0 & 0 \\ -1 & 2 & -1 & 0 & 0 & 0 & 0 \\ 0 & -1 & 2 & -1 & 0 & 0 & 0 \\ 0 & 0 & -1 & 2 & -1 & 0 & 0 \\ 0 & 0 & 0 & -1 & 2 & -1 & 0 \\ 0 & 0 & 0 & 0 & -1 & 2 & -1 \\ 0 & 0 & 0 & 0 & 0 & -1 & 2 \end{pmatrix} \tag{7.3.20}$$

is the Cartan matrix of $\mathfrak{a}_7$.

Given the established representation of the L_{168} group inside the electric group we can now impose the condition of L_{168} invariance on the embedding matrix (7.2.37) which can be expressed in the following way:

$$\begin{aligned} e^{\alpha}_{\Lambda\Sigma}(h,\ell)\,\mathfrak{T}^{-1}\,G_\alpha\,\mathfrak{T} &= \mathscr{T}_{\Lambda\Sigma}{}^{\Delta\Gamma}\,e^{\alpha}_{\Delta\Gamma}(h,\ell)\,G_\alpha \\ e^{\alpha}_{\Lambda\Sigma}(h,\ell)\,\mathfrak{S}^{-1}\,G_\alpha\,\mathfrak{S} &= \mathscr{T}_{\Lambda\Sigma}{}^{\Delta\Gamma}\,e^{\alpha}_{\Delta\Gamma}(h,\ell)\,G_\alpha \\ e^{\alpha}_{\Lambda\Sigma}(h,\ell)\,\mathfrak{R}^{-1}\,G_\alpha\,\mathfrak{R} &= \mathscr{R}_{\Lambda\Sigma}{}^{\Delta\Gamma}\,e^{\alpha}_{\Delta\Gamma}(h,\ell)\,G_\alpha \end{aligned} \tag{7.3.21}$$

Equation (7.3.21) have a unique solution for the parameters h^i, ℓ^a, namely:

$$\ell^i = 0 \quad (i = 1, \ldots, 28) \quad ; \quad \{h_1, h_2, h_3, h_4, h_5, h_6, h_7, h_8\} = \mathbf{e} \times \{1, 1, -1, 1, -1, 1, 1, -1\} \tag{7.3.22}$$

where $\mathbf{e}$ is an arbitrary real parameter that plays the role of gauge coupling constant. Inserting the values (7.3.22) into Eq. (7.2.86) we find that the Killing–Cartan metric $\eta_{W_1 W_2}$ has only negative eigenvalues, namely that all the 28 generators are compact (this conclusion is verified also by looking at eigenvalues of all generators that are all purely imaginary).

Hence when L_{168} is embedded into the Weyl group Weyl$[\mathfrak{a}_7] \subset$ Weyl$[\mathfrak{e}_7]$ the L_{168} invariant gauging is the purely compact SO(8) gauging leading to de Wit Nicolai gauged supergravity.

7.3.1.1 Conformal Speculations

As it is well known SO(8)-gauged supergravity is obtained from d = 11 supergravity compactified on:

$$\mathrm{AdS}_4 \times \mathbb{S}^7 \tag{7.3.23}$$

which is the near horizon geometry of an M2-brane with $\mathbb{R}^8$ transverse space. Indeed $\mathbb{R}^8$ is the metric cone on $\mathbb{S}^7$. The entire Kaluza–Klein spectrum which constitutes the spectrum of BPS operators of the d=3 theory is organized in short representations of the supergroup:

$$\mathrm{Osp}(8|4) \tag{7.3.24}$$

Our discussion leads to the conclusion that we can consider the compactification of supergravity on orbifolds of the following type:

$$\mathbb{C}_\Gamma \mathbb{S}^7 = \frac{\mathbb{S}^7}{\Gamma} \quad ; \quad \Gamma \subset L_{168} \subset \mathrm{Weyl}[\mathfrak{a}_7] \subset \mathrm{SO}(8, \mathbb{Z}) \tag{7.3.25}$$

The corresponding M2-brane solution has the orbifold:

$$\mathbb{C}_\Gamma \mathbb{R}^8 = \frac{\mathbb{R}^8}{\Gamma} \tag{7.3.26}$$

as transverse space.

The massive and massless modes of the Kalauza Klein spectrum are easily worked out from the Osp(8|4) spectrum of the 7-sphere. Indeed since the group Γ is embedded by the above construction into $\mathrm{SO}(8, \mathbb{Z}) \subset \mathrm{SO}(8) \subset \mathrm{Osp}(8|4)$, it suffices to cut the spectrum to the Γ singlets. An observation is particularly important. The 8-dimensional representation of SO(8) is not irreducible under L_{168}. Indeed the vector:

$$\{a, a, a, a, a, a, a, a\} \tag{7.3.27}$$

is invariant under the action of the generators (7.3.7) and it is the only one to be such. The 8-dimensional vector representation is the one to which the gravitinos are assigned. It follows that by means of the L_{168} projection we eliminate 7 out of 8 massless gravitinos. In other words the projected Kaluza–Klein spectrum must be organized into supermultiplets of:

$$\mathrm{Osp}(1|4) \tag{7.3.28}$$

corresponding to $\mathrm{N} = 2$ superconformal symmetry on the brane.

It seems an interesting game to study the formulation of the superconformal field theories dual to supergravity compactified on these peculiar orbifolds.

7.3.2 *Embedding of* L_{168} *into* $\mathrm{SO}(7) \subset \mathrm{SO}(8) \subset \mathrm{SL}(8, \mathbb{R})$ *and the Domain Wall*

We consider next a different embedding of the group L_{168} which leads to a gauged supergravity with domain wall vacua. An embedding with these properties is not into the Weyl group and does not preserve the Cartan subalgebra.

We obtain it with the following simple argument. Let us consider the generators (7.3.18) which are orthogonal with respect to the $\mathfrak{a}_7$ - Cartan matrix and transform them to the basis where such Cartan matrix is reduced to the identity matrix.

We have:

$$Q^T\,\mathscr{C}_{\mathfrak{a}_7}\,Q = \mathbf{1}_{7\times 7} \tag{7.3.29}$$

where

$$Q = \begin{pmatrix} \frac{1}{4} & \frac{1}{4} & \frac{1}{4} & \frac{1}{4} & \frac{1}{4} & -\frac{1}{4} & \frac{1}{\sqrt{2}} \\ 1 & 0 & 0 & 0 & 0 & 0 & \frac{1}{\sqrt{2}} \\ \frac{3}{4} & \frac{3}{4} & -\frac{1}{4} & -\frac{1}{4} & -\frac{1}{4} & \frac{1}{4} & \frac{1}{\sqrt{2}} \\ \frac{1}{2} & \frac{1}{2} & \frac{1}{2} & -\frac{1}{2} & -\frac{1}{2} & \frac{1}{2} & \frac{1}{\sqrt{2}} \\ \frac{1}{4} & \frac{1}{4} & \frac{1}{4} & \frac{1}{4} & -\frac{3}{4} & \frac{3}{4} & \frac{1}{\sqrt{2}} \\ 0 & 0 & 0 & 0 & 0 & 1 & \frac{1}{\sqrt{2}} \\ -\frac{1}{4} & -\frac{1}{4} & -\frac{1}{4} & -\frac{1}{4} & -\frac{1}{4} & \frac{1}{4} & \frac{1}{\sqrt{2}} \end{pmatrix} \tag{7.3.30}$$

Hence if we define:

$$\mathbf{T}_q = Q^{-1}\,\mathbf{T}\,Q \quad ; \quad \mathbf{S}_q = Q^{-1}\,\mathbf{S}\,Q \quad ; \quad \mathbf{R}_q = Q^{-1}\,\mathbf{R}\,Q \tag{7.3.31}$$

we obtain three standard orthogonal 7×7 matrices that generate the group L_{168} inside SO(7). They are not integer valued.

Next let us embed these generators into SO(8) according to the following block-diagonal way:

$$T_q = \left(\begin{array}{c|c} \mathbf{T}_q & \mathbf{0} \\ \hline \mathbf{0} & 1 \end{array}\right) \quad ; \quad S_q = \left(\begin{array}{c|c} \mathbf{S}_q & \mathbf{0} \\ \hline \mathbf{0} & 1 \end{array}\right) \quad ; \quad R_q = \left(\begin{array}{c|c} \mathbf{R}_q & \mathbf{0} \\ \hline \mathbf{0} & 1 \end{array}\right) \tag{7.3.32}$$

Explicitly we obtain

$$T_q = \left(\begin{array}{ccccccc|c} \frac{1}{4} & \frac{1}{4} & \frac{1}{4} & \frac{1}{4} & \frac{1}{4} & -\frac{3}{4} & \frac{1}{2\sqrt{2}} & 0 \\ \frac{3}{4} & -\frac{1}{4} & -\frac{1}{4} & -\frac{1}{4} & -\frac{1}{4} & -\frac{1}{4} & -\frac{1}{2\sqrt{2}} & 0 \\ -\frac{1}{4} & \frac{3}{4} & -\frac{1}{4} & -\frac{1}{4} & -\frac{1}{4} & -\frac{1}{4} & -\frac{1}{2\sqrt{2}} & 0 \\ -\frac{1}{4} & -\frac{1}{4} & \frac{3}{4} & -\frac{1}{4} & -\frac{1}{4} & -\frac{1}{4} & -\frac{1}{2\sqrt{2}} & 0 \\ -\frac{1}{4} & -\frac{1}{4} & -\frac{1}{4} & \frac{3}{4} & -\frac{1}{4} & -\frac{1}{4} & -\frac{1}{2\sqrt{2}} & 0 \\ \frac{1}{4} & \frac{1}{4} & \frac{1}{4} & \frac{1}{4} & -\frac{3}{4} & \frac{1}{4} & \frac{1}{2\sqrt{2}} & 0 \\ -\frac{1}{2\sqrt{2}} & -\frac{1}{2\sqrt{2}} & -\frac{1}{2\sqrt{2}} & -\frac{1}{2\sqrt{2}} & -\frac{1}{2\sqrt{2}} & -\frac{1}{2\sqrt{2}} & \frac{1}{2} & 0 \\ \hline 0 & 0 & 0 & 0 & 0 & 0 & 0 & 1 \end{array}\right) \tag{7.3.33}$$

$$S_q = \left(\begin{array}{ccccccc|c} \frac{3}{4} & \frac{1}{4} & \frac{1}{4} & \frac{1}{4} & \frac{1}{4} & -\frac{1}{4} & \frac{1}{2\sqrt{2}} & 0 \\ \frac{1}{4} & -\frac{1}{4} & -\frac{1}{4} & \frac{3}{4} & -\frac{1}{4} & \frac{1}{4} & -\frac{1}{2\sqrt{2}} & 0 \\ \frac{1}{4} & -\frac{1}{4} & \frac{3}{4} & -\frac{1}{4} & -\frac{1}{4} & \frac{1}{4} & -\frac{1}{2\sqrt{2}} & 0 \\ \frac{1}{4} & -\frac{1}{4} & -\frac{1}{4} & -\frac{1}{4} & -\frac{1}{4} & -\frac{3}{4} & -\frac{1}{2\sqrt{2}} & 0 \\ \frac{1}{4} & -\frac{1}{4} & -\frac{1}{4} & -\frac{1}{4} & \frac{3}{4} & \frac{1}{4} & -\frac{1}{2\sqrt{2}} & 0 \\ -\frac{1}{4} & -\frac{3}{4} & \frac{1}{4} & \frac{1}{4} & \frac{1}{4} & -\frac{1}{4} & \frac{1}{2\sqrt{2}} & 0 \\ -\frac{1}{2\sqrt{2}} & \frac{1}{2\sqrt{2}} & \frac{1}{2\sqrt{2}} & \frac{1}{2\sqrt{2}} & \frac{1}{2\sqrt{2}} & -\frac{1}{2\sqrt{2}} & -\frac{1}{2} & 0 \\ \hline 0 & 0 & 0 & 0 & 0 & 0 & 0 & 1 \end{array}\right) \tag{7.3.34}$$

$$R_q = \left(\begin{array}{ccccccc|c} 0 & 0 & 0 & 0 & 0 & -1 & 0 & 0 \\ 0 & 0 & 1 & 0 & 0 & 0 & 0 & 0 \\ 0 & 1 & 0 & 0 & 0 & 0 & 0 & 0 \\ 0 & 0 & 0 & 0 & 1 & 0 & 0 & 0 \\ 0 & 0 & 0 & 1 & 0 & 0 & 0 & 0 \\ -1 & 0 & 0 & 0 & 0 & 0 & 0 & 0 \\ 0 & 0 & 0 & 0 & 0 & 0 & -1 & 0 \\ \hline 0 & 0 & 0 & 0 & 0 & 0 & 0 & 1 \end{array}\right) \tag{7.3.35}$$

Given these generators we immediately construct a new immersion of the group L_{168} into the 28 representation of $SL(8, \mathbb{R})$ by means of an antisymmetrized tensor product:

$$\begin{aligned} \mathscr{T}_q^{[\Lambda\Sigma]}{}_{[\Delta\Gamma]} &= T_q^{[\Lambda}{}_{[\Delta} T_q^{\Sigma]}{}_{\Gamma]} \\ \mathscr{S}_q^{[\Lambda\Sigma]}{}_{[\Delta\Gamma]} &= S_q^{[\Lambda}{}_{[\Delta} S_q^{\Sigma]}{}_{\Gamma]} \\ \mathscr{R}_q^{[\Lambda\Sigma]}{}_{[\Delta\Gamma]} &= R_q^{[\Lambda}{}_{[\Delta} R_q^{\Sigma]}{}_{\Gamma]} \end{aligned} \tag{7.3.36}$$

This provides also a new immersion of the group L_{168} into the fundamental 56 representation of $E_{7(7)}$ by means of the following definitions:

$$\begin{aligned} \mathfrak{T}_q &= \left(\begin{array}{c|c} \mathscr{T}_1 & 0 \\ \hline 0 & \mathscr{T}_q \end{array}\right) \in E_{7(7)} \\ \mathfrak{S}_q &= \left(\begin{array}{c|c} \mathscr{S} & 0 \\ \hline 0 & \mathscr{S} \end{array}\right) \in E_{7(7)} \\ \mathfrak{R}_q &= \left(\begin{array}{c|c} \mathscr{R}_q & 0 \\ \hline 0 & \mathscr{R}_q \end{array}\right) \in E_{7(7)} \end{aligned} \tag{7.3.37}$$

Differently from the previous case the CSA is not preserved by the new action of the L_{168}-group

$$\begin{aligned}
\mathfrak{T}^{-1}\, H_{\alpha_i}\, \mathfrak{T} &\neq \text{linear combination of } H_{\alpha_j} \\
\mathfrak{S}^{-1}\, H_{\alpha_i}\, \mathfrak{S} &\neq \text{linear combination of } H_{\alpha_j} \\
\mathfrak{R}^{-1}\, H_{\alpha_i}\, \mathfrak{R} &\neq \text{linear combination of } H_{\alpha_j}
\end{aligned} \tag{7.3.38}$$

Nevertheless the group is embedded into the electric group $SL(8, \mathbb{R})$ and this allows us to impose a new condition of L_{168} invariance on the embedding matrix (7.2.37), in the same way as before, namely:

$$\begin{aligned}
e^{\alpha}_{\Lambda\Sigma}(h,\ell)\, \mathfrak{T}_q^{-1}\, G_\alpha\, \mathfrak{T}_q &= \mathscr{T}_{q|\Lambda\Sigma}{}^{\Delta\Gamma}\, e^{\alpha}_{\Delta\Gamma}(h,\ell)\, G_\alpha \\
e^{\alpha}_{\Lambda\Sigma}(h,\ell)\, \mathfrak{S}_q^{-1}\, G_\alpha\, \mathfrak{S}_q &= \mathscr{T}_{q|\Lambda\Sigma}{}^{\Delta\Gamma}\, e^{\alpha}_{\Delta\Gamma}(h,\ell)\, G_\alpha \\
e^{\alpha}_{\Lambda\Sigma}(h,\ell)\, \mathfrak{R}_q^{-1}\, G_\alpha\, \mathfrak{R}_q &= \mathscr{R}_{q|\Lambda\Sigma}{}^{\Delta\Gamma}\, e^{\alpha}_{\Delta\Gamma}(h,\ell)\, G_\alpha
\end{aligned} \tag{7.3.39}$$

Equation (7.3.39) have a two-parameter solution h^i, ℓ^a, namely:

$$\ell^i = 0 \quad (i = 1, \ldots, 28) \quad ; \quad \{h_1, h_2, h_3, h_4, h_5, h_6, h_7, h_8\} = \{x, x, -x, x, -x, x, y, -x\} \tag{7.3.40}$$

Up to rescaling of the gauge coupling constant there are only four cases in the above solution:

SO(8) This gauging obtains for $x = y = \pm 1$. With this choice the Killing metric of the gauge algebra has 28 negative eigenvalues.
SO(1, 7) This gauging obtains for $x = -y = \pm 1$. With this choice the Killing metric of the gauge algebra has 21 negative eigenvalues and 7 positive ones.
CSO(7, 1) This gauging obtains for $x = \pm 1$, $y = 0$. With this choice the Killing metric has 21 negative eigenvalues and 7 vanishing ones.
CSO(1, 7) This gauging obtains for $x = 0$, $y = \pm 1$. With this choice the Killing metric has 28 vanishing eigenvalues. The gauge group is abelian and contains only 7 non vanishing translation generators.

As we see there are four solutions.

The above items provide all the ingredients necessary to calculate the potential reduced to any chosen subset of the 70 scalar fields in any chosen gauging. The algorithm was developed in [33, 54]. First we construct the coset representative in the solvable parameterization explicitly defined as follows

$$\mathbb{L} = \text{LC} \cdot \text{LN}_1 \cdot \text{LN}_2 \cdot \text{LN}_3 \cdot \text{LN}_4 \cdot \text{LN}_5 \cdot \text{LN}_6 \tag{7.3.41}$$

where:

$$\text{LC} = \exp\left[\sum_{i=1}^{7} s_i\, H_{w_i}\right] \tag{7.3.42}$$

and

$$\text{LN}_i = \exp\left[\sum_{\alpha_{i,j} \in \mathbb{D}_i} \phi_{i,j}\, E^{\alpha_{i,j}}\right] \tag{7.3.43}$$

In the above formulae the 7-Cartan fields H_{w_i} are associated with the simple weights w_i of $SL(8, \mathbb{R})$ so that the field s_p is a singlet under the $SL(p, \mathbb{R}) \times SL(8-p, \mathbb{R})$ subalgebra of $SL(8, \mathbb{R})$. Correspondingly it is a singlet under the SO(p) × SO(8 − p) compact subalgebra of the gauge algebra. On the other hand the spaces $\mathbb{D}^i$ correspond to the filtration of the Borel subalgebra of $E_{7(7)}$ into abelian ideals of dimensions:

$$\dim \mathbb{D}^i = 1, 3, 6, 10, 16, 27 \tag{7.3.44}$$

This filtration was introduced in [54].

Once the coset representative $\mathbb{L}$ is constructed in the solvable parametrization in the Dynkin basis, it can be rotated to UspY basis in order to extract the blocks useful to construct the potential. The entire procedure can be automatized on a computer. In 1998 together with Mario Trigiante, the present author constructed a MATHEMATICA package, named **N8potent** that was utilized to obtain the following results.

First within the Cartan subalgebra that contains all the candidate dilaton fields we looked for the scalar that is a singlet of L_{168} according to the embedding presently discussed. To noone's surprise we found that there is only one such singlet s_7 which is actually a singlet with respect to the entire SO(7). Secondly we calculated the form of its kinetic term in the absence of all other nilpotent scalars. We found the lagrangian:

$$\mathrm{L} = \sqrt{-\det g}\left[2\, R[g] + \tfrac{7}{8}\, \partial_\mu s_7\, \partial^\mu s_7 - \mathscr{V}(s_7)\right] \tag{7.3.45}$$

where $\mathscr{V}(s_7)$ is the potential. Thirdly utilizing the package **N8potent** we calculated the potential for the four gaugings enumerated above.

We found

SO(8) This gauging obtains for $x = y = \pm 1$. With this choice the potential is the following one:

$$\mathscr{V}_{\mathrm{SO(8)}}(s_7) = \frac{147}{4} e^{-\frac{s_7}{2}} \left(-2e^{2s_7} + e^{4s_7} + 13\right) \tag{7.3.46}$$

SO(1, 7) This gauging obtains for $x = -y = \pm 1$. With this choice the potential is the following one:

$$\mathscr{V}_{\mathrm{SO(1,7)}}(s_7) = \frac{147}{4} e^{-\frac{s_7}{2}} \left(2e^{2s_7} + e^{4s_7} + 13\right) \tag{7.3.47}$$

CSO(7, 1) This gauging obtains for $x = \pm 1$, $y = 0$. With this choice the potential is the following one:

$$\mathscr{V}_{\mathrm{CSO(7,1)}}(s_7) = \frac{1911}{4} e^{-\frac{s_7}{2}} \tag{7.3.48}$$

CSO(1, 7) This gauging obtains for $x = 0$, $y = \pm 1$. With this choice the potential is the following one:

$$\mathscr{V}_{\mathrm{CSO(1,7)}}(s_7) = \frac{147}{4} e^{\frac{7s_7}{2}} \tag{7.3.49}$$

The renormalization of the scalar field s_7 which reduces its kinetic term to the canonical form is:

$$s_7 = -\frac{2}{\sqrt{7}}\phi \tag{7.3.50}$$

Inserting such a redefinition into the above calculated scalar potentials we obtain interesting domain wall potentials in canonical normalizations.

This is as much as we wanted to say about $\mathcal{N}=8$ supergravity. Our goal was to illustrate how the interaction structure of the lagrangian and in particular all the scalar potentials that can be obtained from its possible gaugings are all structurally encoded in the $\mathfrak{e}_{7(7)}$ Lie algebra, which is the ultimate core of this supergravity theory.

Next we turn to $\mathcal{N}=2$ supergravity models and we consider the mechanism that generates Starobinsky like inflaton potentials in these theories. As announced in the introduction, our main goal is to emphasize the role of the c-map and of the Tits Satake projection in these constructions.

7.4 Abelian Gaugings and General Properties of Their Potentials in the c-Map Framework

As we stressed in the introduction the inclusion into $\mathcal{N}=2$ supergravity obtained in [56] of inflaton potentials such as the Starobinsky potential[8]

$$V_{Starobinsky}(\phi) \equiv (1 - \exp[-\phi])^2 \tag{7.4.1}$$

is not occasional and limited to the case of hypermultiplets lying in $\frac{\mathrm{G}_{(2,2)}}{\mathrm{SU(2)}\times\mathrm{SU(2)}}$, rather it follows a general pattern that can be uncovered and relies on the properties of the c-map. In this way the mechanisms of the [56] can be generalized to larger Quaternionic Kähler manifolds opening a quite interesting new playground for the search of inflaton potentials that can be classified and understood in their geometrical origin.

Let us schematically summarize the main ingredients of the approach pioneered in [56] whose generalization, obtained in [57] we present in this chapter:

(A) The inflaton field ϕ is assumed to belong to the hypermultiplet Quaternionic Kähler manifold $\mathcal{QM}$.

(B) In analogy with the construction in [56], we require the graviphoton not to be minimally coupled to any other field. This condition originally followed from the general argument that in the dual to the $R + R^2$ supergravity the central charge is gauged. This will amount to a constraint on the form of the embedding tensor θ defining the gauge algebra.

(C) The inflaton potential is generated by the gauging of an abelian subalgebra $\mathcal{A} \subset \mathrm{iso}\,[\mathcal{QM}]$ of the isometry algebra of the hypermultiplet manifold.

---

[8] Just as in [56] we mention scalar fields that typically have non canonical kinetic terms.

(D) Since $\mathscr{A}$ is abelian it is not required to have any action on the vector multiplet scalars ω^i which are inert. Actually it is quite desirable that the potential $V_{gauging}$ generated by the gauging allows to fix all the ω^i to their values at some reference point, say $\omega^i = 0$:

$$\frac{\partial}{\partial \omega^i} V_{gauging}\Big|_{\omega^i=0} = 0 \tag{7.4.2}$$

As shown in [56], one can generically guarantee the fixing conditions (7.4.2) if the Special Kähler Geometry of the vector multiplets is chosen to be that of the so named Minimal Coupling, defined below in Eqs. (7.5.1)–(7.5.3).

(E) With the above choice of the vector multiplet geometry, after fixing the scalars ω^i the effective potential reduces to a sum of squares of the tri-holomorphic moment maps $P^x_{\mathscr{A}}$ which still depend on the variables $\{Z, U, a, z^i, \overline{z}^{i^\star}\}$. In order to approach effective potentials recognizable also as $\mathscr{N} = 1$ supergravity potentials one would like to be able to fix all the Heisenberg fields $\mathbf{Z}$ (and possibly also the other fields U and a) to zero, remaining only with the complex fields $(z^i, \overline{z}^{i^\star})$ of the inner Special Kähler manifold. Looking at the general form (4.3.31) of the tri-holomorphic moment map for the Heisenberg algebra generators and (4.3.40) for the tri-holomorphic moment map of the inner Special Kähler isometries we immediately realize that, gauging these isometries *separately*, the condition:

$$\frac{\partial}{\partial \mathbf{Z}^\alpha} \sum_{\mathrm{t}\in\mathscr{A}} (\mathscr{P}_{\mathrm{t}})^2\Big|_{\mathbf{Z}=0} = 0 \tag{7.4.3}$$

is always satisfied. A gauge generator which is a combination of a translation in the Heisenberg algebra and a Special Kähler isometry, yields in general a scalar potential exhibiting linear terms in $\mathbf{Z}$, so that (7.4.3) provides a non-trivial constraint.

The definition of the locus $\mathscr{L}$ involves setting to zero a certain number of fields ϕ^r belonging to $\mathscr{SK}_n$ so that we should also realize the consistency condition:

$$\frac{\partial}{\partial \phi^r} \sum_{\mathrm{t}\in\mathscr{A}} (\mathscr{P}_{\mathrm{t}})^2\Big|_{\substack{\mathbf{Z}=0\\ \phi^r=0}} = 0\,. \tag{7.4.4}$$

As mentioned earlier, the gauging yielding Starobinsky-like potentials need also involve the compact generator $\mathfrak{S}$. As we shall show in the following, if the gauged isometry is a combination of $\mathfrak{S}$ and an $\mathscr{SK}_n$ isometry, (7.4.3) poses no constraint on the gauging.

(F) A favorite, though not mandatory, choice corresponds to looking for abelian generators of iso $[\mathscr{SK}_n]$ such that the locus which satisfies conditions (7.4.4) is defined by setting to zero all the axions p_r, namely all the fields associated with nilpotent generators of the solvable Lie algebra of $\mathscr{SK}_n$. The inclusion of the Starobinsky potential in supergravity was obtained in [56] precisely in this way. In Sect. 7.9.4 we show a generalization of the same mechanism in the case of a bigger manifold $\mathcal{QM}_{4n+4}$, obtaining what can be denominated a multi Starobinsky model.

(G) **The U-problem**. If we use only the type of isometries yielding the tri-holomorphic moment maps (4.3.31), (4.3.35) and (4.3.40) we face a serious problem with the fields U. It appears only through exponentials all of the same sign ($\exp[-2U]$ or $\exp[-U]$ in front of perfect squares. Hence the field U cannot be stabilized unless all such squares are zero which means no residual potential. To overcome such a problem one should have moment maps with the opposite sign of U in the exponential and this can happen only by introducing in the gauging either L^E_- or generators $\mathbf{W}^{2,\alpha}$ this means that such generators should exist, namely the manifold $\mathcal{QM}_{4n+4}$ should be a symmetric space. In [56] the U-problem was solved by adding to a parabolic generator of a $\mathcal{SK}_n$-isometry the universal compact generator (4.3.52). As we have emphasized the Ehlers subalgebra exists in all symmetric spaces and so does the compact generator (4.3.52). This implies that the mechanism leading to the inclusion of the Starobinsky model found in [56] is actually rather universal and can be generalized in several ways.

The above discussion provides a framework for the search of other inflaton potentials.

7.5 Minimal Coupling Special Geometry

In this section we shortly describe the structure of the Minimal Coupling Special Kähler manifold $\mathcal{MSK}_{p+1}$, mostly in order to fix our conventions and to establish our notations. As announced in the introduction, this kind of Special Geometry is our favorite choice for the vector multiplet sector of the $\mathcal{N} = 2$ lagrangian which allows us to construct an entire class of theories where the vector multiplet scalars can be stabilized and the effective potential of an abelian gauging is reduced only to the hypermultiplet sector. In view of such a use of $\mathcal{MSK}_{p+1}$, all items of its Special Geometry will be denoted with a hat, and its complex coordinates will be named ω_i rather than z^i. However it is clear that $\mathcal{MSK}_{p+1}$ might also be used as c-map preimage of a Quaternionic Kähler manifold describing hypermultiplets.

As a manifold $\mathcal{MSK}_{p+1}$ is the following coset:

$$\mathcal{MSK}_{p+1} = \frac{\mathrm{SU(1, p + 1)}}{\mathrm{U(1) \times SU(p + 1)}} \tag{7.5.1}$$

In terms of the complex coordinates ω^i a convenient choice of the $(2p + 4)$-dimensional holomorphic symplect section is the following one:

$$\widehat{\Omega} = \begin{pmatrix} \widehat{X}^{\Lambda} \\ \hline \widehat{F}_{\Sigma} \end{pmatrix} = \begin{pmatrix} 1 \\ \omega^i \\ \hline -\mathrm{i} \\ \mathrm{i}\,\omega^i \end{pmatrix} \quad ; \quad (i = 1, \dots p+1) \tag{7.5.2}$$

which leads to the following Kähler potential:

$$\widehat{\mathscr{K}} = -\log\left[-\mathrm{i}\widehat{\Omega}\,\widehat{\mathbb{C}}\,\overline{\widehat{\Omega}}\right] = -\log\left[2\left(1 - \omega\cdot\overline{\omega}\right)\right] \tag{7.5.3}$$

and to the following Kähler metric:

$$\widehat{g}_{ij^\star} = \partial_i\,\partial_{j^\star}\widehat{\mathscr{K}} = \frac{1}{(1-\omega\cdot\overline{\omega})^2}\left(\delta^{ij}\,(1-\omega\cdot\overline{\omega}) + \overline{\omega}^i\,\omega^j\right) \tag{7.5.4}$$

Defining the Kähler covariant derivatives of the covariantly holomorphic sections as in Eq. (4.2.18) we obtain three results that are very important for the discussion of reduced scalar potentials in the present paper. Firstly we get:

$$\nabla_i\,\widehat{U}_j \equiv \nabla_i\,\nabla_j\widehat{V} = 0 \tag{7.5.5}$$

which compared with Eq. (4.2.19) implies the vanishing of the three-index symmetric tensor $\widehat{C}_{ijk}$. This unique property of the special Kähler manifold $\mathscr{MSK}_{p+1}$ defined by Eq. (7.5.1) is the reason why it has been named the Minimal Coupling Special Geometry, the interpretation of the tensor C_{ijk} in phenomenological applications being that of Yukawa couplings of the gauginos. In Ref. [56] it was shown that the vanishing of $\widehat{C}_{ijk}$ guarantees the consistency (see Eq. (3.10) of the quoted reference) of the truncation of the classical supergravity theory to the hypermultiplet quaternionic scalars by fixing the vector multiplet scalars to the origin of their manifold:

$$\omega^i = 0 \tag{7.5.6}$$

Secondly we evaluate the covariantly symplectic holomorphic section in the origin of the manifold and we obtain:

$$\widehat{V}\Big|_{\omega=0} = \frac{1}{\sqrt{2}}\left\{1\,|\,0\,||-\mathrm{i}\,|\,0\right\} \tag{7.5.7}$$

In the same point we have:

$$\left(\widehat{g}^{ij^\star}\,\nabla_i\widehat{V}^\alpha\,\nabla_{j^\star}\widehat{\overline{V}}\right)\Big|_{\omega=0} = \frac{1}{2}\left(\begin{array}{c|c||c|c} 0 & 0 & 0 & 0 \\ \hline 0 & \mathbf{1}_{(p+1)\times(p+1)} & 0 & \mathrm{i}\,\mathbf{1}_{(p+1)\times(p+1)} \\ \hline 0 & 0 & 0 & 0 \\ \hline 0 & -\mathrm{i}\,\mathbf{1}_{(p+1)\times(p+1)} & 0 & \mathbf{1}_{(p+1)\times(p+1)} \end{array}\right) \tag{7.5.8}$$

7.5.1 Gauging Abelian Isometries of the Hypermultiplets

Relying on these results we see that if the hypermultiplet Quaternionic manifold $\mathscr{QM}_{4m}$ possesses a $p+1$-dimensional abelian Lie algebra of isometries, we can always gauge them by using, for the vector multiplets, the Special Kähler manifold $\mathscr{MSK}_{p+1}$ introducing also the following embedding tensor:

$$\theta^I_M \equiv \left\{\theta^I_\Lambda\,,\,\theta^{\Sigma|I}\right\} = \left\{\theta_0{}^I = 0\,,\,\theta_J{}^I = \delta^I_J\,,\,\theta^{\Sigma|I} = 0\right\}. \tag{7.5.9}$$

Notice that the choice of setting $\theta_0{}^I = 0$ follows from the requirement (B) that the graviphoton should not be gauged. This indeed amounts to requiring:

$$\widehat{V}\Big|^M_{\omega=0}\,\theta_M{}^{\mathscr{A}} = 0 \;\Rightarrow\; \theta_0{}^I = 0\,. \tag{7.5.10}$$

In such a theory the scalar potential has the following general form:

$$\mathscr{V}_{scalar}(\omega,\overline{\omega},q) = 4\,k^u_I k^v_J\,h_{uv}\,\widehat{V}^I\,\widehat{\overline{V}}^J + \left(\widehat{g}^{ij^\star}\,\nabla_i\widehat{V}^I\,\nabla_{j^\star}\widehat{\overline{V}}^J - 3\,\widehat{V}^I\,\widehat{\overline{V}}^J\right)\mathscr{P}^x_I\,\mathscr{P}^x_J \tag{7.5.11}$$

setting $\omega^i = 0$ is a consistent truncation and the reduced potential takes the following universal general form which is positive definite by construction:

$$\mathscr{V}_{scalar}(0,0,q) = \sum_{I=1}^{p+1}\mathscr{P}^x_I(q)\,\mathscr{P}^x_I(q) \tag{7.5.12}$$

In the next Sect. 7.5.2 we reconsider the derivation of the Starobinsky potential obtained in [56] from a parabolic gauging as a master example that can be generalized to bigger manifolds.

7.5.2 The Starobinsky Potential

Recently a great deal of activity was devoted to the inclusion of phenomenologically interesting inflaton potentials into $\mathscr{N}=1$ supergravity. A first wave of investigations considered the possible generation of potentials by means of suitably chosen superpotentials, subsequently, after an important new viewpoint was introduced in [58] and was subsequently developed in [59–64], it became clear that positive definite inflaton potentials can be generated by the gauging of some isometry of the Kähler manifold of scalar multiplets. Such potentials have the form of squares of Kähler moment maps. In [63] this mechanism was applied to the case of constant curvature one-dimensional Kähler manifolds and it was shown that Starobinsky-like potentials [65] emerge from the moment map of a parabolic isometry in $\mathrm{SL}(2,\mathbb{R}) \simeq \mathrm{SU}(1,1)$ with the addition of a Fayet Iliopoulos term. In particular the standard Starobinsky model that is dual to an $R + R^2$ supergravity emerges from gauging the parabolic shift isometry of an $\frac{\mathrm{SU}(1,1)}{\mathrm{U}(1)}$ manifold with Kähler potential $\mathscr{K} = -3\log(z-\bar{z})$ which is precisely the Special Kähler manifold S^3. Let us now consider Eq. (4.3.40) and we can learn an important lesson. If in the c-map image of some $\mathscr{S}\mathscr{K}$ Special Kähler manifold, for instance the S^3 model, we gauge, according to the scheme discussed in Sect. 7.4, some nilpotent Lie algebra element $\mathfrak{N}_+ \in \mathbb{U}_{\mathscr{S}\mathscr{K}} \subset \mathbb{U}_{\mathscr{Q}}$ identical with the parabolic shift generator that we would have gauged in $\mathscr{N}=1$ supergravity, (for instance the generator $L_+ \in \mathfrak{sl}(2,\mathbb{R})$ in the case of the S^3 model), we obtain a moment map that contains precisely the $\mathscr{P}_{\mathbf{I}}$ of the $\mathscr{N}=1$ case, modified by $\mathbf{Z}$ dependent terms. In case the $\mathbf{Z}$ can be stabilized to zero the remaining effective potential is that of the corresponding $\mathscr{N}=1$ theory, apart from the Fayet Iliopoulos term. There are two remaining problems. The generation of a Fayet Iliopoulos term and the stabilization of the U field. They are solved in one stroke by modifying the parabolic generator of the inner Special Kähler isometry with the addition of the universal Ehlers rotation (4.3.52).

Let us see how this works.

With reference to Eq. (7.7.26) let us consider the following generator:

$$\mathfrak{p} = \mathfrak{N}_+ + \kappa\,\mathfrak{S} \tag{7.5.13}$$

where $\mathfrak{N}_+$ is the previously mentioned nilpotent element of the Special Kähler subalgebra ($\mathfrak{N}_+^r = 0$, for some positive integer r) and κ is a parameter. Let us then calculate the tri-holomorphic moment map $\mathscr{P}^x_{\mathfrak{p}}$ according to formula (4.3.54).

Because of the linearity of the momentum map in Lie algebra elements we have:

$$\begin{aligned}
\mathfrak{P}_{\mathfrak{p}} &= \mathfrak{P}_{\mathfrak{N}_+} + \mathfrak{P}_{\mathfrak{S}} \\
\mathfrak{P}_{\mathfrak{N}_+} &= \left(\begin{array}{c|c} \frac{i}{4}\mathscr{P}_{\mathfrak{N}_+} + \mathscr{O}\left(\mathbf{Z}^2\right) & \mathscr{O}\left(\mathbf{Z}\right) \\ \hline \mathscr{O}\left(\mathbf{Z}\right) & -\frac{i}{4}\mathscr{P}_{\mathfrak{N}_+} - \mathscr{O}\left(\mathbf{Z}^2\right) \end{array}\right) \\
\mathfrak{P}_{\mathfrak{S}} &= \left(\begin{array}{c|c} \frac{i}{8}e^{-U}\left(1+a^2+e^{2U}\right) + \mathscr{O}\left(\mathbf{Z}^2\right) & \mathscr{O}\left(\mathbf{Z}\right) \\ \hline \mathscr{O}\left(\mathbf{Z}\right) & -\frac{i}{8}e^{-U}\left(1+a^2+e^{2U}\right) - \mathscr{O}\left(\mathbf{Z}^2\right) \end{array}\right)
\end{aligned} \tag{7.5.14}$$

where $\mathscr{P}_{\mathfrak{N}_+}$ is the Kählerian moment map of the Killing vector associated with the generator $\mathfrak{N}_+$ as defined in Eq. (3.7.22). It is evident by the above completely universal formulae that the potential:

$$V_{gauging} = \text{const}\,\text{Tr}\left[\mathfrak{P}_{\mathrm{p}} \cdot \mathfrak{P}_{\mathrm{p}}\right] \tag{7.5.15}$$

possesses the following universal property:

$$\frac{\partial}{\partial \mathbf{Z}^{\alpha}} V_{gauging}\bigg|_{\mathbf{Z}=0} = 0 \tag{7.5.16}$$

allowing for a consistent truncation of the Heisenberg fields. After such truncation we find:

$$V_{eff}(U, a, z, \overline{z}) = V_{gauging}\big|_{\mathbf{Z}=0} = \text{const} \times \left[\mathscr{P}_{\mathfrak{N}_+} + \frac{\kappa}{2} e^{-U}\left(1 + a^2 + e^{2U}\right)\right]^2 \tag{7.5.17}$$

From Eq. (7.5.17) we further learn that we can consistently truncate the fields a and U setting them to zero since

$$\frac{\partial}{\partial U} V_{eff}\bigg|_{U=a=0} = 0 \quad ; \quad \frac{\partial}{\partial a} V_{eff}\bigg|_{U=a=0} = 0 \tag{7.5.18}$$

We find:

$$V_{infl}(z, \overline{z}) \equiv V_{eff}(0, 0, z, \overline{z}) = \left(\mathscr{P}_{\mathfrak{N}_+} + \kappa\right)^2 \tag{7.5.19}$$

which clearly shows how the universal generator $\mathfrak{S}$ provides, after stabilization of the U field, the mechanism that generates the Fayet Iliopoulos term [66] essential for inflation.

7.6 Examples

As an illustration of the general patterns and mechanisms described in the previous pages we consider two examples of Quaternionic Kähler manifolds $\mathscr{QM}_{4n+4}$ obtained from the c-map of two homogeneous symmetric Special Kähler manifolds $\mathscr{SK}_n$.

1. The manifold $\frac{\mathrm{G}_{(2,2)}}{\mathrm{SU}(2)\times\mathrm{SU}(2)}$ which is the c-map image of the Special Kähler manifold $\frac{\mathrm{SU}(1,1)}{\mathrm{U}(1)}$ with cubic embedding of $\mathrm{SU}(1,1)$ in $\mathrm{Sp}(4,\mathbb{R})$. In this case $n = 1$ and the corresponding coupling of one vector multiplet to supergravity is usually named the S^3 model in the literature. We already used the S^3-model extensively in Chap. 6 as an example for the construction of nilpotent orbits and black-hole solutions. There the key point was the $c^\star$ map of $\frac{\mathrm{SU}(1,1)}{\mathrm{U}(1)}$ to the pseudo-quaternionic manifold $\frac{\mathrm{G}_{(2,2)}}{\mathrm{SU}(1,1)\times\mathrm{SU}(1,1)}$. For the issue of scalar potentials the key point is the c-map of the same special Kähler manifold to the quaternionic Kähler manifold $\frac{\mathrm{G}_{(2,2)}}{\mathrm{SU}(2)\times\mathrm{SU}(2)}$. It is very instructive to make a close comparison between the two cases, lorentzian and Euclidean, respectively.

2. The manifold $\frac{\mathrm{F}_{(4,4)}}{\mathrm{SU}(2)\times\mathrm{USp}(6)}$ which is the c-map image of the Special Kähler manifold $\frac{\mathrm{Sp}(6,\mathbb{R})}{\mathrm{SU}(3)\times\mathrm{U}(1)}$. In this case $n = 6$.

For these two models we provide a full fledged construction of all the geometrical items and in particular we realize the bridge between the algebraic description and the analytic one advocated at the end of Sect. 4.3.4.1. This allows us to discuss a couple of examples of gaugings. In particular in the case of the of the first model which we utilize as a calibration device for our general formulae we retrieve the inclusion of the Starobinsky model first demonstrated in [56].

The detailed construction of the second model is utilized to provide an example of generalization of the results of [56] by means of the inclusion of a multi Starobinsky model.

7.7 The S^3 Model and Its Quaternionic Image $\frac{\mathrm{G}_{(2,2)}}{\mathrm{SU}(2)\times\mathrm{SU}(2)}$

In this which is the simplest example $n = 1$, namely the Special Kähler manifold has complex dimension 1 and it can be identified with the time honored Poincaré Lobachevsky plane:

$$\mathscr{S}\mathscr{K}_1 = \frac{\mathrm{SU}(1,1)}{\mathrm{U}(1)} \tag{7.7.1}$$

The special Kähler structure of this model was exhaustively described in Sect. 6.3 to which we refer for details.

7.7.1 The Matrix $\mathscr{M}_4^{-1}$ and the c-Map

For the S^3 model the matrix $\mathscr{M}_4$ and its inverse have the following explicit appearance:

$$\mathscr{M}_4 = \begin{pmatrix} \frac{4iz\bar{z}\left(z^2+4\bar{z}z+\bar{z}^2\right)}{(z-\bar{z})^3} & -\frac{4i\sqrt{3}z^2\bar{z}^2(z+\bar{z})}{(z-\bar{z})^3} & -\frac{i(z+\bar{z})\left(z^2+10\bar{z}z+\bar{z}^2\right)}{(z-\bar{z})^3} & -\frac{2i\sqrt{3}(z+\bar{z})^2}{(z-\bar{z})^3} \\ -\frac{4i\sqrt{3}z^2\bar{z}^2(z+\bar{z})}{(z-\bar{z})^3} & \frac{8iz^3\bar{z}^3}{(z-\bar{z})^3} & \frac{2i\sqrt{3}z\bar{z}(z+\bar{z})^2}{(z-\bar{z})^3} & \frac{i(z+\bar{z})^3}{(z-\bar{z})^3} \\ -\frac{i(z+\bar{z})\left(z^2+10\bar{z}z+\bar{z}^2\right)}{(z-\bar{z})^3} & \frac{2i\sqrt{3}z\bar{z}(z+\bar{z})^2}{(z-\bar{z})^3} & \frac{4i\left(z^2+4\bar{z}z+\bar{z}^2\right)}{(z-\bar{z})^3} & \frac{4i\sqrt{3}(z+\bar{z})}{(z-\bar{z})^3} \\ -\frac{2i\sqrt{3}(z+\bar{z})^2}{(z-\bar{z})^3} & \frac{i(z+\bar{z})^3}{(z-\bar{z})^3} & \frac{4i\sqrt{3}(z+\bar{z})}{(z-\bar{z})^3} & \frac{8i}{(z-\bar{z})^3} \end{pmatrix} \tag{7.7.2}$$

its inverse being:

$$\mathcal{M}_4^{-1} = \begin{pmatrix} \frac{4i\left(z^2+4\bar{z}z+\bar{z}^2\right)}{(z-\bar{z})^3} & \frac{4i\sqrt{3}(z+\bar{z})}{(z-\bar{z})^3} & \frac{i\left(z^3+11\bar{z}z^2+11\bar{z}^2z+\bar{z}^3\right)}{(z-\bar{z})^3} & -\frac{2i\sqrt{3}z\bar{z}(z+\bar{z})^2}{(z-\bar{z})^3} \\ \frac{4i\sqrt{3}(z+\bar{z})}{(z-\bar{z})^3} & \frac{8i}{(z-\bar{z})^3} & \frac{2i\sqrt{3}(z+\bar{z})^2}{(z-\bar{z})^3} & -\frac{i(z+\bar{z})^3}{(z-\bar{z})^3} \\ \frac{i\left(z^3+11\bar{z}z^2+11\bar{z}^2z+\bar{z}^3\right)}{(z-\bar{z})^3} & \frac{2i\sqrt{3}(z+\bar{z})^2}{(z-\bar{z})^3} & \frac{4iz\bar{z}\left(z^2+4\bar{z}z+\bar{z}^2\right)}{(z-\bar{z})^3} & -\frac{4i\sqrt{3}z^2\bar{z}^2(z+\bar{z})}{(z-\bar{z})^3} \\ -\frac{2i\sqrt{3}z\bar{z}(z+\bar{z})^2}{(z-\bar{z})^3} & -\frac{i(z+\bar{z})^3}{(z-\bar{z})^3} & -\frac{4i\sqrt{3}z^2\bar{z}^2(z+\bar{z})}{(z-\bar{z})^3} & \frac{8iz^3\bar{z}^3}{(z-\bar{z})^3} \end{pmatrix} \tag{7.7.3}$$

Furthermore, in this case a convenient reference point is given by $z_0 = \mathrm{i}$ that can be mapped into any point of the upper complex plane by means of the element:

$$\mathfrak{g}_z = \begin{pmatrix} e^{h/2} & e^{-h/2}y \\ 0 & e^{-h/2} \end{pmatrix} \in \mathrm{SL}(2, \mathbb{R}) \tag{7.7.4}$$

acting by means of fractional linear transformations. The explicit form of the $\Lambda(\mathfrak{g})$ matrix in the **W**-representation was given in Eq. (1.7.19). This provides us with all the necessary information in order to write down the explicit form of the $E^I_{\mathcal{Q}\mathcal{M}}$ vielbein for the S^3 case.

7.7.2 *The Vielbein and the Borellian Maurer Cartan Equations*

They are the following ones:

$$E^I_{\mathcal{Q}\mathcal{M}} = \frac{1}{2}\begin{pmatrix} \mathrm{dU} \\ \sqrt{3}\mathrm{dh} \\ \sqrt{3}\mathrm{dy}e^{-h} \\ e^{-U}\left(\mathrm{da} + \mathrm{dZ}_3\mathrm{Z}_1 + \mathrm{dZ}_4\mathrm{Z}_2 - \mathrm{dZ}_1\mathrm{Z}_3 - \mathrm{dZ}_2\mathrm{Z}_4\right) \\ \sqrt{2}e^{-\frac{h}{2}-\frac{U}{2}}\left(\mathrm{dZ}_1 + y\left(2\mathrm{dZ}_3 - \sqrt{3}y\mathrm{dZ}_4\right)\right) \\ \sqrt{2}e^{-\frac{3h}{2}-\frac{U}{2}}\left(\left(\sqrt{3}\mathrm{dZ}_3 - y\mathrm{dZ}_4\right)y^2 + \sqrt{3}\mathrm{dZ}_1 y + \mathrm{dZ}_2\right) \\ \sqrt{2}e^{\frac{h-U}{2}}\left(\mathrm{dZ}_3 - \sqrt{3}y\mathrm{dZ}_4\right) \\ \sqrt{2}e^{\frac{3h}{2}-\frac{U}{2}}\mathrm{dZ}_4 \end{pmatrix} \tag{7.7.5}$$

Furthermore we find $\mathcal{M}_4^{-1}(\mathrm{i}, -\mathrm{i}) = -\mathbf{1}_{4\times 4}$ so that the quadratic form (4.3.46) is just:

$$\mathfrak{q}_{AB} = \mathrm{diag}\,(1, 1, 1, 1, 1, 1, 1, 1) \tag{7.7.6}$$

The next step consists of calculating the geometry of the space described by the above vielbein and flat metric (7.7.6). To this effect we have first to calculate the contorsion, namely the exterior derivatives of the vielbein and then using such a result the spin connection ω^{IJ}, finally the curvature two-form from which we extract the Riemann and the Ricci tensor.

Addressing the first step, namely the contorsion, we have the first important surprise. The exterior derivatives of the vielbein are expressed in terms of wedge-quadratic products of the same vielbein with constant numerical coefficients. This means that the above constructed vielbein satisfy a set of Maurer Cartan equations describing a Lie algebra, namely[9]:

$$dE^I - \frac{1}{2} f^I_{JK} \, E^I \wedge E^J = 0 \tag{7.7.7}$$

the tensor $f_{BC}{}^A$ being the structure constants of such a Lie algebra. Explicitly for the S^3 model we get:

$$\begin{aligned}
0 &= dE^1 \\
0 &= dE^2 \\
0 &= dE^3 + 2\,\tfrac{E^2\wedge E^3}{\sqrt{3}} \\
0 &= dE^4 + 2\,E^1 \wedge E^4 - 2\,E^5 \wedge E^7 - 2\,E^6 \wedge E^8 \\
0 &= dE^5 + E^1 \wedge E^5 + \tfrac{E^2\wedge E^5}{\sqrt{3}} - \tfrac{4\,E^3\wedge E^7}{\sqrt{3}} \\
0 &= dE^6 + E^1 \wedge E^6 + \sqrt{3}E^2 \wedge E^6 - 2\,E^3 \wedge E^5 \\
0 &= dE^7 + E^1 \wedge E^7 - \tfrac{E^2\wedge E^7}{\sqrt{3}} + 2\,E^3 \wedge E^8 \\
0 &= dE^8 + E^1 \wedge E^8 - \sqrt{3}E^2 \wedge E^8
\end{aligned} \tag{7.7.8}$$

Hence it arises the following question: which Lie algebra is described by such Maurer Cartan equations? Utilizing the standard method of diagonalizing the adjoint action of the two commuting generators $H_{1,2}$ dual to $E^{1,2}$ we find that the eigenvalues are just the positive roots of $\mathfrak{g}_{2,2}$ as given in Eq. (1.6.1). As it is well known the complex Lie algebra $\mathfrak{g}_2(\mathbb{C})$ has rank two and it is defined by the 2×2 Cartan matrix encoded in the following Dynkin diagram:

$\mathfrak{g}_2$  $= \begin{pmatrix} 2 & -3 \\ -1 & 2 \end{pmatrix}$

The real form $\mathfrak{g}_{2,2}$ is the maximally split form of the above complex Lie algebra. With a little bit of more work we can put Eq. (7.7.8) into the standard Cartan Weyl form for the Borel subalgebra of $\mathfrak{g}_{2,2}$, composed by the Cartan generators and by all the positive root step operators. Naming T_J the generators dual to the vielbein E^I such that $E^I(T_J) = \delta^I_J$, we find that the appropriate identifications are the following ones:

[9] Note that here, for simplicity we have dropped the suffix $\mathscr{SK}$. This is done for simplicity since there is no risk of confusion.

$$
\begin{aligned}
T_2 &= 2\,\frac{\mathscr{H}_1}{\sqrt{3}}\\
T_1 &= 2\,\frac{\mathscr{H}_2}{\sqrt{3}}\\
T_3 &= 2\,E^{\alpha_1}\\
T_4 &= 2\,E^{\alpha_6}\\
T_8 &= 2\,E^{\alpha_2}\\
T_7 &= 2\,E^{\alpha_3}\\
T_5 &= 2\,E^{\alpha_4}\\
T_6 &= 2\,E^{\alpha_5}
\end{aligned}
\tag{7.7.9}
$$

We conclude that the manifold on which the metric (4.3.4) is constructed is homeomorphic to the solvable group-manifold $\mathrm{Bor}(\mathfrak{g}_{2,2})$.

7.7.3 The Spin Connection

Next, calculating the Levi-Civita spin connection from its definition, namely the vanishing torsion condition:

$$
0 = dE^I + \omega^{IJ} \wedge E^J \tag{7.7.10}
$$

we find the following result:

$$
\omega^{IJ} = \begin{pmatrix}
0 & 0 & 0 & E^4 & \frac{E^5}{2} & \frac{E^6}{2} & \frac{E^7}{2} & \frac{E^8}{2} \\
0 & 0 & \frac{E^3}{\sqrt{3}} & 0 & \frac{E^5}{2\sqrt{3}} & \frac{1}{2}\sqrt{3}E^6 & -\frac{E^7}{2\sqrt{3}} & -\frac{1}{2}\sqrt{3}E^8 \\
0 & -\frac{E^3}{\sqrt{3}} & 0 & 0 & -\frac{E^6}{2}-\frac{E^7}{\sqrt{3}} & -\frac{E^5}{2} & \frac{E^8}{2}-\frac{E^5}{\sqrt{3}} & \frac{E^7}{2} \\
-E^4 & 0 & 0 & 0 & \frac{E^7}{2} & \frac{E^8}{2} & -\frac{E^5}{2} & -\frac{E^6}{2} \\
-\frac{E^5}{2} & -\frac{E^5}{2\sqrt{3}} & \frac{E^6}{2}+\frac{E^7}{\sqrt{3}} & -\frac{E^7}{2} & 0 & \frac{E^3}{2} & -\frac{E^3}{\sqrt{3}}-\frac{E^4}{2} & 0 \\
-\frac{E^6}{2} & -\frac{1}{2}\sqrt{3}\,E^6 & \frac{E^5}{2} & -\frac{E^8}{2} & -\frac{E^3}{2} & 0 & 0 & -\frac{E^4}{2} \\
-\frac{E^7}{2} & \frac{E^7}{2\sqrt{3}} & \frac{E^5}{\sqrt{3}}-\frac{E^8}{2} & \frac{E^5}{2} & \frac{E^3}{\sqrt{3}}+\frac{E^4}{2} & 0 & 0 & \frac{E^3}{2} \\
-\frac{E^8}{2} & \frac{1}{2}\sqrt{3}E^8 & -\frac{E^7}{2} & \frac{E^6}{2} & 0 & \frac{E^4}{2} & -\frac{E^3}{2} & 0
\end{pmatrix}
\tag{7.7.11}
$$

which can be decomposed in the way we now describe.

7.7.4 Holonomy Algebra and Decompostion of the Spin Connection

Let us introduce two triplets $J^x_{[I]}$ and $J^x_{[II]}$ of 8×8 matrices that can be read off explicitly as the coefficients of α_x and β_x in the following linear combinations:

$$\sum_{x=1}^{3} \alpha_x \, J^x_{[I]} =$$

$$\begin{pmatrix}
0 & 0 & 0 & -\frac{\alpha_1}{2} & -\frac{1}{4}\sqrt{3}\alpha_3 & -\frac{\alpha_2}{4} & \frac{\sqrt{3}\alpha_2}{4} & -\frac{\alpha_3}{4} \\
0 & 0 & \frac{\alpha_1}{2} & 0 & -\frac{\alpha_3}{4} & -\frac{1}{4}\sqrt{3}\alpha_2 & -\frac{\alpha_2}{4} & \frac{\sqrt{3}\alpha_3}{4} \\
0 & -\frac{\alpha_1}{2} & 0 & 0 & -\frac{\alpha_2}{4} & \frac{\sqrt{3}\alpha_3}{4} & \frac{\alpha_3}{4} & \frac{\sqrt{3}\alpha_2}{4} \\
\frac{\alpha_1}{2} & 0 & 0 & 0 & \frac{\sqrt{3}\alpha_2}{4} & -\frac{\alpha_3}{4} & \frac{\sqrt{3}\alpha_3}{4} & \frac{\alpha_2}{4} \\
\frac{\sqrt{3}\alpha_3}{4} & \frac{\alpha_3}{4} & \frac{\alpha_2}{4} & -\frac{1}{4}\sqrt{3}\alpha_2 & 0 & \frac{\sqrt{3}\alpha_1}{4} & -\frac{\alpha_1}{4} & 0 \\
\frac{\alpha_2}{4} & \frac{\sqrt{3}\alpha_2}{4} & -\frac{1}{4}\sqrt{3}\alpha_3 & \frac{\alpha_3}{4} & -\frac{1}{4}\sqrt{3}\alpha_1 & 0 & 0 & \frac{\alpha_1}{4} \\
-\frac{1}{4}\sqrt{3}\alpha_2 & \frac{\alpha_2}{4} & -\frac{\alpha_3}{4} & -\frac{1}{4}\sqrt{3}\alpha_3 & \frac{\alpha_1}{4} & 0 & 0 & \frac{\sqrt{3}\alpha_1}{4} \\
\frac{\alpha_3}{4} & -\frac{1}{4}\sqrt{3}\alpha_3 & -\frac{1}{4}\sqrt{3}\alpha_2 & -\frac{\alpha_2}{4} & 0 & -\frac{\alpha_1}{4} & -\frac{1}{4}\sqrt{3}\alpha_1 & 0
\end{pmatrix} \tag{7.7.12}$$

$$\sum_{x=1}^{3} \beta_x \, J^x_{[II]} =$$

$$\begin{pmatrix}
0 & 0 & 0 & \frac{3\beta_1}{2} & -\frac{1}{4}\sqrt{3}\beta_3 & -\frac{3\beta_2}{4} & -\frac{1}{4}\sqrt{3}\beta_2 & \frac{3\beta_3}{4} \\
0 & 0 & \frac{\beta_1}{2} & 0 & -\frac{\beta_3}{4} & -\frac{3}{4}\sqrt{3}\beta_2 & \frac{\beta_2}{4} & -\frac{3}{4}\sqrt{3}\beta_3 \\
0 & -\frac{\beta_1}{2} & 0 & 0 & \frac{5\beta_2}{4} & \frac{\sqrt{3}\beta_3}{4} & \frac{5\beta_3}{4} & -\frac{1}{4}\sqrt{3}\beta_2 \\
-\frac{3\beta_1}{2} & 0 & 0 & 0 & -\frac{1}{4}\sqrt{3}\beta_2 & \frac{3\beta_3}{4} & \frac{\sqrt{3}\beta_3}{4} & \frac{3\beta_2}{4} \\
\frac{\sqrt{3}\beta_3}{4} & \frac{\beta_3}{4} & -\frac{5\beta_2}{4} & \frac{\sqrt{3}\beta_2}{4} & 0 & \frac{\sqrt{3}\beta_1}{4} & -\frac{5\beta_1}{4} & 0 \\
\frac{3\beta_2}{4} & \frac{3\sqrt{3}\beta_2}{4} & -\frac{1}{4}\sqrt{3}\beta_3 & -\frac{3\beta_3}{4} & -\frac{1}{4}\sqrt{3}\beta_1 & 0 & 0 & -\frac{3\beta_1}{4} \\
\frac{\sqrt{3}\beta_2}{4} & -\frac{\beta_2}{4} & -\frac{5\beta_3}{4} & -\frac{1}{4}\sqrt{3}\beta_3 & \frac{5\beta_1}{4} & 0 & 0 & \frac{\sqrt{3}\beta_1}{4} \\
-\frac{3\beta_3}{4} & \frac{3\sqrt{3}\beta_3}{4} & \frac{\sqrt{3}\beta_2}{4} & -\frac{3\beta_2}{4} & 0 & \frac{3\beta_1}{4} & -\frac{1}{4}\sqrt{3}\beta_1 & 0
\end{pmatrix} \tag{7.7.13}$$

Both triplets form an 8-dimensional representation of the $\mathfrak{su}(2)$ Lie algebra and the two triplets commute with each other:

$$\begin{aligned}
\left[J^x_{[I]} \, , \, J^y_{[I]} \right] &= \varepsilon^{xyz} \, J^y_{[I]} \\
\left[J^x_{[II]} \, , \, J^y_{[II]} \right] &= \varepsilon^{xyz} \, J^y_{[II]} \\
\left[J^x_{[I]} \, , \, J^y_{[II]} \right] &= 0
\end{aligned} \tag{7.7.14}$$

Furthermore all matrices are antisymmetric so that the two Lie algebras $\mathfrak{su}_{\mathrm{I}}(2)$ and $\mathfrak{su}_{\mathrm{II}}(2)$ are both subalgebras of $\mathfrak{so}(8)$. The distinction between these two representations becomes clear when we calculate the Casimir operator for both of them. We obtain:

$$\sum_{x=1}^{3} J^x_{[I]} \cdot J^x_{[I]} = -\frac{3}{4}\mathbf{1} \quad ; \quad \sum_{x=1}^{3} J^x_{[II]} \cdot J^x_{[II]} = -\frac{15}{4}\mathbf{1} \tag{7.7.15}$$

Hence the first $\mathfrak{su}_{\mathrm{I}}(2)$ Lie algebra is realized on the considered eight-dimensional space in the $j = \frac{1}{2}$ representation, while the second $\mathfrak{su}_{\mathrm{II}}(2)$ Lie algebra is realized on the same space in the $j = \frac{3}{2}$. In other words, with respect to both subalgebras of $\mathfrak{so}(8)$, the fundamental representation decomposes as follows:

$$\mathbf{8} \stackrel{\mathfrak{su}_{\mathrm{I}}(2)\oplus\mathfrak{su}_{\mathrm{II}}(2)\subset\mathfrak{so}(8)}{\Longrightarrow} (\mathbf{2}, \mathbf{4}) \tag{7.7.16}$$

By direct calculation we verify that the spin connection displayed in Eq. (7.7.11) has the following structure:

$$\omega = \omega_x^{[I]} J^x_{[I]} \oplus \omega_x^{[II]} J^x_{[II]} \tag{7.7.17}$$

where:

$$\omega_x^{[I]} = \begin{pmatrix} \sqrt{3}\mathrm{E}^3 - \mathrm{E}^4 \\ \sqrt{3}\mathrm{E}^7 - \mathrm{E}^6 \\ -\mathrm{E}^8 - \sqrt{3}\mathrm{E}^5 \end{pmatrix} \quad ; \quad \omega_x^{[II]} = \begin{pmatrix} \frac{\mathrm{E}^4}{2} + \frac{\mathrm{E}^3}{2\sqrt{3}} \\ -\frac{\mathrm{E}^7}{2\sqrt{3}} - \frac{\mathrm{E}^6}{2} \\ \frac{\mathrm{E}^8}{2} - \frac{\mathrm{E}^5}{2\sqrt{3}} \end{pmatrix} \tag{7.7.18}$$

This structure clearly demonstrates the reduced holonomy of the Quaternionic Kähler manifold. Indeed, according to Eq. (7.7.16) the vielbein transforms in the doublet of $\mathfrak{su}_{\mathrm{I}}(2)$ tensored with the fundamental representation of $\mathfrak{sp}(4, \mathbb{R})$. In the present case the symplectic 4×4 matrices are actually reduced to the subalgebra $\mathfrak{su}_{\mathrm{II}}(2) \subset \mathfrak{sp}(4, \mathbb{R})$ with respect to which the fundamental of $\mathfrak{sp}(4, \mathbb{R})$ remains irreducible and coincides with the $j = \frac{3}{2}$ representation of $\mathfrak{su}_{\mathrm{II}}(2)$. The above discussion can be summarized by the statement:

$$\mathfrak{so}(8) \subset \mathfrak{su}(2) \oplus \mathfrak{usp}(4) \subset \mathrm{Hol} = \mathfrak{su}_{\mathrm{I}}(2) \oplus \mathfrak{su}_{\mathrm{II}}(2) \tag{7.7.19}$$

by definition the holonomy algebra being the Lie algebra in which the Levi-Civita spin connection takes values.

7.7.5 Structure of the Isotropy Subalgebra $\mathbb{H}$

It remains to single out the structure of the denominator subalgebra $\mathbb{H} \subset \mathbb{U} \equiv \mathfrak{g}_{2,2}$ in the orthogonal decomposition:

$$\mathbb{U} = \mathbb{H} \oplus \mathbb{K} \quad ; \quad \begin{cases} [\mathbb{H}\,,\,\mathbb{H}] \subset \mathbb{H} \\ [\mathbb{H}\,,\,\mathbb{K}] \subset \mathbb{K} \\ [\mathbb{K}\,,\,\mathbb{K}] \subset \mathbb{H} \end{cases} \tag{7.7.20}$$

Since our quaternionic Kähler manifold is a symmetric space it follows that the Lie algebra $\mathbb{H}$ must be isomorphic with the holonomy algebra $\mathrm{Hol} = \mathfrak{su}_{\mathrm{I}}(2) \oplus \mathfrak{su}_{\mathrm{II}}(2)$ that we have calculated in the previous subsection. By definition the Lie algebra $\mathbb{H}$

is the maximal compact subalgebra which for maximal split algebras has a universal definition in terms of the step operators associated with the positive roots E^{α} and their conjugates $E^{-\alpha}$. In the case of $\mathfrak{g}_{2,2}$ which has six positive roots we can write:

$$\mathbb{H} \equiv \mathrm{span}_{\mathbb{R}} \left\{E^{\alpha_1} - E^{\alpha_1}\,,\, E^{\alpha_2} - E^{\alpha_2}\,,\, E^{\alpha_3} - E^{\alpha_3}\,,\, E^{\alpha_4} - E^{\alpha_4}\,,\, E^{\alpha_5} - E^{\alpha_5}\,,\, E^{\alpha_6} - E^{\alpha_6}\right\} \tag{7.7.21}$$

The structure of (7.7.21) is the following:

$$\mathbb{H} = \mathfrak{su}_{\mathrm{I}}(2) \oplus \mathfrak{su}_{\mathrm{II}}(2) \tag{7.7.22}$$

where the generators of the two subalgebras are:

$$j^x_{[I]} = \begin{pmatrix} \frac{-3\mathrm{E}^{-\alpha_1}+3\mathrm{E}^{\alpha_1}+\sqrt{3}\left(\mathrm{E}^{\alpha_6}-\mathrm{E}^{-\alpha_6}\right)}{6\sqrt{2}} \\ \frac{3\mathrm{E}^{-\alpha_3}-3\mathrm{E}^{\alpha_3}+\sqrt{3}\left(\mathrm{E}^{-\alpha_5}-\mathrm{E}^{\alpha_5}\right)}{6\sqrt{2}} \\ \frac{\sqrt{3}\left(\mathrm{E}^{\alpha_2}-\mathrm{E}^{-\alpha_2}\right)+3\left(\mathrm{E}^{-\alpha_4}-\mathrm{E}^{\alpha_4}\right)}{6\sqrt{2}} \end{pmatrix} \tag{7.7.23}$$

and

$$j^x_{[II]} = \begin{pmatrix} \frac{-\mathrm{E}^{-\alpha_1}+\mathrm{E}^{\alpha_1}+\sqrt{3}\left(\mathrm{E}^{-\alpha_6}-\mathrm{E}^{\alpha_6}\right)}{2\sqrt{2}} \\ \frac{-\mathrm{E}^{-\alpha_3}+\mathrm{E}^{\alpha_3}+\sqrt{3}\left(\mathrm{E}^{-\alpha_5}-\mathrm{E}^{\alpha_5}\right)}{2\sqrt{2}} \\ \frac{\sqrt{3}\left(\mathrm{E}^{-\alpha_2}-\mathrm{E}^{\alpha_2}\right)+\mathrm{E}^{-\alpha_4}-\mathrm{E}^{\alpha_4}}{2\sqrt{2}} \end{pmatrix} \tag{7.7.24}$$

and satisfy among themselves the same relations (7.7.14) as their homologous generators $J^x_{[I]}$ and $J^x_{[II]}$. In Eq. (7.7.22) we have used the same notation as in Eq. (7.7.19) using the obligatory homomorphism between the holonomy algebra Hol and the isotropy subalgebra $\mathbb{H}$. The precise correspondence between generators of one algebra and generators of the other will be established in the next subsection by means of the use of the coset representative.

7.7.6 The Coset Representative

The next step in the development of the coset approach is the construction of the solvable coset representative $\mathbb{L}_{Solv}(\phi)$, advocated in Eqs. (4.3.57) and (4.3.58), namely a coordinate dependent element of the Borel group of $\mathfrak{g}_{(2,2)}$ such that the Maurer Cartan form

$$\Xi = \mathbb{L}_{Solv}(\phi)^{-1}\, \mathrm{d}\mathbb{L}_{Solv}(\phi) \tag{7.7.25}$$

projected along the Borel algebra generators, as given in Eq. (7.7.9), reproduces the vielbein of Eq. (7.7.5). The appropriate coset representative is obtained by exponentiating the Borel Lie algebra and the precise recipe is provided below. First define:

$$
\begin{aligned}
\mathbf{L}_0^E &= \frac{1}{\sqrt{3}}\,\mathscr{H}_2 \quad ; \quad \mathbf{L}_+^E \, = -\sqrt{\frac{2}{3}}\,E^{\alpha_6} \\
\mathbf{L}_0 &= \mathscr{H}_1 \quad ; \quad \mathbf{L}_+ \, = \sqrt{2}E^{\alpha_1} \\
\mathbf{W}^I &= \sqrt{\frac{2}{3}}\,\{E^{\alpha_4}\,,\, E^{\alpha_5}\,,\, -E^{\alpha_3}\,,\, -E^{\alpha_2}\,,\,\}
\end{aligned}
\tag{7.7.26}
$$

and then set:

$$
\begin{aligned}
\mathbb{L} = \exp\left[a\mathbf{L}_+^E\right] \cdot \exp\left[\sqrt{2}\left(Z_1\,\mathbf{W}^1 + Z_3\,\mathbf{W}^3\right)\right] \cdot \\
\cdot \exp\left[\sqrt{2}\left(Z_1\,\mathbf{W}^1 + Z_3\,\mathbf{W}^3\right)\right] \cdot \exp\left[y\mathbf{L}_+\right] \cdot \exp\left[h\mathbf{L}_0\right] \cdot \exp\left[U\mathbf{L}_0^E\right]
\end{aligned}
\tag{7.7.27}
$$

By explicit evaluation we obtain the result displayed in the appendix in formulae (7.12.1) and (7.12.2) and we verify that, if we set:

$$
\mathfrak{T}_I = \left\{\mathbf{L}_0^E\,,\,\mathbf{L}_0\,,\,\mathbf{L}_+^E\,,\,\mathbf{L}_+\,,\,\mathbf{W}_1\,,\,\mathbf{W}_2\,,\,\mathbf{W}_3\,,\,\mathbf{W}_4\right\}
\tag{7.7.28}
$$

upon substitution of (7.12.2) into the Maurer Cartan form (7.7.25) we obtain:

$$
\mathbb{L}_{Solv}(\phi)^{-1}\,\mathrm{d}\mathbb{L}_{Solv}(\phi) = \sum_{I=1}^{8}\mathfrak{T}_I\,E^I_{\mathscr{Q}\mathscr{M}}
\tag{7.7.29}
$$

the forms $E^I_{\mathscr{Q}\mathscr{M}}$ being given in Eq. (7.7.5). Alternatively we can also write:

$$
\mathbb{L}_{Solv}(\phi)^{-1}\,\mathrm{d}\mathbb{L}_{Solv}(\phi) = \sum_{x=1}^{3}\left(\omega_x^{[I]}\,j^x_{[I]} \oplus \omega_x^{[II]}\,j^x_{[II]}\right) \oplus \sum_{I}^{8}\mathbf{T}_I\,E^I_{\mathscr{S}\mathscr{Q}}
\tag{7.7.30}
$$

In the above equation $\omega_x^{[I]}$ and $\omega_x^{[II]}$ are the components of the spin connections given in Eq. (7.7.18), $j^x_{[I]}$ and $j^x_{[II]}$ are the generators of $\mathbb{H}$ defined in Eqs. (7.7.23), (7.7.24) and $\mathbf{T}_I$ denotes a suitable base of generators in the $\mathbb{K}$ subspace of $\mathfrak{g}_{(2,2)}$ defined as:

$$
\begin{aligned}
\mathbb{K} \equiv\,, \mathrm{span}_{\mathbb{R}}\,\{\mathscr{H}_1\,,\,\mathscr{H}_2\,,\,E^{\alpha_1} + E^{\alpha_1}\,,\,E^{\alpha_2} + E^{\alpha_2}\,,\,E^{\alpha_3} + E^{\alpha_3} \\
E^{\alpha_4} + E^{\alpha_4}\,,\,E^{\alpha_5} + E^{\alpha_5}\,,\,E^{\alpha_6} + E^{\alpha_6}\}
\end{aligned}
\tag{7.7.31}
$$

The precise form of the generators $\mathbf{T}_I$ is not relevant to our purposes and we omit it. The key point is instead the identification of the generators $j^x_{[I]}$ of $\mathbb{H}$ with generators $J^x_{[I]}$ of the holonomy algebra. This provides us with the knowledge of the quaternionic complex structures within the algebra $\mathbb{U}_{\mathscr{Q}}$ and allows to calculate the tri-holomorphic moment map of any generator $\mathbf{t} \in \mathbb{U}_{\mathscr{Q}}$ by means of the formula (4.3.54) which in our case reads:

$$\mathscr{P}^x_{\mathbf{t}} = \frac{1}{2}\mathrm{Tr}_{\mathbf{7}}\left(j^x_{[I]}\,\mathbb{L}^{-1}_{Solv}\,\mathbf{t}\,\mathbb{L}_{Solv}\right) \tag{7.7.32}$$

having denoted by **7** the 7-dimensional fundamental representation of $\mathfrak{g}_{(2,2)}$.

7.7.6.1 The Starobinsky Potential

As an immediate application of Eq. (7.7.32) one can retrieve the results of [56] on the inclusion of the Starobinsky potential into supergravity. In Sect. 7.5.2 we presented a general discussion of the gaugings of nilpotent generators in the Special Kähler subalgebra $\mathbb{U}_{\mathscr{SK}} \subset \mathbb{U}_{\mathscr{Q}}$. In the present case where $\mathbb{U}_{\mathscr{SK}} = \mathfrak{sl}(2,\mathbb{R})$ the only available nilpotent operator is L_+ and from the general formula (3.7.22) applied to the case where the metric is given by (6.3.1) and the complex coordinate is parameterized as in Eq. (6.3.7) we find:

$$\mathscr{P}_{L_+} = \text{const} \times \exp[-h] = \text{const} \times (\mathrm{Im}\, z)^{-1} \tag{7.7.33}$$

This result inserted into the general formula (7.5.19) yields

$$V(h) = \text{const} \times (\exp[-h] + \kappa)^2 \tag{7.7.34}$$

which is indeed the Starobinsky potential, since, once expressed in terms of h, the Kähler potential is exactly $\mathscr{K} = 3\,h$. The same result is directly obtained with precise coefficients by inserting in Eq. (7.7.32) the 7-dimensional image of L_+ in the fundamental representation of $\mathfrak{g}_{(2,2)}$.

7.8 The Sp(6, $\mathbb{R}$)/SU(3) x U(1) - Model and Its c-Map Image

Next we consider the Special Kähler manifold

$$\mathscr{M}_{\mathrm{Sp}\,6} = \frac{\mathrm{Sp}(6,\mathbb{R})}{\mathrm{SU}(3)\times\mathrm{U}(1)} \tag{7.8.1}$$

and its c-map image which is the following quaternionic manifold:

$$c\text{-map} \;:\; \mathscr{M}_{\mathrm{Sp}\,6} \mapsto \mathscr{QM}_{F4} \equiv \frac{\mathrm{F}_{(4,4)}}{\mathrm{SU}(2)\times\mathrm{USp}(6)} \tag{7.8.2}$$

$\mathscr{M}_{\mathrm{Sp}6}$ belongs to the magic square of exceptional special Kähler manifolds whose quaternionic c-map is a homogeneous symmetric space having, as it is evident from (7.8.2), an exceptional Lie group as isometry group.

We begin by illustrating some general properties of this remarkable manifold. First of all, in order to discuss them adequately we need to choose a basis for the $\mathfrak{sp}(6,\mathbb{R})$ Lie algebra. Since we are not interested in solving Lax equations we do not choose the basis where the matrices of the Borel subalgebra are upper triangular. We rather use the basis where the symplectic preserved metric is the standard one, namely:

$$\mathbb{C} = \begin{pmatrix} 0 & 0 & 0 & 1 & 0 & 0 \\ 0 & 0 & 0 & 0 & 1 & 0 \\ 0 & 0 & 0 & 0 & 0 & 1 \\ -1 & 0 & 0 & 0 & 0 & 0 \\ 0 & -1 & 0 & 0 & 0 & 0 \\ 0 & 0 & -1 & 0 & 0 & 0 \end{pmatrix} \tag{7.8.3}$$

This traditional choice allows to describe in a simple way other aspects of the manifold geometry that are more relevant to our present purposes.

According to the above choice, an element of the $\mathrm{Sp}(6,\mathbb{R})$ group and an element of the $\mathfrak{sp}(6,\mathbb{R})$ Lie-algebra are matrices respectively fulfilling the following two constraints:

$$\left(\begin{array}{c|c} A & B \\ \hline C & D \end{array}\right)^T \mathbb{C} \left(\begin{array}{c|c} A & B \\ \hline C & D \end{array}\right) = \mathbb{C} \quad ; \quad \left(\begin{array}{c|c} \mathbf{A} & \mathbf{B} \\ \hline \mathbf{C} & \mathbf{D} \end{array}\right)^T \mathbb{C} + \mathbb{C} \left(\begin{array}{c|c} \mathbf{A} & \mathbf{B} \\ \hline \mathbf{C} & \mathbf{D} \end{array}\right) = 0 \tag{7.8.4}$$

where $A, B, C, D, \mathbf{A}, \mathbf{B}, \mathbf{C}, \mathbf{D}$ are 3×3 blocks. By means of the so called Cayley transformation

$$\mathscr{C} = \begin{pmatrix} \frac{1}{\sqrt{2}} & 0 & 0 & \frac{i}{\sqrt{2}} & 0 & 0 \\ 0 & \frac{1}{\sqrt{2}} & 0 & 0 & \frac{i}{\sqrt{2}} & 0 \\ 0 & 0 & \frac{1}{\sqrt{2}} & 0 & 0 & \frac{i}{\sqrt{2}} \\ \frac{1}{\sqrt{2}} & 0 & 0 & -\frac{i}{\sqrt{2}} & 0 & 0 \\ 0 & \frac{1}{\sqrt{2}} & 0 & 0 & -\frac{i}{\sqrt{2}} & 0 \\ 0 & 0 & \frac{1}{\sqrt{2}} & 0 & 0 & -\frac{i}{\sqrt{2}} \end{pmatrix} \tag{7.8.5}$$

a real element of the symplectic group (or algebra) can be mapped into a matrix that is simultaneously symplectic and pseudounitary:

$$\mathscr{S} = \mathscr{C}^\dagger \left(\begin{array}{c|c} A & B \\ \hline C & D \end{array}\right) \mathscr{C} = \begin{pmatrix} U_0 & U_1^\star \\ U_1 & U_0^\star \end{pmatrix} \in \mathrm{Sp}(6,\mathbb{C}) \bigcap \mathrm{SU}(3,3) \tag{7.8.6}$$

The diagonal blocks $U_0 \in \mathrm{U}(3)$ span the H-subgroup of the coset (7.8.1). This allows to introduce a set of projective coordinates that parameterize the points of the manifold (7.8.1) and have a nice fractional linear transformation under the action of the group $\mathrm{Sp}(6,\mathbb{R})$. Given any coset parameterization

$$\left(\begin{array}{c|c} A(\phi) & B(\phi) \\ \hline C(\phi) & D(\phi) \end{array}\right) \in \mathrm{Sp}(6, \mathbb{R}) \tag{7.8.7}$$

namely a family of symplectic group elements depending on 12 parameters ϕ^i such that each different choice of the ϕ^i provides a representative of a different equivalence class in (7.8.1), we can construct the following, *symmetric complex matrix*:

$$Z(\phi) \equiv (A(\phi) - \mathrm{i}\,B(\phi))\,(C(\phi) - \mathrm{i}D(\phi))^{-1} \tag{7.8.8}$$

which has a very simple transformation under the action of the symplectic group. Let us consider the action of any element of Sp(6, ℝ) on the coset representative. We have:

$$\underbrace{\left(\begin{array}{c|c} \hat{A} & \hat{B} \\ \hline \hat{C} & \hat{D} \end{array}\right)}_{=\mathfrak{g}\in \mathrm{Sp}(6,\mathbb{R})} \left(\begin{array}{c|c} A(\phi) & B(\phi) \\ \hline C(\phi) & D(\phi) \end{array}\right) = \left(\begin{array}{c|c} A(\phi') & B(\phi') \\ \hline C(\phi') & D(\phi') \end{array}\right) \mathrm{H}(\phi, \mathfrak{g}) \tag{7.8.9}$$

where ϕ' is the label of a new equivalence class and $\mathrm{H}(\phi, \mathfrak{g}) \in \mathrm{U}(3)$ is a suitable H-compensator. Calculating the matrix $Z(\phi')$ according to the definition (7.8.8) we find that it is related to $Z(\phi)$ by a simple linear fractional transformation (generalized to matrices):

$$Z(\phi') = (AZ(\phi) + B)\,(CZ(\phi) + D)^{-1} \tag{7.8.10}$$

Formula (7.8.10) is of crucial relevance and requires several comments. From a mathematical point of view, (7.8.10) is the well known generalization of the action of the $\mathrm{SL}(2, \mathbb{R}) \simeq \mathrm{Sp}(2, \mathbb{R})$ group on the upper complex plane of Poincaré–Lobachevsky. The complex numbers z with positive imaginary parts ($\mathrm{Im} z > 0$) are replaced by the complex symmetric matrices Z_{ij} whose imaginary part is positive definite. Such matrices constitute the so named **upper Siegel plane**, which indeed is homeomorphic to the coset $\mathrm{Sp}(2n, \mathbb{R})/\mathrm{U(n)}$. From the physical point of view (7.8.10) is just identical to the Gaillard–Zumino formula for the construction of the kinetic matrix $\mathscr{N}_{\Lambda\Sigma}$ which appears in the lagrangian of the vector fields in $\mathscr{N} = 2$ supergravity and is rooted in the structure of special Kähler geometry. Indeed, as we know, for any special Kähler manifold $\mathscr{M}_n$ of complex dimension n that is also a symmetric space G/H, there exists a **W**-representation of G, which is symplectic, has dimension $2n + 2$ and hosts the electric and magnetic field strengths of the model. Such a representation defines a symplectic embedding:

$$\mathrm{G} \rightarrow \mathrm{Sp}\,(2n + 2, \mathbb{R}) \tag{7.8.11}$$

which associates to any coset representative $\mathfrak{g}(\phi) \in \mathrm{G/H}$ its corresponding symplectic $(2n + 2) \times (2n + 2)$ representation $\left(\begin{array}{c|c} A(\phi) & B(\phi) \\ \hline C(\phi) & D(\phi) \end{array}\right)$. From this latter, utilizing the recipe provided by formula (7.8.10) we obtain an $(n + 1) \times (n + 1)$ complex symmetric matrix to be identified with the appropriate $\mathscr{N}$ kinetic matrix largely discussed and utilized in Sect. 4.2.4.

The peculiarity of the $\mathscr{N} = 2$ model under investigation is that the original isometry group G is already symplectic so that we can utilize the Gaillard–Zumino formula (7.8.10) in the fundamental 6 dimensional representation in order to construct a Siegel parametrization of the coset in terms of a symmetric complex 3×3 matrix Z. The **W**-representation is the $\mathbf{14}'$ and this defines the embedding:

$$\mathrm{Sp}(6,\mathbb{R}) \mapsto \mathrm{Sp}(14,\mathbb{R}) \tag{7.8.12}$$

from which we can construct the 7×7 kinetic matrix $\mathscr{N}(\mathrm{Z})$.

7.8.1 *The Transitive Action of* Sp(6, ℝ) *on the Upper Siegel Plane*

Before proceeding with the actual construction of the Lie algebra let us comment on the transitive action of the symplectic group on the Siegel plane. Focusing on the formula (7.8.10), consider the Sp(6, ℝ) parabolic subgroup composed by the following matrices:

$$\mathfrak{g}(B) = \left(\begin{array}{c|c} \mathbf{1}_{3\times3} & B \\ \hline \mathbf{0}_{3\times3} & \mathbf{1}_{3\times3} \end{array}\right) \tag{7.8.13}$$

where B is symmetric and real. By means of such a subgroup we can always map a generic Z matrix into one that has vanishing real part $\mathrm{Re}Z = 0$. Next consider the action on the residual imaginary part of Z of the $\mathrm{GL}(3,\mathbb{R}) \subset \mathrm{Sp}(6,\mathbb{R})$ subgroup composed by the matrices:

$$\mathfrak{g}(B) = \left(\begin{array}{c|c} \mathscr{A} & \mathbf{0}_{3\times3} \\ \hline \mathbf{0}_{3\times3} & \left(\mathscr{A}^T\right)^{-1} \end{array}\right) \quad ; \quad \mathscr{A} \in \mathrm{GL}(3,\mathbb{R}) \tag{7.8.14}$$

We obtain:

$$\mathrm{ImZ} \mapsto \mathscr{A}\, \mathrm{ImZ}\, \mathscr{A}^T \tag{7.8.15}$$

Choosing $\mathscr{A} = (\mathrm{ImZ})^{\frac{1}{2}}$, which is always possible since $\mathrm{Im}Z$ is positive definite we can reduce the imaginary part to the identity matrix. This shows the transitive action of the symplectic group on the Siegel plane and also provides a nice coset parameterization of the coset manifold. Indeed we can introduce the following matrix:

$$\mathfrak{g}(Z) \equiv \left(\begin{array}{c|c} (\mathrm{ImZ})^{\frac{1}{2}} & \mathrm{ReZ}\ (\mathrm{ImZ})^{-\frac{1}{2}} \\ \hline \mathbf{0} & (\mathrm{ImZ})^{-\frac{1}{2}} \end{array}\right) \tag{7.8.16}$$

which maps the origin of the manifold $\mathbf{i1}_{3\times3}$ in the complex symmetric matrix Z.

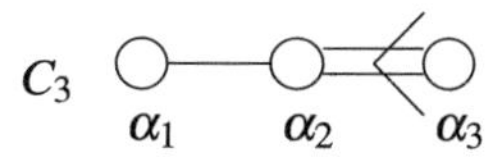

Fig. 7.6 The Dynkin diagram of C_3

7.8.2 *The* $\mathfrak{sp}(6, \mathbb{R})$ *Lie Algebra*

From the point of view of the Dynkin classification the Lie algebra $\mathfrak{sp}(6, \mathbb{R})$ is the maximally split real section of the complex Lie algebra C_3 whose Dynkin diagram is displayed in Fig. 7.6. The root system is composed of 18-roots whose subset of 9 positive ones is displayed here below:

$$\begin{bmatrix} \alpha_1 \\ \alpha_2 \\ \alpha_3 \\ \alpha_4 \\ \alpha_5 \\ \alpha_6 \\ \alpha_7 \\ \alpha_8 \\ \alpha_9 \end{bmatrix} = \begin{bmatrix} \alpha_1 & \{1,-1,0\} \\ \alpha_2 & \{0,1,-1\} \\ \alpha_3 & \{0,0,2\} \\ \alpha_1+\alpha_2 & \{1,0,-1\} \\ \alpha_2+\alpha_3 & \{0,1,1\} \\ \alpha_1+\alpha_2+\alpha_3 & \{1,0,1\} \\ 2\alpha_2+\alpha_3 & \{0,2,0\} \\ \alpha_1+2\alpha_2+\alpha_3 & \{1,1,0\} \\ 2\alpha_1+2\alpha_2+\alpha_3 & \{2,0,0\} \end{bmatrix} \tag{7.8.17}$$

The simple roots are the first three. Of the remaining 6 we have provided both their expression in terms of the simple roots and their realization as three-vectors in $\mathbb{R}^3$. Such a realization is spelled out also for the simple roots. Next we present the basis of 6×6 matrices that fulfill the standard commutation relations of the Lie Algebra in the Cartan Weyl basis.

7.8.2.1 Cartan Generators

The Cartan generators are named $\mathscr{H}^i$ and can be easily read-off from the following formula:

$$\sum_{i=1}^{3} h_i\, \mathscr{H}^i = \begin{pmatrix} h_1 & 0 & 0 & 0 & 0 & 0 \\ 0 & h_2 & 0 & 0 & 0 & 0 \\ 0 & 0 & h_3 & 0 & 0 & 0 \\ 0 & 0 & 0 & -h_1 & 0 & 0 \\ 0 & 0 & 0 & 0 & -h_2 & 0 \\ 0 & 0 & 0 & 0 & 0 & -h_3 \end{pmatrix} \tag{7.8.18}$$

by collecting the coefficient of the parameter h_i.

7.8.2.2 Positive Root Step Operators

The step operator associated with the positive root α_i is named $\mathscr{E}^{\alpha_i}$ and can be easily read-off from the following formula:

$$\sum_{i=1}^{9} a_i\, \mathscr{E}^{\alpha_i} = \begin{pmatrix} 0 & a_1 & a_4 & \sqrt{2}a_9 & a_8 & a_6 \\ 0 & 0 & a_2 & a_8 & \sqrt{2}a_7 & a_5 \\ 0 & 0 & 0 & a_6 & a_5 & \sqrt{2}a_3 \\ 0 & 0 & 0 & 0 & 0 & 0 \\ 0 & 0 & 0 & -a_1 & 0 & 0 \\ 0 & 0 & 0 & -a_4 & -a_2 & 0 \end{pmatrix} \tag{7.8.19}$$

by collecting the coefficient of the parameter a_i.

7.8.2.3 Negative Root Step Operators

The step operator associated with the negative root $-\alpha_i$ is named $\mathscr{E}^{-\alpha_i}$ and can be easily read-off from the following formula:

$$\sum_{i=1}^{9} b_i\, \mathscr{E}^{-\alpha_i} = \begin{pmatrix} 0 & 0 & 0 & 0 & 0 & 0 \\ b_1 & 0 & 0 & 0 & 0 & 0 \\ b_4 & b_2 & 0 & 0 & 0 & 0 \\ \sqrt{2}b_9 & b_8 & b_6 & 0 & -b_1 & -b_4 \\ b_8 & \sqrt{2}b_7 & b_5 & 0 & 0 & -b_2 \\ b_6 & b_5 & \sqrt{2}b_3 & 0 & 0 & 0 \end{pmatrix} \tag{7.8.20}$$

by collecting the coefficient of the parameter b_i.

7.8.3 *The Representation* **14′**

The **14′** representation of $\mathfrak{sp}(6, \mathbb{R})$ which plays the role of **W**-representation for the special manifold under consideration is defined as the representation obeyed by the three-times antisymmetric tensors with vanishing $\mathbb{C}$-traces, namely:

$$\underbrace{t_{ABC}}_{\text{antisymmetric in } A,B,C} \times \mathbb{C}^{BC} = 0 \tag{7.8.21}$$

The generators are constructed in the appendix, Sect. 7.12.2, and displayed in Eqs. (7.12.11)–(7.12.13).

7.8.4 The Holomorphic Symplectic Section and Its Transformation in the $\mathbf{14'}$

In order to construct the special geometry of the manifold (7.8.1) we need to introduce the holomorphic symplectic section that, by definition, should transform in the $14'$ representation of Sp(6, ℝ). To this effect, we choose as special coordinates the components of the symmetric complex matrix defined by Eq. (7.8.10) and we choose a lexicographic order to enumerate its independent components, namely we set:

$$Z = \begin{pmatrix} z_1 & z_2 & z_3 \\ z_2 & z_4 & z_5 \\ z_3 & z_5 & z_6 \end{pmatrix} \tag{7.8.22}$$

Next we introduce the holomorphic prepotential defined by:

$$\begin{aligned} \mathscr{F} &\equiv Z_{a,i}\, Z_{b,j}\, Z_{c,k}\, \varepsilon^{abc}\, \varepsilon^{ijk} \\ &= -6\left(z_6 z_2^2 - 2 z_3 z_5 z_2 + z_3^2 z_4 + z_1\left(z_5^2 - z_4 z_6\right)\right) \end{aligned} \tag{7.8.23}$$

and we can introduce a first ansatz for the symplectic section by writing:

$$\begin{aligned} \tilde{\Omega} &= \left\{1,\, z^I,\, \mathscr{F},\, \frac{\partial \mathscr{F}}{\partial z^J}\right\} \\ &= \left\{1, z_1, z_2, z_3, z_4, z_5, z_6, -6\left(z_6 z_2^2 - 2 z_3 z_5 z_2 + z_3^2 z_4 + z_1\left(z_5^2 - z_4 z_6\right)\right),\right. \\ &\quad 6 z_4 z_6 - 6 z_5^2,\, 12\left(z_3 z_5 - z_2 z_6\right),\, 12\left(z_2 z_5 - z_3 z_4\right), \\ &\quad \left. 6 z_1 z_6 - 6 z_3^2,\, 12\left(z_2 z_3 - z_1 z_5\right),\, 6 z_1 z_4 - 6 z_2^2\right\} \end{aligned} \tag{7.8.24}$$

In order to match the transformation of this holomorphic section with the transformations of the $\mathbf{14'}$ representation as we defined it in Sect. 7.8.4 we still need a change of basis. Consider the following matrix

$$\mathfrak{S} = \begin{pmatrix}
0 & 0 & 0 & 0 & 0 & 0 & \sqrt{2} & 0 & 0 & 0 & 0 & 0 & 0 & 0 \\
0 & 0 & 0 & 0 & 0 & 0 & 0 & 0 & 0 & 0 & 0 & \frac{1}{3\sqrt{2}} & 0 & 0 \\
0 & 0 & 0 & 0 & 0 & 0 & 0 & 0 & -\frac{1}{3\sqrt{2}} & 0 & 0 & 0 & 0 & 0 \\
0 & 0 & 0 & 0 & 0 & 0 & 0 & \frac{1}{3\sqrt{2}} & 0 & 0 & 0 & 0 & 0 & 0 \\
0 & 0 & 0 & 0 & 0 & 0 & 0 & 0 & 0 & 0 & 0 & 0 & \frac{1}{6} & 0 \\
0 & 0 & 0 & 0 & 0 & 0 & 0 & 0 & 0 & 0 & -\frac{1}{6} & 0 & 0 & 0 \\
0 & 0 & 0 & 0 & 0 & 0 & 0 & 0 & 0 & \frac{1}{6} & 0 & 0 & 0 & 0 \\
0 & 0 & 0 & 0 & 0 & 0 & 0 & 0 & 0 & 0 & 0 & 0 & 0 & \frac{1}{3\sqrt{2}} \\
0 & 0 & 0 & 0 & -\sqrt{2} & 0 & 0 & 0 & 0 & 0 & 0 & 0 & 0 & 0 \\
0 & \sqrt{2} & 0 & 0 & 0 & 0 & 0 & 0 & 0 & 0 & 0 & 0 & 0 & 0 \\
\sqrt{2} & 0 & 0 & 0 & 0 & 0 & 0 & 0 & 0 & 0 & 0 & 0 & 0 & 0 \\
0 & 0 & 0 & 0 & 0 & -2 & 0 & 0 & 0 & 0 & 0 & 0 & 0 & 0 \\
0 & 0 & 0 & 2 & 0 & 0 & 0 & 0 & 0 & 0 & 0 & 0 & 0 & 0 \\
0 & 0 & -2 & 0 & 0 & 0 & 0 & 0 & 0 & 0 & 0 & 0 & 0 & 0
\end{pmatrix} \tag{7.8.25}$$

and define:

$$\Omega\,(Z) \,=\, \mathfrak{S}\,\tilde{\Omega}(Z) \,=\, \begin{pmatrix} \sqrt{2}z_6 \\ \sqrt{2}\left(z_1 z_6 - z_3^2\right) \\ \sqrt{2}\left(z_5^2 - z_4 z_6\right) \\ -\sqrt{2}\left(z_6 z_2^2 - 2z_3 z_5 z_2 + z_3^2 z_4 + z_1\left(z_5^2 - z_4 z_6\right)\right) \\ 2z_2 z_3 - 2z_1 z_5 \\ 2z_3 z_4 - 2z_2 z_5 \\ 2z_3 z_5 - 2z_2 z_6 \\ \sqrt{2}\left(z_1 z_4 - z_2^2\right) \\ -\sqrt{2}z_4 \\ \sqrt{2}z_1 \\ \sqrt{2} \\ -2z_5 \\ 2z_3 \\ -2z_2 \end{pmatrix} \tag{7.8.26}$$

Naming $\mathscr{D}_{14}\,[g]$ the 14-dimensional representation of a finite element $g, \in \mathrm{Sp}(6, \mathbb{R})$ of the symplectic group that corresponds to the representation of the algebra as we constructed it above, the holomorphic symplectic section (7.8.26) transforms in the following way:

$$\Omega\left[(A\,Z \,+\, B)\,(C\,Z \,+\, D)^{-1}\right] \,=\, \frac{1}{\mathrm{Det}\,(C\,Z \,+\, D)}\,\mathscr{D}_{14}\left[\left(\begin{array}{c|c} A & B \\ \hline C & D \end{array}\right)\right]\,\Omega[Z] \tag{7.8.27}$$

The formula (7.8.27) can be in particular applied to the case where the original Z is the origin of the coset manifold: $Z_0 \,=\, \mathrm{i}\,\mathbf{1}_{3\times 3}$. In that case, recalling Eq. (7.8.16) we find:

$$\Omega[Z_0] \,=\, \left\{i\sqrt{2}, -\sqrt{2}, \sqrt{2}, -i\sqrt{2}, 0, 0, 0, -\sqrt{2}, -i\sqrt{2}, i\sqrt{2}, \sqrt{2}, 0, 0, 0\right\} \tag{7.8.28}$$

and

$$\Omega[Z_0] \,=\, \sqrt{\mathrm{Det}\,[\mathrm{Im}\,Z]} \,\times\, \mathscr{D}_{14}\left[\left(\begin{array}{c|c} (\mathrm{Im Z})^{\frac{1}{2}} & \mathrm{Re Z}\,\,(\mathrm{Im Z})^{-\frac{1}{2}} \\ \hline \mathbf{0} & (\mathrm{Im Z})^{-\frac{1}{2}} \end{array}\right)\right] \cdot\, \Omega[Z_0] \tag{7.8.29}$$

7.8.5 The Kähler Potential and the Metric

Provided with this information we can now write the explicit form of the Kähler potential and of the Kähler metric for the manifold (7.8.1) according to the rules of special Kähler geometry. We have:

$$\begin{aligned}
\mathscr{K} \equiv & -\log\left(\mathrm{i}\Omega[Z]\,\mathbb{C}_{14}\,\overline{\Omega}[\overline{Z}]\right)\\
= & -\log\Big(2i\Big(-z_6 z_2^2+\bar{z}_6 z_2^2+2z_6\bar{z}_2 z_2-2z_5\bar{z}_3 z_2+2\bar{z}_3\bar{z}_5 z_2-2\bar{z}_2\bar{z}_6 z_2-z_6\bar{z}_2^2-z_4\bar{z}_3^2+\bar{z}_1\bar{z}_5^2+\\
& z_5^2\bar{z}_1-z_4 z_6\bar{z}_1+2z_5\bar{z}_2\bar{z}_3+\bar{z}_3^2\bar{z}_4+z_6\bar{z}_1\bar{z}_4+z_3^2\left(\bar{z}_4-z_4\right)-2z_5\bar{z}_1\bar{z}_5-2\bar{z}_2\bar{z}_3\bar{z}_5\\
& +2z_3\left(-z_5\bar{z}_2+\bar{z}_5\bar{z}_2+z_4\bar{z}_3-\bar{z}_3\bar{z}_4+z_2\left(z_5-\bar{z}_5\right)\right)+\bar{z}_2^2\bar{z}_6+z_4\bar{z}_1\bar{z}_6-\bar{z}_1\bar{z}_4\bar{z}_6\\
& -z_1\left(z_5^2-2\bar{z}_5 z_5+\bar{z}_5^2+z_6\bar{z}_4-\bar{z}_4\bar{z}_6+z_4\left(\bar{z}_6-z_6\right)\right)\Big)\Big)
\end{aligned} \tag{7.8.30}$$

and the line element on the manifold, in terms of the special coordinates z_i takes the standard form:

$$ds_K^2 = \frac{\partial}{\partial z^i}\frac{\partial}{\partial \bar{z}^j}\mathscr{K}\,dz^i \otimes d\bar{z}^j \tag{7.8.31}$$

The explicit form of ds_K^2 in terms of the special coordinate z^i can be worked out by simple derivatives, yet its explicit form is quite lengthy and so much involved that we think it better not to display it. For the purposes that we pursue we rather prefer to write the form of the metric in terms of solvable real coordinates.

7.8.5.1 The Solvable Parametrization

The transition to a solvable parametrization of the coset is rather simple. Let us define the solvable coset representative as the product of the exponentials of all the generators of the Borel subalgebra of $\mathfrak{sp}(6,\mathbb{R})$:

$$\mathbb{L}(h,p) = \prod_{i=1}^{9}\exp\left[p_{10-i}\,\mathscr{E}^{\alpha_{10-i}}\right]\prod_{j=3}^{3}\exp\left[h_j\,\mathscr{H}^j\right] =$$

$$\begin{pmatrix}
e^{h_1} & e^{h_2}p_1 & e^{h_3}p_4 & e^{-h_1}\left(\sqrt{2}p_1p_2p_3p_4+(p_1p_2-p_4)p_6-p_1p_8+\sqrt{2}p_9\right) & e^{-h_2}\left(-\sqrt{2}p_2p_3p_4-p_2p_6+p_8\right) & e^{-h_3}\left(\sqrt{2}p_3p_4+p_6\right)\\
0 & e^{h_2} & e^{h_3}p_2 & e^{-h_1}\left((p_1p_2-p_4)p_5-\sqrt{2}p_1p_7+p_8\right) & e^{-h_2}\left(\sqrt{2}p_7-p_2p_5\right) & e^{-h_3}p_5\\
0 & 0 & e^{h_3} & e^{-h_1}\left(\sqrt{2}p_1p_2p_3-p_1p_5+p_6\right) & e^{-h_2}\left(p_5-\sqrt{2}p_2p_3\right) & \sqrt{2}e^{-h_3}p_3\\
0 & 0 & 0 & e^{-h_1} & 0 & 0\\
0 & 0 & 0 & -e^{-h_1}p_1 & e^{-h_2} & 0\\
0 & 0 & 0 & e^{-h_1}(p_1p_2-p_4) & -e^{-h_2}p_2 & e^{-h_3}
\end{pmatrix} \tag{7.8.32}$$

The real coordinates of the manifold are now the 12 parameters:

$$\text{coordinates} \equiv \{h_1,\ldots,h_3,\ p_1,\ \ldots\ p_9\} \tag{7.8.33}$$

Extracting the complex matrix Z from the symplectic matrix $\mathbb{L}(h,p)$ we find:

$$Z(h,p) =$$
$$\begin{pmatrix} ie^{2h_2}p_1^2 + ie^{2h_1} + \left(\sqrt{2}p_3 + ie^{2h_3}\right)p_4^2 + \sqrt{2}p_9 & ie^{2h_2}p_1 + ie^{2h_3}p_2p_4 + p_8 & \left(\sqrt{2}p_3 + ie^{2h_3}\right)p_4 + p_6 \\ ie^{2h_2}p_1 + ie^{2h_3}p_2p_4 + p_8 & ie^{2h_3}p_2^2 + ie^{2h_2} + \sqrt{2}p_7 & ie^{2h_3}p_2 + p_5 \\ \left(\sqrt{2}p_3 + ie^{2h_3}\right)p_4 + p_6 & ie^{2h_3}p_2 + p_5 & \sqrt{2}p_3 + ie^{2h_3} \end{pmatrix} \tag{7.8.34}$$

which defines the coordinate transformation from the special to the solvable coordinates:

$$\begin{pmatrix} z_1 \\ z_2 \\ z_3 \\ z_4 \\ z_5 \\ z_6 \end{pmatrix} = \begin{pmatrix} \sqrt{2}p_3p_4^2 + i\left(e^{2h_2}p_1^2 + e^{2h_1} + e^{2h_3}p_4^2\right) + \sqrt{2}p_9 \\ i\left(e^{2h_2}p_1 + e^{2h_3}p_2p_4\right) + p_8 \\ ie^{2h_3}p_4 + \sqrt{2}p_3p_4 + p_6 \\ i\left(e^{2h_3}p_2^2 + e^{2h_2}\right) + \sqrt{2}p_7 \\ ie^{2h_3}p_2 + p_5 \\ \sqrt{2}p_3 + ie^{2h_3} \end{pmatrix} \tag{7.8.35}$$

Inserting such a coordinate transformation into the Kähler metric (7.8.31) we obtain its form in terms of the real coordinates (7.8.33). For the explicit form of the metric, we refer the reader to the appendix, Eq. (3.8.80). The complete metric is quite formidable (3.8.80) since it contains a total of 100 terms. It has however quite simple properties when we sit in the neighborhood of the coset origin, in particular at $p_i \sim 0$. In this case it drastically simplifies and becomes diagonal:

$$\begin{aligned} ds_K^2 \overset{p_i \to 0}{\Longrightarrow} & \, \mathrm{dh}_1^2 + \mathrm{dh}_2^2 + \mathrm{dh}_3^2 + \frac{1}{2}e^{2h_2-2h_1}\mathrm{dp}_1^2 + \frac{1}{2}e^{2h_3-2h_2}\mathrm{dp}_2^2 + \frac{1}{2}e^{-4h_3}\mathrm{dp}_3^2 + \frac{1}{2}e^{2h_3-2h_1}\mathrm{dp}_4^2 \\ & + \frac{1}{2}e^{-2h_2-2h_3}\mathrm{dp}_5^2 + \frac{1}{2}e^{-2h_1-2h_3}\mathrm{dp}_6^2 + \frac{1}{2}e^{-4h_2}\mathrm{dp}_7^2 + \frac{1}{2}e^{-2h_1-2h_2}\mathrm{dp}_8^2 + \frac{1}{2}e^{-4h_1}\mathrm{dp}_9^2 \end{aligned} \tag{7.8.36}$$

which shows that it is positive definite as it should be. It is also interesting to note that if the truncation to the Cartan's fields is permitted by the potential, then we just have three dilatons with canonical kinetic terms.

7.8.6 *The Quartic Invariant in the* $\mathbf{14'}$

Of crucial relevance for the analysis of Black Hole charges and in general for the classification of orbits in the **W**-representation is the quartic symplectic invariant. Given a 14-vector

$$\mathcal{Q} = \{q_1, q_2 \ldots, q_{14}\} \tag{7.8.37}$$

the standard form of this invariant can be expressed in the following manifestly $\mathrm{Sp}(6,\mathbb{R})$-invariant form

$$\mathfrak{I}_4(\mathcal{Q}) = -\frac{n_V(2n_V+1)}{6d}(\Lambda_a)_{\alpha\beta}(\Lambda^a)_{\gamma\delta}\,\mathcal{Q}^\alpha\,\mathcal{Q}^\beta\,\mathcal{Q}^\gamma\,\mathcal{Q}^\delta\,, \tag{7.8.38}$$

where in our case $n_V = 7$ and $d = \text{dimSp}(6, \mathbb{R}) = 21$, the symplectic indices are raised and lowered by $\mathbb{C}_{14}^{\alpha\beta}$ and $\mathbb{C}_{14\,\alpha\beta}$ and the index a is raised by the inverse of $\eta_{ab} \equiv \text{Tr}(\Lambda_a\, \Lambda_b)$. The explicit form of $\mathfrak{J}_4(\mathcal{Q})$ reads:

$$
\begin{aligned}
\mathfrak{J}_4(\mathcal{Q}) = & -2q_1q_9q_5^2 + 2q_3q_{11}q_5^2 - 2\sqrt{2}q_6q_7q_{11}q_5 - 2q_1q_8q_{12}q_5 + 2q_2q_9q_{12}q_5 - 2q_3q_{10}q_{12}q_5 \\
& +2q_4q_{11}q_{12}q_5 - 2\sqrt{2}q_7q_9q_{13}q_5 + 2\sqrt{2}q_1q_6q_{14}q_5 + 2\sqrt{2}q_3q_{13}q_{14}q_5 + q_1^2q_8^2 \\
& +q_2^2q_9^2 + q_3^2q_{10}^2 + q_4^2q_{11}^2 + 2q_2q_8q_{12}^2 - 2q_4q_{10}q_{12}^2 - 2q_3q_8q_{13}^2 + 2q_4q_9q_{13}^2 - 2q_2q_3q_{14}^2 \\
& -2q_1q_4q_{14}^2 + 2q_1q_2q_8q_9 + 2q_1q_6^2q_{10} + 2q_1q_3q_8q_{10} - 2q_7^2q_9q_{10} - 2q_2q_3q_9q_{10} \\
& -4q_1q_4q_9q_{10} - 2q_2q_6^2q_{11} - 2q_7^2q_8q_{11} - 4q_2q_3q_8q_{11} \\
& -2q_1q_4q_8q_{11} + 2q_2q_4q_9q_{11} + 2q_3q_4q_{10}q_{11} \\
& +2\sqrt{2}q_6q_7q_{10}q_{12} - 2q_1q_6q_8q_{13} - 2q_2q_6q_9q_{13} + 2q_3q_6q_{10}q_{13} \\
& +2q_4q_6q_{11}q_{13} - 2\sqrt{2}q_7q_8q_{12}q_{13} + 2q_1q_7q_8q_{14} \\
& +2q_2q_7q_9q_{14} + 2q_3q_7q_{10}q_{14} + 2q_4q_7q_{11}q_{14} - 2\sqrt{2}q_2q_6q_{12}q_{14} + 2\sqrt{2}q_4q_{12}q_{13}q_{14}
\end{aligned}
\tag{7.8.39}
$$

7.8.7 Truncation to the STU-Model

Next we analyze how the STU-model is embedded into the Sp(6, ℝ)-model. At the level of the special coordinates the truncation to the STU-model is very simply done. It suffices to set to zero the complex coordinates z_2, z_3, z_5 keeping only z_1, z_4, z_6 that can be identified with the fields S, T, U. When we do so the symplectic section reduces as follows:

$$
\Omega\left[Z\begin{pmatrix} z_1 & 0 & 0 \\ 0 & z_4 & 0 \\ 0 & 0 & z_6 \end{pmatrix}\right] = \begin{pmatrix} \sqrt{2}z_6 \\ \sqrt{2}z_1z_6 \\ -\sqrt{2}z_4z_6 \\ \sqrt{2}z_1z_4z_6 \\ 0 \\ 0 \\ 0 \\ \sqrt{2}z_1z_4 \\ -\sqrt{2}z_4 \\ \sqrt{2}z_1 \\ \sqrt{2} \\ 0 \\ 0 \\ 0 \end{pmatrix}
\tag{7.8.40}
$$

and the Kähler potential reduces to:

$$\mathscr{K} \rightarrow -\log\left[2i\,(z_1 - \overline{z}_1)\,(z_4 - \overline{z}_4)\,(z_6 - \overline{z}_6)\right] \tag{7.8.41}$$

which yields three copies of the Poincaré metric, one for each of the three $\frac{\mathrm{SL}(2,\mathbb{R})}{\mathrm{SO}(2)}$ submanifolds.

The result (7.8.40) is in agreement with the decomposition of the $\mathbf{14}'$ of $\mathfrak{sp}(6,\mathbb{R})$ with respect to the three subalgebras $\mathfrak{sl}(2)$:

$$\mathbf{14}' \stackrel{\mathfrak{sl}(2)\times\mathfrak{sl}(2)\times\mathfrak{sl}(2)}{\Longrightarrow} (\mathbf{2},\mathbf{2},\mathbf{2}) \oplus (\mathbf{2},\mathbf{1},\mathbf{1}) \oplus (\mathbf{1},\mathbf{2},\mathbf{1}) \oplus (\mathbf{1},\mathbf{1},\mathbf{2}) \tag{7.8.42}$$

From (7.8.40) we also learn that the directions $\{1, 2, 3, 4, 8, 9, 10, 11\}$ of the $\mathbf{14}'$ vector space span the representation $(\mathbf{2},\mathbf{2},\mathbf{2})$, while the directions $\{5, 6, 7, 12, 13, 14\}$ of the same space span the representations $(\mathbf{2},\mathbf{1},\mathbf{1}) \oplus (\mathbf{1},\mathbf{2},\mathbf{1}) \oplus (\mathbf{1},\mathbf{1},\mathbf{2})$. The adjoint representation of $\mathfrak{sp}(6,\mathbb{R})$ decomposes instead in the following way:

$$\begin{aligned} \mathrm{adj}\,[\mathfrak{sp}(6,\mathbb{R})] \stackrel{\mathfrak{sl}(2)\times\mathfrak{sl}(2)\times\mathfrak{sl}(2)}{\Longrightarrow} & (\mathbf{3},\mathbf{1},\mathbf{1}) \oplus (\mathbf{1},\mathbf{3},\mathbf{1}) \oplus (\mathbf{1},\mathbf{1},\mathbf{3}) \\ & \oplus (\mathbf{2},\mathbf{2},\mathbf{1}) \oplus (\mathbf{2},\mathbf{1},\mathbf{2}) \oplus (\mathbf{1},\mathbf{2},\mathbf{2}) \end{aligned} \tag{7.8.43}$$

as it is evident by a quick inspection of the roots (7.8.17). In terms of the Cartan–Weyl basis the three $\mathfrak{sl}(2,\mathbb{R})$ subalgebra contains the three Cartan generators $\mathscr{H}_i$ and the step operators $\mathscr{E}^{\pm\alpha_3}$, $\mathscr{E}^{\pm\alpha_7}$, $\mathscr{E}^{\pm\alpha_9}$. The remaining 12 step operators span the representation $(\mathbf{2},\mathbf{2},\mathbf{1}) \oplus (\mathbf{2},\mathbf{1},\mathbf{2}) \oplus (\mathbf{1},\mathbf{2},\mathbf{2})$, namely:

$$(\mathbf{2},\mathbf{2},\mathbf{1}) \oplus (\mathbf{2},\mathbf{1},\mathbf{2}) \oplus (\mathbf{1},\mathbf{2},\mathbf{2}) = \mathrm{span}\left[\mathscr{E}^{\pm\alpha_1}, \mathscr{E}^{\pm\alpha_2}, \mathscr{E}^{\pm\alpha_4}, \mathscr{E}^{\pm\alpha_5}, \mathscr{E}^{\pm\alpha_6}, \mathscr{E}^{\pm\alpha_8}\right] \tag{7.8.44}$$

The explicit form of an $\mathfrak{sp}(6,\mathbb{R})$ Lie algebra element reduced to the $\mathfrak{sl}(2)^3$ subalgebra is the following one:

$$\begin{pmatrix} h_1 & 0 & 0 & b_1 & 0 & 0 \\ 0 & h_2 & 0 & 0 & b_2 & 0 \\ 0 & 0 & h_3 & 0 & 0 & b_3 \\ c_1 & 0 & 0 & -h_1 & 0 & 0 \\ 0 & c_2 & 0 & 0 & -h_2 & 0 \\ 0 & 0 & c_3 & 0 & 0 & -h_3 \end{pmatrix} \in \mathfrak{sl}(2) \otimes \mathfrak{sl}(2) \otimes \mathfrak{sl}(2) \subset \mathfrak{sp}(6,\mathbb{R}) \tag{7.8.45}$$

7.8.8 Reduction of the Charge Vector to the (2, 2, 2)

In order to study the orbits of the charge vectors in the $\mathbf{14}'$ our first step consists of reducing it to normal form, namely to the (2, 2, 2) representation. We claim that for generic charge vectors this is always possible by means of $\mathrm{Sp}(6,\mathbb{R})$ rotations generated by elements of the $(\mathbf{2},\mathbf{2},\mathbf{1}) \oplus (\mathbf{2},\mathbf{1},\mathbf{2}) \oplus (\mathbf{1},\mathbf{2},\mathbf{2})$ subspace. To show this let us consider the six dimensional compact Lie algebra element:

$$\begin{aligned}\mathbb{K}_\psi = \psi_1\left(\mathcal{E}^{\alpha_1}-\mathcal{E}^{-\alpha_1}\right)+\psi_2\left(\mathcal{E}^{\alpha_2}-\mathcal{E}^{-\alpha_2}\right)+\psi_3\left(\mathcal{E}^{\alpha_4}-\mathcal{E}^{-\alpha_4}\right)\\ \psi_4\left(\mathcal{E}^{\alpha_5}-\mathcal{E}^{-\alpha_5}\right)+\psi_5\left(\mathcal{E}^{\alpha_6}-\mathcal{E}^{-\alpha_6}\right)+\psi_6\left(\mathcal{E}^{\alpha_8}-\mathcal{E}^{-\alpha_8}\right)\end{aligned} \quad (7.8.46)$$

and a generic charge vector that has components only in the $(\mathbf{2},\mathbf{2},\mathbf{2})$ subspace.

$$\mathcal{Q}_{2,2,2} = \{\Theta_1,\Theta_2,\Theta_3,\Theta_4,0,0,0,\Theta_5,\Theta_6,\Theta_7,\Theta_8,0,0,0\} \quad (7.8.47)$$

If we apply the $\mathbf{14}'$ representation of $\mathbb{K}_\psi$ to the charge vector $\mathcal{Q}_N$ we obtain:

$$\mathcal{D}_{14}\left(\mathbb{K}_\psi\right)\mathcal{Q}_{2,2,2} = \begin{pmatrix} 0\\ 0\\ 0\\ 0\\ -\sqrt{2}\Theta_2\psi_2+\sqrt{2}\Theta_5\psi_2-\sqrt{2}\Theta_4\psi_4-\sqrt{2}\Theta_7\psi_4\\ -\sqrt{2}\Theta_3\psi_3-\sqrt{2}\Theta_5\psi_3+\sqrt{2}\Theta_4\psi_5-\sqrt{2}\Theta_6\psi_5\\ \sqrt{2}\Theta_2\psi_1+\sqrt{2}\Theta_3\psi_1-\sqrt{2}\Theta_1\psi_6-\sqrt{2}\Theta_4\psi_6\\ 0\\ 0\\ 0\\ 0\\ -\sqrt{2}\Theta_1\psi_2-\sqrt{2}\Theta_6\psi_2+\sqrt{2}\Theta_3\psi_4-\sqrt{2}\Theta_8\psi_4\\ \sqrt{2}\Theta_1\psi_3-\sqrt{2}\Theta_7\psi_3+\sqrt{2}\Theta_2\psi_5+\sqrt{2}\Theta_8\psi_5\\ \sqrt{2}\Theta_6\psi_1+\sqrt{2}\Theta_7\psi_1-\sqrt{2}\Theta_5\psi_6-\sqrt{2}\Theta_8\psi_6 \end{pmatrix} \quad (7.8.48)$$

which clearly shows that the six parameters $\psi_{1,\ldots,6}$ are sufficient to generate arbitrary components {5, 6, 7, 12, 13, 14} of the charge vector starting from vanishing ones. Reverting the path this means that by means of the same rotations, apart from singular orbits that deserve a separate study we can always fix the gauge where the six components {5, 6, 7, 12, 13, 14} vanish.

7.8.8.1 Further Reduction to Normal Form of the Charge Vector

Once the charge vector is reduced to the (2, 2, 2) representation, we can further act on it with the $\mathrm{SL}(2,\mathbb{R})^3$ group in order to further reduce its components. By using the three parameters of the abelian translation group $\mathbb{R}^3$ contained in $\mathrm{SL}(2,\mathbb{R})^3$ we can put to zero three of the eight charges and a possible normal form of the charge vector is the following one:

$$\mathcal{Q}_N = \{0, P_1, P_2, P_3, 0, 0, 0, P_4, 0, 0, P_5, 0, 0, 0\} \quad (7.8.49)$$

The corresponding quartic invariant is:

$$\mathfrak{J}_4\left(\mathcal{Q}_N\right) = P_3^2 P_5^2 - 4P_1P_2P_4P_5 \quad (7.8.50)$$

F_4

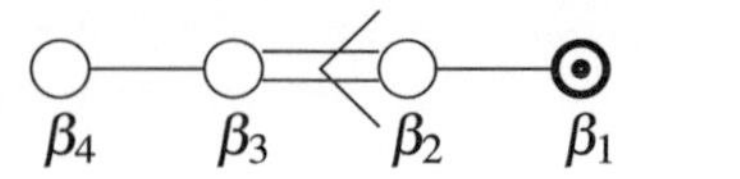

$$\psi = \beta_{24} = 2\beta_1 + 3\beta_2 + 4\beta_3 + 2\beta_4$$
$$(\psi, \beta_1) = 2 \quad ; \quad (\psi, \beta_i) = 0 \quad i \neq 1$$

Fig. 7.7 The Dynkin diagram of $F_{4(4)}$. The only root which is not orthogonal to the highest root is $\beta_V = \beta_1$. The root $\beta_V = \beta_1$ is the highest weight of the **W**-representation of $\mathfrak{sp}(6, \mathbb{R})$

7.9 The $\frac{\mathrm{F}_{(4,4)}}{\mathrm{SU(2)\times USp(6)}}$ Quaternionic Kähler Manifold

Let us now come to the c-map image of the Special Kähler manifold (7.8.1), namely to the quaternionic Kähler manifold (7.8.2). The $F_{(4,4)}$ Lie algebra has rank four and its structure is codified in the Dynkin diagram presented in Fig. 7.7. The complete set of positive roots contains 24 elements that were listed in Table 1.3. In that table the first column is the name of the root, the second column gives its decomposition in terms of simple roots, while the last column provides the component of the root vector in $\mathbb{R}^4$.

The standard Cartan–Weyl form of the Lie algebra is as follows:

$$\left[\mathscr{H}_i\,,\,E^{\pm\beta}\right] = \pm\,\beta^i\,E^{\pm\beta_I} \tag{7.9.1}$$

$$\left[E^{\beta}\,,\,E^{-\beta}\right] = \beta\,\cdot\,\mathscr{H} \tag{7.9.2}$$

$$\left[E^{\beta}\,,\,E^{\gamma}\right] = \begin{cases} N_{\beta\gamma}\,E^{\beta+\gamma} & \text{if } \beta+\gamma \text{ is a root} \\ 0 & \text{if } \beta+\gamma \text{ is not a root} \end{cases} \tag{7.9.3}$$

where $N_{\beta\gamma}$ are numbers that were constructed in Sect. 1.8 and displayed in Eqs. (1.8.8)–(1.8.10). All the rest of the construction can be easily presented in terms of these Weyl generators and this is what we presently do.

7.9.1 *The Maximal Compact Subalgebra* $\mathbb{H} = \mathfrak{su}(2) \oplus \mathfrak{usp}(6)$

The maximal compact subalgebra $\mathbb{H}$ of a maximally split simple Lie algebra such as $F_{(4,4)}$, is just the real span of all the independent compact generators $E^{\beta_i} - E^{-\beta_i}$. In our case we have 24 positive roots and we can write:

$$\mathbb{H} = \mathrm{span}_{\mathbb{R}}\,\{H_1\,,\,H_2\,,\,\ldots\,,\,H_{24}\} \tag{7.9.4}$$

where we have defined:

$$H_i = E^{\beta_i} - E^{-\beta_i} \tag{7.9.5}$$

the positive roots being numbered as in Table 1.3. We know from theory that this maximal compact subalgebra has the structure:

$$\mathbb{H} = \mathfrak{su}(2) \oplus \mathfrak{usp}(6) \tag{7.9.6}$$

It is important to derive an explicit basis of generators satisfying the standard commutation relations of the two simple factors in Eq. (7.9.6) for holonomy calculations of the coset manifold. Particularly important are the three generators J^x of the $\mathfrak{su}(2)$ subalgebra since they will act as quaternionic complex structures in the calculation of the tri-holomorphic moment map. By means of standard techniques of diagonalization of the adjoint action of generators we have retrieved the required basis rearrangement.

7.9.1.1 The $\mathfrak{su}(2)$ Lie Algebra

The three generators J^x have tho following explicit form:

$$\begin{aligned} J^1 &= \frac{H_1 - H_{14} + H_{20} - H_{22}}{4\sqrt{2}} \\ J^2 &= \frac{H_5 + H_{11} - H_{18} + H_{23}}{4\sqrt{2}} \\ J^3 &= -\frac{H_2 - H_9 + H_{16} + H_{24}}{4\sqrt{2}} \end{aligned} \tag{7.9.7}$$

and close the standard commutation relations:

$$\left[J^x \, , \, J^y\right] = \varepsilon^{xyz} \, J^y \tag{7.9.8}$$

7.9.1.2 The $\mathfrak{usp}(6)$ Lie Algebra

The 21 generators of the $\mathfrak{usp}(6)$ Lie algebra are given by the following combinations. First we have three mutually commuting generators (the compact Cartan generators):

$$\left[\mathscr{L}^i \, , \, \mathscr{L}^j\right] = 0 \tag{7.9.9}$$

that are given by the following combinations:

$$\begin{aligned} \mathscr{L}^1 &= -\tfrac{H_2}{2} - \tfrac{H_9}{2} + \tfrac{H_{16}}{2} - \tfrac{H_{24}}{2} \\ \mathscr{L}^2 &= -\tfrac{H_2}{2} + \tfrac{H_9}{2} + \tfrac{H_{16}}{2} + \tfrac{H_{24}}{2} \\ \mathscr{L}^3 &= \tfrac{H_2}{2} + \tfrac{H_9}{2} + \tfrac{H_{16}}{2} - \tfrac{H_{24}}{2} \end{aligned} \tag{7.9.10}$$

Secondly we have 9 pairs of generators $\{X_i\ ,\ Y_i\}$ which are in correspondence with the 9 positive roots of the $\mathfrak{sp}(6, C)$ Lie algebra (see Eq. (7.8.17)). Explicitly we have:

$$\begin{array}{ll} X_1 = H_{10} & ;\ Y_1 = H_7 \\ X_2 = H_4 & ;\ Y_2 = -H_{13} \\ X_3 = H_6 & ;\ Y_3 = -H_3 \\ X_4 = -H_1 + H_{14} + H_{20} - H_{22} & ;\ Y_4 = -H_5 - H_{11} - H_{18} + H_{23} \\ X_5 = H_{21} & ;\ Y_5 = -H_8 \\ X_6 = H_1 + H_{14} + H_{20} + H_{22} & ;\ Y_6 = H_5 - H_{11} - H_{18} - H_{23} \\ X_7 = -H_1 - H_{14} + H_{20} + H_{22} & ;\ Y_7 = H_5 - H_{11} + H_{18} + H_{23} \\ X_8 = H_{17} & ;\ Y_8 = H_{15} \\ X_9 = H_{12} & ;\ Y_9 = H_{19} \end{array} \tag{7.9.11}$$

The commutation relations with the compact Cartan generators are as follows:

$$\left[\mathscr{L}^i\ ,\ X_I\right] = \alpha^i_I\, Y_I \quad ; \quad \left[\mathscr{L}^i\ ,\ Y_I\right] = -\,\alpha^i_I\, X_I \tag{7.9.12}$$

where α_I are the roots of Eq. (7.8.17). The remaining commutation relations mix the Y and the X among themselves and reproduce the Cartan generators.

7.9.2 The Subalgebra $\mathfrak{sl}(2, \mathbb{R})_E \oplus \mathfrak{sp}(6, \mathbb{R})$ and the W-Generators

Of great relevance in all applications of the (pseudo)-quaternionic geometry either in the construction of Black-Hole solutions or in the quest of inflaton potentials by means of the gauging of hypermultiplet isometries is the identification of the subalgebra:

$$\mathfrak{sl}(2, \mathbb{R})_E \oplus \mathfrak{sp}(6, \mathbb{R}) \subset \mathfrak{f}_{(4,4)} \tag{7.9.13}$$

and the recasting of $\mathfrak{f}_{(4,4)}$ in the general form (1.7.10) by means of the identification of the **W**-generators.

To this effect a very powerful tool is provided by the comparison of the $\mathfrak{f}_{(4,4)}$ root system displayed in Eq. (1.3) with the $\mathfrak{sp}(6, \mathbb{R})$ root system displayed in Eq. (7.8.17). The step operators associated with the highest (lowest) root $\pm\beta_{24}$ are the only ones that have a grading ± 2 with respect to the fourth Cartan generator $\mathscr{H}_4$. These three operators close among themselves the Lie algebra $\mathfrak{sl}(2, \mathbb{R})_E$. There are 9 roots that have grading zero with respect to $\mathscr{H}_4$. Projected onto the plane $\mathscr{H}_4 = 0$ these 9 roots form, together with their negatives, a $\mathfrak{sp}(6, \mathbb{R})$ root system. Correspondingly the $\mathfrak{sp}(6, \mathbb{R})$ subalgebra is generated by the step operators associated with these 9 roots (and with their negatives) plus the first 3 Cartan generators. Finally there are 14 positive roots β that have grading 1 with respect to $\mathscr{H}_4$. The step operators associated with these 14 roots form the **W**-generators with index 1 of $\mathrm{SL}(2, \mathbb{R})_E$. Their partners with index 2 are provided by the corresponding negative root step operators.

It is quite important to arrange the generators **W** in such a way that under any element $\mathfrak{g} \in \mathfrak{sp}(6,\mathbb{R}) \subset \mathfrak{f}_{(4,4)}$ they transform exactly with the $\mathscr{D}_{14}(\mathfrak{g})$ matrices defined in Eqs. (7.12.11)–(7.12.13).

The precise definition of all the generators that satisfy the specified requirements is given below.

7.9.2.1 The Ehlers Subalgebra $\mathfrak{sl}(2,\mathbb{R})_E$

The standard commutation relations:

$$\left[L_0^E\,,\,L_\pm^E\right] = \pm\,L_\pm^E \tag{7.9.14}$$

$$\left[L_+^E\,,\,L_-^E\right] = 2\,L_0^E \tag{7.9.15}$$

are satisfied by the following generators:

$$\begin{aligned} L_0^E &= \frac{1}{2}\,\mathscr{H}_4 \\ L_+^E &= \frac{1}{\sqrt{2}}\,\mathscr{E}^{\beta_{24}} \\ L_-^E &= \frac{1}{\sqrt{2}}\,\mathscr{E}^{-\,\beta_{24}} \end{aligned} \tag{7.9.16}$$

7.9.2.2 The Subalgebra $\mathfrak{sp}(6,\mathbb{R})$

The Cartan generators are the following ones:

$$\begin{aligned} \mathscr{H}_1 &= \mathscr{H}_1 \\ \mathscr{H}_2 &= \mathscr{H}_2 \\ \mathscr{H}_3 &= \mathscr{H}_3 \end{aligned} \tag{7.9.17}$$

while the step operators are identified as follows

$$\begin{aligned} \mathscr{E}^{\pm\alpha_1} &= \mathscr{E}^{\pm\beta_4} \\ \mathscr{E}^{\pm\alpha_2} &= \mathscr{E}^{\pm\beta_3} \\ \mathscr{E}^{\pm\alpha_3} &= \mathscr{E}^{\pm\beta_2} \\ \mathscr{E}^{\pm\alpha_4} &= \mathscr{E}^{\pm\beta_7} \\ \mathscr{E}^{\pm\alpha_5} &= -\mathscr{E}^{\pm\beta_6} \\ \mathscr{E}^{\pm\alpha_6} &= \mathscr{E}^{\pm\beta_{10}} \\ \mathscr{E}^{\pm\alpha_7} &= -\mathscr{E}^{\pm\beta_9} \\ \mathscr{E}^{\pm\alpha_8} &= \mathscr{E}^{\pm\beta_{13}} \\ \mathscr{E}^{\pm\alpha_9} &= \mathscr{E}^{\pm\beta_{16}} \end{aligned} \tag{7.9.18}$$

We would like to attract the attention of the reader on the two minus signs introduced in the identifications (7.9.18). Together with the other minus signs that appear below in the identification of the **W**-generators these signs are essential in order for the transformations of the **W**.s to be identical with those given by the previously defined $\mathscr{D}_{14}(\mathfrak{g})$ matrices.

7.9.2.3 The W-Generators

Casting the $\mathfrak{f}_{(4,4)}$ Lie algebra in the general form (1.7.10) is completed by the identification of the **W**-generators. We find:

$$\begin{aligned}
\mathbf{W}^{1,1} &= \mathscr{E}^{\beta_5} \\
\mathbf{W}^{1,2} &= \mathscr{E}^{\beta_{20}} \\
\mathbf{W}^{1,3} &= \mathscr{E}^{\beta_{14}} \\
\mathbf{W}^{1,4} &= -\mathscr{E}^{\beta_{23}} \\
\mathbf{W}^{1,5} &= \mathscr{E}^{\beta_{21}} \\
\mathbf{W}^{1,6} &= \mathscr{E}^{\beta_{19}} \\
\mathbf{W}^{1,7} &= -\mathscr{E}^{\beta_{17}} \\
\mathbf{W}^{1,8} &= -\mathscr{E}^{\beta_{22}} \\
\mathbf{W}^{1,9} &= -\mathscr{E}^{\beta_{11}} \\
\mathbf{W}^{1,10} &= -\mathscr{E}^{\beta_{18}} \\
\mathbf{W}^{1,11} &= -\mathscr{E}^{\beta_{1}} \\
\mathbf{W}^{1,12} &= -\mathscr{E}^{\beta_{8}} \\
\mathbf{W}^{1,13} &= -\mathscr{E}^{\beta_{12}} \\
\mathbf{W}^{1,14} &= -\mathscr{E}^{\beta_{15}}
\end{aligned} \tag{7.9.19}$$

and for all $\mathfrak{g} \in \mathfrak{sp}(6, \mathbb{R}) \subset \mathfrak{f}_{(4,4)}$ we have:

$$\left[\mathfrak{g}\,,\,\mathbf{W}^{1,\alpha}\right] = \mathscr{D}_{14}(\mathfrak{g})^{\alpha}{}_{\gamma}\,\mathbf{W}^{1,\gamma} \tag{7.9.20}$$

The generators $\mathbf{W}^{2,\alpha}$ are then easily obtained from by means of a rotation with the unique compact generator of the Ehlers subalgebra introduced in Eq. (4.3.52):

$$\left[\mathfrak{S}\,,\,\mathbf{W}^{1,\alpha}\right] = \mathbf{W}^{2,\alpha} \tag{7.9.21}$$

7.9.3 The Solvable Coset Representative

The precise constructions of the previous sections enable us to introduce the solvable coset representative $\mathbb{L}_{Solv}\,(a, U, h, p, Z)$ of the manifold (7.8.2) such that the Maurer Cartan form:

$$\Xi \equiv \mathbb{L}_{Solv}^{-1}\,\mathrm{d}\mathbb{L}_{Solv} \tag{7.9.22}$$

decomposed along the generators of the Borel Lie algebra:

$$\begin{aligned}
\Xi &= E^I_{\mathcal{QM}}\, T_I \\
T_I &= \left\{ \underbrace{L^E_0\, ,\, L^E_+}_{2\hookrightarrow Solv[\mathfrak{sl}(2)]}\, ,\, \underbrace{\mathcal{H}^i\, ,\, \mathcal{E}^{\alpha_i}}_{12\hookrightarrow Solv[\mathfrak{sp}(6)]}\, ,\, \underbrace{\mathbf{W}^{1\alpha}}_{14\hookrightarrow \mathbb{H}\mathrm{eis}} \right\}
\end{aligned} \tag{7.9.23}$$

provides the vielbein $E^I_{\mathcal{QM}}$ mentioned in Eq. (4.3.44) and by squaring the metric (4.3.35).

In full analogy with Eqs. (7.7.27) and (7.8.32) we write:

$$\begin{aligned}
\mathbb{L}_{Solv} = \exp\left[a\, L^E_+\right] \cdot \exp\left[\sum_{j=1}^{7} \mathbf{Z}_{2j-1}\, \mathbf{W}^{1,2j-1}\right] \cdot \exp\left[\sum_{j=1}^{7} \mathbf{Z}_{2j}\, \mathbf{W}^{1,2j}\right] \times \\
\times \prod_{i=1}^{9} \exp\left[p_{10-i}\, \mathcal{E}^{\alpha_{10-i}}\right] \cdot \prod_{j=3}^{3} \exp\left[h_j\, \mathcal{H}^j\right] \cdot \exp\left[U\, L^E_0\right]
\end{aligned} \tag{7.9.24}$$

The explicit expression of $\mathbb{L}_{Solv}$ in the fundamental 26-dimensional representation is obviously very large but it can be dealt with by means of an appropriate MATHEMATICA code.

We are finally in the position of calculating the tri-holomorphic moment map of any element $\mathfrak{t} \in \mathfrak{f}_{(4,4)}$ of the isometry Lie algebra of $\mathcal{QM}$ through the formula:

$$\mathcal{P}^x_{\mathfrak{t}} = \mathrm{Tr}_{\mathbf{26}}\left(J^x\, \mathbb{L}^{-1}_{Solv}\, \mathfrak{t}\, \mathbb{L}_{Solv}\right) \tag{7.9.25}$$

7.9.4 The Example of the Inclusion of Multi Starobinsky Models

In Sect. 7.8.7 we studied the truncation of the $\mathfrak{sp}(6,\mathbb{R})$ model to the STU model. There we showed that setting to zero the three complex coordinates z_2, z_3, z_5, the remaining ones z_1, z_4, z_6 span the STU model, namely they parameterize three copies of the Lobachevsky–Poincaré hyperbolic plane. Inspecting Eq. (7.8.35) we also see that the three coordinates z_1, z_4, z_6 are the only ones that survive when all the axions p_i are set to zero. We also recall from Sect. 7.8.7 that the three parabolic generators of the three $SL(2,\mathbb{R})$ groups spanning the STU model are $\mathcal{E}^{\alpha_3}$, $\mathcal{E}^{\alpha_7}$, $\mathcal{E}^{\alpha_9}$ whose identification with $\mathfrak{f}_{(4,4)}$ generators is provided by Eq. (7.9.18). Correspondingly we introduce the following generator:

$$\mathfrak{t}_{STU} = \beta_3\, \mathcal{E}^{\alpha_3} + \beta_2\, \mathcal{E}^{\alpha_7} + \beta_1\, \mathcal{E}^{\alpha_9} - \kappa\, \mathfrak{S} \tag{7.9.26}$$

and we calculate its tri-holomorphic moment map, by means of Eq. (7.9.25). Defining the potential:

$$V_{STU} = \sum_{x=1}^{3} \left(\mathscr{P}^x_{\mathfrak{t}_{STU}}\right)^2 \tag{7.9.27}$$

We can verify that:

$$\begin{aligned} \frac{\partial}{\partial \mathbf{Z}^\alpha} V_{STU}\Big|_{\mathbf{Z}=U=a=0} &= 0 \\ \frac{\partial}{\partial U} V_{STU}\Big|_{\mathbf{Z}=U=a=0} &= 0 \\ \frac{\partial}{\partial a} V_{STU}\Big|_{\mathbf{Z}=U=a=0} &= 0 \end{aligned} \tag{7.9.28}$$

Hence we can consistently truncate U, a and the Heisenberg fields $\mathbf{Z}$. We find:

$$V_{STU}|_{\mathbf{Z}=U=a=0} = \frac{9}{4}\left(2\kappa - \sqrt{2}\sum_{i=1}^{3} \beta_i\, e^{-2h_i}\right)^2 \tag{7.9.29}$$

The above potential can be named a multi-Starobinsky model with three independent dilatons.

First of all let us note that in the above model the absolute value of β_i is irrelevant since we can always reabsorb it by a constant shift $h_i \to h_i - \log|\beta_i|$. The only relevant thing are the signs of β_i including in this notion also zero, namely β_i can be ± 1 or 0. Secondly we observe that when all the non vanishing β_i have the same sign we can make a consistent one field truncation to

$$h_i = h \quad ; \quad \text{for all } i \text{ such that } \beta_i \neq 0 \tag{7.9.30}$$

After this truncation the potential (7.9.29) becomes the following:

$$V_{eff} = \frac{9}{4}\left(2\kappa - \sqrt{2}\, q\, e^{-2h}\right)^2 \tag{7.9.31}$$

where q is the number of equal sign non zero β_i, which obviously can take only three values $q = 1, 2, 3$. In order to compare this result with the definition of α-attractors introduced in [67–69], we just have to compare the potential (7.9.31) with the normalization of the scalar kinetic terms in the lagrangian:

$$\mathscr{L} = \ldots + \frac{1}{4}(\partial U)^2 + (\partial h_1)^2 + (\partial h_2)^2 + (\partial h_3)^2 + \ldots \tag{7.9.32}$$

which follows from Eqs. (4.3.4) and (3.8.80). Renaming $h = \frac{1}{\sqrt{2q}}\,\phi$, so that the new field ϕ has canonical kinetic term $\frac{1}{2}\,(\partial\phi)^2$, we obtain a potential:

$$V_{eff} = \text{const} \times \left(2\kappa - \sqrt{2}\,q\,\exp\left[-\sqrt{\frac{2}{q}}\,\phi\right]\right)^2 \tag{7.9.33}$$

which, in the notation of [67–69], corresponds to $\alpha = \frac{q}{3}$, namely to:

$$\alpha = 1\,,\,\frac{2}{3}\,,\,\frac{1}{3} \tag{7.9.34}$$

The above result has been obtained by gauging only one generator, namely (7.9.26). Correspondingly we have generated Starobinsky-like models with only one massive vector that is the gauge vector associated with the gauged generator. There is another way of obtaining the same potential but with q-massive vectors (one for each constituent Starobinsky model with $q = \frac{1}{3}$). This is very simply understood remarking that the $\mathfrak{f}_{(4,4)}$ algebra contains an $\mathfrak{sl}(2,\mathbb{R})^4$ subalgebra singled out as follows:

$$\mathfrak{f}_{(4,4)} \supset \mathfrak{sl}(2,\mathbb{R})_E \oplus \underbrace{\mathfrak{sl}(2,\mathbb{R})_S \oplus \mathfrak{sl}(2,\mathbb{R})_T \oplus \mathfrak{sl}(2,\mathbb{R})_U}_{\subset\,\mathfrak{sp}(6,\mathbb{R})} \tag{7.9.35}$$

where $\mathfrak{sl}(2,\mathbb{R})_S \oplus \mathfrak{sl}(2,\mathbb{R})_T \oplus \mathfrak{sl}(2,\mathbb{R})_U$ describes the STU model embedded in the Kähler manifold (7.8.1). These four $\mathfrak{sl}(2,\mathbb{R})$ algebras are completely symmetric among themselves and the gauging of their generators produce identical results. So we can introduce the abelian gauge algebra spanned by the following three commuting generators:

$$\begin{aligned}
\mathfrak{t}_S &= \beta_3\,\mathscr{E}^{\alpha_3} - \kappa_3\,\mathfrak{S}\\
\mathfrak{t}_T &= \beta_2\,\mathscr{E}^{\alpha_7} - \kappa_2\,\mathfrak{S}\\
\mathfrak{t}_U &= \beta_1\,\mathscr{E}^{\alpha_9} - \kappa_1\,\mathfrak{S}
\end{aligned} \tag{7.9.36}$$

Gauging with three separate vectors each of the above generators we obtain a new potential:

$$\widehat{V}_{STU} = \sum_{x=1}^{3}\left(\mathscr{P}^x_{\mathfrak{t}_S}\right)^2 + \sum_{x=1}^{3}\left(\mathscr{P}^x_{\mathfrak{t}_T}\right)^2 + \sum_{x=1}^{3}\left(\mathscr{P}^x_{\mathfrak{t}_U}\right)^2 \tag{7.9.37}$$

that has the same property as the potential (7.9.27), namely it allows us to truncate consistently to zero all the axions p_i, all the Heisenberg fields $\mathbf{Z}^\alpha$ and the Taub NUT field a. The reduced potential after such a truncation has the form:

$$\widehat{V}_{red} = \frac{9}{4}\sum_{i=1}^{3}\left(2\kappa_i - \sqrt{2}e^{-2h_i}\beta_i\right)^2 \tag{7.9.38}$$

As we already remarked before, the absolute value of the β_i parameters is irrelevant: what matters is only the relative signs of the β_i with respect to the sign of their corresponding κ_i. If for all non vanishing β_i we have $\frac{\beta_i}{\kappa_i} = 1$, then we can consistently perform the same truncation (7.9.30) as before and we reobtain the potentials (7.9.33) with the same spectrum of α-values (7.9.34). The difference with the previous case is, as we emphasized at the beginning of this discussion, that now the number of massive fields is q, namely as many as the elementary non trivial constituent Starobinsky-like models.

7.9.5 Nilpotent Gaugings and Truncations

Let us now put the above obtained results in the general framework discussed in Sect. 7.5.2. The issue is the classification of orbits of nilpotent operators and the question whether for each of these orbits we can find a consistent one-field reduction that produces a Starobinsky-like model with an appropriate value of α.

To answer this question we follow the algorithm described Chap. 6. As we know, up to conjugation, every nilpotent orbit is associated with a standard triple $\{x, y, h\}$ satisfying the standard commutation relations of the $\mathfrak{sl}(2)$ Lie algebra, namely:

$$[h\,,\,x] = x \quad ; \quad [h\,,\,y] = -\,y \quad ; \quad [x\,,\,y] = 2\,h \tag{7.9.39}$$

Interesting for us is the classification of nilpotent orbits in the Kähler subalgebra $\mathfrak{sp}(6,\mathbb{R})$ and, according to the above mathematical theory, this is just the classification of embeddings of an $\mathfrak{sl}(2)$ Lie algebra in the ambient one, modulo conjugation by the full group $\mathrm{Sp}(6,\mathbb{R})$. The second relevant point emphasized in Sect. 6.6.1 is that embeddings of subalgebras $\mathfrak{h} \subset \mathfrak{g}$ are characterized by the branching law of any representation of $\mathfrak{g}$ into irreducible representations of $\mathfrak{h}$. Clearly two embeddings might be conjugate only if their branching laws are identical. Embeddings with different branching laws necessarily belong to different orbits. In the case of the $\mathfrak{sl}(2) \sim \mathfrak{so}(1,2)$ Lie algebra, irreducible representations are uniquely identified by their spin j, so that the branching law is expressed by listing the angular momenta $\{j_1, j_2, \dots j_n\}$ of the irreducible blocks into which any representation of the original algebra, for instance the fundamental, decomposes with respect to the embedded subalgebra. The dimensions of each irreducible module is $2j+1$ so that an a priori constraint on the labels $\{j_1, j_2, \dots j_n\}$ characterizing an irreducible orbit of $\mathfrak{sp}(6,\mathbb{R})$ is the summation rule:

$$\sum_{i=1}^{n}(2j_i + 1) = 6 = \text{dimension of the fundamental representation} \tag{7.9.40}$$

Therefore we have considered all possible partitions of the number 6 into integers and for each partition we have constructed a candidate h element in the Cartan subalgebra of $\mathfrak{sp}(6, \mathbb{R})$ containing as eigenvalues all the J_3 values of the corresponding $\{j_1, j_2, \dots j_n\}$ representation. To clarify what we mean by this it suffices to consider the example of the first partition $6 = 6$. In this case the 6 dimensional representation of $\mathfrak{sl}(2)$ is the $j = \frac{5}{2}$ and the 6 eigenvalues are $\pm\frac{5}{2}, \pm\frac{3}{2}, \pm\frac{1}{2}$. Having so fixed the so named central element h of the candidate standard triplet we have tried to construct the corresponding x and y. Imposing the standard commutation relations (7.9.39) one obtains quadratic equations on the coefficients of the linear combinations expressing the candidate x and y that may have or may not have solutions. If the solutions exist, then the corresponding standard triple is found, the orbit exists and we have constructed one representative x.

Next, given the existing orbits and the corresponding standard triples, for each of them we have constructed a Lobachevsky complex plane immersed in the Special Kähler manifold $\mathscr{M}_{Sp6}$ defined by Eq. (7.8.1). The construction is very simple. One calculates the group element $\mathfrak{g}(\lambda, \psi) \in \mathfrak{sp}(6, \mathbb{R})$ defined below:

$$\mathfrak{g}(\lambda, \psi) = \exp[\psi\, x] \cdot \exp[\lambda\, h] = \left(\begin{array}{c|c} \mathbf{A}(\lambda, \psi) & \mathbf{B}(\lambda, \psi) \\ \hline \mathbf{C}(\lambda, \psi) & \mathbf{D}(\lambda, \psi) \end{array}\right) \tag{7.9.41}$$

and using Eq. (7.8.8), we write:

$$\begin{aligned} Z(\lambda\,,\, \psi) &= (\mathbf{A}(\lambda, \psi) - \mathbf{iB}(\lambda, \psi)) \cdot (\mathbf{C}(\lambda, \psi) - \mathbf{iD}(\lambda, \psi))^{-1} \\ &\equiv \begin{pmatrix} z_1(\lambda, \psi) & z_2(\lambda, \psi) & z_3(\lambda, \psi) \\ z_2(\lambda, \psi) & z_4(\lambda, \psi) & z_5(\lambda, \psi) \\ z_3(\lambda, \psi) & z_5(\lambda, \psi) & z_6(\lambda, \psi) \end{pmatrix} \end{aligned} \tag{7.9.42}$$

which defines the explicit embedding:

$$\phi\, :\, \frac{\mathrm{SL}(2, \mathbb{R})}{\mathrm{SO}(2)} \rightarrow \frac{\mathrm{Sp}(6, \mathbb{R})}{\mathrm{SU}(3) \times \mathrm{U}(1)} \equiv \mathscr{M}_{Sp6} \tag{7.9.43}$$

of the Lobachevsky plane in $\mathscr{M}_{Sp6}$. Indeed from (7.9.42) we read off the parameterization of the complex coordinates z_i $(i = 1, \dots, 6)$ as functions of $\lambda = \log \mathrm{Im}\, w$ and $\psi = \mathrm{Re}\, w$, the complex variable w being the local variable over the embedded Poincaré–Lobachevsky plane.

The question is whether the field equations of the scalar fields:

$$\partial_i\, \partial_{j^\star}\, \mathscr{K}\, \partial^\mu \partial_\mu\, \overline{z}^{j^\star} + \partial_i\, \partial_{j^\star}\, \partial_{k^\star}\, \mathscr{K}\, \partial_\mu z^{j^\star}\, \partial_\mu z^{k^\star} - \frac{1}{4}\, \partial_i\, V_{gauging}\,(z\,,\, \overline{z}) = 0 \tag{7.9.44}$$

admit first a consistent reduction to the complex scalar field w and then a consistent truncation to a vanishing axion $\psi = 0$. Consistency of the truncation can be verified or disproved in the following simple way. The pull-back on the immersed surface $\phi^\star\left(\frac{\mathrm{SL}(2,\mathbb{R})}{\mathrm{SO}(2)}\right) \subset \mathscr{M}_{Sp6}$ of the twelve field Eq. (7.9.44) (six complex equations) should be consistent among themselves and be identical with the two field equations obtained from the variation of the pull-back $\phi^\star(\mathscr{L})$ on the immersed surface of the Lagrangian $\mathscr{L}$ from which Eq. (7.9.44) derive, namely:

$$\mathscr{L} = 4\,\partial_i\,\partial_{j^\star}\,\mathscr{K}\,\partial_\mu\,z^i\,\partial^\mu\,\overline{z}^{j^\star} - V_{gauging}\,(z\,,\,\overline{z}) \tag{7.9.45}$$

In other words, defining $w = \mathrm{i}\,e^\lambda + \psi$, the truncation is consistent if the following diagram is commutative:

$$\begin{array}{ccc} \mathscr{L}(z,\overline{z}) & \stackrel{\phi^\star}{\Longrightarrow} & \phi^\star\mathscr{L}(w,\overline{w}) \\ \downarrow & & \downarrow \\ \partial^\mu\,\frac{\partial\mathscr{L}}{\partial(\partial_\mu z)} - \frac{\partial\mathscr{L}}{\partial z} & \stackrel{\phi^\star}{\Longrightarrow} & \partial^\mu\,\frac{\partial\phi^\star\mathscr{L}}{\partial(\partial_\mu w)} - \frac{\partial\phi^\star\mathscr{L}}{\partial w} \end{array} \tag{7.9.46}$$

Partition	J.s	Orbit Name	One field reduction
$6=6$	$\left(\frac{5}{2}\right)$	$\mathfrak{O}_1$	NO
$6=5+1$	$(2,0)$	Orbit does not exist	NO
$6=4+2$	$\left(\frac{3}{2},\frac{1}{2}\right)$	$\mathfrak{O}_2$	NO
$6=3+3$	$(1,1)$	$\mathfrak{O}_3$	NO
$6=3+2+1$	$\left(1,\frac{1}{2},0\right)$	Orbit does not exist	NO
$6=3+1+1+1$	$(1,0,0,0)$	Orbit does not exist	NO
$6=2+2+2$	$\left(\frac{1}{2},\frac{1}{2},\frac{1}{2}\right)$	$\mathfrak{O}_4$	YES
$6=2+2+1+1$	$\left(\frac{1}{2},\frac{1}{2},0,0\right)$	$\mathfrak{O}_5$	YES
$6=2+1+1+1+1$	$\left(\frac{1}{2},0,0,0,0\right)$	$\mathfrak{O}_6$	YES

In the above table we have summarized the results of this simple investigation. There is a total of six orbits (up to possible further splitting in Weyl group orbits which we have not analyzed) and for each of them the corresponding immersion formulae in the $\mathscr{M}_{\mathrm{Sp6}}$ manifolds are those described below.

7.9.5.1 Orbit $\mathfrak{O}_1$: $(j = \frac{5}{2})$

$$\begin{aligned}
&\begin{pmatrix} z_1 & z_2 & z_3 \\ z_2 & z_4 & z_5 \\ z_3 & z_5 & z_6 \end{pmatrix} = \\
&\begin{pmatrix} -6\psi^5+10ie^\lambda\psi^4+5ie^{3\lambda}\psi^2+ie^{5\lambda} & \sqrt{5}\psi\left(3\psi^3-4ie^\lambda\psi^2-ie^{3\lambda}\right) & i\sqrt{10}\left(i\psi+e^\lambda\right)\psi^2 \\ \sqrt{5}\psi\left(3\psi^3-4ie^\lambda\psi^2-ie^{3\lambda}\right) & i\left(8i\psi^3+8e^\lambda\psi^2+e^{3\lambda}\right) & \sqrt{2}\psi\left(3\psi-2ie^\lambda\right) \\ i\sqrt{10}\left(i\psi+e^\lambda\right)\psi^2 & \sqrt{2}\psi\left(3\psi-2ie^\lambda\right) & ie^\lambda-3\psi \end{pmatrix} \\
&w = \mathrm{i}\,e^\lambda + \psi
\end{aligned} \tag{7.9.47}$$

The pull-back of the lagrangian is the following one:

$$\phi^\star \mathscr{L} = 35\left(\partial^\mu \psi\, \partial^\mu \psi\, e^{-2\lambda} + \partial^\mu \lambda\, \partial^\mu \lambda\right) - \frac{1}{4} g^2 \left(3\, e^{-\lambda} - \kappa\right)^2 \qquad (7.9.48)$$

The pull-backs of the scalar field equations are inconsistent among themselves and differ from the equations derived from the pull-back of the lagrangian (7.9.48), hence the truncation is not consistent. No Starobinsky-like model can be obtained from this orbit.

One might wonder whether the inconsistency is due to the particularly chosen coset representative (7.9.41) and to the explicit form of the embedding (7.9.47) which turns out to be non-holomorphic. To clarify such a doubt and show that the inconsistency of the equations is an intrinsic property of the orbit, we have addressed the problem from a different view point which leads to a perfectly holomorphic embedding of the Lobachevsky plane associated with the considered orbit into the target Special Kähler manifold (7.8.1).

The argument is the following one. Having fixed the embedding $\mathfrak{sl}(2,\mathbb{R}) \mapsto \mathfrak{sp}(6,\mathbb{R})$ at the level of the fundamental representation **6** it is fixed also in all other representations and we can wonder what is the branching rule of the **W**-representation $\mathbf{14}'$ such an embedding. By direct evaluation of the Casimir we obtain the following branching:

$$\mathbf{14}' \overset{\mathfrak{sl}(2,\mathbb{R})}{\longrightarrow} \left(j = \frac{9}{2}\right) \oplus \left(j = \frac{3}{2}\right) \qquad (7.9.49)$$

This means that the symplectic section (7.8.26) splits into the sum of two vectors, one lying in the 10-dimensional space of the first representation, the other in the 4-dimensional space of the second representation. Imposing the vanishing of the lowest spin representation introduces a set of 4 holomorphic constraints on the six coordinates z_i. By construction these constraints are $\mathfrak{sl}(2,\mathbb{R})$ invariant: therefore the sought for Lobachevsky plane certainly lies in the complex two-folds defined by the vanishing of these constraints. With a little bit of work one can further eliminate one of the two remaining complex coordinates in such a way that the ten entries of the $\left(j = \frac{9}{2}\right)$ representation correspond to all the powers w^r, with $r = 0, 1, \ldots, 9$ of a complex parameter w. Because of this very property w can be interpreted as the local coordinate of the sought for Lobachevsky plane embedded in the Kähler manifold (7.8.1) according to the specified orbit. Indeed if w transforms by fractional linear transformation under some algebra $\mathfrak{sl}(2)$, then the $2j+1$ first powers of w provide a basis for the j-representation of that $\mathfrak{sl}(2)$. Viceversa, if a vector, which is known to transform in the j-representation of a given $\mathfrak{sl}(2)$ (up to an overall function of w), is made by linear combinations of the first $2j+1$ powers of a coordinate w, then that w is the local coordinate on a Lobachevsky plane transitive under the action of that very $\mathfrak{sl}(2)$.

In our case the four holomorphic constraints that express the vanishing of the $j = \frac{3}{2}$ representation inside the $14'$ are the following ones:

$$\begin{aligned}
\sqrt{\frac{2}{7}}\left(\sqrt{5}\left(z_4 z_6 - z_5^2\right) - 2z_2\right) &= 0\\
\frac{8z_3 - \sqrt{10}z_4}{\sqrt{21}} &= 0\\
\sqrt{\frac{2}{7}}\left(\sqrt{5}z_1 + 2z_3 z_5 - 2z_2 z_6\right) &= 0\\
\frac{-\sqrt{10}z_3^2 + 8z_4 z_3 - 8z_2 z_5 + \sqrt{10}z_1 z_6}{\sqrt{21}} &= 0
\end{aligned} \tag{7.9.50}$$

The explicit form of (7.9.50) obviously depends on the standard triple chosen as representative of the orbit, yet for whatever representative the four constraints are holomorphic. The next point consists in solving (7.9.50) in terms of a parameter w so that the complementary set of ten polynomials of the z_i spanning the $j = \frac{9}{2}$ representation provide all the powers of w from 0 to 9. The requested solution is given by:

$$z_1 \to \frac{3w^5}{16},\ z_2 \to \frac{3\sqrt{5}w^4}{16},\ z_3 \to \frac{1}{4}\sqrt{\frac{5}{2}}w^3,\ z_4 \to w^3,\ z_5 \to \frac{3w^2}{2\sqrt{2}},\ z_6 \to \frac{3w}{2} \tag{7.9.51}$$

Implementing the transformation (7.9.51) in the symplectic section (7.8.26) one finds:

$$\Omega[Z] \xrightarrow{\phi} \Omega_{\frac{9}{2}}[w] = \begin{pmatrix} \frac{3w}{\sqrt{2}}\\ \frac{w^6}{4\sqrt{2}}\\ -\frac{3w^4}{4\sqrt{2}}\\ \frac{w^9}{256\sqrt{2}}\\ -\frac{3w^7}{32\sqrt{2}}\\ -\frac{1}{16}\sqrt{\frac{5}{2}}w^6\\ -\frac{3}{16}\sqrt{5}w^5\\ \frac{3w^8}{128\sqrt{2}}\\ -\sqrt{2}w^3\\ \frac{3w^5}{8\sqrt{2}}\\ \sqrt{2}\\ -\frac{3w^2}{\sqrt{2}}\\ \frac{1}{2}\sqrt{\frac{5}{2}}w^3\\ -\frac{3}{8}\sqrt{5}w^4 \end{pmatrix} \tag{7.9.52}$$

which as requested contains all the powers of w and has vanishing projection on the $j = \frac{3}{2}$ representation. Calculating the Kähler potential from such a section we obtain:

$$\mathscr{K}_{\frac{9}{2}} = -\log\left(\overline{\Omega}_{\frac{9}{2}}[\overline{w}]\,\mathbb{C}_{14}\,\Omega_{\frac{9}{2}}[w]\right) = \log\left(-\frac{\mathrm{i}}{256}\,(w-\overline{w})^9\right) \tag{7.9.53}$$

Now the question of consistency can be readdressed in the present context. Implementing the substitution (7.9.51) in the six complex Eq. (7.9.44) (with for instance vanishing potential) do we obtain six consistent equations or not? The answer is no. The six Eq. (7.9.44) are inconsistent and this confirms in a holomorphic set up the same result we had previously obtained in the direct approach of Eqs. (7.9.41)–(7.9.42). Hence the $\mathfrak{sl}(2)$ embedding of orbit $\mathfrak{O}_1$ leads to inconsistent truncations and has to be excluded.

7.9.5.2 Orbit $\mathfrak{O}_2$: ($j_1 = \frac{3}{2}$, $j_2 = \frac{1}{2}$)

For the second orbit, the direct approach (7.9.41) and (7.9.42) leads to:

$$\begin{pmatrix} z_1 & z_2 & z_3 \\ z_2 & z_4 & z_5 \\ z_3 & z_5 & z_6 \end{pmatrix} =$$
$$\begin{pmatrix} \left(e^\lambda - i\psi\right)^2\left(2\psi - ie^\lambda\right) & 0 & \sqrt{3}\psi\left(\psi + ie^\lambda\right) \\ 0 & -\psi - ie^\lambda & 0 \\ \sqrt{3}\psi\left(\psi + ie^\lambda\right) & 0 & -2\psi - ie^\lambda \end{pmatrix}$$
$$w = \mathrm{i}\,e^\lambda + \psi \tag{7.9.54}$$

The pull-back of the lagrangian is the following one:

$$\phi^\star\mathscr{L} = 11\left(\partial^\mu\psi\,\partial^\mu\psi\,e^{-2\lambda} + \partial^\mu\lambda\,\partial^\mu\lambda\right) - \frac{1}{4}g^2\left(3\,e^{-\lambda} - \kappa\right)^2 \tag{7.9.55}$$

Also in this case the pull-back of the scalar field equations yields an inconsistent set and there is no truncation. No Starobinsky-like model can be obtained from this orbit. In a similar way to the previous case we can discuss the same issue in a holomorphic set up. The branching rule of the **14**′ representation in the considered embedding is the following one:

$$\mathbf{14}' \overset{\mathfrak{sl}(2,\mathbb{R})}{\longrightarrow} \left(j = \frac{5}{2}\right) \oplus \left(j = \frac{3}{2}\right) \oplus \left(j = \frac{3}{2}\right) \tag{7.9.56}$$

and we can impose holomorphic constraints that suppress the two lowest spin representations $\left(j = \frac{3}{2}\right)$ leaving only the top one $\left(j = \frac{5}{2}\right)$ spanned by the powers of a parameter w from 0 to 5. Such a holomorphic embedding is given:

$$z_1 \frac{2w^3}{3^{3/4}},\; z_2 \to 0,\; z_3 \to w^2, z_4 \to \frac{w}{\sqrt[4]{3}},\; z_5 \to 0,\; z_6 \to \frac{2w}{\sqrt[4]{3}} \tag{7.9.57}$$

Substitution of Eq. (7.9.57) into the field equations (7.9.44) confirms that their pull-back on this surface is inconsistent.

7.9.5.3 Orbit $\mathfrak{O}_3$: ($j_1 = 1,\ j_2 = 1$)

For the third orbit, the direct approach (7.9.41) and (7.9.42) leads to

$$\begin{pmatrix} z_1 & z_2 & z_3 \\ z_2 & z_4 & z_5 \\ z_3 & z_5 & z_6 \end{pmatrix} = \begin{pmatrix} -ie^{2\lambda} & -\psi^2 & -\sqrt{2}\psi \\ -\psi^2 & -i\left(2\psi^2 + e^{2\lambda}\right) & -i\sqrt{2}\psi \\ -\sqrt{2}\psi & -i\sqrt{2}\psi & -i \end{pmatrix}$$
$$w = \mathrm{i}\, e^{\lambda} + \psi \tag{7.9.58}$$

The pull-back of the lagrangian is the following one:

$$\phi^\star \mathscr{L} = 8\left(\partial^\mu \psi\, \partial^\mu \psi\, e^{-2\lambda} + \partial^\mu \lambda\, \partial^\mu \lambda\right) - \frac{1}{4} g^2 \kappa^2 \tag{7.9.59}$$

while the pull-back of the scalar field equations is an inconsistent set. Hence this truncation is not consistent and no Starobinsky-like model can be obtained from this orbit. As in the previous two cases we can confirm the same result in a holomorphic set up, yet we consider it useless to repeat once more the same type of calculations. What is relevant to mention in view of our subsequent considerations is the branching rule of the $\mathbf{14}'$ representation under this forbidden embedding leading to inconsistent field equations::

$$\mathbf{14}' \stackrel{\mathfrak{sl}(2,\mathbb{R})}{\longrightarrow} (j = 2) \oplus (j = 2) \oplus 4 \times (j = 0) \tag{7.9.60}$$

7.9.5.4 Orbit $\mathfrak{O}_4$: ($j_1 = \frac{1}{2},\ j_2 = \frac{1}{2},\ j_3 = \frac{1}{2}$)

For the fourth orbit, the direct approach (7.9.41) and (7.9.42) leads to

$$\begin{pmatrix} z_1 & z_2 & z_3 \\ z_2 & z_4 & z_5 \\ z_3 & z_5 & z_6 \end{pmatrix} = \begin{pmatrix} ie^{\lambda} - \psi & 0 & 0 \\ 0 & ie^{\lambda} - \psi & 0 \\ 0 & 0 & ie^{\lambda} - \psi \end{pmatrix}$$
$$w = \mathrm{i}\, e^{\lambda} - \psi \tag{7.9.61}$$

The pull-back of the lagrangian is the following one:

$$\phi^\star \mathscr{L} = 3 \left(\partial^\mu \psi \, \partial^\mu \psi \, e^{-2\lambda} + \partial^\mu \lambda \, \partial^\mu \lambda\right) - g^2 \frac{1}{4} \left(3\, e^{-\lambda} - 2\,\kappa\right)^2 \qquad (7.9.62)$$

The pull-back of the scalar field equations produces equations consistent among themselves which coincide with the equations derived from the pull-back of the lagrangian (7.9.62), hence the truncation is consistent. We reobtain the Starobinsky model discussed in the previous section with $q = 3$ and hence with $\alpha = 1$. In this case the consistent truncation is already produced form holomorphic constraints. Indeed Eq. (7.9.61) can be summarized as:

$$z_2 = z_3 = z_5 = 0 \quad ; \quad z_1 = z_4 = z_6 = w \qquad (7.9.63)$$

It is interesting and important for our future consideration to mention the branching rule of the $\mathbf{14}'$ representation under this $\mathfrak{sl}(2)$ subalgebra:

$$\mathbf{14}' \overset{\mathfrak{sl}(2,\mathbb{R})}{\longrightarrow} \left(j = \frac{3}{2}\right) \oplus 5 \times \left(j = \frac{1}{2}\right) \qquad (7.9.64)$$

and the constraints (7.9.63) precisely are the conditions under which the five representations $\left(j = \frac{1}{2}\right)$ vanish and we are left with the representation $\left(j = \frac{3}{2}\right)$ duely spanned by the powers $1, w, w^2, w^3$.

7.9.5.5 Orbit $\mathfrak{O}_5$: $(j_1 = \frac{1}{2},\ j_2 = \frac{1}{2},\ j_3 = 0)$

For the fifth orbit, the direct approach (7.9.41) and (7.9.42) leads to

$$\begin{pmatrix} z_1 & z_2 & z_3 \\ z_2 & z_4 & z_5 \\ z_3 & z_5 & z_6 \end{pmatrix} = \begin{pmatrix} ie^\lambda - \psi & 0 & 0 \\ 0 & ie^\lambda - \psi & 0 \\ 0 & 0 & i \end{pmatrix}$$
$$w = \mathrm{i}\, e^\lambda - \psi \qquad (7.9.65)$$

The pull-back of the lagrangian is the following one:

$$\phi^\star \mathscr{L} = 2 \left(\partial^\mu \psi \, \partial^\mu \psi \, e^{-2\lambda} + \partial^\mu \lambda \, \partial^\mu \lambda\right) - g^2 \left(e^{-\lambda} - \kappa\right)^2 \qquad (7.9.66)$$

The pull-back of the scalar field equations yields a consistent system identical with the field equations derived from the pull-back of the lagrangian (7.9.70), hence the truncation is consistent. We reobtain the Starobinsky-like model discussed in the previous section with $q = 2$ and hence with $\alpha = \frac{2}{3}$.

In this, as in the previous case, the consistent truncation is produced from holomorphic constraints. Indeed Eq. (7.9.61) can be summarized as:

$$z_2 = z_3 = z_5 = 0 \quad ; \quad z_1 = z_4 = w \quad ; \quad z_6 = \mathrm{i} \tag{7.9.67}$$

In this case the branching rule of the $\mathbf{14}'$ representation under the considered $\mathfrak{sl}(2)$ subalgebra is the following one:

$$\mathbf{14}' \xrightarrow{\mathfrak{sl}(2,\mathbb{R})} (j = 1) \oplus (j = 1) \oplus 2 \times \left(j = \frac{1}{2}\right) + 4 \times (j = 0) \tag{7.9.68}$$

and the constraint (7.9.67) guarantees that the singlets and the $\left(j = \frac{1}{2}\right)$ representations are all set to zero.s

7.9.5.6 Orbit $\mathfrak{O}_6$: $(j_1 = \frac{1}{2},\ j_2 = 0,\ j_3 = 0)$

For the sixth orbit, the direct approach (7.9.41) and (7.9.42) leads to

$$\begin{aligned} &\begin{pmatrix} z_1 & z_2 & z_3 \\ z_2 & z_4 & z_5 \\ z_3 & z_5 & z_6 \end{pmatrix} = \\ &\begin{pmatrix} \psi + ie^{\lambda} & 0 & 0 \\ 0 & i & 0 \\ 0 & 0 & i \end{pmatrix} \\ &w = \mathrm{i}\, e^{\lambda} + \psi \end{aligned} \tag{7.9.69}$$

The pull-back of the lagrangian is the following one:

$$\phi^\star \mathscr{L} = \left(\partial^\mu \psi\, \partial^\mu \psi\, e^{-2\lambda} + \partial^\mu \lambda\, \partial^\mu \lambda\right) - \frac{1}{4} g^2 \left(e^\lambda + \kappa\right)^2 \tag{7.9.70}$$

The pull-back of the scalar field equations yields a consistent system coinciding with the equations derived from the pull-back of the lagrangian (7.9.70). So we have a consistent truncation and we reobtain the Starobinsky-like model discussed in the previous section with $q = 1$. It corresponds to $\alpha = \frac{1}{3}$. Equation (7.9.69) can be summarized as:

$$z_2 = z_3 = z_5 = 0 \quad ; \quad z_1 = w \quad ; \quad z_4 = z_6 = \mathrm{i} \tag{7.9.71}$$

The branching of the $\mathbf{14}'$ dimensional representation under this $\mathfrak{sl}(2)$ subalgebra is the following one:

$$\mathbf{14}' \overset{\mathfrak{sl}(2,\mathbb{R})}{\longrightarrow} 5 \times \left(j = \frac{1}{2}\right) \oplus 4 \times (j = 0) \tag{7.9.72}$$

7.9.5.7 Conclusion of the Above Discussion

This concludes our preliminary study of the orbits and shows that the embedded Starobinsky-like models described in Sect. 7.9.4 exhaust the list of possible embeddings, the values of $\alpha = 1, \frac{2}{3}, \frac{1}{3}$ being, apparently the only admissible ones. Next let us observe that the branching rules of the $\mathbf{14}'$ dimensional representation which lead to consistent truncations, namely, (7.9.64), (7.9.68) and (7.9.72) are the only possible ones that we can obtain by embedding:

$$\mathfrak{sl}(2) \mapsto \mathfrak{sl}(2) \times \mathfrak{sl}(2) \times \mathfrak{sl}(2) \tag{7.9.73}$$

if the considered $\mathbf{14}'$ representation of $\mathfrak{sl}(2) \times \mathfrak{sl}(2) \times \mathfrak{sl}(2)$ is the following one:

$$\mathbf{14}' = \left(\frac{1}{2}, \frac{1}{2}, \frac{1}{2}\right) \oplus \left(\frac{1}{2}, 0, 0\right) \oplus \left(0, \frac{1}{2}, 0\right) \oplus \left(0, 0, \frac{1}{2}\right) \tag{7.9.74}$$

This has a profound meaning. It implies that the only consistent truncations occur when the $\mathfrak{sl}(2)$ Lie algebra is embedded in the *sub-Tits-Satake* Lie algebra, which as we discuss in the conclusive part is universal for all $\mathcal{N} = 2$ models. This allows us to make the bold statement that the only values of α one can obtain form the gauging of hypermultiplet isometries in any supergravity theory based on symmetric manifolds is just $\alpha = 1, \frac{2}{3}, \frac{1}{3}$.

7.10 Holomorphic Consistent Truncations and the Sub Tits Satake Algebra

In order to find the deep rationale for the conclusions reached from the above results we need to recall the results on the Tits Satake projection and the universality classes that were derived in Chap. 5, in particular those of Sect. 5.6 concerning the Tits Satake decompositions of the $\mathbf{W}$-representation.

Let us remark that the gauge condition (5.6.38) has another important interpretation if applied to the holomorphic section of special geometry. The key point is the following numerical identity valid for all members of the universality class:

$$\dim \frac{\mathrm{U}_{\mathcal{SK}}}{\mathrm{H}_{\mathcal{SK}}} = \dim \frac{\mathrm{U}^{TS}_{\mathcal{SK}}}{\mathrm{H}^{TS}_{\mathcal{SK}}} \oplus 6 \times \dim \mathcal{D}_{\mathrm{subpaint}} \tag{7.10.1}$$

This means that if we decompose the symplectic section of the big group according to the Tits-Satake subalgebra and we impose on it the condition (5.6.38) we just obtain the right number of holomorphic constraints to project onto the submanifold $\frac{\mathrm{U}^{TS}_{\mathscr{S}\mathscr{K}}}{\mathrm{H}^{TS}_{\mathscr{S}\mathscr{K}}}$. At the level of field equations this is certainly a consistent truncation, since we project onto the singlets of the subpaint group.

On the other hand if we decompose the **W**-representation with respect to the sub-Tits-Satake subalgebra $\mathfrak{sl}(2) \times \mathfrak{sl}(2) \times \mathfrak{sl}(2)$ we have the branching rule:

$$\mathbf{W} \rightarrow (\mathscr{D}_1|\mathbf{2},\mathbf{1},\mathbf{1}) \oplus (\mathscr{D}_2|\mathbf{1},\mathbf{2},\mathbf{1}) \oplus (\mathscr{D}_3|\mathbf{1},\mathbf{1},\mathbf{2}) \oplus (\mathbf{1}|\mathbf{2},\mathbf{2},\mathbf{2}) \tag{7.10.2}$$

where $\mathscr{D}_{1,2,3}$ are three suitable representations of the Paint Group. Imposing on the symplectic section of the big model the constraints:

$$\begin{aligned} (\mathscr{D}_1|\mathbf{2},\mathbf{1},\mathbf{1}) &= 0 \\ (\mathscr{D}_2|\mathbf{1},\mathbf{2},\mathbf{1}) &= 0 \\ (\mathscr{D}_3|\mathbf{1},\mathbf{1},\mathbf{2}) &= 0 \end{aligned} \tag{7.10.3}$$

yields precisely the correct number of holomorphic constraints that restrict the considered Special Kähler manifold to the Special Kähler manifold of the STU-model namely to $\left(\frac{\mathrm{SL}(2,\mathbb{R})}{\mathrm{SO}(2)}\right)^3$. This follows from the numerical identity true for all members of the universality class:

$$\dim \frac{\mathrm{U}_{\mathscr{S}\mathscr{K}}}{\mathrm{H}_{\mathscr{S}\mathscr{K}}} = \sum_{i=1}^{3} 2 \times \dim \mathscr{D}_i + 6 \tag{7.10.4}$$

The reason why the truncation to the STU-model is always a consistent truncation at the level of field equations is obvious in this set up. It corresponds to the truncation to the Paint Group singlets.

7.10.1 W-Representations for the Remaining Models

For the models of type $\mathfrak{sl}(2,\mathbb{R}) \oplus \mathfrak{so}(q,q+p)$ having $\mathfrak{sl}(2,\mathbb{R}) \oplus \mathfrak{so}(q,q+1)$ as Tits Satake subalgebra and $\mathfrak{so}(p-1)$ as subpaint algebra the decomposition of the **W**-representation was presented in Eq. (5.6.40) and the question whether each $\mathfrak{sl}(2,\mathbb{R}) \oplus \mathfrak{so}(q,q+p)$ orbit in the $(\mathbf{2},\mathbf{2q}+\mathbf{p})$ representation intersects the $\mathfrak{sl}(2,\mathbb{R}) \oplus \mathfrak{so}(q,q+1) \oplus \mathfrak{so}(p-1)$-invariant locus (5.6.41) was positively answered.

Relevant for the case of $\mathscr{N}=2$ supersymmetry is the value $q=2$ and, in this case, the sub-Tits-Satake Lie algebra is:

$$\mathbb{G}_{\mathrm{subTS}} = \mathfrak{sl}(2,\mathbb{R}) \oplus \mathfrak{so}(2,2) = \mathfrak{sl}(2,\mathbb{R}) \oplus \mathfrak{sl}(2,\mathbb{R}) \oplus \mathfrak{sl}(2,\mathbb{R}) \tag{7.10.5}$$

namely it is once again the Lie algebra of the STU-model. Reduction to the STU-model is consistent for the same reason as in the other universality classes: it corresponds to truncation to Paint Group singlets.

7.10.2 *Gaugings with Consistent One-Field Truncations*

On the basis of the analysis presented in the previous section we arrive at the following conclusion. By gauging a nilpotent element of the isometry subalgebra of $\mathcal{SK}$ inside $\mathcal{QM}$ we generate a potential. The structure of the theory depends on the nilpotent orbit, namely on the embedding of an $\mathfrak{sl}(2)$ Lie algebra in $\mathbb{U}_{\mathcal{SK}}$ and there are many ways of doing this (the orbits), yet the gauged theory will admit a one-field truncation if and only if the $\mathfrak{sl}(2)$ is embedded into the sub Tits Satake Lie algebra:

$$\mathfrak{sl}(2) \mapsto \mathbb{G}_{\text{subTS}} \subset \mathbb{U}_{\mathcal{SK}} \tag{7.10.6}$$

There are only three different embeddings of $\mathfrak{sl}(2)$ into $(\mathfrak{sl}(2))^3$ and these correspond to the three admissible values $\alpha = 1, \frac{2}{3}, \frac{1}{3}$ in the Starobinsky-like model.

7.11 Conclusions for This Chapter

This chapter tackled the problem of potentials and gaugings in supergravity. The main message that we hope has reached our reader is that while the key item in the construction of black-hole solutions is the $c^{\star}$-map, for the case of gaugings and candidate inflaton potentials the key item is the c-map. In both cases the crucial mathematical structure is encoded in the Tits-Satake projection and in the *sub Tits Satake subalgebra*. Just as we said for the case of black-holes, uplifting the described algebraic mechanisms to a higher degree of abstraction, disentangling them from the details of their physical interpretation might result in a deeper understanding and in new perspectives.

Appendix

7.12 Large Formulae Not Displayed in the Main Text

In this appendix we collect some large formulae that need the landscape format and are not presented in the main text.

7.12.1 *Concerning the* $\mathrm{G}_{(2,2)}$ *Model*

7.12.1.1 The Solvable Coset Representative for $\frac{\mathrm{G}_{(2,2)}}{\mathrm{SU(2)}\times\mathrm{SU(2)}}$

$$\mathbb{L} = \tag{7.12.1}$$

$$\begin{pmatrix}
e^{\frac{h+U}{2}} & e^{\frac{U-h}{2}} y & -\sqrt{\frac{2}{3}}e^{h} Z_3 & -\frac{2Z_1}{\sqrt{3}} - \frac{2yZ_3}{\sqrt{3}} & -\sqrt{\frac{2}{3}}e^{-h} Z_3 y^2 - 2\sqrt{\frac{2}{3}}e^{-h} Z_1 y - \sqrt{2}e^{-h} Z_2 & e^{\frac{h-U}{2}}\left(a - \frac{Z_1 Z_3}{3} - Z_2 Z_4\right) & e^{-\frac{h}{2}-\frac{U}{2}}\left(\frac{2Z_2Z_3}{\sqrt{3}} - \frac{2Z_1^2}{3}\right) + e^{-\frac{h}{2}-\frac{U}{2}} y\left(a - \frac{Z_1Z_3}{3} - Z_2Z_4\right) \\
0 & e^{\frac{U-h}{2}} & -\sqrt{2}e^{h} Z_4 & \frac{2Z_3}{\sqrt{3}} - 2yZ_4 & -\sqrt{2}e^{-h} Z_4 y^2 + 2\sqrt{\frac{2}{3}}e^{-h} Z_3 y + \sqrt{\frac{2}{3}}e^{-h} Z_1 & e^{\frac{h-U}{2}}\left(\frac{2Z_3^2}{3} + \frac{2Z_1Z_4}{\sqrt{3}}\right) & e^{-\frac{h}{2}-\frac{U}{2}} y\left(\frac{2Z_3^2}{3} + \frac{2Z_1Z_4}{\sqrt{3}}\right) + e^{-\frac{h}{2}-\frac{U}{2}}\left(a + \frac{Z_1Z_3}{3} + Z_2Z_4\right) \\
0 & 0 & e^{h} & \sqrt{2}y & e^{-h} y^2 & -\sqrt{\frac{2}{3}}e^{\frac{h-U}{2}} Z_1 & -\sqrt{\frac{2}{3}}e^{-\frac{h}{2}-\frac{U}{2}} yZ_1 - \sqrt{2}e^{-\frac{h}{2}-\frac{U}{2}} Z_2 \\
0 & 0 & 0 & 1 & \sqrt{2}e^{-h} y & \frac{2e^{\frac{h-U}{2}} Z_3}{\sqrt{3}} & \frac{2e^{-\frac{h}{2}-\frac{U}{2}} Z_1}{\sqrt{3}} + \frac{2e^{-\frac{h}{2}-\frac{U}{2}} yZ_3}{\sqrt{3}} \\
0 & 0 & 0 & 0 & e^{-h} & \sqrt{2}e^{\frac{h-U}{2}} Z_4 & \sqrt{2}e^{-\frac{h}{2}-\frac{U}{2}} yZ_4 - \sqrt{\frac{2}{3}}e^{-\frac{h}{2}-\frac{U}{2}} Z_3 \\
0 & 0 & 0 & 0 & 0 & e^{\frac{h-U}{2}} & e^{-\frac{h}{2}-\frac{U}{2}} y \\
0 & 0 & 0 & 0 & 0 & 0 & e^{-\frac{h}{2}-\frac{U}{2}}
\end{pmatrix} \tag{7.12.2}$$

7.12.2 *Explicit Form of Generators for the Lie Algebra* $\mathfrak{sp}(6,\mathbb{R})$ *in the* $\mathbf{14}'$ *Representation*

The $\mathbf{14}'$ representation of $\mathfrak{sp}(6,\mathbb{R})$ which plays the role $\mathbf{W}$-representation for the special manifold under consideration is defined as the representation obeyed by the three-times antisymmetric tensors with vanishing $\mathbb{C}$-traces, namely:

$$\underbrace{t_{ABC}}_{\text{antisymmetric in } A,B,C} \times \mathbb{C}^{BC} = 0 \tag{7.12.3}$$

Let us then consider a lexicographic ordered basis for the 20-dimensional reducible representation provided by the three times antisymmetric tensor:

$$\begin{pmatrix} V_1 \\ V_2 \\ V_3 \\ V_4 \\ V_5 \\ V_6 \\ V_7 \\ V_8 \\ V_9 \\ V_{10} \\ V_{11} \\ V_{12} \\ V_{13} \\ V_{14} \\ V_{15} \\ V_{16} \\ V_{17} \\ V_{18} \\ V_{19} \\ V_{20} \end{pmatrix} \equiv \begin{pmatrix} t_{123} \\ t_{124} \\ t_{134} \\ t_{234} \\ t_{125} \\ t_{135} \\ t_{235} \\ t_{145} \\ t_{245} \\ t_{345} \\ t_{126} \\ t_{136} \\ t_{236} \\ t_{146} \\ t_{246} \\ t_{346} \\ t_{156} \\ t_{256} \\ t_{356} \\ t_{456} \end{pmatrix} \tag{7.12.4}$$

The splitting into the two irreducible subspaces of dimension **14** and **6** respectively can be performed by defining the following new basis vectors:

$$\begin{pmatrix} \Phi_1 \\ \Phi_2 \\ \Phi_3 \\ \Phi_4 \\ \Phi_5 \\ \Phi_6 \\ \Phi_7 \\ \Phi_8 \\ \Phi_9 \\ \Phi_{10} \\ \Phi_{11} \\ \Phi_{12} \\ \Phi_{13} \\ \Phi_{14} \end{pmatrix} = \begin{pmatrix} V_1 \\ V_4 \\ V_6 \\ V_{10} \\ V_{11} \\ V_{15} \\ V_{17} \\ V_{20} \\ V_5 - V_{12} \\ -V_2 - V_{13} \\ V_7 - V_3 \\ V_{16} - V_9 \\ V_8 + V_{19} \\ V_{14} - V_{18} \end{pmatrix} \; ; \; \begin{pmatrix} C_1 \\ C_2 \\ C_3 \\ C_4 \\ C_5 \\ C_6 \end{pmatrix} = \begin{pmatrix} V_5 + V_{12} \\ V_{13} - V_2 \\ -V_3 - V_7 \\ -V_9 - V_{16} \\ V_8 - V_{19} \\ V_{14} + V_{18} \end{pmatrix} \tag{7.12.5}$$

Taking the antisymmetric cubic tensor product and using the splitting (7.12.5), the matrices $\widehat{\mathscr{D}}_{14}(\mathfrak{g})$ representing any element $\mathfrak{g} \in \mathfrak{sp}(6, \mathbb{R})$ of the Lie algebra in the $\mathbf{14}'$ representation can be easily extracted. The so obtained $\widehat{\mathscr{D}}_{14}(\mathfrak{g})$ matrices are symplectic, since, by direct calculation one can determine a unique antisymmetric matrix:

$$
\widehat{\mathbb{C}}_{14} = \begin{pmatrix}
0 & 0 & 0 & 0 & 0 & 0 & 0 & 2 & 0 & 0 & 0 & 0 & 0 & 0 \\
0 & 0 & 0 & 0 & 0 & 0 & -2 & 0 & 0 & 0 & 0 & 0 & 0 & 0 \\
0 & 0 & 0 & 0 & 0 & -2 & 0 & 0 & 0 & 0 & 0 & 0 & 0 & 0 \\
0 & 0 & 0 & 0 & 2 & 0 & 0 & 0 & 0 & 0 & 0 & 0 & 0 & 0 \\
0 & 0 & 0 & -2 & 0 & 0 & 0 & 0 & 0 & 0 & 0 & 0 & 0 & 0 \\
0 & 0 & 2 & 0 & 0 & 0 & 0 & 0 & 0 & 0 & 0 & 0 & 0 & 0 \\
0 & 2 & 0 & 0 & 0 & 0 & 0 & 0 & 0 & 0 & 0 & 0 & 0 & 0 \\
-2 & 0 & 0 & 0 & 0 & 0 & 0 & 0 & 0 & 0 & 0 & 0 & 0 & 0 \\
0 & 0 & 0 & 0 & 0 & 0 & 0 & 0 & 0 & 0 & 0 & 1 & 0 & 0 \\
0 & 0 & 0 & 0 & 0 & 0 & 0 & 0 & 0 & 0 & 0 & 0 & 1 & 0 \\
0 & 0 & 0 & 0 & 0 & 0 & 0 & 0 & 0 & 0 & 0 & 0 & 0 & 1 \\
0 & 0 & 0 & 0 & 0 & 0 & 0 & 0 & -1 & 0 & 0 & 0 & 0 & 0 \\
0 & 0 & 0 & 0 & 0 & 0 & 0 & 0 & 0 & -1 & 0 & 0 & 0 & 0 \\
0 & 0 & 0 & 0 & 0 & 0 & 0 & 0 & 0 & 0 & -1 & 0 & 0 & 0
\end{pmatrix} \tag{7.12.6}
$$

which verifies the relation:

$$
\forall\, \mathfrak{g} \in \mathfrak{sp}(6,\mathbb{R}) \quad : \quad \widehat{\mathscr{D}}_{14}(\mathfrak{g})^T\, \widehat{\mathbb{C}}_{14} + \widehat{\mathbb{C}}_{14}\, \widehat{\mathscr{D}}_{14}(\mathfrak{g}) = 0 \tag{7.12.7}
$$

Unfortunately $\widehat{\mathbb{C}}_{14}$ is not yet the standard symplectic matrix for the Lie algebra $\mathfrak{sp}(14,\mathbb{R})$. Hence we still need to perform a change of basis that brings $\widehat{\mathbb{C}}_{14}$ to its standard form:

$$
\mathbb{C}_{14} \equiv \begin{pmatrix}
0 & 0 & 0 & 0 & 0 & 0 & 0 & 1 & 0 & 0 & 0 & 0 & 0 & 0 \\
0 & 0 & 0 & 0 & 0 & 0 & 0 & 0 & 1 & 0 & 0 & 0 & 0 & 0 \\
0 & 0 & 0 & 0 & 0 & 0 & 0 & 0 & 0 & 1 & 0 & 0 & 0 & 0 \\
0 & 0 & 0 & 0 & 0 & 0 & 0 & 0 & 0 & 0 & 1 & 0 & 0 & 0 \\
0 & 0 & 0 & 0 & 0 & 0 & 0 & 0 & 0 & 0 & 0 & 1 & 0 & 0 \\
0 & 0 & 0 & 0 & 0 & 0 & 0 & 0 & 0 & 0 & 0 & 0 & 1 & 0 \\
0 & 0 & 0 & 0 & 0 & 0 & 0 & 0 & 0 & 0 & 0 & 0 & 0 & 1 \\
-1 & 0 & 0 & 0 & 0 & 0 & 0 & 0 & 0 & 0 & 0 & 0 & 0 & 0 \\
0 & -1 & 0 & 0 & 0 & 0 & 0 & 0 & 0 & 0 & 0 & 0 & 0 & 0 \\
0 & 0 & -1 & 0 & 0 & 0 & 0 & 0 & 0 & 0 & 0 & 0 & 0 & 0 \\
0 & 0 & 0 & -1 & 0 & 0 & 0 & 0 & 0 & 0 & 0 & 0 & 0 & 0 \\
0 & 0 & 0 & 0 & -1 & 0 & 0 & 0 & 0 & 0 & 0 & 0 & 0 & 0 \\
0 & 0 & 0 & 0 & 0 & -1 & 0 & 0 & 0 & 0 & 0 & 0 & 0 & 0 \\
0 & 0 & 0 & 0 & 0 & 0 & -1 & 0 & 0 & 0 & 0 & 0 & 0 & 0
\end{pmatrix} \tag{7.12.8}
$$

Such a change of basis is provided by the matrix:

$$\Lambda = \begin{pmatrix}
0 & 0 & 0 & \frac{1}{\sqrt{2}} & 0 & 0 & 0 & 0 & 0 & 0 & 0 & 0 & 0 & 0 \\
0 & 0 & -\frac{1}{\sqrt{2}} & 0 & 0 & 0 & 0 & 0 & 0 & 0 & 0 & 0 & 0 & 0 \\
0 & -\frac{1}{\sqrt{2}} & 0 & 0 & 0 & 0 & 0 & 0 & 0 & 0 & 0 & 0 & 0 & 0 \\
\frac{1}{\sqrt{2}} & 0 & 0 & 0 & 0 & 0 & 0 & 0 & 0 & 0 & 0 & 0 & 0 & 0 \\
0 & 0 & 0 & 0 & 0 & 0 & 0 & \frac{1}{\sqrt{2}} & 0 & 0 & 0 & 0 & 0 & 0 \\
0 & 0 & 0 & 0 & 0 & 0 & 0 & 0 & \frac{1}{\sqrt{2}} & 0 & 0 & 0 & 0 & 0 \\
0 & 0 & 0 & 0 & 0 & 0 & 0 & 0 & 0 & \frac{1}{\sqrt{2}} & 0 & 0 & 0 & 0 \\
0 & 0 & 0 & 0 & 0 & 0 & 0 & 0 & 0 & 0 & \frac{1}{\sqrt{2}} & 0 & 0 & 0 \\
0 & 0 & 0 & 0 & 1 & 0 & 0 & 0 & 0 & 0 & 0 & 0 & 0 & 0 \\
0 & 0 & 0 & 0 & 0 & 1 & 0 & 0 & 0 & 0 & 0 & 0 & 0 & 0 \\
0 & 0 & 0 & 0 & 0 & 0 & 1 & 0 & 0 & 0 & 0 & 0 & 0 & 0 \\
0 & 0 & 0 & 0 & 0 & 0 & 0 & 0 & 0 & 0 & 0 & 1 & 0 & 0 \\
0 & 0 & 0 & 0 & 0 & 0 & 0 & 0 & 0 & 0 & 0 & 0 & 1 & 0 \\
0 & 0 & 0 & 0 & 0 & 0 & 0 & 0 & 0 & 0 & 0 & 0 & 0 & 1
\end{pmatrix} \tag{7.12.9}$$

Indeed we have:

$$\begin{aligned}
\Lambda^T \, \widehat{\mathbb{C}}_{14} \Lambda^T &= \mathbb{C}_{14} \\
\mathscr{D}_{14}(\mathfrak{g}) &\equiv \Lambda^{-1} \, \widehat{\mathscr{D}}_{14}(\mathfrak{g}) \, \Lambda \\
0 &= \mathscr{D}_{14}(\mathfrak{g})^T \mathbb{C}_{14} + \mathbb{C}_{14} \, \mathscr{D}_{14}(\mathfrak{g})
\end{aligned} \tag{7.12.10}$$

We do not present the intermediate matrices $\widehat{\mathscr{D}}_{14}(\mathfrak{g})$. We go directly to the presentation of the final ones $\mathscr{D}_{14}(\mathfrak{g})$ that are standard symplectic matrices of the $\mathfrak{sp}(14, \mathbb{R})$ Lie algebra.

7.12.2.1 Cartan Generators in the 14′

The Cartan generators are named $\mathscr{H}^i_{14} \equiv \mathscr{D}_{14}\left(\mathscr{H}^i\right)$ and can be easily read-off from the following formula:

$$\sum_{i=1}^{3} h_i \, \mathscr{H}^i_{14} =$$

$$\left(\begin{array}{ccccccc}
-h_1 - h_2 + h_3 & 0 & 0 & 0 & 0 & 0 & 0 \\
0 & h_1 - h_2 + h_3 & 0 & 0 & 0 & 0 & 0 \\
0 & 0 & -h_1 + h_2 + h_3 & 0 & 0 & 0 & 0 \\
0 & 0 & 0 & h_1 + h_2 + h_3 & 0 & 0 & 0 \\
0 & 0 & 0 & 0 & h_1 & 0 & 0 \\
0 & 0 & 0 & 0 & 0 & h_2 & 0 \\
0 & 0 & 0 & 0 & 0 & 0 & h_3 \\
0 & 0 & 0 & 0 & 0 & 0 & 0 \\
0 & 0 & 0 & 0 & 0 & 0 & 0 \\
0 & 0 & 0 & 0 & 0 & 0 & 0 \\
0 & 0 & 0 & 0 & 0 & 0 & 0 \\
0 & 0 & 0 & 0 & 0 & 0 & 0 \\
0 & 0 & 0 & 0 & 0 & 0 & 0 \\
0 & 0 & 0 & 0 & 0 & 0 & 0
\end{array}\right.$$

$$
\left.\begin{array}{ccccccc}
0 & 0 & 0 & 0 & 0 & 0 & 0 \\
0 & 0 & 0 & 0 & 0 & 0 & 0 \\
0 & 0 & 0 & 0 & 0 & 0 & 0 \\
0 & 0 & 0 & 0 & 0 & 0 & 0 \\
0 & 0 & 0 & 0 & 0 & 0 & 0 \\
0 & 0 & 0 & 0 & 0 & 0 & 0 \\
0 & 0 & 0 & 0 & 0 & 0 & 0 \\
h_1+h_2-h_3 & 0 & 0 & 0 & 0 & 0 & 0 \\
0 & -h_1+h_2-h_3 & 0 & 0 & 0 & 0 & 0 \\
0 & 0 & h_1-h_2-h_3 & 0 & 0 & 0 & 0 \\
0 & 0 & 0 & -h_1-h_2-h_3 & 0 & 0 & 0 \\
0 & 0 & 0 & 0 & -h_1 & 0 & 0 \\
0 & 0 & 0 & 0 & 0 & -h_2 & 0 \\
 & 0 & 0 & 0 & 0 & 0 & 0 \; -h_3
\end{array}\right)
\tag{7.12.11}
$$

by collecting the coefficient of the parameter h_i.

7.12.2.2 Positive Root Step Operators in the 14′

The step operator associated with the positive root α_i is named $\mathscr{E}_{14}^{\alpha_i} \equiv \mathscr{D}_{14}\left(\mathscr{E}^{\alpha_i}\right)$ and can be easily read-off from the following formula:

$$
\sum_{i=1}^{9} a_i\, \mathscr{E}_{14}^{\alpha_i} =
$$

$$
\left(\begin{array}{cccccccccccccc}
0 & 0 & 0 & 0 & 0 & 0 & 0 & 0 & 0 & 0 & \sqrt{2}a_3 & 0 & 0 & 0 \\
\sqrt{2}a_9 & 0 & 0 & 0 & 0 & 0 & -\sqrt{2}a_1 & 0 & 0 & \sqrt{2}a_3 & 0 & 0 & -\sqrt{2}a_6 & 0 \\
-\sqrt{2}a_7 & 0 & 0 & 0 & 0 & 0 & 0 & 0 & \sqrt{2}a_3 & 0 & 0 & -\sqrt{2}a_5 & 0 & 0 \\
0 & \sqrt{2}a_7 & -\sqrt{2}a_9 & 0 & \sqrt{2}a_5 & -\sqrt{2}a_6 & \sqrt{2}a_8 & \sqrt{2}a_3 & 0 & 0 & 0 & 0 & 0 & 0 \\
0 & -\sqrt{2}a_2 & 0 & 0 & 0 & a_1 & -a_4 & 0 & 0 & -\sqrt{2}a_5 & 0 & \sqrt{2}a_9 & a_8 & -a_6 \\
0 & 0 & -\sqrt{2}a_4 & 0 & 0 & 0 & a_2 & 0 & -\sqrt{2}a_6 & 0 & 0 & a_8 & \sqrt{2}a_7 & a_5 \\
-\sqrt{2}a_8 & 0 & \sqrt{2}a_1 & 0 & 0 & 0 & 0 & 0 & 0 & 0 & 0 & -a_6 & a_5 & \sqrt{2}a_3 \\
0 & 0 & 0 & 0 & -\sqrt{2}a_2 & \sqrt{2}a_4 & 0 & 0 & -\sqrt{2}a_9 & \sqrt{2}a_7 & 0 & 0 & 0 & \sqrt{2}a_8 \\
0 & 0 & 0 & 0 & 0 & 0 & 0 & 0 & 0 & 0 & -\sqrt{2}a_7 & \sqrt{2}a_2 & 0 & 0 \\
0 & 0 & 0 & 0 & 0 & 0 & 0 & 0 & 0 & 0 & \sqrt{2}a_9 & 0 & \sqrt{2}a_4 & -\sqrt{2}a_1 \\
0 & 0 & 0 & 0 & 0 & 0 & 0 & 0 & 0 & 0 & 0 & 0 & 0 & 0 \\
-\sqrt{2}a_2 & 0 & 0 & 0 & 0 & 0 & 0 & 0 & 0 & 0 & -\sqrt{2}a_5 & 0 & 0 & 0 \\
\sqrt{2}a_4 & 0 & 0 & 0 & 0 & 0 & 0 & 0 & 0 & 0 & \sqrt{2}a_6 & -a_1 & 0 & 0 \\
0 & 0 & 0 & 0 & 0 & 0 & 0 & 0 & \sqrt{2}a_1 & 0 & -\sqrt{2}a_8 & a_4 & -a_2 & 0
\end{array}\right)
\tag{7.12.12}
$$

by collecting the coefficient of the parameter a_i.

7.12.2.3 Negative Root Step Operators in the 14′

The step operator associated with the negative root $-\alpha_i$ is named $\mathscr{E}_{14}^{-\alpha_i} \equiv \mathscr{D}_{14}\left(-\mathscr{E}^{\alpha_i}\right)$ and can be easily read-off from the following formula:

$$\sum_{i=1}^{9} b_i\, \mathscr{E}_{14}^{-\alpha_i} = \left(\begin{array}{cccccccccccccc}
0 & \sqrt{2}b_9 & -\sqrt{2}b_7 & 0 & 0 & 0 & -\sqrt{2}b_8 & 0 & 0 & 0 & 0 & -\sqrt{2}b_2 & \sqrt{2}b_4 & 0 \\
0 & 0 & 0 & \sqrt{2}b_7 & -\sqrt{2}b_2 & 0 & 0 & 0 & 0 & 0 & 0 & 0 & 0 & 0 \\
0 & 0 & 0 & -\sqrt{2}b_9 & 0 & -\sqrt{2}b_4 & \sqrt{2}b_1 & 0 & 0 & 0 & 0 & 0 & 0 & 0 \\
0 & 0 & 0 & 0 & 0 & 0 & 0 & 0 & 0 & 0 & 0 & 0 & 0 & 0 \\
0 & 0 & 0 & \sqrt{2}b_5 & 0 & 0 & 0 & -\sqrt{2}b_2 & 0 & 0 & 0 & 0 & 0 & 0 \\
0 & 0 & 0 & -\sqrt{2}b_6 & b_1 & 0 & 0 & \sqrt{2}b_4 & 0 & 0 & 0 & 0 & 0 & 0 \\
0 & -\sqrt{2}b_1 & 0 & \sqrt{2}b_8 & -b_4 & b_2 & 0 & 0 & 0 & 0 & 0 & 0 & 0 & 0 \\
0 & 0 & 0 & \sqrt{2}b_3 & 0 & 0 & 0 & 0 & 0 & 0 & 0 & 0 & 0 & 0 \\
0 & 0 & \sqrt{2}b_3 & 0 & 0 & -\sqrt{2}b_6 & 0 & -\sqrt{2}b_9 & 0 & 0 & 0 & 0 & 0 & \sqrt{2}b_1 \\
0 & \sqrt{2}b_3 & 0 & 0 & -\sqrt{2}b_5 & 0 & 0 & \sqrt{2}b_7 & 0 & 0 & 0 & 0 & 0 & 0 \\
\sqrt{2}b_3 & 0 & 0 & 0 & 0 & 0 & 0 & 0 & -\sqrt{2}b_7 & \sqrt{2}b_9 & 0 & -\sqrt{2}b_5 & \sqrt{2}b_6 & -\sqrt{2}b_8 \\
0 & 0 & -\sqrt{2}b_5 & 0 & \sqrt{2}b_9 & b_8 & -b_6 & 0 & \sqrt{2}b_2 & 0 & 0 & 0 & -b_1 & b_4 \\
0 & -\sqrt{2}b_6 & 0 & 0 & b_8 & \sqrt{2}b_7 & b_5 & 0 & 0 & \sqrt{2}b_4 & 0 & 0 & 0 & -b_2 \\
0 & 0 & 0 & 0 & -b_6 & b_5 & \sqrt{2}b_3 & \sqrt{2}b_8 & 0 & -\sqrt{2}b_1 & 0 & 0 & 0 & 0
\end{array}\right) \tag{7.12.13}$$

by collecting the coefficient of the parameter b_i.

7.12.2.4 The Metric in Solvable Coordinates

The explicit form of the metric in solvable coordinates reads:

$$\begin{aligned}
ds_K^2 =\; & \mathrm{dh}_1^2 + \mathrm{dh}_2^2 + \mathrm{dh}_3^2 + \frac{1}{2}e^{2h_2-2h_1}\mathrm{dp}_1^2 + \frac{1}{2}e^{2h_3-2h_2}\mathrm{dp}_2^2 \\
& + \frac{1}{2}e^{-4h_3}\mathrm{dp}_3^2 + \frac{1}{2}e^{2h_3-2h_1}\mathrm{dp}_4^2 + \frac{1}{2}e^{-2h_2-2h_3}\mathrm{dp}_5^2 + \frac{1}{2}e^{-2h_1-2h_3}\mathrm{dp}_6^2 + \frac{1}{2}e^{-4h_2}\mathrm{dp}_7^2 \\
& + \frac{1}{2}e^{-2h_1-2h_2}\mathrm{dp}_8^2 - \sqrt{2}e^{-2h_1-2h_2}\mathrm{dp}_7 p_1 \mathrm{dp}_8 + \frac{1}{2}e^{-4h_1}\mathrm{dp}_9^2 \\
& - e^{2h_3-2h_1}\mathrm{dp}_2\mathrm{dp}_4 p_1 - e^{-2h_1-2h_3}\mathrm{dp}_5\mathrm{dp}_6 p_1 \\
& + \frac{1}{2}e^{2h_3-2h_1}\mathrm{dp}_2^2 p_1^2 + \frac{1}{2}e^{-2h_1-2h_3}\mathrm{dp}_5^2 p_1^2 + e^{-2h_1-2h_2}\mathrm{dp}_7^2 p_1^2 + e^{-4h_1}\mathrm{dp}_8^2 p_1^2 \\
& - \sqrt{2}e^{-4h_1}\mathrm{dp}_8\mathrm{dp}_9 p_1 \\
& + \frac{1}{2}e^{-4h_1}\mathrm{dp}_7^2 p_1^4 - \sqrt{2}e^{-4h_1}\mathrm{dp}_7\mathrm{dp}_8 p_1^3 + e^{-4h_1}\mathrm{dp}_7\mathrm{dp}_9 p_1^2 - \sqrt{2}e^{-2h_2-2h_3}\mathrm{dp}_3\mathrm{dp}_5 p_2 \\
& - \sqrt{2}e^{-4h_2}\mathrm{dp}_5\mathrm{dp}_7 p_2 \\
& - e^{-2h_1-2h_2}\mathrm{dp}_6\mathrm{dp}_8 p_2 + \sqrt{2}e^{-2h_1-2h_3}\mathrm{dp}_3\mathrm{dp}_6 p_1 p_2 + \sqrt{2}e^{-2h_1-2h_2}\mathrm{dp}_6\mathrm{dp}_7 p_1 p_2 \\
& + 2e^{-2h_1-2h_2}\mathrm{dp}_5\mathrm{dp}_8 p_1 p_2 \\
& + \sqrt{2}e^{-4h_1}\mathrm{dp}_6\mathrm{dp}_9 p_1 p_2 \\
& + \sqrt{2}e^{-4h_1}\mathrm{dp}_6\mathrm{dp}_7 p_2 p_1^3 - \sqrt{2}e^{-2h_1-2h_3}\mathrm{dp}_3\mathrm{dp}_5 p_2 p_1^2 - 2\sqrt{2}e^{-2h_1-2h_2}\mathrm{dp}_5\mathrm{dp}_7 p_2 p_1^2 \\
& - 2e^{-4h_1}\mathrm{dp}_6\mathrm{dp}_8 p_2 p_1^2 - \sqrt{2}e^{-4h_1}\mathrm{dp}_5\mathrm{dp}_9 p_2 p_1^2 \\
& - \sqrt{2}e^{-4h_1}\mathrm{dp}_5\mathrm{dp}_7 p_2 p_1^4 + \sqrt{2}e^{-4h_1}\mathrm{dp}_6\mathrm{dp}_7 p_2 p_1^3 + 2e^{-4h_1}\mathrm{dp}_5\mathrm{dp}_8 p_2 p_1^3 \\
& + e^{-2h_2-2h_3}\mathrm{dp}_3^2 p_2^2 \\
& + e^{-4h_2}\mathrm{dp}_5^2 p_2^2 + \frac{1}{2}e^{-2h_1-2h_2}\mathrm{dp}_6^2 p_2^2 \\
& + e^{-2h_1-2h_3}\mathrm{dp}_3^2 p_1^2 p_2^2 + 2e^{-2h_1-2h_2}\mathrm{dp}_5^2 p_1^2 p_2^2 + e^{-4h_2}\mathrm{dp}_3\mathrm{dp}_7 p_2^2 \\
& - 2e^{-2h_1-2h_2}\mathrm{dp}_5\mathrm{dp}_6 p_1 p_2^2 - \sqrt{2}e^{-2h_1-2h_2}\mathrm{dp}_3\mathrm{dp}_8 p_1 p_2^2 \\
& - 2e^{-4h_1}\mathrm{dp}_5\mathrm{dp}_6 p_2^2 p_1^3 - \sqrt{2}e^{-4h_1}\mathrm{dp}_3\mathrm{dp}_8 p_2^2 p_1^3 + 2e^{-2h_1-2h_2}\mathrm{dp}_5^2 p_2^2 p_1^2 \\
& + e^{-4h_1}\mathrm{dp}_6^2 p_2^2 p_1^2 \\
& + 2e^{-2h_1-2h_2}\mathrm{dp}_3\mathrm{dp}_7 p_2^2 p_1^2 + e^{-4h_1}\mathrm{dp}_3\mathrm{dp}_9 p_2^2 p_1^2 \\
& + e^{-4h_1}\mathrm{dp}_5^2 p_2^2 p_1^4 + e^{-4h_1}\mathrm{dp}_3\mathrm{dp}_7 p_2^2 p_1^4 - 2\sqrt{2}e^{-2h_1-2h_2}\mathrm{dp}_3\mathrm{dp}_5 p_2^3 p_1^2 \\
& + \sqrt{2}e^{-2h_1-2h_2}\mathrm{dp}_3\mathrm{dp}_6 p_2^3 p_1 - \sqrt{2}e^{-4h_2}\mathrm{dp}_3\mathrm{dp}_5 p_2^3
\end{aligned}$$

$$
\begin{aligned}
&+\frac{1}{2}e^{-4h_1}dp_3^2p_2^4p_1^4-\sqrt{2}e^{-4h_1}dp_3dp_5p_2^3p_1^4+\sqrt{2}e^{-4h_1}dp_3dp_6p_2^3p_1^3\\
&+e^{-2h_1-2h_2}dp_3^2p_2^4p_1^2\\
&-2\sqrt{2}e^{-2h_1-2h_2}dp_3dp_5p_2^3p_1^2+\frac{1}{2}e^{-4h_2}dp_3^2p_2^4\\
&+\sqrt{2}e^{-2h_1-2h_3}dp_4dp_6p_3-\sqrt{2}e^{-2h_1-2h_3}dp_4dp_5p_1p_3\\
&-\sqrt{2}e^{-2h_1-2h_2}dp_4dp_8p_2p_3+2e^{-2h_1-2h_3}dp_3dp_4p_1p_2p_3+2e^{-2h_1-2h_2}dp_4dp_7p_1p_2p_3\\
&+2e^{-4h_1}dp_4dp_7p_2p_3p_1^3-2\sqrt{2}e^{-4h_1}dp_4dp_8p_2p_3p_1^2\\
&-2\sqrt{2}e^{-2h_1-2h_2}dp_4dp_5p_2^2p_3p_1+2e^{-4h_1}dp_4dp_9p_2p_3p_1+\sqrt{2}e^{-2h_1-2h_2}dp_4dp_6p_2^2p_3\\
&+2e^{-4h_1}dp_3dp_4p_2^3p_3p_1^3-2\sqrt{2}e^{-4h_1}dp_4dp_5p_2^2p_3p_1^3+2\sqrt{2}e^{-4h_1}dp_4dp_6p_2^2p_3p_1^2\\
&+2e^{-2h_1-2h_2}dp_3dp_4p_2^3p_3p_1+e^{-2h_1-2h_3}dp_4^2p_3^2\\
&+e^{-2h_1-2h_2}dp_4^2p_2^2p_3^2+2e^{-4h_1}dp_4^2p_1^2p_2^2p_3^2-e^{-2h_1-2h_2}dp_5dp_8p_4\\
&-\sqrt{2}e^{-4h_1}dp_6dp_9p_4+\sqrt{2}e^{-2h_1-2h_2}dp_5dp_7p_1p_4\\
&+\sqrt{2}e^{-4h_1}dp_5dp_7p_4p_1^3-\sqrt{2}e^{-4h_1}dp_6dp_7p_4p_1^2-2e^{-4h_1}dp_5dp_8p_4p_1^2\\
&+2e^{-4h_1}dp_6dp_8p_4p_1+\sqrt{2}e^{-4h_1}dp_5dp_9p_4p_1\\
&-2e^{-4h_1}dp_5^2p_2p_4p_1^3+4e^{-4h_1}dp_5dp_6p_2p_4p_1^2-2e^{-2h_1-2h_2}dp_5^2p_2p_4p_1\\
&-2e^{-4h_1}dp_6^2p_2p_4p_1+e^{-2h_1-2h_2}dp_5dp_6p_2p_4\\
&\sqrt{2}e^{-4h_1}dp_3dp_5p_2^2p_4p_1^3-\sqrt{2}e^{-4h_1}dp_3dp_6p_2^2p_4p_1^2+\sqrt{2}e^{-2h_1-2h_2}dp_3dp_5p_2^2p_4p_1\\
&-2\sqrt{2}e^{-4h_1}dp_4dp_6p_2p_3p_4p_1+\sqrt{2}e^{-2h_1-2h_2}dp_4dp_5p_2p_3p_4\\
&+e^{-4h_1}dp_5^2p_4^2p_1^2+2\sqrt{2}e^{-4h_1}dp_4dp_5p_2p_3p_4p_1^2-2e^{-4h_1}dp_5dp_6p_4^2p_1\\
&+\frac{1}{2}e^{-2h_1-2h_2}dp_5^2p_4^2+e^{-4h_1}dp_6^2p_4^2
\end{aligned}
\tag{7.12.14}
$$

References

1. E. Cremmer, B. Julia, The SO(8) supergravity. Nucl. Phys. B **159**, 141–212 (1979)
2. B. De Wit, H. Nicolai, N=8 supergravity with local SO(8) × SU(8) invariance. Phys. Lett. **108B**, 285 (1982)
3. B. de Wit, H. Nicolai, N=8 supergravity. Nucl. Phys. B **208**, 323 (1982)
4. P.G.O. Freund, M.A. Rubin, Dynamics of dimensional reduction. Phys. Lett. B **97**, 233–235 (1980)
5. L. Castellani, R. D'Auria, P. Fré, SU(3) ⊗ SU(2) ⊗ U(1) from D=11 supergravity. Nucl. Phys. B **239**, 610–652 (1984)
6. F. Englert, Spontaneous compactification of eleven-dimensional supergravity. Phys. Lett. B **119**, 339–342 (1982)
7. M.A. Awada, M.J. Duff, C.N. Pope, $\mathcal{N}=8$ supergravity breaks down to $\mathcal{N}=1$. Phys. Rev. Lett. **50**, 294 (1983)
8. R. D'Auria, P. Fré, P. Van Nieuwenhuizen, $\mathcal{N}=2$ matter coupled supergravity from compactification on a coset G/H possessing an additional killing vector. Phys. Lett. B **136**, 347–353 (1984)

9. L. Castellani, L.J. Romans, $N = 3$ and $N = 1$ supersymmetry in a new class of solutions for $d = 11$ supergravity. Nucl. Phys. B **238**, 683–701 (1984)
10. L. Castellani, L.J. Romans, N.P. Warner, A classification of compactifying solutions for D=11 supergravity. Nucl. Phys. B **241**, 429–462 (1984)
11. R. D'Auria, P. Fré, On the fermion mass-spectrum of Kaluza-Klein supergravity. Ann. Phys. **157**, 1–100 (1984)
12. D.Z. Freedman, H. Nicolai, Multiplet shortening in Osp(N|4). Nucl. Phys. B **237**, 342–366 (1984)
13. A. Ceresole, P. Fré, H. Nicolai, Multiplet structure and spectra of $\mathcal{N} = 2$ supersymmetric compactifications. Class. Quantum Gravity **2**, 133 (1985)
14. A. Casher, F. Englert, H. Nicolai, M. Rooman, The mass spectrum of supergravity on the round seven sphere. Nucl. Phys. B **243**, 173 (1984)
15. M. Duff, B. Nilsson, C. Pope, Kaluza klein supergravity. Phys. Rep. **130**, 1 (1986)
16. C.M. Hull, Noncompact gaugings of $N = 8$ supergravity. Phys. Lett. **142B**, 39 (1984)
17. C.M. Hull, More gaugings of $N = 8$ supergravity. Phys. Lett. **148B**, 297–300 (1984)
18. C.M. Hull, The construction of new gauged $N = 8$ supergravities. Physica **15D**, 230 (1985)
19. C.M. Hull, A new gauging of $N = 8$ supergravity. Phys. Rev. D **30**, 760 (1984)
20. C.M. Hull, N.P. Warner, The potentials of the gauged $N = 8$ supergravity theories. Nucl. Phys. B **253**, 675–686 (1985)
21. C.M. Hull, N.P. Warner, The structure of the gauged $N = 8$ supergravity theories. Nucl. Phys. B **253**, 650–674 (1985)
22. C.M. Hull, The minimal couplings and scalar potentials of the gauged $N = 8$ supergravities. Class. Quant. Grav. **2**, 343 (1985)
23. J. Maldacena, The large-N limit of superconformal field theories and supergravity. Int. J. Theor. Phys. **4**, 1113–1133 (1999). https://doi.org/10.1023/A:1026654312961 [hep-th/9711200]
24. R. Kallosh, A. Van Proeyen, Conformal symmetry of supergravities in AdS spaces. Phys. Rev. D **60**, 026001 (1999). https://doi.org/10.1103/PhysRevD.60.026001 [hep- th/9804099]
25. S. Ferrara, C. Fronsdal, A. Zaffaroni, On $\mathcal{N} = 8$ supergravity in AdS_5 and $\mathcal{N} = 4$ superconformal Yang-Mills theory. Nucl. Phys. B **532**, 153–162 (1998). https://doi.org/10.1016/S0550-3213(98)00444-1 [hep-th/9802203]
26. S. Ferrara, A. Zaffaroni, Superconformal field theories, multiplet shortening, and the AdS(5) / SCFT(4) correspondence, in *Conference Moshe Flato Dijon, France*, September 5–8, (1999), pp. 177–188
27. S. Ferrara, M. Porrati, A. Zaffaroni, N=6 supergravity on AdS(5) and the SU(2,2/3) superconformal correspondence. Lett. Math. Phys. **47**, 255–263 (1999)
28. A. Ceresole, G. Dall'Agata, R. D'Auria, S. Ferrara, Spectrum of type IIB supergravity on $AdS_5 \times T^{11}$: predictions on $\mathcal{N} = 1$ SCFT's. Phys. Rev. D **61**, 066001 (2000). [hep-th/9905226]
29. D. Fabbri, P. Fré, L. Gualtieri, P. Termonia, Osp(N|4) supermultiplets as conformal superfields on ∂AdS_4 and the generic form of $\mathcal{N} = 2$, D=3 gauge theories. Class. Quantum Gravity **17**, 55 (2000). [hep-th/9905134]
30. D. Fabbri, P. Fré, L. Gualtieri, C. Reina, A. Tomasiello, A. Zaffaroni, A. Zampa, 3D superconformal theories from Sasakian seven-manifolds: new non-trivial evidences for AdS_4/CFT_3. Nucl. Phys. B, **577**, 547–608 (2000). [hep- th/9907219]
31. M. Billó, D. Fabbri, P. Fré, P. Merlatti, A. Zaffaroni, Shadow multiplets in AdS_4/CFT_3 and the super-Higgs mechanism: hints of new shadow supergravities. Nucl. Phys. B **591**, 139–194 (2000). [hep-th/0005220]
32. D. Fabbri, P. Fré, L. Gualtieri, P. Termonia, M-theory on $AdS_4 \times M^{1,1,1}$: the complete $Osp(2|4) \times SU(3) \times SU(2)$ spectrum from harmonic analysis. Nucl. Phys. B **560**, 617–682 (1999). [hep-th/9903036]
33. F. Cordaro, P. Fre, L. Gualtieri, P. Termonia, M. Trigiante, N=8 gaugings revisited: an exhaustive classification. Nucl. Phys. B **532**, 245–279 (1998)
34. B. de Wit, H. Samtleben, M. Trigiante, On Lagrangians and gaugings of maximal supergravities. Nucl. Phys. B **655**, 93–126 (2003)

35. B. De Wit, H. Samtleben, M. Trigiante, Gauging maximal supergravities. Fortsch. Phys. **52**, 489–496 (2004)
36. B. De Wit, H. Samtleben, M. Trigiante, The maximal D=4 supergravities. JHEP **06**, 049 (2007)
37. M. Trigiante, Gauged supergravities. Phys. Rept. **680**, 1–175 (2017)
38. H. Samtleben, Lectures on gauged supergravity and flux compactifications. Class. Quant. Grav. **25**, 214002 (2008)
39. P. Fre, M. Trigiante, A. Van Proeyen, Stable de Sitter vacua from N=2 supergravity. Class. Quant. Grav. **19**, 4167–4194 (2002)
40. S. Ferrara, P. van Nieuwenhuizen, Noether coupling of massive gravitinos to $N = 1$ supergravity. Phys. Lett. **127B**, 70–74 (1983)
41. M. de Roo, Gauged N=4 matter couplings. Phys. Lett. **156B**, 331–334 (1985)
42. S. Cecotti, L. Girardello, M. Porrati, Some features of SUSY breaking in $N = 2$ supergravity. Phys. Lett. **151B**, 367–371 (1985)
43. S. Ferrara, P. Fre, L. Girardello, Spontaneously broken N=3 supergravity. Nucl. Phys. B **274**, 600–618 (1986)
44. YuM Zinovev, Spontaneous symmetry breaking in N=2 supergravity with matter. Int. J. Mod. Phys. A **7**, 7515–7539 (1992)
45. S. Ferrara, L. Girardello, M. Porrati, Minimal Higgs branch for the breaking of half of the supersymmetries in N=2 supergravity. Phys. Lett. B **366**, 155–159 (1996)
46. P. Fre, L. Girardello, I. Pesando, M. Trigiante, Spontaneous N=2 -¿ N=1 local supersymmetry breaking with surviving compact gauge group. Nucl. Phys. B **493**, 231–248 (1997)
47. L. Girardello, M. Porrati, A. Zaffaroni, Spontaneously broken N=2 supergravity without light mirror fermions. Nucl. Phys. B **505**, 272–290 (1997)
48. S. Cecotti, L. Girardello, M. Porrati, Ward identities of local supersymmetry and spontaneous breaking of extended supergravity, in *Johns Hopkins Workshop on Current Problems in Particle Theory 9: New Trends in Particle Theory Florence, Italy*, June 5–7 (1985)
49. S. Ferrara, L. Maiani, An introduction to supersymmetry breaking in extended supergravity, in *5th SILARG Symposium on Relativity, Supersymmetry and Cosmology. Bariloche, Argentina*, January 4–11, (1985), p. 349
50. M. De Roo, P. Wagemans, Partial supersymmetry breaking in $N = 4$ supergravity. Phys. Lett. **B177**, 352 (1986)
51. M. de Roo, P. Wagemans, Gauge matter coupling in $N = 4$ supergravity. Nucl. Phys. B **262**, 644 (1985)
52. P.C.C. Wagemans, Aspects of N=4 supergravity. Ph.D. thesis, Groningen U (1990)
53. G. Arcioni, A. Ceresole, F. Cordaro, R. D'Auria, P. Fre, L. Gualtieri, M. Trigiante, N=8 BPS black holes with 1/2 or 1/4 supersymmetry and solvable Lie algebra decompositions. Nucl. Phys. B **542**, 273–307 (1999)
54. L. Andrianopoli, R. D'Auria, S. Ferrara, P. Fre, M. Trigiante, E(7)(7) duality, BPS black hole evolution and fixed scalars. Nucl. Phys. B **509**, 463–518 (1998)
55. L. Castellani, R. D'Auria, P. Fré, *Supergravity and Superstrings: A Geometric Perspective*, vol. 1,2,3 (World Scientific, Singapore, 1991)
56. A. Ceresole, G. Dall'Agata, S. Ferrara, M. Trigiante, A. Van Proeyen, A search for an $\mathcal{N} = 2$ inflaton potential. Fortsch. Phys. **62**, 584–606 (2014)
57. P. Fré, A.S. Sorin, M. Trigiante, The *c*- map, Tits Satake subalgebras and the search for $\mathcal{N} = 2$ inflaton potentials. Fortsch. Phys. **63**, 198–258 (2015)
58. S. Ferrara, R. Kallosh, A. Linde, M. Porrati, Minimal supergravity models of inflation. Phys. Rev. D **88**, 085038 (2013)
59. P. Fré, A.S. Sorin, Inflation and integrable one-field cosmologies embedded in rheonomic supergravity. Fortsch. Phys. **62**, 4–25 (2014)
60. P. Fré, A.S. Sorin, Axial symmetric Kahler manifolds, the D -map of inflaton potentials and the Picard-Fuchs equation. Fortsch. Phys. **62**, 26–77 (2014)
61. S. Ferrara, R. Kallosh, A. Van Proeyen, On the supersymmetric completion of $R + R^2$ gravity and cosmology. JHEP **11**, 134 (2013)

62. S. Ferrara, R. Kallosh, A. Linde, M. Porrati, Higher order corrections in minimal supergravity models of inflation. JCAP **1311**, 046 (2013)
63. S. Ferrara, P. Fré, A.S. Sorin, On the topology of the inflaton field in minimal supergravity models. JHEP **04**, 095 (2014)
64. S. Ferrara, P. Fre, A.S. Sorin, On the Gauged Khler isometry in minimal supergravity models of inflation. Fortsch. Phys. **62**, 277–349 (2014)
65. A.A. Starobinsky, A new type of isotropic cosmological models without singularity. Phys. Lett. **91B**, 99–102 (1980)
66. P. Fayet, J. Iliopoulos, Spontaneously broken supergauge symmetries and goldstone spinors. Phys. Lett. **51B**, 461–464 (1974)
67. R. Kallosh, A. Linde, D. Roest, Superconformal inflationary α -attractors. JHEP **11**, 198 (2013)
68. R. Kallosh, A. Linde, D. Roest, Large field inflation and double α -attractors. JHEP **08**, 052 (2014)
69. R. Kallosh, A. Linde, D. Roest, The double attractor behavior of induced inflation. JHEP **09**, 062 (2014)

Chapter 8
(Hyper)Kähler Quotients, ALE-Manifolds and $\mathbb{C}^n/\Gamma$ Singularities

Quando chel cubo con le cose appresso
Se agguaglia a qualche numero discreto
Trouan dui altri differenti in esso.

Niccoló Tartaglia

8.1 Historical and Conceptual Introduction

In this last chapter we turn to the analysis of important developments in complex geometry which took place in the 1980–1990s, directly motivated by supersymmetry and supergravity and completely inconceivable outside such a framework. Notwithstanding their roots in the theoretical physics of the superworld, such developments constitute, by now, the basis of some of the most innovative and alive research directions of contemporary geometry.

8.1.1 *From Supersymmetric σ-Models to (Hyper)Kähler Quotients*

As we remarked in previous chapters, an entire new life was contributed to Geometry by the problems posed by the coupling of matter multiplets to supergravity or by the description of their self-interaction in rigid supersymmetry. This was the cradle of *special geometries* whose theory gained momentum by the end of the 1980s and the beginning of the 1990s. In connection with supersymmetry a basic problem which was to reveal his deep geometrical implications is that of *gauging*: namely how to promote global symmetries of supersymmetric lagrangians to *local gauge* ones. In that context one crucial geometrical item happens to be the moment-map

P. G. Frè, *Advances in Geometry and Lie Algebras from Supergravity*, Theoretical and Mathematical Physics,
https://doi.org/10.1007/978-3-319-74491-9_8

we already discussed in previous chapters. Indeed the hamiltonian functions $\Sigma_A(\phi)$ associated with the generators T_A of any Lie isometry group play a distinctive role in supersymmetric field theories: they are the on-shell value taken by the so named *auxiliary fields* and appear in the supersymmetry transformation rules of the fermion members of the supermultiplets: spin $\frac{1}{2}$ or spin $\frac{3}{2}$ fields. Furthermore, according with a general scheme touched upon in Chap. 7, these hamiltonian functions, or *moment maps*, are also the building blocks of the *scalar potential* generated by the *gauging*.

By the end of the 1980s the geometrical characterization of the scalar manifolds appearing in $\mathcal{N} = 2$ field theories in $D = 4$ or $\mathcal{N} = 4$ in $D = 3$ was universally clear and the notion of *HyperKähler manifolds*, well established both in Theoretical Physics and in Mathematics, was attracting a lot of interest in both communities. The prototype of compact *HyperKähler manifolds* were the torus T^4 and the Kummer surface K3, largely utilized in supergravity and string compactifications. From the mathematical point of view the main interest was focused on the identification and on the construction of new examples, compact or non compact of *HyperKählerian spaces*: supersymmetry came to aid.

In the 1980s, with the presence of Peter van Nieuwenhuizen, one of the three founders of supergravity, and the contiguity to a Department of Mathematics of very high level, the Institute of Theoretical Physics (ITP) of New York State University at Stony Brook had become a very prominent center of Mathematical Physics, particularly active in those geometrical directions that are more closely related to supersymmetry. Several young researchers from Europe that extensively contributed to the topics discussed in this book spent research stages in Stony Brook in various capacities, either as post-doctoral fellows or as visiting scientists (see Fig. 8.1).

In 1987 a milestone paper for the history of HyperKähler geometry was written by four authors, three of which were or had been associated with Stony Brook (see Fig. 8.2). The mentioned paper, entitled *HyperKähler metrics and supersymmetry* authored by Anders Karlhede, Nigel Hitchin, Ulf Lindstrom and Martin Rocek [1] grew out from two different cultural traditions turning out to be extremely influential both in Physics and in Mathematics.

The British author Hitchin, former student of Sir Michael Atiyah and presently his successor on the Savilian Chair of Geometry in Oxford, brought in the distinguished geometrical and topological tradition of the Cambridge school, whose roots can be traced back to Hodge and which is responsible for such other milestones as, for instance, the index theorem. Martin Rocek, Anders Karlhede and Ulf Lindstrom, together with Marc Grisaru and Jim Gates, were among the early founders of superspace formalism for supersymmetric theories and had a deep working knowledge of the latter. From the inbreeding of these two traditions arose a quite powerful new mathematical vision, that of *HyperKähler quotient*.

The guiding line was provided by the lagrangian realization of a supersymmetric field theory encompassing hypermultiplets that span a flat HyperKähler manifold $\mathcal{S}$ and are coupled to gauge vector multiplets which promote a group $\mathcal{G}$ of global isometries of the space $\mathcal{S}$ to local symmetries of the lagrangian. If the kinetic terms of these vector multiplets $\mathbf{V}$ are omitted, the latter can be integrated away by means of a gaussian integration. The result of this functional integration yields, as a remnant, a set of constraints. The systematic solution of such constraints provides the geometrical

Fig. 8.1 The first picture dating 1979 is the historical one taken during the first international conference on Supergravity, held at Stony Brook ITP. The second picture dating 1982 shows Peter van Nieuwenhuizen, the present author and Riccardo D'Auria in front of the Stony Brook house they were sharing during a one month stay of the two Italians for collaboration with van Nieuwenhuizen. The third and the fourth pictures were taken in November 2001 during the conference *Supergravity at 25* held in Stony Brook ITP. In the second picture one sees Leonardo Castellani, the present author, Peter van Nieuwenhuizen and Alberto Lerda. The last picture is the group photo of all participants to the workshop. In the 1980s the scientific relations between Torino University and Stony Brook were particularly intense and fruitful. Equally important were the relations of Stony Brook with Leuven in Belgium, Utrecht in the Netherlands and the École Normale Superiéure in Paris

construction of a new non trivial, yet smaller, HyperKähler manifold, namely the HyperKähler quotient $\mathscr{S}//\mathscr{G}$.

The great value of paper [1] was the clear cut axiomatization of this procedure which, extracted from field theory, was recast in pure mathematical terms as a self contained mathematical construction.

In the following years the HyperKähler quotient was adopted by mathematicians as a preferred constructive algorithm for new HyperKähler manifolds.

A very important instance of such constructions was provided a couple of years after the publication of [1] by Kronheimer, who succeeded in showing that all asymptotically flat gravitational instantons, the so named ALE manifolds, can be realized as HyperKähler quotients [2, 3]. To ALE spaces and to the Kronheimer construction we devote several sections of the present chapter. We anticipate here that the classification of ALE manifolds is a new incarnation of the ADE classification of simply laced Lie algebras, finite subgroups of SU(2) and of singularities. It clearly encodes a very deep connection between fundamental issues of Geometry and Physics.

Fig. 8.2 From the left to the right: Martin Rocek, Anders Karlhede (1952), Nigel J. Hitchin (1946), Ulf Lindstrom (1947), finally a view of the campus of New York University at Stony Brook. Martin Rocek is currently Professor of Theoretical Physics at Stony Brook and a member of the C. N. Yang Institute for Theoretical Physics. He received A.B. and Ph.D. degrees from Harvard University in 1975 and 1979. He did post-doctoral research at the University of Cambridge and Caltech before becoming a professor at Stony Brook. Anders Karlhede is currently Vice Rector of Stockholm University and a member of the Swedish Academy of Sciences. Nigel Hitchin is currently Savilian Professor of Geometry, Oxford, a position previously held by his doctoral supervisor (and later research collaborator) Sir Michael Atiyah. Hitchin is responsible, together with Atiyah for the index theorem and for the ADHM construction of instantons. Ulf Lindstrom is currently chairman of the theoretical physics department at the University of Upsala. He originally graduated from Stockholm University. Lindstrom and Hitchin have both contributed to the development of the notion of generalized complex geometry. In 1987 when their fundamental paper on HyperKähler quotients was written, three of the above four authors (Karlhede, Lindstrom and Rocek) were working at the ITP of Stony Brook

Many current research lines in geometry related with manifolds of *restricted holonomy*, *spin(7) manifolds* and the like are intimately related with the idea of the HyperKähler quotient. Similarly *quiver* constructions in brane physics and most of the geometrical constructions in the CFT/gauge correspondence are off-springs of the HyperKähler quotient algorithm.

8.1.2 Further Geometrical Visions from p-Branes and the Gauge/Gravity Correspondence

The immensely fertile field of the *gauge/gravity correspondence*, originally viewed as the AdS/CFT correspondence, has its starting point in November 1997 with the publication on the ArXive of a paper by Juan Maldacena [4] on the large N limit of gauge theories.

In this book we do not address the multifaceted history of this important subject that has generated an extremely large corpus of physically relevant results. We rather concentrate on its more geometrical aspects and group-theoretical foundations.

The central idea of the gauge/gravity correspondence, frequentely referred to as the *holographic principle*, envisages that fundamental informations on the quantum behavior of fields leaving on some boundary of a larger space-time can be obtained from the classical gravitational dynamics of fields leaving in the bulk of that space-time. This can be regarded as a modern mathematical reformulation of Plato's myth of the shadows on the walls of the cavern (the myth of the *antrum platonicum*). In such a framework geometrical issues are the central focus of attention.

The group theoretical foundations of the AdS/CFT correspondence were explored in many papers and important contributions were given by Ferrara, Fronsdal, Zaffaroni, Kallosh and Van Proeyen in [5–8]. As everything important and profound, the AdS/CFT correspondence has a relative simple origin which, however, is extremely rich in ramified and powerful consequences. The key point is the double interpretation of any anti de Sitter group $\mathrm{SO}(2, \mathrm{p}+1)$ as the isometry group of $\mathrm{AdS}_{\mathrm{p}+2}$ space or as the conformal group on the $p+1$-dimensional boundary $\partial\mathrm{AdS}_{\mathrm{p}+2}$. Such a double interpretation is inherited by the supersymmetric extensions of $\mathrm{SO}(2, \mathrm{p}+1)$. This is what leads to consider *superconformal field theories* on the boundary. Two cases are of particular relevance because of concurrent reasons which are peculiar to them: from the algebraic side the essential use of one of the low rank sporadic isomorphisms of orthogonal Lie algebras, from the supergravity side the existence of a spontaneous compactification of the Freund–Rubin type [9]. The two cases are:

(A) The case $p = 3$ which leads to AdS_5 and to its 4-dimensional boundary. Here the sporadic isomorphism is $\mathrm{SO}(2, 4) \sim \mathrm{SU}(2, 2)$ which implies that the list of superconformal algebras is given by the superalgebras $\mathfrak{su}(2, 2 \mid \mathcal{N})$ for $1 \leq \mathcal{N} \leq 4$. On the other hand in Type IIB Supergravity, there is a self-dual five-form field strength. Giving a v.e.v to this latter ($F_{a_1a_2a_3a_4a_5} \ltimes \varepsilon_{a_1a_2a_3a_4a_5}$), one splits the ambient ten-dimensional space into $5 \oplus 5$ where the first 5 stands for the

AdS$_5$ space, while the second 5 stands for any compact 5-dimensional Einstein manifold $\mathcal{M}_5$. The holonomy of $\mathcal{M}_5$ decides the number of supersymmetries and on the 4-dimensional boundary ∂AdS$_5$ we have a superconformal Yang-Mills gauge theory.

(B) The case $p = 2$ which leads which leads to AdS$_4$ and to its 3-dimensional boundary. Here the sporadic isomorphism is $\mathrm{SO}(2,3) \sim \mathrm{Sp}(4,\mathbb{R})$ which implies that the list of superconformal algebras is given by the superalgebras $\mathrm{Osp}(\mathcal{N} \mid 4)$ for $\mathcal{N} = 1, 2, 3, 6, 8$. On the other hand in $D = 11$ Supergravity, there is a a four-form field strength. Giving a v.e.v to this latter ($F_{a_1a_2a_3a_4} \ltimes \varepsilon_{a_1a_2a_3a_4}$), one splits the ambient ten-dimensional space into $4 \oplus 7$ where 4 stands for the AdS$_4$ space, while 7 stands for any compact 7-dimensional Einstein manifold $\mathcal{M}_7$. The holonomy of $\mathcal{M}_7$ decides the number of supersymmetries and on the 3-dimensional boundary ∂AdS$_4$ we should have a superconformal gauge theory.

From the p-brane point of view, case (A) is associated with $D3$-branes, while case (B) is associated with $M2$-branes.

The former case was that mostly explored at the beginning of the AdS/CFT tale in 1998 and in successive years. Yet the existence of the second case was immediately evident to anyone who had experience in supergravity and particularly to those who had worked in Kaluza–Klein supergravity in the years 1982–1985. Thus in a series of papers [10–16], mostly produced by the Torino Group but in some instances in collaboration with the SISSA Group, the AdS$_4$/CFT$_3$ correspondence was proposed and intensively developed in the spring and in the summer of the year 1999.

One leading idea, motivating this outburst of activity, was that the entire corpus of results on Kaluza–Klein mass-spectra [9, 17–32], which had been derived in the years 1982–1986, could now be recycled in the new superconformal interpretation. Actually it was immediately clear that the Kaluza–Klein towers of states, in particularly those corresponding to short representations of the superalgebra $\mathrm{Osp}(\mathcal{N} \mid 4)$, provided an excellent testing ground for the AdS$_4$/CFT$_3$ correspondence.[1] One had to conceive candidate superconformal field theories living on the boundary, that were able to reproduce all the infinite towers of Kaluza Klein multiplets as corresponding towers of composite operators with the same quantum numbers.

In the case the manifold $\mathcal{M}_7$ was a coset manifold $\mathcal{G}/\mathcal{H}$, an exhaustive list of cases was known since the middle eighties, thanks to the work of Castellani, Romans and Warner [32]. The supersymmetric cases form an even shorter sublist of the main list in [32] and were also classified by the same authors (see Tables 8.1 and 8.2).

Since it was clear that the theory on the boundary had to be a matter coupled gauge-theory, in three papers [12, 13, 15], the general form of matter coupled $\mathcal{N} = 2, 3$ non abelian gauge theories in D = 3, with both a canonical kinetic term for the gauge fields and a Chern Simons one, were constructed using auxiliary fields and the rheonomic approach.

[1]The unitary induced representations of the $\mathrm{Osp}(\mathcal{N} \mid 4)$ superalgebra in their double interpretation as gravitational multiplets or as multiplets of superconformal fields were discussed in [12] and have been systematically reviewed in Chap. 12 of the book [33].

Table 8.1 The homogeneous 7-manifolds that admit at least 2 Killing spinors are all sasakian or tri-sasakian. This is evident from the fibration structure of the 7-manifold, which is either a fibration in circles $\mathbb{S}^1$ for the $\mathcal{N}=2$ cases or a fibration in $\mathbb{S}^3$ for the unique $\mathcal{N}=3$ case corresponding to the N^{010} manifold. Since this latter is also an $N=2$ manifold, there is in addition the $\mathbb{S}^1$ fibration.

$\mathcal{N}$	Name	Coset	Holon. $\mathfrak{so}(8)$bundle	Fibration
8	$\mathbb{S}^7$	$\frac{SO(8)}{SO(7)}$	1	$\begin{cases} \mathbb{S}^7 \stackrel{\pi}{\Longrightarrow} \mathbb{P}^3 \\ \forall p \in \mathbb{P}^3 ; \pi^{-1}(p) \sim \mathbb{S}^1 \end{cases}$
2	M^{111}	$\frac{SU(3)\times SU(2)\times U(1)}{SU(2)\times U(1)\times U(1)}$	SU(3)	$\begin{cases} M^{111} \stackrel{\pi}{\Longrightarrow} \mathbb{P}^2 \times \mathbb{P}^1 \\ \forall p \in \mathbb{P}^2 \times \mathbb{P}^1 ; \pi^{-1}(p) \sim \mathbb{S}^1 \end{cases}$
2	Q^{111}	$\frac{SU(2)\times SU(2)\times SU(2)\times U(1)}{U(1)\times U(1)\times U(1)}$	SU(3)	$\begin{cases} Q^{111} \stackrel{\pi}{\Longrightarrow} \mathbb{P}^1 \times \mathbb{P}^1 \times \mathbb{P}^1 \\ \forall p \in \mathbb{P}^1 \times \mathbb{P}^1 \times \mathbb{P}^1 ; \pi^{-1}(p) \sim \mathbb{S}^1 \end{cases}$
2	$V^{5,2}$	$\frac{SO(5)}{SO(2)}$	SU(3)	$\begin{cases} V^{5,2} \stackrel{\pi}{\Longrightarrow} M_a \sim \text{quadric in } \mathbb{P}^4 \\ \forall p \in M_a ; \pi^{-1}(p) \sim \mathbb{S}^1 \end{cases}$
3	N^{010}	$\frac{SU(3)\times SU(2)}{SU(2)\times U(1)}$	SU(2)	$\begin{cases} N^{010} \stackrel{\pi}{\Longrightarrow} \mathbb{P}^2 \\ \forall p \in \mathbb{P}^2 ; \pi^{-1}(p) \sim \mathbb{S}^3 \\ N^{010} \stackrel{\pi}{\Longrightarrow} \frac{SU(3)}{U(1)\times U(1)} \\ \forall p \in \frac{SU(3)}{U(1)\times U(1)} ; \pi^{-1}(p) \sim \mathbb{S}^1 \end{cases}$

Table 8.2 The homogeneous 7-manifolds that admit just one Killing spinors are the squashed 7-sphere and the infinite family of N^{pqr} manifolds for $pqr \neq 010$.

$\mathcal{N}$	Name	Coset	Holon. $\mathfrak{so}(8)$ bundle
1	$\mathbb{S}^7_{squashed}$	$\frac{SO(5)\times SO(3)}{SO(3)\times SO(3)}$	$SO(7)^+$
1	N^{pqr}	$\frac{SU(3))\times U(1)}{U(1)\times U(1)}$	$SO(7)^+$

In the series of papers [10, 12–16], it was also conjectured that the gauge theories *dual to the supergravity backgrounds* of type $AdS_4 \times \mathcal{M}_7$ have an infrared fixed point where the Yang Mills coupling constant goes to infinity. In this limit the kinetic terms are removed for all the fields in the gauge multiplet. These latter become auxiliary fields and, with the exception of the non abelian gauge one-forms, they can be integrated away leaving, as remnant, a pure Chern Simons gauge theory with a very specific form, that was discussed in the quoted papers.

The question remains how to fill the blackbox of matter multiplets in the general Chern Simons lagrangian constructed in the way sketched above.

8.1.3 The Sasakian Structure and the Metric Cone

It is in the resolution of this problem that the interplay between the geometry of the compactification manifold $\mathcal{M}_7$ and the structure of the $d = 3$ superconformal field theory becomes evident.

In paper [14] the authors introduced a systematic bridge between the geometry of $\mathcal{M}_7$ and the structure of the boundary gauge theory based on the crucial observation that all the 7-dimensional cosets with at least two Killing spinors of AdS-type are sasakian manifolds or tri-sasakian manifolds.

What sasakian means is visually summarized in the following table. First of all

base of the fibration	projection		7-manifold	metric cone
$\mathcal{B}_6$	$\overset{\pi}{\longleftarrow}$		$\mathcal{M}_7$	$\mathcal{C}(\mathcal{M}_7)$
$\Updownarrow$	$\forall p \in \mathcal{B}_6$	$\pi^{-1}(p) \sim \mathbb{S}^1$	$\Updownarrow$	$\Updownarrow$
Kähler K_3			sasakian	Kähler Ricci flat K_4

the $\mathcal{M}_7$ manifold must admit an $\mathbb{S}^1$-fibration over a complex Kähler three-fold K_3:

$$\pi \quad : \quad \mathcal{M}_7 \overset{\mathbb{S}^1}{\longrightarrow} K_3 \tag{8.1.1}$$

Calling z^i the three complex coordinates of K_3 and ϕ the angle spanning $\mathbb{S}^1$, the fibration means that the metric of $\mathcal{M}_7$ admits the following representation:

$$ds^2_{\mathcal{M}_7} = (d\phi - \mathcal{A})^2 + g_{ij^\star}\, dz^i \otimes d\bar{z}^{j^\star} \tag{8.1.2}$$

where the one-form $\mathcal{A}$ is some suitable connection one-form on the U(1)-bundle (8.1.1).

Secondly the metric cone $\mathcal{C}(\mathcal{M}_7)$ over the manifold $\mathcal{M}_7$ defined by the direct product $\mathbb{R}_+ \otimes \mathcal{M}_7$ equipped with the following metric :

$$ds^2_{\mathcal{C}(\mathcal{M}_7)} = dr^2 + 4\, e^2\, r^2\, ds^2_{\mathcal{M}_7} \tag{8.1.3}$$

should also be a Ricci-flat complex Kähler 4-fold. In the above equation e just denotes a constant scale parameter with the dimensions of an inverse length $[e] = \ell^{-1}$.

Altogether the Ricci flat Kahler manifold K_4, which plays the role of transverse space to the M2-branes, is a rank-one holomorphic vector bundle over the base manifold K_3 associated to a corresponding principal fibre-bundle over the same base with $\mathbb{C}^\star$ structural group (i.e. a line-bundle over the base manifold K_3):

$$\begin{array}{lcl} \pi & : & K_4 \longrightarrow K_3 \\ \forall p \in K_3 & & \pi^{-1}(p) \sim \mathbb{C} \end{array} \tag{8.1.4}$$

All the manifolds listed in Table 8.1 are sasakian in the sense described above. The $\mathfrak{so}(8)$-holonomy mentioned in this table is the holonomy of the Levi-Civita connection of the metric cone $\mathscr{C}(\mathscr{M}_7)$ which can be easily calculated from that of the $\mathscr{M}_7$-manifold relying on the following one-line construction. Define the vielbein of $\mathscr{C}(\mathscr{M}_7)$ in terms of the vielbein of $\mathscr{M}_7$ in the following way:

$$V^I = \begin{cases} V^0 = dr \\ V^\alpha = e\, r\, \mathscr{B}^\alpha \end{cases} \quad r \in \mathbb{R}_+ \tag{8.1.5}$$

where $ds^2_{\mathscr{M}_7} = \sum_{\alpha=1}^7 \mathscr{B}^\alpha \otimes \mathscr{B}^\alpha$. The torsion equation:

$$dV^I + \Omega^{IJ} \wedge V^J = 0 \tag{8.1.6}$$

where Ω^{IJ} is the spin-connection of the metric cone, is solved by:

$$\begin{aligned} \Omega^{\alpha\beta} &= \mathscr{B}^{\alpha\beta} \\ \Omega^{0\beta} &= -2\, e\, r\, \mathscr{B}^\beta \end{aligned} \tag{8.1.7}$$

having denoted by $\mathscr{B}^{\alpha\beta}$ the spin-connection of $\mathscr{M}_7$, namely $d\mathscr{B}^\alpha + \mathscr{B}^{\alpha\beta} \wedge \mathscr{B}^\beta = 0$. According to the summary of Kaluza–Klein supergravity presented in [34], Ω^{IJ} is the $\mathfrak{so}(8)$-connection whose holonomy decides the number of Killing spinor admitted by the $AdS_4 \times \mathscr{M}_7$ compactification of M-theory. When this holonomy vanishes we have the maximal number of preserved supersymmetries. When it is $SU(3) \subset SO(8)$ we have $\mathscr{N} = 2$. When it is $SU(2) \subset SO(8)$ we might in principle expect $\mathscr{N} = 4$, but we actually have only $\mathscr{N} = 3$, as firstly remarked by Castellani, Romans and Warner in 1985.

In [14], it was emphasized that the fundamental geometrical clue to the field content of the *superconformal gauge theory* on the boundary is provided by the construction of the Kähler manifold K_4 as a holomorphic algebraic variety in some higher dimensional affine or projective space $\mathbb{V}_q$, plus a Kähler quotient. The equations identifying the algebraic locus in $\mathbb{V}_q$ are related with the superpotential W appearing in the $d = 3$ lagrangian, while the Kähler quotient is related with the D-terms appearing in the same lagrangian. The coordinates u, v of the space $\mathbb{V}_q$ are the scalar fields of the *superconformal gauge theory*, whose vacua, namely the set of extrema of its scalar potential, should be in one-to-one correspondence with the points of K_4. Going from one to multiple M2-branes just means that the coordinate u, v of $\mathbb{V}_q$ acquire color indices under a proper set of color gauge groups and are turned into matrices. In this way we obtain *quivers*.

All these conceptual and algorithmic points were enumerated in the set of papers [10, 12–14, 16], where the cases $Q^{1,1,1}$, $M^{1,1,1}$ and $N^{0,1,0}$ were worked out in detail, finding the algebraic embedding, defining the superpotential and the quiver. Finally the Kaluza–Klein spectrum of supergravity compactified on each of these three spaces was matched with the spectrum of composite conformal operators in the corresponding boundary superconformal theory.

The subject of the AdS_4/CFT_3 correspondence received new powerful momentum in 2007–2009 by the work presented in papers [35–37] which stirred a great interest in the scientific community and obtained a very large number of citations.

The ABJM-construction of [36] is very clear and the attentive reader, making the required changes of notations and names of the objects, can verify that the $\mathcal{N} = 3$ lagrangian presented there is just the same as that one obtains from the lagrangian constructed in papers [10, 12] by letting the Yang-Mills coupling constant go to infinity. What is really new and extremely important in ABJM is the relative quantization of the Chern Simons levels $k_{1,2}$ of the two gauge groups and its link to a quotiening of the seven sphere by means of a cyclic group $\mathbb{Z}_k$. Indeed the theories presented in [36] pertain to the first case in Table 8.1, modified by a *finite group quotiening*.

8.1.4 Finite Group Quotiening

As we emphasized the key guiding item in the construction of the d = 3 gauge theory is the K_4 manifold and its representation as an algebraic locus in some $\mathbb{V}_q$. We can extract the logic which underlies [36], by means of the following arguments. First consider the following projections and embeddings pertaining to the case where $\mathcal{M}_7$ is a smooth coset manifold

$$K_3 \stackrel{\pi}{\longleftarrow} \mathcal{G}/\mathcal{H} \stackrel{C}{\hookrightarrow} K_4 \stackrel{A}{\hookrightarrow} \mathbb{V}_q \tag{8.1.8}$$

In the above formula $\stackrel{C}{\hookrightarrow}$ is the embedding map into the metric cone, while $\stackrel{A}{\hookrightarrow}$ denotes the algebraic embedding into an affine of projective variety by means of a suitable set of algebraic equations.

For instance in the case of the seven sphere $\mathcal{G}/\mathcal{H} = \mathrm{SO}(8)/\mathrm{SO}(7)$, we have $K_3 = \mathbb{P}^3$ and $K_4 = \mathbb{C}^4 \sim \mathbb{R}^8$. Then the algebraic map $\stackrel{A}{\hookrightarrow}$ is just the identity map since $\mathbb{V}_q = \mathbb{C}^4$.

On the contrary, in the case N^{010}, the base manifold $K_3 = \frac{\mathrm{SU}(3)}{\mathrm{U}(1)\times\mathrm{U}(1)}$ is just the $\mathfrak{su}(3)$ flag manifold and K_4 is obtained as the Kähler quotient of an algebraic locus cutout in $\mathbb{V}_q = \mathbb{C}^6$ by a quadric equation. In this particular case the entire procedure how to go from $\mathbb{C}^6$ to K_4 can be seen as a HyperKähler quotient with respect to the triholomorphic action of a U(1) group:

$$K_4 = \mathbb{C}^6//_H \mathrm{U}(1) \tag{8.1.9}$$

The quadric constraint is traced back to the vanishing of the holomorphic part of the triholomorphic moment map, while the Kähler quotient encodes the constraint coming from the real part of the same moment map.

Next we consider some finite group $\Gamma \subset \mathcal{G}$ and in Eq. (8.1.8) we replace the homogeneous space $\mathcal{G}/\mathcal{H}$ with the orbifold $\frac{\mathcal{G}/\mathcal{H}}{\Gamma}$. The finite group quotient extends both to the projection map and to the metric cone enlargement. Thus Eq. (8.1.8) is replaced by:

$$\frac{K_3}{\Gamma} \xleftarrow{\pi} \frac{\mathscr{G}/\mathscr{H}}{\Gamma} \overset{C}{\hookrightarrow} \frac{K_4}{\Gamma} \overset{A}{\hookrightarrow} \mathbb{V}_q \tag{8.1.10}$$

Typically the quotient $\frac{K_4}{\Gamma}$ is a singular manifold. We need a resolution of the singularities by means of an appropriate resolving map:

$$X^{\text{res}} \rightarrow \frac{K_4}{\Gamma} \tag{8.1.11}$$

which typically leads to an affine variety $X^{\text{res}} \overset{A}{\hookrightarrow} \mathbb{C}^q$ embedded by suitable algebraic equations into some $\mathbb{C}^q$.

The final outcome is that the coordinates of $\mathbb{C}^q$ are the matter fields in the $d = 3$ conformal field theory, while the embedding equations should determine the superpotential W. The gauging is instead dictated by the final Kähler quotient of the resolved algebraic variety X^{res} which produces the resolved metric cone K_4^{res}.

New contributions to this algebro-geometric approach to the gauge theories dual to $M2$-brane solutions of $D = 11$ supergravity have been recently given in [38, 39]. The rest of this chapter is essentially based on these two papers and on the much earlier paper [40] of 1994 where the Kronheimer construction was firstly applied to $2D$ Conformal Field Theories.

8.1.5 *Crepant Resolution of Gorenstein Singularities*

It appears from the above discussion that the most fundamental question at stake is a classical problem of algebraic geometry, namely the resolution of singularities, in particular of the quotient singularities. For this there is a well established set of results that were all obtained by the mathematical community at the beginning of 1990s, under the stimulus of string and supergravity theory.

First of all we fix sum vocabulary.

Definition 8.1.1 The **canonical line bundle** $K_{\mathbb{V}}$ over a complex algebraic variety $\mathbb{V}$ of complex dimension n is the bundle of holomorphic $(n, 0)$-forms $\Omega^{(n,0)}$ defined over $\mathbb{V}$.

Definition 8.1.2 An orbifold $\mathbb{V}/\Gamma$ of an algebraic variety modded by the action of a finite group is named **Gorenstein** if the isotropy subgroup $H_p \subset \Gamma$ of every point $p \in \mathbb{V}$ has a trivial action on the canonical bundle $K_{\mathbb{V}}$.

Definition 8.1.3 A resolution of singularities $\pi : \mathbb{W} \rightarrow \mathbb{X} \equiv \mathbb{V}/\Gamma$ is named **crepant**, if $K_{\mathbb{W}} = \pi^\star K_{\mathbb{X}}$. In particular this implies that the first Chern class of the resolved variety vanishes ($c_1(T\mathbb{W}) = 0$), if it vanishes for the orbifold, namely if $c_1(T\mathbb{X}) = 0$.

In the case $\mathbb{V} = \mathbb{C}^n$, a resolution of quotient singularity:

$$\pi : \mathbb{W} \to \mathbb{C}^n/\Gamma \tag{8.1.12}$$

is crepant if the resolved variety $\mathbb{W}$ has vanishing first Chern class, namely it is a Calabi-Yau q-fold.

The Gorenstein condition plus the request that there should be a crepant resolution restricts the possible Γs to be subgroups of SL(n, $\mathbb{C}$).

Concerning the crepant resolution of Gorenstein singularities $\mathbb{C}^n/\Gamma$, what was established in the early 1990s is what follows:

1. For $n = 2$ the classification of Gorenstein singularities boils down to the classification of finite Kleinian subgroups $\Gamma \subset \mathrm{SU}(2)$. This latter is just the A-D-E classification and the crepant resolution of singularities is done in one stroke by the Kronheimer construction of ALE-manifolds [2, 3] via an HyperKähler quotient of a flat HyperKähler manifold $\mathbb{H}_\Gamma$, whose dimension and structure depends on the group Γ.
2. For $n = 3$ the classification of finite subgroups of $\mathrm{SL}(3, \mathbb{C})$ was performed at the very beginning of the XX century [41–43] and it is summarized in [44]. As stressed by Markushevich in [45] in that list there are only two types of groups, either solvable groups or the simple group PSL(2,7) of order 168. For this reason the same Markushevich studied the resolution of the Gorenstein orbifold:

$$\mathscr{O}_{168} \equiv \frac{\mathbb{C}^3}{\mathrm{PSL}(2, \mathbb{Z}_7)} \tag{8.1.13}$$

   which corresponds to a unique truely new case. There are several other physical motivations for the study of orbifolds with respect to

$$\mathrm{L}_{168} \equiv \mathrm{PSL}(2, \mathbb{Z}_7) \tag{8.1.14}$$

   or one of its maximal subgroups.
3. For $n > 3$ essentially nothing is known with the exception of those cases that can be reduced to singularities in $n = 2, 3$.

8.1.6 *The Complex Hopf Fibration of $\mathbb{S}^7$ and Quotient Singularities $\mathbb{C}^4/\Gamma$*

In order to arrive at what is for us most interesting, namely quotient singularities of the type $\mathbb{C}^4/\Gamma$ we start from the first of the cases listed in Table 8.1, namely the complex Hopf fibration of the seven sphere:

$$\begin{aligned} \pi : \mathbb{S}^7 &\rightarrow \mathbb{CP}^3 \\ \forall y \in \mathbb{CP}^3 : \pi^{-1}(y) &\sim \mathbb{S}^1 \end{aligned} \tag{8.1.15}$$

We want to establish the following important conclusion. Writing the metric cone over the seven sphere as $\mathbb{C}^4$, namely:

$$\mathscr{C}(\mathbb{S}^7) = \mathbb{R}^8 \sim \mathbb{C}^4 \tag{8.1.16}$$

the homogeneous coordinates Z^i of $\mathbb{CP}^3$ can be identified with the standard affine coordinates of $\mathbb{C}^4$ defined above.

To this purpose we consider the standard definition of the $\mathbb{CP}^3$ manifold as the set of quadruplets $\left\{Z^1, \ldots, Z^4\right\}$ modulo an overall complex factor:

$$\left\{Z^1, \ldots, Z^4\right\} \sim \lambda \left\{Z^1, \ldots, Z^4\right\} \quad , \quad \forall \lambda \in \mathbb{C}^\star \tag{8.1.17}$$

On the other hand we define the 7-sphere as the locus in $\mathbb{C}^4$ cut out by the following constraint:

$$|\mathbf{Z}|^2 \equiv \sum_{i=1}^{4} |Z^i|^2 = 1 \tag{8.1.18}$$

Let us define the Kähler metric on the $\mathbb{CP}^3$ in terms of the homogeneous coordinates:

$$ds^2_{\mathbb{CP}^3} = \frac{d\mathbf{Z} \cdot d\bar{\mathbf{Z}}}{|\mathbf{Z}|^2} - \frac{(\mathbf{Z} \cdot d\bar{\mathbf{Z}})\,(\bar{\mathbf{Z}} \cdot d\mathbf{Z})}{|\mathbf{Z}|^4} \tag{8.1.19}$$

That the above is indeed a metric on $\mathbb{CP}^3$ is verified in the following way: if in Eq. (8.1.19) $\mathbf{Z}$ is replaced by $\lambda\mathbf{Z}$ all the factors λ and all their differentials cancel identically. If we fix the λ-gauge by setting $Z_4 = 1$ and we rename $Z_{1,2,3} = Y_{1,2,3}$, then we find that the above metric is identical with the Kähler metric obtained from the Fubini–Study Kähler potential:

$$\mathscr{K}_{\mathbb{CP}^3}(\mathbf{Y}) = \log\left(1 + |\mathbf{Y}|^2\right) \tag{8.1.20}$$

On the other hand if we consider the pull-back of the flat Kähler metric of C^4 on the locus (8.1.17) we obtain the metric of the seven sphere:

$$ds^2_{\mathbb{S}^7} = d\mathbf{Z} \cdot d\bar{\mathbf{Z}}\,|_{|\mathbf{Z}|^2=1} \tag{8.1.21}$$

Let us next consider the following 1-form:

$$\Omega\,(\mathbf{Z}) = \frac{\mathrm{i}}{2\,|\mathbf{Z}|^2}\left(\mathbf{Z} \cdot d\bar{\mathbf{Z}} - \bar{\mathbf{Z}} \cdot d\mathbf{Z}\right) \tag{8.1.22}$$

and perform the following two calculations. If we replace $\mathbf{Z} \rightarrow \lambda\mathbf{Z}$, we obtain:

$$\Omega\left(\lambda \mathbf{Z}\right) = \frac{\mathrm{i}}{2}\left(\lambda d\bar{\lambda} - \bar{\lambda} d\lambda\right) + \Omega\left(\mathbf{Z}\right) \tag{8.1.23}$$

In particular if $\lambda = e^{\mathrm{i}\theta}$ we get:

$$\Omega\left(e^{\mathrm{i}\theta}\mathbf{Z}\right) = d\theta + \Omega\left(\mathbf{Z}\right) \tag{8.1.24}$$

This shows that Ω is a U(1)-connection on the principal $U(1)$-bundle that has $\mathbb{CP}^3$ as base manifold and which can be identified with the 7-sphere. The curvature of this connection is just the Kähler 2-form on $\mathbb{CP}^3$.

On the other hand we have:

$$ds^2_{\mathbb{S}} \equiv d\Omega^2 + ds^2_{\mathbb{CP}^3} = \frac{d\mathbf{Z}\cdot d\bar{\mathbf{Z}}}{|\mathbf{Z}|^2} - \frac{\left(\mathbf{Z}\cdot d\bar{\mathbf{Z}} + \bar{\mathbf{Z}}\cdot d\mathbf{Z}\right)^2}{|\mathbf{Z}|^4} \tag{8.1.25}$$

If we restrict the above line element to the locus (8.1.17) we find:

$$ds^2_{\mathbb{S}}\mid_{|\mathbf{Z}|^2=1} = d\mathbf{Z}\cdot d\bar{\mathbf{Z}}\mid_{|\mathbf{Z}|^2=1} = ds^2_{\mathbb{S}^7} \tag{8.1.26}$$

In this way we have obtained the desired result: the metric cone over the 7-sphere is described by the homogeneous coordinates of CP^3 interpreted as affine ones on $\mathbb{C}^4$:

$$ds^2_{\mathscr{C}} = dr^2 + r^2 ds^2_{\mathbb{S}^7} = d\mathbf{Z}\cdot d\bar{\mathbf{Z}} \tag{8.1.27}$$

Another way of stating the same result is the following one. We can regard $\mathbb{C}^4$ as the total space of a rank = 1 holomorphic vector bundle over $\mathbb{CP}^3$, with structural group $\mathbb{C}^\star$:

$$\begin{aligned} \pi : \mathbb{C}^4 &\rightarrow \mathbb{CP}^3 \\ \forall y \in \mathbb{CP}^3 &: \pi^{-1}(y) \sim \mathbb{C} \end{aligned} \tag{8.1.28}$$

The form Ω is a connection on this line-bundle.

The consequence of this discussion is that if we have a finite subgroup $\Gamma \subset \mathrm{SU}(4)$, which obviously is an isometry of $\mathbb{CP}^3$ we can consider its action both on $\mathbb{CP}^3$ and on the seven sphere so that we have:

$$\mathrm{AdS}_4 \times \frac{\mathbb{S}^7}{\Gamma} \rightarrow \partial\mathrm{AdS}_4 \times \frac{\mathbb{C}^4}{\Gamma} \tag{8.1.29}$$

This is the setup for the theory of M2-branes that probe the singularity $\frac{\mathbb{C}^4}{\Gamma}$.

8.2 From Singular Orbifolds to Smooth Resolved Manifolds

The next point which provides an important orientation in addressing mathematical questions comes from physics in view, as we stressed above, of the final use of the considered mathematical lore in connection with M2-brane solutions of $D = 11$ supergravity and later on in the construction of quantum gauge theories dual to such M2-solutions of supergravity.

Let us start once again from

$$K_3 \stackrel{\pi}{\longleftarrow} \mathscr{M}_7 \stackrel{Cone}{\hookrightarrow} K_4 \stackrel{\mathscr{A}}{\hookrightarrow} \mathbb{V}_q \tag{8.2.1}$$

namely from Eq. (8.1.8) that we are going to rewrite in slightly more general terms. The AdS_4 compactification of $D = 11$ supergravity is obtained by utilizing as complementary 7-dimensional space a manifold $\mathscr{M}_7$ which occupies the above displayed position in the inclusion-projection diagram (8.2.1). The metric cone $\mathscr{C}(\mathscr{M}_7)$ enters the game when, instead of looking at the vacuum:

$$\mathrm{AdS}_4 \times \mathscr{M}_7 \tag{8.2.2}$$

we consider the more general M2-brane solutions of $\mathrm{D} = 11$ supergravity, where the $\mathrm{D} = 11$ metric is of the following form:

$$ds^2_{11} = H(y)^{-\frac{2}{3}} \left(d\xi^\mu \otimes d\xi^\nu \eta_{\mu\nu}\right) - H(y)^{\frac{1}{3}} \left(ds^2_{\mathscr{M}_8}\right) \tag{8.2.3}$$

$\eta_{\mu\nu}$ being the constant Lorentz metric of $\mathrm{Mink}_{1,2}$ and:

$$ds^2_{\mathscr{M}_8} = dy^I \otimes dy^J g_{IJ}(y) \tag{8.2.4}$$

being a Ricci-flat metric on an asymptotically locally Euclidean 8-manifold $\mathscr{M}_8$. In Eq. (8.2.3) the symbol $H(y)$ denotes a harmonic function over the manifold $\mathscr{M}_8$, namely:

$$\Box_g H(y) = 0 \tag{8.2.5}$$

Equation (8.2.5) is the only differential constraint required in order to satisfy all the field equations of $D = 11$ supergravity in presence of the standard M2-brain ansatz for the 3-form field:

$$\mathbf{A}^{[3]} \propto H(y)^{-1} \left(d\xi^\mu \wedge d\xi^\nu \wedge d\xi^\rho \, \varepsilon_{\mu\nu\rho}\right) \tag{8.2.6}$$

In this more general setup the manifold $\mathscr{M}_8$ is what substitutes the metric cone $\mathscr{C}(\mathscr{M}_7)$. To see the connection between the two viewpoints it suffices to introduce the radial coordinate $r(y)$ by means of the position:

$$H(y) = 1 - \frac{1}{r(y)^6} \tag{8.2.7}$$

The asymptotic region where $\mathcal{M}_8$ is required to be locally Euclidean is defined by the condition $r(y) \to \infty$. In this limit the metric (8.2.4) should approach the flat Euclidean metric of $\mathbb{R}^8 \simeq \mathbb{C}^4$. The opposite limit where $r(y) \to 0$ defines the near horizon region of the M2-brane solution. In this region the metric (8.2.3) approaches that of the space (8.2.2), the manifold $\mathcal{M}_7$ being a codimension one submanifold of $\mathcal{M}_8$ defined by the limit $r \to 0$.

To be mathematically more precise let us consider the harmonic function as a map:

$$\mathfrak{H} \quad : \quad \mathcal{M}_8 \to \mathbb{R}_+ \tag{8.2.8}$$

This viewpoint introduces a foliation of $\mathcal{M}_8$ into a one-parameter family of 7-manifolds:

$$\forall h \in \mathbb{R}_+ \quad : \quad \mathcal{M}_7(h) \equiv \mathfrak{H}^{-1}(h) \subset \mathcal{M}_8 \tag{8.2.9}$$

In order to have the possibility of residual supersymmetries we are interested in cases where the Ricci flat manifold $\mathcal{M}_8$ is actually a Ricci-flat Kähler 4-fold.

In this way the appropriate rewriting of Eq. (8.1.8)–(8.2.1) is as follows:

$$K_3 \underbrace{\stackrel{\pi}{\Longleftarrow}}_{\text{if it applies}} \mathcal{M}_7 \stackrel{\mathfrak{H}^{-1}}{\longleftarrow} K_4 \stackrel{\mathscr{A}}{\hookrightarrow} \mathbb{V}_q \tag{8.2.10}$$

Next we recall the general pattern laid down in [38] that will be our starting point.

The $\mathcal{N} = 8$ Case with no Singularities.

The prototype of the above inclusion-projection diagram is provided by the case of the M2-brane solution with all preserved supersymmetries. In this case we have:

$$\mathbb{CP}^3 \stackrel{\pi}{\longleftarrow} \mathbb{S}^7 \stackrel{Cone}{\hookrightarrow} \mathbb{C}^4 \stackrel{\mathscr{A}=\mathrm{Id}}{\hookrightarrow} \mathbb{C}^4 \tag{8.2.11}$$

On the left we just have the projection map of the Hopf fibration of the 7-sphere. On the right we have the inclusion map of the 7 sphere in its metric cone $\mathscr{C}(\mathbb{S}^7) \equiv \mathbb{R}^8 \sim \mathbb{C}^4$. The last algebraic inclusion map is just the identity map, since the algebraic variety $\mathbb{C}^4$ is already smooth and flat and needs no extra treatment.

The Singular Orbifold Cases.

The next orbifold cases are those of interest to us here. Let $\Gamma \subset \mathrm{SU}(4)$ be a finite discrete subgroup of $\mathrm{SU}(4)$. Then Eq. (8.2.11) is replaced by the following one:

$$\frac{\mathbb{CP}^3}{\Gamma} \stackrel{\pi}{\longleftarrow} \frac{\mathbb{S}^7}{\Gamma} \stackrel{Cone}{\hookrightarrow} \frac{\mathbb{C}^4}{\Gamma} \stackrel{\mathscr{A}=?}{\hookrightarrow} \; ? \tag{8.2.12}$$

The consistency of the above quotient is guaranteed by the inclusion $\mathrm{SU}(4) \subset \mathrm{SO}(8)$. The question marks can be removed only by separating the two cases:

(A) Case: $\Gamma \subset \mathrm{SU}(2) \subset \mathrm{SU}(2)_{\mathrm{I}} \otimes \mathrm{SU}(2)_{\mathrm{II}} \subset \mathrm{SU}(4)$. Here we obtain:

$$\frac{\mathbb{C}^4}{\Gamma} \simeq \mathbb{C}^2 \times \frac{\mathbb{C}^2}{\Gamma} \tag{8.2.13}$$

and everything is under full control for the Kleinian $\frac{\mathbb{C}^2}{\Gamma}$ singularities and their resolution à la Kronheimer in terms of hyperKähler quotients.

(B) Case: $\Gamma \subset \mathrm{SU}(3) \subset \mathrm{SU}(4)$. Here we obtain:

$$\frac{\mathbb{C}^4}{\Gamma} \simeq \mathbb{C} \times \frac{\mathbb{C}^3}{\Gamma} \tag{8.2.14}$$

and the study and resolution of the singularity $\frac{\mathbb{C}^3}{\Gamma}$ in a physicist friendly way is the main issue of this chapter. The comparison of case (B) with the well known case (A) will provide us with many important hints.

Let us begin by erasing the question marks in case (A). Here we can write:

$$\frac{\mathbb{CP}^3}{\Gamma} \overset{\pi}{\longleftarrow} \frac{\mathbb{S}^7}{\Gamma} \overset{Cone}{\hookrightarrow} \mathbb{C}^2 \times \frac{\mathbb{C}^2}{\Gamma} \overset{\mathrm{Id}\times\mathscr{A}_W}{\hookrightarrow} \mathbb{C}^2 \times \mathbb{C}^3 \tag{8.2.15}$$

In the first inclusion map on the right, Id denotes the identity map $\mathbb{C}^2 \to \mathbb{C}^2$ while $\mathscr{A}_W$ denotes the inclusion of the orbifold $\frac{\mathbb{C}^2}{\Gamma}$ as a singular variety in $\mathbb{C}^3$ cut out by a single polynomial constraint:

$$\begin{aligned} \mathscr{A}_W \; &: \frac{\mathbb{C}^2}{\Gamma} \to \mathbf{V}(\mathscr{I}_\Gamma^W) \subset \mathbb{C}^3 \\ \mathbb{C}\left[\mathbf{V}(\mathscr{I}_\Gamma)\right] \; &= \frac{\mathbb{C}[u, w, z]}{W_\Gamma(u, w, z)} \end{aligned} \tag{8.2.16}$$

where by $\mathbb{C}\left[\mathbf{V}(\mathscr{I}_\Gamma)\right]$ we denote the *coordinate ring* of the algebraic variety $\mathbf{V}$. As we recall in more detail in next sections, the variables u, w, z are polynomial Γ-invariant functions of the coordinates z_1, z_2 on which Γ acts linearly. The unique generator $W_\Gamma(u, w, z)$ of the ideal $\mathscr{I}_\Gamma^W$ which cuts out the singular variety isomorphic to $\frac{\mathbb{C}^2}{\Gamma}$ is the unique algebraic relation existing among such invariants. In the next sections we discuss the relation between this algebraic equation and the embedding in higher dimensional algebraic varieties associated with the McKay quiver and the hyperKähler quotient.

Let us now consider the case (B). Up to this level things go in a quite analogous way with respect to case (A). Indeed we might write

$$\frac{\mathbb{CP}^3}{\Gamma} \overset{\pi}{\longleftarrow} \frac{\mathbb{S}^7}{\Gamma} \overset{Cone}{\hookrightarrow} \mathbb{C} \times \frac{\mathbb{C}^3}{\Gamma} \overset{\mathrm{Id}\times\mathscr{A}_{\mathscr{W}}}{\hookrightarrow} \mathbb{C} \times \mathbb{C}^4 \tag{8.2.17}$$

In the last inclusion map on the right, Id denotes the identity map $\mathbb{C} \to \mathbb{C}$ while $\mathscr{A}_{\mathscr{W}}$ denotes the inclusion of the orbifold $\frac{\mathbb{C}^3}{\Gamma}$ as a singular variety in $\mathbb{C}^4$ cut out by a single polynomial constraint:

$$\begin{aligned} \mathscr{A}_{\mathscr{W}} \ &: \frac{\mathbb{C}^3}{\Gamma} \to \mathbf{V}(\mathscr{I}_\Gamma) \subset \mathbb{C}^4 \\ \mathbb{C}\left[\mathbf{V}\left(\mathscr{I}_\Gamma\right)\right] \ &\sim \frac{\mathbb{C}[u_1, u_2, u_3, u_4]}{\mathscr{W}_\Gamma(u_1, u_2, u_3, u_4)} \end{aligned} \tag{8.2.18}$$

For the case $\Gamma = \mathrm{L}_{168}$ discussed by Markushevich the variables u_1, u_2, u_3, u_4 are polynomial Γ-invariant functions of the coordinates z_1, z_2, z_3 on which Γ acts linearly. The unique generator $\mathscr{W}_\Gamma(u_1, u_2, u_3, u_4)$ of the ideal $\mathscr{I}_\Gamma$ which cuts out the singular variety isomorphic to $\frac{\mathbb{C}^3}{\Gamma}$ is the unique algebraic relation existing among such invariants.

The simple representation of the orbifold as a hypersurface in $\mathbb{C}^4$ is no longer true for the subgroups of L_{168}. For instance for the maximal subgroup $\mathrm{G}_{21} \subset \mathrm{L}_{168}$ the orbifold $\frac{\mathbb{C}^3}{\mathrm{G}_{21}}$ is an affine algebraic variety in $\mathbb{C}^5$, the corresponding ideal being generated by two polynomials.[2]

As for the relation of this algebraic equation with the embedding in higher dimensional algebraic varieties associated with the McKay quiver, things are now more complicated.

In the years 1990s up to 2010s there has been an intense activity in the mathematical community on the issue of the crepant resolutions of $\mathbb{C}^3/\Gamma$ (see for [44–46, 51]) that has gone on almost unnoticed by physicists since it was mostly formulated in the abstract language of algebraic geometry, providing few clues to the applicability of such results to gauge theories and branes. Yet, once translated into more explicit terms, by making use of coordinate patches, and equipped with some additional ingredients of Lie group character, these mathematical results turn out to be not only useful, but rather of outmost relevance for the physics of M2-branes. In the present paper we aim at drawing the consistent, systematic scheme which emerges in this context upon a proper interpretation of the mathematical constructions.

So let us consider the case of smooth resolutions. In case (A) the smooth resolution is provided by a manifold ALE_Γ and we obtain the following diagram:

$$\mathscr{M}_7 \ \overset{\mathfrak{H}^{-1}}{\longleftarrow} \ \mathbb{C}^2 \times ALE_\Gamma \ \overset{\mathrm{Id}\times qK}{\longleftarrow} \ \mathbb{C}^2 \times \mathbb{V}_{|\Gamma|+1} \ \overset{\mathscr{A}_{\mathscr{P}}}{\hookrightarrow} \ \mathbb{C}^2 \times \mathbb{C}^{2|\Gamma|} \tag{8.2.19}$$

In the above equation the map $\overset{\mathfrak{H}^{-1}}{\longleftarrow}$ denotes the inverse of the harmonic function map on $\mathbb{C}^2 \times ALE_\Gamma$ that we have already discussed. The map $\overset{\mathrm{Id}\times qK}{\longleftarrow}$ is instead the product of the identity map $\mathrm{Id} : \mathbb{C}^2 \to \mathbb{C}^2$ with the Kähler quotient map:

$$qK \ : \ \mathbb{V}_{|\Gamma|+1} \longrightarrow \mathbb{V}_{|\Gamma|+1} /\!\!/_K \mathscr{F}_{|\Gamma|-1} \simeq ALE_\Gamma \tag{8.2.20}$$

[2]This result was derived in private conversations of the author with Dimitry Markushevich.

of an algebraic variety of complex dimension $|\Gamma| + 1$ with respect to a suitable Lie group $\mathscr{F}_{|\Gamma|-1}$ of real dimension $|\Gamma| - 1$. Finally the map $\stackrel{\mathscr{A}_{\mathscr{P}}}{\hookrightarrow}$ denotes the inclusion map of the variety $\mathbb{V}_{|\Gamma|+1}$ in $\mathbb{C}^{2|\Gamma|}$. Let $y_1, \ldots, y_{2|\Gamma|}$ be the coordinates of $\mathbb{C}^{2|\Gamma|}$. The variety $\mathbb{V}_{|\Gamma|+1}$ is defined by an ideal generated by $|\Gamma| - 1$ quadratic generators:

$$\mathbb{V}_{|\Gamma|+1} = \mathbf{V}\left(\mathscr{I}_{\Gamma}\right)$$
$$\mathbb{C}\left[\mathbf{V}\left(\mathscr{I}_{\Gamma}\right)\right] = \frac{\mathbb{C}\left[y_1, \ldots, y_{2|\Gamma|}\right]}{\left(\mathscr{P}_1(y), \mathscr{P}_2(y), \ldots, \mathscr{P}_{|\Gamma|-1}(y)\right)} \tag{8.2.21}$$

Actually the $|\Gamma| - 1$ polynomials $\mathscr{P}_i(y)$ are the holomorphic part of the triholomorphic moment maps associated with the triholomorphic action of the group $\mathscr{F}_{|\Gamma|-1}$ on $\mathbb{C}^{2|\Gamma|}$ and the entire procedure from $\mathbb{C}^{2|\Gamma|}$ to ALE_{Γ} can be seen as the hyperKähler quotient:

$$ALE_{\Gamma} = \mathbb{C}^{2|\Gamma|} /\!\!/_{HK} \mathscr{F}_{|\Gamma|-1} \tag{8.2.22}$$

yet we have preferred to split the procedure into two steps in order to compare case (A) with case (B) where the two steps are necessarily distinct and separated.

Indeed in case (B) we can write the following diagram:

$$\mathscr{M}_7 \stackrel{\mathfrak{H}^{-1}}{\longleftarrow} \mathbb{C} \times Y_{\Gamma} \stackrel{\mathrm{Id} \times qK}{\longleftarrow} \mathbb{C} \times \mathbb{V}_{|\Gamma|+2} \stackrel{\mathrm{Id} \times \mathscr{A}_{\mathscr{P}}}{\hookrightarrow} \mathbb{C} \times \mathbb{C}^{3|\Gamma|} \tag{8.2.23}$$

In this case, just as in the previous one, the intermediate step is provided by the Kähler quotient but the map on the extreme right $\stackrel{\mathscr{A}_{\mathscr{P}}}{\hookrightarrow}$ denotes the inclusion map of the variety $\mathbb{V}_{|\Gamma|+2}$ in $\mathbb{C}^{3|\Gamma|}$. Let $y_1, \ldots, y_{3|\Gamma|}$ be the coordinates of $\mathbb{C}^{3|\Gamma|}$. The variety $\mathbb{V}_{|\Gamma|+2}$ is defined as the principal branch of a set of quadratic algebraic equations that are group-theoretically defined. Altogether the mentioned construction singles out the holomorphic orbit of a certain group action to be discussed in detail in the sequel. So we anticipate:

$$\mathbb{V}_{|\Gamma|+2} = \mathscr{D}_{\Gamma} \equiv \mathrm{Orbit}_{\mathscr{G}_{\Gamma}}\left(L_{\Gamma}\right) \tag{8.2.24}$$

where both the set L_{Γ} and the complex group $\mathscr{G}_{\Gamma}$ are completely defined by the discrete group Γ defining the quotient singularity.

8.3 Generalities on $\frac{\mathbb{C}^3}{\Gamma}$ Singularities

Recalling what we summarized above we conclude that the singularities relevant to our goals are of the form:

$$X = \frac{\mathbb{C}^3}{\Gamma} \tag{8.3.1}$$

where the finite group $\Gamma \subset \mathrm{SU}(3)$ has a holomorphic action on $\mathbb{C}^3$. For this case, as we mentioned above, there is a series of general results and procedures developed in algebraic geometry that we want to summarize in the perspective of their use in physics.

To begin with let us observe the schematic diagram sketched here below:

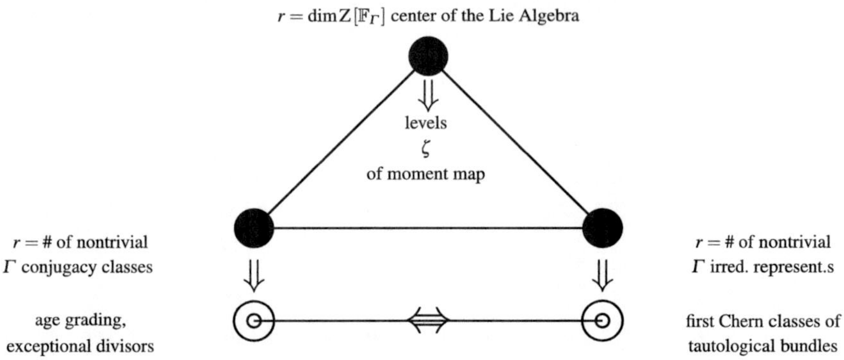

(8.3.2)

The fascination of the mathematical construction lying behind the desingularization process, which has a definite counterpart in the structure of the Chern-Simons gauge theories describing M2-branes at the $\mathbb{C}^3/\Gamma$ singularity, is the triple interpretation of the same number r which alternatively yields:

- The number of nontrivial conjugacy classes of the finite group Γ,
- The number of irreducible representations of the finite group Γ,
- The center of the Lie algebra $\zeta\,[\mathbb{F}_\Gamma]$ of the compact gayge group $\mathscr{F}_\Gamma$, whose structure, as we will see, is:

$$\mathscr{F}_\Gamma = \bigotimes_{i=1}^{r} \mathrm{U(n_i)} \tag{8.3.3}$$

The levels ζ_I of the moment maps are the main ingredient of the singularity resolution. At level$\zeta^I = 0$ we have the singular orbifold $\mathscr{M}_0 = \frac{\mathbb{C}^3}{\Gamma}$, while at $\zeta^i \neq 0$ we obtain a smooth manifold $\mathscr{M}_\zeta$ which develops a nontrivial homology and cohomology. In physical parlance the levels ζ^I are the Fayet-Iliopoulos parameters appearing in the lagrangian, while $\mathscr{M}_\zeta$ is the manifold of vacua of the theory, namely of extrema of the potential, as we already emphasized.

Quite generally, we find that each of the gauge factors $\mathrm{U(n_i)}$ is the structural group of a holomorphic vector bundle of rank n_i:

$$\mathfrak{V}_i \xrightarrow{\pi} \mathscr{M}_\zeta \tag{8.3.4}$$

whose first Chern class is a nontrivial (1,1)-cohomology class of the resolved smooth manifold:

$$c_1(\mathfrak{V}_i) \in H^{1,1}(\mathscr{M}_\zeta) \tag{8.3.5}$$

On the other hand a very deep theorem originally proved in the nineties by Reid and Ito [46] relates the dimensions of the cohomology groups $H^{q,q}\left(\mathscr{M}_\zeta\right)$ to the conjugacy classes of Γ organized according to the grading named *age*. So named *junior classes* of $age = 1$ are associated with $H^{1,1}\left(\mathscr{M}_\zeta\right)$ elements, while the so-named *senior classes* of $age = 2$ are associated with $H^{2,2}\left(\mathscr{M}_\zeta\right)$ elements.

The link that pairs irreps with conjugacy classes is provided by the relation, well-known in algebraic geometry, between *divisors* and *line bundles*. The conjugacy classes of γ can be put into correspondence with the exceptional divisors created in the resolution $\mathscr{M}_\zeta \xrightarrow{\zeta\to 0} \frac{\mathbb{C}^3}{\Gamma}$ and each divisor defines a line bundle whose first Chern class is an element of the $H^{1,1}\left(\mathscr{M}_\zeta\right)$ cohomology group.

These line bundles labeled by conjugacy classes have to be compared with the line bundles created by the Kähler quotient procedure that are instead associated with the irreps, as we have sketched above. In this way we build the bridge between conjugacy classes and irreps.

Finally there is the question whether the divisor is compact or not. In the first case, by Poincaré duality, we obtain nontrivial $H^{2,2}\left(\mathscr{M}_\zeta\right)$ elements. In the second case we have no new cohomology classes. The age grading precisely informs us about the compact or noncompact nature of the divisors. Each senior class corresponds to a cohomology class of degree 4, thus signaling the existence of a non-trivial closed (2,2) form, and via Poincaré duality, it also corresponds to a compact component of the exceptional divisor.

The physics-friendly illustration of this general beautiful scheme, is the main goal of the present chapter. We begin with the concept of age grading.

8.3.1 The Concept of Aging for Conjugacy Classes of the Discrete Group Γ

According to the above quoted theorem that we shall explain below, the *age grading* of Γ conjugacy classes allows to predict the Dolbeault cohomology of the resolved algebraic variety. It goes as follows.

Suppose that Γ (a finite group) acts in a linear way on $\mathbb{C}^n$. Consider an element $\gamma \in \Gamma$ whose action is the following:

$$\gamma.\mathbf{z} = \underbrace{\begin{pmatrix} \cdots & \cdots & \cdots \\ \vdots & \vdots & \vdots \\ \cdots & \cdots & \cdots \end{pmatrix}}_{\mathscr{D}(\gamma)} \cdot \begin{pmatrix} z_1 \\ \vdots \\ z_n \end{pmatrix} \tag{8.3.6}$$

Since in a finite group all elements have a finite order, there exists $r \in \mathbb{N}$, such that $\gamma^r = \mathbf{1}$. We define the age of an element in the following way. Let us diagonalize $D(\gamma)$, namely compute its eigenvalues. They will be as follows:

$$(\lambda_1, \ldots, \lambda_n) = \exp\left[\frac{2\pi\, i}{r}\, a_i\right] \quad ; \quad r > a_i \in \mathbb{N} \quad i = 1, \ldots, n \tag{8.3.7}$$

We define:

$$\text{age}\,(\gamma) = \frac{1}{r}\sum_{i=1}^{n} a_i \tag{8.3.8}$$

Clearly the age is a property of the conjugacy class of the element, relative to the considered three-dimensional complex representation.

8.3.2 The Fundamental Theorem

In [46] Y. Ito and M. Reid proved the following fundamental theorem:

Theorem 8.3.1 *Let $Y \to \mathbb{C}^3/\Gamma$ be a crepant [3] resolution of a Gorenstein [4] singularity. Then we have the following relation between the de-Rham cohomology groups of the resolved smooth variety Y and the ages of Γ conjugacy classes:*

$$\dim H^{2k}\,(Y) = \#\mathit{of\ age\ k conjugacy\ classes\ of}\ \Gamma$$

On the other hand it happens that all odd cohomology groups are trivial:

$$\dim H^{2k+1}\,(Y) = 0 \tag{8.3.9}$$

This is true also in the case of $\mathbb{C}^2/\Gamma$ singularities, yet in $n = 2, 3$ the consequences of the same fact are drastically different. In all complex dimensions n the deformations of the Kähler class are in one-to-one correspondence with the harmonic forms $\omega^{(1,1)}$, while those of the complex structure are in correspondence with the harmonic forms $\omega^{(n-1,1)}$. In $n = 2$ the harmonic $\omega^{(1,1)}$ forms play the double role of Kähler class deformations and complex structure deformations. This is the reason why we can do a hyperKähler quotient and we have both moduli parameters in the Kähler potential and in the polynomials cutting out the smooth variety. Instead in $n = 3$ Eq. (8.3.9) implies that the polynomials constraints cutting the singular locus have no deformation parameters. The parameters of the resolution occur only at the level of the Kähler quotient and are the levels of the Kählerian moment maps.

Given an algebraic representation of the variety Y as the vanishing locus of certain polynomials $W(x) = 0$, the algebraic $2k$-cycles are the $2k$-cycles that can be holomorphically embedded in Y. The following statement in $n = 3$ is elementary:

---

[3] A resolution of singularities $X \to Y$ is crepant when the canonical bundle of X is the pullback of the canonical bundle of Y.

[4] A variety is Gorenstein when the canonical divisor is a Cartier divisor, i.e., a divisor corresponding to a line bundle.

Statement 8.3.1 *The Poincaré dual of any algebraic* $2k$*-cycle is necessarily of type* (k, k)

Its converse is known as the Hodge conjecture.

Taking the above for granted we conclude that the so named *junior conjugacy classes* (age $= 1$) are in a one-to-one correspondence with $\omega^{(1,1)}$-forms that span $H^{1,1}$, while *conjugacy classes of age* 2 are in one-to-one correspondence with $\omega^{(2,2)}$-forms that span $H^{2,2}$.

8.3.3 *Ages for* $\Gamma \subset \mathrm{L}_{168}$

For the holomorphic action on $\mathbb{C}^3$ of the group L_{168} we have calculated the ages of the various conjugacy classes, starting from the construction of the irreducible three-dimensional complex representation discussed in Sect. 1.3.4.

In order to be able to compare with Markushevich's paper [45], we sometimes utilize its basis for the generators. It is important to note that the form given by Markusevich of the generators which he calls τ, χ and ω, respectively of order 7, 3 and 2, does not correspond to the standard generators in the presentation of the group L_{168} utilized in Sect. 1.3. Yet there is no problem since we have a translation vocabulary. Setting:

$$R = \omega.\chi \quad ; \quad S = \chi.\tau \quad ; \quad T = \chi^2.\omega \tag{8.3.10}$$

these new generators satisfy the standard relations of the presentation displayed in Eq. (1.3.7). For practical convenience we distinguish the abstract description of the group, from its concrete realization in terms of matrices, by rewriting Eq. (1.3.7) in terms of abstract generators denoted by the corresponding greek letters:

$$\mathrm{L}_{168} = \left(\rho, \sigma, \tau \parallel \rho^2 = \sigma^3 = \tau^7 = \rho.\sigma.\tau = (\tau.\sigma.\rho)^4 = \varepsilon\right) \tag{8.3.11}$$

In this way we can give an exhaustive enumeration of all the group elements as words in the three symbols ρ,σ,τ.

We begin by constructing explicitly the group L_{168} in Markusevich basis substituting the analytic form of the generators which follows from the identification (8.3.10). We find

$$\varepsilon \to \begin{pmatrix} 1 & 0 & 0 \\ 0 & 1 & 0 \\ 0 & 0 & 1 \end{pmatrix}$$

$$\rho \to \begin{pmatrix} -\frac{2\mathrm{Cos}[\frac{\pi}{14}]}{\sqrt{7}} & -\frac{2\mathrm{Cos}[\frac{3\pi}{14}]}{\sqrt{7}} & \frac{2\mathrm{Sin}[\frac{\pi}{7}]}{\sqrt{7}} \\ -\frac{2\mathrm{Cos}[\frac{3\pi}{14}]}{\sqrt{7}} & \frac{2\mathrm{Sin}[\frac{\pi}{7}]}{\sqrt{7}} & -\frac{2\mathrm{Cos}[\frac{\pi}{14}]}{\sqrt{7}} \\ \frac{2\mathrm{Sin}[\frac{\pi}{7}]}{\sqrt{7}} & -\frac{2\mathrm{Cos}[\frac{\pi}{14}]}{\sqrt{7}} & -\frac{2\mathrm{Cos}[\frac{3\pi}{14}]}{\sqrt{7}} \end{pmatrix}$$

$$
\begin{aligned}
\sigma &\to \begin{pmatrix} 0 & 0 & -(-1)^{1/7} \\ (-1)^{2/7} & 0 & 0 \\ 0 & (-1)^{4/7} & 0 \end{pmatrix} \\
\tau &\to \begin{pmatrix} \frac{i+(-1)^{13/14}}{\sqrt{7}} & -\frac{(-1)^{1/14}\left(-1+(-1)^{2/7}\right)}{\sqrt{7}} & \frac{(-1)^{9/14}\left(1+(-1)^{1/7}\right)}{\sqrt{7}} \\ \frac{(-1)^{11/14}\left(-1+(-1)^{2/7}\right)}{\sqrt{7}} & \frac{i+(-1)^{5/14}}{\sqrt{7}} & \frac{(-1)^{3/14}\left(1+(-1)^{3/7}\right)}{\sqrt{7}} \\ -\frac{(-1)^{11/14}\left(1+(-1)^{1/7}\right)}{\sqrt{7}} & -\frac{(-1)^{9/14}\left(1+(-1)^{3/7}\right)}{\sqrt{7}} & -\frac{-i+(-1)^{3/14}}{\sqrt{7}} \end{pmatrix}
\end{aligned}
\tag{8.3.12}
$$

We remind the reader that ρ, σ, τ are the abstract names for the generators of L_{168} whose 168 elements are written as words in these letters (modulo relations). Substituting these letters with explicit matrices that satisfy the defining relation of the group one obtains an explicit representation of the latter. In the present case the substitution (8.3.12) produces the irreducible 3-dimensional representation DA_3.

8.3.3.1 The Case of the Full Group $\Gamma = \mathrm{L}_{168}$

Utilizing this explicit representation it is straightforward to calculate the age of each conjugacy class and we obtain the result displayed in the following table.

Conjugacy class of L_{168}	$\mathscr{C}_1$	$\mathscr{C}_2$	$\mathscr{C}_3$	$\mathscr{C}_4$	$\mathscr{C}_5$	$\mathscr{C}_6$
representative of the class	**e**	R	S	TSR	T	SR
order of the elements in the class	1	2	3	4	7	7
age	0	1	1	1	1	2
number of elements in the class	1	21	56	42	24	24

(8.3.13)

8.3.3.2 The Case of the Maximal Subgroup $\Gamma = \mathrm{G}_{21} \subset \mathrm{L}_{168}$

In order to obtain the ages for the conjugacy classes of the maximal subgroup G_{21}, we just need to obtain the explicit three-dimensional form of its generators $\mathscr{X}$ and $\mathscr{Y}$ satisfying the defining relations (1.3.35). This latter is determined by the above explicit form of the L_{168} generators, by recalling the embedding relations:

$$
\mathscr{Y} = \rho\,\sigma\,\tau^3\,\sigma\,\rho \quad ; \quad \mathscr{X} = \sigma\,\rho\,\sigma\,\rho\,\tau^2 \tag{8.3.14}
$$

In this way we obtain the following explicit result:

$$
\begin{aligned}
\mathscr{Y} &\to \mathrm{Y} = \begin{pmatrix} -(-1)^{3/7} & 0 & 0 \\ 0 & (-1)^{6/7} & 0 \\ 0 & 0 & -(-1)^{5/7} \end{pmatrix} \\
\mathscr{Y} &\to \mathrm{X} = \begin{pmatrix} 0 & 1 & 0 \\ 0 & 0 & 1 \\ 1 & 0 & 0 \end{pmatrix}
\end{aligned}
\tag{8.3.15}
$$

Hence, for the action on $\mathbb{C}^3$ of the maximal subgroup $G_{21} \subset L_{168}$ we obtain the following ages of its conjugacy classes:

Conjugacy Class of G_{21}	C_1	C_2	C_3	C_4	C_5
representative of the class	e	$\mathscr{Y}$	$\mathscr{X}^2\mathscr{Y}\mathscr{X}\mathscr{Y}^2$	$\mathscr{Y}\mathscr{X}^2$	$\mathscr{X}$
order of the elements in the class	1	7	7	3	3
age	0	2	1	1	1
number of elements in the class	1	3	3	7	7

(8.3.16)

8.3.3.3 The Case of the Two Maximal Octahedral Subgroups

For the other two maximal subgroups O_{24A} and O_{24B} we find instead an identical result. This latter is retrieved from the two embedding conditions of the generators S and T.

Subgroup O_{24A}

$$T = \rho\,\sigma\,\rho\,\tau^2\,\sigma\,\rho\,\tau \quad ; \quad S = \tau^2\,\sigma\,\rho\,\tau\,\sigma^2 \tag{8.3.17}$$

Subgroup O_{24B}

$$T = \rho\,\tau\,\sigma\,\rho\,\tau^2\,\sigma\,\rho\,\tau \quad ; \quad S = \sigma\,\rho\,\tau\,\sigma\,\rho\,\tau \tag{8.3.18}$$

In this way we get:

Conjugacy Class of the O_{24A}	C_1	C_2	C_3	C_4	C_5
representative of the class	e	T	$STST$	S	ST
order of the elements in the class	1	3	2	2	4
age	0	1	1	1	1
number of elements in the class	1	8	3	6	6

(8.3.19)

and

Conjugacy Class of the O_{24B}	C_1	C_2	C_3	C_4	C_5
representative of the class	e	T	$STST$	S	ST
order of the elements in the class	1	3	2	2	4
age	0	1	1	1	1
number of elements in the class	1	8	3	6	6

(8.3.20)

8.3.3.4 The Case of the Cyclic Subgroups $\mathbb{Z}_3$ and $\mathbb{Z}_7$

Last we consider the age grading for the quotient singularities $\mathbb{C}^3/\mathbb{Z}_3$ and $\mathbb{C}^3/\mathbb{Z}_7$. As generators of the two cyclic groups we respectively choose the matrices X and Y displayed in Eq. (8.3.15). In other words we utilize either one of the two generators of the maximal subgroup $G_{21} \subset L_{168}$.

The $\Gamma = \mathbb{Z}_3$ Case

The first step consists of diagonalizing the action of the generator X. Introducing the unitary matrix:

$$\mathfrak{q} = \begin{pmatrix} \frac{1}{\sqrt{3}} & \frac{1}{\sqrt{3}} & \frac{1}{\sqrt{3}} \\ \frac{-1+i\sqrt{3}}{2\sqrt{3}} & \frac{-1-i\sqrt{3}}{2\sqrt{3}} & \frac{1}{\sqrt{3}} \\ \frac{-1-i\sqrt{3}}{2\sqrt{3}} & \frac{-1+i\sqrt{3}}{2\sqrt{3}} & \frac{1}{\sqrt{3}} \end{pmatrix} \tag{8.3.21}$$

we obtain:

$$\widetilde{\mathrm{X}} \equiv \mathfrak{q}^\dagger \,\mathrm{X}\, \mathfrak{q} = \begin{pmatrix} e^{\frac{2i\pi}{3}} & 0 & 0 \\ 0 & e^{-\frac{2i\pi}{3}} & 0 \\ 0 & 0 & 1 \end{pmatrix} \tag{8.3.22}$$

This shows that the quotient singularity $\mathbb{C}^3/\mathbb{Z}_3$ is actually of the form $\mathbb{C}^2/\mathbb{Z}_3 \otimes \mathbb{C}$ since it suffices to change basis of $\mathbb{C}^3$ by introducing the new complex coordinates:

$$\tilde{z}_a = \mathfrak{q}_a{}^b\, z_b \tag{8.3.23}$$

It follows that in the resolution of the singularity we will obtain:

$$ALE_{\mathbb{Z}_3} \otimes \mathbb{C} \rightarrow \frac{\mathbb{C}^3}{\mathbb{Z}_3} \tag{8.3.24}$$

Yet, as we discuss more extensively below, the starting setup $\mathbb{C}^3/\Gamma$ produces a special type of ALE-manifold where all the holomorphic moment map levels are frozen to zero and only the Kähler quotient parameters are switched on.

Equation (8.3.22) corresponds also to the decomposition of the three-dimensional representation of $\mathbb{Z}_3$ into irreducible representations of $\mathbb{Z}_3$. From the diagonalized form (8.3.22) of the generator we immediately obtain the ages of the conjugacy classes:

Conjugacy Class of $\mathbb{Z}_3$	C_1	C_2	C_3
representative of the class	e	X	X^2
order of the elements in the class	1	3	3
age	0	1	1
number of elements in the class	1	1	1

(8.3.25)

The $\Gamma = \mathbb{Z}_7$ Case

In the $\mathbb{Z}_7$ case, the generator Y is already diagonal and, as we see none of the three complex coordinates is invariant under the action of the group. Hence differently from the previous case we obtain:

$$\mathscr{M}_{\mathbb{Z}_7} \rightarrow \frac{\mathbb{C}^3}{\mathbb{Z}_7} \tag{8.3.26}$$

where the resolved smooth manifold is not the direct product of $\mathbb{C}$ with an ALE-manifold:

$$\mathcal{M}_{\mathbb{Z}_7} \neq ALE_{\mathbb{Z}_7} \otimes \mathbb{C} \tag{8.3.27}$$

From the explicit diagonal form (8.3.15) of the generator we immediately obtain the ages of the conjugacy classes:

Conjugacy Class of $\mathbb{Z}_7$	C_1	C_2	C_3	C_4	C_5	C_6	C_7
representative of the class	e	Y	Y^2	Y^3	Y^4	Y^5	Y^6
order of the elements in the class	1	7	7	7	7	7	7
age	0	2	2	1	2	1	1
number of elements in the class	1	1	1	1	1	1	1

(8.3.28)

8.4 ALE Manifolds and the Resolution of $\mathbb{C}^2/\Gamma$ Singularities

ALE manifolds are interesting *per se* since they are gravitational instantons; they also provide a very important item of comparison for $\mathbb{C}^3/\Gamma$ singularities, since they happen to be the crepant resolution of $\mathbb{C}^2/\Gamma$ singularities $\Gamma \subset \mathrm{SU}(2)$ being a kleinian group.

8.4.1 ALE Manifolds

ALE means asymptotically locally Euclidean. This means that ALE manifolds are smooth 4-manifolds with Euclidean signature and a metric leading to a self-dual curvature two-form:

$$\mathfrak{R}^{ab}_{ALE} = \frac{1}{2}\,\varepsilon^{abcd}\,\mathfrak{R}^{cd}_{ALE} \tag{8.4.1}$$

which, for large distances from a core, approaches the flat Euclidean metric.

Actually ALE manifolds are all Ricci flat and constitute vacuum solutions of Einstein equations after Wick rotation. In this sense ALE-manifolds are gravitational instantons in the same way as the connections with a self dual field strength are gauge instantons.

The first instance of an ALE manifold was found by Eguchi and Hanson [52] in 1979 and its explicit form will be discussed in Sect. 8.7.

The fascination of ALE manifolds is that they happen to be in one-to-one correspondence with the finite subgroups $\Gamma \subset \mathrm{SU}(2)$ and are similarly classified by the ADE classification of simply-laced Lie algebras.

Fig. 8.3 Peter Benedict Kronheimer (born 1963) is a British mathematician, known for his work on gauge theory and its applications to 3- and 4-dimensional topology. He is currently William Caspar Graustein Professor of Mathematics at Harvard University. He completed his PhD at Oxford University under the direction of Sir Michael Atiyah. Kronheimer's early work was on gravitational instantons, in particular the classification of HyperKähler four manifolds with asymptotical locally Euclidean geometry (ALE spaces) leading to the papers *The construction of ALE spaces as hyper-Kahler quotients* and *A Torelli-type theorem for gravitational instantons*. He also contributed extensively to the topology of 4-manifolds and to theory of Donaldson invariants. He and Nakajima gave a construction of instantons on ALE spaces generalizing the Atiyah-Hitchin-Drinfeld-Manin construction

In 1989 Peter Kronheimer (see Fig. 8.3) succeeded in constructing all of them as HyperKähler quotients of suitably chosen flat HyperKähler manifolds dictated by the structure of the finite group Γ to which each of them corresponds.

The association between ALE manifolds, ADE singularities and subgroups $\Gamma \subset \mathrm{SU}(2)$ is not a superficial matter rather it is a very deep and structural one. The topological properties of the ALE four-manifold are identified with intrinsic numbers of the corresponding Lie algebra; for instance the Hirzebruch signature τ of the ALE coincides with the rank of the corresponding Lie Algebra $\mathbb{G}$ and with the dimension of the chiral ring $\mathscr{R}_\Gamma$ associated with the singular potential W_Γ. On the other hand the same number is also that of the non trivial conjugacy classes of Γ, apart of the identity class.

The catch of all this is encoded in a surprising correspondence between extended Dynkin diagrams and irreducible representations of the finite groups Γ that had been discovered years before Kronheimer by McKay [53]. Without any doubt the McKay correspondence provided Kronheimer with an essential guideline for his construction.

Fig. 8.4 Gary William Gibbons (born 1946) is a British theoretical physicist. Gibbons was born in Coulsdon, Surrey. He was educated at Purley County Grammar School and the University of Cambridge, where in 1969 he became a research student under the supervision of Dennis Sciama. When Sciama moved to the University of Oxford, he became a student of Stephen Hawking, obtaining his PhD from Cambridge in 1973. Gibbons became a full professor in 1997, a Fellow of the Royal Society in 1999, and a Fellow of Trinity College, Cambridge in 2002. He has given outstanding contributions to the theory of quantum black holes and to the theory of gravitational instantons. His special interests in geometry in all of its aspects led him to contribute to many issues in string and M-theory compactifications

A very important basis for Kronheimer work was encoded in the work on gravitational instantons previously conducted by Gibbons and Hawking [47] (see Fig. 8.4) and by Hitchin [48] (see also [49, 50]).

In the next subsections we begin the discussion of ALE manifolds and of their topology. Kronheimer construction will be presented in Sect. 8.5.1.

8.4.2 ALE Manifolds and Their Relation with the ADE Singularities

ALE spaces are non-compact manifolds that have originally emerged in the literature as gravitational instantons. Indeed they are Riemannian 4-manifolds with an (anti)selfdual curvature 2-form and a metric that approaches the Euclidean metric at infinity. In polar coordinates (r, Θ) on $\mathbb{R}^4$, we have $g_{\mu\nu}(r, \Theta) = \delta_{\mu\nu} + O(r^{-4})$. This corresponds to the intuitive concept of an instanton as a defect which is localized in a finite region of space-time. This picture, however, is verified only modulo an additional subtlety that is of utmost relevance in the present geometrical construction. The base manifold of the gravitational instanton has a boundary at infinity which, rather than a pure 3-sphere is:

$$\mathbb{S}^3/\Gamma \tag{8.4.2}$$

$\Gamma \subset \mathrm{SU}(2)$ being a finite subgroup of $\mathrm{SU}(2) \sim \mathbb{S}^3$. Therefore, outside the core of the instanton, rather than $\mathbb{R}^4$, the manifold looks like the quotient singularity $\mathbb{R}^4/\Gamma$. This is the reason for the name given to these spaces: the asymptotic behaviour is Euclidean only *locally*.

For the sake of our purposes the most important aspect of ALE spaces is that they are complex 2-folds endowed with a HyperKähler structure and a trivial canonical bundle $c_1\,(ALE_\Gamma) = 0$. This makes ALE spaces the *non-compact* analogues of the $K3$ surface which, apart from the T^4 torus is the only compact Calabi–Yau 2-fold. Indeed viewed as a complex manifold, outside the core of the instanton, the ALE space looks like the quotient singularity

$$ALE_\Gamma \sim \mathbb{C}^2/\Gamma \quad ; \quad \Gamma \subset \mathrm{SU}(2) \tag{8.4.3}$$

where Γ is the above mentioned finite subgroup of SU(2). In this way we have explained the rationale for the subindex Γ attached to the symbol denoting an ALE space. Indeed it can be shown that the choice of the identification group at infinity completely fixes the *topological type* of the ALE manifold. These types are in one-to-one correspondence with the finite groups Γ which admit an ADE classification, like simple Lie algebras and simple singularities. The correspondence between the ADE classification of ALE spaces and that of simple singularities will be discussed below. For the moment we note that the remaining ambiguity, once the identification group Γ has been fixed is given by the moduli of the self dual metric (i.e. of the HyperKähler structure) at fixed topological type.

In the HyperKähler quotient construction of the ALE spaces the complete set of the HyperKähler structure moduli can be seen as the *levels* of the *quaternionic momentum map*.

Let us summarize this viewpoint. In this approach the 4-dimensional HyperKähler ALE space is obtained from a flat HyperKähler manifold $\mathcal{S}$ of real dimension $4n$ invariant under the action of a compact Lie group $\mathcal{G}_\Gamma$ whose generators are vector fields $\mathbf{X}$ holomorphic with respect to the three complex structures of $\mathcal{S}$. In this construction the compact Lie group $\mathcal{G}_\Gamma$ is fully determined by the choice of the discrete group Γ. Because of their triholomorphicity the vector fields $\mathbf{X}$ preserve also the HyperKähler 2-forms Ω^i (i = 1, 2, 3):

$$0 = \mathcal{L}_{\mathrm{X}}\Omega^i = i_{\mathrm{X}}d\Omega^i + d(i_{\mathrm{X}}\Omega^i) = d(i_{\mathrm{X}}\Omega^i)\,. \tag{8.4.4}$$

If $\mathcal{S}$ is simply connected, $d(i_{\mathbf{X}}\Omega^i) = 0$ implies the existence of three functions $\mu_i^{\mathbf{X}}$ such that $d\mu_i^{\mathbf{X}} = i_{\mathrm{X}}\Omega^i$. The functions $\mu_i^{\mathbf{X}}$ are defined up to a constant, which can be arranged so as to make them equivariant: $\mathbf{X}\mu_i^{\mathbf{Y}} = \mu_i^{[\mathbf{X},\mathbf{Y}]}$. The $\{\mu_i^{\mathbf{X}}\}$ constitute the *triholomorphic momentum map* discussed in Sect. 3.7.2, namely a map $\mu : \mathcal{S} \to \mathbb{R}^3 \otimes \mathbb{G}_\Gamma^\star$, where $\mathbb{G}_\Gamma^\star$ denotes the dual of the Lie algebra $\mathbb{G}_\Gamma$ of the group $\mathcal{G}_\Gamma$. Indeed let $x \in \mathbb{G}_\Gamma$ be the element corresponding to the Killing vector $\mathbf{X}$; then for a given $m \in \mathcal{F}, \mu_i(m) : x \longmapsto \mu_i^{\mathbf{X}}(m) \in \mathbb{C}$ is a linear functional on $\mathcal{G}_\Gamma$. Let $\mathcal{Z} \subset \mathbb{G}_\Gamma^\star$ be the dual of the centre of $\mathbb{G}_\Gamma$. For each $\boldsymbol{\ell} \in \mathbb{R}^3 \otimes \mathcal{Z}$ the level set of the momentum map

$$\mathscr{N} \equiv \bigcap_{\mathbf{X}\in\mathscr{Z}} \left(\mu_i^{\mathbf{X}}\right)^{-1} (\ell_i^{\mathbf{X}}) \subset \mathscr{F}, \tag{8.4.5}$$

which has dimension $\dim \mathscr{N} = \dim \mathscr{S} - 3 \dim \mathscr{G}_\Gamma$, is left invariant by the action of $\mathscr{G}_\Gamma$, due to the equivariance of μ. Thus we can take the quotient

$$ALE_\Gamma = \mathscr{N}/\mathscr{G}_\Gamma. \tag{8.4.6}$$

which is a manifold of dimension $\dim ALE_\Gamma = \dim\mathscr{S} - 4\dim\mathscr{G}_\Gamma$ as long as the action of $\mathscr{G}_\Gamma$ on $\mathscr{N}$ has no fixed points. The triplet $\hat{\Omega}^i$ of 2-forms on ALE_Γ, defined via the restriction of Ω^i to $\mathscr{N} \subset \mathscr{S}$ and the quotient projection from $\mathscr{N}$ to ALE_Γ, are the HyperKähler forms on ALE_Γ. It is important to note that, once the third complex structure $\mathscr{J}^3$ is chosen as the preferred complex structure, the momentum maps $\mu_\pm^{\mathbf{X}} = \mu_1^{\mathbf{X}} \pm i\mu_2^{\mathbf{X}}$ are holomorphic (resp. antiholomorphic) functions. For $\forall \mathbf{X} \in \mathscr{Z}$ the level parameter $\boldsymbol{\ell}^{\mathbf{X}}$ is a 3-vector, corresponding to a unit quaternion:

$$\boldsymbol{\ell}^{\mathbf{X}} = \begin{pmatrix} \ell_3^{\mathbf{X}} & \mathrm{i}(\ell_1^{\mathbf{X}} - \mathrm{i}\ell_2^{\mathbf{X}}) \\ \mathrm{i}(\ell_1^{\mathbf{X}} + \mathrm{i}\ell_2^{\mathbf{X}}) & -\ell_3^{\mathbf{X}} \end{pmatrix} \tag{8.4.7}$$

and the complex combinations:

$$\ell_\pm^{\mathbf{X}} = \ell_1^{\mathbf{X}} \pm \mathrm{i}\ell_2^{\mathbf{X}} \tag{8.4.8}$$

can be regarded as moduli of the complex structure deformations. This goes as follows.

As a complex manifold, the ALE space equipped with the HyperKähler metric of moduli $\boldsymbol{\ell}^{\mathbf{X}}$ can be identified with the zero-locus in $\mathbb{C}^3$ of the following polynomial:

$$\begin{aligned} W_\Gamma^{ALE}(u, w, z;\, \mathbf{t}) &= W_\Gamma(u, w, z) + \sum_{i=1}^{r} t_i\, \mathscr{P}^{(i)}(u, w, z) \\ r &\equiv \dim \mathscr{R}_\Gamma \end{aligned} \tag{8.4.9}$$

where

1. $W_\Gamma(u, w, z)$ is the simple singularity polynomial corresponding to the finite subgroup $\Gamma \subset \mathrm{SU}(2)$
2. $\mathscr{P}^{(i)}(u, w, z)$ is a basis spanning the chiral ring

$$\mathscr{R}_\Gamma = \frac{\mathbb{C}[u, w, z]}{\partial W_\Gamma} \tag{8.4.10}$$

   of polynomials in u, w, z that do not vanish upon use of the vanishing relations $\partial_u W_\Gamma = \partial_w W_\Gamma = \partial_z W_\Gamma = 0$.
3. The complex parameters t^i are the complex structure moduli and they are in one-to-one correspondence with the set of complex level parameters $\ell_+^{\mathbf{X}}$. As an

illustration, in a later section the explicit relation between these two sets will be worked out for the case of the A_k finite groups. Alternatively the parameters t^i can be seen as the moduli for the resolution of the quotient singularity $\mathbb{C}^2/\Gamma$

It is a matter of fact that the dimension of the chiral ring $r \equiv |\mathscr{R}_\Gamma|$ is precisely equal to the number of non-trivial conjugacy classes (or of non trivial irreducible representations) of the finite group Γ. From the geometrical point of view this implies an identification between the number of complex structure deformations of the ALE manifold and the number r of non-trivial conjugacy classes just mentioned. As we recall below this implies that $\tau = r$, where τ is the Hirzebruch signature of the manifold ALE_Γ. In the language of algebraic geometry the singular orbifold $\mathbb{C}^2/\Gamma$, which is in one-to-one correspondence with the vanishing locus Z_0 of the potential $W_\Gamma(u, w, z)$ admits an equivariant minimal resolutions of singularity $Z \xrightarrow{\lambda} Z_0$, where Z is a smooth variety, λ is an isomorphism outside the singular point $\{0\} \in Z_0$ and it is a proper map such that $\lambda^{-1}(Z_0 - 0)$ is dense in Z. The fundamental fact is that the exceptional divisor $\lambda^{-1}(0) \subset Z$ consists of a set of irreducible curves $c_\alpha, \alpha = 1, \ldots, r$ one for each vertex of the Dynkin diagram of the simple Lie Algebra associated with Γ in the ADE classification of finite rotation groups. Each c_α is isomorphic to a copy of $\mathbb{P}^1$; the intersection matrix of these non-trivial two-cycles is the negative of the Cartan matrix:

$$c_\alpha \cap c_\beta = -C_{\alpha\beta}\,. \tag{8.4.11}$$

Kronheimer construction, reviewed in Sect. 8.5.1, shows that the base manifold ALE_Γ of an ALE space is diffeomorphic to the space Z supporting the resolution of the orbifold $Z_0 \sim \mathbb{C}^2/\Gamma$. Therefore Eq. (8.4.11) applies to the generators of the second homology group of ALE_Γ. In particular we see that

$$\begin{aligned}\tau &= dim H^2_{\mathbf{c}}(ALE_\Gamma) = dim H_2(ALE_\Gamma) = \\ &= \text{rank of the corresponding Lie Algebra} = \\ &= \#\text{ of non trivial conj. classes in } \Gamma = |\mathscr{R}_\Gamma|\,. \end{aligned}\tag{8.4.12}$$

For a proper illustration of (8.4.12) let us recall that on a non-compact manifold it is worth considering the "compact-support" cohomology groups, which coincide with the relative cohomology groups of forms vanishing on the boundary at infinity of the manifold:

$$H^p_{\mathbf{c}} = \frac{\{\mathbf{L}^2 \text{ integrable, closed } p - \text{forms}\}}{\{\mathbf{L}^2 \text{ integrable, exact } p - \text{forms}\}} = H^p(ALE_\Gamma, \partial ALE_\Gamma),$$

of dimensions $b^p_{\mathbf{c}}$. Analogously we will consider the compact support Dolbeault cohomology groups $H^{p,q}_{\mathbf{c}}$, of dimensions $h^{p,q}_{\mathbf{c}}$. The Poincaré duality provides an isomorphism $H_p(ALE_\Gamma) \sim H^{4-p}_{\mathbf{c}}(ALE_\Gamma)$, where $H_p(ALE_\Gamma)$ are the homology groups. Call b_p their dimensions (the *Betti numbers*); then $b_p = b^{4-p}_{\mathbf{c}}$. The fundamental topological invariants characterizing gravitational instantons were recognized long time ago to be the Euler characteristic χ and the Hirzebruch signature τ of the base manifold. The Euler characteristic is the alternating sum of the Betti numbers:

$$\chi = \sum_{p=0}^{4}(-1)^p b_p = \sum_{p=0}^{4}(-1)^p b_{\mathbf{c}}^{4-p} = \sum_{p=0}^{4}(-1)^p b_{\mathbf{c}}^{p}. \tag{8.4.13}$$

The Hirzebruch signature is the difference between the number of positive and negative eigenvalues of the quadratic form on $H_{\mathbf{c}}^2(ALE_\Gamma)$ given by the cup product $\int_{ALE_\Gamma} \alpha \wedge \beta$, with $\alpha, \beta \in H_{\mathbf{c}}^2(ALE_\Gamma)$. That is, if $b_{\mathbf{c}}^{2(+)}$ and $b_{\mathbf{c}}^{2(-)}$ are the number of selfdual and anti-selfdual 2-forms with compact support, $\tau = b_{\mathbf{c}}^{2(+)} - b_{\mathbf{c}}^{2(-)}$. At this point, we need two observations.

1. The HyperKähler forms Ω^3, $\Omega^\pm$, being covariantly constant, cannot be $\mathbf{L}^2$ if the space is non-compact
2. In the compact case, for instance for the $K3$ manifold, they are the unique antiselfdual 2-forms, so that $b^{2(-)} = 3$, $b^{2(+)} = \tau + 3$. Indeed from the expression of the Hirzebruch signature in terms of the Hodge numbers, $\tau = \sum_{p+q=0\mathrm{mod}2}(-1)^p h^{p,q}$, using the Calabi–Yau condition $c_1(K3) = 0$, which implies $h^{2,0} = h^{0,2} = 1$, and the fact that $h^{0,0} = h^{2,2} = 1$ we obtain $h^{1,1} = \tau + 4$. Hence the cohomology in degree two splits as follows:

$$\begin{array}{ccc} h^{2,0} & h^{1,1} & h^{0,2} \\ 1 & 1+(\tau+3) & 1 \end{array}$$

   This leads to the conclusion that $\Omega^3 \in H^{1,1}$ and $\Omega^\pm \in H^{2,0}$ (resp. $H^{0,2}$) are the unique antiselfdual two-forms.

In the non compact case, by the observation (1) the HyperKähler two-forms are deleted from the compact support cohomology groups. However the Hirzebruch signature is what it is, hence also other three selfdual two-forms have to be deleted as being non square-integrable, in order to maintain the value of τ. The "Hodge diamonds" for the usual and $\mathbf{L}^2$ Dolbeault cohomology groups are respectively given by:

$$\text{tot. cohom.} = \begin{array}{ccccc} & & 1 & & \\ & 0 & & 0 & \\ 1 & & \tau+4 & & 1 \\ & 0 & & 0 & \\ & & 0 & & \end{array} \qquad \text{comp. cohom.} = \begin{array}{ccccc} & & 0 & & \\ & 0 & & 0 & \\ 0 & & \tau & & 0 \\ & 0 & & 0 & \\ & & 1 & & \end{array} \tag{8.4.14}$$

Note that, from (8.4.13), $\chi = \tau + 1$.

The list of the simple singularity potentials in association with the various finite groups $\Gamma \subset \mathrm{SU}(2)$ and the corresponding ALE manifolds is given in Table 8.3. We show later on how the undeformed singularity potentials can be derived from consideration of Γ-invariants.

Table 8.3 FINITE SU(2) SUBGROUP versus ALE MANIFOLD properties

Γ.	$W_\Gamma(u,w,z)$	$\mathscr{R} = \frac{\mathbb{C}[u,w,z]}{\partial W}$	$\lvert\mathscr{R}\rvert$	#c.c.	$\tau \equiv \chi - 1$
A_k	$u^2+w^2-z^{k+1}$	$\{1, z, ..$ $.., z^{k-1}\}$	k	$k+1$	k
D_{k+2}	u^2+w^2z+ z^{k+1}	$\{1, w, z, w^2,$ $z^2, ..., z^{k-1}\}$	$k+2$	$k+3$	$k+2$
$E_6 = \mathscr{T}$	$u^2+w^3+z^4$	$\{1, w, z,$ $wz, z^2, wz^2\}$	6	7	6
$E_7 = \mathscr{O}$	$u^2+w^3+wz^3$	$\{1, w, z, w^2,$ $z^2, wz, w^2z\}$	7	8	7
$E_8 = \mathscr{I}$	$u^2+w^3+z^5$	$\{1, w, z, z^2, wz,$ $z^3, wz^2, wz^3\}$	8	9	8

8.4.3 ALE Manifolds as Algebraic Loci

The ALE manifold is the algebraic locus cut out in $\mathbb{C}^3$ by the polynomial constraint:

$$W_\Gamma^{ALE}\,(u,w,z,\mathbf{t}) = 0 \tag{8.4.15}$$

The complex parameters $\mathbf{t}$ parameterize the complex structure of the considered ALE manifold.

An example of this construction corresponds to the following choices:

$$\begin{aligned} \Gamma &= A_k \\ W_{A_k}\,(u,w,z,\zeta) &= u^2+w^2+z^{k+1} \\ \sum_{i=1}^{|\mathscr{R}_\Gamma|} t^i\,\mathscr{P}^i(u,w,z) &= t_k\,z^{k-1}\,t_{k-1}\,z^{k-2}\,+\cdots+\,t_2z\,+\,t_1 \end{aligned} \tag{8.4.16}$$

In the case we consider we can rewrite:

$$\begin{aligned} W_{A_k}^{ALE}\,(u,w,z,\mathbf{t}) &= u^2+w^2+P(z,\mathbf{t}) \\ P(z,\mathbf{t}) &\equiv z^{k+1}+t_k\,z^{k-1}\,+\,t_{k-1}\,z^{k-2}\,+\cdots+\,t_2\,z\,+\,t_1 \end{aligned} \tag{8.4.17}$$

The order $k+1$ polynomial $P(x,\mathbf{t})$ can always be factorized according to its roots a_i and we can write:

$$P(z,\mathbf{t}) = \prod_{i=1}^{k+1}(z-a_i) \tag{8.4.18}$$

This parametrization allows a simple characterization of the k homology cycles c_α with intersection matrix given by (8.4.11).

8.4.4 Explicit Form of the Homology 2-Cycles on ALE Manifolds

Consider the particular case where the ALE moduli relative to the complex structure, namely the roots a_i introduced in (8.4.18) are all real and ordered on the real line as:

$$a_i \in \mathbb{R} \qquad a_1 > a_2 > \cdots > a_k \tag{8.4.19}$$

we can introduce the following k maps of the compact 2-sphere

$$\mathbb{S}^2 = \{0 \leq \theta \leq \pi \quad ; \quad 0 \leq \varphi \leq 2\pi\} \tag{8.4.20}$$

into the ALE manifold:

$$(\theta, \varphi) \longrightarrow c_\alpha \equiv \begin{cases} u = A_\alpha\,(\theta)\ \sin\varphi \\ w = A_\alpha\,(\theta)\ \cos\varphi \\ z = \frac{a_\alpha + a_{\alpha+1}}{2} + \cos\theta\ \left(\frac{a_\alpha - a_{\alpha+1}}{2}\right) \end{cases} \tag{8.4.21}$$

where the function $A_\alpha\,(\theta)$ is to be determined in such a way that the ALE equation is satisfied, that is:

$$u^2 + w^2 + \prod_{i=1}^{k+1} (z - a_i) \; = \; 0 \tag{8.4.22}$$

We immediately obtain:

$$A_\alpha\,(\theta) = \frac{a_\alpha - a_{\alpha+1}}{2}\ \sin\theta\,\sqrt{\Delta_\alpha\,(\theta)} \tag{8.4.23}$$

where

$$\Delta_\alpha\,(\theta) = \prod_{j \neq \alpha, \alpha+1} \left[\frac{a_\alpha + a_{\alpha+1} - 2a_j}{2} + \cos\theta \left(\frac{a_\alpha - a_{\alpha+1}}{2} \right) \right] \tag{8.4.24}$$

The 2-spheres do touch in one point on the real z-axis since for $\theta = 0, \pi$ we have $A_\alpha(0) = A_\alpha(\pi) = 0$ implying $u = w = 0$, while:

$$z_\alpha(\pi) = a_\alpha = z_{\alpha+1}(0) \tag{8.4.25}$$

This shows that the intersection matrix of these cycles is indeed the negative of the Cartan matrix for the A_k Lie algebra:

$$c_\alpha \cap c_\beta = -\begin{pmatrix} 2 & -1 & 0 & 0 & \dots & 0 \\ -1 & 2 & -1 & 0 & \dots & 0 \\ \dots & \dots & \dots & \dots & \dots & \dots \\ 0 & \dots & 0 & 0 & -1 & 2 \end{pmatrix} \tag{8.4.26}$$

Next we can observe that for any value of the moduli away from the singularities, namely where all the roots a_i are distinct, the ALE manifold, viewed as a real manifold, is always the same manifold in the same way as for any non singular value of the modulus parameter τ the complex torus

$$z \sim z + n + m\tau \qquad n, m \in \mathbb{Z} \tag{8.4.27}$$

is always the same real torus. Hence the homology basis we have constructed in a specific complex structure is a homology basis for the underlying *real* ALE-manifold.

8.4.5 Periods of the $\Omega^{(2,0)}_{ALE}$ Form on ALE Spaces

Given these preliminaries we can calculate the periods of the $\Omega^{(2,0)}_{ALE}$ holomorphic 2-form on the ALE manifolds. For an arbitrary ALE-space the number of homology 2-cycles is:

$$\begin{aligned} \#\text{2-cycles} &= |\mathscr{R}| = \text{dimension of chiral ring} \\ &= \#\text{ of irreps of } \Gamma - 1 \\ &= \tau = \text{Hirzebruch signature of ALE} \\ &= k = \text{rank of the Lie algebra } \leftrightarrow \ \Gamma \end{aligned} \tag{8.4.28}$$

and we can introduce the following $k = \tau$-dimensional vector of periods:

$$\Theta^{ALE}\left(\mathbf{t}\right) \equiv= \begin{pmatrix} \int_{c_1} \Omega^{(2,0)}_{ALE} \\ \int_{c_2} \Omega^{(2,0)}_{ALE} \\ \dots \\ \int_{c_k} \Omega^{(2,0)}_{ALE} \end{pmatrix} \tag{8.4.29}$$

8.4.5.1 Explicit Calculation of the Periods in the A_k Case

In the case of the ALE manifolds encoded in Eq. (8.4.17) and (8.4.18) it is fairly simple to calculate explicitly the periods (8.4.29).

As holomorphic 2-form we can choose:

$$\Omega^{(2,0)}_{ALE} = \frac{du \wedge dz}{\partial_w W^{ALE}_{A_k}} = \frac{1}{2}\,\frac{du \wedge dz}{w} \tag{8.4.30}$$

Specializing $\Omega^{(2,0)}_{ALE}$ to the homology two-cycles we have:

$$du = A'_\alpha(\theta)\,\sin\varphi\,d\theta + A_\alpha(\theta)\,\cos\varphi\,d\varphi \tag{8.4.31}$$

$$dz = \frac{a_\alpha - a_{\alpha+1}}{2}\,\sin\theta\,d\theta \tag{8.4.32}$$

$$w = 2A_\alpha(\theta)\,\cos\varphi \tag{8.4.33}$$

Hence we obtain:

$$\begin{aligned}\int_{C_\alpha} \Omega^{(2,0)}_{ALE} &= \int_{S^2} c^\star_\alpha\left[\Omega^{(2,0)}_{ALE}\right] \\ &= \frac{1}{2}\,\frac{a_\alpha - a_{\alpha+1}}{2} \int d\varphi \wedge \sin\theta\,d\theta \\ &= \pi\,(a_\alpha - a_{\alpha+1})\end{aligned} \tag{8.4.34}$$

The ALE manifold develops a singularity when some of the periods (volumes) of the cohomology 2-cycles shrinks to zero and this happens when two contiguous roots of the polynomial (8.4.18) coincide. In particular all cycles shrink to zero and we are at the orbifold singular point when all roots coincide (Fig. 8.5).

8.4.6 Comparison with the $\mathbb{C}^3/\Gamma$ Case

Let us compare the above predictions for the case (B) of $\mathbb{C}^3/\Gamma$ singularities with the well known case (A) of $\mathbb{C}^2/\Gamma$ where Γ is a Kleinian subgroup of SU(2) and the resolution of the singularity leads to an ALE manifold. As we already stressed above this latter can be explicitly constructed by means of a HyperKähler quotient, according with Kronheimer's construction.

Table 8.3 encodes the main properties about $Y \to X = \mathbb{C}^2/\Gamma$ which we have been discussing throughout the present section. They can be summarized as follows:

1. As an affine variety the singular orbifold X is described by a single polynomial equation $W_\Gamma(u, w, z) = 0$ in $\mathbb{C}^3$. This equation is simply given by a relation existing among the invariants of Γ as we anticipated in the previous section. Note that this is the case also for $X = \frac{\mathbb{C}^3}{\mathrm{L}_{168}}$, as Markushevich has shown. As we already recalled, he has found one polynomial constraint $W_{\mathrm{L}_{168}}(u_1, u_2, u_3, u_4) = 0$ of degree 10 in $\mathbb{C}^4$ which describes X. Generically this is not rue for other groups including subgroups of L_{168}.

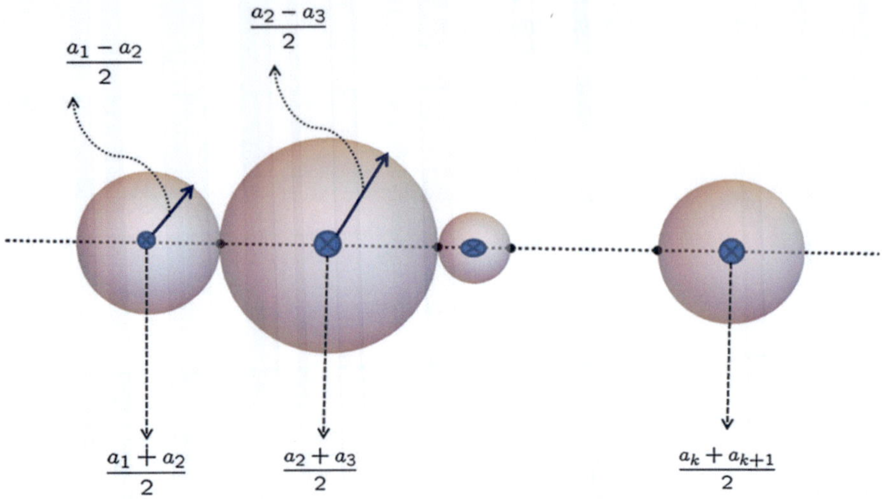

Fig. 8.5 The homology 2-cycles of an ALE manifold of type A_k. Naming a_i the $k+1$ roots of the polynomial $P(x, \mathbf{t})$, the radii of the homology two spheres are $r_\alpha = \frac{a_\alpha - a_{\alpha+1}}{2}$ and the centers are $\frac{a_\alpha + a_{\alpha+1}}{2}$

2. The resolved locus Y in the case of ALE manifolds is obtained by a deformed equation:

$$W_\Gamma^{ALE}(u, w, z; \mathbf{t}) = W_\Gamma(u, w, z) + \sum_{i=1}^{r} t_i \, \mathscr{P}^{(i)}(u, w, z)$$
$$r \equiv \dim \mathscr{R}_\Gamma \tag{8.4.35}$$

where

(a) $W_\Gamma(u, w, z)$ is the simple singularity polynomial corresponding to the finite subgroup $\Gamma \subset \mathrm{SU}(2)$

(b) $\mathscr{P}^{(i)}(u, w, z)$ is a basis spanning the chiral ring

$$\mathscr{R}_\Gamma = \frac{\mathbb{C}[u, w, z]}{\partial W_\Gamma} \tag{8.4.36}$$

of polynomials in u, w, z that do not vanish upon use of the vanishing relations $\partial_u W_\Gamma = \partial_w W_\Gamma = \partial_z W_\Gamma = 0$.

(c) The complex parameters t^i are the complex structure moduli and they are in one-to-one correspondence with the set of complex level parameters $\ell_+^{\mathbf{X}}$.

3. According with the general view put forward in the previous section, for ALE manifolds we have:

$$\dim H^{(1,1)} = r \equiv \# \text{ non trivial conjugacy classes of } \Gamma \tag{8.4.37}$$

We also have:

$$\dim \mathscr{R}_\Gamma = r \tag{8.4.38}$$

as one sees from Table 8.3. From the point of view of complex differential geometry this is the consequence of a special coincidence, already stressed in the previous section, which applies only to the case of complex dimension 2. As one knows, for Calabi-Yau n-folds complex structure deformations are associated with $\omega^{n-1,1} \in H^{(n-1,1)}$ harmonic forms, while Kähler structure deformations, for all n, are associated with $\omega^{1,1} \in H^{(1,1)}$ harmonic forms. Hence when $n = 2$, the $(1,1)$-forms play a double role as complex structure deformations and as Kähler structure deformations. For instance, this is well known in the case of $K3$. Hence in the $n = 2$ case the number of *non trivial conjugacy classes* of the group Γ coincides both with the number of Kähler moduli and with number of complex structure moduli of the resolved variety.

4. In the case of $Y \to X = \frac{\mathbb{C}^3}{\Gamma}$ the number of $(1,1)$-forms and hence of Kähler moduli is still related with r = *#junior conjugacy classes of Γ* but there are no complex-structure deformations.

8.5 The McKay Correspondence for $\mathbb{C}^2/\Gamma$

Next we address the McKay correspondence and we show how it leads, according to Kronheimer, to the explicit construction of ALE-manifolds as HyperKähler quotients.

The character table of any finite group γ allows to reconstruct the decomposition coefficients of any representation along the irreducible representations:

$$D = \bigoplus_{\mu=1}^{r} a_\mu \, D_\mu$$

$$a_\mu = \frac{1}{g} \sum_{i=1}^{r} g_i \, \chi_i^{(D)} \, \chi_i^{(\mu)\star} \tag{8.5.1}$$

For the finite subgroups $\Gamma \subset \mathrm{SU}(2)$ a particularly important case is the decomposition of the tensor product of an irreducible representation D_μ with the defining 2-dimensional representation $\mathscr{Q}$. It is indeed at the level of this decomposition that the relation between these groups and the simply laced Dynkin diagrams becomes explicit and it is named the McKay correspondence. This decomposition plays a crucial role in the explicit construction of ALE manifolds according to Kronheimer. Setting

$$\mathscr{Q} \otimes D_\mu = \bigoplus_{\nu=0}^{r} A_{\mu\nu}\, D_\nu \tag{8.5.2}$$

where D_0 denotes the identity representation, one finds that the matrix $\bar{c}_{\mu\nu} = 2\delta_{\mu\nu} - A_{\mu\nu}$ is the *extended Cartan matrix* relative to the *extended Dynkin diagram* corresponding to the given group. We remind the reader that the extended Dynkin diagram of any simply laced Lie algebra is obtained by adding to the *dots* representing the *simple roots* $\{\alpha_1 \ldots\ldots \alpha_r\}$ an *additional dot* (marked black in Figs. 8.6, 8.7) representing the negative of the highest root $\alpha_0 = \sum_{i=1}^{r} n_i\, \alpha_i$ (n_i are the Coxeter numbers). Thus we see a correspondence between the non-trivial conjugacy classes $\mathscr{C}_i$ (or equivalently the non-trivial irrepses) of the group $\Gamma(\mathbb{G})$ and the simple roots of $\mathbb{G}$. In this correspondence the extended Cartan matrix provides the Clebsch–Gordon coefficients (8.5.2), while the Coxeter numbers n_i express the dimensions of the irreducible representations. All these informations are summarized in Figs. 8.6, 8.7 where the numbers n_i are attached to each of the dots: the number 1 is attached to the extra dot since it stands for the identity representation.

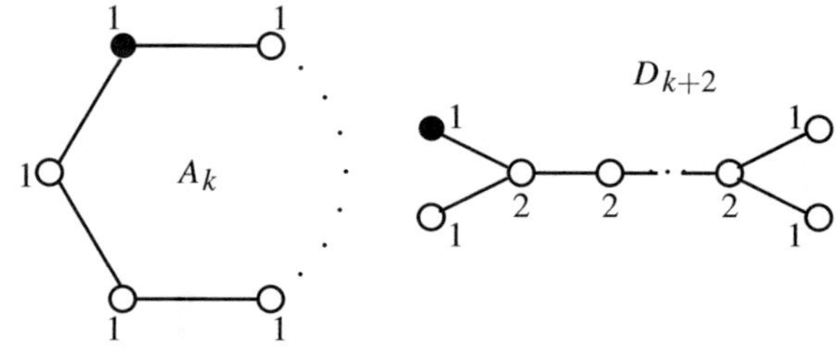

Fig. 8.6 Extended Dynkin diagrams of the infinite series

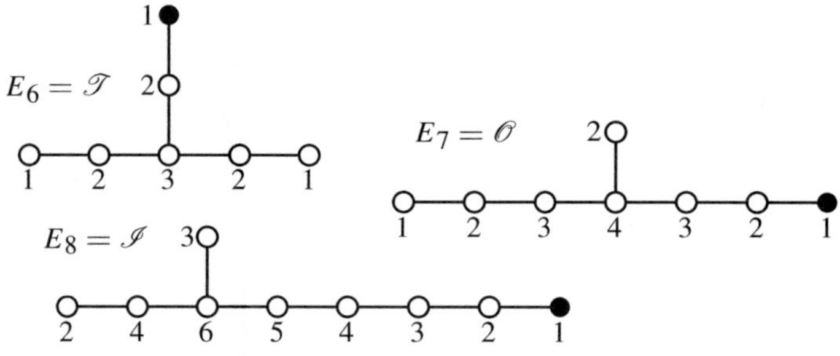

Fig. 8.7 Exceptional extended Dynkin diagrams

8.5.1 Kronheimer's Construction

Given any finite subgroup of $\Gamma \subset \mathrm{SU}(2)$, we consider a space $\mathscr{P}$ whose elements are two-vectors of $|\Gamma| \times |\Gamma|$ complex matrices: $p \in \mathscr{P} = (A, B)$. The action of an element $\gamma \in \Gamma$ on the points of $\mathscr{P}$ is the following:

$$\begin{pmatrix} A \\ B \end{pmatrix} \xrightarrow{\gamma} \begin{pmatrix} u_\gamma & i\,\bar{v}_\gamma \\ i\,v_\gamma & \bar{u}_\gamma \end{pmatrix} \begin{pmatrix} R(\gamma)\,A\,R(\gamma^{-1}) \\ R(\gamma)\,B\,R(\gamma^{-1}) \end{pmatrix} \tag{8.5.3}$$

where the two-dimensional matrix on the right hand side is the realization of γ inside the defining two-dimensional representation $\mathscr{Q} \subset \mathrm{SU}(2)$, while $R(\gamma)$ is the regular, $|\Gamma|$-dimensional representation. The basis vectors in R named e_γ are in one-to-one correspondence with the group elements $\gamma \in \Gamma$ and transform as follows:

$$R(\gamma)\,e_\delta = e_{\gamma\cdot\delta} \qquad \forall\gamma\,,\,\delta \in \Gamma \tag{8.5.4}$$

In mathematical notation the space $\mathscr{P}$ is named as:

$$\mathscr{P} \simeq \mathrm{Hom}\,(R, \mathscr{Q} \otimes R) \tag{8.5.5}$$

Next we introduce the space $\mathscr{S}$, which by definition is the subspace of Γ-invariant elements in $\mathscr{P}$:

$$\mathscr{S} \equiv \{p \in \mathscr{P}/\forall\gamma \in \Gamma, \gamma \cdot p = p\} \tag{8.5.6}$$

Explicitly the invariance condition reads as follows:

$$\begin{pmatrix} u_\gamma & i\,\bar{v}_\gamma \\ i\,v_\gamma & \bar{u}_\gamma \end{pmatrix} \begin{pmatrix} A \\ B \end{pmatrix} = \begin{pmatrix} R(\gamma^{-1})\,A\,R(\gamma) \\ R(\gamma^{-1})\,B\,R(\gamma) \end{pmatrix} \tag{8.5.7}$$

The decomposition (8.5.2) is very useful in order to determine the Γ-invariant flat space (8.5.6).

A two-vector of matrices can be thought of also as a matrix of two-vectors: that is, $\mathscr{P} = \mathscr{Q} \otimes \mathrm{Hom}(R, R) = \mathrm{Hom}(R, \mathscr{Q} \otimes R)$. Decomposing the regular representation, $R = \bigoplus_{\nu=0}^{r} n_\mu D_\mu$ into irrepses, using Eq. (8.5.2) and Schur's lemma, we obtain:

$$\mathscr{S} = \bigoplus_{\mu,\nu} A_{\mu,\nu} \mathrm{Hom}(\mathbb{C}^{n_\mu}, \mathbb{C}^{n_\nu})\ . \tag{8.5.8}$$

The dimensions of the irrepses, n_μ are displayed in Figs. 8.6, 8.7. From Eq. (8.5.8) the real dimension of $\mathscr{S}$ follows immediately: $\dim \mathscr{S} = \sum_{\mu,\nu} 2A_{\mu\nu} n_\mu n_\nu$ implies, recalling that $A = 2 \times \mathbf{1} - \bar{c}$ [see Eq. (8.5.2)] and that for the extended Cartan matrix $\bar{c} n = 0$:

$$\dim_{\mathbb{C}} \mathscr{S} = 2 \sum_\mu n_\mu^2 = 2|\Gamma|\ . \tag{8.5.9}$$

In mathematical notation the space $\mathscr{S}$ is denoted as follows:

$$\mathscr{S} \simeq \mathrm{Hom}_{\Gamma}\,(R, \mathscr{Q} \otimes R) \tag{8.5.10}$$

So we can summarize the discussion by saying that:

$$\dim_{\mathbb{C}}\,[\mathrm{Hom}_{\Gamma}\,(R, \mathscr{Q} \otimes R)] = 2\,|\Gamma| \tag{8.5.11}$$

The quaternionic structure of the flat manifolds $\mathscr{P}$ and $\mathscr{S}$ can be seen by simply writing their elements as follows:

$$p = \begin{pmatrix} A & iB^{\dagger} \\ iB & A^{\dagger} \end{pmatrix} \in \mathrm{Hom}\,(R, \mathscr{Q} \otimes R) \qquad A, B \in \mathrm{End}(R)\ .$$

Then the HyperKähler forms and the HyperKähler metric are defined by the following formulae:

$$\begin{aligned} \Theta = \mathrm{Tr}(\mathrm{d}p^{\dagger} \wedge \mathrm{d}p) &= \begin{pmatrix} \mathrm{i}\,\mathbf{K} & \mathrm{i}\overline{\boldsymbol{\Omega}} \\ \mathrm{i}\,\boldsymbol{\Omega} & -\mathrm{i}\,\mathbf{K} \end{pmatrix} \\ ds^2 \times \mathbf{1} &= \mathrm{Tr}(\mathrm{d}p^{\dagger} \otimes \mathrm{d}p) \end{aligned} \tag{8.5.12}$$

In the above equations the trace is taken over the matrices belonging to $\mathrm{End}(R)$ in each entry of the quaternion. From Eq. (8.5.12) we extract the explicit expressions for the Kähler 2-form $\mathbf{K}$ and the holomorphic 2-form $\boldsymbol{\Omega}$ of the flat HyperKähler manifold $\mathrm{Hom}\,(R, \mathscr{Q} \otimes R)$. We have:

$$\begin{aligned} \mathbf{K} &= -\mathrm{i}\left[\mathrm{Tr}\left(\mathrm{d}A^{\dagger} \wedge \mathrm{d}A\right) + \mathrm{Tr}\left(\mathrm{d}B^{\dagger} \wedge \mathrm{d}B\right)\right] \equiv \mathrm{i}g_{\alpha\bar{\beta}}\,\mathrm{d}q^{\alpha} \wedge \mathrm{d}q^{\bar{\beta}} \\ ds^2 &= g_{\alpha\bar{\beta}}\,\mathrm{d}q^{\alpha} \otimes \mathrm{d}q^{\bar{\beta}} \\ \boldsymbol{\Omega} &= 2\mathrm{Tr}\,(\mathrm{d}A \wedge \mathrm{d}B) \equiv \Omega_{\alpha\beta}\,\mathrm{d}q^{\alpha} \wedge \mathrm{d}q^{\beta} \end{aligned} \tag{8.5.13}$$

Starting from the above written formulae, by means of an elementary calculation one verifies that both the metric and the HyperKähler forms are invariant with respect to the action of the discrete group Γ defined in Eq. (8.5.3). Hence one can consistently reduce the space $\mathrm{Hom}\,(R, \mathscr{Q} \otimes R)$ to the invariant space $\mathrm{Hom}_{\Gamma}\,(R, \mathscr{Q} \otimes R)$ defined in Eq. (8.5.6). The HyperKähler 2-forms and the metric of the flat space $\mathscr{S}$, whose real dimension is $4|\Gamma|$ are given by Eq. (8.5.13) where the matrices A, B satisfy the invariance condition Eq. (8.5.7).

8.5.1.1 Solution of the Invariance Constraint in the Case of the Cyclic Groups A_k

The space $\mathscr{S}$ can be easily described when Γ is the cyclic group A_k. The order of A_k is $k + 1$; the abstract multiplication table is that of $\mathbb{Z}_{k+1}$. We can immediately read it off from the matrices of the regular representation. Obviously, it is sufficient to consider the representative of the first element e_1, as $R(e_j) = (R(e_1))^j$.

One has:

$$R(e_1) = \begin{pmatrix} 0\,0 \cdots 0\,1 \\ 1\,0 \cdots 0\,0 \\ 0\,1 \cdots 0\,0 \\ \vdots\,\vdots \ddots \vdots\,\vdots \\ 0\,0 \cdots 1\,0 \end{pmatrix} \tag{8.5.14}$$

Actually, the invariance condition Eq. (8.5.7) is best solved by changing basis so as to diagonalize the regular representation, realizing explicitly its decomposition in terms of the k unidimensional irrepses. Let $\nu = e^{\frac{2\pi i}{k+1}}$, be a $(k+1)$th root of unity so that $\nu^{k+1} = 1$. The looked for change of basis is performed by means of the matrix:

$$\begin{aligned} S_{ij} &= \frac{1}{\sqrt{k+1}}\,\nu^{ij} \quad ; \quad i, j = 0, 1, 2, \ldots, k \\ \left(S^{-1}\right)_{ij} = \left(S^{\dagger}\right)_{ij} &= \frac{1}{\sqrt{k+1}}\,\nu^{k+1-ij} \end{aligned} \tag{8.5.15}$$

In the new basis we find:

$$\begin{aligned} \widehat{R}(e_0) \equiv S^{-1}\,R(e_0)\,S &= \begin{pmatrix} 1\,0 & \ldots & 0\,0 \\ 0\,1\,0 & \ldots & 0\,0 \\ \vdots\,\vdots & \ddots & \vdots\,\vdots \\ 0\,0 & \ldots & 1\,0 \\ 0\,0 & \ldots & 0\,1 \end{pmatrix} \\ \widehat{R}(e_1) \equiv S^{-1}\,R(e_1)\,S &= \begin{pmatrix} 1\,0 & \ldots & 0 & 0 \\ 0\,\nu\,0 & \ldots & 0 & 0 \\ \vdots\,\vdots & \ddots & \vdots & \vdots \\ 0\,0 & \ldots & \nu^{k-1} & 0 \\ 0\,0 & \ldots & 0 & \nu^k \end{pmatrix} \end{aligned} \tag{8.5.16}$$

Equation (8.5.16) displays on the diagonal the representatives of e_j in the one-dimensional irrepses.

In the above basis, the explicit solution of Eq. (8.5.7) is given by

$$A = \begin{pmatrix} 0 & u_0 & 0 & \cdots & 0 \\ 0 & 0 & u_1 & \cdots & 0 \\ \vdots & \vdots & \vdots & \ddots & \vdots \\ \vdots & \vdots & \vdots & & u_{k-1} \\ u_k & 0 & 0 & \cdots & 0 \end{pmatrix} \quad ; \quad B = \begin{pmatrix} 0 & 0 & \cdots & \cdots & v_k \\ v_0 & 0 & \cdots & \cdots & 0 \\ 0 & v_1 & \cdots & \cdots & 0 \\ \vdots & \vdots & \ddots & & \vdots \\ 0 & 0 & \cdots & v_{k-1} & 0 \end{pmatrix} \tag{8.5.17}$$

We see that these matrices are parameterized in terms of $2k+2$ complex, i.e. $4(k+1) = |A_k|$ real parameters. In the D_{k+2} case, where the regular representation is $4k$-dimensional, choosing appropriately a basis, one can solve analogously Eq. (8.5.7); the explicit expressions are too large, so we do not write them. The essential point is that the matrices A and B no longer correspond to two distinct set of parameters, the group being non-abelian.

8.5.2 The Gauge Group for the Quotient and Its Moment Maps

The next step in the Kronheimer construction of the ALE manifolds is the determination of the group $\mathscr{F}$ of triholomorphic isometries with respect to which we will perform the quotient. We borrow from physics the nomenclature *gauge group* since in a would be $\mathscr{N}=3,4$ rigid three-dimensional gauge theory where the space $\text{Hom}_\Gamma(R, \mathscr{Q}\otimes R)$ is the flat manifold of hypermultiplet scalars, the triholomorphic moment maps of $\mathscr{F}$ emerge as scalar dependent non derivative terms in the hyperino supersymmetry transformation rules generated by the *gauging* of the group $\mathscr{F}$.

Consider the action of $\text{SU}(|\Gamma|)$ on $\text{Hom}(R, \mathscr{Q}\otimes R)$ given, using the quaternionic notation for the elements of $\text{Hom}(R, \mathscr{Q}\otimes R)$, by

$$\forall g \in \text{SU}(|\Gamma|) \quad , \quad g \quad : \quad \begin{pmatrix} A & i\,B^\dagger \\ i\,B & A^\dagger \end{pmatrix} \longmapsto \begin{pmatrix} gAg^{-1} & i\,g\,B^\dagger\,g^{-1} \\ i\,g\,B\,g^{-1} & g\,A^\dagger\,g^{-1} \end{pmatrix} \tag{8.5.18}$$

It is easy to see that this action is a triholomorphic isometry of $\text{Hom}(R, \mathscr{Q}\otimes R)$. Indeed both the HyperKähler forms Θ and the metric ds^2 are invariant.

Let $\mathscr{F}\subset \text{SU}(|\Gamma|)$ be the subgroup of the above group which *commutes with the action of Γ on the space* $\text{Hom}(R, \mathscr{Q}\otimes R)$, action which was defined in Eq. (8.5.3). Then the action of $\mathscr{F}$ descends to $\text{Hom}_\Gamma(R, \mathscr{Q}\otimes R)\subset \text{Hom}(R, \mathscr{Q}\otimes R)$ to give a *triholomorphic isometry*: indeed the metric and the HyperKähler forms on the space $\text{Hom}_\Gamma(R, \mathscr{Q}\otimes R)$ are just the restriction of those on $\text{Hom}(R, \mathscr{Q}\otimes R)$. Therefore one can take the HyperKähler quotient of $\text{Hom}_\Gamma(R, \mathscr{Q}\otimes R)$ with respect to $\mathscr{F}$.

Let $\{f_A\}$ be a basis of generators for $\mathbb{F}$, the Lie algebra of $\mathscr{F}$. Under the infinitesimal action of $f = \mathbf{1} + \lambda^A f_A \in \mathbb{F}$, the variation of $p\in \text{Hom}_\Gamma(R, \mathscr{Q}\otimes R)$ is $\delta p = \lambda^A \delta_A p$, with

$$\delta_A p = \begin{pmatrix} [f_A, A] & i[f_A, B^\dagger] \\ i[f_A, B] & [f_A, A^\dagger] \end{pmatrix}$$

The components of the momentum map are then given by

$$\mu_A = \text{Tr}\,(q^\dagger\,\delta_A p) \quad \equiv \quad \text{Tr}\begin{pmatrix} f_A\,\mu_3(p) & f_A\,\mu_-(p) \\ f_A\,\mu_+(p) & f_A\,\mu_3(p) \end{pmatrix}$$

so that the real and holomorphic maps μ_3 : Hom $_\Gamma(R, \mathcal{Q} \otimes R) \to \mathbb{F}^*$ and μ_+ : Hom $_\Gamma(R, \mathcal{Q} \otimes R) \to \mathbb{C} \times \mathbb{F}^*$ can be represented as matrix-valued maps:

$$\begin{aligned} \mu_3(p) &= -i\left([A, A^\dagger] + [B, B^\dagger]\right) \\ \mu_+(p) &= ([A, B]) \ . \end{aligned} \tag{8.5.19}$$

In this way we get:

$$\mu_A = \begin{pmatrix} \mathfrak{P}^3_A & \mathfrak{P}^-_A \\ \mathfrak{P}^+_A & -\mathfrak{P}^3_A \end{pmatrix} \tag{8.5.20}$$

where:

$$\begin{aligned} \mathfrak{P}^3_A &= -\mathrm{i}\left[\mathrm{Tr}\left(\left[A\,,\,A^\dagger\right] f_A\right) + \mathrm{Tr}\left(\left[B^\dagger\,,\,B\right] f_A\right)\right] \\ \mathfrak{P}^+_A &= \mathrm{Tr}\left([A\,,\,B]\ f_A\right) \end{aligned} \tag{8.5.21}$$

Let $\mathfrak{Z}^\star$ be the dual of the center of $\mathbb{F}$.

In correspondence with a level $\zeta = \{\zeta^3, \zeta^+\} \in \mathbf{R}^3 \otimes \mathfrak{Z}^\star$ we can form the Hyper-Kähler quotient:

$$\mathcal{M}_\zeta \equiv \mu^{-1}(\zeta) //_{HK} \mathcal{F} \tag{8.5.22}$$

Varying ζ and Γ all ALE manifolds can be obtained as $\mathcal{M}_\zeta$.

First of all, it is not difficult to check that $\mathcal{M}_\zeta$ is four-dimensional. Let us see how this happens. There is a nice characterization of the group $\mathcal{F}$ in terms of the extended Dynkin diagram associated with Γ. We have

$$\mathcal{F} = \bigotimes_{\mu=1}^{r+1} \mathrm{U}(\mathrm{n}_\mu) \bigcap \mathrm{SU}(|\Gamma|) \tag{8.5.23}$$

where the sum is extended to all the irreducible representations of the group Γ and n_μ are their dimensions. One should also take into account that the determinant of all the elements must be one, since $\mathcal{F} \subset \mathrm{SU}(|\Gamma|)$. Pictorially the group $\mathcal{F}$ has a $\mathrm{U}(\mathrm{n}_\mu)$ factor for each dot of the diagram, n_μ being associated with the dots as in Figs. 8.6, 8.7. $\mathcal{F}$ acts on the various *components* of Hom $_\Gamma(R, \mathcal{Q} \otimes R)$ that are in correspondence with the edges of the diagram, see Eq. (8.5.8), as dictated by the diagram structure. From Eq. (8.5.23) it is immediate to derive:

$$\dim \mathcal{F} = \sum_\mu n_\mu^2 - 1 = |\Gamma| - 1 \tag{8.5.24}$$

It follows that

$$\dim_{\mathbb{R}} \mathcal{M}_\zeta = \dim_{\mathbb{R}} \mathrm{Hom}_\Gamma(R, \mathcal{Q} \otimes R) - 4 \dim_{\mathbb{R}} \mathcal{F} = 4|\Gamma| - (4|\Gamma| - 1) = 4 \tag{8.5.25}$$

Analyzing the construction we see that there are two steps. In the first step, by setting the holomorphic part of the moment map to its level ζ, we define an algebraic locus in a larger space. Next the Kähler quotient further reduces such a locus to the necessary complex dimension 2. The two steps are united in one because of the triholomorphic character of the isometries. As we are going to stress in a subsequent section, in complex dimension 3 there is only holomorphicity; hence the two steps are separated. There must be another principle that leads to impose those constraints that cut out the algebraic locus $\mathbb{V}_{|\Gamma|+2}$ of which we perform the Kähler quotient in the next step (see Eq. (8.2.24)). The main question is to spell out such principles. As anticipated, equation $\mathbf{p} \wedge \mathbf{p} = 0$ is the one that does the job. We are not able to reduce the $3|\Gamma|^2$ quadrics on $3|\Gamma|$ variables to an ideal with $2|\Gamma| - 2$ generators, yet we know that such reduction must exist. Indeed, by means of another argument that utilizes Lie group orbits we can show that there is a variety of complex dimension 3, named $\mathscr{D}^0_\Gamma$ which is in the kernel of equation $\mathbf{p} \wedge \mathbf{p} = 0$.

8.5.2.1 The Triholomorphic Moment Maps in the A_k Case of Kronheimer Construction

The structure of $\mathscr{F}$ and the momentum map for its action are very simply worked out in the A_k case. An element $f \in \mathscr{F}$ must commute with the action of A_k on $\mathscr{P}$, Eq. (8.5.3), where the two-dimensional representation in the l.h.s. is given by:

$$\Gamma(A_k) \ni \gamma_\ell = \mathscr{Q}_\ell \equiv \begin{pmatrix} e^{2\pi i \ell/(k+1)} & 0 \\ 0 & e^{-2\pi i \ell/(k+1)} \end{pmatrix} \quad ; \quad \{\ell = 1, \ldots.., k+1\}$$

Then f must have the form

$$f = \mathrm{diag}(e^{i\varphi_0}, e^{i\varphi_1}, \ldots, e^{i\varphi_k}) \quad ; \quad \sum_{i=0}^{k} \varphi_i = 0 \,. \tag{8.5.26}$$

Thus $\mathbb{F}$ is just the algebra of diagonal traceless $k+1$-dimensional matrices, which is k-dimensional. Choose a basis of generators for $\mathbb{F}$, for instance:

$$\begin{aligned} f_1 &= \mathrm{diag}(1, -1, 0, \ldots, 0) \\ f_2 &= \mathrm{diag}(1, 0, -1, 0, \ldots, 0) \\ \ldots &= \ldots \\ f_k &= \mathrm{diag}(1, 0, 0, \ldots, 0, -1) \end{aligned} \tag{8.5.27}$$

From Eq. (8.5.21) we immediately obtain the components of the momentum map:

$$\begin{aligned} \mathfrak{P}^3_A &= |u^0|^2 - |u^k|^2 - |v_0|^2 + |v_k|^2 + \left(|u^{A-1}|^2 - |u^A|^2 - |v_{A-1}|^2 + |v_A|^2\right) \\ \mathfrak{P}^+_A &= u^0 v_0 - u^k v_k + \left(u^{A-1} v_{A-1} \; - \; u^A \, v_A\right) \quad , \quad (A = 1, \ldots, k) \end{aligned} \tag{8.5.28}$$

8.5.3 Level Sets and Weyl Chambers

In order for $\mathscr{M}_\zeta$ to be a smooth manifold, it is

necessary that $\mathscr{F}$ acts freely on $\mu^{-1}(\zeta)$. This happens or does not happen depending on the value of ζ. A simple characterization of $\mathfrak{Z}$ can be given in terms of the simple Lie algebra $\mathbb{G}$ associated with Γ. There exists an isomorphism between $\mathfrak{Z}$ and the Cartan subalgebra $\mathscr{H}_{CSA} \subset \mathbb{G}$. Thus we have

$$\begin{aligned}\dim \mathfrak{Z} = \dim \mathscr{H}_{CSA} &= \text{rank}\,\mathbb{G} \\ &= \#\text{of non trivial conj. classes in } Gamma \end{aligned} \tag{8.5.29}$$

The space $\mathscr{M}_\zeta$ turns out to be singular when, under the above identification $\mathfrak{Z} \sim \mathscr{H}_{CSA}$, any of the level components $\zeta^i \in \mathbf{R}^3 \otimes \mathfrak{Z}$ lies on the walls of a Weyl chamber. In particular, as the point $\zeta^i = 0$, $(i = 1, \ldots, r)$ is identified with the origin of the root space, which lies of course on all the walls of the Weil chambers, *the space $\mathscr{M}_0$ is singular*. Not too surprisingly we will see in a moment that $\mathscr{M}_0$ corresponds to the *orbifold limit* $\mathbb{C}^2/\Gamma$ of a family of ALE manifolds with boundary at infinity $\mathbb{S}^3/\Gamma$.

To verify this statement in general let us choose the natural basis for the regular representation R, in which the basis vectors e_δ transform as in Eq. (8.5.4). Define the space $L \subset \mathscr{S}$ as follows:

$$L = \left\{ \begin{pmatrix} C \\ D \end{pmatrix} \in \mathscr{S} \,/\, C, D \text{ are diagonal in the basis } \{e_\delta\} \right\} \tag{8.5.30}$$

For every element $\gamma \in \Gamma$ there is a pair of numbers (c_γ, d_γ) given by the corresponding entries of C, D: $C \cdot e_\gamma = c_\gamma e_\gamma$, $D \cdot e_\gamma = d_\gamma e_\gamma$. Applying the invariance condition Eq. (8.5.7), which is valid since $L \subset \mathscr{S}$, we obtain:

$$\begin{pmatrix} c_{\gamma\cdot\delta} \\ d_{\gamma\cdot\delta} \end{pmatrix} = \begin{pmatrix} u_\gamma & i\bar{v}_\gamma \\ iv_\gamma & \bar{u}_\gamma \end{pmatrix} \begin{pmatrix} c_\delta \\ d_\delta \end{pmatrix} \tag{8.5.31}$$

We can identify L with $\mathbb{C}^2$ associating for instance $(C, D) \in L \longmapsto (c_0, d_0) \in \mathbb{C}^2$. Indeed all the other pairs (c_γ, d_γ) are determined in terms of Eq. (8.5.31) once (c_0, d_0) are given. By Eq. (8.5.31) the action of Γ on L induces exactly the action of Γ on $\mathbb{C}^2$ provided by the its two-dimensional defining representation inside SU(2). It is quite easy to show the following fundamental fact: *each orbit of $\mathscr{F}$ in $\mu^{-1}(0)$ meets L in one orbit of Γ*. Because of the above identification between L and $\mathbb{C}^2$, this leads to conclude that *$\mu^{-1}(0)/\mathscr{F}$ is isometric to $\mathbb{C}^2/\Gamma$*. Instead of reviewing the formal proof of these statements devised by Kronheimer, we will verify them explicitly in the case of the cyclic groups, giving a description which sheds some light on the *deformed* situation; that is we show in which way a non-zero level ζ^+ for the holomorphic momentum map puts $\mu^{-1}(\zeta)$ in correspondence with the affine hypersurface in $\mathbb{C}^3$ cut out by the polynomial constraint (8.4.35) which is a deformation of that describing the $\mathbb{C}^2/\Gamma$ orbifold, obtained for $\zeta^+ = 0$.

8.5.3.1 Retrieving the Polynomial Constraint from the HyperKähler Quotient in the $\Gamma = A_k$ Case

We can directly realize $\mathbb{C}^2/\Gamma$ as an affine algebraic surface in $\mathbb{C}^3$ by expressing the coordinates x, y and z of $\mathbb{C}^3$ in terms of the matrices $(C, D) \in L$. The explicit parametrization of the matrices in $\mathscr{S}$ in the A_k case, which was given in Eq. (8.5.17) in the basis in which the regular representation R is diagonal, can be conveniently rewritten in the *natural basis* $\{e_\gamma\}$ via the matrix S^{-1} defined in Eq. (8.5.15). The subset L of diagonal matrices (C, D) is given by:

$$C = c_0 \,\mathrm{diag}(1, \nu, \nu^2, \ldots, \nu^k), \qquad D = d_0 \,\mathrm{diag}(1, \nu^k, \nu^{k-1}, \ldots, \nu), \tag{8.5.32}$$

This is nothing but the fact that $\mathbb{C}^2 \sim L$. The set of pairs $\begin{pmatrix} \nu^m c_0 \\ \nu^{k-m} d_0 \end{pmatrix}$, $m = 0, 1, \ldots, k$ is an orbit of Γ in $\mathbb{C}^2$ and determines the corresponding orbit of Γ in L. To describe $\mathbb{C}^2/A_k$ one needs to identify a suitable set of invariants $(u, w, z) \in \mathbb{C}^3$ such that

$$0 = W_\Gamma(u, w, z) \equiv u^2 + w^2 - z^{k+1} \tag{8.5.33}$$

To this effect we define:

$$u = \tfrac{1}{2}(x + y) \quad ; \quad w = -\mathrm{i}\tfrac{1}{2}(x - y) \quad \Leftrightarrow \quad xy = u^2 + w^2 \tag{8.5.34}$$

and we make the following ansatz:

$$x = \det C \quad ; \quad y = \det D, \quad ; \quad z = \frac{1}{k+1} \mathrm{Tr}\, CD. \tag{8.5.35}$$

This guess is immediately confirmed by the study of the deformed surface. We know that there is a one-to-one correspondence between the orbits of $\mathscr{F}$ in $\mu^{-1}(0)$ and those of Γ in L. Let us realize this correspondence explicitly.

Choose the basis where R is diagonal. Then $(A, B) \in \mathscr{S}$ have the form of Eq. (8.5.17). The relation $xy = z^{k+1}$ holds also true when, in Eq. (8.5.35), the pair $(C, D) \in L$ is replaced by an element $(A, B) \in \mu^{-1}(0)$.

To see this, let us describe the elements $(A, B) \in \mu^{-1}(0)$. We have to equate the right hand sides of Eq. (8.5.19) to zero. We note that:

$$[A, B] = 0 \quad \Rightarrow \quad v_i = \frac{u_0 v_0}{u_i} \quad \forall i \tag{8.5.36}$$

Secondly,

$$[A, A^\dagger] + [B, B^\dagger] = 0 \quad \Rightarrow \quad |u_i| = |u_j| \text{ and } |v_i| = |v_j| \quad \forall i, j \tag{8.5.37}$$

From the previous two equations we conclude that:

$$u_j = |u_0| e^{i\phi_j} \quad ; \quad v_j = |v_0| e^{i\psi_j} \tag{8.5.38}$$

Finally:

$$[A, B] = 0 \quad \Rightarrow \quad \psi_j = \Phi - \phi_j \quad \forall j \tag{8.5.39}$$

where Φ is an arbitrary overall phase.

In this way, we have characterized $\mu^{-1}(0)$ and we immediately check that the pair $(A, B) \in \mu^{-1}(0)$ satisfies $xy = z^{k+1}$ if $x = \det A$, $y = \det B$ and $z = 1/(k+1)\,\mathrm{Tr}\,AB$ as we have proposed in Eq. (8.5.35).

After this explicit solution of the momentum map constraint has been implemented we are left with $k+4$ parameters, namely the $k+1$ phases ϕ_j, $j = 0, 1, \dots k$, plus the absolute values $|u_0|$ and $|v_0|$ and the overall phase Φ. So we have:

$$\dim \mu^{-1}(0) = \dim \mathscr{S} - 3 \dim \mathscr{F} = 4|\Gamma| - 3(|\Gamma| - 1) = |\Gamma| + 3 \tag{8.5.40}$$

where $|\Gamma| = k + 1$.

Now we perform the quotient of $\mu^{-1}(0)$ with respect to $\mathscr{F}$. Given a set of phases f_i such that $\sum_{i=0}^{k} f_i = 0 \bmod 2\pi$ and given $f = \mathrm{diag}(e^{i f_0}, e^{i f_1}, \dots, e^{i f_k}) \in \mathscr{F}$, the orbit of $\mathscr{F}$ in $\mu^{-1}(0)$ passing through $\begin{pmatrix} A \\ B \end{pmatrix}$ has the form $\begin{pmatrix} fAf^{-1} \\ fBf^{-1} \end{pmatrix}$.

Choosing $f_j = f_0 + j\psi + \sum_{n=0}^{j-1} \phi_n$, $j = 1, \dots, k$, with $\psi = -\frac{1}{k} \sum_{n=0}^{k} \phi_n$, and f_0 determined by the condition $\sum_{i=0}^{k} f_i = 0 \bmod 2\pi$, one obtains

$$fAf^{-1} = a_0 \begin{pmatrix} 0 & 1 & 0 & \dots & 0 \\ 0 & 0 & 1 & \dots & 0 \\ & \dots & & \dots & \\ 0 & 0 & \dots & 0 & 1 \\ 1 & 0 & 0 & \dots & 0 \end{pmatrix}, \qquad fBf^{-1} = b_0 \begin{pmatrix} 0 & 0 & \dots & 0 & 1 \\ 1 & 0 & 0 & \dots & 0 \\ 0 & 1 & 0 & \dots & 0 \\ & \dots & & \dots & \\ 0 & \dots & 0 & 1 & 0 \end{pmatrix} \tag{8.5.41}$$

where $a_0 = |u_0| e^{i\psi}$ and $b_0 = |v_0| e^{i(\Phi - \psi)}$. Since the phases ϕ_j are determined modulo 2π, it follows that ψ is determined modulo $\frac{2\pi}{k+1}$. Thus we can say $(a_0, b_0) \in \mathbb{C}^2/\Gamma$. This is the one-to-one correspondence between $\mu^{-1}(0)/\mathscr{F}$ and $\mathbb{C}^2/\Gamma$.

Next we derive the deformed relation between the invariants x, y, z. It fixes the correspondence between the resolution of the singularity performed in the momentum map approach and the resolution performed on the hypersurface $xy = z^{k+1}$ in $\mathbb{C}^3$. To this purpose, we focus on the holomorphic part of the momentum map, i.e. on the equation:

$$[A, B] = \Lambda_0 = \mathrm{diag}(\lambda_0, \lambda_1, \lambda_2, \dots, \lambda_k) \in \mathfrak{Z} \otimes \mathbb{C} \tag{8.5.42}$$

$$\lambda_0 = -\sum_{i=1}^{k} \lambda_i \tag{8.5.43}$$

Let us recall the expression (8.5.17) for the matrices A and B. Naming $a_i = u_i v_i$, Eq. (8.5.42) implies:

$$a_i = a_0 + \lambda_i \quad ; \quad i = 1, \dots, k \tag{8.5.44}$$

Let $\Lambda = \mathrm{diag}(\lambda_1, \lambda_2, \dots, \lambda_k)$. We have

$$xy = \det A \, \det B = a_0 \, \Pi_{i=1}^k (a_0 + \lambda_i) = a_0^{k+1} \, \det\left(1 + \frac{1}{a_0}\Lambda\right) = \sum_{i=0}^{k} a_0^{k+1-i} S_i(\Lambda) \tag{8.5.45}$$

The $S_i(\Lambda)$ are the symmetric polynomials in the eigenvalues of Λ. They are defined by the relation $\det(1+\Lambda) = \sum_{i=0}^k S_i(\Lambda)$ and are given by:

$$S_i(\Lambda) = \sum_{j_1 < j_2 < \cdots < j_i} \lambda_{j_1} \lambda_{j_2} \cdots \lambda_{j_i} \tag{8.5.46}$$

In particular, $S_0 = 1$ and $S_1 = \sum_{i=1}^k \lambda_i$. Define $S_{k+1}(\Lambda) = 0$, so that we can rewrite:

$$xy = \sum_{i=0}^{k+1} a_0^{k+1-i} S_i(\Lambda) \tag{8.5.47}$$

and note that

$$z = \frac{1}{k+1} \mathrm{Tr} AB = a_0 + \frac{1}{k+1} S_1(\Lambda). \tag{8.5.48}$$

Then the desired deformed relation between x, y and z is obtained by substituting $a_0 = z - \frac{1}{k} S_1$ in (8.5.45), thus obtaining

$$\begin{aligned} xy &= \sum_{m=0}^{k+1} \sum_{n=0}^{k+1-m} \binom{k+1-m}{n} z^n \left(-\frac{1}{k+1} S_1\right)^{k+1-m-n} S_m \, z^n = \sum_{n=0}^{k+1} t_{n+1} \, z^n \\ \Longrightarrow \quad t_{n+1} &= \sum_{m=0}^{k+1-n} \binom{k+1-m}{n} \left(-\frac{1}{k+1} S_1\right)^{k+1-m-n} \end{aligned} \tag{8.5.49}$$

Note in particular that $t_{k+2} = 1$ and $t_{k+1} = 0$, i.e.

$$xy = z^{k+1} + \sum_{n=0}^{k} t_{n+1} z^n \tag{8.5.50}$$

which means that the deformation proportional to z^k is absent. This establishes a clear correspondence between the momentum map construction and the polynomial ring $\frac{\mathbb{C}[x,y,z]}{\partial W}$ where $W(x, y, z) = xy - z^{k+1}$. Moreover, note that we have only used one of the momentum map equations, namely $[A, B] = \Lambda_0$. The equation

$[A, A^\dagger]+[B, B^\dagger] = \Sigma$ has been completely ignored. This means that the deformation of the complex structure is described by the parameters Λ, while the parameters Σ describe the deformation of the Kähler structure. The relation (8.5.49) can also be written in a simple factorized form, namely

$$xy = \Pi_{i=0}^{k}(z - \mu_i), \tag{8.5.51}$$

where

$$\begin{aligned}\mu_i &= \frac{1}{k}(\lambda_1 + \lambda_2 + \cdots + \lambda_{i-1} - 2\lambda_i + \lambda_{i+1} + \cdots + \lambda_k), \quad i = 1, \ldots, k-1\\ \mu_0 &= -\sum_{i=1}^{k}\mu_i = \frac{1}{k}S_1.\end{aligned} \tag{8.5.52}$$

8.6 Generalization of the Correspondence: McKay Quivers for $\mathbb{C}^3/\Gamma$ Singularities

One can generalize the extended Dynkin diagrams obtained in the above way by constructing McKay quivers, according to the following definition:

Definition 8.6.1 Let us consider the quotient $\mathbb{C}^n/\Gamma$, where Γ is a finite group that acts on $\mathbb{C}^n$ by means of the complex representation $\mathcal{Q}$ of dimension n and let D_i, $(i = 1, \ldots, r+1)$ be the set of irreducible representations of Γ having denoted by $r+1$ the number of conjugacy classes of Γ. Let the matrix $\mathcal{A}_{ij}$ be defined by:

$$\mathcal{Q} \otimes \mathrm{D}_i = \bigoplus_{j=1}^{r+1} \mathcal{A}_{ij}\,\mathrm{D}_j \tag{8.6.1}$$

To such a matrix we associate a quiver diagram in the following way. Every irreducible representation is denoted by a circle labeled with a number equal to the dimension of the corresponding irrep. Next we write an oriented line going from circle i to circle j if D_j appears in the decomposition of $\mathcal{Q} \otimes \mathrm{D}_i$, namely if the matrix element $\mathcal{A}_{ij}$ does not vanish.

The analogue of the extended Cartan matrix discussed in the case of $\mathbb{C}^2/\Gamma$ is defined below:

$$\bar{c}_{ij} = n\,\delta_{ij} - \mathcal{A}_{ij} \tag{8.6.2}$$

and it has the same property, namely, it admits the vector of irrep dimensions

$$\mathbf{n} \equiv \{1, n_1, \ldots, n_r\} \tag{8.6.3}$$

as a null vector:

$$\bar{c}.\mathbf{n} = \mathbf{0} \tag{8.6.4}$$

Typically the McKay quivers encode the information determining the interaction structure of the dual gauge theory on the brane world volume. Indeed the bridge between Mathematics and Physics is located precisely at this point. In the case of a single $M2$-brane, the $n|\Gamma|$ complex coordinates ($n = 2$, or 3) of the flat Kähler manifold $Hom_\Gamma(R, Q \otimes R)$ are the scalar fields of the Wess-Zumino multiplets, the unitary group $\mathscr{F}$ commuting with the action of Γ is the *gauge group*, the moment maps of $\mathscr{F}$ enter the definition of the potential, according to the standard supersymmetry formulae and the holomorphic constraints defining the $\mathbf{V}_{|\Gamma|+2}$ variety have to be related with the superpotential $\mathfrak{W}$ of the $\mathcal{N} = 2$ theories in $\mathrm{d} = 3$ i.e. the $n = 3$ case where the singular space is $\mathbb{C} \times \mathbb{C}^3/\Gamma$). In the case of $\mathcal{N} = 4$ theories, also in $\mathrm{d} = 3$, (i.e. the n=2 case where the singular space is $\mathbb{C}^2 \times \mathbb{C}^2/\Gamma$), the holomorphic constraints $\mathscr{P}_i(y)$ are identified with the holomorphic part of the tri-holomorphic moment map. When one goes to the case of multiple $M2$-branes the gauge group is enlarged by color indices. This is another story. The first step is to understand the case of one $M2$-brane and here the map between Physics and Mathematics is one-to-one.

8.6.1 Representations of the Quivers and Kähler Quotients

Let us now follow the same steps of the Kronheimer construction and derive the representations of the $\mathbb{C}^3/\Gamma$ quivers. The key point is the construction of the analogues of the spaces $\mathscr{P}_\Gamma$ in Eq. (8.5.5) and of its invariant subspace $\mathscr{S}_\Gamma$ in Eq. (8.5.6). To this effect we introduce three matrices $|\Gamma| \times |\Gamma|$ named A, B, C and set:

$$p \in \mathscr{P}_\Gamma \equiv \mathrm{Hom}\,(R, \mathscr{Q} \otimes R) \Rightarrow p = \begin{pmatrix} A \\ B \\ C \end{pmatrix} \tag{8.6.5}$$

The action of the discrete group Γ on the space $\mathscr{P}_\Gamma$ is defined in full analogy with the Kronheimer case:

$$\forall \gamma \in \Gamma : \quad \gamma \cdot p \equiv \mathscr{Q}(\gamma) \begin{pmatrix} R(\gamma)\, A\, R(\gamma^{-1}) \\ R(\gamma)\, B\, R(\gamma^{-1}) \\ R(\gamma)\, C\, R(\gamma^{-1}) \end{pmatrix} \tag{8.6.6}$$

where $\mathscr{Q}(\gamma)$ denotes the three-dimensional complex representation of the group element γ, while $R(\gamma)$ denotes its $|\Gamma| \times |\Gamma|$-matrix image in the regular representation.

In complete analogy with Eq. (8.5.6) the subspace $\mathscr{S}_\Gamma$ is obtained by setting:

$$\mathscr{S}_\Gamma \equiv \mathrm{Hom}_\Gamma\,(R, Q \otimes R) = \{p \in \mathscr{P}_\Gamma / \forall \gamma \in \Gamma, \gamma \cdot p = p\} \tag{8.6.7}$$

Just as in the previous case a three-vector of matrices can be thought as a matrix of three-vectors: that is, $\mathscr{P}_\gamma = \mathscr{Q} \otimes \mathrm{Hom}(R, R) = \mathrm{Hom}(R, \mathscr{Q} \otimes R)$. Decomposing the regular representation, $R = \bigoplus_{i=0}^{r} n_i D_i$ into irreps, using Eq. (8.6.1) and Schur's lemma, we obtain:

$$\mathscr{S}_\Gamma = \bigoplus_{i,j} A_{i,j} \mathrm{Hom}(\mathbb{C}^{n_i}, \mathbb{C}^{n_j}) \tag{8.6.8}$$

The properties (8.6.2)–(8.6.4) of the matrix A_{ij} associated with the quiver diagram guarantee, in perfect analogy with Eq. (8.5.9)

$$\dim_{\mathbb{C}} \mathscr{S}_\Gamma \simeq \mathrm{Hom}_\Gamma (R, \mathscr{Q} \otimes R) = 3 \sum_i n_i^2 = 3|\Gamma| \,. \tag{8.6.9}$$

8.6.2 The Quiver Lie Group, Its Maximal Compact Subgroup and the Kähler Quotient

We address now the most important point, namely the reduction of the $3|\Gamma|$-dimensional complex manifold $\mathrm{Hom}_\Gamma (R, \mathscr{Q} \otimes R)$ to a $|\Gamma| + 2$-dimensional subvariety of which we will perform the Kähler quotient in order to obtain the final 3-dimensional (de-singularized) smooth manifold that provides the crepant resolution. The inspiration about how this can be done is provided by comparison with the $\mathbb{C}^2/\Gamma$ case, *mutatis mutandis*. The key formulae to recall are the following ones: Eqs. (8.5.19), (8.5.23) and (8.5.30).

From Eq. (8.5.19) we see that the analytic part of the triholomorphic moment map is provided by the projection onto the gauge group generators of the commutator $[A\,,\ B]$. When the level parameters are all zero (namely when the locus equation is not perturbed by the elements of the chiral ring) the outcome of the moment map equation is simply the condition $[A\,,\ B] = 0$. In the case of $\mathbb{C}^3/\Gamma$ we already know that there are no deformations of the complex structure and that the analogue of the holomorphic moment map constraint has to be a rigid parameterless condition. Namely the ideal that cuts out the $\mathbb{V}_{|\Gamma|+2}$ variety should be generated by a list of quadric polynomials $\mathscr{P}_i(y)$ fixed once and for all in a parameterless way. It is reasonable to guess that these equations should be a generalization of the condition $[A\,,\ B] = 0$. In the $\mathbb{C}^3/\Gamma$ case we have three matrices A, B, C and the obvious generalization is given below:

$$\mathbf{p} \wedge \mathbf{p} = 0 \tag{8.6.10}$$

where:

$$\mathbf{p} = \begin{pmatrix} A \\ B \\ C \end{pmatrix} \in \mathrm{Hom}_\Gamma (R, \mathscr{Q} \otimes R)$$
$$p_1 = A \;\; ; \quad p_2 = B \;\; ; \quad p_3 = C \tag{8.6.11}$$

This is a short-hand for the following explicit equations

$$\begin{aligned} 0 &= \varepsilon^{ijk}\mathbf{p}_i \cdot \mathbf{p}_j \\ &\Updownarrow \\ 0 &= [A, B] = [B, C] = [C, A] \end{aligned} \tag{8.6.12}$$

Equation (8.6.10) is the very same equation numbered (1.18) in Craw's doctoral thesis [51]. We will see in a moment that it is indeed the correct equation reducing $\mathrm{Hom}_\Gamma(R, \mathscr{Q} \otimes R)$ to a $|\Gamma| + 2$-dimensional subvariety. The way to understand it goes once again through a detailed comparison with the Kronheimer case.

One has to discuss the construction of the gauge group and to recall the identification of the singular orbifold $\mathbb{C}^2/\Gamma$ with the subspace named L defined by Eq. (8.5.30). Both constructions have a completely parallel analogue in the $\mathbb{C}^3/\Gamma$ case and these provide the key to understand why (8.6.10) is the right choice.

Before we do that let us provide the main link between the here considered mathematical constructions and the Physics of three-dimensional Chern-Simons gauge theories. To this purpose let us go back to the results of [38]. For those special $\mathscr{N} = 2$ Chern-Simons gauge theories that are actually $\mathscr{N} = 3$, the superpotential $\mathscr{W}$ has the form displayed below:

$$\mathscr{W} = -\frac{1}{8\alpha}\,\mathscr{P}_+^\Lambda\,\mathscr{P}_+^\Sigma\,\kappa_{\Lambda\Sigma} \tag{8.6.13}$$

where $\mathscr{P}_\Lambda^+$ denote the holomorphic parts of the triholomorphic moment maps and $\kappa_{\Lambda\Sigma}$ is the Killing metric of the gauge Lie algebra. When looking for extrema at $V = 0$ of the scalar potential, namely for classical vacua of the gauge theory, taking into account the positive definiteness of the scalar metric $g^\alpha\beta^\star$ of the Killing metric $\kappa_{\Lambda\Sigma}$ and of the matrix $\mathbf{m}^{\Lambda\Sigma}$ one obtains the following conditions:

$$\mathscr{P}_\Lambda^3 = \zeta_\Lambda^3 \tag{8.6.14}$$

$$\mathscr{P}_\Lambda^+ = \zeta_\Lambda^+ \tag{8.6.15}$$

In mathematical language, the above equations just define the level set $\mu^{-1}(\zeta)$ utilized in the hyperKähler quotient.

The same field theoretic mechanism is realized in a gauge theory whose scalar fields span the space $\mathscr{S}_\Gamma$ for a $\mathbb{C}^3/\Gamma$ singularity, if we introduce the following superpotential:

$$\mathscr{W} = \mathrm{Tr}\left[p_x\, p_y\, p_z\right]\,\varepsilon^{xyz} \tag{8.6.16}$$

With this choice the conditions for the vanishing of the scalar potential are indeed the Kähler moment map equations that we are going to discuss and Eq. (8.6.10).

8.6.2.1 Quiver Lie Groups

We are interested in determining the subgroup

$$\mathscr{G}_\Gamma \subset \mathrm{SL}(|\Gamma|, \mathbb{C}) \tag{8.6.17}$$

made by those elements that commute with the group Γ.

$$\mathscr{G}_\Gamma = \left\{ g \in \mathrm{SL}(|\Gamma|, \mathbb{C}) \mid \forall \gamma \in \Gamma : \left[\mathrm{D_R}(\gamma) , \mathrm{D_{def}}(g) \right] = 0 \right\} \tag{8.6.18}$$

In the above equation $D_R()$ denotes the regular representation while $\mathrm{D_{def}}$ denotes the defining representation of the complex linear group. The two representations, by construction, have the same dimension and this is the reason why equation (8.6.18) makes sense.

It is sufficient to impose the defining constraint for the generators of the group on a generic matrix depending on $|\Gamma|^2$ parameters: this reduces it to a specific matrix depending on $|\Gamma|$-parameters. The further condition that the matrix should have determinant one, reduces the number of free parameters to $|\Gamma| - 1$. In more abstract terms we can say that the group $\mathscr{G}_\Gamma$ has the following general structure:

$$\mathscr{G}_\Gamma = \bigotimes_{\mu=1}^{r+1} \mathrm{GL}(\mathrm{n}_\mu, \mathbb{C}) \bigcap \mathrm{SL}(|\Gamma|, \mathbb{C}) \tag{8.6.19}$$

This is a perfectly analogous result to that displayed in Eq. (8.5.23) for the Kronheimer case. The difference is that there we had unitary groups while here we are talking about general linear complex groups with a holomorphic action on the quiver coordinates. The reason is that we have not yet introduced a Kähler structure on the quiver space $\mathrm{Hom}_\Gamma(R\,,\,\mathscr{Q}\otimes R)$: we do it presently and we shall realize that isometries of the constructed Kähler metric will be only those elements of $\mathscr{G}_\Gamma$ that are contained in the unitary subgroup mentioned below:

$$\mathscr{F}_\Gamma \equiv \bigotimes_{\mu=1}^{r+1} \mathrm{U}(\mathrm{n}_\mu) \bigcap \mathrm{SU}(|\Gamma|) \subset \mathscr{G}_\Gamma \tag{8.6.20}$$

8.6.2.2 The Holomorphic Quiver Group and the Reduction to $V_{|\Gamma|+2}$

Yet the group $\mathscr{G}_\Gamma$ plays an important role in understanding the rationale of the holomorphic constraint (8.6.10). The key item is the coset $\mathscr{G}_\Gamma/\mathscr{F}_\Gamma$.

Let us introduce some notations. Relaying on Eq. (8.6.5) we define the diagonal embedding:

$$\mathbb{D} \quad : \quad \mathrm{GL}(|\Gamma|, \mathbb{C}) \to \mathrm{GL}(3|\Gamma|, \mathbb{C}) \tag{8.6.21}$$

$$\forall M \in \mathrm{GL}(|\Gamma|, \mathbb{C}) \quad ; \quad \mathbb{D}[M] \equiv \left(\begin{array}{c|c|c} M & 0 & 0 \\ \hline 0 & M & 0 \\ \hline 0 & 0 & M \end{array}\right) \tag{8.6.22}$$

In this notation, the invariance condition that defines $\mathscr{S}_\Gamma = \mathrm{Hom}_\Gamma(R, \mathscr{Q} \times R)$ can be rephrased as follows:

$$\forall \gamma \in \Gamma \quad : \quad \mathscr{Q}[\gamma]\,\mathbf{p} = \mathbb{D}[R_\gamma^{-1}]\,\mathbf{p}\,\mathbb{D}[R_\gamma] \tag{8.6.23}$$

It is clear that any $|\Gamma| \times |\Gamma|$ - matrix M that commutes with R_γ realizes an automorphism of the space $\mathscr{S}_\Gamma$, namely it maps it into itself. The group $\mathscr{G}_\Gamma$ is such an automorphism group. In particular equation (8.6.10) or alternatively (8.6.12) is invariant under the action of $\mathscr{G}_\Gamma$. Hence the locus:

$$\begin{aligned} &\mathscr{D}_\Gamma \subset \mathscr{S}_\Gamma \\ &\mathscr{D}_\Gamma \equiv \{\mathbf{p} \in \mathscr{S}_\Gamma \mid [A, B] = [B, C] = [C, A] = 0\} \end{aligned} \tag{8.6.24}$$

is invariant under the action of $\mathscr{G}_\Gamma$. A priori the locus $\mathscr{D}_\Gamma$ might be empty, but this is not so because there exists an important solution of the constraint (8.6.10) which is the obvious analogue of the space L_Γ defined for the $\mathbb{C}^2/\Gamma$-case in Eq. (8.5.30). In full analogy we set:

$$\mathscr{S}_\Gamma \supset L_\Gamma \equiv \left\{ \begin{pmatrix} A_0 \\ B_0 \\ C_0 \end{pmatrix} \in \mathscr{S}_\Gamma \mid A_0, B_0, C_0 \text{ are diagonal in the natural basis of R} : \{e_\delta\} \right\} \tag{8.6.25}$$

Obviously diagonal matrices commute among themselves and they do the same in any other basis where they are not diagonal, in particular in the *split basis*. By definition we name in this way the basis where the regular representation R is split into irreducible representations. A general result in finite group theory tells us that every n_i-dimensional irrep $\boldsymbol{D}_i$ appears in R exactly n_i-times:

$$R = \bigoplus_{i=0}^{r} n_i\, \boldsymbol{D}_i \quad ; \quad \mathrm{dim}\boldsymbol{D}_i \equiv n_i \tag{8.6.26}$$

In the split basis every element $\gamma \in \Gamma$ is given by a block diagonal matrix of the following form:

$$
R(\gamma) = \left(\begin{array}{c|c|c|c|c|c}
\mathbf{1} & \mathbf{0} & \cdots & \cdots & \mathbf{0} & \mathbf{1} \\
\hline
\mathbf{0} & \begin{matrix} a_{1,1} & \cdots & a_{1,n_1} \\ \vdots & \cdots & \vdots \\ a_{n_1,1} & \cdots & a_{n_1,n_1} \end{matrix} & \mathbf{0} & \cdots & \cdots & \mathbf{0} \\
\hline
\vdots & \cdots & \cdots & \cdots & \cdots & \vdots \\
\hline
\vdots & \cdots & \cdots & \cdots & \cdots & \vdots \\
\hline
\mathbf{0} & \cdots & \cdots & \mathbf{0} & \begin{matrix} b_{1,1} & \cdots & b_{1,n_{r-1}} \\ \vdots & \cdots & \vdots \\ b_{n_{r-1},1} & \cdots & b_{n_{r-1},n_{r-1}} \end{matrix} & \mathbf{0} \\
\hline
\mathbf{0} & \cdots & \cdots & \cdots & \mathbf{0} & \begin{matrix} c_{1,1} & \cdots & c_{1,n_r} \\ \vdots & \cdots & \vdots \\ c_{n_r,1} & \cdots & c_{n_r,n_r} \end{matrix}
\end{array}\right) \tag{8.6.27}
$$

In analogy to what was noticed for the Kronheimer case, the space L_Γ has complex dimension three (in Kronheimer case it was two):

$$
\dim_{\mathbb{C}} L_\Gamma = 3 \tag{8.6.28}
$$

Indeed if we fix the first diagonal entry of each of the three matrices, the invariance condition (8.6.23) determines all the other ones uniquely. In any other basis the number of parameters remains three. Let us call them (a_0, b_0, c_0). Because of the above argument and, once again, in full analogy with the Kronheimer case, we can conclude that the space L_Γ is isomorphic to the singular orbifold $\mathbb{C}^3/\Gamma$, the Γ-orbit of a triple (a_0, b_0, c_0) representing a point in $\mathbb{C}^3/\Gamma$.

The existence of the solution of the constraint (8.6.10) provided by the complex three-dimensional space L_Γ shows that we can construct a variety of dimension $|\Gamma| + 2$ which is in the kernel of the constraint (8.6.10). This is just the orbit, under the action of $\mathscr{G}_\Gamma$ of L_Γ. We set:

$$
\mathscr{D}_\Gamma \equiv \text{Orbit}_{\mathscr{G}_\Gamma}(L_\Gamma) \tag{8.6.29}
$$

The counting is easily done.

1. A generic point in L_Γ has the identity as stability subgroup in $\mathscr{G}_\Gamma$.
2. The group $\mathscr{G}_\Gamma$ has complex dimension $|\Gamma| - 1$, hence we get:

$$
\dim_{\mathbb{C}}(\mathscr{D}_\Gamma) = |\Gamma| - 1 + 3 = |\Gamma| + 2 \tag{8.6.30}
$$

In the sequel we define the variety $V_{|\Gamma|+2}$ to be equal to $\mathscr{D}^0_\Gamma$.

8.6.2.3 The Coset $\mathscr{G}_\Gamma/\mathscr{F}_\Gamma$ and the Kähler Quotient

It is now high time to introduce the Kähler potential of the original $3|\Gamma|$-dimensional complex flat manifold $\mathscr{S}_\Gamma$. We set:

$$\mathscr{K}_{\mathscr{S}_\Gamma} \equiv \mathrm{Tr}\left(\mathbf{p}^\dagger \, \mathbf{p}\right) = \mathrm{Tr}\left(A^\dagger \, A\right) + \mathrm{Tr}\left(B^\dagger \, B\right) + \mathrm{Tr}\left(C^\dagger \, C\right) \qquad (8.6.31)$$

Using the matrix elements of A, B, C as complex coordinates of the manifold and naming λ_i the independent parameters from which they depend in a given explicit solution of the invariance constraint, the Kähler metric is defined, as usual, by:

$$ds^2_{\mathscr{S}_\Gamma} = g_{\ell\bar{m}} \, d\lambda^\ell \otimes d\bar{\lambda}^{\bar{m}} \qquad (8.6.32)$$

where:

$$g_{\ell\bar{m}} = \partial_\ell \, \bar{\partial}_{\bar{m}} \, \mathscr{K} \qquad (8.6.33)$$

From Eq. (8.6.31) we easily see that the Kähler potential is invariant under the unitary subgroup of the quiver group defined by:

$$\mathscr{F}_\Gamma = \left\{ M \in \mathscr{G}_\Gamma \mid M \, M^\dagger = \mathbf{1} \right\} \qquad (8.6.34)$$

whose structure was already mentioned in Eq. (8.6.20). The center $\mathfrak{Z}$ of the Lie algebra $\mathbb{F}_\Gamma$ has dimension r, namely the same as the number of nontrivial conjugacy classes of Γ and it has the following structure:

$$\mathfrak{Z} = \underbrace{\mathfrak{u}(1) \oplus \mathfrak{u}(1) \oplus \cdots \oplus \mathfrak{u}(1)}_{r} \qquad (8.6.35)$$

In the appendices we provide the explicit form of $\mathbb{F}_\Gamma$ while working out examples.

Since $\mathscr{F}_\Gamma$ acts as a group of isometries on the space $\mathscr{S}_\Gamma$ we might construct the Kähler quotient of the latter with respect to the former, yet we may do better.

In the case of an abelian $|\Gamma|$ the center $\mathfrak{z}[\mathbb{F}] = \mathbb{F}$ coincides with the entire gauge algebra. We discuss in detail these cases in the sequel.

Let us consider the inclusion map of the variety $\mathscr{D}_\Gamma$ into $\mathscr{S}_\Gamma$:

$$\iota = \mathscr{D}_\Gamma \to \mathscr{S}_\Gamma \qquad (8.6.36)$$

and let us define as Kähler potential and Kähler metric of the locus $\mathscr{D}_\Gamma$ the pull backs of the Kähler potential (8.6.31) and of metric (8.6.32) of $\mathscr{S}_\Gamma$, namely let us set:

$$\mathscr{K}_{\mathscr{D}_\Gamma} \equiv \iota^\star \, \mathscr{K}_{\mathscr{S}_\Gamma} \qquad (8.6.37)$$

$$ds^2_{\mathscr{D}_\Gamma} = \iota^\star \, ds^2_{\mathscr{S}_\Gamma} \qquad (8.6.38)$$

By construction, the isometry group $\mathscr{F}_\Gamma$ is inherited by the pullback metric on $\mathscr{D}_\Gamma$ and we can consider the Kähler quotient:

$$\mathscr{M}_\zeta \equiv \mathscr{D}_\Gamma /\!/^{\zeta}_{\mathscr{F}_\Gamma} \tag{8.6.39}$$

Let f_I be a basis of generators of $\mathscr{F}_\Gamma$ ($I = 1, \ldots, |\Gamma| - 1$) and let Z_i ($i = 1, \ldots, |\Gamma| + 2$) be a system of complex coordinates spanning the points of $\mathscr{D}_\Gamma$. By means of the inclusion map we have:

$$\forall Z \in \mathscr{D}_\Gamma \quad : \quad \iota(Z) = \mathbf{p}(Z) = \begin{pmatrix} A(Z) \\ B(Z) \\ C(Z) \end{pmatrix} \tag{8.6.40}$$

The action of the *gauge group* $\mathscr{F}_\Gamma$ on $\mathscr{D}_\Gamma$ is implicitly defined by:

$$\mathbf{p}(\delta_I Z) = \delta_I \mathbf{p}(Z) = \begin{pmatrix} [f_I\,,\, A(Z)] \\ [f_I\,,\, B(Z)] \\ [f_I\,,\, C(Z)] \end{pmatrix} \tag{8.6.41}$$

and the corresponding real moment maps are easily calculated:

$$\mu_I(Z,\bar{Z}) = \mathrm{Tr}\left(f_I\left[A(Z), A^\dagger(\bar{Z})\right]\right) + \mathrm{Tr}\left(f_I\left[B(Z), B^\dagger(\bar{Z})\right]\right) + \mathrm{Tr}\left(f_I\left[C(Z), C^\dagger(\bar{Z})\right]\right) \tag{8.6.42}$$

One defines the level sets by means of the equation:

$$\mu^{-1}(\zeta) = \left\{Z \in \mathscr{D}_\Gamma \;\|\; \mu_I(Z,\bar{Z}) = 0 \;\; \text{if} f_I \notin \mathfrak{Z} \;\; ; \;\; \mu_I(Z,\bar{Z}) = \zeta_I \;\; \text{if} f_I \in \mathfrak{Z}\right\} \tag{8.6.43}$$

which, by construction, are invariant under the gauge group $\mathscr{F}_\Gamma$ and we can finally set:

$$\mathscr{M}_\zeta \equiv \mu^{-1}(\zeta) /\!/ \mathscr{F}_\Gamma \equiv \mathscr{D}_\Gamma /\!/^{\zeta}_{\mathscr{F}_\Gamma} \tag{8.6.44}$$

The real and complex dimensions of $\mathscr{M}_\zeta$ are easily calculated. We start from $|\Gamma|+2$ complex dimensions, namely from $2|\Gamma|+4$ real dimensions. The level set equation imposes $|\Gamma|-1$ real constraints, while the quotiening by the group action takes other $|\Gamma|-1$ parameters away. Altogether we remain with 6 real parameters that can be seen as 3 complex ones. Hence the manifolds $\mathscr{M}_\zeta$ are always complex three-folds that, for generic values of ζ, are smooth: supposedly the crepant resolutions of the singular orbifold. For $\zeta = 0$ the manifold $\mathscr{M}_0$ degenerates into the singular orbifold $\mathbb{C}^3/\Gamma$, since the solution of the moment map equation is given by the $\mathscr{F}_\Gamma$ orbit of the locus L_Γ, namely:

$$\mu^{-1}(0) = \mathrm{Orbit}_{\mathscr{F}_\Gamma}(L_\Gamma) \tag{8.6.45}$$

Comparing Eq. (8.6.29) with Eq. (8.6.45) we are led to consider the direct sum of the Lie algebra:

$$\mathbb{G}_\Gamma = \mathbb{F}_\Gamma \oplus \mathbb{K}_\Gamma \tag{8.6.46}$$

$$[\mathbb{F}_\Gamma\,,\,\mathbb{F}_\Gamma] \subset \mathbb{F}_\Gamma \quad ; \quad [\mathbb{F}_\Gamma\,,\,\mathbb{K}_\Gamma] \subset \mathbb{K}_\Gamma \quad ; \quad [\mathbb{K}_\Gamma\,,\,\mathbb{K}_\Gamma] \subset \mathbb{F}_\Gamma \tag{8.6.47}$$

where $\mathbb{F}_\Gamma$ is the maximal compact subalgebra.

A special feature of all the quiver Groups and Lie Algebras is that $\mathbb{F}_\Gamma$ and $\mathbb{K}_\Gamma$ have the same real dimension $|\Gamma|-1$ and one can choose a basis of hermitian generators T_I such that:

$$\begin{array}{l} \forall \boldsymbol{\Phi} \in \mathbb{F}_\Gamma : \boldsymbol{\Phi} = \mathrm{i} \times \sum_{I=1}^{|\Gamma|-1} c_I T^I \; ; \; c_I \in \mathbf{R} \\ \forall \boldsymbol{K} \in \mathbb{K}_\Gamma : \boldsymbol{K} = \quad \sum_{I=1}^{|\Gamma|-1} b_I T^I \quad ; \; b_I \in \mathbf{R} \end{array} \tag{8.6.48}$$

Correspondingly a generic element $g \in \mathscr{G}_\Gamma$ can be split as follows:

$$\forall g \in \mathscr{G}_\Gamma \quad : \quad g = \mathscr{U}\,\mathscr{H} \quad ; \quad \mathscr{U} \in \mathscr{F}_\Gamma \quad ; \quad \mathscr{H} \in \exp\left[\mathbb{K}_\Gamma\right] \tag{8.6.49}$$

Using the above property we arrive at the following parametrization of the space $\mathscr{D}_\Gamma$

$$\mathscr{D}_\Gamma \;=\; \mathrm{Orbit}_{\mathscr{F}_\Gamma}\left(\exp\left[\mathbb{K}_\Gamma\right] \cdot L_\Gamma\right) \tag{8.6.50}$$

where, by definition, we have set:

$$p \in \exp\left[\mathbb{K}_\Gamma\right] \cdot L_\Gamma \Rightarrow p = \left\{ \begin{array}{l} \exp\left[-\boldsymbol{K}\right]\, A_0 \exp\left[\boldsymbol{K}\right] \\ \exp\left[-\boldsymbol{K}\right]\, B_0 \exp\left[\boldsymbol{K}\right] \\ \exp\left[-\boldsymbol{K}\right]\, C_0 \exp\left[\boldsymbol{K}\right] \end{array} \right\} \tag{8.6.51}$$

$$\{A_0,\, B_0,\, C_0\} \;\in\; L_\Gamma \tag{8.6.52}$$

$$\boldsymbol{K} \;=\; \mathbb{K}_\Gamma \tag{8.6.53}$$

Relying on this, in the Kähler quotient we can invert the order of the operations. First we quotient the action of the compact gauge group $\mathscr{F}_\Gamma$ and then we implement the moment map constraints. We have:

$$\mathscr{D}_\Gamma /\!\!/ _{\mathscr{F}_\Gamma} \;=\; \exp\left[\mathbb{K}_\Gamma\right] \cdot L_\Gamma \tag{8.6.54}$$

Calculating the moment maps on $\exp\left[\mathbb{K}_\Gamma\right] \cdot L_\Gamma$ and imposing the moment map constraint we find:

$$\mu^{-1}\left(\zeta\right) /\!\!/ _{\mathscr{F}_\Gamma} \;=\; \left\{ Z \in \exp\left[\mathbb{K}_\Gamma\right] \cdot L_\Gamma \;\|\; \mu_I(Z,\bar{Z}) \;=\; \left\{ \begin{array}{l} 0 \;\; \mathrm{if} f_I \notin \mathfrak{Z} \\ \zeta_I \;\; \mathrm{if} f_I \in \mathfrak{Z} \end{array} \right\| \right\} \tag{8.6.55}$$

Equation (8.6.55) provides an explicit algorithm to calculate the Kähler potential of the final resolved manifold if we are able to solve the constraints in terms of an appropriate triple of complex coordinates. Furthermore for each level parameter ζ_a we have to find the appropriate one-parameter subgroup of $\mathscr{G}_\Gamma$ that lifts the corresponding moment map from the 0-value to the generic value ζ. Indeed we recall that the Kähler potential of the resolved variety is given by the celebrated formula:

$$\mathscr{K}_{\mathscr{M}} = \pi^{\star} \mathscr{K}_{\mathscr{N}} + \zeta_I \mathfrak{C}^{IJ} \boldsymbol{\Phi}_J \tag{8.6.56}$$

where, by definition:

$$\pi : \mathscr{N} \to \mathscr{M} \tag{8.6.57}$$

is the quotient map and $\exp[\zeta_I \mathfrak{C}^{IJ} \boldsymbol{\Phi}_I] \in \exp[\mathbb{K}_\Gamma] \subset \mathscr{G}_\Gamma$ is the element of the quiver group which lifts the moment maps from zero to the values ζ_I, while $\mathfrak{C}^{IJ}$ is a constant matrix whose definition we discuss later on. Indeed the rationale behind formula (8.6.56) requires a careful discussion, originally due to Hitchin, Karlhede, Lindström and Roček [1] which we shall review in the next sections.

8.7 The Example of the Eguchi–Hanson Space

In order to give a concrete illustrative example of the Kronheimer construction we focus on the simplest and oldest known ALE manifold, namely on the Eguchi–Hanson space [52] (see Fig. 8.8).

To this effect we begin by introducing a set of Maurer Cartan forms on the three sphere $\mathbb{S}^3 \sim \mathrm{SU}(2)$:

Fig. 8.8 On the left Tohru Eguchi (1948), on the right Andrew J. Hanson. Eguchi is currently emeritus professor of the University of Tokyo, Yukawa Institute. He held positions at SLAC and at the Enrico Fermi Institute of Chicago University. Andrew J. Hanson received the BA degree in chemistry and physics from Harvard College in 1966 and the PhD degree in theoretical physics from MIT in 1971. He is an Emeritus Professor of Computer Science in the School of Informatics and Computing at Indiana University, Bloomington. He worked in theoretical physics from 1971 until 1980, when he began working in machine vision, graphics, and visualization, first with the perception research group at the SRI Artificial Intelligence Center, and then at Indiana University from 1989 until his retirement in 2012. The Eguchi Hanson metric was derived by the two authors in 1978 when both of them were in California, the first in Stanford, the second in Berkeley

$$\begin{aligned}
\sigma_1 &= -\frac{1}{2}(d\theta\cos(\psi) + d\phi\sin(\theta)\sin(\psi)) \\
\sigma_2 &= \frac{1}{2}(d\theta\sin(\psi) - d\phi\sin(\theta)\cos(\psi)) \\
\sigma_3 &= -\frac{1}{2}(d\phi\cos(\theta) + d\psi)
\end{aligned} \tag{8.7.1}$$

which depend on three Euler angles θ, ϕ, ψ and satisfy the Maurer Cartan equations in the form:

$$d\sigma_i = \varepsilon_{ijk}\,\sigma_j \wedge \sigma_k \tag{8.7.2}$$

Furthermore, let us introduce a radial coordinate $m \le r \le +\infty$ and the following function:

$$G(r) = \sqrt{1 - \left(\frac{m}{r}\right)^2} \tag{8.7.3}$$

The Eguchi Hanson metric is given by the following expression:

$$\begin{aligned}
ds^2_{EH} &= G(r)^{-2}\,dr^2 + r^2\left(\sigma_1^2 + \sigma_2^2\right) + r^2\,G(r)^2\,\sigma_3^2 \\
&= \frac{1}{4}\left(\frac{\left(r^4 - a^4\right)(d\phi\cos(\theta) + d\psi)^2}{r^2} + \frac{4dr^2}{1 - \frac{a^4}{r^4}} + r^2\left(d\phi^2\sin^2(\theta) + d\theta^2\right)\right)
\end{aligned} \tag{8.7.4}$$

Calculating the curvature two-form of the above metric, we find that it is self-dual, while its Ricci tensor vanishes. Hence the Eguchi–Hanson metric is an Euclidean vacuum solution of Einstein equations and it describes a gravitational instanton. As $r \to \infty$ the Eguchi–Hanson metric approaches the flat Euclidean metric:

$$ds^2_{EH} \overset{r\to\infty}{\Longrightarrow} \frac{1}{2}r^2 d\psi d\phi\cos(\theta) + \frac{1}{4}r^2 d\theta^2 + \frac{1}{4}r^2 d\psi^2 + \frac{1}{4}r^2 d\phi^2 + dr^2 \tag{8.7.5}$$

Next we can show that the Eguchi–Hanson space is a complex manifold $\mathcal{M}_{EH}$ and that the Eguchi Hanson metric $\widehat{ds^2_{EH}}$ is a Kähler metric on $\mathcal{M}_{EH}$. To this effect let us introduce the following two complex coordinates:

$$\begin{aligned}
Z^1 &= \left(r^4 - m^4\right)^{\frac{1}{4}} \frac{\left(e^{i(\theta+\phi)} + ie^{i\theta} + e^{i\phi} - i\right)e^{-\frac{1}{2}i(\theta-\psi+\phi)}}{2\sqrt{2}} \\
Z^2 &= \left(r^4 - m^4\right)^{\frac{1}{4}} \frac{\left(e^{i(\theta+\phi)} - ie^{i\theta} + e^{i\phi} + i\right)e^{-\frac{1}{2}i(\theta-\psi+\phi)}}{2\sqrt{2}}
\end{aligned} \tag{8.7.6}$$

By direct calculation we can verify that:

$$ds^2_{EH} = \frac{\partial}{\partial Z^i}\frac{\bar{\partial}}{\partial \bar{Z}^{j^\star}}\mathcal{K}_{EH}\,dZ^i \otimes d\bar{Z}^{j^\star} \tag{8.7.7}$$

where:

$$\mathcal{K}_{EH} = \sqrt{\tau^2 + m^4} - m^2 \log\left(\sqrt{\tau^2 + 1} + m^4\right) + m^2 \log(\tau)$$
$$\tau \equiv |Z^1|^2 + |Z^2|^2 \tag{8.7.8}$$

Having derived the form of the Kähler potential for the Eguchi–Hanson metric we can now connect it to the Kronheimer construction of the ALE manifolds by recalling Eq. (8.5.28) and rewriting them in the case $k = 1$ for which the group $\mathcal{F} = \mathrm{U}(1)$ so that there is only one component of the triholomorphic moment map:

$$\mathfrak{P}^3 = |u_0|^2 - |v_0|^2 + |v_1|^2 - |u_1|^2 \tag{8.7.9}$$
$$\mathfrak{P}^+ = u^0\, v_0 - u^1\, v_1 \tag{8.7.10}$$

In this case it is convenient to redefine:

$$U = \{u_0, v_1\} \tag{8.7.11}$$
$$V = \{v_0, u_1\} \tag{8.7.12}$$

so that Eq. (8.7.10) can be rewritten as follows:

$$\mathfrak{P}^3 = \mathcal{P}^3(U, V) \equiv \sum_{i=1}^{2} |U_i|^2 - \sum_{i=1}^{2} |V_i|^2 \tag{8.7.13}$$
$$\mathfrak{P}^+ = \mathcal{P}^+(U, V) \equiv \sum_{i=1}^{2} U_i\, V_i \tag{8.7.14}$$

Furthermore the action of the group $\mathcal{F}_{\mathbb{Z}_2} = \mathrm{U}(1)$ on the complex coordinates U, V is the following one:

$$\mathrm{U}(1) \quad : \quad (U, V) \quad \Longrightarrow \quad \left(e^{i\varphi} U \,,\, e^{-i\varphi} V\right) \tag{8.7.15}$$

Considering the quiver group $\mathcal{G}_{\mathbb{Z}_2}$ which is just the complexification of $\mathcal{F}_{\mathbb{Z}_2}$ we obtain the transformation:

$$\mathcal{G}_{\mathbb{Z}_2} \quad : \quad (U, V) \quad \Longrightarrow \quad \left(e^{-\Phi} U \,,\, e^{\Phi} V\right) \tag{8.7.16}$$

Relying on these preliminaries we are ready to perform the algebro-geometric Hyper-Kähler quotient. Introducing the level parameters we have to solve the equations:

$$\ell = \mathcal{P}^3\left(e^{-\Phi} U, e^{\Phi} V\right)$$
$$\mathfrak{s} = \mathcal{P}^+\left(e^{-\Phi} U, e^{\Phi} V\right) = \mathcal{P}^+(U, V) \tag{8.7.17}$$

As stated several times and recalled in the second line of the above equation the holomorphic part of the moment-map is invariant under the action of the quiver group. This is very useful for the solution of the constraints. Indeed we can just choose a gauge condition like the following one:

$$U_1 = V_2 \tag{8.7.18}$$

Furthermore, in the case $k = 1$ the holomorphic level parameter $\mathfrak{s}$ can be just set equal to zero without loss of generality since it simply amounts to a change of coordinates. In this way we arrive at:

$$U_1 = V_2 \equiv \tfrac{1}{2}Z^1 \quad ; \quad U_2 = V_1 \equiv \tfrac{1}{2}Z^2 \tag{8.7.19}$$

and the first of Eq. (8.7.17) is solved by:

$$\begin{aligned} \boldsymbol{\Phi} &= -\log\left[\frac{\ell \pm \sqrt{\ell^2 + 4|U|^2\,|V^2|}}{2|V^2|}\right] = -\log\left[\frac{\ell \pm \sqrt{\ell^2 + |\mathbf{Z}|^4}}{2|\mathbf{Z}|^2}\right] \\ |\mathbf{Z}|^2 &\equiv |Z^1|^2 + |Z^2|^2 \end{aligned} \tag{8.7.20}$$

The restriction to the level surface of the ambient Kähler potential is calculated in an equally easy fashion:

$$\mathscr{K}|_{\mathscr{N}} = e^{-2\boldsymbol{\Phi}}\,|U|^2 + e^{2\boldsymbol{\Phi}}\,|V|^2 = \sqrt{\ell^2 + |\mathbf{Z}|^4} \tag{8.7.21}$$

Choosing one branch of the solution (8.7.20) and applying the general formula (8.6.56) to the case under consideration we obtain the Kähler potential of the manifold $\mathscr{M}$:

$$\mathscr{K}_{\mathscr{M}} = \sqrt{\ell^2 + |\mathbf{Z}|^4} - \ell \log\left[\frac{\ell \pm \sqrt{\ell^2 + |\mathbf{Z}|^4}}{2|\mathbf{Z}|^2}\right] \tag{8.7.22}$$

For $\ell = m^2$, we see that the Kähler potential (8.7.22) obtained by means of the HyperKähler quotient advocated in the Kronheimer construction coincides with that of the Eguchi–Hanson manifold displayed in Eq. (8.7.8). This concludes the proof that the Eguchi–Hanson manifold is the smooth resolution of the singularity $\mathbb{C}^2/\mathbb{Z}_2$.

8.7.1 The Algebraic Equation of the Locus and the Exceptional Divisor

First we consider the algebraic equation of the locus in $\mathbb{C}^3$ that corresponds to the Eguchi Hanson manifold. According to the discussion following Eq. (8.5.35) such an equation is provided by the relation between the Γ invariants:

$$x \equiv \text{Det}\, A \quad ; \quad \text{Det}\, B \quad ; \quad z \equiv \tfrac{1}{2}\, \text{Tr}\, (A\, B) \tag{8.7.23}$$

Upon use of the gauge condition (8.7.18) and of the solution of the holomorphic moment map constraint (8.7.19) we have:

$$A = \begin{pmatrix} 0 & \frac{1}{2}Z_1 \\ \frac{1}{2}Z_1 & 0 \end{pmatrix} \quad ; \quad B = \begin{pmatrix} 0 & \frac{1}{2}Z_2 \\ \frac{1}{2}Z_2 & 0 \end{pmatrix} \tag{8.7.24}$$

so that:

$$x = -\tfrac{1}{4}\, Z_1\, Z_2 \quad ; \quad y = -\tfrac{1}{4} Z_1\, Z_2 \quad ; \quad z = \tfrac{1}{4}\, Z_1\, Z_2 \tag{8.7.25}$$

and the equation of the orbifold locus $\mathbb{C}^2/\Gamma$:

$$xy = z^2 \tag{8.7.26}$$

remains unmodified. This happens because the holomorphic moment map has not been lifted away from zero and similarly will happen in all the resolutions of the $\mathbb{C}^3/\Gamma$ singularities since, as we stressed, there we have no complex structure deformations and the analogue of the holomorphic moment map equation $[A\,,\,B] = [B\,,\,C] = [C\,,\,A]$ obtains no deformation. Yet we know that by lifting the level of the real moment-map we obtain the smooth Eguchi Hanson manifold which has a non trivial homology 2-cycle, as foreseen by the general Theorem 8.3.1. In quasi polar coordinates these homology cycle is the two-sphere spanned by angles θ and ϕ when we set $r = m$ and we disregard the angle ψ. Such a homology cycle disappears when $m \to 0$ hence it is the exceptional divisor generated by the minimal resolution of the singularity. Hence it is interesting to see where such an exceptional divisor is located in the complex description of the Eguchi Hanson manifold obtained from the Kronheimer construction. To this effect it is convenient to recall the relation between divisors and line-bundles.

8.7.1.1 Divisors and Line Bundles

A *prime divisor* in a complex manifold or algebraic variety X is an irreducible closed codimension one subvariety of X. A divisor $\mathfrak{D}$ is a locally finite formal linear combination

$$\mathfrak{D} = \sum_i a_i\, \mathfrak{D}_i \tag{8.7.27}$$

where the a_i are integers, and the $\mathfrak{D}_i$ are prime divisors. A prime divisor $\mathfrak{D}$ can be described by a collection $\{(U_\alpha,\, f_\alpha)\}$, where $\{U_\alpha\}$ is an open cover of X, and the $\{f_\alpha\}$ are holomorphic functions on U_α such that $f_\alpha = 0$ is the equation of $\mathfrak{D} \cap U_\alpha$ in U_α.

As a consequence, the functions $g_{\alpha\beta} = f_\alpha/f_\beta$ are holomorphic nowhere vanishing functions

$$g_{\alpha\beta} : U_\alpha \cap U_\beta \to \mathbb{C}^*$$

that on triple intersections $U_\alpha \cap U_\beta \cap U_\gamma$ satisfy the cocycle condition

$$g_{\alpha\beta} g_{\beta\gamma} = g_{\alpha\gamma}$$

and therefore define a line bundle $\mathscr{L}(\mathfrak{D})$. If $\mathfrak{D}$ is a divisor as in (8.7.27) then one sets

$$\mathscr{L}(\mathfrak{D}) = \bigotimes_i \mathscr{L}(\mathfrak{D}_i)^{a_i}.$$

The inverse correspondence (from line bundles to divisors) is described as follows. If s is a nonzero meromorphic section of a line bundle $\mathscr{L}$, and V is a codimension one subvariety of X over which s has a zero or a pole, denoted by $\mathrm{ord}_V(s)$ the order of the zero, or minus the order of the pole; then

$$\mathfrak{D} = \sum_V \mathrm{ord}_V(s) \cdot V$$

is a divisor, whose associated line bundle $\mathscr{L}(\mathfrak{D})$ is isomorphic to $\mathscr{L}$.

8.7.1.2 The Exceptional Divisor

It is easy to work out the exceptional divisor in the Eguchi–Hanson case by performing the following holomorphic coordinate transformation:

$$Z^1 \to (1 - \xi_1)\, \xi_2 \quad ; \quad Z^2 \to -(1 + \xi_1)\, \xi_2 \tag{8.7.28}$$

Upon the substitution (8.7.28) and the identification $\ell = m^2$ the Kähler potential (8.7.22) becomes:

$$\begin{aligned}
\mathscr{K}_{EH} &= \mathscr{K}_0 + m^2 \left(\mathscr{K}_\mathrm{E} + \log |W|^2\right) \\
\mathscr{K}_0 &= \sqrt{m^4 + 4\left(1 + |\xi_1|^2\right)^2 |\xi_2|^4} - m^2 \log\left(m^2 + \sqrt{m^4 + 4\left(1 + |\xi_1|^2\right)^2 |\xi_2|^4}\right) \\
\mathscr{K}_\mathrm{E} &= \log\left(1 + |\xi_1|^2\right) \\
W &\equiv \sqrt{2}\, \xi_2
\end{aligned} \tag{8.7.29}$$

Inspecting Eq. (8.7.29) we realize that $\mathscr{K}_\mathrm{E}$ is the standard Kähler potential of a $\mathbb{P}^1$ written in the affine coordinate ξ_1. This suggests that the Eguchi–Hanson manifold is covered by two open charts:

$$U_N = \{\xi_1^N, \xi_2^N\}$$
$$U_S = \{\xi_1^S, \xi_2^S\} \tag{8.7.30}$$

with transition function:

$$\{\xi_1^N, \xi_2^N\} = \left\{\frac{1}{\xi_1^S}, \xi_2^S\, \xi_1^S\right\} \tag{8.7.31}$$

Under the transformation (8.7.30) the function $\mathscr{K}_0$ is invariant, while $\mathscr{K}_{\mathrm{E}}$ transforms as follows:

$$\mathscr{K}_{\mathrm{E}}\left(\xi^N, \bar{\xi^N}\right) = \mathscr{K}_{\mathrm{E}}\left(\xi^S, \bar{\xi^S}\right) - \log|\xi_1^S|^2 \tag{8.7.32}$$

Therefore we can introduce a line bundle $\mathscr{L} \xrightarrow{\pi} \mathscr{M}_{EH}$ defined by two local trivializations, one based on U_N, the other on U_S with transition function:

$$g_{NS} \quad : \quad W_N(\xi^N) = \xi_1^S W_S(\xi^S) \tag{8.7.33}$$

a fiber metric on such a bundle is obtained by defining the following invariant norm for the bundle sections:

$$\| W \|^2 \equiv e^{\mathscr{K}_{\mathrm{E}}} |W|^2 \tag{8.7.34}$$

The corresponding first Chern class is:

$$c_1(\mathscr{L}) = \omega^{(1,1)} \equiv \frac{i}{2\pi}\partial\bar{\partial}\log \| W \|^2 \xrightarrow{W\to 0} \frac{i}{2\pi}\frac{d\xi_1 \wedge d\bar{\xi}_1}{\left(1+|\xi_1|^2\right)^2} \equiv \omega_{\mathscr{D}}^{(1,1)} \tag{8.7.35}$$

The divisor $\mathfrak{D}$ related with this line bundle is obviously obtained as the vanishing locus of the global section $W = \xi_2 = 0$. The cohomology class of $\omega^{(1,1)}$ is that of the Poincaré dual $\omega_{\mathscr{D}}^{(1,1)}$ of the vanishing section W, namely of the divisor $\mathfrak{D}$:

$$\left[\omega^{(1,1)}\right] = \left[\omega_{\mathfrak{D}}^{(1,1)}\right] \tag{8.7.36}$$

What we have discussed so far is just an explicit illustration of the well known fact that the Eguchi–Hanson manifold is the total space of the fiber bundle $\mathscr{O}_{\mathbb{P}^1}(-2)$.

Since the function $\mathscr{K}_0$ is invariant, it is clear that its contribution $\partial\bar{\partial}\mathscr{K}_0$ to the Kähler 2-form is cohomologous to zero which implies:

$$[\mathbb{K}_{EH}] = m^2\left[\omega_{\mathfrak{D}}^{(1,1)}\right] \tag{8.7.37}$$

Finally it is instructive to compare the above complex description of the Eguchi–Hanson space with its description in terms of quasi polar coordinates. To this effect it suffices to rewrite the coordinate transformation (8.7.6) in terms of the x_i coordinates. We have:

$$\xi_1 = e^{i\phi} \cot\left(\frac{\theta}{2}\right), \; \xi_2 = \frac{1}{2}\sqrt{1-\cos(\theta)}\sqrt[4]{r^4 - m^4} e^{\frac{1}{2}i(\psi-\phi)} \tag{8.7.38}$$

As we see the locus $\xi_2 = 0$ corresponds to $r = m$ and $\psi =$ any value.

8.8 Gibbons Hawking Metrics and the Resolution of $\mathbb{C}^2/\Gamma$ Singularities

As an exercise that exemplifies the generalized Kronheimer construction which resolves the quotient singularities $\mathbb{C}^3/\Gamma$, we intend to discuss the abelian cases $\Gamma = \mathbb{Z}_3$ and $\mathbb{Z}_7$. In this way, by steps of increasing complexity, we approach the discussion of the non abelian cases. When $\Gamma = \mathbb{Z}_3$ we already pointed out that the singularity is actually of the type mentioned in Eq. (8.3.26). This is quite useful for our purposes since the $ALE_{\mathbb{Z}_k}$ manifolds admit another well known representation with which we can compare the Kronheimer construction in order to get orientation in our main task of understanding the cohomology of the resolved Kähler manifold. The representation we are alluding to is that of the Gibbons Hawking multicenter metrics that are known to be HyperKählerian and indeed equivalent to $ALE_{\mathbb{Z}_k}$. The comparison between these two forms of the same metrics is very useful in order to get queues about the mechanisms by means of which the moment map parameters blow up the singularities in the purely Kählerian case. Hence let us start with the general form of the GH-metrics.

Let the x, y, z be the real coordinates of $\mathbb{R}^3$ to which we adjoin an angle τ spanning a circle $\mathbb{S}^1$. A general GH-metric has the following form:

$$ds^2_{\text{GH}} = \frac{(\text{d}\,\tau + \omega)^2}{\mathscr{V}} + \mathscr{V}\left(\text{dx}^2 + \text{dy}^2 + \text{dz}^2\right) \tag{8.8.1}$$

where $\mathscr{V} = \mathscr{V}(x, y, z)$ is a harmonic function on $\mathbb{R}^3$:

$$\frac{\partial^2 \mathscr{V}}{\partial x^2} + \frac{\partial^2 \mathscr{V}}{\partial y^2} + \frac{\partial^2 \mathscr{V}}{\partial z^2} = 0 \tag{8.8.2}$$

and

$$\omega = \omega_x \, dx + \omega_y \, dy + \omega_z \, dz \tag{8.8.3}$$

is a one-form whose external derivative is requested to be Hodge dual, in the flat metric $ds^2_{\mathbb{R}^3} = \text{dx}^2 + \text{dy}^2 + \text{dz}^2$ of $\mathbb{R}^3$, to the gradient of $\mathscr{V}$:

$$\star_{\mathbb{R}^3} \, \text{d}\omega = \text{d}\mathscr{V} \tag{8.8.4}$$

Without loss of generality we can choose an axial gauge for the connection ω by setting:

$$\omega_z = 0 \tag{8.8.5}$$

The four-dimensional Riemannian space $\mathscr{M}_{\text{GH}}$, whose metric is (8.8.1), is a U(1)-bundle over $\mathbb{R}^3$. Actually we can easily prove that $\mathscr{M}_{\text{GH}}$ is Kählerian by means of the following argument. Consider the following two-form:

$$\mathbb{K}_{\text{GH}} = 2\left((\text{d}\tau + \omega) \wedge \text{d}z - \mathscr{V}\,\text{d}x \wedge \text{d}y\right) \tag{8.8.6}$$

which is closed in force of Eqs. (8.8.2) and (8.8.3):

$$\text{d}\mathbb{K}_{\text{GH}} = 0 \tag{8.8.7}$$

From Eq. (8.8.1) we easily workout the components of the metric in the x, y, z, τ coordinate basis:

$$g_{ij} = \begin{pmatrix} \mathscr{V} + \frac{\omega_x^2}{\mathscr{V}} & \frac{\omega_x \omega_y}{\mathscr{V}} & 0 & \frac{\omega_x}{\mathscr{V}} \\ \frac{\omega_x \omega_y}{\mathscr{V}} & \mathscr{V} + \frac{\omega_y^2}{\mathscr{V}} & 0 & \frac{\omega_y}{\mathscr{V}} \\ 0 & 0 & \mathscr{V} & 0 \\ \frac{\omega_x}{\mathscr{V}} & \frac{\omega_y}{\mathscr{V}} & 0 & \frac{1}{\mathscr{V}} \end{pmatrix} \tag{8.8.8}$$

and of its inverse:

$$g^{ij} = \begin{pmatrix} \frac{1}{\mathscr{V}} & 0 & 0 & -\frac{\omega_x}{\mathscr{V}} \\ 0 & \frac{1}{\mathscr{V}} & 0 & -\frac{\omega_y}{\mathscr{V}} \\ 0 & 0 & \frac{1}{\mathscr{V}} & 0 \\ -\frac{\omega_x}{\mathscr{V}} & -\frac{\omega_y}{\mathscr{V}} & 0 & \frac{\mathscr{V}^2+\omega_x^2+\omega_y^2}{\mathscr{V}} \end{pmatrix} \tag{8.8.9}$$

Similarly, from Eq. (8.8.6) we work out the components of the form $\mathbb{K}_{\text{GH}}$:

$$K_{ij} = \begin{pmatrix} 0 & -\mathscr{V} & \omega_x & 0 \\ \mathscr{V} & 0 & \omega_y & 0 \\ -\omega_x & -\omega_y & 0 & -1 \\ 0 & 0 & 1 & 0 \end{pmatrix} \tag{8.8.10}$$

Raising the second index of the antisymmetric tensor K_{ij} with the inverse metric $g^{j\ell}$ we obtain a mixed tensor

$$J_i^{\ \ell} \equiv K_{ij}\, g^{j\ell} = \begin{pmatrix} 0 & -1 & \frac{\omega_x}{\mathscr{V}} & \omega_y \\ 1 & 0 & \frac{\omega_y}{\mathscr{V}} & -\omega_x \\ 0 & 0 & 0 & -\mathscr{V} \\ 0 & 0 & \frac{1}{\mathscr{V}} & 0 \end{pmatrix} \tag{8.8.11}$$

which satisfies the property:

$$J_i^{\ \ell}\, J_\ell^{\ \text{m}} = -\,\delta_i^{\text{m}} \tag{8.8.12}$$

Hence J is a *quasi-complex structure* which is proved to be a *complex structure* by verifying that its Nienhuis tensor vanishes:

$$N^{\ell}_{ij} \equiv \partial_{[i} J_{j]}{}^{\ell} - J_i{}^{\mathrm{m}} J_j{}^{\mathrm{n}} \partial_{[\mathrm{m}} J_{\mathrm{n}]}{}^{\ell} = 0 \tag{8.8.13}$$

It follows that $\mathcal{M}_{\mathrm{GH}}$ is a complex manifold, the metric (8.8.6) being hermitian with respect to J since the matrix $K_{ij} \equiv J_i{}^{\ell} g_{\ell j}$ is by construction antisymmetric and, as such, it defines a Kähler 2-form. Thus we have a Kähler form which is closed and this, by definition, implies that the complex manifold $\mathcal{M}_{\mathrm{GH}}$ is a Kähler manifold.

8.8.1 Integration of the Complex Structure and the Issue of the Kähler Potential

The first task in order to put the Kähler metric of a $2n$-dimensional real manifold into a standard complex form derived from a Kähler potential is that of deriving a suitable set of complex coordinates Z_μ that are eigenstates of the complex structure. This means to find a complete set of n complex solutions of the following differential equation:

$$J_i{}^{\ell} \partial_\ell Z = i\, \partial_i Z \tag{8.8.14}$$

In the case of the complex structure in Eq. (8.8.11) a basis of the eigenspace pertaining to the eigenvalue i is easily provided by the following two vectors

$$\begin{aligned} \mathbf{v}_1 &= \left\{-i\omega_y, i\omega_x, i\mathcal{V}, 1\right\} \\ \mathbf{v}_2 &= \{i, 1, 0, 0\} \\ J\,\mathbf{v}_{1,2} &= i\mathbf{v}_{1,2} \end{aligned} \tag{8.8.15}$$

The second eigenvector $\mathbf{v}_2$ inserted into Eq. (8.8.14) immediately singles out one of the two complex coordinates:

$$\mathfrak{z} \equiv y + i\,x \tag{8.8.16}$$

In order to integrate Eq. (8.8.14) utilizing the first eigenvector $\mathbf{v}_1$, a very useful tool is provided by a recent observation made by Ortin et al. in [54] who pointed out that a convenient way of automatically realizing conditions (8.8.2) and (8.8.4) is obtained by setting:

$$\omega_x = \frac{\partial^2 \mathcal{F}}{\partial y \partial z} \quad ; \quad \omega_y = -\frac{\partial^2 \mathcal{F}}{\partial x \partial z} \quad ; \quad \mathcal{V} = \frac{\partial^2 \mathcal{F}}{\partial z^2} \tag{8.8.17}$$

where $\mathcal{F}(x, y, z)$ is a harmonic prepotential:

$$\frac{\partial^2 \mathscr{F}}{\partial x^2} + \frac{\partial^2 \mathscr{F}}{\partial y^2} + \frac{\partial^2 \mathscr{F}}{\partial z^2} = 0 \tag{8.8.18}$$

Using the prepotential $\mathscr{F}$ the differential equation to be satisfied by the searched for complex coordinate $\mathfrak{w}$ is the following one:

$$\left\{ i\,\frac{\partial^2 \mathscr{F}}{\partial z \partial z}, i\,\frac{\partial^2 \mathscr{F}}{\partial y \partial z}, i\,\frac{\partial^2 \mathscr{F}}{\partial z^2}, 1 \right\} = \left\{ \partial_x \mathfrak{w}, \partial_y \mathfrak{w}, \partial_z \mathfrak{w}, \partial_\tau \mathfrak{w} \right\} \tag{8.8.19}$$

In view of Eq. (8.8.17) we can set:

$$\mathscr{F}(x, y, z) = \int \mathrm{d}z \int \mathrm{d}z\, \mathscr{V}(x, y, z) \tag{8.8.20}$$

and the differential Eq. (8.8.19) is immediately integrated by setting:

$$\mathfrak{w} = \tau + i\, \partial_z \mathscr{F} = \tau + i \int \mathrm{d}z\, \mathscr{V} \tag{8.8.21}$$

Obviously, whenever a complex coordinate has been found, any holomorphic function of the same is an equally good complex coordinate. Hence in addition to $\mathfrak{z}$, defined in Eq. (8.8.16), we choose the second complex coordinate as follows:

$$\mathfrak{h} = \exp\left[i \mathfrak{w}\right] = e^{i\tau} \rho \quad ; \quad \rho = \exp\left[- \int \mathscr{V}\, \mathrm{d}z \right] \tag{8.8.22}$$

Using the above implicit definition of the complex coordinates one can transform the Kähler 2-form (8.8.6) to the complex coordinates obtaining:

$$\mathbb{K}_{\mathrm{GH}} = K_{\mathfrak{h}\bar{\mathfrak{h}}}\, d\mathfrak{h} \wedge d\bar{\mathfrak{h}} + K_{\mathfrak{h}\bar{\mathfrak{z}}}\, d\mathfrak{h} \wedge d\bar{\mathfrak{z}} + K_{\mathfrak{z}\bar{\mathfrak{h}}}\, d\mathfrak{z} \wedge d\bar{\mathfrak{h}} + K_{\mathfrak{z}\bar{\mathfrak{z}}}\, d\mathfrak{z} \wedge d\bar{\mathfrak{z}} \tag{8.8.23}$$

where

$$\begin{aligned} K_{\mathfrak{h}\bar{\mathfrak{h}}} &= i\,\frac{1}{\mathfrak{h}\,\bar{\mathfrak{h}}\,\mathscr{V}} = \partial_{\mathfrak{h}} \bar{\partial}_{\bar{\mathfrak{h}}} \mathscr{K} \\ K_{\mathfrak{h}\bar{\mathfrak{z}}} &= \frac{i\omega_x + \omega_y}{\mathfrak{h}\,\mathscr{V}} = \partial_{\mathfrak{h}} \bar{\partial}_{\bar{\mathfrak{z}}} \mathscr{K} \\ K_{\mathfrak{z}\bar{\mathfrak{h}}} &= \frac{i\omega_x - \omega_y}{\bar{\mathfrak{h}}\,\mathscr{V}} = \partial_{\mathfrak{z}} \bar{\partial}_{\bar{\mathfrak{h}}} \mathscr{K} \\ K_{\mathfrak{z}\bar{\mathfrak{z}}} &= i\,\frac{\omega_x^2 + \omega_y^2 + \mathscr{V}^2}{\mathscr{V}} = \partial_{\mathfrak{z}} \bar{\partial}_{\bar{\mathfrak{z}}} \mathscr{K} \end{aligned} \tag{8.8.24}$$

The problem of deriving the Kähler potential $\mathscr{K}(\mathfrak{h}, \mathfrak{z}, \bar{\mathfrak{h}}, \bar{\mathfrak{z}})$ corresponding to the GH-metric is reduced to the inverting the coordinate transformation encoded in Eqs. (8.8.21) and (8.8.16) and then solving the system of coupled differential equations encoded in Eq. (8.8.24). Typically this is far from being an easy task, but in some simple cases it can be done. The primary illuminating example is provided by the Eguchi Hanson metric corresponding to $ALE_{\mathbb{Z}_2}$.

8.8.2 Identification of the Eguchi–Hanson Space with the Two-Center GH-Manifold

We derive the map between the manifold with a two center GH metric and the Eguchi Hanson space. We begin with a conceptual discussion about the parameters of GH-metrics.

The Gibbons Hawking multi-center metrics have a number of parameters that can be counted in the following way. Let n be the number of centers. Each center has 3-coordinates, hence a priori we have $3n$ parameters. Yet, using the Euclidean group of translations and rotations, which is a symmetry of the $3d$ laplacian, we can always bring a center to a reference point, say the origin $\boldsymbol{x} = 0$. So we are left with $3(n-1)$ parameters. Furthermore, once a center is fixed, another center lies somewhere on a two-sphere around the first center and we can use the rotation group to bring it to a preferred direction. This kills two other parameters. In this way we have:

$$\#\text{ of effective parameters in a GH metric} = 3n - 5 \tag{8.8.25}$$

From the point of view of the Kronheimer construction, the n-center metric corresponds to the resolution $Y \to \frac{\mathbb{C}^2}{\mathbb{Z}_m}$ via a HyperKähler quotient. In this case the gauge group is $\mathrm{U}(1)^{n-1}$ and we have indeed $3(n-1)$ parameters. Two parameters corresponding to one complex moment-map level can be disposed of by a redefinition of the complex coordinates for the resolved manifold Y. Hence also on the side of the HyperKähler quotient we have:

$$\#\text{ of effective parameters in a HyperKähler quotient resolution of } \frac{\mathbb{C}^2}{\mathbb{Z}_n} = 3n - 5 \tag{8.8.26}$$

In the Eguchi–Hanson case $n = 2$ and there is only one effective parameter on both sides of the correspondence, namely the parameter m^2 that we have associated with the real moment map level. The level of the holomorphic moment map corresponds to the two parameters that can be disposed of by a coordinate transformation and was set to zero.

From the GH-side, the removal of the spurious parameters can be conventionally performed by aligning the two centers on the z-axis at symmetrical positions with respect to the origin $z = 0$. Hence referring to Eqs. (8.8.1) and (8.8.2) we set:

$$\mathcal{V}_{EH} = \frac{1}{\sqrt{\left(\frac{m^2}{8}+z\right)^2+x^2+y^2}} + \frac{1}{\sqrt{\left(z-\frac{m^2}{8}\right)^2+x^2+y^2}} \tag{8.8.27}$$

and we obtain the following connection one-form :

$$\begin{aligned}\omega_{EH} = &\left(m^2\left(\frac{1}{\sqrt{(m^2-8z)^2+64\,(x^2+y^2)}} - \frac{1}{\sqrt{(m^2+8z)^2+64\,(x^2+y^2)}}\right)\right.\\ &-\frac{8z}{\sqrt{(m^2-8z)^2+64\,(x^2+y^2)}}\\ &\left.-\frac{8z}{\sqrt{(m^2+8z)^2+64\,(x^2+y^2)}}+2\right)\times\frac{y\,\mathrm{d}x - x\,\mathrm{d}y}{x^2+y^2}\end{aligned} \tag{8.8.28}$$

which satisfies with $\mathcal{V}_{EH}$ the duality relation (8.8.4). The one-form ω_{EH} agrees with Eq. (8.8.17) if we set:

$$\begin{aligned}\partial_z\mathcal{F}_{EH} &= \int dz\,\mathcal{V}_{EH}\\ &= \log\left(\sqrt{\left(z-\frac{m^2}{8}\right)^2+x^2+y^2}-\frac{m^2}{8}+z\right)\\ &\quad+\log\left(\sqrt{\left(\frac{m^2}{8}+z\right)^2+x^2+y^2}+\frac{m^2}{8}+z\right)\end{aligned} \tag{8.8.29}$$

The metric:

$$ds^2_{two-center} = \frac{1}{\mathcal{V}_{EH}}\,(\mathrm{d}\tau+\omega_{EH})^2+\mathcal{V}_{EH}\left(\mathrm{d}x^2+\mathrm{d}y^2+\mathrm{d}z^2\right) \tag{8.8.30}$$

is exactly mapped into the Eguchi–Hanson metric (8.7.4) by the following coordinate transformation:

$$\begin{aligned}&x\to\frac{1}{8}\sin(\theta)\sqrt{r^4-m^4}\cos(\psi)\quad,\quad y\to\frac{1}{8}\sin(\theta)\sqrt{r^4-m^4}\sin(\psi)\\ &z\to\frac{1}{8}r^2\cos(\theta)\quad,\quad \tau\to 2\psi+2\phi\end{aligned} \tag{8.8.31}$$

It is also interesting to work out the explicit form, in the present case of the complex coordinates $\mathfrak{h}$ and $\mathfrak{z}$ introduced in Eqs. (8.8.16) and (8.8.22) within the framework of the general discussion. After some algebra one finds:

$$\mathfrak{h} = \frac{64\, e^{2i(\psi+\phi)}}{(\cos(\theta)+1)^2\left(r^4-m^4\right)} \quad , \quad \mathfrak{z} = \frac{1}{8}\, i\, e^{-i\psi}\sin(\theta)\sqrt{r^4-m^4} \tag{8.8.32}$$

As one realizes, both these coordinates are singular on the exceptional divisor $r = m$ and they are not convenient to describe it. The relation with the good coordinates $\xi_{1,2}$ is actually antiholomorphic and it would be difficult to be guessed a priori:

$$\bar{\xi}_1 = -\frac{i}{\mathfrak{z}\sqrt{\mathfrak{h}}} \quad , \quad \bar{\xi}_2 = -\frac{\sqrt{2}\,\mathfrak{z}\,\sqrt[4]{\mathfrak{h}}}{m} \tag{8.8.33}$$

In terms of the GH-coordinates, by inspecting Eq. (8.8.31) we readily retrieve the image of exceptional divisor inside the GH space. It is given by the locus:

$$D_E = \left\{x = y = 0\,,\ -\frac{m^2}{8} \le z \le \frac{m^2}{8}\,,\ 0 \le \tau \le 2\pi\right\} \tag{8.8.34}$$

namely the tensor product of the segment joining the two centers on the z-axis with the circle spanned by the τ-angle. This observation is useful in order to find the exceptional divisors in the more complicated multi-center cases.

8.9 The Generalized Kronheimer Construction for $\frac{\mathbb{C}^3}{\Gamma}$ and the Tautological Bundles

In the present section we aim at extracting a general scheme from the detailed discussions presented in the previous sections. Our final goal is to establish all the algorithmic steps that give a precise meaning to each of the lines appearing in the conceptual diagram of Eq. (8.3.2).

8.9.1 Construction of the Space $\mathcal{N}_\zeta \equiv \mu^{-1}(\zeta)$

Summarizing the points of our construction we have the following situation. We have considered the moment map

$$\mu : \mathcal{S}_\Gamma \longrightarrow \mathbb{F}_\Gamma{}^* \tag{8.9.1}$$

where $\mathbb{F}_\Gamma{}^*$ is the dual of the Lie algebra of the maximal compact subgroup $\mathcal{F}_\Gamma$ of the quiver group $\mathcal{G}_\Gamma$. Next we have considered the center of the above Lie algebra

$\mathfrak{z}[\mathbb{F}_\Gamma]\subset\mathbb{F}_\Gamma$ and its dual $\mathfrak{Z}[\mathbb{F}_\Gamma]^*$. The moment map can be restricted to the subspace:

$$\mathscr{D}_\Gamma \subset \mathscr{S}_\Gamma \quad ; \quad \mathscr{D}_\Gamma \equiv \{p \in \mathscr{S}_\Gamma | \, p \wedge p = 0\} \tag{8.9.2}$$

which is just the orbit, with respect to the quiver group $\mathscr{G}_\Gamma$, of a locus $\mathscr{E}_\Gamma \subset \mathscr{S}_\Gamma$ of complex dimension three obtained in the following way.

Consider the following subspace of $\mathscr{S}_\Gamma^{[0,0]} \subset \mathscr{S}_\Gamma$

$$\mathscr{S}_\Gamma^{[0,0]} = \{p \in \mathscr{S}_\Gamma | \, p \wedge p = 0 \quad ; \quad \mu(p) = 0\} \tag{8.9.3}$$

whose elements are triples of $|\Gamma| \times |\Gamma|$ complex matrices (A,B,C) satisfying, by the above definition, in addition to the invariance constraint (8.6.6), (8.6.7) also the following two ones:

$$\begin{array}{l} [A, B] = [B, C] = [C, A] = 0 \\ \mathrm{Tr}\left[T_I \left(\left[A, A^\dagger\right] + \left[B, B^\dagger\right] + \left[C, C^\dagger\right]\right)\right] = 0 \; ; \; I = 1, \ldots, |\Gamma| - 1 \end{array} \tag{8.9.4}$$

Since the action of the compact group $\mathscr{F}_\Gamma$ leaves both the first and the second constraint invariant, it follows that it maps the locus $\mathscr{S}_\Gamma^{[0,0]}$ into itself

$$\mathscr{F}_\Gamma \quad : \quad \mathscr{S}_\Gamma^{[0,0]} \to \mathscr{S}_\Gamma^{[0,0]} \tag{8.9.5}$$

The locus $\mathscr{E}_\Gamma$ is defined as the quotient:

$$\mathscr{E}_\Gamma \equiv \frac{\mathscr{S}_\Gamma^{[0,0]}}{\mathscr{F}_\Gamma} \tag{8.9.6}$$

which turns out to be of complex dimension three and to be isomorphic to the singular orbifold :

$$\frac{\mathscr{S}_\Gamma^{[0,0]}}{\mathscr{F}_\Gamma} \simeq \frac{\mathbb{C}^3}{\Gamma} \tag{8.9.7}$$

Choosing a representative in each equivalence class $\frac{\mathscr{S}_\Gamma^{[0,0]}}{\mathscr{F}_\Gamma}$ simply amounts to a choice of local coordinates on $\frac{\mathbb{C}^3}{\Gamma}$ which will be promoted to a system of local coordinates on the manifold $\mathscr{M}_\zeta$ of the final resolved singularity.

We have a canonical algorithm to construct a canonical coordinate system for $\mathscr{E}_\Gamma$ which originates from Kronheimer and from the 1994 paper by Anselmi, Billò, Frè, Girardello and Zaffaroni on ALE manifolds and conformal field theories [40]. The construction is the following. We begin with the locus $L_\Gamma \subset \mathscr{S}_\Gamma$ defined as the set of triples (A_d, B_d, C_d) such that the invariance constraint (8.6.7) is satisfied with respect to Γ and they are diagonal in the natural basis of the regular representation. We have shown on the basis of several examples that :

$$\mathscr{D}_\Gamma = \mathrm{Orbit}_{\mathscr{G}_\Gamma}\left(L_\Gamma\right) \tag{8.9.8}$$

We obtain an explicit parameterization of the locus $\mathscr{E}_\Gamma$ by solving the algebraic equation for the hermitian matrix $\mathscr{V}_0 \in \exp[\mathbb{K}_\Gamma]$, such that

$$\forall p \in L_\Gamma \quad : \quad \mu(\mathscr{V}_0.p) = 0 \tag{8.9.9}$$

The important thing is that the solution for the above equation is a constant matrix $\mathscr{V}_0$, independent from the point $p \in L_\Gamma$. Then we fix the coordinates of our manifold by parameterizing

$$p \in \mathscr{E}_\Gamma \Rightarrow p = \begin{pmatrix} A_0 \\ B_0 \\ C_0 \end{pmatrix} = \begin{pmatrix} \mathscr{V}_0{}^{-1} A_d \mathscr{V}_0 \\ \mathscr{V}_0{}^{-1} B_d \mathscr{V}_0 \\ \mathscr{V}_0{}^{-1} C_d \mathscr{V}_0 \end{pmatrix} \text{ where } \begin{pmatrix} A_d \\ B_d \\ C_d \end{pmatrix} \in L_\Gamma \tag{8.9.10}$$

It follows that Eq. (8.9.8) can be substituted by

$$\mathbb{V}_{|\Gamma|+2} \equiv \mathscr{D}_\Gamma = \text{Orbit}_{\mathscr{G}_\Gamma}(\mathscr{E}_\Gamma) \tag{8.9.11}$$

We can also introduce a subspace $\mathscr{D}_\Gamma{}^0 \subset \mathbb{V}_{|\Gamma|+2}$ which is the orbit of $\mathscr{E}_\Gamma$ under the compact subgroup $\mathscr{F}_\Gamma$:

$$\mathscr{D}_\Gamma{}^0 = \text{Orbit}_{\mathscr{F}_\Gamma}(\mathscr{E}_\Gamma) \tag{8.9.12}$$

This being the case we consider the restriction of the moment map to $\mathscr{D}_\Gamma$

$$\mu : \mathscr{D}_\Gamma \longrightarrow \mathbb{F}_\Gamma{}^* \tag{8.9.13}$$

and given an element

$$\zeta \in \mathfrak{z}[\mathbb{F}_\Gamma]^* \tag{8.9.14}$$

we define:

$$\mathscr{N}_\zeta \equiv \mu^{-1}(\zeta) \subset \mathscr{D}_\Gamma \quad : \quad \mathscr{N}_\zeta = \{p \in \mathscr{D}_\Gamma \,|\, \mu(p) = \zeta\} \tag{8.9.15}$$

Obviously we have:

$$\mathscr{N}_0 \equiv \mu^{-1}(0) = \mathscr{D}_\Gamma{}^0 \tag{8.9.16}$$

8.9.2 The Space $\mathscr{N}_\zeta$ as a Principal Fiber Bundle

The space $\mathscr{N}_\zeta$ has a natural structure of an $\mathscr{F}_\Gamma$ principal line bundle over the quotient $\mathscr{M}_\zeta$:

$$\mathscr{N}_\zeta \xrightarrow{\pi} \mathscr{M}_\zeta \equiv \mathscr{N}_\zeta /\!/ \mathscr{F}_\Gamma \tag{8.9.17}$$

On the tangent space to the total space of the $\mathscr{F}_\Gamma$-bundle $T\mathscr{N}_\zeta$ we have a metric induced, as the pullback, by the inclusion map:

$$\iota : \mathscr{N}_\zeta \longrightarrow \mathscr{S}_\Gamma \tag{8.9.18}$$

of the flat metric g on $\mathscr{S}_\Gamma$

$$g_{\mathscr{N}} = \iota^* \left(g_{\mathscr{S}_\Gamma} \right) \tag{8.9.19}$$

Since the metric $g_{\mathscr{S}_\Gamma}$ is Kählerian we have a Kähler potential $\mathscr{K}_{\mathscr{S}_\Gamma}$ from which it derives and we define the function

$$\mathscr{K}_{\mathscr{N}} \equiv \iota^* \left(\mathscr{K}_{\mathscr{S}_\Gamma} \right) \tag{8.9.20}$$

This function is not the Kähler potential of $\mathscr{N}_\zeta$ which is not even Kählerian (it has odd dimensions) but it will be related to the Kähler potential of the final quotient $\mathscr{M}_\zeta$ by means of an argument due to [1], that we spell out a few lines below. Let us denote by $p \in \mathscr{M}_\zeta$ a point of the base manifold and by $\pi^{-1}(p)$ the $\mathscr{F}_\Gamma$-fiber over that point.

8.9.2.1 The Natural Connection and the Tautological Bundles

We can determine a natural connection on the principal bundle (8.9.17) through the following steps. As it is observed in Eq. (2.7) of the paper by Degeratu and Walpuski [55], which agrees with the formulae of the present paper, the quiver group has always the following form:

$$\mathscr{G}_\Gamma = \prod_{i=1}^{r} \mathrm{GL}\left(\mathbb{C}^{\dim[\boldsymbol{D}_i]} \right) \tag{8.9.21}$$

where $\boldsymbol{D}_i$ are the nontrivial irreducible representations of the finite group Γ, with the exclusion of $\boldsymbol{D}_0$, the identity representation. It also follows that the compact gauge subgroup $\mathscr{F}_\Gamma$ has the corresponding following structure

$$\mathscr{F}_\Gamma = \prod_{i=1}^{r} \mathrm{U}\left(\dim\left[D_i \right] \right) \tag{8.9.22}$$

Consequently, the principal bundle (8.9.17) induces holomorphic vector bundles of rank $\dim[\boldsymbol{D}_i]$ on which the compact structural group acts non-trivially only with its component $\mathrm{U}(\dim[\boldsymbol{D}_i])$. A natural connection on these bundles is obtained as it follows

$$\mathbb{A} = \frac{\mathrm{i}}{2} \left(\mathscr{H}^{-1} \partial \mathscr{H} - \mathscr{H} \bar{\partial} \mathscr{H}^{-1} \right) + g^{-1} dg \in \bigoplus_{i=1}^{r} \mathbf{u}\left(\dim\left[D_i \right] \right) \tag{8.9.23}$$

where $\mathscr{H}$ is a hermitian fiber-metric on the direct sum of the tautological vector bundles defined below:

$$\mathscr{R} \equiv \bigoplus_{i=1}^{r} \mathscr{R}_i \quad ; \quad \mathscr{R}_i \xrightarrow{\pi} \mathscr{M}_\zeta \quad ; \quad \forall p \in \mathscr{M}_\zeta \quad : \quad \pi^{-1}(p) \simeq \mathbb{C}^{\dim[D_i]} \tag{8.9.24}$$

By definition the matrix $\mathscr{H}$ must be of dimension

$$\dim[\mathscr{H}] = n \times n \qquad \text{where} \qquad n = \sum_{i=1}^{r} \dim[D_i] = \sum_{i=1}^{r} n_i \tag{8.9.25}$$

In order to find the hermitian matrix $\mathscr{H}$, we argue in the following way. First we observe that in the regular representation R each irreducible representation $\boldsymbol{D}_i$ is contained exactly $\dim[\boldsymbol{D}_i]$ times, so that the form of the matrix $\mathscr{V}$ corresponding to the hermitian parameterization of the coset $\frac{\mathscr{G}_\Gamma}{\mathscr{F}_\Gamma}$ has always the following form:

$$\mathscr{V} = \begin{pmatrix} \mathfrak{H}_0 & 0 & 0 & \cdots & 0 \\ 0 & \mathfrak{H}_1 \otimes \mathbf{1}_{n_1 \times n_1} & 0 & \cdots & \vdots \\ 0 & 0 & \mathfrak{H}_2 \otimes \mathbf{1}_{n_2 \times n_2} & \cdots & \vdots \\ \vdots & \cdots & \cdots & \cdots\cdots & \vdots \\ \vdots & \cdots & \cdots & \cdots & 0 \\ 0 & \cdots & \cdots & 0 & \mathfrak{H}_r \otimes \mathbf{1}_{n_r \times n_r} \end{pmatrix} \tag{8.9.26}$$

where n_i denotes the dimension of the ith nontrivial representation of the discrete group Γ and from this we extract the block diagonal matrix:

$$\mathscr{H} \equiv \begin{pmatrix} \mathfrak{H}_1 & 0 & \cdots & \cdots & 0 \\ 0 & \mathfrak{H}_2 & \cdots & \cdots & \vdots \\ \vdots & \cdots\cdots & \cdots\cdots & \cdots & \vdots \\ \vdots & \cdots\cdots & \cdots & \mathfrak{H}_{r-1} & 0 \\ 0 & \cdots\cdots & \cdots & 0 & \mathfrak{H}_r \end{pmatrix} \tag{8.9.27}$$

The hermitian matrix $\mathscr{H}$ is the fiber metric on the direct sum:

$$\mathscr{R} = \bigoplus_{i=1}^{r} \mathscr{R}_i \tag{8.9.28}$$

of the r tautological bundles that, by construction, are holomorphic vector bundles with rank equal to the dimension of the r irreducible representations of Γ:

$$\mathscr{R}_i \xrightarrow{\pi} \mathscr{M}_\zeta \quad ; \quad \forall p \in \mathscr{M}_\zeta \quad : \quad \pi^{-1}(p) \approx \mathbb{C}^{n_i} \tag{8.9.29}$$

The compatible connection[5] on the holomorphic vector bundle $\mathscr{R}$ is given by:

$$\begin{aligned} \vartheta &= \mathscr{H}^{-1}\,\partial\mathscr{H} \;=\; \bigoplus_{i=1}^{r} \theta_i \\ \theta_i &= \mathfrak{H}_i^{-1}\,\partial\mathfrak{H}_i \;\in\; \mathbb{GL}(n_i, \mathbb{C}) \end{aligned} \tag{8.9.30}$$

where $\mathbb{GL}(n_i, \mathbb{C})$ is the Lie algebra of $\mathrm{GL}(n_i, \mathbb{C})$ which is structural group of the ith tautological vector bundle. The natural connection of the $\mathscr{F}_\Gamma$ principal bundle, mentioned in Eq. (8.9.23) is just, according to a universal scheme, the imaginary part of the holomorphic connection ϑ.

8.9.2.2 The Tautological Bundles from the Irrep Viewpoint and the Kähler Potential

From the analysis of the above section we have reached a very elegant conclusion. Once the matrix $\mathscr{V}$ is calculated as function of the level parameters ζ and of the base-manifold coordinates $(z_m, \bar{z}_m)$ $(m = 1, 2, 3)$, we also have the block diagonal hermitian matrix $\mathscr{H}$ which encodes the hermitian fiber metrics $\mathfrak{H}_i(\zeta, z, \bar{z})$ on each of the tautological holomorphic bundles $\mathfrak{V}_i$ whose ranks are equal, one by one, to the dimensions n_i of the irreps of Γ. The first Chern classes of these bundles are represented by the differential $(1, 1)$ forms:

$$\omega_i^{(1,1)} = \frac{i}{2\pi}\,\bar{\partial}\partial \mathrm{Log}\,[\mathrm{Det}\,[\mathfrak{H}_i]] \tag{8.9.31}$$

Let us recall another remarkable group theoretical fact. The number r of nontrivial irreps of Γ is equal to the number r of nontrivial conjugacy classes and to the number r of generators of the center of the compact Lie algebra $\mathbb{F}_\Gamma$, hence also to the number r of level parameters ζ and to number r of holomorphic tautological bundles. Now we are in a position to derive in full generality the formula for the Kähler potential and, hence, for the Kähler metric of the resolved manifold $\mathscr{M}_\zeta$ that we anticipated in (8.6.56) . In view of the above discussion, we rewrite the latter as it follows:

$$\mathscr{K}_{\mathscr{M}_\zeta} = \mathscr{K}_{\mathscr{S}_\Gamma}\mid_{\mathscr{N}_\zeta} + \zeta^i \mathfrak{C}_{\mathrm{ij}} \mathrm{Log}\left[\mathrm{Det}\left[\mathfrak{H}_j\right]\right] \tag{8.9.32}$$

where $\mathscr{K}_{\mathscr{S}_\Gamma}$ is the Kähler potential of the flat space $\mathscr{S}_\Gamma$ and $|_{\mathscr{N}_\zeta}$ denotes its restriction to the level surface $\mathscr{N}_\zeta$, while $\mathfrak{C}_{\mathrm{IJ}}$ is an r×r constant matrix whose structure we will define and determine below. Why the matrix defined there yields the appropriate

[5] Following standard mathematical nomenclature, we call compatible connection on a holomorphic vector bundle, one whose $(0, 1)$ part is the Cauchy Riemann operator of the bundle.

Kähler potential is what we will now explain starting from an argument introduced in 1987 by Hitchin, Karlhede, Lindström and Roček.

The HKLR Differential Equation and Its Solution

Before explaining the origin of the matrix $\mathfrak{C}_{IJ}$, we would like to stress that, conceptually it encodes a pairing between the level parameters (= generators of the Lie algebra center) and the tautological bundles (= irreps). If we could understand the relation between conjugacy classes with their ages and cohomology classes, then we would have a relation between irreps and conjugacy classes and we could close the three-sided relation diagram among the center $\mathfrak{z}\left[\mathbb{F}_\Gamma\right]$ and the other two items, irreps and conjugacy classes. As we are going to show, this side of the relation is based on the concept of weighted blowup. On the other hand, understanding the matrix $\mathfrak{C}_{IJ}$, is a pure Lie algebra theory issue, streaming from the HKLR argument.

Hence, continuing such an argument, let us consider the flat Kähler manifold $\mathscr{S}_\Gamma$ and its Kähler potential

$$\mathscr{K} = \sum_{i=1}^{3} \mathrm{Tr}\left[A_i A_i^\dagger\right] \text{ where we have defined } A_i = \{A, B, C\} \tag{8.9.33}$$

The exponential of the Kähler potential is also, by definition, the hermitian metric on the Hodge line bundle:

$$\begin{aligned} &\mathscr{L}_{\text{Hodge}} \xrightarrow{\pi} \mathscr{S}_\Gamma \quad \text{where} \quad \forall p \in \mathscr{S}_\Gamma \quad : \quad \pi^{-1}(p) \approx \mathbb{C}^* \\ &\| W \|^2 \equiv e^{\mathscr{K}_\mathscr{S}}\, W \bar{W} \end{aligned} \tag{8.9.34}$$

Indeed, the second line of the above equation $\| W \|^2$ defines the invariant norm of any section of $\mathscr{L}_{\text{Hodge}}$.

Let us know consider the action of the quiver group on $\mathscr{S}_\Gamma$ and its effect on the fiber metric $\mathrm{h} = e^{\mathscr{K}}$. The maximal compact subgroup $\mathscr{F}_\Gamma$ is an isometry group for the Kähler metric defined by (8.9.33). Hence we focus on the orthogonal (with respect to the Killing form) complement of $\mathscr{F}_\Gamma$. Let

$$\boldsymbol{\Phi} \in \mathbb{K}_\Gamma \tag{8.9.35}$$

be an element of the orthogonal subspace to the maximal compact subalgebra

$$\mathbb{G}_\Gamma = \mathbb{F}_\Gamma \oplus \mathbb{K}_\Gamma \tag{8.9.36}$$

consider the one parameter subgroup generated by this Lie algebra element

$$g(\lambda) \equiv e^{\lambda \boldsymbol{\Phi}} \tag{8.9.37}$$

The action of this group on the Kähler potential is easily calculated

$$\mathcal{K}_{\mathcal{S}}(\lambda) = \sum_{i=1}^{3} \text{Tr}\left[A_i e^{2\lambda \boldsymbol{\Phi}} A_i^{\dagger} e^{-2\lambda \boldsymbol{\Phi}}\right] \tag{8.9.38}$$

Performing the derivative with respect to λ we obtain

$$\partial_\lambda \mathcal{K}_{\mathcal{S}}(\lambda) \mid_{\lambda=0} = \sum_{i=1}^{3} \text{Tr}\left(\boldsymbol{\Phi}\left[A_i, A_i^{\dagger}\right]\right) \tag{8.9.39}$$

Then we utilize the fact that each element $\boldsymbol{\Phi} \in \mathbb{K}_\Gamma$ is just equal to i× $\boldsymbol{\Phi}_c$ where $\boldsymbol{\Phi}_c$ denotes an appropriate element of the compact subalgebra. Hence the above equation becomes

$$\partial_\lambda \mathcal{K}_{\mathcal{S}}(\lambda) \mid_{\lambda=0} = \mathrm{i} \times \sum_{i=1}^{3} \text{Tr}\left(\boldsymbol{\Phi}_c\left[A_i, A_i^{\dagger}\right]\right) = i\mathfrak{P}_{\boldsymbol{\Phi}_c} \tag{8.9.40}$$

Let us decompose the moment map along the standard basis of compact generators. We obtain:

$$\begin{aligned}
\mathfrak{P}_{\Phi} &= \sum_{I=1}^{|\Gamma|-1} \boldsymbol{\Phi}^I \text{Tr}\left(\mathfrak{K}_I^c\left[A_i, A_i^{\dagger}\right]\right) \\
&= \mathrm{i} \sum_{I=1}^{|\Gamma|-1} \boldsymbol{\Phi}_c^I \mathfrak{P}_I(p) = \sum_{I=1}^{|\Gamma|-1} \boldsymbol{\Phi}^I \mathfrak{P}_I(p) = \sum_{I=1}^{|\Gamma|-1} \boldsymbol{\Phi}^I \text{Tr}\left(\mathfrak{K}_I\left[A_i, A_i^{\dagger}\right]\right)
\end{aligned} \tag{8.9.41}$$

where p $\in \mathcal{D}_\Gamma \subset \mathcal{S}_\Gamma$ denotes the arbitrary point in the ambient space described by the triple of matrices A_i, $\mathfrak{K}_I = \mathrm{i}\ \mathfrak{K}_I^c$ are the $|\Gamma|$-1 noncompact generators of the quiver group $\mathcal{G}_\Gamma$ that, by construction, are just as many as the compact generators $\mathfrak{K}_I^c$ of the maximal compact subgroup $\mathcal{F}_\Gamma$. Formally integrating the above differential equation it follows that the fiber of the metric Hodge line bundle (8.9.34)

$$h(p) \equiv \text{Exp}\left[\mathcal{K}_{\mathcal{S}}(p)\right] \tag{8.9.42}$$

transforms in the following way under the action of the quiver group

$$\forall g \in \mathcal{G}_\Gamma \; g : h(p) \longrightarrow h^g(p) \equiv h\left(e^{\text{Log}[g]} p\right) = e^{c(g,p)} h(p) \tag{8.9.43}$$

where

$$\text{Log}[g] \in \mathbb{G}_\Gamma \tag{8.9.44}$$

is an element of the quiver group Lie algebra and as such can be decomposed along a complete basis of generators

$$\mathrm{Log}[g] = \sum_{I=1}^{7} \boldsymbol{\Phi}^I \mathfrak{K}_I + \boldsymbol{\Phi}_c{}^I \mathfrak{K}^c{}_I \tag{8.9.45}$$

and the anomaly $c(g, p)$introduced in Eq. (8.9.43) has, in force of the differential equation discussed above the following form:

$$c(g, p) = \sum_{I=1}^{7} \left(\boldsymbol{\Phi}^I + i\boldsymbol{\Phi}_c{}^I\right) \mathfrak{P}_I(p) \tag{8.9.46}$$

where $\mathfrak{P}_I(p)$ are the moment maps at point p.

Next consider the diagram

$$\mathscr{S}_\Gamma \;\stackrel{\iota}{\longleftarrow}\; \mathscr{N}_\zeta \;\stackrel{\pi}{\longrightarrow}\; \mathscr{M}_\zeta \;\equiv\; \mathscr{N}_\zeta/\mathscr{F}_\Gamma \tag{8.9.47}$$

where $\mathscr{N}_\zeta$ is the level surface and $\mathscr{M}_\zeta$ the final Kähler threefold with its associated Hodge line bundle whose curvature is the Kähler form $\mathbf{K}_\mathscr{M}$

$$\mathbf{K}_\mathscr{M} \equiv \frac{i}{2\pi}\bar{\partial}\partial\mathscr{K}_\mathscr{M} = \frac{i}{2\pi}\bar{\partial}\left(\frac{1}{h_\mathscr{M}}\partial h_\mathscr{M}\right) \tag{8.9.48}$$

$\mathscr{K}_\mathscr{M}$ being the Kähler potential of the resolved variety. Following HKLR, we require that

$$\pi^*\mathbf{K}_\mathscr{M} = \iota^*\mathbf{K}_{\mathscr{S}_\Gamma} \tag{8.9.49}$$

where $\mathbf{K}_{\mathscr{S}_\Gamma}$ is the Kähler form of the flat Kähler manifold $\mathscr{S}_\Gamma = \mathrm{Hom}_\Gamma(\mathrm{Q}\otimes\mathrm{R},\mathrm{R})$. At the level of fiber metric on the associated Hodge line bundles, Eq. (8.9.49) amounts to stating that

$$\forall p \in \mathscr{M}_\zeta \quad : \quad h_\mathscr{M}\,(p) = h^g_{\mathscr{S}_\Gamma}(p) = h_{\mathscr{S}_\Gamma}(g.p) = e^{c(g,p)}\, h_{\mathscr{S}_\Gamma}(p) \tag{8.9.50}$$

where g is an element of the quiver group that brings the point $\mathrm{p} \in \mathscr{N}_\zeta$ on the level surface of level ζ to the reference level surface $\mathscr{N}_0$ which corresponds to the singular orbifold $\frac{\mathbb{C}^3}{\Gamma}$. Applying this to Eq. (8.9.46) we obtain:

$$c(g, p) = \zeta^I \boldsymbol{\Phi}_I(p) = \zeta^I * \mathrm{Tr}\left[\mathfrak{K}_I \mathrm{Log}[g]\right] = \sum_{i=1}^{r} \zeta^I * \mathrm{Tr}\left[\mathfrak{K}_I^{\mathrm{central}} \mathrm{Log}[g]\right] \tag{8.9.51}$$

since the only non-vanishing levels are located in the Lie Algebra center. On the other hand we have $\mathrm{g} = \mathscr{H}$:

$$\mathrm{Tr}\left[\mathfrak{K}_I^{\mathrm{central}} \mathrm{Log}[\mathscr{H}]\right] \equiv \sum_{J=1}^{r} \mathfrak{C}_{IJ} \mathrm{Log}\left[\mathrm{Det}\left[\mathfrak{H}_J\right]\right] \tag{8.9.52}$$

The above formula defines the constant matrix $\mathfrak{C}_{IJ}$ and justifies the final formula (8.9.32). In the case of cyclic Γ the center of the Lie Algebra $\mathbb{F}_\Gamma$ coincides with the algebra itself and the matrix $\mathfrak{C}_{IJ}$ is just diagonal and essentially trivial.

Dolbeault Cohomology

The objects we are dealing with are Dolbeault cohomology classes of the final resolved manifold $\mathscr{M}_\zeta$ which is Kähler as a result of its Kähler quotient construction.

When we say that $\omega^{p,q}$ is a harmonic representative of a nontrivial cohomology class in $H^{1,1}\left(\mathscr{M}_\zeta\right)$ we are stating that:

- The form is ∂-closed and $\overline{\partial}$-closed

$$\partial\omega^{p,q} = \overline{\partial}\omega^{p,q} = 0$$

- There do not exist forms $\phi^{p-1,q}$ and $\phi^{p,q-1}$ such that:

$$\omega^{p,q} = \partial\phi^{p-1,q} = \overline{\partial}\phi^{p,q-1}$$

The reason why the $\omega_i^{(1,1)}$ are nontrivial representatives of $(1,1)$ cohomology classes is that they are obtained as $\overline{\partial}$ of connection one-forms $\theta^{(1,0)}$ that are not globally defined. Indeed if we introduce the curvatures and the first Chern classes of the tautological vector bundles we have the elegant formula anticipated in Eq. (8.9.31):

$$\begin{aligned} \Theta_i &= \overline{\partial}\theta_i \\ \omega_i^{(1,1)} &\equiv c_1(\mathscr{R}_i) = \mathrm{Tr}(\Theta_i) = \overline{\partial}\,\partial \log\left[\mathrm{Det}\left(\mathfrak{H}_i\right)\right] \end{aligned} \quad (8.9.53)$$

Comparing now with the definition of Dolbeault cohomology we see that $\omega_i^{(1,1)}$ are nontrivial cohomology classes because either

$$\theta^{(1,0)} \equiv \partial \log\left[\mathrm{Det}\left(\mathfrak{H}_i\right)\right] \quad \text{or} \quad \theta^{(0,1)} \equiv \overline{\partial} \log\left[\mathrm{Det}\left(\mathfrak{H}_i\right)\right] \quad (8.9.54)$$

are non-globally defined 1-forms on the base manifold. This is so because they transform nontrivially from one local trivialization of the bundle to the next one. The transition functions on the connections are determined by the transition functions on the metric $\mathscr{H}$, namely on the coset representative. Here comes the delicate point.

Where from in the Kronheimer-like construction do we know that there are different local trivializations, otherwise that the tautological bundles are nontrivial? Computationally we solve the algebraic equations for $\mathscr{H}$ in terms of the coordinates z_i $(i = 1, 2, 3)$ parameterizing the locus L_Γ, which is equivalent to the singular locus $\frac{\mathbb{C}^3}{\Gamma}$ and we find $\mathscr{H} = \mathscr{H}(\zeta, z)$ where ζ are the level parameters. In order to conclude that the tautological bundle is nontrivial we should divide the locus L_Γ into patches and find the transition functions of the connections θ_i from one patch to the other. Obviously the transition function must be an element of the quiver group $g \in \mathscr{G}_\Gamma$. At the first sight it is not clear how to implement such a program, since we do not know how we should partition the locus L_Γ. Clearly the actual solution of the algebraic equations is complicated and, as we very well know, we are able to implement it only by means of a power series in ζ, yet it is obvious that this is not a case by case study. As everything else in the Kronheimer-like construction, it must be based on first principles and it is precisely these first principles that we are going to find out. It is at this level that the issue of ages is going to come into play in an algorithmic way. We begin by inspecting the only case where the final analytic form of all the construction items is available in closed form, namely the Eguchi–Hanson case.

8.9.3 What We See in the Eguchi–Hanson Case

Let us briefly summarize what we have verified in the EH case. The space $\mathscr{N}_\zeta$ has a natural structure of principal U(1)-bundle over the quotient $\mathscr{M}_\zeta$, as the maximal compact subgroup of the quiver group $\mathscr{F}_\Gamma \subset \mathscr{G}_\Gamma$ in this case is just U(1).

$$\mathscr{N}_\zeta \xrightarrow{\pi} \mathscr{M}_\zeta \equiv \mathscr{N}_\zeta /\!\!/ \mathscr{F}_\Gamma \tag{8.9.55}$$

As $\mathscr{N}_\zeta$ is a closed submanifold of $\mathscr{S}_\Gamma$ it has an induced metric g. The vertical tangent bundle to $\mathscr{N}_\zeta$ is locally generated by the vector field

$$V_{\text{vert}} = \frac{\partial}{\partial \phi} \tag{8.9.56}$$

Pointwise we can consider the space $T\mathscr{N}_{\text{hor}}$ orthogonal to the vertical tangent bundle

$$T\mathscr{N}_{\text{hor}} = \left\{ X \in T\mathscr{N}_\zeta \middle| < X, \frac{\partial}{\partial \phi} > \equiv g\left(X, \frac{\partial}{\partial \phi}\right) = 0 \right\} \tag{8.9.57}$$

This assignment of a complement to the vertical tangent spaces is smooth and U(1)-invariant, and therefore defines a connection on the principal bundle $\mathscr{N}_\zeta$, whose connection form $\boldsymbol{A}$ satisfies

$$\forall X \in T\mathscr{N}_{\text{hor}} : \boldsymbol{A}(X) = 0\,;\, \boldsymbol{A}\left(\frac{\partial}{\partial \phi}\right) = 1 \tag{8.9.58}$$

In the chosen coordinates we find:

$$\boldsymbol{A} = \mathrm{d}\phi - \frac{\zeta\,\mathrm{d}\theta_1 \rho_1^2}{2\left(1+\rho_1^2\right)\sqrt{\zeta^2 + 64\left(1+\rho_1^2\right)^2 \rho_2^4}} - \frac{\zeta\,\mathrm{d}\theta_2}{2\sqrt{\zeta^2 + 64\left(1+\rho_1^2\right)^2 \rho_2^4}} \tag{8.9.59}$$

where:

$$z_{1,2} = \exp\left[\mathrm{i}\,\theta_{1,2}\right] \rho_{1,2} \tag{8.9.60}$$

are the standard complex coordinates labeling the points of the locus L_Γ, namely parametrizing the two matrices A, B that solve the invariance constraint of Γ, defining $\mathrm{Hom}_\Gamma(\mathscr{Q} \times R, R)$, and are also diagonal in the natural basis of the regular representation. In the split basis they turn out to be antidiagonal:

$$A = \begin{pmatrix} 0 & Z_1 \\ Z_1 & 0 \end{pmatrix} \quad ; \quad B = \begin{pmatrix} 0 & Z_2 \\ Z_2 & 0 \end{pmatrix} \tag{8.9.61}$$

By means of the usual correspondence between U(1) bundles and line bundles we conclude that this connection $\boldsymbol{A}$ is the imaginary part of the connection theta of the corresponding bundle and we write the equation:

$$\theta = \mathfrak{H}^{-1}\partial\mathfrak{H} \tag{8.9.62}$$

where the explicit solution of the algebraic moment map equations yields:

$$\mathfrak{H} = \frac{\sqrt[4]{\frac{\zeta+\sqrt{\zeta^2+16\left|Z_1^2\right|^2+\left|Z_2^2\right|^2}}{|Z_1|^2+|Z_2|^2}}}{\sqrt{2}} \tag{8.9.63}$$

Curvature of the Line Bundle

In this way we find that the tautological bundle has the following curvature:

$$\Theta = \bar{\partial}\partial\mathrm{Log}[\mathfrak{H}] \tag{8.9.64}$$

Θ is the first Chern class of the tautological line bundle implicitly defined by the above construction

$$\begin{aligned} &\mathscr{L} \xrightarrow{\pi} \mathscr{M}_\zeta \\ &c_1(\mathscr{L}) = \left[\frac{i}{2\pi}\Theta\right] \in H^{1,1}\left(\mathscr{M}_\zeta\right) \end{aligned} \tag{8.9.65}$$

where $H^{1,1}\left(\mathscr{M}_\zeta\right)$ is the (1,1) cohomology group of the manifold $\mathscr{M}_\zeta$. On the other hand the very space of Eguchi–Hanson $\mathscr{M}_\zeta$ is a line bundle over $\mathbb{P}_1$:

$$\mathscr{M}_\zeta \xrightarrow{\pi_0} \mathbb{P}_1 \tag{8.9.66}$$

There is a (1,1)-form ω over $\mathbb{P}_1$ which is the the first Chern class of the bundle $\mathscr{M}_\zeta$.

$$c_1\left(\mathscr{M}_\zeta\right) = \omega \in H^{1,1}\left(\mathbb{P}_1\right) \tag{8.9.67}$$

We find that, as usual the pullback $\pi_0{}^*$of the projection π_0 works in particular as follows:

$$\pi_0{}^* : T_{(1,1)}{}^*\mathbb{P}_1 \longrightarrow T_{(1,1)}{}^*\mathscr{M}_\xi \tag{8.9.68}$$

We find that the (1,1)-form Θ which is defined over the whole $\mathcal{M}_\zeta$ is the pullback image of the first Chern class of the line bundle $\mathcal{M}_\zeta$.

$$\pi_0{}^* \left[c_1\left(\mathcal{M}_\zeta\right)\right] = c_1(\mathcal{L}) \tag{8.9.69}$$

The line bundle $\mathcal{M}_\zeta \xrightarrow{\pi_0} \mathbb{P}_1$ is by definition the one associated with the vanishing locus of the section ξ_2.

What We Have Learned from this Explicit Case?

The above detailed analysis reveals that, according to general lore, the cohomology classes constructed as first Chern classes of the tautological holomorphic vector bundles defined by the Kähler quotient via hermitian matrices $\mathfrak{H}_i$, are naturally associated with the components of the exceptional divisor. This latter is defined as the vanishing locus of a global holomorphic section $W(p)$ of a line bundle:

$$\begin{aligned} &\mathcal{L}_{\mathfrak{D}} \xrightarrow{\pi} \mathcal{M}_\zeta \\ &\mathfrak{D} \subset \mathcal{M}_\zeta \quad ; \quad \mathfrak{D} = \left\{p \in \mathcal{M}_\zeta \mid W(p) = 0 \quad \text{where} \quad W \in \Gamma\left(\mathcal{L}_{\mathfrak{D}}\right)\right\} \end{aligned} \tag{8.9.70}$$

The line bundle $\mathcal{L}_{\mathfrak{D}}$ is singled out by the divisor $\mathfrak{D}$ and for this reason it is labeled by it. Its first Chern class $\omega_{\mathfrak{D}}^{(1,1)}$ is certainly a cohomology class and so it must be a linear combination of the first Chern classes $\omega_i^{(1,1)}$ created by the Kähler quotient and associated with the hermitian matrices $\mathfrak{H}_i(\zeta, p)$:

$$\left[\omega_{\mathfrak{D}}^{(1,1)}\right] = \mathrm{S}_{\mathfrak{D},i} \left[\omega_i^{(1,1)}\right] \tag{8.9.71}$$

The question is to know which is which and to determine the constant matrix $\mathrm{S}_{\mathfrak{D},i}$.

Another point revealed by the analysis of the Eguchi–Hanson case is that, at least locally, the entire space $\mathcal{M}_\zeta$ can be viewed as the total space of a line bundle over the divisor $\mathfrak{D}$:

$$\begin{aligned} &\mathcal{M}_\zeta \xrightarrow{\pi_d} \mathfrak{D} \\ &\forall p \in \mathfrak{D} \quad ; \quad \pi_d^{-1}(p) \simeq \mathbb{C}^\star \end{aligned} \tag{8.9.72}$$

Furthermore the matrix $\mathfrak{H}_i$ can be viewed as the invariant norm of a section of the appropriate line bundle:

$$\mathfrak{H}_i(\zeta, z, \bar{z}) = H_i(\xi, \bar{\xi}, W, \bar{W}) \, |W|^2 \tag{8.9.73}$$

where ξ denote the two coordinates spanning the divisor $\mathfrak{D}$ and W (as in Fig. 8.9) spans the vertical fibers out of the divisor. The projection π_d corresponds to setting $W \to 0$ and obtaining:

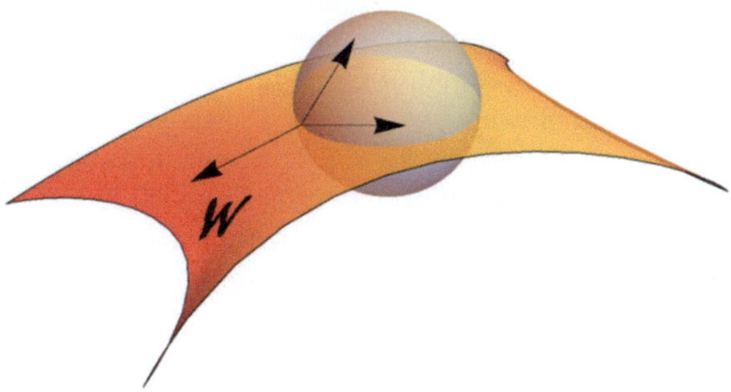

Fig. 8.9 In the Eguchi–Hanson case the exceptional divisor is a submanifold $\mathfrak{D} \subset \mathcal{M}_\zeta$ of codimension one that is mapped into the singular point by the resolving morphism $\mathcal{M}_\zeta \longrightarrow \frac{\mathbb{C}^2}{\Gamma}$. There is a projection operation $\mathcal{M}_\zeta \xrightarrow{\pi} \mathfrak{D}$ that makes $\mathcal{M}_\zeta$ the total space of a line bundle over the divisor. Accordingly we can choose a local coordinate system for $\mathcal{M}_\zeta$ such that two coordinates span the divisor while the third, named W, transforms as if it were a section of the mentioned line bundle

$$\pi_d \;:\; H(\xi, \bar{\xi}, W, \bar{W}) \longrightarrow h(\xi, \bar{\xi}) \equiv H(\xi, \bar{\xi}, 0, 0) \tag{8.9.74}$$

Just as in the case of Eguchi–Hanson, we expect that the two (1,1)-forms:

$$\begin{aligned} \Omega_i &= \overline{\partial}\partial H_i(\xi, \bar{\xi}, W, \bar{W}) \\ \widehat{\Omega}_i &= \overline{\partial}\partial h(\xi, \bar{\xi}) \end{aligned} \tag{8.9.75}$$

should be cohomologous:

$$[\Omega_i] = \left[\widehat{\Omega}_i\right] \tag{8.9.76}$$

The form $\widehat{\Omega}_i$ is the first Chern class of the line bundle (8.9.72) while Ω_i is the first Chern class of the line bundle (8.9.70) that defines the divisor.

Divisors and Conjugacy Classes Graded by Age

Hence the question boils down to the following: *What are the components of the exceptional divisor of a crepant resolution of the singularity $\mathbb{C}^3/\Gamma$, and how many are they?* The answer is provided by Theorem 8.3.1 (Theorem 1.6 in [46]); they are the inverse images via the blowdown morphism of the irreducible components of the fixed locus of the action of Γ on $\mathbb{C}^3$, and are in a one-to-one correspondence with the junior conjugacy classes of Γ. The irreducible components of the exceptional divisor may be compact (corresponding to a component of the fixed locus which is just the origin of $\mathbb{C}^3$) or noncompact (corresponding to fixed loci of higher dimensions, i.e., curves).

Let us consider the case of a cyclic group Γ, with only the origin as fixed locus, and choose a generator γ of Γ of order r. As in Eq. (8.3.7), we can write $\gamma = \frac{1}{r}(a_1, a_2, a_3)$. As described in [46], Sects. 2.3 and 2.4, the resolution of singularities is obtained by iterating the following construction, which uses toric geometry (a general reference for toric geometry, which in particular explains how to perform a toric blowup by subdividing the fan of the toric variety one wants to blowup, is [56]). The fan of the toric variety $\mathbb{C}^3$ is the first octant of $\mathbb{R}^3$, with all its faces; by adding the ray $\frac{1}{r}(a_1, a_2, a_3)$ we perform a blowup $\mathbb{B}_{[a_1,a_2,a_3]} \to \mathbb{C}^3$ whose exceptional divisor F is the weighted projective space $\mathbb{WP}[a_1, a_2, a_3]$. The same procedure applied to $\mathbb{C}^3/\Gamma$ produces a partial desingularization $W_\gamma \to \mathbb{C}^3/\Gamma$ which is the base of a cyclic covering $\mathbb{B}_{[a_1,a_2,a_3]} \to W_\gamma$, ramified along the exceptional divisor E of $W_\gamma \to \mathbb{C}^3/\Gamma$. The situation is summarised by the following diagram

$$\begin{array}{ccccc} F & \hookrightarrow & \mathbb{B}_{[a_1,a_2,a_3]} & \xrightarrow{\text{weighted blowup}} & \mathbb{C}^3 \\ \downarrow & & \downarrow & & \downarrow \\ E & \hookrightarrow & W_\gamma & \xrightarrow{\text{crepant resolution}} & \mathbb{C}^3/\Gamma \end{array} \quad . \tag{8.9.77}$$

The full desingularization is obtained by performing a multiple toric blowup, adding all rays corresponding to junior conjugacy classes.

8.10 Analysis of the (1,1)-Forms: Irreps Versus Conjugacy Classes that is Cohomology Versus Homology

In the present section we plan to analyze in full detail, within the scope of a one junior class model, the relation between the above extensively discussed $\omega_\alpha^{(1,1)}$ forms ($\alpha = 1, \ldots, r = \#$ of nontrivial irreps), with the exceptional divisors generated by the blowup of the singularity, together with the other predictions of the fundamental Theorem 8.3.1 which associates cohomology classes of $\mathscr{M}_\zeta$ with conjugacy classes of Γ. The number of nontrivial conjugacy classes and the number of nontrivial irreps are equal to each other so that we use r in both cases, yet what is the actual pairing is not clear a priori and it is not intrinsic to group theory, as we have stressed several times. In this section we want to explore this pairing and to do that in an explicit way we need explicit calculable examples. These are very few because of the bottleneck constituted by the solution of the moment map equations, that are algebraic of higher degree and only seldom admit explicit analytic solutions. For this reason we introduce here the full-fledged construction of one of those rare examples, where the moment map equations are solved in terms of radicals. As anticipated above this model has the additional nice feature that the number of junior conjugacy classes is just one. It will be the master model for our explicit analysis.

It is also important to stress that aim of the Kronheimer-like construction is not only the calculation of cohomology but also the actual determination of the Kähler potential (yielding the Kähler metric), which is encoded in formula (8.9.32). From this point of view one of the $\mathrm{Det}\mathfrak{H}_i$ may lead to a corresponding $\omega_i^{(1,1)} = \frac{i}{2\pi}\bar{\partial}\partial\mathrm{Det}\mathfrak{H}_i$ that is either exact or cohomologous to another one, yet its contribution to the Kähler potential, which is very important in physical applications, cannot be neglected. It is only the cohomology class of the Kähler 2-form that is affected by the triviality of one or more of the $\omega_i^{(1,1)}$; the contributions to the Kähler potential that give rise to exact form deformations of the Kähler 2-form are equally important as others.

Having anticipated these general considerations we turn to our master model.

8.10.1 The Master Model $\frac{\mathbb{C}^3}{\Gamma}$ with Generator $\{\xi, \xi, \xi\}$

In this section we develop all the calculations for the Kähler quotient resolution of the quotient singularity $\frac{\mathbb{C}^3}{\mathbb{Z}_3}$ in the case where the generator Y of $\mathbb{Z}_3$ is of the following form:

$$Y = \begin{pmatrix} \xi & 0 & 0 \\ 0 & \xi & 0 \\ 0 & 0 & \xi \end{pmatrix} \tag{8.10.1}$$

ξ being a primite cubic root of unity $\xi^3 = 1$.

The equation $p \wedge p = 0$ which is a set of quadrics has solutions arranged in various branches. There is a unique, principal branch of the solution that has maximal dimension $\mathscr{D}_\Gamma^0$ and is indeed isomorphic to the $\mathscr{G}_\Gamma$ orbit of the singular locus L_Γ. This principal branch is the algebraic variety $\mathbb{V}_{|\Gamma|+2}$ mentioned in Eq. (8.2.24), of which we perform the Kähler quotient with respect to the group $\mathscr{F}_\Gamma$

$$\mathscr{F}_\Gamma = \bigotimes_{\mu=1}^{r+1} \mathrm{U}\left(n_\mu\right) \cap \mathrm{SU}(|\Gamma|) = \mathrm{U}(1) \otimes \mathrm{U}(1) \tag{8.10.2}$$

in order to obtain the crepant resolution together with its Kähler metric. In the above formula $n_\mu = \{1,1,1\}$ are the dimensions of the irreducible representations of $\Gamma = \mathbb{Z}_3$ and $\mathrm{r} + 1 = 3$ is the number of conjugacy classes of the group (r is the number of nontrivial representations).

To make a long story short, exactly as in the Kronheimer case we are able to retrieve the algebraic equation of the singular locus from traces and determinants of the quiver matrices restricted to L_Γ. Precisely for the $\mathbb{Z}_3$ case under consideration we obtain

$$\mathscr{I}_1 = \mathrm{Det}\left[A_o\right]; \; \mathscr{I}_2 = \mathrm{Det}\left[B_o\right]; \; \mathscr{I}_3 = \mathrm{Det}\left[C_o\right]; \; \mathscr{I}_4 = \frac{1}{3}\mathrm{Tr}\left[A_o B_o C_o\right] \tag{8.10.3}$$

and we find the relation

$$\mathscr{I}_1 \mathscr{I}_2 \mathscr{I}_3 = \mathscr{I}_4{}^3 \tag{8.10.4}$$

which reproduces the $\mathbb{C}^3$ analogue of Eqs. (8.5.33)–(8.5.35) applying to the $\mathbb{C}^2$ case of Kronheimer and Arnold.

The main difference, as we have several time observed, is that now the same equations remain true, with no deformation for the entire $\mathscr{G}_\Gamma = \mathbb{C}^* \times \mathbb{C}^*$, orbit of the locus L_Γ, namely for the entire $\mathbb{V}_{|\Gamma|+2} = \mathbb{V}_5$ variety of which we construct the Kähler quotient with respect to the compact subgroup $\mathrm{U}(1) \times \mathrm{U}(1) \subset \mathbb{C}^* \times \mathbb{C}^*$. This is in line with the many times emphasized feature that in the $\mathbb{C}^3$ case there is no deformation of the complex structure.

8.10.1.1 The Actual Calculation of the Kähler Quotient and of the Kähler Potential

The calculation of the final form of the Kähler potential is reduced to the solution of a set of two algebraic equations. The solutions of such equations are accessible in this particular case since they reduce to a single cubic for which we have Cardano's formula. For this reason the present case is the three-dimensional analogue of the Eguchi–Hanson space where everything is explicitly calculable and all theorems admit explicit testing and illustration.

By calculating the ages we determine the number of $\omega^{(q,q)}$ harmonic forms (where $q = 1, 2$). According to Theorem 8.3.1 all these forms (and their dual cycles in homology) should be in one-to-one correspondence with the r nontrivial conjugacy classes of Γ. On the other hand the Kähler quotient construction associates one level parameter ζ to each generator of the center $\mathfrak{z}(\mathscr{F}_\Gamma)$ of the group $\mathscr{F}_\Gamma$, two ζs in this case, that are in one-to-one correspondence with the r nontrivial irreducible representation of Γ. The number is the same, but what is the pairing between **irreps** and **conjugacy classes**? More precisely how do we see the homology cycles that are created when each of the r level parameters ζ departs from its original zero value? Using the explicit expression of the functions $\mathfrak{H}_{1,2}$ defined in Eqs. (8.9.26)–(8.9.31) we arrive at the calculation of the $\omega^{(1,1)}{}_{i=1,2}$ forms that encode the first Chern classes of the two tautological bundles. The expectation from the age argument is that these two 2-forms should be cohomologous corresponding to just the unique predicted class of type (1,1) since $h^{1,1}$=1. On the other hand we should be able to construct an $\omega^{(2,2)}$ form representing the unique class that is Poincarè dual to the exceptional divisor.

In this case we can successfully answer both questions and this is very much illuminating.

Ages

Indeed taking the explicit generator

$$Y = \begin{pmatrix} (-1)^{2/3} & 0 & 0 \\ 0 & (-1)^{2/3} & 0 \\ 0 & 0 & (-1)^{2/3} \end{pmatrix} \tag{8.10.5}$$

we easily calculate the $\{a_1, a_2, a_3\}$ vectors respectively associated to each of the three conjugacy classes and we obtain:

$$a - \text{vectors} = \{\{0, 0, 0\}, \tfrac{1}{3}\{1, 1, 1\}, \tfrac{1}{3}\{2, 2, 2\}\} \tag{8.10.6}$$

from which we conclude that, apart from the class of the identity, there is just one junior and one senior class.

Hence we conclude that the Hodge numbers of the resolved variety should be $h^{(0,0)} = 1$; $h^{(1,1)} = 1$; $h^{(2,2)} = 1$.

If we follow the weighted blowup procedure described in [39] using the weights of the unique junior class $\{1, 1, 1\}$, we see that the bundle projection π yields

$$\pi : \mathbb{B}_{(1,1,1)} \longrightarrow \mathbb{WP}_{(1,1,1)} \sim \mathbb{P}^2 \tag{8.10.7}$$

So the blowup replaces the singular point $0 \in \mathbb{C}^3$ with a $\mathbb{P}^2$, which is compact. As a result, also the exceptional divisor in the resolution $\mathscr{M}_\zeta$ is compact. By Poincaré duality this entrains the existence of a harmonic (2,2)-form associated with the unique senior class.

8.10.1.2 The Quiver Matrix

In this case, the quiver matrix defined by Eq. (8.6.1) is the following one :

$$A_{ij} = \begin{pmatrix} 0 & 3 & 0 \\ 0 & 0 & 3 \\ 3 & 0 & 0 \end{pmatrix} \tag{8.10.8}$$

and it has the graphical representation displayed in Fig. 8.10

8.10.1.3 The Space $\mathscr{S}_\Gamma = \mathrm{Hom}_\Gamma(\mathscr{Q} \otimes R, R)$ in the Natural Basis

Solving the invariance constraints (8.6.7) in the natural basis of the regular representation we find the triples of matrices {A,B,C} spanning the locus $\mathscr{S}_\Gamma$. They are as follows:

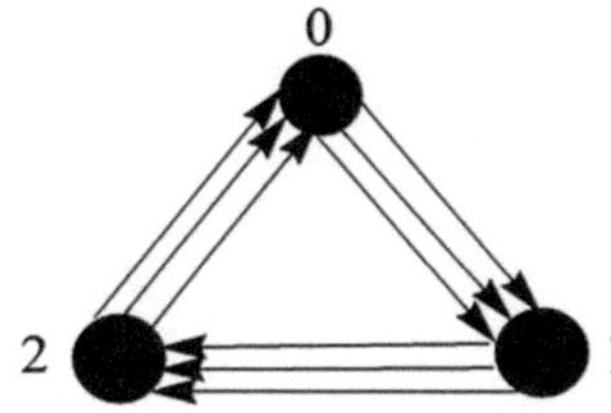

Fig. 8.10 The quiver diagram of the cyclic group with generator $Y = \mathrm{diag}\{\xi, \xi.\xi\}$

$$A = \begin{pmatrix} \alpha_{1,1} & \alpha_{1,2} & \alpha_{1,3} \\ (-1)^{2/3}\alpha_{1,3} & (-1)^{2/3}\alpha_{1,1} & (-1)^{2/3}\alpha_{1,2} \\ -(-1)^{1/3}\alpha_{1,2} & -(-1)^{1/3}\alpha_{1,3} & -(-1)^{1/3}\alpha_{1,1} \end{pmatrix}$$
$$B = \begin{pmatrix} \beta_{1,1} & \beta_{1,2} & \beta_{1,3} \\ (-1)^{2/3}\beta_{1,3} & (-1)^{2/3}\beta_{1,1} & (-1)^{2/3}\beta_{1,2} \\ -(-1)^{1/3}\beta_{1,2} & -(-1)^{1/3}\beta_{1,3} & -(-1)^{1/3}\beta_{1,1} \end{pmatrix}$$
$$C = \begin{pmatrix} \gamma_{1,1} & \gamma_{1,2} & \gamma_{1,3} \\ (-1)^{2/3}\gamma_{1,3} & (-1)^{2/3}\gamma_{1,1} & (-1)^{2/3}\gamma_{1,2} \\ -(-1)^{1/3}\gamma_{1,2} & -(-1)^{1/3}\gamma_{1,3} & -(-1)^{1/3}\gamma_{1,1} \end{pmatrix} \quad (8.10.9)$$

The Locus L_Γ

The locus $L_\Gamma \subset \mathscr{S}_\Gamma$ is easily described by the equation:

$$A_0 = \begin{pmatrix} \alpha_{1,1} & 0 & 0 \\ 0 & (-1)^{2/3}\alpha_{1,1} & 0 \\ 0 & 0 & -(-1)^{1/3}\alpha_{1,1} \end{pmatrix}$$
$$B_0 = \begin{pmatrix} \beta_{1,1} & 0 & 0 \\ 0 & (-1)^{2/3}\beta_{1,1} & 0 \\ 0 & 0 & -(-1)^{1/3}\beta_{1,1} \end{pmatrix}$$
$$C_0 = \begin{pmatrix} \gamma_{1,1} & 0 & 0 \\ 0 & (-1)^{2/3}\gamma_{1,1} & 0 \\ 0 & 0 & -(-1)^{1/3}\gamma_{1,1} \end{pmatrix} \quad (8.10.10)$$

8.10.1.4 The Space $\mathscr{S}_\Gamma$ in the Split Basis

Solving the invariance constraints in the split basis of the regular representation we find another representation of the triples of matrices $\{A, B, C\}$ that span the space $\mathscr{S}_\Gamma$. They are as follows:

$$A = \begin{pmatrix} 0 & 0 & m_{1,3} \\ m_{2,1} & 0 & 0 \\ 0 & m_{3,2} & 0 \end{pmatrix}$$
$$B = \begin{pmatrix} 0 & 0 & n_{1,3} \\ n_{2,1} & 0 & 0 \\ 0 & n_{3,2} & 0 \end{pmatrix}$$
$$C = \begin{pmatrix} 0 & 0 & r_{1,3} \\ r_{2,1} & 0 & 0 \\ 0 & r_{3,2} & 0 \end{pmatrix} \quad (8.10.11)$$

8.10.1.5 The Equation $p \wedge P = 0$ and the Characterization of the Variety $\mathbb{V}_5 = \mathscr{D}_\Gamma$

Here we are concerned with the solution of Eq. (8.6.12) and the characterization of the locus $\mathscr{D}_\Gamma$.

Differently from the more complicated cases of larger groups, in the present abelian case of small order, we can explicitly solve the quadratic equations provided by the commutator constraints and we discover that there is a principal branch of the solution, named $\mathscr{D}^0_\Gamma$ that has indeed dimension $5 = |\Gamma|+2$. In addition however there are several other branches with smaller dimension. These branches describe different components of the locus $\mathscr{D}_\Gamma$. Actually as already pointed out they are all contained in the $\mathscr{G}_\Gamma$ orbit of the subspace L_Γ defined above. The quadratic equations defining $\mathscr{D}_\Gamma$ have a set of 14 different solutions realized by a number n_i of constraints on the 9 parameters. Hence there are 14 branches $\mathscr{D}^i_\Gamma$(i = 0, 1, …, 16) of dimensions:

$$\dim_{\mathbb{C}} \mathscr{D}^i_\Gamma = 9 - n_i \tag{8.10.12}$$

The full dimension table of these branches is displayed below

$$\{5, 4, 4, 4, 4, 3, 3, 3, 3, 3, 3, 3, 2, 2\}$$

As we see, there is a unique branch that has the maximal dimension $5 = |\mathbb{Z}_3| + 2$. This is the principal branch $\mathscr{D}^0_\Gamma$. It can be represented by the substitution:

$$n_{2,1} \to \frac{m_{2,1}n_{1,3}}{m_{1,3}}, \quad n_{3,2} \to \frac{m_{3,2}n_{1,3}}{m_{1,3}}, \quad r_{2,1} \to \frac{m_{2,1}r_{1,3}}{m_{1,3}}, \quad r_{3,2} \to \frac{m_{3,2}r_{1,3}}{m_{1,3}} \tag{8.10.13}$$

In this way we have reached a complete resolution of the following problem. We have an explicit parametrization of the variety $V_{|\Gamma|+2}$. This variety is described by the following three matrices depending on the 5 complex variables ω_i (i = 1, …, 5):

$$A = \begin{pmatrix} 0 & 0 & \omega_1 \\ \omega_2 & 0 & 0 \\ 0 & \omega_3 & 0 \end{pmatrix}$$
$$B = \begin{pmatrix} 0 & 0 & \omega_4 \\ \frac{\omega_2\omega_4}{\omega_1} & 0 & 0 \\ 0 & \frac{\omega_3\omega_4}{\omega_1} & 0 \end{pmatrix}$$
$$C = \begin{pmatrix} 0 & 0 & \omega_5 \\ \frac{\omega_2\omega_5}{\omega_1} & 0 & 0 \\ 0 & \frac{\omega_3\omega_5}{\omega_1} & 0 \end{pmatrix} \tag{8.10.14}$$

8.10.1.6 The Quiver Group

Our next point is the derivation of the group $\mathcal{G}_\Gamma$ defined in Eqs. (8.6.17) and (8.6.18), namely:

$$\mathcal{G}_\Gamma = \left\{ g \in \mathrm{SL}(|\Gamma|, \mathbb{C}) \,\middle|\, \forall \gamma \in \Gamma : \left[D_R(\gamma), D_{\mathrm{def}}(g) \right] = 0 \right\} \tag{8.10.15}$$

Let us proceed to this construction. In the diagonal basis of the regular representation this is a very easy task, since the group is simply given by the diagonal 3×3 matrices with determinant one. We introduce such matrices

$$\mathfrak{g} \in \mathcal{G}_\Gamma \quad : \quad \mathfrak{g} = \begin{pmatrix} a_1 & 0 & 0 \\ 0 & a_2 & 0 \\ 0 & 0 & a_3 \end{pmatrix} \tag{8.10.16}$$

8.10.1.7 $\mathbb{V}_5$ as the Orbit Under $\mathcal{G}_\Gamma$ of the Locus L_Γ

In this section we want to verify and implement Eq. (8.2.24), namely we aim at showing that $\mathbb{V}_5 = \mathcal{D}_\Gamma = \mathrm{Orbit}_{\mathcal{G}_\Gamma}(L_\Gamma)$. To this effect we rewrite the locus L_Γ in the diagonal split basis of the regular representation. The change of basis is performed by the character table of the cyclic group $\mathbb{Z}_3$. The result is displayed below:

$$\begin{aligned} A_0 &= \begin{pmatrix} 0 & 0 & \alpha_{1,1} \\ \alpha_{1,1} & 0 & 0 \\ 0 & \alpha_{1,1} & 0 \end{pmatrix} \\ B_0 &= \begin{pmatrix} 0 & 0 & \beta_{1,1} \\ \beta_{1,1} & 0 & 0 \\ 0 & \beta_{1,1} & 0 \end{pmatrix} \\ C_0 &= \begin{pmatrix} 0 & 0 & \gamma_{1,1} \\ \gamma_{1,1} & 0 & 0 \\ 0 & \gamma_{1,1} & 0 \end{pmatrix} \end{aligned} \tag{8.10.17}$$

Eventually the complex parameters

$$z_1 \equiv \alpha_{1,1}; \quad z_2 \equiv \beta_{1,1}; \quad z_3 \equiv \gamma_{1,1} \tag{8.10.18}$$

will be utilized as complex coordinates of the resolved variety when the level parameters $\zeta_{1,2}$ are switched on. Starting from the above the orbit is given by:

$$\mathrm{Orbit}_{\mathcal{G}_\Gamma} \equiv \left\{ \left\{ \mathfrak{g}\, A_0\, \mathfrak{g}^{-1}, \mathfrak{g}\, B_0\, \mathfrak{g}^{-1}, \mathfrak{g}\, C_0\, \mathfrak{g}^{-1} \right\} \;\middle|\; \begin{array}{c} \forall\, \mathfrak{g} \in \mathcal{G}_\Gamma \\ \forall \{A_0, B_0, C_0\} \in L_\Gamma \end{array} \right\} \supset \mathcal{D}^0_\Gamma \tag{8.10.19}$$

and the correspondence between the parameters of the principal branch $\mathscr{D}^0_\Gamma$ and the parameters spanning $\mathscr{G}_\Gamma$ and L_Γ is provided below:

$$\begin{aligned} a_1 &\to \frac{\omega_2^{1/3}}{\omega_1^{1/3}},\ a_2 \to \frac{\omega_3^{1/3}}{\omega_2^{1/3}},\ a_3 \to \frac{\omega_1^{1/3}}{\omega_3^{1/3}} \\ z_1 &\to \omega_1^{1/3}\omega_2^{1/3}\omega_3^{1/3} \\ z_2 &\to \frac{\omega_2^{1/3}\omega_3^{1/3}\omega_4}{\omega_1^{2/3}},\ z_3 \to \frac{\omega_2^{1/3}\omega_3^{1/3}\omega_5}{\omega_1^{2/3}} \end{aligned} \tag{8.10.20}$$

Branches of smaller dimension of the solution are all contained in the $\text{Orbit}_{\mathscr{G}_\Gamma}(L_\Gamma)$ and correspond to the orbits of special points of L_Γ where some of the z_i vanish or satisfy special relations among themselves. Hence, indeed we have:

$$\text{Orbit}_{\mathscr{G}_\Gamma} = \mathscr{D}_\Gamma$$

8.10.1.8 The Compact Gauge Group $\mathscr{F}_\Gamma = \mathrm{U}(1)^2$

We introduce a basis for the generators of the compact subgroup $\mathrm{U}(1)^2 = \mathscr{F}_\Gamma \subset \mathscr{G}_\Gamma$ provided by the set of two generators displayed here below

$$T^1 = \begin{pmatrix} i & 0 & 0 \\ 0 & -i & 0 \\ 0 & 0 & 0 \end{pmatrix} \quad ; \quad T^2 = \begin{pmatrix} 0 & 0 & 0 \\ 0 & i & 0 \\ 0 & 0 & -i \end{pmatrix} \tag{8.10.21}$$

whose trace-normalization is the A_2 Cartan matrix

$$\mathrm{Tr}\left(T^i T^j\right) = \mathfrak{C}^{ij} = \begin{pmatrix} 2 & -1 \\ -1 & 2 \end{pmatrix} \tag{8.10.22}$$

8.10.1.9 Calculation of the Kähler Potential and of the Moment Maps

Naming Δ_i the moduli of the coordinates z_i and θ_i their phases according to $z_i = e^{i\theta_i}\Delta_i$ and considering a generic element $\mathfrak{g}_R$ of the quiver group that is real and hence is a representative of a coset class in $\frac{\mathscr{G}_\Gamma}{\mathscr{F}_\Gamma}$:

$$\mathfrak{g}_R = \begin{pmatrix} e^{\lambda_1} & 0 & 0 \\ 0 & e^{-\lambda_1+\lambda_2} & 0 \\ 0 & 0 & e^{-\lambda_2} \end{pmatrix} \quad ; \quad \lambda_{1,2} \in \mathbb{R} \tag{8.10.23}$$

The triple of matrices {A,B,C}=$\left\{\mathfrak{g}_R A_0 \mathfrak{g}_R{}^{-1}, \mathfrak{g}_R B_0 \mathfrak{g}_R{}^{-1}, \mathfrak{g}_R C_0 \mathfrak{g}_R{}^{-1}\right\}$ have the following explicit appearance:

$$A = \begin{pmatrix} 0 & 0 & e^{i\theta_1-\lambda_1-\lambda_2}\Delta_1 \\ e^{i\theta_1+2\lambda_1-\lambda_2}\Delta_1 & 0 & 0 \\ 0 & e^{i\theta_1-\lambda_1+2\lambda_2}\Delta_1 & 0 \end{pmatrix}$$
$$B = \begin{pmatrix} 0 & 0 & e^{i\theta_2-\lambda_1-\lambda_2}\Delta_2 \\ e^{i\theta_2+2\lambda_1-\lambda_2}\Delta_2 & 0 & 0 \\ 0 & e^{i\theta_2-\lambda_1+2\lambda_2}\Delta_2 & 0 \end{pmatrix}$$
$$C = \begin{pmatrix} 0 & 0 & e^{i\theta_3-\lambda_1-\lambda_2}\Delta_3 \\ e^{i\theta_3+2\lambda_1-\lambda_2}\Delta_3 & 0 & 0 \\ 0 & e^{i\theta_3-\lambda_1+2\lambda_2}\Delta_3 & 0 \end{pmatrix} \quad (8.10.24)$$

Calculating the Kähler potential we find

$$\begin{aligned} \mathscr{K}_{\mathscr{S}}|_{\mathscr{D}} &= \left(\mathrm{Tr}\left[A\ A^\dagger\right] + \mathrm{Tr}\left[B\ B^\dagger\right] + \mathrm{Tr}\left[C\ C^\dagger\right]\right) \\ &= e^{-2(\lambda_1+\lambda_2)}\left(1 + e^{6\lambda_1} + e^{6\lambda_2}\right)\left(\Delta_1^2 + \Delta_2^2 + \Delta_3^2\right) \end{aligned} \quad (8.10.25)$$

We have used the above notation since $\mathrm{Tr}\left[A, A^\dagger\right] + \mathrm{Tr}\left[B, B^\dagger\right] + \mathrm{Tr}\left[C, C^\dagger\right]$ is the Kähler potential of the ambient space $\mathscr{S}_\Gamma$ restricted to the orbit $\mathscr{D}_\Gamma$. Indeed since $\mathscr{F}_\Gamma$ is an isometry of $\mathscr{S}_\Gamma$, the dependence in $\mathscr{K}_{\mathscr{S}}|_{\mathscr{D}}$ is only on the real part of the quiver group, namely on the real factors $\lambda_{1,2}$. Just as it stands, $\mathscr{K}_{\mathscr{S}}|_{\mathscr{D}}$ cannot work as Kähler potential of a complex Kähler metric. Yet, when the real factors $\lambda_{1,2}$ will be turned into functions of the complex coordinates z_i, then $\mathscr{K}_{\mathscr{S}}|_{\mathscr{D}}$ will be enabled to play the role of a contribution to the Kähler potential of the resolved manifold $\mathscr{M}_\zeta$.

Next we calculate the moment maps according to the formulas:

$$\begin{aligned} \mathfrak{P}^1 &\equiv -i\,\mathrm{Tr}\left[T^1\left(\left[A, A^\dagger\right] + \left[B, B^\dagger\right] + \left[C, C^\dagger\right]\right)\right] \\ &= e^{-2(\lambda_1+\lambda_2)}\left(1 - 2e^{6\lambda_1} + e^{6\lambda_2}\right)\left(\Delta_1^2 + \Delta_2^2 + \Delta_3^2\right) \\ \mathfrak{P}^2 &\equiv -i\,\mathrm{Tr}\left[T^2\left(\left[A, A^\dagger\right] + \left[B, B^\dagger\right] + \left[C, C^\dagger\right]\right)\right] \\ &= e^{-2(\lambda_1+\lambda_2)}\left(1 + e^{6\lambda_1} - 2e^{6\lambda_2}\right)\left(\Delta_1^2 + \Delta_2^2 + \Delta_3^2\right) \end{aligned} \quad (8.10.26)$$

8.10.1.10 Solution of the Moment Map Equations

In order to solve the moment map equations it is convenient to introduce the new variables

$$\Upsilon_{1,2} = \exp\left[2\,\lambda_{1,2}\right] \quad (8.10.27)$$

and to redefine the moment maps with indices lowered by means of the inverse of the Cartan matrix mentioned above

$$\mathfrak{P}_i = \left(\mathfrak{C}^{-1}\right)_{\mathrm{ij}}\mathfrak{P}^j \quad (8.10.28)$$

In this way imposing the level condition

$$\mathfrak{P}_i = -\zeta_i \tag{8.10.29}$$

where $\zeta_{1,2} > 0$ are the two level parameters, we obtain the final pair of algebraic equations for the factors $\Upsilon_{1,2}$

$$\left\{ \frac{\Sigma \left(-1+\Upsilon_1^3\right)}{\Upsilon_1 \Upsilon_2}, \frac{\Sigma \left(-1+\Upsilon_2^3\right)}{\Upsilon_1 \Upsilon_2} \right\} = \{\zeta_1, \zeta_2\} \tag{8.10.30}$$

where we have introduced the shorthand notation:

$$\Sigma = \sum_{i=1}^{3} |z_i|^2 \tag{8.10.31}$$

The above algebraic system composed of two cubic equations is simple enough in order to find all of its nine roots by means of Cardano's formula. The very pleasant property of these solutions is that one and only one of the nine branches satisfies the correct boundary conditions, namely provides real $\Upsilon_i(\zeta, \Sigma)$ that are positive for all values of Σ and ζ and reduce to 1 when $\zeta \to 0$.

The complete solution of the algebraic equations can be written in the following way. For the first factor we have:

$$\Upsilon_1 = \frac{1}{6^{1/3}} \left(\frac{N}{\Sigma^3 \Pi^{\frac{1}{3}}} \right)^{\frac{1}{3}} \tag{8.10.32}$$

where

$$\begin{aligned} N &= 2 \times 2^{1/3} \zeta_1^3 \zeta_2^2 + 6 \Sigma^3 \Pi^{\frac{1}{3}} + 2\zeta_1^2 \left(3 \times 2^{1/3} \Sigma^3 + \zeta_2 \Pi^{\frac{1}{3}}\right) \\ &\quad + \zeta_1 \left(6 \times 2^{1/3} \Sigma^3 \zeta_2 + 2^{2/3} \Pi^{\frac{2}{3}}\right) \\ \Pi &= 27\, \Sigma^6 + 9\, \Sigma^3\, \zeta_1^2 \zeta_2 + 9\, \Sigma^3\, \zeta_1 \zeta_2^2 + 2\, \zeta_1^3 \zeta_2^3 + 3\sqrt{3}\, \Sigma^3\, \mathfrak{R} \\ \mathfrak{R} &= \sqrt{27\, \Sigma^6 + 6\, \Sigma^3 \zeta_1 \zeta_2^2 - \zeta_1^4 \zeta_2^2 - 4\, \Sigma^3 \zeta_2^3 + \zeta_1^3 \left(-4\, \Sigma^3 + 2\zeta_2^3\right) + \zeta_1^2 \left(6\, \Sigma^3 \zeta_2 - \zeta_2^4\right)} \end{aligned} \tag{8.10.33}$$

For the second factor we have

$$\Upsilon_2 = \frac{-\frac{M^{8/3}}{\Sigma^5} + \frac{18 M^{5/3}}{\Sigma^2} - 72\, M^{2/3} \Sigma + 36 \left(\frac{M}{\Sigma^3}\right)^{2/3} \zeta_1^3 - 36 \left(\frac{M}{\Sigma^3}\right)^{2/3} \zeta_1^2 \zeta_2 + 6 \left(\frac{M}{\Sigma^3}\right)^{5/3} \zeta_1^2 \zeta_2}{36 \times 6^{2/3}\, \Sigma^2 \zeta_1} \tag{8.10.34}$$

where

$$M = \frac{6\, \Sigma^3 \Pi^{1/3} + 2^{2/3} \Pi^{2/3} \zeta_1 + 6 \times 2^{1/3} \Sigma^3\, \zeta_1^2 + 6 \times 2^{1/3} \Sigma^3\, \zeta_1 \zeta_2 + 2 \Pi^{1/3} \zeta_1^2 \zeta_2 + 2 \times 2^{1/3} \zeta_1^3 \zeta_2^2}{\Omega^{1/3}} \tag{8.10.35}$$

8.10.2 Discussion of Cohomology in the Master Model

Since the two scale factors $\Upsilon_{1,2}$ are functions only of Σ, the two (1,1)-forms, relative to the two tautological bundles, respectively associated with the first and second nontrivial irreps of the cyclic group, defined in Eq. (8.9.31) take the following general appearance:

$$\begin{aligned} \omega_{1,2}^{(1,1)} &= \frac{i}{2\pi}\left(\frac{d}{\mathrm{d}\,\Sigma}\mathrm{Log}\left[\Upsilon_{1,2}(\Sigma)\right] d\bar{z}^i \wedge \mathrm{dz}^i + \frac{d^2}{\mathrm{d}\,\Sigma^2}\mathrm{Log}\left[\Upsilon_{1,2}(\Sigma)\right] z^j \bar{z}^i \mathrm{dz}^i \wedge d\bar{z}^j\right) \\ &= \frac{i}{2\pi}\left(f_{1,2}\Theta + g_{1,2}\Psi\right) \end{aligned} \tag{8.10.36}$$

where we have introduced the short hand notation

$$\Theta = \sum_{i=1}^{3} d\bar{z}^i \wedge \mathrm{dz}^i \quad ; \quad \Psi = \sum_{i,j=1}^{3} z^j\, \bar{z}^i dz^i \wedge d\bar{z}^j \tag{8.10.37}$$

Indeed in the present case the fiber metrics $\mathfrak{H}_{1,2}$ are one-dimensional and given by $\mathfrak{H}_{1,2} = \sqrt{\Upsilon_{1,2}}$. The most relevant point is that the two functions $f_{1,2}$ and $g_{1,2}$ being the derivatives (first and second) of $\Upsilon_{1,2}$ depend only on the variable Σ.

It follows that a triple wedge product of the two-forms $\omega_a^{(1,1)}$ (a=1,2) has always the following structure:

$$\omega_a^{(1,1)} \wedge \omega_b^{(1,1)} \wedge \omega_b^{(1,1)} = \left(\frac{i}{2\,\pi}\right)^3 (f_a\, f_b\, f_c + 2\,\Sigma\,(g_a\, f_b\, f_c + g_b\, f_c\, f_a + g_c\, f_a\, f_b)) \times \mathrm{Vol} \tag{8.10.38}$$

where

$$\mathrm{Vol} = dz_1 \wedge dz_2 \wedge dz_3 \wedge d\bar{z}_1 \wedge d\bar{z}_2 \wedge d\bar{z}_3 \tag{8.10.39}$$

This structure enables us to calculate intersection integrals of the considered forms very easily. It suffices to change variables as we explain below. The equations

$$\Sigma = \sum_{i=1}^{3} |z_i|^2 = \rho^2 \tag{8.10.40}$$

define 5-spheres of radius ρ. Introducing the standard Euler angle parametrization of a 5-sphere, the volume form (8.10.39) reduces to:

$$\mathrm{Vol} = 8i\rho^5 \cos^4(\theta_1) \cos^3(\theta_2) \cos^2(\theta_3) \cos(\theta_4) \prod_{i=1}^{5} d\theta_i \tag{8.10.41}$$

The integration on the Euler angles can be easily performed and we obtain:

$$\prod_{i}^{4} \int_{\frac{\pi}{2}}^{\frac{\pi}{2}} d\theta_i \int_0^{2\pi} d\theta_5 \left(8 i \rho^5 \cos^4(\theta_1) \cos^3(\theta_2) \cos^2(\theta_3) \cos(\theta_4)\right) = 8 i \pi^3 \rho^5 \tag{8.10.42}$$

Hence defining the intersection integrals:

$$\mathscr{I}_{abc} = \int_{\mathscr{M}} \omega_a^{(1,1)} \wedge \omega_b^{(1,1)} \wedge \omega_b^{(1,1)} \tag{8.10.43}$$

we arrive at

$$\begin{aligned} \mathscr{I}_{\text{abc}} &= \left(\frac{i}{2\pi}\right)^3 \times 8 i \pi^3 \times \int_0^\infty \left(6\rho^5 f_a f_b f_c + 2\rho^7 (f_b f_c g_a + f_a f_c g_b + f_a f_b g_c)\right) d\rho \\ &= \int_0^\infty \left(6\rho^5 f_a f_b f_c + 2\rho^7 (f_b f_c g_a + f_a f_c g_b + f_a f_b g_c)\right) d\rho \end{aligned} \tag{8.10.44}$$

We have performed the numerical integration of these functions and we have found the following results

$$\begin{array}{l} (\zeta_1 > 0, \zeta_2 = 0) : \mathscr{I}_{111} = \frac{1}{8} \\ (\zeta_1 = 0, \zeta_2 \geq 0) : \mathscr{I}_{111} = 0 \\ (\zeta_1 > 0, \zeta_2 > 0) : \mathscr{I}_{111} = 1 \end{array} \tag{8.10.45}$$

From this we reach the following conclusion. Since the corresponding integral is nonzero it follows that:

$$\omega_S^{(2,2)} \equiv \omega_1^{(1,1)} \wedge \omega_1^{(1,1)} \tag{8.10.46}$$

is closed but not exact and by Poincaré duality it is the Poincaré dual of some cycle $S \in H_2(\mathscr{M})$ such that:

$$\int_S \iota^* \omega_1^{(1,1)} = \int_{\mathscr{M}} \omega_1^{(1,1)} \wedge \omega_S^{(2,2)} \tag{8.10.47}$$

where

$$\iota : S \longrightarrow \mathscr{M} \tag{8.10.48}$$

is the inclusion map. Since $H_c^2(\mathscr{M}) = H^2(\mathscr{M})$ and both have dimension 1 it follows that $\dim H_2(\mathscr{M}) = 1$, so that every nontrivial cycle S is proportional, (as homology class) via some coefficient α to a single cycle $\mathscr{C}$, namely we have $S = \alpha\, \mathscr{C}$. Then we can interpret Eq. (29) as follows

$$\int_{\alpha\mathscr{C}} \iota^* \omega_1^{(1,1)} = \alpha \int_{\mathscr{M}} \omega_1^{(1,1)} \wedge \omega_{\mathscr{C}}^{(2,2)} \tag{8.10.49}$$

If we choose as fundamental cycle, that one for which

$$\int_{\mathscr{C}} \iota^* \omega_1^{(1,1)} = 1 \tag{8.10.50}$$

we conclude that

$$\alpha = \begin{cases} 1 & \text{case } \{\zeta_1 > 0, \zeta_2 > 0\} \\ \frac{1}{8} & \text{case } \{\zeta_1 > 0, \zeta_2 = 0\} \end{cases} \tag{8.10.51}$$

Next we have calculated the intersection integral $\mathscr{I}_{211}$ and we have found:

$$\begin{array}{l} (\zeta_1 > 0, \zeta_2 = 0) : \mathscr{I}_{211} = 0 \\ (\zeta_1 = 0, \zeta_2 \geq 0) : \mathscr{I}_{211} = 0 \\ (\zeta_1 > 0, \zeta_2 > 0) : \mathscr{I}_{211} = 1 \end{array} \tag{8.10.52}$$

Conclusions on Cohomology

We have two cases.

case $\{\zeta_1 > 0, \zeta_2 > 0\}$. The first Chern classes of the two tautological bundles are cohomologous:

$$\left[\omega_1^{(1,1)}\right] = \left[\omega_2^{(1,1)}\right] = \left[\omega^{(1,1)}\right] \tag{8.10.53}$$

case $\{\zeta_1 > 0, \zeta_2 = 0\}$.] The first Chern class of the first tautological bundle is nontrivial and generates $H_c^{(1,1)}(\mathscr{M}) = H^{1,1}(\mathscr{M})$.

$$\left[\omega_1^{(1,1)}\right] = \text{nontrivial} \tag{8.10.54}$$

The first Chern class of the second tautological bundle is trivial, namely

$$\omega_2^{(1,1)} = \text{exact form} \tag{8.10.55}$$

Obviously since there is symmetry in the exchange of the first and second scale factors, exchanging $\zeta_1 \leftrightarrow \zeta_2$, the above conclusion is reversed in the case $\{\zeta_1 = 0, \zeta_2 > 0\}$.

In passing we have also proved that the unique (2,2)-class is just the square of the unique (1,1)-class

$$\left[\omega^{(2,2)}\right] = \left[\omega^{(1,1)}\right] \wedge \left[\omega^{(1,1)}\right] \tag{8.10.56}$$

8.10.2.1 The Exceptional Divisor

Finally let us discuss how we retrieve the exceptional divisor $\mathbb{P}^2$ predicted by the weighted blowup argument. As we anticipated in Eqs. (8.9.73)–(8.9.74), replacing the three coordinates z_i with

$$z_1 = W \quad ; \quad z_2 = W\,\xi_1 \quad ; \quad z_3 = W\,\xi_2 \tag{8.10.57}$$

which is the appropriate change for one of the three standard open charts of $\mathbb{P}^2$, we obtain

$$\mathfrak{H}_1(\Sigma) = \frac{1}{|W|^2 H_1(\xi, \bar{\xi}, W, \bar{W})} \quad (8.10.58)$$

where the function $H_1(\xi, \bar{\xi}, W, \bar{W})$ has the property that:

$$\lim_{W \to 0} \log[H_1(\xi, \bar{\xi}, W, \bar{W})] = -\log[1 + |\xi_1|^2 + |\xi_2|^2] + \log[\text{const}] \quad (8.10.59)$$

From the above result we conclude that the exceptional divisor $\mathfrak{D}^{(E)}$ is indeed the locus $W = 0$ and that on this locus the first Chern class of the first tautological bundle reduces to the Kähler 2-form of the Fubini–Study Kähler metric on $\mathbb{P}^2$. Indeed we can write:

$$c_1(\mathscr{L}_1)\,|_{\mathfrak{D}^{(E)}} = -\frac{i}{2\pi}\,\bar{\partial}\,\partial\,\log[1 + |\xi_1|^2 + |\xi_2|^2] \quad (8.10.60)$$

From this point of view this master example is the perfect three-dimensional analogue of the Eguchi–Hanson space, the $\mathbb{P}^1$ being substituted by a $\mathbb{P}^2$.

With this we conclude our long and detailed exposition of the Kronheimer construction for the resolution of $\mathbb{C}^2/\Gamma$ and $\mathbb{C}^3/\Gamma$ singularities. By now the reader should have accumulated enough insight in the simplicity of the principles and the calculation complexity of this beautiful branch of modern geometry.

8.11 Conclusion

In this chapter we have seen that the 2400 year old classification problem of platonic solids is still alive and able to produce very challenging modern fruits. We started in Chap. 1 with the diophantine equation that provides the ADE classification of finite rotation groups. In Sect. 1.5 we retrieved, via Dynkin diagrams, the same classification in terms of simply laced Lie algebras. In the present chapter we found a third incarnation of the same classification under the form of gravitational instantons associated with the resolution of singularities [57–59].

The relation between finite groups, Lie algebras and complex geometry have in the topics discussed in this chapter a most exciting illustration. Furthermore the profound role played by supersymmetry in bringing to the surface deep and unexpected connections is exemplified by the contents of the present chapter in a paradigmatic way.

Indeed one of the most fundamental question at stake in many problems of supergravity and superstring, in particular related with compactifications and with the AdS/CFT correspondence, is just the classical algebraic geometry problem of resolving quotient singularities. Under the inspired stimulus of supersymmetric theories a rich set of results were obtained by the mathematical community at the beginning of 1990s, those reviewed in this chapter being just the first ones in such a list.

Entering a more circumstantial analysis I have tried to emphasize where the catch of a such a stimulus is. The example of (Hyper)Kähler quotients is indeed paradigmatic. The whole story began from the physical interpretation of the mathematical notion of moment-map. Identifying the moment-maps with the auxiliary fields of supersymmetric gauge-theories new scenarios opened up. Extremization of the scalar potential, namely the physical problem of searching for classical vacua of a field-theory naturally produced the notion of (Hyper)Kähler quotient. It was once again a physical problem, that of instantons extended from gauge-theories to gravity, what motivated the consideration of ALE manifolds. Yet their construction as HyperKähler quotients would not have been possible without the further ingredient of the McKay correspondence. This latter came neither from physics nor from the solution of some mathematical problem posed in a standard way. It just came from that type of Interrogation of Mathematics rather than of Nature which I discussed in the twin book [60]. It is looking for traces of unexpected correspondences that sometimes we uncover the deeper nature of certain mathematical structures we have known for long time. As a result of such discoveries we usually open new scenarios not only for Mathematics but also for Physics, where identifications, such as that of the moment-maps with the auxiliary fields, become possible with far reaching consequences of the type highlighted above.

From these considerations it is evident how wide and deep is the extent of the fertilizing influence exerted by *supersymmetry* on the development of modern Geometry. I believe that this latter has entered a new season of expansion and progress that can, in the long run, lead to new conceptions in Physics.

References

1. N.J. Hitchin, A. Karlhede, U. Lindström, M. Roček, Hyperkahler metrics and supersymmetry. Commun. Math. Phys. **108**, 535 (1987)
2. P.B. Kronheimer, The construction of ALE spaces as hyper-Kähler quotients. J. Differ. Geom. **29**, 665–683 (1989)
3. P.B. Kronheimer, A Torelli-type theorem for gravitational instantons. J. Differ. 1 Geom. **29**, 685–697 (1989)
4. J. Maldacena, The large-N limit of superconformal field theories and supergravity. Int. J. Theor. Phys. **4**, 1113–1133 (1999). https://doi.org/10.1023/A:1026654312961, arXiv:hep-th/9711200
5. S. Ferrara, C. Fronsdal, Conformal maxwell theory as a singleton field theory on AdS_5, IIB 3-branes and duality. Class. Quantum Gravity **15**, 2153 (1998). https://doi.org/10.1088/0264-9381/15/8/004, arXiv:hep-th/9712239
6. S. Ferrara, C. Fronsdal, A. Zaffaroni, On $\mathcal{N} = 8$ supergravity in AdS_5 and $\mathcal{N} = 4$ superconformal Yang-Mills theory. Nucl. Phys. B **532**, 153–162 (1998). https://doi.org/10.1016/S0550-3213(98)00444-1, arXiv:hep-th/9802203
7. R. Kallosh, A. Van Proeyen, Conformal symmetry of supergravities in AdS spaces. Phys. Rev. D **60**, 026001 (1999). https://doi.org/10.1103/PhysRevD.60.026001, arXiv:hep-th/9804099
8. S. Ferrara, C. Fronsdal, Gauge fields as composite boundary excitations. Phys. Lett. B **433**, 19–28 (1998). https://doi.org/10.1016/S0370-2693(98)00664-9, arXiv:hep-th/9802126

9. P.G.O. Freund, M.A. Rubin, Dynamics of dimensional reduction. Phys. Lett. B **97**, 233–235 (1980)
10. D. Fabbri, P. Fré, L. Gualtieri, P. Termonia, M-theory on $AdS_4 \times M^{1,1,1}$: the complete $Osp(2|4) \times SU(3) \times SU(2)$ spectrum from harmonic analysis. Nucl. Phys. B **560**, 617–682 (1999). arXiv:hep-th/9903036
11. A. Ceresole, G. Dall'Agata, R. D'Auria, S. Ferrara, Spectrum of type IIB supergravity on $AdS_5 \times T^{11}$: predictions on $\mathcal{N} = 1$ SCFT's. Phys. Rev. D **61**, 066001 (2000). arXiv:hep-th/9905226
12. D. Fabbri, P. Fré, L. Gualtieri, P. Termonia, Osp(N|4) supermultiplets as conformal superfields on ∂AdS_4 and the generic form of $N = 2$, D=3 gauge theories. Class. Quantum Gravity **17**, 55 (2000). arXiv:hep-th/9905134
13. P. Fré, L. Gualtieri, P. Termonia, The structure of $\mathcal{N} = 3$ multiplets in AdS_4 and the complete $Osp(3|4) \times SU(3)$ spectrum of M-theory on $AdS_4 \times N^{0,1,0}$. Phys. Lett. B **471**, 27–38 (1999). arXiv:hep-th/9909188
14. D. Fabbri, P. Fré, L. Gualtieri, C. Reina, A. Tomasiello, A. Zaffaroni, A. Zampa, 3D superconformal theories from Sasakian seven-manifolds: new non-trivial evidences for AdS_4/CFT_3. Nucl. Phys. B **577**, 547–608 (2000). arXiv:hep-th/9907219
15. M. Billó, D. Fabbri, P. Fré, P. Merlatti, A. Zaffaroni, Rings of short $\mathcal{N} = 3$ superfields in three dimensions and M-theory on $AdS_4 \times N^{010}$. Class. Quantum Gravity **18**, 1269 (2001). https://doi.org/10.1088/0264-9381/18/7/310, arXiv:hep-th/0005219
16. M. Billó, D. Fabbri, P. Fré, P. Merlatti, A. Zaffaroni, Shadow multiplets in AdS_4/CFT_3 and the super-Higgs mechanism: hints of new shadow supergravities. Nucl. Phys. B **591**, 139–194 (2000). arXiv:hep-th/0005220
17. M.J. Duff, C.N. Pope, Kaluza-Klein supergravity and the seven sphere. Supersymmetry and supergravity (1983), p. 183; ICTP/82/83-7, Lectures given at September school on supergravity and supersymmetry, Trieste, Italy, Sep 6–18, 1982. Published in Trieste Workshop 1982:0183 (QC178:T7:1982)
18. R. D'Auria, P. Fré, Spontaneous generation of Osp(4|8) symmetry in the spontaneous compactification of D = 11 supergravity. Phys. Lett. B **121**, 141–146 (1983)
19. M.A. Awada, M.J. Duff, C.N. Pope, $\mathcal{N} = 8$ supergravity breaks down to $\mathcal{N} = 1$. Phys. Rev. Lett. **50**, 294 (1983)
20. B. Biran, F. Englert, B. de Wit, H. Nicolai, Gauged N = 8 supergravity and its breaking from spontaneous compactification. Phys. Lett. B **124**, 45–50 (1983)
21. M. Günaydin, N.P. Warner, Unitary supermultiplets of Osp(8|4, R) and the spectrum of the S^7 compactification of 11-dimensional supergravity. Nucl. Phys. B **272**, 99–124 (1986)
22. E. Witten, Search for a realistic Kaluza-Klein theory. Nucl. Phys. B **186**, 412–428 (1981)
23. L. Castellani, R. D'Auria, P. Fré, $SU(3) \otimes SU(2) \otimes U(1)$ from D = 11 supergravity. Nucl. Phys. B **239**, 610–652 (1984)
24. R. D'Auria, P. Fré, On the spectrum of the $\mathcal{N} = 2$ $SU(3) \otimes SU(2) \otimes U(1)$ gauge theory from D = 11 supergravity. Class. Quantum Gravity **1**, 447 (1984)
25. A. Ceresole, P. Fré, H. Nicolai, Multiplet structure and spectra of $\mathcal{N} = 2$ supersymmetric compactifications. Class. Quantum Gravity **2**, 133 (1985)
26. D.N. Page, C.N. Pope, Stability analysis of compactifications of D = 11 supergravity with $SU(3) \times SU(2) \times U(1)$ symmetry. Phys. Lett. B **145**, 337–341 (1984)
27. R. D'Auria, P. Fré, P. Van Nieuwenhuizen, $\mathcal{N} = 2$ matter coupled supergravity from compactification on a coset G/H possessing an additional killing vector. Phys. Lett. B **136**, 347–353 (1984)
28. D.N. Page, C.N. Pope, Which compactifications of D = 11 supergravity are stable? Phys. Lett. B **144**, 346–350 (1984)
29. R. D'Auria, P. Fré, Universal Bose-Fermi mass-relations in Kaluza-Klein supergravity and harmonic analysis on coset manifolds with Killing spinors. Ann. Phys. **162**, 372–412 (1985)
30. L. Castellani, R. D'Auria, P. Fré, K. Pilch, P. van Nieuwenhuizen, The bosonic mass formula for Freund-Rubin solutions of D = 11 supergravity on general coset manifolds. Class. Quantum Gravity **1**, 339 (1984)

31. D.Z. Freedman, H. Nicolai, Multiplet shortening in Osp($N|4$). Nucl. Phys. B **237**, 342–366 (1984)
32. L. Castellani, L.J. Romans, N.P. Warner, A classification of compactifying solutions for D = 11 supergravity. Nucl. Phys. B **241**, 429–462 (1984)
33. P.G. Fré, A. Fedotov, *Groups and Manifolds: Lectures for Physicists with Examples in Mathematica* (De Gruyter, Berlin, 2018)
34. P. Fré, P.A. Grassi, Constrained supermanifolds for AdS M-theory backgrounds. JHEP **01**, 036 (2008)
35. D. Gaiotto, X. Yin, Notes on superconformal Chern-Simons-matter theories. J. High Energy Phys. **2007**, 056 (2007). https://doi.org/10.1088/1126-6708/2007/08/056, arXiv:0704.3740 [hep-th]
36. O. Aharony, O. Bergman, D.L. Jafferis, J. Maldacena, N = 6 superconformal Chern-Simons-matter theories, M2-branes and their gravity duals. J. High Energy Phys. **2008**, 091 (2008). https://doi.org/10.1088/1126-6708/2008/10/091, arXiv:0806.1218 [hep-th]
37. J. Bagger, N. Lambert, Comments on multiple M2-branes. J. High Energy Phys. **2008**, 105 (2008). https://doi.org/10.1088/1126-6708/2008/02/105, arXiv:0712.3738 [hep-th]
38. P. Fré, P.A. Grassi, The integral form of D = 3 Chern-Simons theories probing $\mathbb{C}^n/\Gamma$ singularities (2017). arXiV:hep-th1705.00752
39. U. Bruzzo, A. Fino, P. Fr, The Kähler quotient resolution of $\mathscr{C}^3/\Gamma$ singularities, the McKay correspondence and D = 3 $\mathscr{N}$ = 2 Chern-Simons gauge theories (2017)
40. D. Anselmi, M. Billó, P. Fré, L. Girardello, A. Zaffaroni, ALE manifolds and conformal field theories. Int. J. Mod. Phys. A **9**, 3007–3058 (1994)
41. H.F. Blichfeldt, On the order of linear homogeneous groups. IV. Trans. Am. Math. Soc. **12**, 39–42 (1911)
42. H.F. Blichfeldt, Blichfeldt's finite collineation groups. Bull. Am. Math. Soc. **24**, 484–487 (1918)
43. G.A. Miller, H.F. Blichfeldt, L.E. Dickson, *Theory and Applications of Finite Groups* (Dover Publications Inc., New York, 1961)
44. S.-S. Roan, Minimal resolutions of Gorenstein orbifolds in dimension three. Topology **35**, 489–508 (1996)
45. D. Markushevich, Resolution of C^3/H_{168}. Math. Ann. **308**, 279–289 (1997)
46. Y. Ito, M. Reid, The McKay correspondence for finite subgroups of SL(3, **C**), in *Higher-dimensional Complex Varieties (Trento, (1994)* (de Gruyter, Berlin, 1996), pp. 221–240
47. G.W. Gibbons, S.W. Hawking, Classification of gravitational instanton symmetries. Commun. Math. Phys. **66**, 291–310 (1979)
48. N.J. Hitchin, Polygons and gravitons. Math. Proc. Camb. Philos. Soc. **85**, 465–476 (1979)
49. E. Witten, Phases of N = 2 theories in two-dimensions. Nucl. Phys. **B403**, 159–222 (1993) [AMS/IP Stud. Adv. Math. **1**, 143 (1996)]
50. E. Calabi, Metriques kähleriennes et fibres holomorphes. Ann. Scie. Ec. Norm. Sup. **12**, 269 (1979)
51. A. Craw, The McKay correspondence and representations of the McKay quiver. Ph.D. thesis, Warwick University, United Kingdom, 2001
52. T. Eguchi, A.J. Hanson, Selfdual Solutions to Euclidean Gravity. Ann. Phys. **120**, 82 (1979)
53. J. McKay, Graphs, singularities and finite groups. Proc. Symp. Pure Math. **37**, 183 (1980)
54. S. Chimento, T. Ortin, On 2-dimensional Kaehler metrics with one holomorphic isometry (2016). arXiv:hep-th1610.02078
55. A. Degeratu, T. Walpuski, Rigid HYM connections on tautological bundles over ALE crepant resolutions in dimension three. SIGMA Symmetry Integr. Geom. Methods Appl. **12**, Paper No. 017, 23 (2016)
56. W. Fulton, *Introduction to Toric Varieties*. Volume 131 of Annals of Mathematics Studies (Princeton University Press, Princeton, 1993) (The William H. Roever Lectures in Geometry)
57. V. Arnold, S. Gusein-Zade, A. Varchenko, *Singularities of Differentiable Maps* (Birkhäuser, 1975)
58. E. Brieskorn, Unknown, Actes Congrés Intern. Math. Ann. t, 2 (1970)

59. P. Slodowy, *Simple Singularities and Simple Algebraic Groups*. Lecture Notes in Mathematics, vol. 815 (1980)
60. P.G. Fré, *A Conceptual History of Symmetry from Plato to the Superworld* (Springer, Berlin, 2018)

Chapter 9
Epilogue

> *M. Poincaré a écrit que dans les Sciences mathématiques une bonne notation a la même importance philosophique qu'une bonne classification dans les Sciences naturelles. Évidemment, et même avec plus de raison, on peut en dire autant des méthodes, car c'est bien de leur choix que dépend la possibilité de forcer (pour nous servir encore des paroles de lillustre géomètre francais) une multitude de faits sans aucun lien apparent à se grouper suivant leurs affinités naturelles.*
>
> Ricci and Levi-Civita

What happened in mathematics since the mid thirties of the XX century to the early eighties of the same century is deeply characterized, in my opinion, by the following two highly momentous developments, one intrinsic to the mathematical community the other forced on it by the new visions of theoretical physics. These developments are the following ones:

(a) Starting with the monumental work of Cartan on symmetric spaces the *theory of symmetry*, meaning group theory, Lie Algebra theory and associated topics merged more tightly with the *theory of geometry*, meaning manifolds and fibre-bundles, their isometries, their holonomies and their topology.
(b) With the advent of supersymmetry and of its obligatory consequences, namely supergravity, superstrings and branes, what in geometry was so far generic, for instance the dimensions D of the space-time manifold or the possible scalar potentials ceased to be such and started being determined within finite ranges of choices that are dictated by a superior structure, at the same time very restrictive and surprisingly rich in its power to relate so far uncorrelated mathematical objects.

P. G. Frè, *Advances in Geometry and Lie Algebras from Supergravity*, Theoretical and Mathematical Physics,
https://doi.org/10.1007/978-3-319-74491-9_9

Before supersymmetry D might be any number, after supersymmetry it took the fixed values either $D = 11$, or $D = 10$, related to each other by a deep mechanism named duality. Before supersymmetry, all Riemannian spaces were equally interesting, after supersymmetry *special geometries* occupied the scene introducing new exciting mathematical structures that I have described at length in several chapters of this book. Before supersymmetry, exceptional Lie algebras were mathematical curiosities mostly disregarded by physicists, after supersymmetry all the exceptional Lie algebras fell into appropriate boxes specially prepared for them in a grandiose fresco which almost unexpectedly started revealing itself.

Looking at matters from a distance and with a mathematical attitude one gets the impression that supersymmetry played the role of that critical tile in a puzzle, putting which into its proper place, all the other tiles almost automatically find their way to their correct positions. Many examples can be made but one spectacular one might suffice to clarify this point.

The possible holonomy groups of Riemannian manifolds were classified before supersymmetry and fill a very short list. Generic manifolds have holonomy SO(n) in $d = n$ dimensions. In even dimensions $d = 2n$, manifolds with holonomy $\mathrm{U(n)} \subset \mathrm{SO(2n)}$ are the complex manifolds. Those among the complex manifolds that have holonomy $\mathrm{SU(n)} \subset \mathrm{U(n)} \subset \mathrm{SO(2n)}$ are the Kähler manifolds and here we meet with $\mathcal{N} = 1$ supersymmetry, as my attentive reader already knows. In $d = 4n$, manifolds with holonomy $\mathrm{USp(2n)} \subset \mathrm{U(2n)} \subset \mathrm{SO(4n)}$ are the HyperKähler manifolds while those with holonomy $\mathrm{USp(2n)} \times \mathrm{SU(2)} \subset \mathrm{U(2n)} \subset \mathrm{SO(4n)}$ are the quaternionic Kähler manifolds. In both cases we meet here with $\mathcal{N} = 2$ supersymmetry, rigid in the first case, local in the second one. The list contained two more exceptional cases, the mysterious 7-dimensional manifolds with $\mathrm{G}_{2(-14)}$ holonomy and the 8-dimensional manifolds with $\mathrm{Spin(7)} \subset \mathrm{SO(8)}$ holonomy. Both cases were decoded by supergravity. The first was decoded by observing that $d = 7$ is the complement of $d = 4$ in compactifications of $D = 11$ supergravity and that $\mathrm{G}_{2(-14)}$ holonomy is the condition for a residual $\mathcal{N} = 1$ supersymmetry of the compactified vacuum. The second case was decoded considering M2-branes in $D = 11$ space-time, Spin(7)-holonomy of the 8-manifold transverse to the M2-brane being the condition for its $\mathcal{N} = 1$ supersymmetry.

Not only known mathematics found its interpretation within the framework of supersymmetry and supergravity but new entire chapters of geometry were constructed under the stimulus of supergravity. Most notable among them are some of the topics extensively discussed in this book, namely:

1. Special Kähler Geometry.
2. The c and $c^\star$ maps from Special Kähler Geometry to quaternionic or pseudo quaternionic geometry.
3. The relations of the above constructions with the *Tits Satake projection*.
4. The systematics of Kähler and HyperKähler quotients leading, for instance, to the classifications and construction of all ALE manifolds.
5. The σ-model approach to supergravity black-holes and the refinement of the theory of nilpotent orbits.

One more item in this already large conceptual landscape needs proper emphasis. This is the compound of physical-mathematical ideas and conceptual frameworks that has been stirred by the AdS/CFT correspondence in its various declinations.

The original emphasis in these developments was on the holographic principle which is an intrinsic property of anti de Sitter space, liable to be extended to more complicated space-time geometries, and on the cheered by physicists opportunity of calculating exact quantum field-theoretic correlators by means of classical evaluations of geometrical nature. Yet, in my humble opinion, there is an even more important physical-mathematical conception that has emerged within the AdS/CFT framework and which was discussed with a certain amplitude in the last chapter of the present book.

This is the dynamical interpretation of *geometry* as a low-energy effect in field theories with a much larger set of coordinate-fields endowed with canonical kinetic terms. This is the lagrangian description, within gauged–coupled σ-models, of the (Hyper)Kähler quotient algorithm able to construct ALE–manifolds and other *crepant resolutions* of $\mathbb{C}^n/\Gamma$ singularities, Γ being a discrete group.

From a philosophical point of view that above is a new variant of Einstein guiding principle that space-time geometry is created by gravitating energy. In the brane setup the geometry of the space transverse to the brane world volume is just a collective phenomenon due to the gauge interactions of microscopic coordinates spanning a much larger flat world. From a physical-mathematical viewpoint the crucial item is the appearance of a discrete group Γ and the relation of the field-theoretical setup with the issue of quotient singularity resolutions.

In this respect the ADE classification of ALE-manifolds as resolutions of the singularities $\mathbb{C}^2$/*Platonic Group* plays a fundamental paradigmatic role. Its generalization to higher dimensional resolutions, in particular $\mathbb{C}^3/\Gamma$ brings in new challenging actors, the simple Hurwitz group $L_{168} = \mathrm{PSL}(2, \mathbb{Z}_7)$ and the generalized McKay correspondence.

In my humble opinion, further investigation of all the implications, physical, geometrical and philosophical of the vast panorama only very partially unveiled by the above considerations is mandatory.

Indeed the same general conclusions put forward in the epilogue of the twin historical book [1] fit equally well in the epilogue of the present one.

After the spectacular detection of gravitational waves emitted from the coalescence of two massive black holes and the numerical verification of Einstein field equations, the points (A)–(E) introduced in the preface are firmly established, at least within our Western Analytical System of Thought. The choices of symmetries, bundles and potentials within such a framework have to be made in a way enlightened by the lesson of supersymmetry. It is not yet clear whether supersymmetry is realized in Nature in the way we think and it might take a quite long time before we are able to answer such a question in an experimental way, yet we cannot ignore the geometrical structures and the miraculous relations among them that supersymmetry has brought to the front stage. We have to continue the exploration of the new mathematics introduced by supergravity and superstrings to find new hidden clues, so far not yet observed.

From generic choices we have been instructed to look at special structures, restricted holonomy, for instance, exceptional Lie algebras, hyperbolic algebras, sporadic simple groups and the like, searching for new corners where other tiles of the mathematical puzzle might find their proper place. At the end of a long day it might happen that supersymmetry is only the tip of an iceberg and that in the deep waters under the cold sea surface there lies another mathematical logic able to lead us to a new physical vision and to new far reaching conclusions. Yet the tip is there, it was observed and one cannot avoid to explore further what lies underneath the surface of the sea.

Reference

1. P.G. Fré, *A Conceptual History of Symmetry from Plato to the Superworld* (Springer, Berlin, 2018)

MIX
Papier aus verantwortungsvollen Quellen
Paper from responsible sources
FSC® C105338

If you have any concerns about our products,
you can contact us on
ProductSafety@springernature.com

In case Publisher is established outside the EU,
the EU authorized representative is:
Springer Nature Customer Service Center GmbH
Europaplatz 3, 69115 Heidelberg, Germany

Printed by Libri Plureos GmbH
in Hamburg, Germany